Distributions for Modeling Location, Scale, and Shape

Chapman & Hall/CRC
The R Series

Recently Published Titles

Basics of Matrix Algebra for Statistics with R
Nick Fieller

Introductory Fisheries Analyses with R
Derek H. Ogle

Statistics in Toxicology Using R
Ludwig A. Hothorn

Spatial Microsimulation with R
Robin Lovelace, Morgane Dumont

Extending R
John M. Chambers

Using the R Commander: A Point-and-Click Interface for R
John Fox

Computational Actuarial Science with R
Arthur Charpentier

bookdown: Authoring Books and Technical Documents with R Markdown
Yihui Xie

Testing R Code
Richard Cotton

R Primer, Second Edition
Claus Thorn Ekstrøm

Flexible Regression and Smoothing: Using GAMLSS in R
Mikis D. Stasinopoulos, Robert A. Rigby, Gillian Z. Heller, Vlasios Voudouris, and Fernanda De Bastiani

The Essentials of Data Science: Knowledge Discovery Using R
Graham J. Williams

blogdown: Creating Websites with R Markdown
Yihui Xie, Alison Presmanes Hill, Amber Thomas

Handbook of Educational Measurement and Psychometrics Using R
Christopher D. Desjardins, Okan Bulut

Displaying Time Series, Spatial, and Space-Time Data with R, Second Edition
Oscar Perpinan Lamigueiro

Reproducible Finance with R
Jonathan K. Regenstein, Jr.

R Markdown
The Definitive Guide
Yihui Xie, J.J. Allaire, Garrett Grolemund

Practical R for Mass Communication and Journalism
Sharon Machlis

Analyzing Baseball Data with R, Second Edition
Max Marchi, Jim Albert, Benjamin S. Baumer

Spatio-Temporal Statistics with R
Christopher K. Wikle, Andrew Zammit-Mangion, and Noel Cressie

Statistical Computing with R, Second Edition
Maria L. Rizzo

Geocomputation with R
Robin Lovelace, Jakub Nowosad, Jannes Münchow

Dose-Response Analysis with R
Christian Ritz, Signe M. Jensen, Daniel Gerhard, Jens C. Streibig

Advanced R, Second Edition
Hadley Wickham

For more information about this series, please visit: https://www.crcpress.com/go/the-r-series

Distributions for Modeling Location, Scale, and Shape

Using GAMLSS in R

By

Robert A. Rigby

Mikis D. Stasinopoulos

Gillian Z. Heller

Fernanda De Bastiani

CRC Press is an imprint of the
Taylor & Francis Group, an informa business

CRC Press
Taylor & Francis Group
6000 Broken Sound Parkway NW, Suite 300
Boca Raton, FL 33487-2742

First issued in paperback 2021

Printed on acid-free paper

ISBN-13: 978-0-367-27884-7 (hbk)
ISBN-13: 978-1-03-208942-3 (pbk)

Please visit Library of Congress for this book's Cataloging-in-Publication Data

Visit the Taylor & Francis Web site at
http://www.taylorandfrancis.com

and the CRC Press Web site at
http://www.crcpress.com

To
Marcel,
Nikos, Jill, Harry, and Melissa,
Steven, Ilana, and Monique,
Nelson, Angélica, and Felipe.

Contents

III Reference guide 361

Preface

This is a wide-ranging book about statistical *distributions*, their properties, and how they can be used in practice to model the dependence of the distribution of a response variable on explanatory variables. It will be especially useful to applied statisticians in a wide range of application areas, and also to those interested in understanding the properties of distributions.

This book describes over one hundred distributions, which are available in the **R** package **gamlss.dist**, and also demonstrates how more distributions can be created by transformation, truncation, or censoring, using the GAMLSS packages. The distributions are continuous, discrete, and mixed (i.e. continuous-discrete). Many real data applications are included, where the response variable is modeled using a Generalized Additive Model for Location, Scale, and Shape (GAMLSS), [Rigby and Stasinopoulos, 2005], which allows any distribution for the response variable and the modeling of all parameters of the distribution using explanatory variables. It is a distributional regression model.

This book follows an earlier book, 'Flexible Regression and Smoothing: Using GAMLSS in R' [Stasinopoulos et al., 2017], which focuses on the GAMLSS model and software implementation using the GAMLSS packages in **R**. This book focuses on distributions for modeling a response variable and applications. It will be a useful resource for users of the GAMLSS software, and statistical modelers in general.

Historically the normal and Poisson distributions were commonly used for modeling continuous and discrete count response variables, respectively. Unfortunately, especially with the large data sets often encountered more recently, these distributions are often found to be inappropriate or to provide inadequate fits to the observed response variable distribution in many practical situations.

In practice, for a continuous response variable, the shape of its distribution can be highly positively or negatively skewed and/or highly platykurtic or leptokurtic (e.g. lighter or heavier tails than the normal distribution). Continuous variables can also have different ranges from that of the normal distribution, i.e. $(-\infty, \infty)$. The ranges $(0, \infty)$ and $(0, 1)$ are especially common.

Also in practice a discrete count response variable can have a distribution which is overdispersed or underdispersed relative to the Poisson distribution, and/or highly positively (or negatively) skewed and/or leptokurtic (or platykurtic) and/or have an excess (or reduced) incidence of zero values.

There are also occasions where a response variable requires a mixture of a continuous and a discrete distribution, i.e. a continuous distribution with additional specific values with point probabilities. For example, in insurance claims, the claim amount can be modeled with a continuous distribution on $(0, \infty)$ if a claim is made, and with a point probability at zero if no claim is made. In this case the distribution should allow values in the range $[0, \infty)$, which includes the value zero. Another important example is a proportion (i.e. percentage or fractional) response variable which has a continuous distribution on $(0, 1)$ with additional point probabilities at zero and one, giving range $[0, 1]$.

All of the above cases are dealt with in this book, together with analysis of real data examples.

This book is divided into three parts:

I Parametric distributions and the GAMLSS family of distributions

II Advanced topics

III Distributions reference guide.

The aims of Part I 'Parametric distributions and the GAMLSS family of distributions' are to introduce:

- the basic ideas and properties of distributions,
- the distributions implemented in the **gamlss** packages, in particular:
  - to investigate continuous distributions on ranges $(-\infty, \infty)$, $(0, \infty)$, and $(0, 1)$, including highly positively or negatively skewed and/or highly platykurtic or leptokurtic,
  - to investigate discrete distributions for counts, including overdispersed or underdispersed and/or highly positively (or negatively) skewed and/or leptokurtic (or platykurtic) and/or an excess (or reduced) incidence of zero values,
  - to investigate discrete distributions for 'binomial counts',
  - to investigate mixed distributions with range $[0, \infty)$ including exact value zero, or range $[0, 1), (0, 1]$, or $[0, 1]$, i.e. the interval from zero to one including the exact values zero, one, or both.

Real data are used throughout Part I to emphasize the practical considerations when choosing a distribution.

The aims of Part II 'Advanced topics' are to:

- explain basic ideas of statistical inference by emphasizing the difference between population, sample, and model,
- investigate the robustness of parameter estimation to outlier contamination,
- describe different methods for generating new distributions,
- discuss skewness and kurtosis,
- compare continuous distributions based on their moment and centile skewness and kurtosis,
- investigate and classify the heaviness of tails of distributions.

Note that some of the material covered in part II is original, and therefore potentially controversial. Our aim is to provoke further discussion and further research.

The aims of Part III 'Distributions reference guide' are to provide a reference guide to the continuous, discrete, and mixed distributions available in the **gamlss.dist** package. In particular, tables and plots of the explicitly defined distributions are provided. The tables contain (most or all of) the following:

- the range of the response variable,
- the range of the parameters of the distribution,
- the probability (density) function,
- the median and the mode,
- the mean, variance, moment skewness, and moment excess kurtosis,
- the moment or probability generating function,
- the cumulative distribution function,
- the inverse cumulative distribution function, i.e. the quantile (or centile) function.

Motivation

A theoretical distribution for a response variable is a representation (usually an approximation) of its population distribution. The theoretical distribution may be parametric, and this generally provides a simple parsimonious (and usually smooth) representation of the population distribution. It can be used for inference (e.g. estimation) for the centiles (or quantiles) and moments (e.g. mean) of the population distribution and other population measures. A better theoretical distribution model for a population distribution should provide better inference for population centiles and moments. Note however inference for the moments can be particularly sensitive to the use of a theoretical distribution model which misspecifies the population distribution, and especially if it misspecifies the heaviness of the tail(s) of the population distribution.

In a regression situation a theoretical parametric conditional distribution for a response variable, given the values of the explanatory variables, may provide a simple parsimonious representation (usually an approximation) of the corresponding conditional population distribution. This allows a potentially simple interpretation of how changing values of the explanatory variables affects the conditional distribution of the response variable. A better theoretical conditional distribution should provide better inference for the centiles and moments of the population conditional distribution, and potentially better inference regarding the relationship of the response variable population conditional distribution to explanatory variables. Furthermore a flexible theoretical conditional distribution, with say four distribution parameters, potentially allows modeling of changes in the location, scale, skewness, and kurtosis of the population conditional distribution of the response variable with explanatory variables.

An alternative approach for conditional population centile estimation, to the approach using a theoretical parametric conditional distribution, is quantile regression [Koenker, 2017]. The advantages and disadvantages of the two approaches are discussed in Rigby and Stasinopoulos [2013] and Stasinopoulos et al. [2017, p451-452].

An alternative approach for estimation of the conditional population mean (and variance), to the approach using a theoretical parametric conditional distribution, is generalized estimating equations (GEE) [Ziegler, 2011].

Software information

The software used in this book is almost entirely within the GAMLSS **R** packages.

1. The original **gamlss** package for fitting a GAMLSS model contains the main function `gamlss()` for fitting a GAMLSS model, and methods for dealing with fitted `gamlss` objects. It automatically downloads **gamlss.dist** and **gamlss.data**.
2. The package **gamlss.dist** contains all the explicitly defined distributions. Its function `gen.Family()` can generate 'log' and 'logit' distributions with ranges $(0, \infty)$ and $(0, 1)$, respectively, from any **gamlss.dist** continuous distribution with range $(-\infty, \infty)$.
3. The **gamlss.data** package contains most of the data sets used in this book, and is automatically loaded with **gamlss**. In the data sets in **gamlss.data** categorical variables are, in general, declared as factors within the data frame. [This can be checked using the function `str()`.]
4. The package **gamlss.tr** can take any **gamlss.dist** distribution and generate left, right or both sides truncated distributions.
5. The package **gamlss.cens**, designed for censored data, can be used to fit generalized Tobit models, as well as discretized continuous distributions for count data. Examples of the use of these features are given in Chapter 3 and throughout this book.
6. The package **gamlss.countKinf** is for count response variables where one of its values (not necessarily zero) occurs more often than expected from a standard count distribution, see Chapter 7.
7. The **gamlss.add** package provides extra additive terms for fitting a parameter of the response variable distribution. This is mainly achieved by providing interfaces with other **R** packages.
8. For other GAMLSS **R** packages see Section 2.2 in Stasinopoulos et al. [2017].

The majority of data sets used are contained either within **gamlss.data** or alternative **R** packages. In a few exceptions the data can be readily accessed on the internet.

More about latest developments, further examples, and exercises using GAMLSS can be found on `www.gamlss.org`. The GAMLSS **R** packages are available from The Comprehensive R Archive Network: `https://cran.r-project.org`.

There also several packages in **R** designed either for fitting distributions to a single variable, or more generally for distributional regression models. For example the package **fitdistrplus** [Delignette-Muller and Dutang, 2015] is designed to fit distributions to data. The package **Newdistns** [Nadarajah and Rocha, 2016] computes the pdf, cdf, quantile and random generator functions of nineteen families of distributions. For regression models the package **bamlss** [Umlauf et al., 2017] is a Bayesian version of GAMLSS. The package **gamboostLSS** [Hofner et al., 2016] is designed for boosting with a GAMLSS distribution. The package **mgcv**, automatically distributed with **R**, is designed for generalized additive models, but is increasingly being extended to include GAMLSS features, see Wood et al. [2017]. The package **VGAM** [Yee, 2019] contains a large number of different parametric distributions whose parameters can be fitted with additive smoothing terms. The package **countreg**, Zeileis et al. [2008], can be used for modeling count response variable. The package **GJRM** [Marra and Radice, 2019] for generalized joint regression modeling contains its own `gamlss()` function. The package **distreg.vis** [Stadlmann, 2019] is designed to visually help the interpretation of distributional regression models, including GAMLSS.

Notation used in this book

In this book we distinguish between statistical models, **R** packages, and **R** functions. We use capital letters for models, bold characters for packages, and code type characters (with extra brackets) for functions. For example

- GAMLSS refers to the statistical model,
- **gamlss** refers to the **R** package, and
- `gamlss()` refers to the **R** function for fitting a GAMLSS model,
- **gamlss.dist** refers to the **R** package that includes the fitting functions for all the explicit distributions in GAMLSS,
- `gamlss.family()` is an S3 class of objects in **R** for fitting a GAMLSS distribution to a model.

Within the book, italic text is used for emphasis or for an item being defined.

Vectors in general will be represented in a lower case bold letters, e.g. $\mathbf{x} = (x_1, x_2, \ldots, x_n)^\top$ and matrices in an upper case bold letter, e.g. $\mathbf{X}$. Scalar random variables are represented by upper case (Y) and the observed value by lower case (y).

The following table shows the notation that is used throughout this book.

	Distributions
Y	a univariate response variable
y	a single value of the response variable Y
$\mathbf{y}$	the vector of observed values of the response variable Y, i.e. $(y_1, y_2, \ldots, y_n)^\top$
n	total number of observations
K	the number of parameters in the distribution of Y
$f(y)$	theoretical probability function (pf) or probability density function (pdf) or mixed probability function (mpf) of the random variable Y (`d` function in **R**). Occasionally, for clarity the subscript Y is used, i.e. $f_Y(y)$.

$F(y)$	$\mathrm{P}(Y \leq y)$, cumulative distribution function (cdf) of Y (`p` function in **R**)
y_p	$F^{-1}(p)$, inverse cumulative distribution (or quantile) function of Y (`q` function in **R**)
$S(y)$	$\mathrm{P}(Y > y)$, survival function of Y
$h(y)$	hazard function of Y
$\mathcal{D}(\cdot)$	generic distribution
R_Y	range of Y, i.e. an interval, or discrete list, for the set of all possibles values of Y
$\mathcal{E}(\cdot)$	exponential family of distributions
$Y \sim \mathcal{D}$	Y has distribution $\mathcal{D}$
$Y \stackrel{\text{ind}}{\sim} \mathcal{D}(\boldsymbol{\mu}, \boldsymbol{\sigma}, \boldsymbol{\nu}, \boldsymbol{\tau})$	$Y_1, Y_2, \ldots, Y_n$ are independently distributed with distributions $\mathcal{D}(\mu_i, \sigma_i, \nu_i, \tau_i)$ for $i = 1, 2, \ldots, n$
$f_{Y\mid X}(\cdot)$	conditional probability (density) function of the random variable Y given X
$\mathcal{N}(\mu, \sigma^2)$	normal distribution with mean μ and variance σ^2
$\phi(\cdot)$	pdf of standard normal distribution `NO`$(0,1)$
$\Phi(\cdot)$	cdf of standard normal distribution `NO`$(0,1)$
	Moment measures and functions
$\mathrm{E}(Y)$	Expected value (or mean) of random variable Y
$\mathrm{Var}(Y)$	Variance of random variable Y
$\mathrm{SD}(Y)$	Standard deviation of random variable Y
${\mu_k}'$	$\mathrm{E}(Y^k)$, kth population moment of Y about zero, for $k = 1, 2, 3 \ldots$
μ_k	$\mathrm{E}\{[Y - \mathrm{E}(Y)]^k\}$, kth population central moment (or kth moment about the mean) of Y, for $k = 2, 3, \ldots$
$M_Y(t)$	$\mathrm{E}(e^{tY})$, moment generating function (MGF) of Y
$K_Y(t)$	$\log[M_Y(t)]$, cumulant generating function (CGF) of Y
$G_Y(t)$	$\mathrm{E}(t^Y)$, probability generating function (PGF) of a discrete random variable Y
γ_1	$\sqrt{\beta_1}$, moment skewness
γ_2	$\beta_2 - 3$, moment excess kurtosis
β_2	moment kurtosis
γ_{1t}	$\gamma_1/(1 + \lvert\gamma_1\rvert)$, transformed moment skewness
γ_{2t}	$\gamma_2/(1 + \lvert\gamma_2\rvert)$, transformed moment kurtosis
	Centile (or quantile) measures and functions
Q_1	$y_{0.25} = F^{-1}(0.25)$, first quartile of Y
m	$y_{0.5} = F^{-1}(0.5)$, median of Y
Q_3	$y_{0.75} = F^{-1}(0.75)$, third quartile of Y
IR	$Q_3 - Q_1$, interquartile range
SIR	$(Q_3 - Q_1)/2$, semi interquartile range
s_p	$\dfrac{(y_p + y_{1-p})/2 - y_{0.5}}{(y_{1-p} - y_p)/2}$, centile skewness function of Y, $s_p(Y)$, for $0 < p < 0.5$
γ	$s_{0.25} = [(Q_3 + Q_1)/2 - m]/\mathrm{SIR}$, Galton's centile skewness measure (also called central centile skewness measure)
$s_{0.01}$	tail centile skewness measure
$S(p)$	$y_{1-p} - y_p$, spread function of Y, for $0 < p < 0.5$

k_p	$(y_{1-p} - y_p)/\text{IR}$, centile kurtosis function of Y, $k_p(Y)$, for $0 < p < 0.5$
δ	$k_{0.01} = (y_{0.99} - y_{0.01})/\text{IR}$, Andrew's centile kurtosis measure measure of Y
$sk_p(Y)$	$k_p(Y)/k_p(Z)$, standardized kurtosis function of Y, for $0 < p < 0.5$, where $Z \sim \mathcal{N}(0,1)$
$ek_{0.01}$	$k_{0.01} - 3.449$, centile excess kurtosis
$tk_{0.01}$	$ek_{0.01}/(1 + \lvert ek_{0.01} \rvert)$, transformed centile kurtosis
	Intervals
$(-\infty, \infty)$	interval from minus infinity to infinity (also called the real line $\mathbb{R}$)
$(0, \infty)$	interval from zero to infinity, excluding zero (also called the positive real line $\mathbb{R}_+$)
$[0, \infty)$	interval from zero to infinity, including zero
$(0, 1)$	interval from zero to one, excluding both zero and one
$[0, 1)$	interval from zero to one, including zero but excluding one
$(0, 1]$	interval from zero to one, excluding zero but including one
$[0, 1]$	interval from zero to one, including both zero and one
$\{0, 1, 2, \ldots, n\}$	integer values from zero to n (called binomial values)
$\{0, 1, 2, \ldots\}$	non-negative integer values from zero to infinity
	Distribution parameters
θ_k	the kth distribution parameter, where $\theta_1 = \mu$, $\theta_2 = \sigma$, $\theta_3 = \nu$ and $\theta_4 = \tau$
$\boldsymbol{\theta}_k$	a vector of length n of the kth distribution parameter, e.g. $\boldsymbol{\theta}_1 = \boldsymbol{\mu}$, $\boldsymbol{\theta}_2 = \boldsymbol{\sigma}$
μ	the first parameter of the distribution (usually location)
σ	the second parameter of the distribution (usually scale)
ν	the third parameter of the distribution (usually shape, e.g. skewness)
τ	the fourth parameter of the distribution (usually shape, e.g. kurtosis)
λ	a hyperparameter
$\boldsymbol{\lambda}$	the vector of all hyperparameters in the model
	Likelihood and information criteria
L	likelihood function
ℓ	log-likelihood function
Λ	generalized likelihood ratio test statistic
$\boldsymbol{i}$	Fisher's expected information matrix
I	observed information matrix
GDEV	global deviance, i.e. minus twice the fitted log-likelihood
GAIC	generalized Akaike information criterion: GDEV $+ \kappa \cdot$ df
df	total (effective) degrees of freedom used in the model
κ	penalty for each degree of freedom used in the model
	Residuals
$\mathbf{u}$	vector of (randomized) quantile residuals
$\mathbf{r}$	vector of normalized (randomized) quantile residuals
$\boldsymbol{\varepsilon}$	vector of (partial) residuals
Q	Q statistic calculated from the residuals
Z	Z-statistic calculated from the residuals

	GAMLSS model components
$\mathcal{M}$	a GAMLSS model containing $\{\mathcal{D},\mathcal{G},\mathcal{T},\mathcal{L}\}$
$\mathcal{D}$	the specification of the distribution of the response variable
$\mathcal{G}$	the link functions $g_k(\cdot)$ where $g_k(\boldsymbol{\theta}_k)=\boldsymbol{\eta}_k$ for $k=1,2,\ldots,K$
$\mathcal{T}$	the explanatory variable terms in the model
$\mathcal{L}$	the specification of the smoothing parameters
	Systematic part of the GAMLSS model
p_k	the number of columns in the design matrix $\mathbf{X}_k$ for parameter $\boldsymbol{\theta}_k$
J_k	the total number of smoothers for $\boldsymbol{\theta}_k$
q_{kj}	the dimension of the random effect vector $\boldsymbol{\gamma}_{kj}$
$\mathbf{x}_{kj}$	the jth explanatory variable vector for $\boldsymbol{\theta}_k$
$\mathbf{X}_k$	an $n\times p_k$ fixed effects design matrix for $\boldsymbol{\theta}_k$
$\boldsymbol{\beta}_k$	a vector of fixed effect parameters for $\boldsymbol{\theta}_k$ of length $J_K^{'}$ i.e. $(\beta_{k1},\beta_{k2},\ldots,\beta_{kJ_K^{'}})^\top$
$\boldsymbol{\beta}$	a vector of all fixed effects parameters, i.e. $(\boldsymbol{\beta}_1^\top,\boldsymbol{\beta}_2^\top,\ldots,\boldsymbol{\beta}_k^\top)^\top$
$\boldsymbol{\gamma}_{kj}$	the jth random effect parameter vector for $\boldsymbol{\theta}_k$, of length q_{kj}
$\boldsymbol{\gamma}$	a vector of all random effects parameters, i.e. $(\boldsymbol{\gamma}_{11}^\top,\ldots,\boldsymbol{\gamma}_{KJ_K}^\top)^\top$
$\mathbf{Z}_{kj}$	an $n\times q_{kj}$ random effect design matrix for the jth smoother of $\boldsymbol{\theta}_k$
$\mathbf{G}_{kj}$	an $q_{kj}\times q_{kj}$ penalty matrix for $\boldsymbol{\gamma}_{kj}$
$g_k(\cdot)$	link function applied to model $\boldsymbol{\theta}_k$
$\boldsymbol{\eta}_k$	the predictor for $\boldsymbol{\theta}_k$ i.e. $\boldsymbol{\eta}_k=g_k(\boldsymbol{\theta}_k)$
$\mathbf{H}_k$	the hat matrix for $\boldsymbol{\theta}_k$
$\mathbf{z}_k$	the adjusted dependent variable for $\boldsymbol{\theta}_k$
$s_{kj}(\cdot)$	the jth nonparametric or nonlinear function in $\boldsymbol{\eta}_k$
$\mathbf{W}$	an $n\times n$ diagonal matrix of weights
$\mathbf{w}$	an n-dimensional vector of weights (the diagonal elements of $\mathbf{W}$)
$\mathbf{S}_{kj}$	the jth smoothing matrix for $\boldsymbol{\theta}_k$
	Functions
$\Gamma(a)$	$\int_0^\infty t^{a-1}e^{-t}dt$, gamma function, for $a>0$. (Note $\Gamma(a)$ can also be defined for $a<0$, see Johnson et al. [2005] p 6-7)
$\gamma(a,x)$	$\int_0^x t^{a-1}e^{-t}dt$, incomplete gamma function, for $a>0$ and $x>0$.
$\Gamma(a,x)$	$\int_x^\infty t^{a-1}e^{-t}dt$, complement of the incomplete gamma function, for $a>0$ and $x>0$. Note $\Gamma(a,x)=\Gamma(a)-\gamma(a,x)$.
$\Psi(a)$	$\frac{d}{da}\log\Gamma(a)$, psi (or digamma) function
$\Psi^{(r)}(a)$	$\frac{d^{(r)}}{da^{(r)}}\Psi(a)$, rth derivative of $\Psi(a)$
$B(a,b)$	$\int_0^1 t^{a-1}(1-t)^{b-1}dt=\frac{\Gamma(a)\Gamma(b)}{\Gamma(a+b)}$, beta function, for $a>0$ and $b>0$
$B(a,b,x)$	$\int_0^x t^{a-1}(1-t)^{b-1}dt$, incomplete beta function, for $a>0$, $b>0$ and $0<x<1$
$(a)_j$	$a(a+1)\ldots(a+j+1)$, Pochhammer's symbol
${}_2F_1(a,b;c;x)$	$\sum_{j=0}^\infty \dfrac{(a)_j(b)_jx^j}{(c)_jj!}$, Gaussian hypergeometric function, for $c\neq 0,-1,-2,\ldots$
$\text{logit}(a)$	$\log[a/(1-a)]$, logit function, for $0<a<1$

$\text{logit}^{-1}(a)$	$1/[1+\exp(-a)]$, inverse logit function, for $-\infty < a < \infty$
$K_\lambda(t)$	$\frac{1}{2}\int_0^\infty x^{\lambda-1}\exp\{-\frac{1}{2}t(x+x^{-1})\}\,dx$, modified Bessel function of the second kind
$K_{\lambda+1}(t)$	$(2\lambda/t)K_\lambda(t)+K_{\lambda-1}(t)$
$K_{-1/2}(t)$	$[\pi/(2t)]^{1/2}e^{-t}=K_{1/2}(t)$
$\xi(a)$	$\sum_{i=1}^\infty i^{-a}$, Riemann zeta function
$\sinh(x)$	$(e^x-e^{-x})/2$, sinh function
$\sinh^{-1}(x)$	$\log[(x^2+1)^{1/2}+x]$, inverse sinh function, or arcsinh function
$\cosh(x)$	$(e^x+e^{-x})/2$, cosh function
$d[f_1(y),f_2(y)]$	$\int[\log f_1(y)-\log f_2(y)]\,f_1(y)\,dy$, Kullback-Leibler distance
$\text{IC}(y,\hat{\theta})$	influence function (or curve) for maximum likelihood estimator $\hat{\theta}$ of parameter θ
z_p	$\Phi^{-1}(p)$, inverse cdf of $\mathcal{N}(0,1)$, for $0<p<1$
$t_{p,\nu}$	$F_T^{-1}(p)$, inverse cdf of $T\sim t_\nu$, t distribution with ν degrees of freedom (t_ν), for $0<p<1$
	Other notation
$\lfloor y \rfloor$	largest integer less than or equal to y (the floor function)
$\lceil y \rceil$	smallest integer greater than or equal to y (the ceiling function)
C	Euler's constant, $C=0.57722$ (to 5 decimal places)
$F_1 <_2 F_2$	distribution of Y_2 is more van Zwet skew to the right than the distribution of Y_1
$F_1 <_S F_2$	distribution of Y_2 is more Balanda-MacGillivray kurtotic than the distribution of Y_1
$\Rightarrow$	implies
$\Longleftrightarrow$	if and only if (i.e. implies both ways, $\Rightarrow$, $\Leftarrow$)
o	order less than, i.e. $f_1(y)=o[f_2(y)] \Leftrightarrow f_1(y)/f_2(y)\to 0$, as $y\to\infty$
$\mathcal{O}$	order equal to, i.e. $f_1(y)=O[f_2(y)] \Leftrightarrow f_1(y)/f_2(y)\to c$, where c is a finite constant, as $y\to\infty$
$\sim$	asymptotically equivalent to, i.e. $f_1(y)\sim f_2(y) \Leftrightarrow f_1(y)/f_2(y)\to 1$, as $y\to\infty$

Acknowledgements

The GAMLSS packages, which are dedicated to fitting GAMLSS models, have been constantly developing over the last 20 years. We therefore would like to thank all those people who during those years helped us to improve our knowledge and the software by directly contributing or sending us corrections. In particular we like to thank Robert Gilchrist, for his continuous support, Abu Hossain for his asssistance with the development of the inflated distributions in Chapter 9, Simon Wood especially for his bivariate smoothing functions used in Chapter 5, Chris Jones for information provided to us,

Athanasios Lopatatzidis for introducing us to the `GG` distribution, Peter Lane for introducing us to response variables defined on $[0, 1]$, Dan Rosenheck for introducing us to the discretized Burr 12 distribution, and Gauss Cordereio for references on methods of generating distributions. We would like to thank Jim Lindsey for his pioneering work on distributional regression, and Tim Cole and Peter Green for their LMS method which helped the development of centile fitting in GAMLSS. We would like to thank Christian Kleiber and Achim Zeileis especialy for their `rootogram()` function in the **R** package **countreg**. We would also like to thank Calliope Akantziliotou, Vlasios Voudouris, Raydonal Ospina, Nicoletta Motpan, Fiona McElduff, Majid Djennad, Marco Enea, Luiz Nakamura, Alexios Ghalanos, Christos Argyropoulos, Almond Stocker, Jens Lichter, and Stanislaus Stadlmann for directly helping with **gamlss.dist**. For corrections in the software we would like to thank Almond Stöcker, Stephen Cromie, Francimário Alves de Lima, Felix Hunschede, and others we have inadvertently omitted. Finally we would like to thank Harry Stasinopoulos for providing us with the cover figure of this book.

Part I

Parametric distributions and the GAMLSS family of distributions

1

Types of distributions

CONTENTS

This chapter provides:

1. the definition of a statistical distribution; and
2. the implementation of distributions in **R**.

1.1 Introduction

The first two chapters are an introduction to general theoretical parametric distributions. In this chapter we explore the different types of parametric distributions. In Chapter 2, properties and summary measures of parametric distributions are considered. Chapter 3 introduces the GAMLSS family of distributions.

Statistical modeling is the art of building parsimonious statistical models for a better understanding of the response variables of interest. A statistical model deals with uncertainties and therefore it contains a *stochastic* or *random* part, that is, a component which describes the uncertain nature of the response variable. The main instrument to deal with uncertainty in mathematics is the concept of *probability* and all statistical models incorporate some probabilistic assumptions by assuming that response variables of interest have a probabilistic distribution. This probabilistic or stochastic part of a statistical model usually arises from the assumption that the data we observe are a sample from a larger (unknown) population whose properties we wish to study. The statistical model is a representation (usually an approximation) of the population and its behaviour. Essential to understanding the basic concepts of statistical modeling are the ideas of

- the *population*,
- the *sample* and
- the *model.*

These concepts are explained in detail in Chapter 10.

An *'experiment'*, in the statistical sense, is any action which has an outcome that cannot be predicted with certainty. For example, "toss a die and record the score" is an experiment with possible outcomes $S = \{1, 2, 3, 4, 5, 6\}$. The set of all possible outcomes S is called the sample space of the experiment. Outcomes of the experiment are grouped into *events* (denoted as E). Events that correspond to a single outcome are called elementary events, while those corresponding to more than one outcome are called composite events. For example, the event "even score" corresponds to the outcomes $E = \{2, 4, 6\}$ and is a composite event. Probabilities, denoted here as $\mathrm{P}(\cdot)$, are functions defined on the events and have the properties:

- $\mathrm{P}(E) \geqslant 0$
- $\mathrm{P}(E_1 \cup E_2) = P(E_1) + P(E_2)$ where E_1 and E_2 are events which have no common outcomes (i.e. are disjoint)
- $\mathrm{P}(S) = 1$.

That is, probabilities take values from zero to one. If an event has a probability of zero then this event will never occur, while an event with probability of one is a certain event.

A *random variable* Y assigns a value to an outcome determined by chance. The *range* of Y, denoted R_Y, is the set of all possible values of Y. For example, let us consider the simple action of tossing a coin and observing whether the result is a head or a tail. If we assign $Y = 0$ for a tail and $Y = 1$ for a head, then the random variable Y can take values in the set $\{0, 1\}$. We say that Y has *range* $R_Y = \{0, 1\}$.

A *continuous* random variable Y has a range that is an interval, or the union of two or more intervals, of the real line. The most common continuous ranges R_Y are:

- *the real line* $\mathbb{R} = (-\infty, \infty)$
- *the positive real line* $\mathbb{R}_+ = (0, \infty)$
- *values between zero and one* $\mathbb{R}_0^1 = (0, 1)$.

A *discrete* random variable Y has a range that is a distinct set of values. The number of distinct values may be either finite or countably infinite. The most common discrete ranges R_Y are:

- the binary values : $\{0, 1\}$

- the binomial values : $\{0, 1, 2, \ldots, n\}$
- the non-negative integer values : $\{0, 1, 2, \ldots\}$.

Note that the binary values are a special case of the binomial values.

1.2 Probability (density) function

1.2.1 Discrete random variable

The distribution of a discrete random variable Y is defined by the probabilities for all possible values of Y, and is called a *probability function (pf)*. The probability function $\mathrm{P}(\cdot)$ assigns the probability that Y takes each particular value in its range. For example, in the coin toss, $\mathrm{P}(Y = 0) = \mathrm{P}(Y = 1) = 0.5$ if the coin is fair.

For a discrete random variable Y, let $f(y) = \mathrm{P}(Y = y)$, that is $f(y)$[a] represents the probability that Y is equal to specific value y. The function $f(y)$ is said to be a proper probability function of the discrete random variable Y if $f(y)$ is positive for all values of y within its range R_Y and if

$$\sum_{y \in R_Y} f(y) = \sum_{y \in R_Y} \mathrm{P}(Y = y) = 1 .$$

For convenience, let $f(y) = \mathrm{P}(Y = y) = 0$ for all y that are in the real line $\mathbb{R} = (-\infty, \infty)$, but are not in R_Y.

1.2.2 Continuous random variable

The distribution of a continuous random variable Y is defined by a *probability density function (pdf)*, $f(y)$.

Let Y be a continuous random variable. The function $f(y)$ is said to be a proper probability density function of the continuous variable Y if $f(y)$ is positive for all values of y within its range R_Y and if

$$\int_{R_Y} f(y)\, dy = 1 .$$

For convenience let $f(y) = 0$ for all y in the real line $\mathbb{R} = (-\infty, \infty)$ not in R_Y. Then $f(y)$ is defined on $y \in \mathbb{R}$. Probabilities are given by areas under $f(y)$:

$$\mathrm{P}(a \leq Y \leq b) = \int_a^b f(y)\, dy .$$

[a]There are two notational comments needed to be made here: (i) the notation $\mathrm{P}(Y = y)$ is preferable to $f(y)$ for discrete random variables, since it emphasizes the discrete nature of the variable involved. In this book we will use both notations since it is sometimes convenient to use the same notation for discrete and continuous random variables; (ii) it is common practice to use the notation $f_Y(y)$ to emphasize that the f function refers to the random variable Y. We generally drop Y from the notation $f_Y(y)$ for simplicity, but use it when needed for clarification, as for example when more than one random variable is involved.

Note here the peculiarity that $P(Y = a) = \int_a^a f(y)\, dy = 0$, for any arbitrary value a. That is, the probability of a continuous random variable Y being exactly equal to any specific value is equal to zero. (This can be circumvented by defining the probability on a small interval $(a - \Delta y, a + \Delta y)$ around a, where Δy has a small value. Then $P(Y \in (a - \Delta y, a + \Delta y)) = \int_{a-\Delta y}^{a+\Delta y} f(y)\, dy$ is properly defined.)

1.2.3 Mixed random variable

A mixed (continuous-discrete) random variable Y has a distribution which is a mixture of a continuous and a discrete distribution with ranges R_1 and R_2, respectively, i.e. a continuous distribution with additional specific values with point probabilities. The distribution of a mixed random variable Y is defined by a *mixed probability function*, $f(y)$, which is said be proper if $f(y)$ is positive for all values of y within its range R_Y and if

$$\int_{R_1} f(y)\, dy + \sum_{y \in R_2} f(y) = 1 \; .$$

Most commonly the mixed distributions that we encounter as response distributions have either:

- $R_1 = (0, \infty)$ and a single probability mass at zero, i.e. $R_2 = \{0\}$. This is called a *zero-adjusted* distribution and has range $R_Y = [0, \infty)$ including exact value $Y = 0$; or
- $R_1 = (0, 1)$, and $R_2 = \{0\}$, $R_2 = \{1\}$ or $R_2 = \{0, 1\}$, called an *inflated* distribution with ranges $R_Y = [0, 1)$, $R_Y = (0, 1]$ and $R_Y = [0, 1]$, respectively, i.e. the interval from 0 to 1 including exact values $Y = 0$, 1, or both 0 and 1, respectively.

In the case of a zero-adjusted distribution, we write the mixed probability function as, for example,

$$f(y \,|\, \mu, \sigma, \nu) = \begin{cases} \nu & \text{if } y = 0 \\ (1 - \nu) f_W(y \,|\, \mu, \sigma) & \text{if } y > 0 \; , \end{cases} \tag{1.1}$$

where $P(Y = 0) = \nu$, $0 < \nu < 1$ and $f_W(\cdot)$ is the pdf of the continuous part of the distribution with parameters μ and σ. The mixed probability function of an inflated distribution is similar.

For examples of mixed distributions see Section 3.3.3 and Chapter 9.

1.2.4 Continuous example: exponential distribution

The function

$$f(y) = e^{-y} \qquad \text{for } y > 0$$

is a proper pdf since $e^{-y} > 0$ for $y \in (0, \infty)$ and $\int_0^\infty e^{-y} dy = 1$. Its flexibility is enhanced by introducing a single parameter in the definition of the pdf, i.e.

$$f(y) = \theta e^{-\theta y} \qquad \text{for } y > 0 \; , \tag{1.2}$$

where $\theta > 0$. Since $\int_0^\infty \theta e^{-\theta y} dy = 1$, the new function is still a pdf. Because $f(y)$ now contains the parameter θ, it is called a *parametric probability density function*. The key

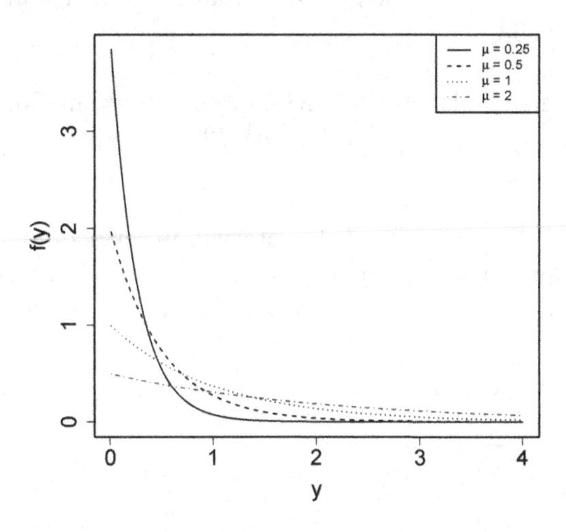

FIGURE 1.1: The exponential pdf with different values of μ

advantage of introducing the parameter θ is that it makes the shape of the distribution more flexible in modeling a set of observations. Finding a suitable estimated value for parameter θ for a given set of data is an example of *statistical inference*, see Chapters 10 and 11.

In order to emphasize the fact that the pdf depends on θ we write it as $f(y \mid \theta)$[b]. The notation $f(y \mid \theta)$ now represents, for different values of θ, a *family* of pdf's. Note that if instead of parameter θ we choose any other one-to-one function of θ, e.g. $\mu = 1/\theta$, the pdf family remains the same. For example, (1.2) and

$$f(y \mid \mu) = \frac{1}{\mu} e^{-y/\mu} \qquad \text{for } y > 0 \tag{1.3}$$

where $\mu > 0$, define the same pdf family (the *exponential distribution*). Figure 1.1 shows $f(y \mid \mu)$ plotted against y, for each of four different values of the parameter μ, i.e. $\mu = (0.25, 0.5, 1, 2)$. As y tends to zero, $f(y \mid \mu) \to \frac{1}{\mu}$ which decreases as μ increases. Note that (1.2) and (1.3) are different parameterizations of the same family. We shall see later that in practice some parameterizations are preferable to others, as far as model interpretation is concerned. The exponential distribution with pdf given by (1.3) is denoted by `EXP`(μ) in the **gamlss.dist** package.

The following is an example of the use of the exponential distribution in practice.

[b]For simplicity in general we will drop the conditioning on θ unless the condition is important, for example in statistical inference.

R data vector: `aircond` in package **gamlss.data** of dimension 24×1
source: Proschan [1963]
variables

`aircond` : the intervals, in service-hours, between failures of the air-conditioning equipment in a Boeing 720 aircraft.

Proschan [1963] reports observations of the interval, in service-hours, between failures of the air-conditioning equipment in ten Boeing 720 aircraft. Here we use 24 observations from one of the aircraft. These data are also available in the **rpanel** package [Bowman et al., 2007].

```
data(aircond)
aircond

## [1]  50  44 102  72  22  39   3  15 197 188  79  88  46   5
## [15]   5  36  22 139 210  97  30  23  13  14
```

A histogram of the data is shown in Figure 1.2. All observations are positive so we require a distribution defined on the positive real line, such as the exponential distribution. In the **R** commands below we plot a histogram of the data and then fit an exponential distribution using the **gamlss** function `histDist()`. The exponential distribution is specified by the argument `family = EXP`. Note that the fitting (i.e. estimation of parameter μ in (1.3)) is performed using maximum likelihood estimation, see Chapter 11.

```
m1 <- histDist(aircond, family = EXP, main="")
fitted(m1, "mu")[1]

## [1] 64.125
```

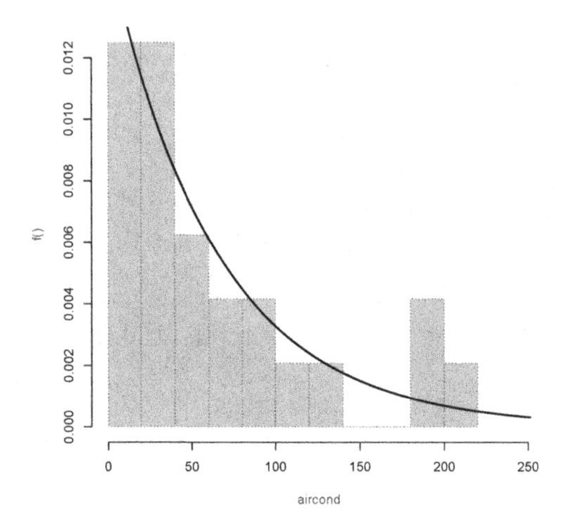

FIGURE 1.2: Histogram of the `aircond` data together with the fitted exponential distribution.

The **gamlss** implementation of the exponential distribution uses parameterization (1.3) and the fitted μ is given by the `fitted(m1, "mu")` command as $\hat{\mu} = 64.125$.

1.2.5 Discrete example: Poisson distribution

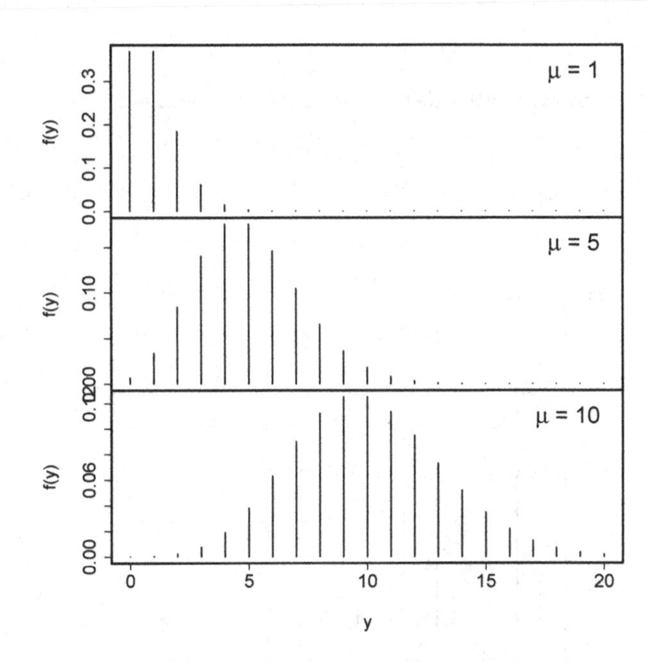

FIGURE 1.3: The Poisson probability function for $\mu = 1, 5, 10$.

The function

$$P(Y = y \mid \mu) = \frac{e^{-\mu}\mu^y}{y!}, \qquad \text{for } y = 0, 1, 2, \ldots \tag{1.4}$$

is the probability function of one of the most popular discrete distributions, the *Poisson distribution*, which is discussed in detail in Section 7.1.1. Figure 1.3 plots $P(Y = y \mid \mu)$ against y for each of the values $\mu = 1, 5, 10$, showing how the shape of the distribution changes with μ. The Poisson distribution with probability function given by (1.4) is denoted by PO(μ) in the **gamlss.dist** package.

An example of the use of the Poisson distribution in practice is given below.

R data file: `prussian` in package **pscl** of dimension 280×3
source: von Bortkiewicz [1898]
variables
- `y` : count of deaths
- `year` : year of observation
- `corp` : corps of Prussian Army

In 1898 the Russian economist Ladislaus von Bortkiewicz published a book entitled 'Das gesetz der kleinen zahlen', in which he included a data example that became famous for illustrating the use of the Poisson distribution. The response variable is the number of deaths by kicks of a horse (`y`) in different years and different corps (regiments) of the Prussian Army. Here we use the data with no explanatory variables and fit a Poisson distribution to the number of deaths. For the fitting we use the `histDist()` function. The Poisson distribution is specified by the argument `family = PO`, which uses parameterization (1.4).

```
library(pscl)
h1 <- histDist(y,  family = PO, data = prussian, main="")
fitted(h1)[1]

## [1] 0.7
```

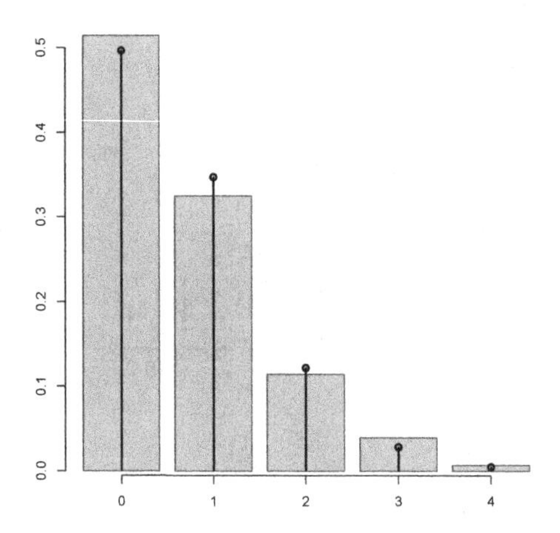

FIGURE 1.4: Barplot of the `prussian` data together with the fitted Poisson distribution (vertical lines).

The estimated parameter is given by $\hat{\mu} = 0.7$, which is the estimated mean of the distribution. Note that no value of Y larger than 4 is observed, but because the Poisson distribution is defined on $\{0, 1, 2, \ldots\}$ we are able to estimate probabilities at values larger than 4. For example the estimated probability that $Y = 5$ is given by

```
dPO(5, mu=fitted(h1,"mu")[1])

## [1] 0.0006955091
```

which is a very small probability.

1.2.6 Summary

- Any non-negative function $f(y) = \text{P}(Y = y)$ which sums over a discrete range to one can be a probability function (pf).

- Any non-negative function $f(y)$ which integrates over a continuous range to one can be a probability density function (pdf).
- Probability (density) functions may depend on more than one parameter. In general, we write a probability (density) function depending on K parameters $(\theta_1, \ldots, \theta_K) = \boldsymbol{\theta}^\top$ as $f(y \mid \theta_1, \ldots, \theta_K)$, or $f(y \mid \boldsymbol{\theta})$. In the **gamlss.dist** distributions, K can be 1, 2, 3 or 4, and parameters are denoted as $\theta_1 = \mu$, $\theta_2 = \sigma$, $\theta_3 = \nu$, $\theta_4 = \tau$, i.e. $\boldsymbol{\theta}^\top = (\mu, \sigma, \nu, \tau)$.
- A one-to-one reparameterization of the parameters does not change the pf or pdf family.
- Parameters can affect the shape of the distribution.

1.3 Cumulative distribution function, cdf

The *cumulative distribution function* (*cdf*), $F(y)$, is the probability of observing a value less than or equal to a specified value y. It is defined as

$$F(y) = \mathrm{P}(Y \le y) = \begin{cases} \sum_{w \le y} f(w) = \sum_{w \le y} \mathrm{P}(Y = w) & \text{for discrete } Y \\ \int_{-\infty}^{y} f(w)\, dw & \text{for continuous } Y \end{cases},$$

for $y \in (-\infty, \infty)$, where in the discrete case the sum is over all $w \in R_Y$ for which $w \le y$.

The cdf is a nondecreasing function, with $\lim_{y \to -\infty} F(y) = 0$ and $\lim_{y \to \infty} F(y) = 1$. For a continuous random variable Y, $F(y)$ is a continuous function, often S-shaped; for discrete Y, $F(y)$ is a step function, with jumps at the distinct values $y \in R_Y$.

1.3.1 Continuous example: exponential distribution

For the exponential distribution (1.3), we have for $y > 0$,

$$F(y) = \mathrm{P}(Y \le y) = \int_0^y \frac{1}{\mu} e^{-w/\mu}\, dw = 1 - e^{-y/\mu} ,$$

giving the cdf

$$F(y) = \begin{cases} 0 & \text{for } y \le 0 \\ 1 - e^{-y/\mu} & \text{for } y > 0 . \end{cases} \tag{1.5}$$

Figure 1.5 plots $f(y)$ and $F(y)$ given by (1.3) and (1.5), respectively, against y, for $\mu = 0.5$, demonstrating the link between $f(y)$ and $F(y)$. For example, $F(0.7) = \mathrm{P}(Y \le 0.7) = 0.75$, i.e. the area under function $f(y)$ for $0 < y < 0.7$ is 0.75.

The next question arising here is how the empirical (i.e. sample) cumulative distribution

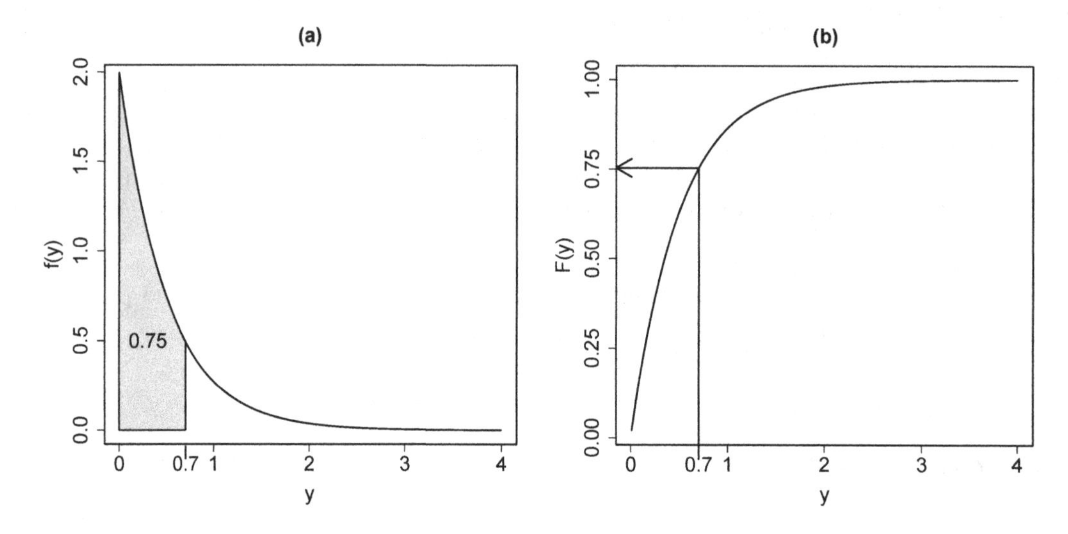

FIGURE 1.5: (a) The probability density function $f(y)$ and (b) the cumulative distribution function $F(y)$, of the exponential distribution with $\mu = 0.5$.

function compares with the fitted cdf. In general the fitted cdf is a smoother function than the empirical cdf. We see this in Figure 1.6, where the empirical cdf of the `aircond` data is plotted together with the fitted cdf of the exponential distribution (i.e. with $\hat{\mu} = 64.125$). Note that this empirical cdf is a step function with steps of height f/n (i.e. $f/24$) where f is the frequency at each distinct value of Y and n is the total number of observations.

```
Ecdf <- ecdf(aircond) # the ecdf
plot(Ecdf, ylab="F(y)", xlab="y", main="")      # plot ecdf
lines(pEXP(0:250, mu=64.125)) # plot the fitted cdf
```

1.3.2 Discrete example: Poisson distribution

The Poisson cdf is

$$F(y) = \mathrm{P}(Y \le y) = \begin{cases} 0 & \text{for } y < 0 \\ \displaystyle\sum_{w=0}^{\lfloor y \rfloor} \frac{e^{-\mu}\mu^w}{w!} & \text{for } y \ge 0 \ , \end{cases} \tag{1.6}$$

where $\lfloor y \rfloor$ denotes the largest integer less than or equal to y (the floor function). Figure 1.7 plots $\mathrm{P}(Y = y)$ and $F(y)$, given by (1.4) and (1.6), respectively, against y, for $\mu = 2$, and shows

$$F(1.5) = \mathrm{P}(Y \le 1.5) = \mathrm{P}(Y = 0) + \mathrm{P}(Y = 1) = 0.406 \ .$$

It is of interest to compare the empirical cdf of the `prussian` data set with the fitted Poisson cdf (i.e. with $\mu = 0.7$). Figure 1.8 shows the two cumulative distribution functions which have very similar values, indicating that the Poisson distribution model provides an adequate fit to the data.

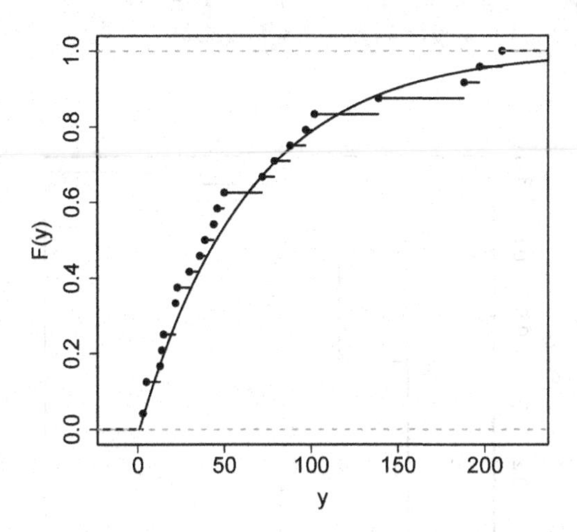

FIGURE 1.6: The fitted cumulative distribution function of the exponential distribution with $\hat{\mu} = 64.125$ (the continuous line), and the empirical (i.e. sample) cumulative distribution function (the step function) for the `aircond` data.

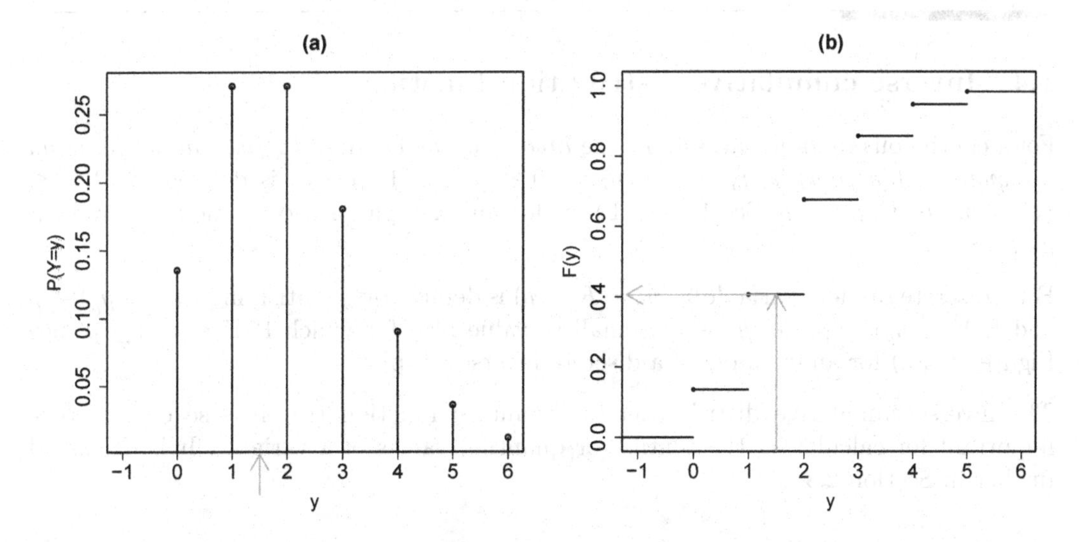

FIGURE 1.7: (a) The probability function $\text{P}(Y = y)$ and (b) the cumulative distribution function $F(y)$, for the Poisson distribution (with $\mu = 2$), showing $F(1.5) = 0.406$. Note that while the probability of observing 1.5 is zero, i.e. $\text{P}(Y = 1.5) = 0$, the cumulative distribution $F(1.5) = \text{P}(Y \leq 1.5) = 0.406$ is positive.

```
Ecdf <- ecdf(prussian$y) # the ecdf
plot(Ecdf, col="gray", main="")    # plot ecdf
cdf <- stepfun(0:6, c(0,pPO(0:6, mu=0.7)), f = 0)
plot(cdf, add=T)              # plot fitted Poisson
```

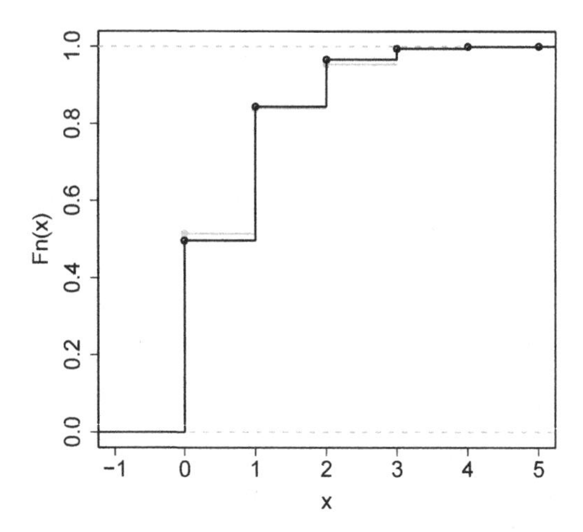

FIGURE 1.8: The empirical cdf, (gray line), and fitted cdf, (black line), for the Poisson distribution (with $\mu = 0.7$), for the `prussian` data.

1.4 Inverse cumulative distribution function

For a continuous random variable Y, the *inverse of the cumulative distribution function* (or *quantile function*) is $y_p = F^{-1}(p)$ for $0 < p < 1$. Hence y_p is defined by $\text{P}(Y \leq y_p) = p$, i.e. $F(y_p) = p$. See Figure 1.9(c) for an example of a continuous inverse cdf plot.

For a discrete random variable Y, $y_p = F^{-1}(p)$ is defined by y_p satisfying $\text{P}(Y < y_p) < p$ and $\text{P}(Y \leq y_p) \geq p$, i.e. y_p is the smallest value of y for which $\text{P}(Y \leq y_p) \geq p$. See Figure 1.10(c) for an example of a discrete inverse cdf plot.

The inverse cumulative distribution (or quantile) function (i.e. inverse cdf) is very important for calculating the centile (or quantile) values of a variable. It is discussed further in Section 2.3.1.

1.5 Survival and hazard functions

The *survival function* $S(y)$ is the probability of Y 'surviving' beyond y, i.e.

$$S(y) = \mathrm{P}(Y > y) = 1 - \mathrm{P}(Y \leq y) = 1 - F(y) \ .$$

Alternative names for the survival function are the *complementary cumulative distribution function* (ccdf), the *tail distribution*, *exceedance*, or the *reliability function* (common in engineering).

In **R** the survival function can be obtained from the cdf (see the `p` function in Section 1.6) by using the argument `lower.tail = FALSE` . For example

```
curve(pEXP(x, mu=1, lower.tail = TRUE), 0,10)
```

plots the cdf of the exponential distribution with mean $\mu = 1$, while

```
curve(pEXP(x, mu=1, lower.tail = FALSE), 0,10)
```

plots the survival function.

For a continuous random variable Y, the *hazard function* $h(y)$ is the instantaneous likelihood of ('dying at') y given survival up to y, i.e.

$$h(y) = \lim_{\delta y \to 0} \frac{\mathrm{P}\left(y < Y \leq y + \delta y \,|\, Y > y\right)}{\delta y} = \frac{f(y)}{S(y)} \ .$$

Both survival and hazard functions play important roles in medical statistics and reliability theory. The survival function is also important in the analysis of tails of distributions, see Chapter 17.

The **gamlss.dist** package provides the facility of taking any `gamlss.family` distribution, see Chapter 3, and creating its hazard function using the functions `gen.hazard()` or `hazardFun()`. For example, to generate the hazard function `hEXP` for the exponential distribution use one of the following:

```
hEXP<-hazardFun(family = "EXP")
gen.hazard("EXP")
```

The hazard function of the exponential distribution has the characteristic that it is flat (i.e. constant). That is, the instantaneous likelihood of 'dying' in the exponential distribution is constant, which is called the 'no memory' property. This can be verified by plotting the created hazard function `hEXP()`.

```
curve(hEXP(x, mu=0.7), 0,3)
```

TABLE 1.1: Standard distributions in **R**.

Continuous distributions

R name	Distribution
`beta`	beta
`cauchy`	Cauchy
`chisq`	chi-squared
`exp`	exponential
`f`	F
`gamma`	gamma
`lnorm`	log-normal
`logis`	logistic
`norm`	normal
`t`	student's t
`unif`	uniform
`weibull`	Weibull

Discrete distributions

R name	Distribution
`binom`	binomial
`geom`	geometric
`hyper`	hypergeometric
`multinom`	multinomial
`nbinom`	negative binomial
`pois`	Poisson

1.6 Distributions in R: the d, p, q, and r functions

The **R** statistical environment contains several popular distributions. **R** uses the `d`, `p`, `q`, and `r` convention where

`d` is the probability (density) function;

`p` is the cumulative distribution function (cdf);

`q` is the quantile (or inverse cdf) function; and

`r` generates random numbers from the distribution.

For example `dnorm()`, `pnorm()`, `qnorm()`, and `rnorm()` are the pdf, cdf, quantile, and random number generating functions of the normal distribution, respectively. Table 1.1 gives the distributions which are provided in the **R base** package.

This book is concerned with distributions in the **gamlss.dist** package. This package follows the `d`, `p`, `q`, and `r` convention, but parameterizations may be different from those used in the standard **R** distributions. The main reason is that **gamlss.dist** is built to support the GAMLSS regression framework. In a regression model, interpretability of the distribution parameters, which are modeled as functions of explanatory variables, is important. For example, the gamma distribution in **gamlss.dist** is parameterized as `GA`(μ, σ), where μ is the mean and σ is the coefficient of variation. The gamma distribution in **R** is `gamma`(a, s), where a is the 'shape' and s is the 'scale', with mean as and coefficient of variation $a^{-0.5}$.

Graphical representation of the `d`, `p`, `q`, and `r` functions can be accomplished easily. The following **R** code produces Figure 1.9, a plot of the pdf, cdf, and quantile function, and a histogram of a randomly generated sample from a gamma, `GA`(μ, σ), distribution, with $\mu = 3$ and $\sigma = 0.5$, using the **gamlss.dist** functions `dGA()`, `pGA()`, `qGA()`, and `rGA()`, respectively.

```
mu=3
sigma=.5
curve(dGA(y, mu, sigma), 0.01, 10, xname="y",ylab="f(y)", main="(a)") # pdf
curve(pGA(y, mu, sigma), 0.01, 10, xname="y", ylab="F(y)", main="(b)")# cdf
curve(qGA(p, mu, sigma), 0.01, 1, xname="p",
      ylab=expression(y[p]=={F^{-1}} (p)), main="(c)") # icdf
set.seed(125)
y<-rGA(1000, mu, sigma) # random sample
hist(y,col="lightgray",main="(d)")
```

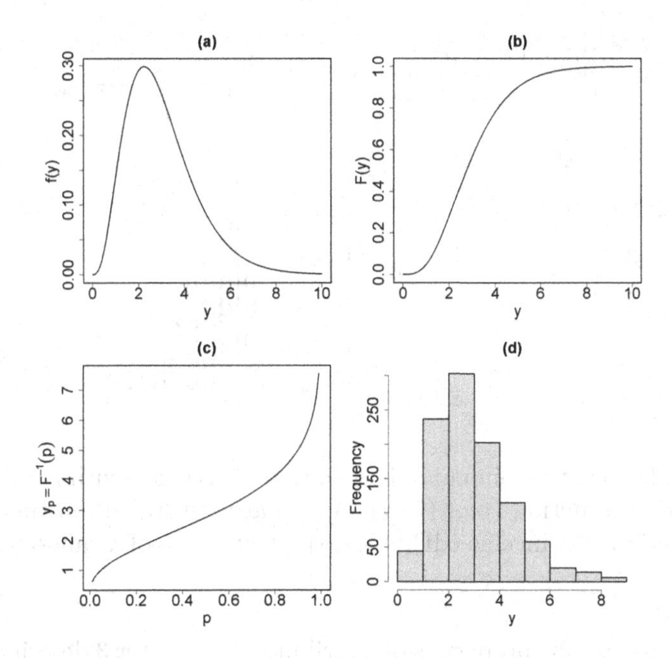

FIGURE 1.9: The gamma distribution GA(μ, σ) with $\mu = 3$ and $\sigma = 0.5$: (a) the probability density function (pdf), (b) the cumulative distribution function (cdf), (c) the quantile function (i.e. inverse cdf), and (d) a histogram of a random sample of 1000 observations.

An example of a discrete distribution, the negative binomial, is given in Figure 1.10. The **gamlss.dist** functions `dNBI()`, `pNBI()`, `qNBI()`, and `rNBI()` are used:

```
mu=8
sigma=.25
plot(function(y) dNBI(y, mu, sigma), from=0, to=30, n=30+1,
     type="h",xlab="y",ylab="f(y)", main="(a)")
cdf <- stepfun(0:29, c(0,pNBI(0:29,mu, sigma)), f = 0)
plot(cdf, xlab="y", ylab="F(y)", verticals=FALSE,
     cex.points=.8, pch=16, main="(b)")
invcdf <- stepfun(seq(0.01,.99,length=39),
                  qNBI(seq(0.1,.99,length=40),mu, sigma), f = 0)
plot(invcdf, ylab=expression(y[p]=={F^{-1}} (p)), do.points = FALSE,
     verticals=TRUE, cex.points=.8, pch=16, main="(c)",
```

```
    xlab="p")
y <- rNBI(1000, mu, sigma)
barplot(table(y), xlab="y", ylab="Frequency", main="(d)")
```

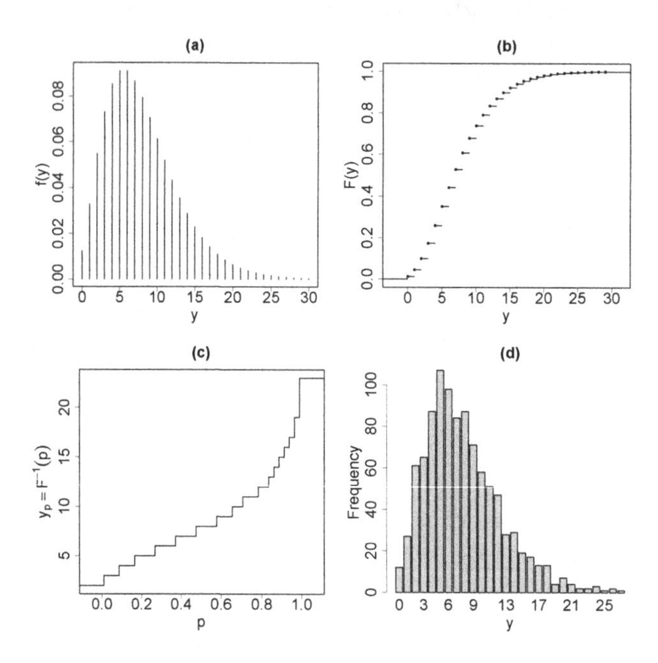

FIGURE 1.10: The negative binomial distribution, NBI(μ, σ), with $\mu = 8$ and $\sigma = 0.25$: (a) The probability function (pf), (b) the cumulative distribution function (cdf), (c) the quantile function (i.e. inverse cdf), and (d) a histogram of a random sample of 1000 observations.

In Chapter 2 we discuss the properties of distributions. Chapter 3 describes all GAMLSS family distributions currently available in the **gamlss.dist** package and how more distributions can be created by transformation, truncation and censoring using GAMLSS packages.

1.7 Bibliographic notes

The material in this chapter is an introduction to distributions. Further information can be found online or in introductory books on probability or statistics. For continuous distributions, classic references are Johnson et al. [1994] and Johnson et al. [1995]. For discrete distributions a classic reference is Johnson et al. [2005]. A comprehensive thesaurus of 750 discrete distributions is given by Wimmer and Altmann [1999]. Other books on distributions are Forbes et al. [2011], Krishnamoorthy [2016] and Balakrishnan and Nevzorov [2003]. Stahl [2006] presents the evolution of the normal distribution. Generating new distributions has attracted theoretical and applied statisticians; see, for example Tahir and Nadarajah [2015] and Cordeiro et al. [2017].

1.8 Exercises

1. Let Y represent the variable `aircont` in Section 1.2.4. The fitted exponential distribution for Y has mean $\mu = 64.125$.

   (a) Use the cdf of the fitted exponential distribution to find (i) $P(Y < 50)$, (ii) $P(Y > 150)$ and (iii) $P(35 < Y < 100)$.

   (b) Plot the histogram and pdf, cdf, and quantile function (i.e. inverse cdf) for the fitted exponential distribution.

   (c) Plot the survival function and the hazard function of the fitted exponential distribution.

2. Let Y represent the count of deaths in the Prussian data set in Section 1.2.5. The fitted Poisson distribution for Y has parameter $\mu = 0.7$.

   (a) Use the fitted distribution to calculate (i) $P(Y = 0)$, (ii) $P(Y \leq 3)$, and (iii) $P(0 < Y \leq 4)$.

   (b) Plot the pf (with a bar plot of the data), cdf, and quantile function for the fitted Poisson distribution.

3. Plot the pdf, cdf, and quantile function, and a histogram of 1000 randomly generated values from each of the following continuous distributions: normal, $\texttt{NO}(0, 1)$; inverse Gaussian, $\texttt{IG}(1, 2)$; Gumbel, $\texttt{GU}(1, 2)$; and reverse Gumbel, $\texttt{RG}(1, 2)$.

4. Plot the pf, cdf, and quantile function, and a histogram of 1000 randomly generated values from each of the following discrete distributions: Poisson inverse Gaussian, $\texttt{PIG}(5, 1)$; Sichel, $\texttt{SICHEL}(5, 2, 1)$; and double Poisson, $\texttt{DPO}(5, 2)$.

2

Properties of distributions

CONTENTS

This chapter provides an introduction to properties of the distributions:

1. moment based properties;
2. moment, cumulant, and probability generating functions; and
3. centile based properties.

2.1 Introduction

In this chapter we are concerned with some general distributional properties, including measures of location, scale, skewness, and kurtosis. For parametric distributions it is common that these measures are related to some or all of the parameters.

It is assumed that the response variable Y comes from a population distribution which can be modeled by a theoretical probability (density) function $f(y \mid \boldsymbol{\theta})$, where the dimension of the parameter vector $\boldsymbol{\theta}$ has no restriction, but in practice is generally limited to a maximum of four, denoted as $\boldsymbol{\theta}^\top = (\mu, \sigma, \nu, \tau)$. Limiting $\boldsymbol{\theta}$ to four dimensions is not usually a serious restriction. The parameters μ, σ, ν, and τ usually represent location, scale, skewness, and kurtosis parameters, respectively, although more generally they can be any parameters of a distribution.

A *location* parameter usually represents the 'center' of the distribution, and is often:

- the *mean*, the average of Y, or expected value of Y, or
- the *median*, the value of Y which cuts the distribution in two halves, each with probability 0.50, or
- the *mode*, the value of Y which has the highest value of the probability (density) function.

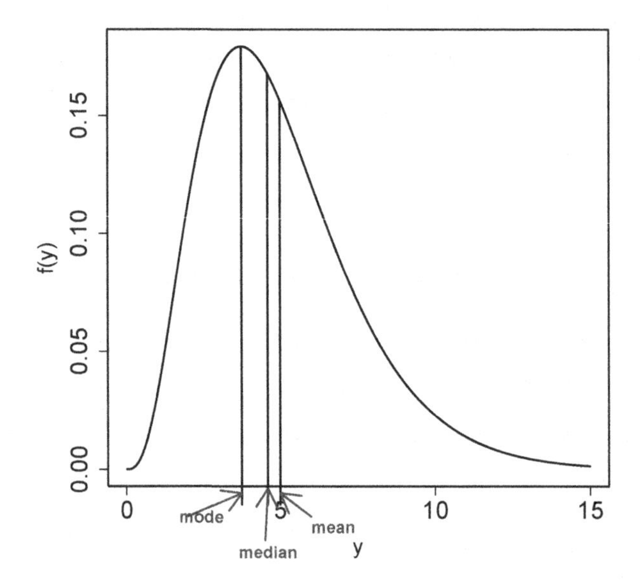

FIGURE 2.1: The mean, median, and mode of a positively skewed distribution (GA$(5, 0.5)$).

For symmetric unimodal distributions the three location measures coincide, but this is not the case for an asymmetric distribution, as Figure 2.1 shows.

A *scale* parameter usually represents or is related to the 'spread' of the distribution. Occasionally it is the standard deviation or the coefficient of variation of the distribution.

The *skewness* is a measure of distributional asymmetry. Informally, a distribution with a heavier tail to the right than the left usually has positive skewness, while one with a heavier tail to the left usually has negative skewness, and a symmetric distribution has zero skewness. An example of a distribution that can be symmetric, positively or negatively skewed, is the skew normal type 2 distribution (SN2, Section 18.3.5), shown in Figure 2.2(a). For a discussion of skewness see Chapter 14.

Kurtosis is a measure primary of heavy tails. Informally a distribution with heavy (i.e. fat) tails (relative to a normal distribution) will usually have high kurtosis (*leptokurtosis*), while a distribution with light (i.e. thin) tails (relative to a normal distribution)

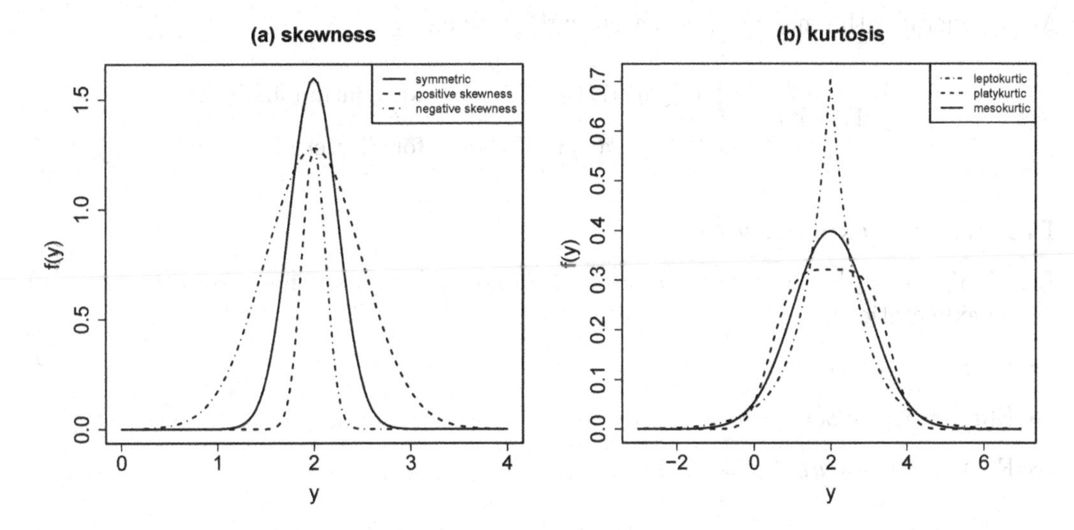

FIGURE 2.2: (a) Skew-normal type 2 distribution: symmetric, positively skewed, and negatively skewed; (b) Power exponential distribution: leptokurtic, platykurtic, and mesokurtic

will usually have low kurtosis (*platykurtosis*). Note that leptokurtic, and platykurtic distributions are judged by comparison to the normal distribution, which is called *mesokurtic*. Figure 2.2(b) shows an example of a distribution which can be leptokurtic, mesokurtic, or platykurtic. This is the power exponential (PE) distribution, see Section 18.3.3. For a discussion of kurtosis see Chapter 15.

Skewness and kurtosis measures can be either moment or centile based, see Sections 2.2 and 2.3, respectively.

2.2 Mean, variance, and moment measures of skewness and kurtosis

In this section we omit conditioning on the parameter vector $\boldsymbol{\theta}$, for simplicity of notation. So, for example, we use $f(y)$ throughout, which has the meaning $f(y \mid \boldsymbol{\theta})$. The mean, variance, and other distributional properties are also conditional on $\boldsymbol{\theta}$, but this conditioning is not explicitly denoted.

2.2.1 Mean or expected value

The population mean or expected value of Y is denoted by $\text{E}(Y)$ and is given by

$$\text{E}(Y) = \begin{cases} \int_{-\infty}^{\infty} y f(y)\, dy & \text{for a continuous random variable } Y \\ \sum_{y \in R_Y} y \, \text{P}(Y = y) & \text{for a discrete random variable } Y \,. \end{cases}$$

More generally, the mean of a function $g(Y)$ is given by

$$\mathrm{E}\left[g(Y)\right] = \begin{cases} \int_{-\infty}^{\infty} g(y) f(y)\, dy & \text{for continuous } Y \\ \sum_{y \epsilon R_Y} g(y)\, \mathrm{P}(Y=y) & \text{for discrete } Y\ . \end{cases}$$

Properties of expectations

Let Y, Y_1 and Y_2 be random variables and a and b constants. Then (from the linearity of the expectation):

- $\mathrm{E}(a) = a$
- $\mathrm{E}(aY + b) = a\mathrm{E}(Y) + b$
- $\mathrm{E}(aY_1 + bY_2) = a\mathrm{E}(Y_1) + b\mathrm{E}(Y_2)$.

2.2.2 Variance and standard deviation

The population variance of Y is defined as

$$\mathrm{Var}(Y) = \mathrm{E}\left\{[Y - \mathrm{E}(Y)]^2\right\}.$$

Hence

$$\mathrm{Var}(Y) = \begin{cases} \int_{-\infty}^{\infty} \left[y - \mathrm{E}(Y)\right]^2 f(y)\, dy & \text{for continuous } Y \\ \sum_{y \in R_Y} \left[y - \mathrm{E}(Y)\right]^2 \mathrm{P}(Y=y) & \text{for discrete } Y\ . \end{cases}$$

The standard deviation of Y is $\mathrm{SD}(Y) = \sqrt{\mathrm{Var}(Y)}$. Both variance and standard deviation give a measure of the spread of the distribution about the mean. The variance is the average squared distance of Y to the mean of Y.

Properties of variances

Let, Y, Y_1 and Y_2 be random variables and a and b constants. Then:

- $\mathrm{Var}(Y) = \mathrm{E}\left(Y^2\right) - \left[\mathrm{E}\left(Y\right)\right]^2$
- $\mathrm{Var}(a) = 0$
- $\mathrm{Var}(aY + b) = \mathrm{Var}(aY) = a^2\mathrm{Var}(Y)$
- If Y_1 and Y_2 are independent then $\mathrm{Var}(aY_1 + bY_2) = a^2\mathrm{Var}(Y_1) + b^2\mathrm{Var}(Y_2)$.

2.2.3 Moments

The kth population raw moment (or kth moment about zero) of Y is given by

$$\mu_k{}' = \mathrm{E}\left(Y^k\right) \qquad \text{for } k = 1, 2, 3, \ldots$$

Hence $\mu_1{}' = \mathrm{E}\left(Y\right)$.

The kth central moment (or kth moment about the mean) of Y is

$$\mu_k = \mathrm{E}\left\{[Y - \mathrm{E}(Y)]^k\right\} \qquad \text{for } k = 2, 3, \ldots .$$

Hence $\mu_2 = \mathrm{Var}(Y)$. For symmetric distributions, all odd central moments are zero (if they exist), i.e. $\mu_k = 0$ for $k = 1, 3, 5, \ldots$. Note that $\mu_1' = \mathrm{E}(Y)$ should not be confused with the distribution parameter μ, which may or may not represent the population mean of Y.

Relationship between central moments and moments about zero

$$\begin{aligned} \mu_2 &= {\mu_2}' - {\mu_1}'^2 \\ \mu_3 &= {\mu_3}' - 3{\mu_2}'{\mu_1}' + 2{\mu_1}'^3 \\ \mu_4 &= {\mu_4}' - 4{\mu_3}'{\mu_1}' + 6{\mu_2}'{\mu_1}'^2 - 3{\mu_1}'^4 \end{aligned} \qquad (2.1)$$

Hence

$$\begin{aligned} \mathrm{Var}(Y) &= \mathrm{E}(Y^2) - [\mathrm{E}(Y)]^2 \\ \mu_3 &= {\mu_3}' - 3\mathrm{Var}(Y)\mathrm{E}(Y) - [\mathrm{E}(Y)]^3 \\ \mu_4 &= {\mu_4}' - 4{\mu_3}'\mathrm{E}(Y) + 6\mathrm{Var}(Y)[\mathrm{E}(Y)]^2 + 3[\mathrm{E}(Y)]^4 \end{aligned} \qquad (2.2)$$

Relationship between moments about zero and central moments

$$\begin{aligned} {\mu_2}' &= \mu_2 + {\mu_1}'^2 \\ {\mu_3}' &= \mu_3 + 3\mu_2{\mu_1}' + {\mu_1}'^3 \\ {\mu_4}' &= \mu_4 + 4\mu_3{\mu_1}' + 6\mu_2{\mu_1}'^2 + {\mu_1}'^4 \end{aligned} \qquad (2.3)$$

2.2.4 Moment skewness and kurtosis

The population moment skewness is defined as

$$\gamma_1 = \sqrt{\beta_1} = \frac{\mu_3}{{\mu_2}^{1.5}} .$$

For symmetric distributions, $\mu_3 = 0$ and so $\gamma_1 = 0$, provided that μ_3 exists. Distributions with negative moment skewness have $\gamma_1 < 0$, and conversely distributions with positive moment skewness $\gamma_1 > 0$.

The population moment kurtosis is defined as

$$\beta_2 = \frac{\mu_4}{{\mu_2}^2} .$$

For the normal distribution, $\beta_2 = 3$, which leads to the definition of the *moment excess kurtosis*:

$$\gamma_2 = \frac{\mu_4}{{\mu_2}^2} - 3 ,$$

which measures the moment excess kurtosis relative to the normal distribution.

One of the problems with moment based properties of a distribution is the fact that for some distributions, moments do not exist (because the integrals used to define them do not exist). This is common for very heavy-tailed distributions, for example the Cauchy distribution, which is a t distribution with one degree of freedom. The centile based properties described in Section 2.3 avoid this problem because they always exist.

2.3 Centiles and centile measures of spread, skewness, and kurtosis

2.3.1 Centile and quantile

Continuous random variable For a continuous random variable Y, the $100p$th centile (or p quantile) of Y is the value y_p such that

$$P(Y \leq y_p) = p \ ,$$

i.e. $F(y_p) = p$ and hence $y_p = F^{-1}(p)$, where $F^{-1}(\cdot)$ is the *inverse cumulative distribution function*, evaluated at $0 < p < 1$. Hence the centile or quantile function is the inverse cdf. For example the 5th centile, $y_{0.05}$, is the value of Y for which the probability of being at or below $y_{0.05}$ is 0.05 (i.e. 5%) and is defined by $F(y_{0.05}) = 0.05$, i.e. $y_{0.05} = F^{-1}(0.05)$.

Using the exponential distribution as an example, to find y_p we solve $F(y_p) = p$, for $0 < p < 1$, i.e. using equation (1.5),

$$1 - \exp(-y_p/\mu) = p \ ,$$

so

$$y_p = -\mu \log(1-p) \ .$$

The cdf of an exponential distribution with $\mu = 0.5$ is obtained from (1.5) and plotted in Figure 2.3 (a). To compute the 60th centile (or 0.6 quantile) of the exponential distribution with $\mu = 0.5$ use:

```
qEXP(0.6, mu=0.5)
## [1] 0.4581454
```

Hence the 60th centile (or 0.6 quantile) of the exponential distribution with $\mu = 0.5$ is $y_{0.6} = 0.4581$ and is displayed in Figure 2.3 (a).

Discrete random variable

For a discrete random variable Y the $100p$th centile (or p quantile) y_p is defined by

$$P(Y < y_p) < p \text{ and } P(Y \leq y_p) \geq p, \text{ i.e.}$$

$$y_p = \text{smallest value of } Y \text{ for which } F(y) \geq p \ . \qquad (2.4)$$

For a discrete random variable Y, strictly speaking mathematically the inverse cdf $y_p = F^{-1}(p)$ does not exist for all p. For example, the cdf of a Poisson distribution with

$\mu = 2$, graphed in Figure 2.3(b) has values (shown up to y just above 6):

$$F(y) = \begin{cases} 0 & y < 0 \\ 0.135 & 0 \leq y < 1 \\ 0.406 & 1 \leq y < 2 \\ 0.677 & 2 \leq y < 3 \\ 0.857 & 3 \leq y < 4 \\ 0.947 & 4 \leq y < 5 \\ 0.983 & 5 \leq y < 6 \\ 0.995 & 6 \leq y < 7 \, . \end{cases}$$

The corresponding probability function $\mathrm{P}(Y = y)$ is plotted in Figure 1.7(a).

If, for example, we wanted the 60th centile (or 0.6 quantile), this would be $y_{0.6} = F^{-1}(0.6)$, i.e. the value of y at which $F(y) = 0.6$, which does not exist. Hence, $y_{0.6}$ is defined as the smallest value of y for which $F(y)$ equates or exceeds 0.6, which is 2. To compute the 60th centile (or 0.6 quantile) of a Poisson distribution with $\mu = 2$:

```
qPO(.60, mu=2)

## [1] 2
```

Hence the 60th centile (or 0.6 quantile) of a Poisson distribution with $\mu = 2$ is $y_{0.6} = 2$ and is displayed in Figure 2.3 (b).

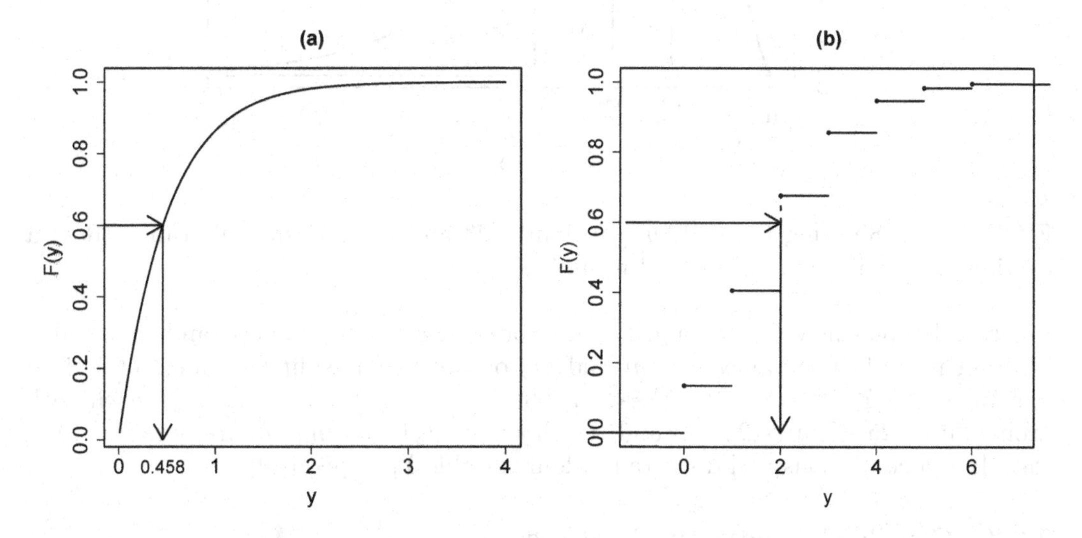

FIGURE 2.3: (a) The 60th centile of the exponential ($\mu = 0.5$) distribution, $y_{0.6} = 0.458$ and (b) the 60th centile of the Poisson ($\mu = 2$) distribution, $y_{0.6} = 2$.

2.3.2 Median, first quartile, and third quartile

For a continuous random variable Y, the median m is the value of Y for which $\mathrm{P}(Y \leq m) = 0.5$. Similarly the first and third quartiles, Q_1 and Q_3, are given by $\mathrm{P}(Y \leq Q_1) =$

0.25 and $P(Y \leq Q_3) = 0.75$. Hence

$$\begin{aligned} Q_1 &= y_{0.25} = F^{-1}(0.25) \\ m &= y_{0.5} = F^{-1}(0.5) \\ Q_3 &= y_{0.75} = F^{-1}(0.75) \ . \end{aligned}$$

For a discrete random variable Y, the definitions of m, Q_1 and Q_3 follow (2.4), i.e.

$$m = y_{0.5} = \text{smallest value of } y \text{ for which } F(y) \geq 0.5 \ ,$$

with corresponding definitions for $Q1$ and $Q3$.

The *interquartile* range is $\text{IR} = Q_3 - Q_1$.

The *semi-interquartile* range, $\text{SIR} = (Q_3 - Q_1)/2$, is a measure of spread.

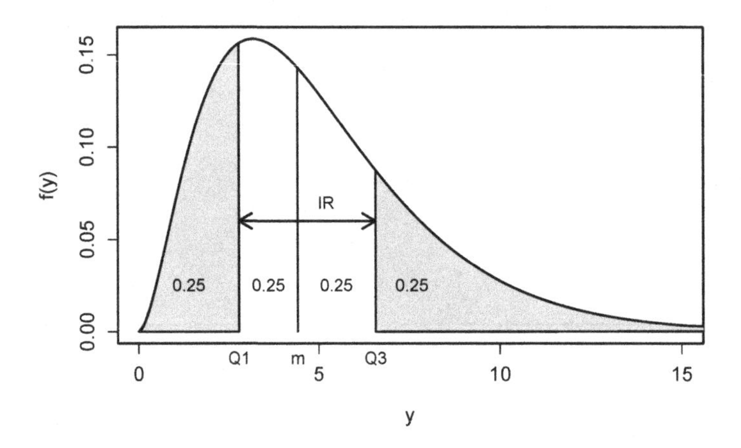

FIGURE 2.4: Showing how $Q1$, m (median), $Q3$ and the interquartile range IR of a continuous distribution are derived from $f(y)$.

Figure 2.4 shows how Q_1, m, and Q_3 are obtained for a continuous random variable Y, from its pdf. The values Q_1, m and Q_3 of variable Y split the range of Y into four intervals $(Y < Q_1)$, $(Q_1 < Y < m)$, $(m < Y < Q_3)$, and $(Y > Q_3)$, each with probability 0.25. Figures 2.5 (a) and (b) show how $Q1$, m, and $Q3$ are obtained from the cdf of a continuous and discrete random variable Y, respectively.

2.3.3 Centile skewness and kurtosis

Centile measures of skewness and kurtosis are given by the following.

Galton's centile skewness measure γ is defined as

$$\gamma = \frac{(Q_3 + Q_1)/2 - m}{\text{SIR}} \ . \tag{2.5}$$

Hence γ is the mid-quartile minus the median, divided by the semi-interquartile range.

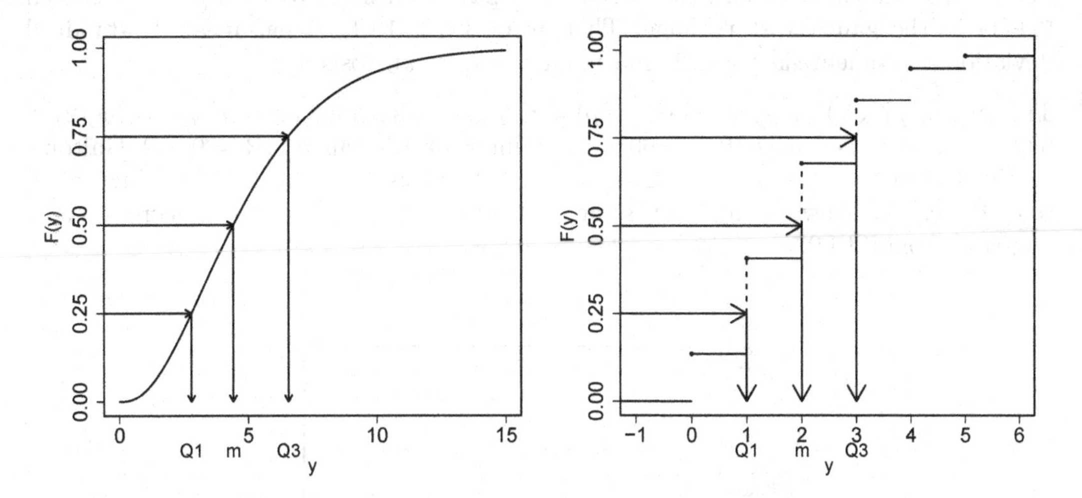

FIGURE 2.5: Showing how Q_1, m and Q_3 are derived for (a) a continuous and (b) a discrete distribution, from the cdf.

Andrew's centile kurtosis measure is given by Andrews et al. [1972], denoted here as δ, is defined as

$$\delta = \frac{y_{0.99} - y_{0.01}}{\text{IR}} .$$

Hence δ is the length of a central 98% interval, divided by the interquartile range (a central 50% interval).

Centile skewness function

A *centile skewness function* of Y is given by MacGillivray [1986]:

$$s_p = s_p(Y) = \frac{(y_p + y_{1-p})/2 - y_{0.5}}{(y_{1-p} - y_p)/2} \qquad \text{for } 0 < p < 0.5 \;, \tag{2.6}$$

i.e. the midpoint of a central $100(1-2p)\%$ interval minus the median, divided by the half length of the central $100(1-2p)\%$ interval. Note that $-1 \le s_p \le 1$.

One important case is $p = 0.25$, giving Galton's centile skewness measure (2.5) i.e. $\gamma = s_p(0.25)$. This can be considered as a central *centile skewness measure* since it focuses on the skewness within the interquartile range.

A second important case is $p = 0.01$, giving

$$s_{0.01} = \frac{(y_{0.01} + y_{0.99})/2 - y_{0.5}}{(y_{0.99} - y_{0.01})/2} \tag{2.7}$$

i.e. the midpoint of a central 98% interval minus the median, divided by the half length of the central 98% interval. This can be considered as a *tail centile skewness measure* since it focuses on skewness within a central 98% interval. A third important case is $p = 0.001$ which is an *extreme tail centile skewness measure.*

Example 1

Let Y have the exponential distribution with pdf given by (1.3) with $\mu = 1$, denoted EXP(1) in the **gamlss.dist** package. Then from Table 19.1, Y has mean 1, standard deviation 1, moment skewness 2, and moment excess kurtosis 6.

The inverse cdf of Y is $y_p = -\log(1-p)$ and hence Y has median $m = y_{0.5} = \log(2) = 0.693$, interquartile range IR $= 1.0986$, semi-interquartile range SIR $= 0.549$, Galton's centile skewness $\gamma = s_{0.25} = 0.262$, and centile kurtosis $\delta = k_{0.01} = 4.183$. Figure 2.6 plots the centile skewness function s_p against p for $0 < p < 0.5$. Note s_p decreases as p increases from 0 to 0.5.

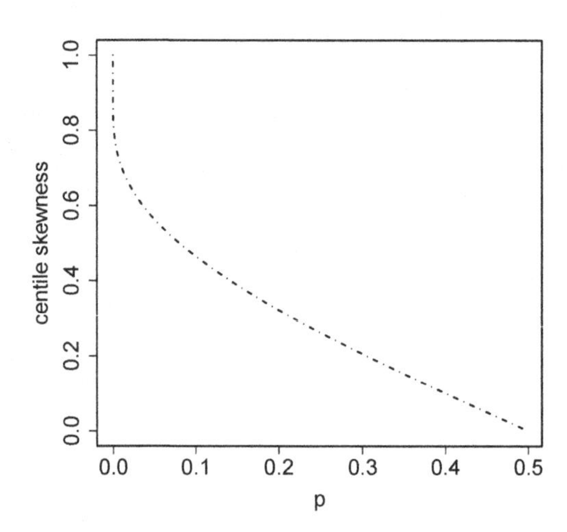

FIGURE 2.6: Centile skewness function s_p against p for $0 < p < 0.5$ for the exponential distribution, EXP(1).

Centile kurtosis function

Following Balanda and MacGillivray [1988], a *centile kurtosis function* of Y is given by Andrews et al. [1972]:

$$k_p = k_p(Y) = \frac{y_{1-p} - y_p}{Q_3 - Q_1} \qquad \text{for } 0 < p < 0.5 \ , \tag{2.8}$$

i.e. the length of a central $100(1-2p)\%$ interval divided by its interquartile range. An important case is $p = 0.01$, giving Andrews et al. [1972] centile kurtosis measure, i.e. $\delta = k_{0.01}$. This has been standardized relative to a normal distribution for which $k_{0.01} = 3.449$, giving

$$sk_{0.01} = \frac{y_{0.99} - y_{0.01}}{3.449\,(Q_3 - Q_1)} \ , \tag{2.9}$$

Rosenberger and Gasko [1983]. Hence a normal distribution has $sk_{0.01} = 1$.

More generally, the centile kurtosis function $k_p(Y)$ for Y has been standardized relative to a normal distribution giving the *standardized centile kurtosis function* $sk_p(Y)$:

$$sk_p(Y) = \frac{k_p(Y)}{k_p(Z)} \ , \tag{2.10}$$

for $0 < p < 0.5$, where Z has a standard normal distribution, denoted $Z \sim \texttt{NO}(0,1)$ in the **gamlss.dist** package. Hence $sk_p(Y)$ gives the ratio of the lengths of the central $100(1-2p)\%$ intervals of Y and Z, when both are scaled to have the same SIR.

Example 2

Let $Z \sim \texttt{NO}(0,1)$, then, from Table 18.3, Z has mean 0, standard deviation 1, moment skewness 0, and moment excess kurtosis 0. The inverse cdf of Z is denoted $z_p = \Phi^{-1}(p)$ and hence Z has median $m = 0$, interquartile range IR=1.349, SIR=0.674, Galton's centile skewness $\gamma = s_{0.25} = 0$, and Andrew's centile kurtosis $\delta = k_{0.01} = 3.449$. Hence the length of the central 98% interval for Z is 3.449 times its interquartile range (the central 50% interval).

Let T have a t distribution with 5 degrees of freedom, denoted $T \sim \texttt{TF}(0,1,5)$ in the **gamlss.dist** package, then from Table 18.11, T has mean 0, standard deviation 1.291, moment skewness 0, and moment excess kurtosis 6. Also T has median $m = 0$, interquartile range IR=1.453, SIR=0.727, Galton's centile skewness $\gamma = 0$, and Andrew's centile kurtosis $\delta = 4.631$. Hence the length of the central 98% interval for T is 4.631 times its interquartile range. Figure 2.7 (a) plots the centile kurtosis function k_p against p, for $0 < p < 0.5$, for the standard normal distribution, $\texttt{NO}(0,1)$, and t distribution with 5 degrees of freedom, $\texttt{TF}(0,1,5)$.

The standardized centile kurtosis of T at $p = 0.01$ is

$$sk_{0.01}(T) = k_{0.01}(T)/k_{0.01}(Z) = 4.631/3.449 = 1.343 \ .$$

Hence the ratio of the lengths of central 98% intervals for T and Z, when both are scaled to have the same SIR, is 1.343. Figure 2.7 (b) plots the standardized centile kurtosis function $sk_p(T)$ against p, for $0 < p < 0.5$, for $T \sim \texttt{TF}(0,1,5)$. Note that for very small p, $sk_p(T)$ becomes very large.

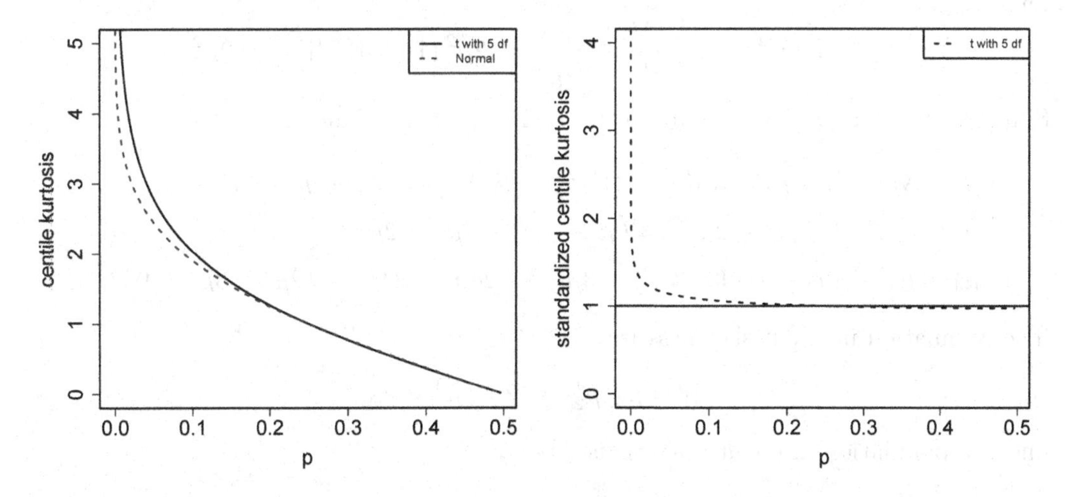

FIGURE 2.7: (a) Centile kurtosis function k_p against p, for $\texttt{NO}(0,1)$ and $\texttt{TF}(0,1,5)$, (b) standardized centile kurtosis function sk_p against p for $\texttt{TF}(0,1,5)$.

2.4 Moment, cumulant, and probability generating functions

2.4.1 Moment generating function

The moment generating function of Y is given by

$$M_Y(t) = \mathrm{E}\left(e^{tY}\right) \tag{2.11}$$

provided it exists. It is called the moment generating function because setting $t = 0$ in its rth derivative with respect to t gives $\mathrm{E}\left(Y^r\right)$, i.e.

$$\mathrm{E}\left(Y^r\right) = \left.\frac{d^{(r)} M_Y(t)}{dt^{(r)}}\right|_{t=0} . \tag{2.12}$$

Example

Let $Y \sim$ `EXP`(μ) with pdf (1.3). The moment generating function of Y is given by:

$$\begin{aligned} M_Y(t) = \mathrm{E}\left(e^{tY}\right) &= \int_0^\infty e^{ty} \frac{1}{\mu} e^{-y/\mu}\, dy \\ &= \frac{1}{\mu} \int_0^\infty e^{-y\left(\frac{1}{\mu}-t\right)}\, dy \\ &= (1-\mu t)^{-1} , \end{aligned}$$

provided $t < 1/\mu$. Hence from (2.12),

$$\mu_1{}' = \mathrm{E}(Y) = \left.\frac{dM_Y(y)}{dt}\right|_{t=0} = \left.\mu\,(1-\mu t)^{-2}\right|_{t=0} = \mu$$

and

$$\mu_2{}' = \mathrm{E}\left(Y^2\right) = \left.\frac{d^{(2)} M_Y(y)}{dt^{(2)}}\right|_{t=0} = \left.2\mu^2\,(1-\mu t)^{-3}\right|_{t=0} = 2\mu^2 .$$

Similarly $\mu_3{}' = \mathrm{E}\left(Y^3\right) = 6\mu^3$ and $\mu_4{}' = \mathrm{E}\left(Y^4\right) = 24\mu^4$. Hence from (2.1),

$$\begin{aligned} \mu_2 &= \mathrm{Var}(Y) = \mu_2{}' - \mu_1{}'^2 = \mathrm{E}\left(Y^2\right) - \left[\mathrm{E}(Y)\right]^2 = 2\mu^2 - \mu^2 = \mu^2 \\ \mu_3 &= \mu_3{}' - 3\mu_2{}'\mu_1{}' + 2\mu_1{}'^3 = 6\mu^3 - 6\mu^3 + 2\mu^3 = 2\mu^3 \\ \mu_4 &= \mu_4{}' - 4\mu_3{}'\mu_1{}' + 6\mu_2{}'\mu_1{}'^2 - 3\mu_1{}'^4 = 24\mu^4 - 24\mu^4 + 12\mu^4 - 3\mu^4 = 9\mu^4 . \end{aligned}$$

The population moment skewness is

$$\gamma_1 = \mu_3/\mu_2^{1.5} = 2\mu^3/\mu^3 = 2 ,$$

and the population moment kurtosis is

$$\beta_2 = \mu_4/\mu_2^2 = 9\mu^4/\mu^4 = 9$$

with moment excess kurtosis $\gamma_2 = \beta_2 - 3 = 6$.

This is a more advanced topic and can be omitted by more practical readers.

2.4.2 Cumulant generating function

The cumulant generating function of Y is given by

$$\mathcal{K}_Y(t) = \log\left[M_Y(t)\right]$$

provided it exists. It is called the cumulant generating function because setting $t = 0$ in its rth derivative with respect to t gives κ_r, the kth cumulant of Y:

$$\kappa_r = \left.\frac{d^{(r)}\mathcal{K}_Y(t)}{dt^{(r)}}\right|_{t=0} . \tag{2.13}$$

Relationship between cumulants and moments

$$\begin{aligned} \kappa_1 &= \mu_1{}' = \mathrm{E}\left(Y\right) \\ \kappa_2 &= \mu_2 = \mathrm{Var}(Y) \\ \kappa_3 &= \mu_3 \\ \kappa_4 &= \mu_4 - 3\mu_2^2 . \end{aligned} \tag{2.14}$$

Hence the population moment skewness and moment excess kurtosis of Y are $\gamma_1 = \kappa_3/\kappa_2^{1.5}$ and $\gamma_2 = \kappa_4/\kappa_2^2$, respectively.

Example

Let $Y \sim \texttt{EXP}(\mu)$ then $\mathcal{K}_Y(t) = \log[(M_Y(t))] = -\log\left(1 - \mu t\right)$, provided $t < 1/\mu$. Hence from (2.13) and (2.14),

$$\begin{aligned} \kappa_1 &= \left.\mu\left(1-\mu t\right)^{-1}\right|_{t=0} = \mu, \text{ so } \mathrm{E}\left(Y\right) = \mu \\ \kappa_2 &= \left.\mu^2\left(1-\mu t\right)^{-2}\right|_{t=0} = \mu^2, \text{ so } \mathrm{Var}\left(Y\right) = \mu^2 \\ \kappa_3 &= \left.2\mu^3\left(1-\mu t\right)^{-3}\right|_{t=0} = 2\mu^3, \text{ so } \mu_3 = 2\mu^3 \\ \kappa_4 &= \left.6\mu^4\left(1-\mu t\right)^{-4}\right|_{t=0} = 6\mu^4, \text{ so } \mu_4 = 9\mu^4 . \end{aligned}$$

The population moment skewness is $\gamma_1 = \kappa_3/\kappa_2^{1.5} = 2$ and the moment excess kurtosis is $\gamma_2 = \kappa_4/\kappa_2^2 = 6$.

2.4.3 Probability generating function

The probability generating function (PGF) of a discrete random variable Y is given by

$$G_Y(t) = \mathrm{E}\left(t^Y\right) \tag{2.15}$$

provided it exists. It is called the probability generating function because setting $t = 0$ in its rth derivative with respect to t gives

$$\mathrm{P}(Y = r) = \left.\frac{1}{r!}\frac{d^{(r)}G_Y(t)}{dt^{(r)}}\right|_{t=0} . \tag{2.16}$$

Note that $G_Y(t) = M_Y(\log t)$ and $M_Y(t) = G_Y(e^t)$.

Example

Let $Y \sim$ PO(μ), with probability function (1.4). The PGF of Y is given by

$$\begin{aligned} G_Y(t) = \mathrm{E}\left(t^Y\right) &= \sum_{y=0}^{\infty} t^y \frac{e^{-\mu}\mu^y}{y!} \\ &= e^{-\mu} \sum_{y=0}^{\infty} \frac{(t\mu)^y}{y!} \\ &= e^{-\mu} e^{\mu t} = e^{\mu(t-1)}. \end{aligned}$$

Hence from (2.16),

$$\begin{aligned} \mathrm{P}(Y=0) &= G_Y(t)\big|_{t=0} = e^{\mu(t-1)}\Big|_{t=0} = e^{-\mu} \\ \mathrm{P}(Y=1) &= \frac{dG_Y(t)}{dt}\bigg|_{t=0} = \mu e^{\mu(t-1)}\Big|_{t=0} = e^{-\mu}\mu \\ \mathrm{P}(Y=2) &= \frac{1}{2!}\frac{d^{(2)}G_Y(t)}{dt^{(2)}}\bigg|_{t=0} = \frac{1}{2!}\mu^2\, e^{\mu(t-1)}\Big|_{t=0} = \frac{e^{-\mu}\mu^2}{2!}. \end{aligned}$$

2.4.4 Properties of MGF and PGF

Let Y, Y_1 and Y_2 be random variables and a and b constants.

- Let $Y = a + bY_1$, then $M_Y(t) = e^{at} M_{Y_1}(bt)$.
- Let $Y = aY_1 + bY_2$ where Y_1 and Y_2 are independent random variables. Then $M_Y(t) = M_{Y_1}(at) M_{Y_2}(bt)$ for values of t for which $M_{Y_1}(at)$ and $M_{Y_1}(bt)$ exist. Also $G_Y(t) = G_{Y_1}(t^a)\, G_{Y_2}(t^b)$ for values of t for which $G_{Y_1}(t^a)$ and $G_{Y_2}(t^b)$ exist. In particular, if $Y = Y_1 + Y_2$ and Y_1 and Y_2 are independent, then $M_Y(t) = M_{Y_1}(t) M_{Y_2}(t)$ and $G_Y(t) = G_{Y_1}(t)\, G_{Y_2}(t)$.
- The MGF and PGF are useful for finding the distribution of finite, countably infinite and continuous mixture distributions.
- The PGF is useful for finding the distribution of a stopped sum.

Example 1
Let $Y_1 \sim$ EXP(μ), then $Y = bY_1$ for $b > 0$ has MGF given by $M_Y(t) = M_{Y_1}(bt) = (1 - b\mu t)^{-1}$ for $t < 1/b\mu$. Hence $Y \sim$ EXP($b\mu$).

Example 2
Let $Y_1 \sim$ PO(μ_1) and $Y_2 \sim$ PO(μ_2) where Y_1 and Y_2 are independent. Then $Y = Y_1 + Y_2$ has PGF given by $G_Y(t) = G_{Y_1}(t)\, G_{Y_2}(t) = e^{\mu_1(t-1)} e^{\mu_2(t-1)} = e^{(\mu_1+\mu_2)(t-1)}$. Hence $Y \sim$ PO($\mu_1 + \mu_2$).

2.5 Bibliographic notes

For properties of univariate discrete distributions see Johnson et al. [2005], and for univariate continuous distributions Johnson et al. [1994] and Johnson et al. [1995]. van Zwet [1964a] presents many properties of distributions in a sophisticated mathematical background. For properties of common probability distributions see Balakrishnan and Nevzorov [2003], Forbes et al. [2011], and Krishnamoorthy [2016].

Pearson [1895] first presented a moment measure of skewness. MacGillivray [1986] discusses measures and ordering of skewness and asymmetry. MacGillivray and Balanda [1988] and Balanda and MacGillivray [1990] discuss measures and ordering for kurtosis.

Smith [1995] presents recursive formulas that translate moments to cumulants and vice versa, in the univariate and multivariate situations.

2.6 Exercises

1. Let X, Y, and Z be independent random variables with finite variance. Show that

$$\mathrm{Var}(XY) = \mathrm{Var}(X)\mathrm{Var}(Y) + \mathrm{E}^2(X)\mathrm{Var}(Y) + \mathrm{E}^2(Y)\mathrm{Var}(X)\ .$$

2. Let $X_1, \ldots, X_n$ be independent random variables with finite variance. Show that $\mathrm{Var}(X_1 + \ldots + X_n) = \mathrm{Var}(X_1) + \ldots + \mathrm{Var}(X_n)$.

3. Let X, Y and Z be independent random variables with common distribution $U(0, 1)$. Calculate the expectation and variance of $W = (X + Y)Z$.

4. Prove the following theorem (the cumulant/moment connection). Let Y be a random variable with r moments about zero $\mu'_1, \mu'_2, \ldots, \mu'_r$. Then Y has r cumulants $\kappa_1, \kappa_2, \ldots, \kappa_r$, and

$$\mu'_r = \sum_{j=0}^{r-1} \binom{r-1}{j} \kappa_{r-j}\mu'_j$$

[Johnson et al., 1995, p 54.]

5. The probability density function of the Pareto type 1 distribution, `PARETO1o`(μ, σ) is given by

$$f(y) = \frac{\mu^\sigma}{y^{\sigma+1}}, \text{ for } y > \mu,\ \mu > 0 \text{ and } \sigma > 0.$$

   (a) Determine the mean and variance of a Pareto distributed random variable, `PARETO1o`(μ, σ).

   (b) For which values of μ and σ are the mean and the variance finite?

6. Let Y have a gamma, `GA`(μ, σ) distribution with pdf given by (19.2).

   (a) Find the moment generating function for Y.

(b) Hence, or otherwise, show that Y has mean μ, variance $\sigma^2\mu^2$, moment skewness 2σ, and moment excess kurtosis $6\sigma^2$.

7. The **gamlss.dist** functions `theoCentileSK()` and `plotCentileSK()` can be used to check the centile skewness and centile kurtosis of a given distribution.

   The function `theoCentileSK()` produces as output the IR, SIR, Galton's centile skewness $\gamma = s_{0.25}$, the tail centile skewness $s_{0.01}$, Andrews's centile kurtosis $\delta = k_{0.01}$, and the standardized centile kurtosis $sk_{0.01}$.

   The function `plotCentileSK()` can be used to plot centile skewness and centile kurtosis functions. For example Figure 2.6 in Section 2.3.3 can be created using:

```
plotCentileSK("EXP", "skew", mu=1)
```

   while Figures 2.7(a) and (b) can be generated using:

```
# (a)
plotCentileSK("NO", "kurt", mu=0, sigma=1)
plotCentileSK("TF", "kurt", mu=0, sigma=1, nu=5, add=T, lty=2)
# (b)
plotCentileSK("TF", "standKurt", mu=0, sigma=1,nu=5)
```

   (a) Find the centile measures for the gamma distribution, GA(μ, σ), with $\mu = 1$ and $\sigma = 2, 1, 0.5$, using for example `theoCentileSK("GA", mu=1,sigma=2)`.

   Plot the centile skewness and centile kurtosis functions of gamma distributions with $\mu = 1$ and $\sigma = 2, 1, 0.5$.

   (b) Repeat (a) for the Box-Cox t distribution, BCT(μ, σ, ν, τ), with $\mu = 1$, $\sigma = 0.1$, $\nu = -1, 1$, and $\tau = 5, 1$.

   (c) Comment on the results.

3

The GAMLSS family of distributions

CONTENTS

This chapter provides:

1. a definition of GAMLSS;
2. an introduction to different types of distributions within GAMLSS and in particular: continuous, discrete, and mixed distributions;
3. information on how to visualize the distributions; and
4. information on link functions in the **gamlss** package.

3.1 Introduction to the GAMLSS model

Generalized Additive Models for Location Scale, and Shape (GAMLSS) are distributional regression models where the response variable can have any distribution and all the parameters of the distribution can depend on explanatory variables. In this section we briefly explain the GAMLSS model, which was introduced by Rigby and Stasinopoulos [2005] and is discussed in Chapter 3 of Stasinopoulos et al. [2017]. For more information on the use of the model in practice see Stasinopoulos et al. [2017]. In the following sections and chapters we focus on the types of distributions for the response variable of the model.

The semiparametric form of the GAMLSS model is defined as follows. Let $Y_1, Y_2, \ldots, Y_n$ be independent response variable observations with $Y_i \sim \mathcal{D}(\mu_i, \sigma_i, \nu_i, \tau_i)$, for $i = 1, 2, \ldots, n$, where $\mathcal{D}$ is any distribution with (up to) four distribution parameters. This is written in vector form as:

$$\mathbf{Y} \overset{\text{ind}}{\sim} \mathcal{D}(\boldsymbol{\mu}, \boldsymbol{\sigma}, \boldsymbol{\nu}, \boldsymbol{\tau})$$

where

$$\begin{aligned}
g_1(\boldsymbol{\mu}) &= \boldsymbol{\eta}_1 = \mathbf{X}_1\boldsymbol{\beta}_1 + s_{11}(\mathbf{x}_{11}) + \ldots + s_{1J_1}(\mathbf{x}_{1J_1}) \\
g_2(\boldsymbol{\sigma}) &= \boldsymbol{\eta}_2 = \mathbf{X}_2\boldsymbol{\beta}_2 + s_{21}(\mathbf{x}_{21}) + \ldots + s_{2J_2}(\mathbf{x}_{2J_2}) \\
g_3(\boldsymbol{\nu}) &= \boldsymbol{\eta}_3 = \mathbf{X}_3\boldsymbol{\beta}_3 + s_{31}(\mathbf{x}_{31}) + \ldots + s_{3J_3}(\mathbf{x}_{3J_3}) \\
g_4(\boldsymbol{\tau}) &= \boldsymbol{\eta}_4 = \mathbf{X}_4\boldsymbol{\beta}_4 + s_{41}(\mathbf{x}_{41}) + \ldots + s_{4J_4}(\mathbf{x}_{4J_4}) \ .
\end{aligned} \tag{3.1}$$

- The vectors $\boldsymbol{\eta}_1$, $\boldsymbol{\eta}_2$, $\boldsymbol{\eta}_3$, and $\boldsymbol{\eta}_4$ are called the predictors of $\boldsymbol{\mu}$, $\boldsymbol{\sigma}$, $\boldsymbol{\nu}$, and $\boldsymbol{\tau}$, respectively;
- The $g_k(\cdot)$ are known monotonic link functions for $k = 1, 2, 3, 4$. See Table 3.1 for the link functions available in the **gamlss** package;
- $\mathbf{X}_k$ are the design matrices incorporating the linear additive terms in the model and $\boldsymbol{\beta}_k$ are the linear coefficient parameters. See Chapter 8 of Stasinopoulos et al. [2017];
- $s_{kj}(\mathbf{x}_{kj})$ represent smoothing functions, for explanatory variables $\mathbf{x}_{kj}$, for $k = 1, 2, 3, 4$ and $j = 1, 2, \ldots, J_k$. See Chapters 9 and 10 of Stasinopoulos et al. [2017].

Note that the quantitative explanatory variables in the $\mathbf{X}$'s can be the same or different from the $\mathbf{x}_{kj}$ in the smoothers.

Most of the smooth functions used within GAMLSS can be written in the form $s(\mathbf{x}) = \mathbf{Z}\boldsymbol{\gamma}$, where $\mathbf{Z}$ is the basis matrix which depends on the explanatory variable $\mathbf{x}$. (See Section 8.2 of Stasinopoulos et al. [2017] for the definition of a basis.) $\boldsymbol{\gamma}$ is a parameter vector to be estimated, subject to a quadratic penalty of the form $\lambda\boldsymbol{\gamma}^\top\mathbf{G}\boldsymbol{\gamma}$, where $\mathbf{G} = \mathbf{D}^\top\mathbf{D}$ is a known matrix and the hyperparameter λ regulates the amount of smoothing needed for the fit. We shall refer to functions in this form as *penalized smooth functions* (or *penalized smoothers*). Penalized smoothers are the subject of Chapter 9 of Stasinopoulos et al. [2017], where it is shown that different formulations for the $\mathbf{Z}$'s

TABLE 3.1: Link functions available within the **gamlss** packages, (with the usual range of the distribution parameters).

Parameter θ range	Link function name	Formula for $g(\theta)$
$-\infty$ to ∞	`identity`	θ
0 to ∞	`log`	$\log(\theta)$
	`sqrt`	$\sqrt{\theta}$
	`inverse`	$1/\theta$
	`'1/mu^2'`	$1/\theta^2$
	`'mu^2'`	θ^2
0 to 1	`logit`	$\log[\theta/(1-\theta)]$
	`probit`	$\Phi^{-1}(\theta)$
	`cauchit`	$\tan(\pi(\theta - 0.05))$
	`cloglog`	$\log(-\log(1-\theta))$
1 to ∞	`logshiftto1`	$\log(\theta - 1)$
2 to ∞	`logshiftto2`	$\log(\theta - 2)$
0.00001 to ∞	`logshiftto0` or `Slog`[a]	$\log(\theta - 0.00001)$
-1 to 1	`'[-1,1]'`[b]	$\log((\theta-1)/(1-\theta))$
0 to 2	`'(0,2]'`[b]	$\log(\theta/(2-\theta))$

[a] This function was created to avoid the value of positive parameters too close to zero.

[b] These two functions were created for the 'Stable' distribution see Section 13.8.2.

and the $\mathbf{D}$'s lead to different types of smoothing functions with different statistical properties.

The model (3.1) can be generalized and written as the random effects GAMLSS model:

$$\begin{aligned}
\mathbf{Y} &\overset{\text{ind}}{\sim} \mathcal{D}(\boldsymbol{\mu}, \boldsymbol{\sigma}, \boldsymbol{\nu}, \boldsymbol{\tau}) \\
g_1(\boldsymbol{\mu}) &= \boldsymbol{\eta}_1 = \mathbf{X}_1\boldsymbol{\beta}_1 + \mathbf{Z}_{11}\boldsymbol{\gamma}_{11} + \ldots + \mathbf{Z}_{1J_1}\boldsymbol{\gamma}_{1J_1} \\
g_2(\boldsymbol{\sigma}) &= \boldsymbol{\eta}_2 = \mathbf{X}_2\boldsymbol{\beta}_2 + \mathbf{Z}_{21}\boldsymbol{\gamma}_{21} + \ldots + \mathbf{Z}_{2J_2}\boldsymbol{\gamma}_{2J_2} \\
g_3(\boldsymbol{\nu}) &= \boldsymbol{\eta}_3 = \mathbf{X}_3\boldsymbol{\beta}_3 + \mathbf{Z}_{31}\boldsymbol{\gamma}_{31} + \ldots + \mathbf{Z}_{3J_3}\boldsymbol{\gamma}_{3J_3} \\
g_4(\boldsymbol{\tau}) &= \boldsymbol{\eta}_4 = \mathbf{X}_4\boldsymbol{\beta}_4 + \mathbf{Z}_{41}\boldsymbol{\gamma}_{41} + \ldots + \mathbf{Z}_{4J_4}\boldsymbol{\gamma}_{4J_4}
\end{aligned} \tag{3.2}$$

where the 'betas' are fixed effects parameters:

$$\boldsymbol{\beta} = (\boldsymbol{\beta}_1^\top, \boldsymbol{\beta}_2^\top, \boldsymbol{\beta}_3^\top, \boldsymbol{\beta}_4^\top)^\top$$

and the 'gammas' are random effect parameters:

$$\boldsymbol{\gamma} = (\boldsymbol{\gamma}_{11}^\top, \ldots, \boldsymbol{\gamma}_{1J_1}^\top, \boldsymbol{\gamma}_{21}^\top, \ldots, \boldsymbol{\gamma}_{4J_4}^\top)^\top$$

and $\mathbf{X}_k$ and $\mathbf{Z}_{kj}$ are design matrices for the fixed and random effects, respectively, for $k = 1, 2, 3, 4$ and $j = 1, 2, \ldots, J_k$. For most smooth functions, model (3.1) can be written in the form of model (3.2). Assume in (3.2) that the $\boldsymbol{\gamma}_{kj}$'s are independent of each other, each with (prior) distribution

$$\boldsymbol{\gamma}_{kj} \sim \mathcal{N}\left(\mathbf{0}, [\mathbf{G}_{kj}(\boldsymbol{\lambda}_{kj})]^{-1}\right) \tag{3.3}$$

where $[\mathbf{G}_{kj}(\boldsymbol{\lambda}_{kj})]^{-1}$ is the (generalized) inverse of a $q_{kj} \times q_{kj}$ symmetric matrix $\mathbf{G}_{kj}(\boldsymbol{\lambda}_{kj})$ which may depend on a vector of hyperparameters $\boldsymbol{\lambda}_{kj}$. If $\mathbf{G}_{kj}(\boldsymbol{\lambda}_{kj})$ is singular, then $\boldsymbol{\gamma}_{kj}$ is understood to have an improper density function proportional to $\exp\left(-\frac{1}{2}\boldsymbol{\gamma}^\top \mathbf{G}_{kj}(\boldsymbol{\lambda}_{kj})\boldsymbol{\gamma}\right)$. An important special case of (3.3) is given when $\mathbf{G}_{kj}(\boldsymbol{\lambda}_{kj}) = \lambda_{kj}\mathbf{G}_{kj}$ and $\mathbf{G}_{kj}$ is a known matrix, for all k, j.

If there are no random effects in the model (3.2) it simplifies to:

$$
\begin{aligned}
\mathbf{Y} &\overset{\text{ind}}{\sim} \mathcal{D}(\boldsymbol{\mu}, \boldsymbol{\sigma}, \boldsymbol{\nu}, \boldsymbol{\tau}) \\
g_1(\boldsymbol{\mu}) &= \boldsymbol{\eta}_1 = \mathbf{X}_1\boldsymbol{\beta}_1 \\
g_2(\boldsymbol{\sigma}) &= \boldsymbol{\eta}_2 = \mathbf{X}_2\boldsymbol{\beta}_2 \\
g_3(\boldsymbol{\nu}) &= \boldsymbol{\eta}_3 = \mathbf{X}_3\boldsymbol{\beta}_3 \\
g_4(\boldsymbol{\tau}) &= \boldsymbol{\eta}_4 = \mathbf{X}_4\boldsymbol{\beta}_4 \ .
\end{aligned}
\tag{3.4}
$$

We refer to (3.4) as the *parametric* GAMLSS model, and to (3.2) as the *random effects* GAMLSS model.

Fitting the parametric model (3.4) requires only estimates for the 'betas' $\boldsymbol{\beta}$. Fitting the random effects GAMLSS model (3.2) requires estimates for the 'betas' $\boldsymbol{\beta}$, the 'gammas' $\boldsymbol{\gamma}$, and also the 'lambdas' $\boldsymbol{\lambda}$ (see (3.6) below), where:

$$\boldsymbol{\lambda} = (\boldsymbol{\lambda}_{11}^\top, \ldots, \boldsymbol{\lambda}_{1J_1}^\top, \boldsymbol{\lambda}_{21}^\top, \ldots, \boldsymbol{\lambda}_{4J_4}^\top)^\top \ .$$

Within **gamlss**, model (3.4) is fitted by maximum likelihood estimation with respect to $\boldsymbol{\beta}$, while the more general random effects model (3.2) is fitted by maximum penalized likelihood estimation, or equivalently posterior mode or maximum a posteriori (MAP) estimation with respect to $\boldsymbol{\beta}$ and $\boldsymbol{\gamma}$ for fixed $\boldsymbol{\lambda}$. (See Rigby and Stasinopoulos [2005].) The log-likelihood function for model (3.4) is given by

$$\ell = \sum_{i=1}^{n} \log f(y_i \mid \mu_i, \sigma_i, \nu_i, \tau_i) \ . \tag{3.5}$$

The penalized log-likelihood function for model (3.2) is given by

$$\ell_p = \ell - \frac{1}{2}\sum_{k=1}^{4}\sum_{j=1}^{J_k} \boldsymbol{\gamma}_{kj}^\top \mathbf{G}_{kj}(\lambda_{kj})\,\boldsymbol{\gamma}_{kj} \ . \tag{3.6}$$

The two basic algorithms for fitting the parametric model (3.4) with respect to $\boldsymbol{\beta}$, and the random effects model (3.2) with respect to $\boldsymbol{\beta}$ and $\boldsymbol{\gamma}$ for fixed $\boldsymbol{\lambda}$, are the RS and CG algorithms described in detail in Chapter 3 of Stasinopoulos et al. [2017]. For the parametric model these produce maximum likelihood estimators for $\boldsymbol{\beta}$, whereas for the random effects model they produce MAP estimators for $\boldsymbol{\beta}$ and $\boldsymbol{\gamma}$ for fixed $\boldsymbol{\lambda}$. Estimation of $\boldsymbol{\lambda}$ is discussed in Stasinopoulos et al. [2017]. Section 3.4.

3.2 Types of distribution within the GAMLSS family

One of the great advantages of the GAMLSS framework, and its associated distributions in package **gamlsss.dist**, is its ability to fit a variety of different distributions to the response variable, so that an appropriate distribution can be chosen among different alternatives. Within the GAMLSS framework it is assumed that the probability (density) function is $f(y \mid \boldsymbol{\theta})$, where $\boldsymbol{\theta}$ is s a vector of up to four distribution parameters, i.e. $\boldsymbol{\theta} = (\mu, \sigma, \nu, \tau)$. There are over one hundred explicit distributions available in **gamlss.dist**, but more distributions can be generated automatically, as we will see in this chapter.

In addition, to create a new explicit distribution is relatively easy, see for example Section 6.4 of Stasinopoulos et al. [2017] describing how to construct a new distribution. The only restriction that the **gamlss** package has for specifying a new distribution is that $f(y \mid \boldsymbol{\theta})$ and its first derivatives with respect to each of the parameters in $\boldsymbol{\theta}$ must be computable. Explicit derivatives are preferable, but numerical derivatives can be used (resulting in reduced computational speed).

Note that for consistency of notation with the **gamlss** package, we denote the $\boldsymbol{\theta}$ parameters as μ, σ, ν, and τ. As we have mentioned in the previous chapter, the parameters μ, σ, ν, and τ represent in many distributions the location, scale, skewness, and kurtosis parameters, respectively, if such a one-to-one relationship exists. More generally μ, σ, ν, and τ are just convenient symbols for the parameters. The relationship between the distribution parameters (μ, σ, ν, and τ) and moment-based measures (mean, variance, skewness, and kurtosis) is related to the properties of the specific distribution. For example, for the gamma distribution, `GA`(μ, σ), as parameterized in **gamlss.dist**, the mean is given by the distribution parameter μ, while the variance is given by $\sigma^2\mu^2$ and therefore the standard deviation is given by $\text{SD}(Y) = \sigma\mu$ and hence the coefficient of variation is σ. It is important to note that although μ is often a location parameter, it is not in general the mean of the distribution. Similarly although σ is often a scale parameter, it is not in general the standard deviation of the distribution.

This chapter introduces all the available distributions within the current implementation of GAMLSS in **R**. Some of those distributions are *explicit* in the sense that their implementations exist within **gamlss.dist**, while others are *generated* from an existing explicit distribution. We refer to both types of distributions as the *GAMLSS family*, to be consistent with the **R** implementation, where the class of all these distributions is defined as the `gamlss.family`. The next section describes the explicit and generated distributions within the `gamlss.family`.

Note there also many important distributions which are special cases of `gamlss.family` distributions and can be fitted by fixing one (or more) parameters at specific values. For example the Cauchy distribution is a special case of the *t* family distribution, `TF`(μ, σ, ν), where $\nu = 1$. This can be fitted using the arguments, `family = TF`, `nu.fix = TRUE` and `nu.start = 1` in the `gamlss()` function.

3.3 Explicit GAMLSS family distributions

The explicit `gamlss.family` distributions (i.e. within **gamlss.dist**) are subdivided into three distinct types according to the type of random variable modeled as a response. These are:

1. continuous `gamlss.family` distributions,
2. discrete `gamlss.family` distributions, and
3. mixed `gamlss.family` distributions.

In the GAMLSS model, Y has probability (density) function $f(y \mid \boldsymbol{\theta})$ with up to four parameters, i.e. $\boldsymbol{\theta} = \mu$ or (μ, σ) or (μ, σ, ν) or (μ, σ, ν, τ).

3.3.1 Continuous distributions in GAMLSS

Table 3.2 shows all the explicit continuous distributions currently available in the `gamlss.family`. Link functions were introduced by Nelder and Wedderburn [1972] for Generalized Linear Models (GLM), but are appropriate for all regression models since they can guarantee that parameter estimates remain within the appropriate range. A link function relates a parameter θ to a predictor η, which may depend on explanatory variables and in general has range $-\infty < \eta < \infty$. For example if a parameter θ has range $0 < \theta < \infty$, then the logarithmic transformation

$$\log(\theta) = \eta$$

produces $-\infty < \eta < \infty$. In parameter estimation, if the logarithmic link is used, η is estimated and transformed back to θ as

$$\theta = e^{\eta},$$

so that θ is guaranteed to be in the range $(0, \infty)$ whatever the value of η. For the logarithmic link, $\log(\theta)$ is the link function and e^{η} is the inverse link function. In general, the link function is denoted as $g(\theta) = \eta$, and the inverse link as $g^{-1}(\eta) = \theta$. Link functions $g(\cdot)$ have to be monotonic and differentiable. Table 3.1 gives the current available link functions existing in **gamlss.dist**. Section 6.5 of of Stasinopoulos et al. [2017] provides advice on how to amend and create new link functions.

For any **gamlss** distribution, each model parameter has its own link function, appropriate to its range[a], and these are denoted as

$$\begin{aligned} g_1(\mu) &= \eta_1 \\ g_2(\sigma) &= \eta_2 \\ g_3(\nu) &= \eta_3 \\ g_4(\tau) &= \eta_4 \ . \end{aligned}$$

[a]Exceptions are the `BCCG`, `BCPE`, and `BCCG` distributions where, for conformity with the LMS method for centile estimation, μ has an identity link function rather than a log link function.

TABLE 3.2: Continuous distributions implemented within the **gamlss.dist** package, with default link functions. (ident = `identity`, log-2 = `logshiftto2`)

Distribution	**gamlss** name	Parameter link functions μ	σ	ν	τ
		$R_Y = (-\infty, \infty)$			
exponential Gaussian	`exGAUS`	ident	log	log	-
exponential gen beta 2	`EGB2`	ident	log	log	log
generalized *t*	`GT`	ident	log	log	log
Gumbel	`GU`	ident	log	-	-
Johnson's original SU	`JSUo`	ident	log	ident	log
Johnson's SU repar	`JSU`	ident	log	ident	log
logistic	`LO`	ident	log	-	-
NET	`NET`	ident	log	fixed	fixed
normal	`NO`	ident	log	-	-
normal 2	`NO2`	ident	log	-	-
normal family	`NOF`	ident	log	ident	-
power exponential	`PE`	ident	log	log	-
power exponential 2	`PE2`	ident	log	log	-
reverse Gumbel	`RG`	ident	log	-	-
sinh-arcsinh	`SHASH`	ident	log	log	log
sinh-arcsinh original	`SHASHo`	ident	log	ident	log
sinh-arcsinh original 2	`SHASHo2`	ident	log	ident	log
skew normal type 1	`SN1`	ident	log	ident	-
skew normal type 2	`SN2`	ident	log	log	-
skew power exp type 1	`SEP1`	ident	log	ident	log
skew power exp type 2	`SEP2`	ident	log	ident	log
skew power exp type 3	`SEP3`	ident	log	log	log
skew power exp type 4	`SEP4`	ident	log	log	log
skew *t* type 1	`ST1`	ident	log	ident	log
skew *t* type 2	`ST2`	ident	log	ident	log
skew *t* type 3	`ST3`	ident	log	log	log
skew *t* type 3 repar	`SST`	ident	log	log	log-2
skew *t* type 4	`ST4`	ident	log	log	log
skew *t* type 5	`ST5`	ident	log	ident	log
t family	`TF`	ident	log	log	-
t family repar	`TF2`	ident	log	log-2	-
		$R_Y = (0, \infty)$			
Box-Cox Cole Green	`BCCG`	ident	log	ident	-
Box-Cox Cole Green orig	`BCCGo`	log	log	ident	-
Box-Cox power exponential	`BCPE`	ident	log	ident	log
Box-Cox power expon orig	`BCPEo`	log	log	ident	log
Box-Cox *t*	`BCT`	ident	log	ident	log
Box-Cox *t* orig	`BCTo`	log	log	ident	log
exponential	`EXP`	log	-	-	-
gamma	`GA`	log	log	-	-
gamma family	`GAF`	log	log	ident	-
generalized beta type 2	`GB2`	log	log	log	log
generalized gamma	`GG`	log	log	ident	-
generalized inv. Gaussian	`GIG`	log	log	ident	-

inverse Gamma	IGAMMA	log	log	-	-
inverse Gaussian	IG	log	log	-	-
log normal	LOGNO	ident	log	-	-
log normal 2	LOGNO2	log	log	-	-
log normal (Box-Cox)	LNO	ident	log	fixed	-
Pareto 2	PARETO2	log	log	-	-
Pareto 2 original	PARETO2o	log	log	-	-
Weibull	WEI	log	log	-	-
Weibull (PH)	WEI2	log	log	-	-
Weibull (μ the mean)	WEI3	log	log	-	-
	$R_Y = (0, 1)$				
beta	BE	logit	logit	-	-
beta original	BEo	log	log	-	-
generalized beta type 1	GB1	logit	logit	log	log
logit normal	LOGITNO	logit	log	-	-
simplex	SIMPLEX	logit	log	-	-
	$R_Y = (\mu, \infty)$				
Pareto 1	PARETO1o	fixed	log	-	-

There are three distinct ranges for continuous distributions: $(-\infty, \infty)$, $(0, \infty)$, and $(0, 1)$. Chapter 4 provides information and examples for the explicit distributions on $(-\infty, \infty)$, while Chapter 18 provides plots and reference tables for them. Information and examples for the explicit continuous distributions on $(0, \infty)$ are given in Chapter 5 and the plots and reference tables of the distributions are in Chapter 19. Information and examples for the explicit distributions on the range $(0, 1)$ are given in Chapter 6 and the plots and reference tables in Chapter 21.

3.3.2 Discrete distributions in GAMLSS

Table 3.3 contains all the explicit discrete distributions currently available in the `gamlss.family`. There are two types of discrete distribution according to their range:

- the *count* distributions with range $\{0, 1, 2, \ldots\}$ or $\{1, 2, \ldots\}$, and
- the *binomial* type distributions with range $\{0, 1, \ldots, n\}$.

The basic count distribution is the Poisson, from which most of the other count data distributions are derived. Chapter 7 provides more details and examples for count distributions, and plots and reference tables are given in Chapter 22.

The binomial type distributions include the binomial, beta binomial, their zero-inflated and zero-adjusted versions, and the double binomial. These are discussed in Chapter 8, and their plots and reference tables are given in Chapter 23.

TABLE 3.3: Discrete distributions implemented within **gamlss.dist**, with default link functions. (inf=inflated, adj=adjusted, inv=inverse)

Distribution	**gamlss** name	Parameter link function μ	σ	ν	τ
	$R_Y = \{0,1,2,\ldots\}$				
beta neg binomial	`BNB`	log	log	log	-
Delaporte	`DEL`	log	log	logit	-
discrete Burr XII	`DBURR12`	log	log	log	-
double Poisson	`DPO`	log	log	-	-
generalized Poisson	`GPO`	log	log	-	-
geometric	`GEOM`	log	-	-	-
geometric (original)	`GEOMo`	logit	-	-	-
negative binomial type I	`NBI`	log	log	-	-
negative binomial type II	`NBII`	log	log	-	-
neg binomial family	`NBF`	log	log	log	-
Poisson	`PO`	log	-	-	-
Poisson-inv Gaussian	`PIG`	log	log	-	-
Poisson-inv Gaussian 2 [a]	`PIG2`	log	log	-	-
Poisson shifted GIG[a]	`PSGIG`	log	log	ident	logit
Sichel	`SI`	log	log	ident	-
Sichel (μ the mean)	`SICHEL`	log	log	ident	-
Waring (μ the mean)	`WARING`	log	log	-	-
Yule (μ the mean)	`YULE`	log	-	-	-
zero-adj beta neg binom	`ZABNB`	log	log	log	logit
zero-adj logarithmic	`ZALG`	logit	logit	-	-
zero-adj neg binomial	`ZANBI`	log	log	logit	-
zero-adj PIG	`ZAPIG`	log	log	logit	-
zero-adj Sichel	`ZASICHEL`	log	log	ident	logit
zero-adj Poisson	`ZAP`	log	logit	-	-
zero-adj Zipf	`ZAZIPF`	log	logit	-	-
zero-inf beta neg binom	`ZIBNB`	log	log	log	logit
zero-inf neg binomial	`ZINBI`	log	log	logit	-
zero-inf neg binom fam[a]	`ZINBF`	log	log	log	logit
zero-inf Poisson	`ZIP`	log	logit	-	-
zero-inf Poisson (μ the mean)	`ZIP2`	log	logit	-	-
zero-inf PIG	`ZIPIG`	log	log	logit	-
zero-inf Sichel	`ZISICHEL`	log	log	ident	logit
	$R_Y = \{0, 1, \ldots, n\}$				
binomial	`BI`	logit	-	-	-
beta binomial	`BB`	logit	log	-	-
double binomial	`DBI`	logit	log	-	-
zero-adj beta binomial	`ZABB`	logit	log	logit	-
zero-adj binomial	`ZABI`	logit	logit	-	-
zero-inf beta binomial	`ZIBB`	logit	log	logit	-
zero-inf binomial	`ZIBI`	logit	logit	-	-
	$R_Y = \{1, 2, \ldots\}$				
logarithmic	`LG`	logit	-	-	-
Zipf	`ZIPF`	log	-	-	-

3.3.3 Mixed distributions in GAMLSS

Mixed distributions, which are a special case of finite mixture distributions, are described in Chapter 9. They are mixtures of continuous and discrete distributions, i.e. continuous distributions in which the range of Y has been expanded to include some discrete values with non-zero probabilities. Table 3.4 shows the explicit mixed distributions currently available in `gamlss.family`.

TABLE 3.4: Mixed distributions implemented within the **gamlss.dist** package, with default link functions.

Distribution	**gamlss**	Range	Parameter link functions			
	name	R_Y	μ	σ	ν	τ
zero-inflated beta	`BEINF0`	$[0, 1)$	logit	logit	log	-
zero-inflated beta	`BEOI`	$[0, 1)$	logit	log	logit	-
one-inflated beta	`BEINF1`	$(0, 1]$	logit	logit	log	-
one-inflated beta	`BEZI`	$(0, 1]$	logit	log	logit	-
zero- and one-inflated beta	`BEINF`	$[0, 1]$	logit	logit	log	log
zero-adjusted gamma	`ZAGA`	$[0, \infty)$	log	log	logit	-
zero-adjusted inv Gaussian	`ZAIG`	$[0, \infty)$	log	log	logit	-

Note that in the range R_Y in Table 3.4, a square bracket indicates the endpoint is included in the interval, while a round bracket indicates the endpoint is not included. (This is consistent with conventional mathematical notation.) Hence, for example, $[0, 1)$ includes 0 but not 1.

The zero-adjusted distributions, `ZAGA` and `ZAIG`, are useful for modeling a response variable on the interval $[0, \infty)$, such as amount of insurance claim in a year. Most of the policies do not have any claims and therefore there is a high probability of the claim amount being zero, but for policies that do have claims, the distribution of the claim amount is defined on the positive real line $(0, \infty)$. The zero-adjusted gamma (`ZAGA`) distribution is shown in Figure 3.1.

Beta inflated distributions are useful for modeling a proportion (or fractional) response variable on the interval 0 to 1, including 0 or 1 or both. Other mixed distributions may be defined using the package **gamlss.inf**. More details on mixed distributions are given in Chapter 9.

3.4 Generated GAMLSS family distributions

There are several ways to generate a new `gamlss.family` distribution from an existing one. This can be achieved in the following ways:

[a]Currently under development

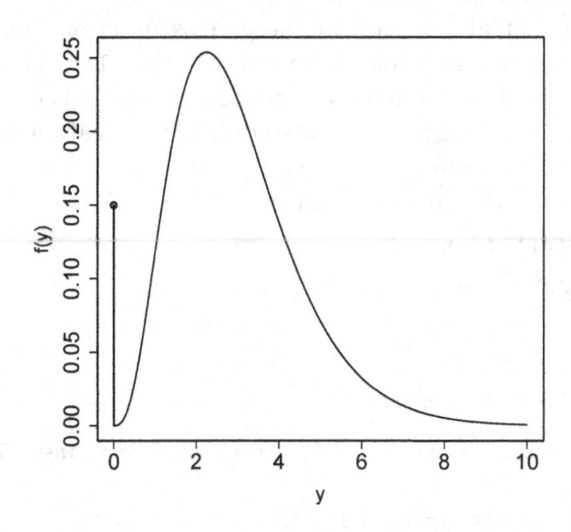

FIGURE 3.1: The zero-adjusted gamma, ZAGA(μ, σ, ν), distribution with $\mu = 3$, $\sigma = 0.5$ and $\nu = 0.15$, an example of a mixed distribution.

- take a continuous distribution defined on $(-\infty, \infty)$ and create a log version with range $(0, \infty)$ (Sections 3.4.1, 5.6.1, and 13.2.1);
- take a continuous distribution defined on $(-\infty, \infty)$ and create a logit version with range $(0, 1)$ (Sections 3.4.2, 6.3.1, and 13.2.2);
- take any continuous or discrete distribution and truncate its range. This can be "left", "right", or "both" truncation (Sections 3.4.3, 5.6.2, 6.3.2, and 13.7);
- take any continuous distribution defined on $(0, \infty)$ and by interval censoring create a discrete count distribution defined on $\{0, 1, 2, \ldots\}$ (Sections 3.4.4 and 7.3.4);
- take a continuous distribution defined on $(-\infty, \infty)$ or $(0, \infty)$ and create by left, right or interval censoring a generalized Tobit model (Sections 3.4.5 and 9.2.3);
- take any continuous distribution defined on $(0, \infty)$ and zero-adjust it to create a mixed distribution on $[0, \infty)$ (Sections 3.4.6 and 9.2.2);
- take any continuous distribution defined on $(0, 1)$ and zero- and/or one-inflate it to create a mixed distribution on $[0, 1)$, $(0, 1]$ or $[0, 1]$ (Sections 3.4.7 and 9.3);
- mix different `gamlss.family` distributions to create a new finite mixture distribution. Chapter 7 of Stasinopoulos et al. [2017] extensively covers this case.

Here we will deal with all cases except the last.

3.4.1 Log distributions on $(0, \infty)$

Any continuous random variable Z defined on $(-\infty, \infty)$ can be transformed to a random variable Y defined on $(0, \infty)$, by the exponential transformation $Y = \exp(Z)$. A well-known example of this is the log normal distribution, which is defined by $Y = \exp(Z)$ where Z is a normally distributed random variable. The log normal distribution is represented in the explicit continuous family distributions of Table 3.2 by LOGNO and LOGNO2. Here we demonstrate the use of the **gamlss** function gen.Family() with the option type="log" to generate a log power exponential (logPE) distribution. Starting with $Z \sim \texttt{PE}(\mu, \sigma, \nu)$, which is defined on $(-\infty, \infty)$, we apply the exponential transform $Y = \exp(Z)$ to give $Y \sim \log \texttt{PE}(\mu, \sigma, \nu)$. In so doing, we create a logPE distribution on $(0, \infty)$. We then generate a random sample of 200 observations from the distribution and finally fit the distribution to the generated data.

```
# generate the distribution
gen.Family("PE", type="log")

## A  log  family of distributions from PE has been generated
##  and saved under the names:
##  dlogPE plogPE qlogPE rlogPE logPE

#generate 200 observations with mu=1, sigma=0.6 and nu=2.5
set.seed(1434)
Y<- rlogPE(200, mu=1, sigma=.6, nu=2.5)
# fit the distribution
h1 <- histDist(Y, family=logPE, nbins=30, ylim=c(0,.30),
               line.col=1, line.wd=2.5, main="(a)")
```

The resulting plot is given in Figure 3.2(a). See Sections 5.6.1 and 13.2.1 for further information.

3.4.2 Logit distributions on $(0, 1)$

Any continuous random variable Z on $(-\infty, \infty)$ can be transformed to a random variable Y on $(0, 1)$ by the inverse logit transformation $Y = 1/(1 + \exp(-Z))$. For example let Z have a t family distribution, $Z \sim \texttt{TF}(\mu, \sigma, \nu)$, and apply the inverse logit transformation to give $Y \sim \texttt{logitTF}(\mu, \sigma, \nu)$, i.e. a logit-$t$ family distribution on $(0, 1)$:

```
gen.Family("TF", type="logit")

## A  logit  family of distributions from TF has been generated
##  and saved under the names:
##  dlogitTF plogitTF qlogitTF rlogitTF logitTF

#generate 200 observations with mu=0, sigma=1, nu=1.5
Y<- rlogitTF(200, mu=0, sigma=1, nu=1.5)
# fit the distribution
h1 <- histDist(Y, family=logitTF, nbins=20, ylim=c(0,2), xlim=c(0,1),
               line.col=1,line.wd=2.5, main="(b)")
```

The resulting plot is given in Figure 3.2(b). See Sections 6.3.1, and 13.2.2.

3.4.3 Truncated distributions

Truncated distributions are appropriate when the range of a response variable Y is a subset of the range of the original distribution. The package **gamlss.tr** allows the user to take any continuous or discrete gamlss.family distribution and restrict its range to obtain a truncated distribution. The generation of a new truncated distribution is

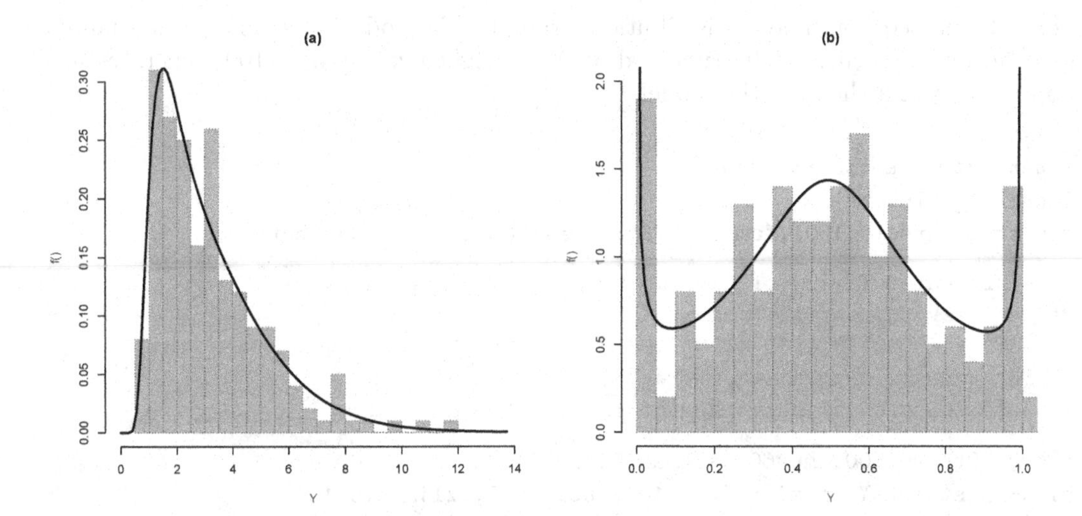

FIGURE 3.2: (a) `logPE` and (b) `logitTF` distributions, fitted to 200 simulated observations.

achieved by the function `gen.trun()`, which takes as arguments: (i) the points of truncation, `par`, (ii) the `gamlss.family`, e.g. `family=NO`, (iii) the new name of the distribution, e.g. `name="truncatedNO"` and (iv) the type of truncation, i.e. `type="left"`, `type="right"`, or `type="both"`. If the type is `"left"` or `"right"`, the argument `par` expects one value, e.g. `par=0`; if it is `type="both"` then `par` expects two values, e.g. `par=c(0,100)`. The function `gen.trun()` generates the functions `d`, `p`, `q`, `r` and the appropriate fitting functions for the truncated distribution. An important point to remember is that truncation works differently for discrete and continuous distributions. For discrete distributions,

- "left" truncation at the integer a means that the truncated random variable can take values $\{a+1, a+2, \ldots\}$;
- "right" truncation at the integer b means that the truncated random variable can take values up to but not including b, i.e. $\{\ldots, b-2, b-1\}$.
- "both" truncation at the integer interval (a, b) means the truncated random variable can take values $\{a+1, a+2, \ldots, b-1\}$.

For a continuous random variable Y, $\text{P}(Y = a) = 0$ for any constant a and therefore the inclusion or not of an endpoint in the truncated range is not an issue. Hence "left" truncation at value a results in truncated variable $Y \geq a$, "right" truncation at b results in $Y \leq b$, and "both" truncation results in $a \leq Y \leq b$. More theoretical detail about truncation of distributions is given in Sections 5.6.2, 6.3.2, and 13.7.

Let us assume that we would like to analyze student marks which take values from 0 to 100. If the marks are recorded on a discrete scale we can take a discrete count distribution and right truncate it at 101. If the marks contain decimal places then, provided there are no values at exactly 0 or 100, we have two options: (i) divide the marks by 100 and use a distribution defined on $(0, 1)$ or (ii) take a continuous distribution defined on either $(0, \infty)$ or $(-\infty, \infty)$ and truncate it to the range $(0, 100)$.

For a truncated continuous distribution example, the code below creates a t-family distribution, TF(μ, σ, ν), left-truncated at 0 and right-truncated at 100, generates 500 observations and then fits the model.

```
# generate the distribution
library(gamlss.tr)
gen.trun(par=c(0,100),family="TF", name="0to100", type="both")

## A truncated family of distributions from TF has been generated
##  and saved under the names:
##  dTF0to100 pTF0to100 qTF0to100 rTF0to100 TF0to100
## The type of truncation is both
##  and the truncation parameter is 0 100

Y <-rTF0to100(500, mu=60 ,sigma=10, nu=5)
h1 <- histDist(Y, family=TF0to100, nbins=20, xlim=c(0,100),
               line.col=gray(.2),line.wd=2.5,
               main="(a)")
```

For a truncated discrete distribution example we generate a negative binomial distribution, NBI(μ, σ), right-truncated at 101, generate 500 observations, and then fit the model.

```
gen.trun(par=101,family="NBI", name="0to100", type="right")

## A truncated family of distributions from NBI has been generated
##  and saved under the names:
##  dNBI0to100 pNBI0to100 qNBI0to100 rNBI0to100 NBI0to100
## The type of truncation is right
##  and the truncation parameter is 101

 y <- rNBI0to100(500, mu=60, sigma=0.1)
h1 <- histDist(y, family=NBI0to100, nbins=100, xlim=c(0,100),
               line.col=gray(.2),line.wd=2.5,
               main= "(b)" )
```

The resulting plots are given in Figure 3.3.

Section 5.6.2 gives an example of the four-parameter SST distribution defined on $(-\infty, \infty)$ truncated to $(0, \infty)$, while Section 6.3.2 gives the SST distribution truncated to $(0, 1)$. Practical examples using truncation are given in Sections 6.4.1 and 9.2.4. See also Section 13.7.

3.4.4 Interval censoring for generation of discrete distributions

The package **gamlss.cens** is useful if the response variable is left-, right-, or interval-censored. Censoring occurs when we do not observe the exact value of the response variable, but we know in which interval it occurred. Section 6.2.2 of Stasinopoulos et al. [2017] gives an example of interval censoring. Here we demonstrate the use of the **gamlss.cens** package to fit a discretized continuous distribution to count data. In the **R** code below we generate a sample from a negative binomial distribution (NBI) and fit an NBI model. Then we use **gamlss.cens** to generate an interval gamma family function,

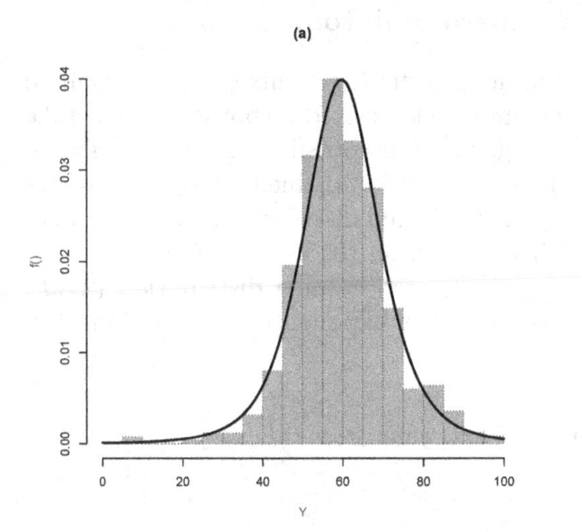

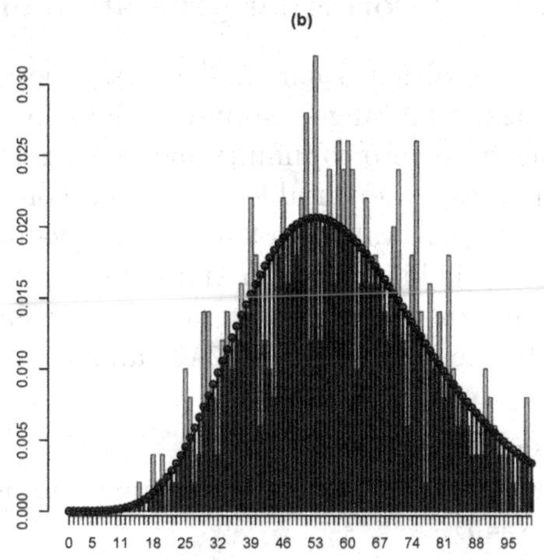

FIGURE 3.3: (a) Truncated t distribution on $(0, 100)$, fitted to 500 continuous simulated observations; (b) negative binomial distribution right truncated at 101 and fitted to 500 discrete simulated observations.

correct for zeros in the response (since gamma does not accept zeros), and then fit a discretized gamma distribution by using the function `Surv()` as response. Note that values from 0 to 1 in the gamma distribution will be treated as 0 in the discretized gamma distribution, values from 1 to 2 as 1, etc.

```
library(gamlss.cens)

## Loading required package: survival

set.seed(342)
Y <- rNBI(500, mu=1, sigma=1)  # generate a NBI sample
m1 <- gamlss(Y~1, family=NBI, trace=F) # fit the NBI model
# generate a GA interval family
gen.cens("GA", type="interval", name="D")

## A censored family of distributions from GA has been generated
##  and saved under the names:
##  dGAiD pGAiD qGAiD GAiD
## The type of censoring is interval

y <- ifelse(Y==0, 0.000001, Y) # taking care of zeros
# fit the discretized GA
m2 <- gamlss(Surv(y, y+1, type="interval2")~1, family=GAiD, trace=F)
AIC(m1,m2)

##    df      AIC
## m1  2 1388.818
## m2  2 1388.842
```

The two models have almost identical fits. Discretized continuous distributions are explained in more detail in Section 7.3.4.

3.4.5 Censoring for generation of a generalized Tobit model

The generalized Tobit model is a general method for fitting a mixed distribution to any restricted range response variable. For example, let a mixed response variable take values from zero to infinity including zero, i.e. $[0, \infty)$. One possible way to fit a model to this response variable is to use a continuous distribution defined on $(-\infty, \infty)$, and treat the zero values as left-censored values, giving a mixed distribution with a discrete point probability at zero and a continuous distribution on $(0, \infty)$. The following is an example in which we generate data from a zero-adjusted gamma distribution (ZAGA) and fit the data with both ZAGA and a left-censored power exponential distribution (PE).

```
set.seed(342)
Y <- rZAGA(500, mu = 1, sigma = 1, nu = 0.1)# generate the sample
sum(Y==0)

## [1] 39

m1 <- gamlss(Y~1, family=ZAGA, trace=F) # fit the ZAGA model
gen.cens("PE", type="left", name="T")

## A censored family of distributions from PE has been generated
##  and saved under the names:
##  dPE1T pPE1T qPE1T PE1T
## The type of censoring is left

m2 <- gamlss(Surv(Y, Y!=0, type="left")~1, family=PE1T, trace=F)
AIC(m1,m2)

##     df      AIC
## m1   3 1179.418
## m2   3 1325.644
```

Note that in m1 model above, $Y \sim 1$ fits a constant for μ in ZAGA(μ, σ, ν), while constants are fitted for σ and ν by default.

Note that in m2 above, Surv() creates from Y a left censored at zero response variable. Unsurprisingly the ZAGA model m1 (from which the response variable was generated) fits better than the generalized Tobit PE model m2. The generalized Tobit model is explained in more detail in Section 9.2.3, and Sections 9.2.4 and 9.3.4 give examples of its use.

3.4.6 Zero-adjusted distributions on [0,∞)

A zero-adjusted distribution is a mixed distribution (introduced in Section 1.2.3) consisting of two components: a continuous distribution defined on $(0, \infty)$, and a probability at zero. The zero-adjusted gamma, ZAGA(μ, σ, ν), and zero-adjusted inverse Gaussian, ZAIG(μ, σ, ν), distributions both have μ and σ as location and scale parameters, while ν models the probability at zero. The package **gamlss.inf** has the capability of extending the zero-adjusted distributions by extending any gamlss.family distribution defined on $(0, \infty)$, including generated distributions, to a zero-adjusted distribution by adding a probability at zero.

To demonstrate the use of the **gamlss.inf** package, we use the data generated in the previous section from a `ZAGA` distribution and fitted as model `m1`. We refit the zero-adjusted gamma model using the `gamlssZadj()` function, and also fit the `BCTo` zero-adjusted model. The models are compared using AIC.

```
library(gamlss.inf)
m3 <- gamlssZadj(Y, mu.formula=~1, family=GA) #
m4 <- gamlssZadj(Y, mu.formula=~1, family=BCTo) #
GAIC(m1,m3,m4)

##    df      AIC
## m1  3 1179.418
## m3  3 1179.418
## m4  5 1185.459

fitted(m1,"nu")[1]

## [1] 0.078

fitted(m3,"xi0")[1]

## [1] 0.078
```

There is a difference in the use of formulae in fitting the model using `gamlss()` or `gamlssZadj()`. In the former the response variable is declared with the μ formula, while in the latter it is specified separately. Note also that for the original `ZAGA` model the zero probability is the parameter ν (`nu`), while in the `gamlssZadj()` formulation it is ξ_0 (`xi0`).

Chapter 9 and specifically Section 9.2 give further information on zero-adjusted distributions.

3.4.7 Inflated distributions on $[0, 1]$

We use the term 'inflated distribution' for a mixed distribution with continuous range defined on $(0, 1)$ and probability masses at zero, one or both, i.e. for range $[0, 1)$, $(0, 1]$, or $[0, 1]$. They are useful for modeling a response variable which is a proportion that may include values 0 or 1, or both. The only explicit `gamlss.family` inflated distributions are based on the beta distribution. However any `gamlss.family` distribution on $(0, 1)$ can be inflated to generate an inflated distribution on $[0, 1)$, $(0, 1]$, or $[0, 1]$, using the function `gen.Inf0to1()`.

As an example, we generate a zero- and one-inflated simplex distribution and create a random sample from it. We then fit a model to the data and display the fitted model in Figure 3.4.

```
library(gamlss.inf)
gen.Inf0to1("SIMPLEX",  type.of.Inflation ="Zero&One")

## A  0to1 inflated SIMPLEX distribution has been generated
##  and saved under the names:
##  dSIMPLEXInf0to1 pSIMPLEXInf0to1 qSIMPLEXInf0to1 rSIMPLEXInf0to1
##  plotSIMPLEXInf0to1

Y<-rSIMPLEXInf0to1(500, mu = 0.2, sigma = 2, xi0=0.1, xi1 = 0.15)
```

```
i1 <- gamlssInf0to1(Y, mu.formula=~1, family=SIMPLEX)
op <- par(cex.lab=1.7,cex.axis=1.7,cex.main=1.7,
       lwd=2,mar=c(5, 5, 4, 1) + 0.1)
plotSIMPLEXInf0to1(mu=fitted(i1,"mu")[1], sigma=fitted(i1, "sigma")[1],
                   xi0=fitted(i1,"xi0")[1], xi1=fitted(i1,"xi1")[1])
```

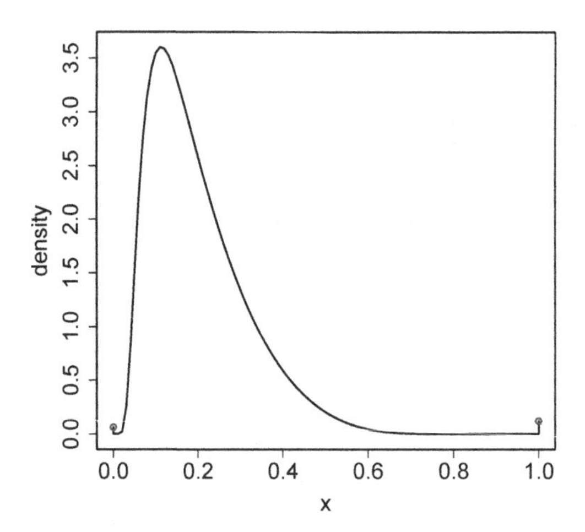

FIGURE 3.4: A fitted inflated simplex distribution. The distribution was fitted to data generated from an inflated simplex distribution with $\mu = 0.1$, $\sigma = 2$, $\xi_0 = 0.1$, and $\xi_1 = 0.15$.

Further information about inflated distributions can be found in Chapter 9 and specifically Section 9.3.

3.5 Displaying GAMLSS family distributions

Each `gamlss.family` distribution has five functions: the "fitting" function, which is used in the argument `family` of the `gamlss()` function, and the usual four **R** functions, `d`, `p`, `q`, and `r`, for the probability (density) function (pdf), cumulative distribution function (cdf), inverse cdf (quantile function), and random generating function, respectively. For example, within the `gamlss.family` the fitting function, pdf, cdf, inverse cdf, and random generating functions of the gamma (`GA`) distribution are `GA`, `dGA`, `pGA`, `qGA`, and `rGA`, respectively.

There are several ways to explore the `gamlss.family` distributions. Here we suggest a few.

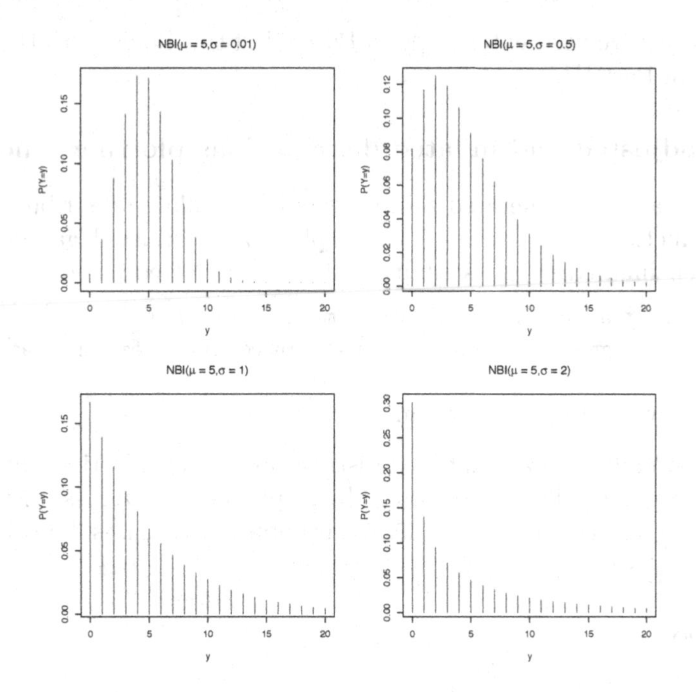

FIGURE 3.5: Plotting the negative binomial distribution, `NBI`(μ, σ), using the `pdf.plot()` function for $\mu = 5$ and $\sigma = 0.01, 0.5, 1$, and 2.

3.5.1 Using the `pdf.plot()` function

A method of graphically displaying the probability (density) function is to use the `pdf.plot()` function. The following code produces Figure 3.5, which plots the negative binomial, `NBI`(μ, σ), distribution for $\mu = 5$, and $\sigma = 0.01, 0.5, 1$, and 2:

```
pdf.plot(family=NBI,mu=5, sigma=c(0.01,0.5,1,2), min=0, max=20)
```

3.5.2 Plotting the `d`, `p`, `q`, and `r` functions

Examples of plotting using the `d`, `p`, `q`, and `r` functions are given in Figures 1.9 and 1.10.

3.5.3 Specific plotting functions

Functions `contR_2_12()`, `contR_3_11()`, and `contR_4_13()` plot continuous distributions with range $(-\infty, \infty)$ and two, three, and four parameters, respectively. (The number after the first underscore indicates the number of parameters in the distribution; the two numbers after the second underscore indicate how many rows and columns are in the plot.) Similarly, `contRplus_2_11()`, `contRplus_3_13()`, and `contRplus_4_33()` plot continuous distributions with range $(0, \infty)$ and two, three, and four parameters, respectively. For count distributions, we have the functions `count_1_31()`, `count_1_22()`, `count_2_32()`, `count_2_32R()`, `count_2_33()`, `count_3_32()`, and `count_3_33()`; and for binomial type distributions `binom_1_31()`, `binom_2_33()`, and `binom_3_33()`. All of the above functions are used extensively in Part III of this book, with the **R** code given in Appendix A of Part III.

3.5.4 Zero-adjusted and inflated distributions plotting functions

Both the explicit and the generated zero-adjusted and inflated distributions have their own plotting functions. For example, the explicit zero-adjusted gamma distribution ZAGA and explicit inflated beta distribution BEINF are plotted using:

```
plotZAGA(mu=1, sigma=.5, nu=0.1, to=5, main="(a) ZAGA")
plotBEINF(mu=.5, sigma=.5 , nu=.5, tau=1, main="(b) beta inflated")
```

The plots are shown in Figure 3.6(a) and (b).

For generated zero-adjusted and inflated distributions, their plotting functions may be used. For example, the following code produces plots of the zero-adjusted Box-Cox t (BCT) distribution and the inflated SIMPLEX distribution (Figures 3.6 (c) and (d)).

```
library(gamlss.inf)
## BCT
gen.Zadj("BCT")

## A zero adjusted BCT distribution has been generated
##  and saved under the names:
##  dBCTZadj pBCTZadj qBCTZadj rBCTZadj
##  plotBCTZadj

plotBCTZadj(mu=1, sigma=0.3, nu=2, tau=5, xi0=.3,
            to= 3)
title("(c) zero-adjusted BCT")
## simplex
gen.Inf0to1("SIMPLEX",  type.of.Inflation ="Zero&One")

## A  0to1 inflated SIMPLEX distribution has been generated
##  and saved under the names:
##  dSIMPLEXInf0to1 pSIMPLEXInf0to1 qSIMPLEXInf0to1 rSIMPLEXInf0to1
##  plotSIMPLEXInf0to1

plotSIMPLEXInf0to1(mu=.5, sigma=2, xi0=.4, xi1=.1)
title("(d) inflated simplex")
```

3.6 Bibliographic notes

For fitting distributions to a random sample, the **R** package **fitdistrplus** of Delignette-Muller and Dutang [2015] is available. We are not aware of any **R** package besides **gamlss** with the facility for generating and fitting distributions, in a regression framework, as described in this chapter. The **VGAM** package [Yee, 2019] is a regression framework which includes several of the explicit response distributions present in **gamlss**.

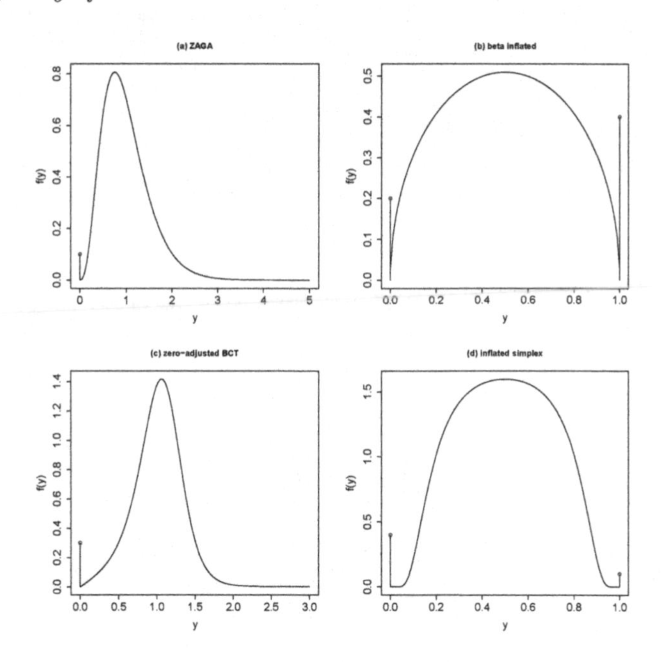

FIGURE 3.6: Plots of (a) an explicit zero-adjusted gamma distribution, ZAGA(μ, σ, ν), (b) an explicit beta inflated distribution, BEINF(μ, σ, ν, τ), (c) a generated zero-adjusted BCT distribution, BCTZadj($\mu, \sigma, \nu, \tau, \xi_0$), and (d) a generated inflated simplex distribution, SIMPLEXInf0to1($\mu, \sigma, \xi_0, \xi_1$).

3.7 Exercises

1. **Explicit distributions:** Explain the different types of distributions existing within the `gamlss.family` and for each type give, with reason, an example of an appropriate response variable.

2. **Generated distributions:**

   (a) Generate a 'log' t family distribution (using the name `logTF`) and randomly generate 1000 observations from this distribution using the parameter values $\mu = 1, \sigma = 1$, and $\nu = 10$. Fit the distribution to the data using the `histDist()` function. The `logTF` distribution is a very heavy right-tailed distribution. See what happens if the value of ν is decreased.

   (b) Write down the pdf of the `logitTF`(μ, σ, ν) distribution obtained in the example given in Section 3.4.2, which transforms a `TF`(μ, σ, ν) distribution on $(-\infty, \infty)$ to a logitTF(μ, σ, ν) distribution on $(0, 1)$, using the logit transformation. [Hint: see equations (13.4) and (18.11)].

4

Continuous distributions on $(-\infty, \infty)$

CONTENTS

This chapter provides:

1. an explanation of continuous distributions defined on $(-\infty, \infty)$ within the GAMLSS family,
2. a definition of a location and scale family,
3. an explanation of how the GAMLSS family distributions can model skewness and kurtosis, and
4. a demonstration of how the distributions are used in practice.

4.1 Introduction

As discussed in Chapter 2, continuous distributions can be symmetric, negatively or positively skewed, and also mesokurtic, leptokurtic or platykurtic. In this chapter we study the GAMLSS family of continuous distributions with range $(-\infty, \infty)$ in more detail. In particular we discuss the shapes of the distributions, and their flexibility in modeling a response variable. Table 4.1 provides a list of the distributions. Many of these can be generated by one (or more) of the methods described in Chapter 13. The definitions, tables of information on distributional properties, plots and references of all explicit `gamlss.family` distributions discussed in this chapter are in Part III, Chapter 18.

Within Table 4.1, positive (negative) skewness is taken to indicate that the moment skewness $\gamma_1 > 0$ ($\gamma_1 < 0$), if γ_1 exists, leptokurtic (platykurtic) indicates that the moment excess kurtosis $\gamma_2 > 0$ ($\gamma_2 < 0$), if γ_2 exists, while mesokurtic is taken to indicate $\gamma_2 = 0$. The measures γ_1 and γ_2, are defined in Section 2.2.4. In the skewness column, '+ve', '-ve', and 'sym' indicate positive moment skewness, negative moment skewness, and symmetry, respectively, while 'both' indicates that the distribution can have positive or negative moment skewness. In the kurtosis column 'lepto', 'platy', and 'meso' indicate moment leptokurtic, moment platykurtic, and moment mesokurtic, respectively, while 'both' indicates that the distribution can be moment platykurtic or moment leptokurtic. Brackets in the skewness (or kurtosis) column indicates that it cannot be modeled independently of the location and scale for a two-parameter distribution (or independently of the location, scale, and either skewness or kurtosis for a three-parameter distribution). Skewness and kurtosis are discussed in Chapters 14 and 15, respectively.

Location-scale families

All of the distributions in Table 4.1, with the exception of the `exGAUS` and `NOF` distributions, are *location-scale* families of distributions with *location shift* parameter μ and *scaling* parameter σ (for fixed ν and τ). Section 18.1 gives a formal definition for a location-scale family of distribution. For these distributions, if

$$Y \sim \mathcal{D}(\mu, \sigma, \nu, \tau)$$

then

$$Z = (Y - \mu)/\sigma \sim \mathcal{D}(0, 1, \nu, \tau),$$

i.e. $Y = \mu + \sigma Z$, so Y is a scaled and shifted version of the random variable Z (where the distribution of Z does not depend on μ or σ). Hence

$$Y_1 = a + bY \sim \mathcal{D}(a + b\mu, b\sigma, \nu, \tau).$$

The location and scale parameters are called "location shift" and "scaling" parameters in the tables in Chapter 18. There are two major advantages of a distribution being a location-scale family.

The first advantage of a location-scale family is that the fitted GAMLSS model is *invariant* to a location and scale change in the unit of measurement of the response variable (e.g. temperature from °F to °C), if an identity link is used for μ and a log link for σ, and if the predictor models for μ and $\log \sigma$ include a constant term. The model is invariant in the sense that the fitted model for Y_1 is the same as the fitted model for Y, when it is converted back to the original unit of measurement.

The second advantage of a location-scale family is that the shape of the distribution does not depend on μ or σ. The location shift parameter μ will shift the distribution to the left or to the right as in Figure 4.1(a), while the scaling parameter σ will contract (or expand) the distribution towards (or away from) μ, as in Figure 4.1(b). The shape of the distribution (for fixed ν and τ, if those parameters exist) will be the same. This

TABLE 4.1: Continuous distributions on $\mathbb{R} = (-\infty, \infty)$ implemented in the **gamlss.dist** package. Note that $\mathbb{R}_+ = (0, \infty)$, $\mathbb{R}_2 = (2, \infty)$; in the distribution name 'gen', 'repar', and 'exp' are abbreviations for 'generalized', 'reparameterized' and 'exponential'; while sym, meso and lepto are abbreviations for symmetric, mesokurtotic and leptokurtotic.

Distribution	**gamlss** name	p	Parameter range μ	σ	ν	τ	skewness	kurtosis
exponential Gaussian	`exGAUS`	3	$\mathbb{R}$	$\mathbb{R}_+$	$\mathbb{R}_+$	-	+ve	(lepto)
exponential gen beta 2	`EGB2`	4	$\mathbb{R}$	$\mathbb{R}_+$	$\mathbb{R}_+$	$\mathbb{R}_+$	both	lepto
generalized t	`GT`	4	$\mathbb{R}$	$\mathbb{R}_+$	$\mathbb{R}_+$	$\mathbb{R}_+$	(sym)	both
Gumbel	`GU`	2	$\mathbb{R}$	$\mathbb{R}_+$	-	-	(-ve)	(lepto)
Johnson's original SU	`JSUo`	4	$\mathbb{R}$	$\mathbb{R}_+$	$\mathbb{R}$	$\mathbb{R}_+$	both	lepto
Johnson's SU repar	`JSU`	4	$\mathbb{R}$	$\mathbb{R}_+$	$\mathbb{R}$	$\mathbb{R}_+$	both	lepto
logistic	`LO`	2	$\mathbb{R}$	$\mathbb{R}_+$	-	-	(sym)	(lepto)
NET	`NET`	2	$\mathbb{R}$	$\mathbb{R}_+$	fixed	fixed	(sym)	lepto
normal	`NO`	2	$\mathbb{R}$	$\mathbb{R}_+$	-	-	(sym)	(meso)
normal	`NO2`	2	$\mathbb{R}$	$\mathbb{R}_+$	-	-	(sym)	(meso)
normal family	`NOF`	3	$\mathbb{R}$	$\mathbb{R}_+$	$\mathbb{R}$	-	(sym)	(meso)
power exponential	`PE`	3	$\mathbb{R}$	$\mathbb{R}_+$	$\mathbb{R}_+$	-	(sym)	both
power exponential	`PE2`	3	$\mathbb{R}$	$\mathbb{R}_+$	$\mathbb{R}_+$	-	(sym)	both
reverse Gumbel	`RG`	2	$\mathbb{R}$	$\mathbb{R}_+$	-	-	(+ve)	(lepto)
sinh-arcsinh	`SHASH`	4	$\mathbb{R}$	$\mathbb{R}_+$	$\mathbb{R}_+$	$\mathbb{R}_+$	both	both
sinh-arcsinh original	`SHASHo`	4	$\mathbb{R}$	$\mathbb{R}_+$	$\mathbb{R}$	$\mathbb{R}_+$	both	both
sinh-arcsinh original 2	`SHASHo2`	4	$\mathbb{R}$	$\mathbb{R}_+$	$\mathbb{R}$	$\mathbb{R}_+$	both	both
skew normal type 1	`SN1`	3	$\mathbb{R}$	$\mathbb{R}_+$	$\mathbb{R}$	-	both	(lepto)
skew normal type 2	`SN2`	3	$\mathbb{R}$	$\mathbb{R}_+$	$\mathbb{R}_+$	-	both	(lepto)
skew exp power 1	`SEP1`	4	$\mathbb{R}$	$\mathbb{R}_+$	$\mathbb{R}$	$\mathbb{R}_+$	both	both
skew exp power 2	`SEP2`	4	$\mathbb{R}$	$\mathbb{R}_+$	$\mathbb{R}$	$\mathbb{R}_+$	both	both
skew exp power 3	`SEP3`	4	$\mathbb{R}$	$\mathbb{R}_+$	$\mathbb{R}_+$	$\mathbb{R}_+$	both	both
skew exp power 4	`SEP4`	4	$\mathbb{R}$	$\mathbb{R}_+$	$\mathbb{R}_+$	$\mathbb{R}_+$	both	both
skew t type 1	`ST1`	4	$\mathbb{R}$	$\mathbb{R}_+$	$\mathbb{R}$	$\mathbb{R}_+$	both	lepto [a]
skew t type 2	`ST2`	4	$\mathbb{R}$	$\mathbb{R}_+$	$\mathbb{R}$	$\mathbb{R}_+$	both	lepto [a]
skew t type 3	`ST3`	4	$\mathbb{R}$	$\mathbb{R}_+$	$\mathbb{R}_+$	$\mathbb{R}_+$	both	lepto
skew t type 3 repar	`SST`	4	$\mathbb{R}$	$\mathbb{R}_+$	$\mathbb{R}_+$	$\mathbb{R}_2$	both	lepto
skew t type 4	`ST4`	4	$\mathbb{R}$	$\mathbb{R}_+$	$\mathbb{R}_+$	$\mathbb{R}_+$	both	lepto [a]
skew t type 5	`ST5`	4	$\mathbb{R}$	$\mathbb{R}_+$	$\mathbb{R}$	$\mathbb{R}_+$	both	lepto [a]
t Family	`TF`	3	$\mathbb{R}$	$\mathbb{R}_+$	$\mathbb{R}_+$	-	(sym)	lepto
t Family type 2	`TF2`	3	$\mathbb{R}$	$\mathbb{R}_+$	$\mathbb{R}_2$	-	(sym)	lepto

[a] Conjecture i.e. believed but not proven.

makes the interpretation of a model with a location-scale family easier, since the effect of changing μ or σ is simple. Also when we plot a location-scale family distribution, say $\mathcal{D}(\mu, \sigma, \nu, \tau)$, we can set $\mu = 0$ and $\sigma = 1$ and study the shape of $\mathcal{D}(0, 1, \nu, \tau)$ by changing ν and τ (if they exist), as we do in Chapter 18.

For details about the exceptions, the `exGAUS` and `NOF` distributions, see Sections 18.3.1 and 18.3.2, respectively.

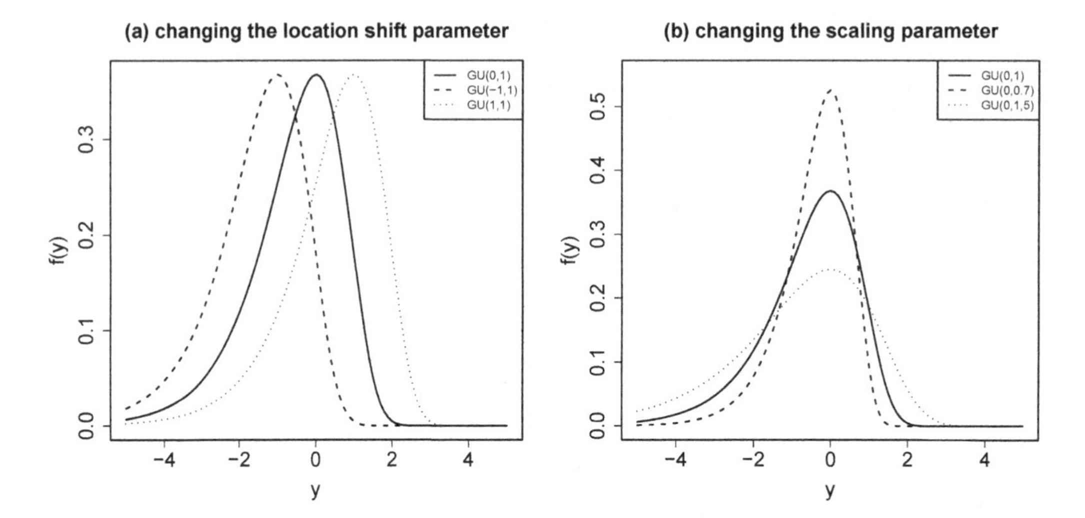

FIGURE 4.1: Plots illustrating how changing the location shift parameter (μ) and the scaling parameters (σ) affect a typical location-scale family distribution (the Gumbel distribution): (a) changing the location shift parameter μ moves the distribution to the left or to the right; (b) changing the scaling parameter σ contracts or expands the distribution around μ.

We now discuss the two-, three-, and four-parameter distributions listed in Table 4.1. More details are given in Chapter 18.

4.2 Two-parameter distributions on $(-\infty, \infty)$

There are four two-parameter continuous distributions on $(-\infty, \infty)$ in the `gamlss.family`:

Gumbel:[a] `GU`(μ, σ), a left-skewed distribution sometimes referred to as the Gumbel distribution for the minimum;

logistic: `LO`(μ, σ) a symmetric leptokurtic distribution;

Gaussian: `NO`(μ, σ) and `NO2`(μ, σ), the normal distribution, symmetric and mesokurtic;

reverse Gumbel: `RG`(μ, σ) a right-skewed distribution, a reflection of `GU`(μ, σ), sometimes referred to as the Gumbel distribution for the maximum.

Two-parameter distributions are only able to model the location and scale of the distribution independently, while skewness and/or kurtosis are defined implicitly from those

[a]Emil Julius Gumbel (1891-1966) was a German mathematician and political writer whose anti-Nazi articles in the 1920s led him to leave Germany in 1932 for Paris and later the USA. He was instrumental in developing extreme value theory along with Leonard Tippett and Ronald Fisher. In 1958 he published the book on 'Statistics of Extremes' [Gumbel, 1958].

two parameters. For example, the normal distribution `NO`(μ, σ) has a location shift parameter μ (its mean) and a scaling parameter σ (its standard deviation). It is a symmetric and mesokurtic distribution, so both the moment skewness and the moment excess kurtosis are zero. Hence the skewness and kurtosis cannot be modeled in the normal distribution. The logistic distribution, `LO`(μ, σ), is also symmetric but has a higher kurtosis than the normal, as shown in Figure 4.2(a). For the Gumbel, `GU`(μ, σ), and reverse Gumbel, `RG`(μ, σ), distributions, the moment skewness is fixed at a negative and a positive value, respectively, and cannot be modeled as a function of explanatory variables. The `GU`(μ, σ), distribution is negatively skewed, appropriate for modeling minimum values of a random variable; while the `RG`(μ, σ) is positively skewed, appropriate for modeling maximum values of a random variable. Figure 4.2(b) compares the `NO`$(0, 1)$, `GU`$(0, 1)$, and `RG`$(0, 1)$ distributions.

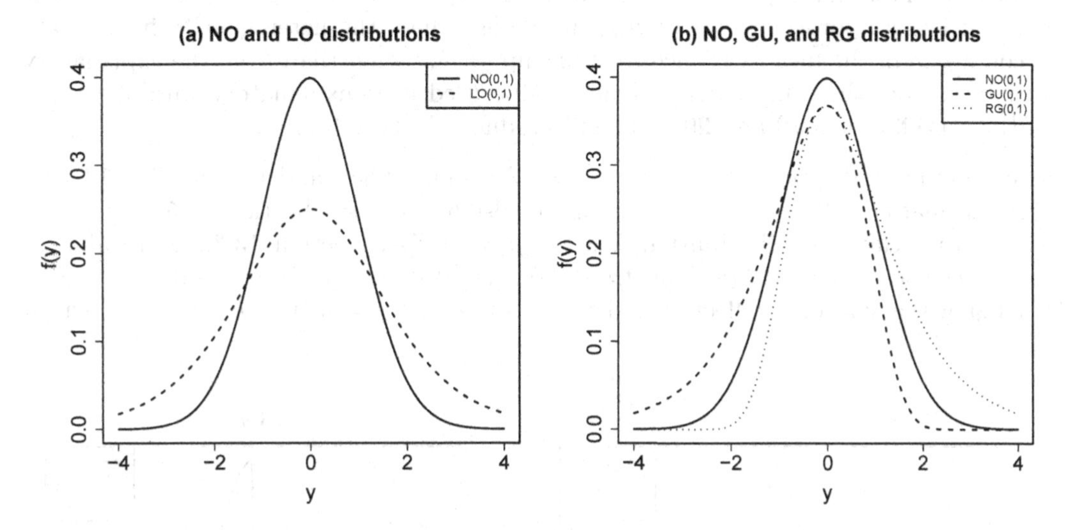

FIGURE 4.2: Plots of (a) normal `NO`(0,1), and logistic `LO`(0,1), (b) normal, `NO`(0,1), Gumbel, `GU`(0,1) and reverse Gumbel `RG`(0,1) distributions.

4.3 Three-parameter distributions on $(-\infty, \infty)$

The following three-parameter continuous distributions on $(-\infty, \infty)$ are in the `gamlss.family`:

exponential Gaussian: `exGAUS`(μ, σ, ν) for modeling a right-skewed response variable, popular with psychologists;

normal family: `NOF`(μ, σ, ν) for modeling power variance-to-mean relationships, i.e. $\text{Var}(Y) = \sigma^2 \mu^\nu$ in a symmetric response variable;

power exponential: `PE`(μ, σ, ν) and `PE2`(μ, σ, ν) for modeling a symmetric response variable which can be leptokurtic or platykurtic;

***t* family:** TF(μ, σ, ν) and TF2(μ, σ, ν) for modeling a symmetric response variable that is leptokurtic;

skew normal: SN1(μ, σ, ν) and SN2(μ, σ, ν) for modeling a response variable which can be positively or negatively skewed.

Three-parameter distributions are able to model either skewness or kurtosis in addition to location and scale. The exception here is the normal family, NOF(μ, σ, ν), for which the parameter ν exists to model the power variance-to-mean relationship. The NOF(μ, σ, ν), is a symmetric mesokurtic distribution. Note that NOF(μ, σ, ν) = NO($\mu, \sigma^2\mu^\nu$). In NOF(μ, σ, ν) the mean is equal to μ and the variance is

$$\text{Var}(Y) = \sigma^2\mu^\nu \ . \tag{4.1}$$

This type of relationship is an instance of Taylor's power law, see Enki et al. [2017]. The distribution is appropriate for regression models where the heterogeneity in the data is coming from the mean-variance relationship (4.1), rather than from the explanatory variables. Note that NOF(μ, σ, ν) should not be used if no explanatory variables exist. In this case NOF(μ, σ, ν) and NO(μ, σ) will produce identical fits.

The skew normal type 1, SN1(μ, σ, ν), (see Azzalini [1985], and also Section 13.4.1), skew normal type 2, SN2(μ, σ, ν), (a spliced distribution, see Section 13.5.2), and the exponential Gaussian distribution, exGAUS(μ, σ, ν), (see Section 13.3.2), are able to model skewness. Figure 4.3(a) plots the SN1($0, 1, \nu$) distribution for $\nu = -100, -3, -1, 0$. Changing the sign of ν reflects the distribution about $y = \mu$, i.e. $y = 0$, as shown in Figure 4.3(b).

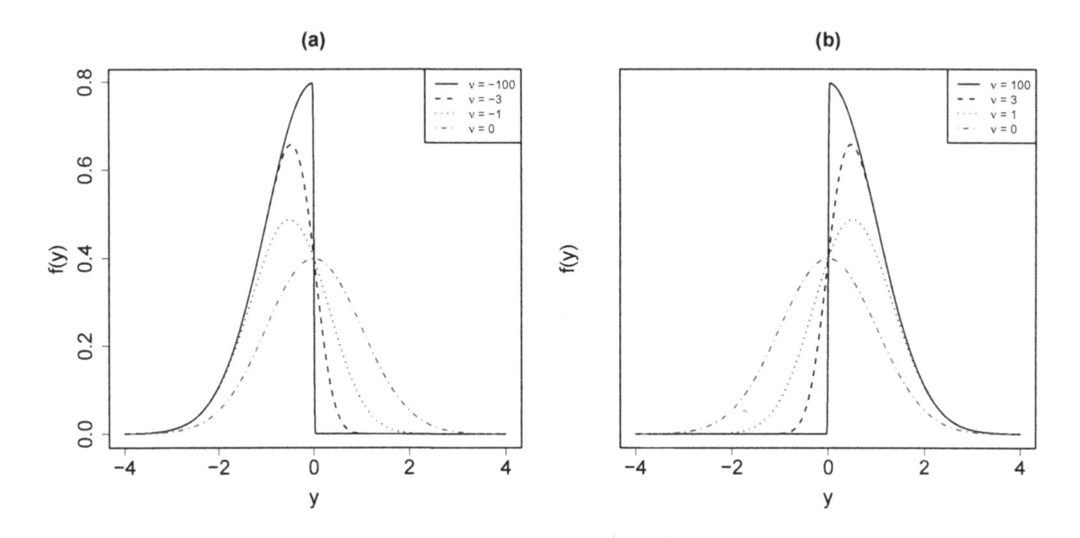

FIGURE 4.3: The skew normal type 1 distribution, SN1($0, 1, \nu$), (a) for $\nu = 0, -1, -3, -100$, and (b) for $\nu = 0, 1, 3, 100$.

The Student[b] *t*-distribution family, TF(μ, σ, ν), is symmetric but able to model leptokur-

[b]The *t*-distribution was developed by William Sealy Gosset in the paper 'The Probable Error of a Mean', which was published under the pseudonym Student, to deal with the problem of inference for the mean of a normally distributed population in situations where the sample size is small and the population standard deviation is unknown.

tosis. The parameter ν is known in the literature as the *degrees of freedom* of the t-distribution. For small values of ν the distribution becomes highly leptokurtic. When $\nu = 1$, we have the Cauchy distribution which is used as the typical example of a distribution in which the first two moments, i.e. mean and variance, do not exist. For values of $\nu > 100$ the distribution is almost indistinguishable from the normal distribution. The `TF2`(μ, σ, ν) is a reparameterization of `TF`(μ, σ, ν) in such a way that μ is the mean, σ is the standard deviation, and ν is always greater than 2. Therefore the distribution always has finite mean and standard deviation. Figure 4.4(a) shows the t-family distribution with $\nu = 1000, 5, 1$. The distribution at $\nu = 1$ is the Cauchy distribution while for $\nu = 1000$ is almost identical to the normal distribution.

The power exponential distribution, `PE`(μ, σ, ν), is symmetric and has mean μ and standard deviation σ. Figure 4.4(b) plots `PE`$(0, 1, \nu)$ for $\nu = 1, 2, 10, 1000$. The power exponential includes as a special cases (i) the Laplace (a two-sided exponential) when $\nu = 1$, (ii) the normal when $\nu = 2$, and (iii) the uniform distribution as a limiting case as $\nu \to \infty$. For $\nu < 2$ the distribution is leptokurtic, while for $\nu > 2$ it is platykurtic. Note that the power exponential can be more kurtotic than the Laplace when $0 < \nu < 1$. The power exponential type 2 distribution, `PE2`(μ, σ, ν), is a reparameterization of `PE`(μ, σ, ν).

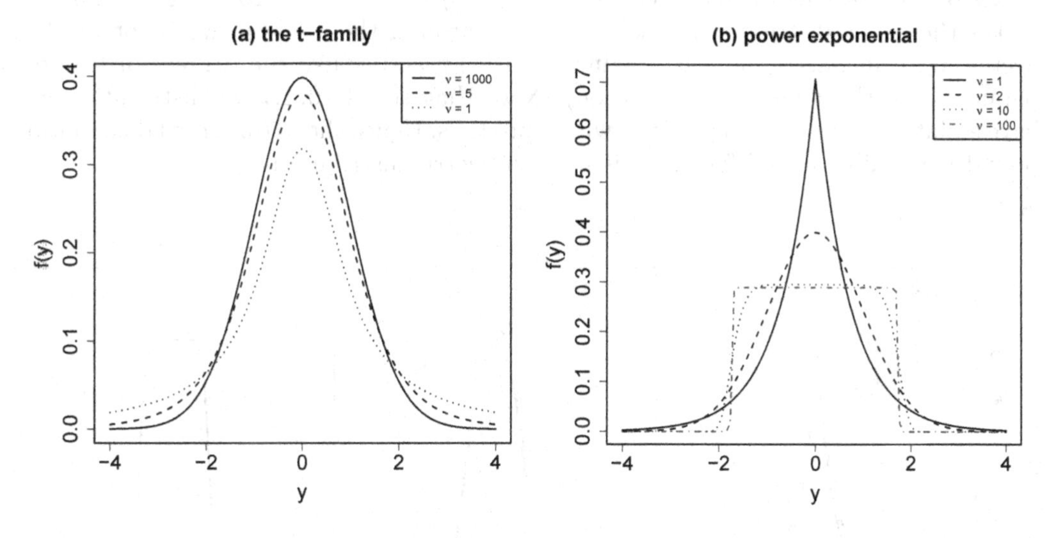

FIGURE 4.4: Plots of (a) t family, TF$(0, 1, \nu)$, for ν=1000, 5, and 1, and (b) the power exponential distributions, PE$(0, 1, \nu)$, for ν=1, 2, 10, and 1000.

4.4 Four-parameter distributions on $(-\infty, \infty)$

Four-parameter distributions are able to model both skewness and kurtosis in addition to location and scale. The only exceptions are the `NET` and `GT` distributions, which are symmetric with ν and τ both modeling kurtosis. The following four-parameter continuous distributions on $(-\infty, \infty)$ are in **gamlss.dist**:

exponential generalized beta type 2: EGB2(μ, σ, ν, τ) for modeling skewness and leptokurtosis;

generalized *t*: GT(μ, σ, ν, τ) a symmetric distribution capable of modeling kurtosis only;

Johnson's SU: JSU(μ, σ, ν, τ) and JSUo(μ, σ, ν, τ) for modeling skewness and leptokurtosis;

normal-exponential-*t*: NET(μ, σ, ν, τ), a symmetric distribution used to model location and scale parameters robustly;

skew exponential power: SEP1(μ, σ, ν, τ), SEP2(μ, σ, ν, τ), SEP3(μ, σ, ν, τ), and also SEP4(μ, σ, ν, τ) model skewness and both leptokurtosis and platykurtosis;

sinh-arcsinh: SHASH(μ, σ, ν, τ), SHASHo(μ, σ, ν, τ), and SHASHo2(μ, σ, ν, τ) model skewness and both leptokurtosis and platykurtosis;

skew *t*: ST1(μ, σ, ν, τ), ST2(μ, σ, ν, τ), ST3(μ, σ, ν, τ), ST4(μ, σ, ν, τ), ST5(μ, σ, ν, τ), and SST(μ, σ, ν, τ) model skewness and leptokurtosis.

Figure 4.5 shows the standardized generalized t distribution, GT$(0, 1, \nu, \tau)$, for different values of the two kurtosis parameters $\nu = (1, 2, 10)$ and $\tau = (1, 3, 10)$. The τ parameter makes the shape distribution spiky or smooth around the mode μ, while for fixed τ, decreasing ν increases the tail heaviness (with $\nu\tau$ acting like the degrees of freedom parameter of the t family distribution). Note that the GT$(0, 1, \nu, \tau)$ distribution can have a platy-shaped central part but with lepto tails. Hence the GT$(0, 1, \nu, \tau)$ distribution provides a flexible model for kurtosis in a symmetric distribution.

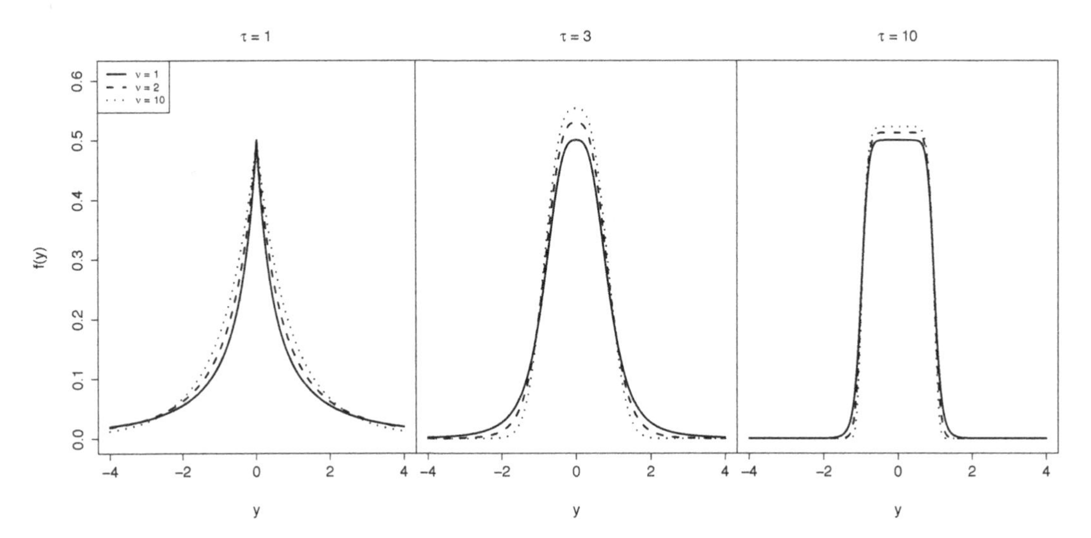

FIGURE 4.5: The generalized t distribution, GT(μ, σ, ν, τ), with $\mu = 0$, $\sigma = 1$, $\nu = 1, 2, 10$, and $\tau = 1, 3, 10$. Both ν and τ are kurtosis parameters.

Chapter 13 explains how some of the distributions presented in this chapter are generated. For example EGB2, JSU, SHASH, and ST5 are generated by transformation from a single random variable, see Section 13.2. JSUo is the original Johnson's SU distribution presented by Johnson [1949], while JSU is a reparameterized version in which μ is the mean and σ the standard deviation. There are three sinh-arcsinh distributions: the SHASH was presented in Jones [2005] and SHASHo in Jones and Pewsey [2009], together with its more stable parameterization SHASHo2. EGB2 was created by McDonald [1996], and ST5 by Jones and Faddy [2003].

The distributions SEP1, SEP2, ST1, and ST2 were generated using Azzalini's methods [Azzalini, 1986, Azzalini and Capitanio, 2003], as explained in Section 13.4.

The distributions SEP3, SEP4, ST3, and ST4 are generated by splicing, as explained in Section 13.5. The SST is a reparameterization of ST3 where μ is the mean and σ the standard deviation, and the condition $\tau > 2$ ensures the distribution has finite mean and standard deviation.

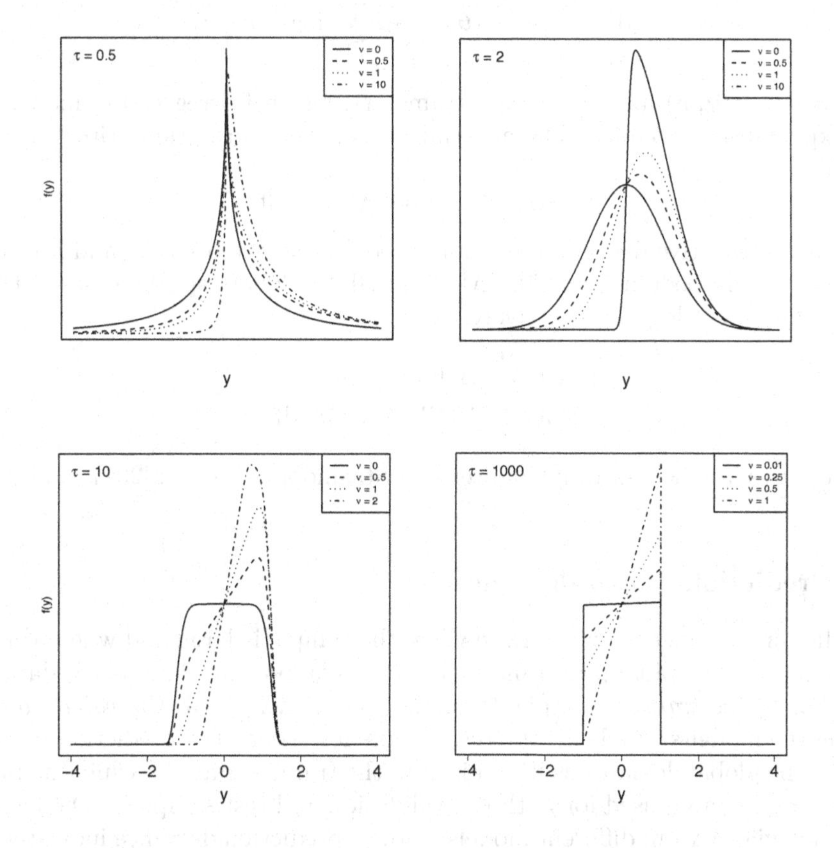

FIGURE 4.6: The skew exponential power type 1 distribution, $\texttt{SEP1}(0, 1, \nu, \tau)$.

Figure 4.6 shows the $\texttt{SEP1}(0, 1, \nu, \tau)$ distribution. Changing the sign of ν reflects the distributions about the origin, changing the skewness from positive to negative or vice versa. Setting $\nu = 0$ gives a symmetric distribution (which is a reparameterized power exponential distribution). This illustrates the great flexibility of skew exponential power distributions, which are able to model both skewness and kurtosis.

4.5 Choosing an appropriate distribution

The choice of an appropriate response distribution is based on how well the distribution fits the data as judged by (i) the generalized Akaike information criterion GAIC, (ii) the prediction global deviance VDEV, (iii) the residuals, or (iv) other criteria.

4.5.1 Generalized Akaike information criterion (GAIC)

The fitted global deviance is defined as minus twice the fitted log likelihood (defined in Section 10.2):

$$\text{GDEV} = -2\hat{\ell}(\hat{\boldsymbol{\theta}}) = -2\sum_{i=1}^{n} \log f(y_i \,|\, \hat{\boldsymbol{\theta}}) \tag{4.2}$$

where $\hat{\boldsymbol{\theta}} = (\hat{\boldsymbol{\mu}}, \hat{\boldsymbol{\sigma}}, \hat{\boldsymbol{\nu}}, \hat{\boldsymbol{\tau}})$ are the fitted parameters, which for regression models depend on the explanatory variables. The generalized Akaike information criterion, GAIC, is defined as

$$\text{GAIC}(\kappa) = \text{GDEV} + \kappa \cdot \text{df} \tag{4.3}$$

where df is the effective degrees of freedom used for the fitted model and κ is a penalty for each degree of freedom used. The AIC and BIC/SBC are special cases of the GAIC when $\kappa = 2$ and $\kappa = \log n$, respectively, i.e.

$$\begin{aligned} \text{AIC} &= \text{GDEV} + 2\,\text{df} \\ \text{SBC} &= \text{GDEV} + \log n \cdot \text{df}\,. \end{aligned}$$

These measures are discussed in Chapter 3 of Stasinopoulos et al. [2017] and in Section 11.5.5.

4.5.2 Prediction global deviance

The prediction deviance is appropriate when the sample is large and where the original observations can be split up into more than one component. The usual 'data mining' split is into: (i) the *training* sample, for fitting the model(s), (ii) the *validation* sample, for tuning the model(s), and (iii) the *test* sample for testing the prediction power of the model(s). The global deviance is minimized in the training sample, while the prediction global deviance can be used for both the validation and test samples. When comparing prediction deviance from different models, a lower prediction deviance indicates a better fitted model for predicting the behavior of the (independent) validation or test sample.

The prediction global deviance is defined as:

$$\text{VDEV} = \text{TDEV} = -2\,\tilde{\ell}(\tilde{\boldsymbol{\theta}}) = -2\sum_{i=1}^{n_P} \log f(\tilde{y}_i \,|\, \tilde{\boldsymbol{\theta}}) \tag{4.4}$$

where $\tilde{\mathbf{y}}$ indicates the response variable values at the validation (or test) sample, $\tilde{\boldsymbol{\theta}} = (\tilde{\boldsymbol{\mu}}, \tilde{\boldsymbol{\sigma}}, \tilde{\boldsymbol{\nu}}, \tilde{\boldsymbol{\mu}})$ are the fitted parameters for the validation (or test) sample (which in **gamlss** can be obtained using the functions `predict()`, or `predictAll()` with argument `newdata`). Note that the fitted parameters $\tilde{\boldsymbol{\theta}}$ depend on both the values of the explanatory variables in the validation (or test) sample, and on the fitted parameters of the model which in GAMLSS are $\hat{\boldsymbol{\beta}}$, $\hat{\boldsymbol{\gamma}}$ and $\hat{\boldsymbol{\lambda}}$, see Section 3.1 and Stasinopoulos et al. [2017, Chapter 3]. Sections 11.7 and 11.8 of Stasinopoulos et al. [2017] explain the **gamlss** functions for obtaining the validation (or test) deviance.

4.5.3 Residuals

The residuals used in **gamlss** (for a continuous response variable) are the normalized quantile residuals as defined in Stasinopoulos et al. [2017, Section 12.2]. If the GAMLSS model (including the distribution) is adequate for the data then the residuals approximate a random sample from the standard normal distribution. Stasinopoulos et al. [2017, Chapter 12] explains some of the GAMLSS diagnostics such as the worm plots of van Buuren and Fredriks [2001] and the Q-statistics of Royston and Wright [2000].

4.5.4 Other criteria

It is worth mentioning at this point that for fitting a specific distribution in **gamlss** the algorithm requires the `d` function which gives the pdf of the distribution (which defines the likelihood function), and (at least) the first (and optionally the expected second) derivatives with respect to the distribution parameters μ, σ, ν, and τ. The `p` function, the cdf of the distribution, is required at the end of the fitting algorithm in order to calculate the normalized quantile residuals. Therefore the cdf is used only once after fitting, while the pdf is used repeatedly during the fitting. The computation of fitted quantiles (or centiles) needs the `q` function, i.e. the inverse cdf. The `q` function is also needed for random variable generation used by the `r` function. Simulation of random variables when the `q` function is not given explicitly but calculated numerically can be slow.

Where more than one distribution fits the data adequately, the choice of distribution may be based on other criteria, depending on the properties of the particular distribution. For example a simple explicit[c] formula for the mean, median, or mode of Y may be desirable in a particular application.

The following are properties of the distribution that may be relevant in choosing the response variable distribution:

- The range of the distribution.
- Explicit pdf, cdf, and inverse cdf. As mentioned above, the explicit pdf is crucial for the speed of the fitting algorithm.
- Explicit (and simple) moment measures of location, scale, skewness, and kurtosis (i.e. population mean, standard deviation, γ_1 and γ_2 respectively), see Section 2.2 and the tables of Chapter 18. The explicit moments could help the interpretation of a model. (For example if μ is the mean of the distribution this helps interpretation of how explanatory variables affect this location measure.)
- Explicit centiles and centile-based measures of location, scale, skewness, and kurtosis

[c]The term explicit indicates that the particular function or measure can be obtained using closed-form mathematical functions available in **R**, i.e. not requiring additional numerical integration or numerical solution.

(i.e. median, semi-interquartile range, γ and δ, see Section 2.3). If the emphasis is on the quantiles of the distribution this is important.

- Continuity of $f(y \mid \mu, \sigma, \nu, \tau)$ and its derivatives with respect to y.
- Continuity of the derivatives of $f(y \mid \mu, \sigma, \nu, \tau)$ with respect to μ, σ, ν, and τ. This is important for convergence of the algorithm.
- Flexibility in modeling skewness and kurtosis. This is especially important when the emphasis in modeling is placed on the tails of the distribution rather than the center.
- Robustness of the MLE of the parameters of the distribution to the presence of outliers, see Chapter 12.
- The interpretation of the location parameter is especially important in regression models. Parameter μ is the mean for the `SST` and `JSU` distributions, μ is the median for the `SHASH` distribution, μ is the mode for the `SEP3`, `SEP4`, `ST3`, `ST4`, `SN2`, `GU`, and `RG` distributions and μ is the mean, median, and mode for all symmetric distributions, i.e. `GT`, `NET`, `TF`, `TF2`, `PE`, `PE2`, `NOF`, `NO`, `NO2`, and `LO`.

Many of the distributions of Table 4.1 can be generated by one or more of the methods described in Chapter 13, including univariate transformation, Azzalini type methods and splicing. Distributions generated by univariate transformation often satisfy all the desirable properties above, except perhaps the flexibility in modeling skewness and kurtosis.

An important disadvantage of distributions generated by Azzalini type methods (`SEP1`, `SEP2`, `ST1`, `ST2`, `SN1`) is that their cdf is not explicitly available, but requires numerical integration. Their inverse cdf requires a numerical search and many integrations. Consequently both functions can be slow, particularly for large data sets. Centiles and centile measures (e.g. the median) are not explicitly available. Moment measures are usually complicated, if available. However they can be very flexible in modeling skewness and kurtosis, see for example Figure 4.6.

An important disadvantage of distributions generated by splicing (`SEP3`, `SEP4`, `ST3`, `ST4`, `SN2`, `NET`) is sometimes a lack of continuity of the second derivatives of the pdf with respect to y and μ at the splicing point. However their mode and moment-based measures are explicit and they can be very flexible in modeling skewness and kurtosis.

Another important criterion of choosing a distribution is the robustness of the parameter estimation to outliers. Robustness in this context is how the maximum likelihood estimates of the distribution parameters are affected by extreme outlier values in the response variable. Robustness is discussed in more detail in Chapter 12. For distributions on $(-\infty, \infty)$, robustness to outliers means how the MLE of a specific parameter will be affected if an observation y tends to plus or minus infinity.

4.6 Example: DAX returns data

We use as illustration the DAX returns data from 1991 to 1998. The original DAX index is one of the four financial indices given in the **R** data set `EuStockMarkets`. Returns are calculated as the first difference of the natural logarithm of the series, i.e. $R_t = \log(Y_t/Y_{t-1})$. The objective is to fit a distribution to the DAX returns.

R data file: `EuStockMarkets` in the **R** package **datasets** of dimension 1860×4.
variables
`DAX` Germany DAX index,
`SMI` Switzerland SMI index,
`CAC` France CAC index,
`FTSE` : UK FTSE index.
purpose: to fit a GAMLSS family distribution with range $(-\infty, \infty)$.

We read the data, calculate the returns, plot the data over time, and construct a histogram of the sample of the `DAX` returns.

```
dax <- EuStockMarkets[,"DAX"]
Rdax<-diff(log(dax))
plot(Rdax, col=gray(.2)); title("(a)")
library(MASS)
truehist(Rdax, col=gray(.7)); title("(b)")
```

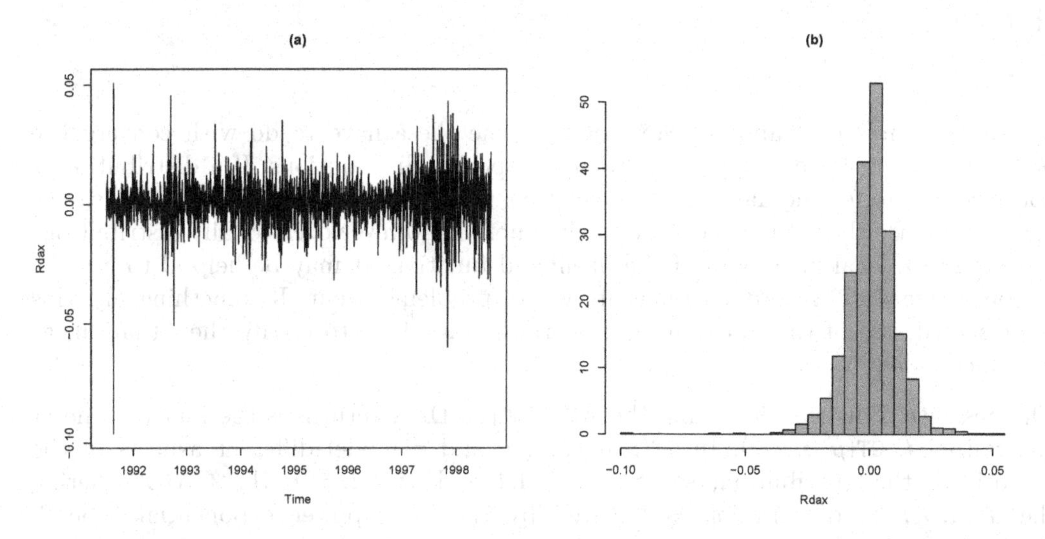

FIGURE 4.7: Plots of the DAX returns: (a) time series plot, (b) histogram.

Figure 4.7(a) shows a typical time series pattern for returns. The mean of the returns is close to zero, while the variance (i.e. variation around the mean) appears to be different over time.[d] Figure 4.7(a) indicates that there are some unusually large positive and negative observations, therefore kurtosis may be an issue in the returns. Here we will ignore the time series nature of the returns and attempt to fit a distribution to the returns shown in Figure 4.7(b). Since returns can take both negative and positive values, a continuous distribution on the real line is appropriate.

4.6.1 Selecting the distribution

`fitDist()`

The **gamlss** function `fitDist()` fits multiple distributions using maximum likelihood and automatically chooses the best fitting distribution according to the GAIC, with default penalty $\kappa = 2$ (i.e. the AIC). The resulting **R** object `f1` contains the 'final best' model, all the fitted distribution names and their corresponding GAIC values in increasing order stored in `f1$fits`, and the failed fits (if any) stored in `f1$failed`.

```
f1 <- fitDist(Rdax, type="realline")
f1$fits

##        GT      SEP2       PE2        PE       JSU      JSUo
## -11967.68 -11965.53 -11962.46 -11962.46 -11961.48 -11961.48
##      SEP1      SEP4      SEP3       TF2        TF       ST4
## -11961.31 -11961.27 -11960.82 -11960.64 -11960.64 -11960.15
##       ST1      EGB2       ST5       ST2       ST3       SST
## -11959.83 -11959.70 -11959.48 -11959.28 -11958.87 -11958.87
##     SHASH   SHASHo2    SHASHo        LO       NET       SN1
## -11957.54 -11956.24 -11956.24 -11932.11 -11919.18 -11759.87
##       SN2        NO    exGAUS        GU        RG
## -11739.13 -11733.21 -11733.17 -11262.12 -10249.88

f1$failed

## list()
```

Several warnings may appear. Most of the time these have to do with convergence problems that occurred during the fitting of specific distributions. If a distribution is not appropriate for the data set, the fact that it failed to converge is of no relevance, and in this case the warnings can be safely ignored. However occasionally distributions converge to a local maximum of the likelihood function. It may be helpful to use the option `trace=TRUE` in order to identify where problems occur. If something like this is suspected, use of the function `chooseDist()` may help to clarify the situation, as described below.

The best fitted distribution using the AIC for the DAX returns is the four-parameter generalized t, GT(μ, σ, ν, τ). In GT(μ, σ, ν, τ), ν and τ model different aspects of the kurtosis of the distribution; see Sections 4.4 and 18.4.2. In the GAIC ordering the GT(μ, σ, ν, τ) distribution is followed by the skew power exponential type 2,

[d] The variance of the time series process is called `volatility` in finance. Robert F. Engle was one of the first researchers to model volatility [Engle, 1982] by introducing the Autoregressive Conditional Heteroscedastic (ARCH) model. The ARCH model and its generalization the GARCH model [Bollerslev, 1986] are the standard techniques in finance for modeling volatility in returns. Engle was one of the winners of the 2003 Nobel Memorial Prize in Economic Sciences.

SEP2(μ, σ, ν, τ), and the power exponential, PE(μ, σ, ν). Note that the two power exponential distributions, PE2(μ, σ, ν) and PE(μ, σ, ν), have identical GAIC. This is the result of fitting reparameterized distributions when no explanatory variables are used. The fitted distributions in these cases are identical (and therefore have identical deviances), but the fitted parameters and their interpretations differ. If explanatory variables are fitted then PE2(μ, σ, ν) and PE(μ, σ, ν) would in general lead to different models.

All of the above 'best' distributions are able to model kurtosis, which of course does not come as a surprise, given the unusual observations in the tails of the returns. Note that the 'best' fitted distribution given by `fitDist()` is for a specific penalty κ with default $\kappa = 2$, i.e. the AIC. For different penalties the final 'best' model can be different. Next we will use the function `chooseDist()`, which is similar to but more general than `fitDist()`.

`chooseDist()`

`fitDist()` can be applied only to fitting distributions when no explanatory variables are in the model. The function `chooseDist()` can be used for any GAMLSS model. Additionally `chooseDist()` provides results for more than one penalty κ, simultaneously. `chooseDist()` requires a fitted **gamlss** model object as its first argument. This can be any conditional or marginal distribution model, (i.e. with or without explanatory variables respectively), fitted using `gamlss()` or `gamlssML()`. The following is an example of the use of `chooseDist()` to fit a variety of distributions to `Rdax`. A normal distribution model is fitted first to start the process.

```
m1 <- gamlssML(Rdax, family = NO)
t1 <- chooseDist(m1, type = "realline")
t1
```

```
## minimum GAIC(k= 2 ) family: GT
## minimum GAIC(k= 3.84 ) family: GT
## minimum GAIC(k= 7.53 ) family: PE2
##                  2       3.84       7.53
## NO       -11733.21 -11729.53 -11722.15
## GU       -11262.12 -11258.44 -11251.06
## RG       -10249.88 -10246.20 -10238.82
## LO       -11932.11 -11928.43 -11921.05
## . . .
## SST      -11958.87 -11951.51 -11936.75
```

The output of `chooseDist()` is a matrix `t1`, with rows for the different fitted distributions and columns giving the corresponding GAICs. In this case the distributions fitted are all the `gamlss.family` distributions on the real line, i.e. with range $(-\infty, \infty)$. The different GAICs are determined by the penalty argument `k`, with default values 2, 3.84 and $\log(n)$, i.e. AIC, χ^2_1 test, and BIC/SBC, respectively. The function `getOrder()` provides an ordering of a specified column according to the relevant GAIC. Below column 1 of `t1` is used, i.e. $GAIC(\kappa = 2) = AIC$.

```
getOrder(t1,1)[1:6]
```

```
## GAIG with k= 2
##          GT         SEP2           PE2             PE            JSU           JSUo
```

```
## -11967.68 -11965.53 -11962.46 -11962.46 -11961.48 -11961.48
```

No final fitted model is provided by `chooseDist()`, unlike `fitDist()`. The final model can be refitted using the `update()` function.

```
mf <- update(m1, family="GT")
```

4.6.2 Interpreting the fitted distribution

The `summary()` function is used to display the fitted coefficients and their standard errors.

```
summary(mf)

## ******************************************************************
## Family:  c("GT", "Generalized t")
##
## Call:
## gamlssML(formula = Rdax, family = "GT", data = sys.parent())
##
##
## Fitting method: "nlminb"
##
##
## Coefficient(s):
##                 Estimate   Std. Error    t value   Pr(>|t|)
## eta.mu       0.000683804  0.000196709    3.47623 0.00050852
## eta.sigma   -4.644880913  0.058193368  -79.81805 < 2.22e-16
## eta.nu       1.785097387  0.487622502    3.66082 0.00025141
## eta.tau      0.348678566  0.107100405    3.25562 0.00113144
##
## eta.mu     ***
## eta.sigma  ***
## eta.nu     ***
## eta.tau    **
## ---
## Signif. codes:
## 0 '***' 0.001 '**' 0.01 '*' 0.05 '.' 0.1 ' ' 1
##
##  Degrees of Freedom for the fit: 4 Residual Deg. of Freedom   1855
## Global Deviance:       -11975.7
##                 AIC:   -11967.7
##                 SBC:   -11945.6
```

There are two important points worth mentioning here. The first is the interpretation of the coefficients. Since the parameters μ, σ, ν, and τ of the GT distribution were fitted using default link functions: 'identity', 'log', 'log', and 'log', respectively (see Table 4.1), the coefficients shown: $\hat{\eta}_\mu = 0.000684$, $\hat{\eta}_\sigma = -4.644881$, $\hat{\eta}_\nu = 1.785097$, and $\hat{\eta}_\tau = 0.348679$ correspond to fitted parameters $\hat{\mu} = \hat{\eta}_\mu = 0.000684$, $\hat{\sigma} = \exp(\hat{\eta}_\sigma) = 0.00961$, $\hat{\nu} = \exp(\hat{\eta}_\nu) = 5.96016$ and $\hat{\tau} = \exp(\hat{\eta}_\tau) = 1.41719$, respectively. The value of the fitted parameters can also be obtained using:

```
fitted(f1,"mu")[1]

## [1] 0.0006838042

fitted(f1,"sigma")[1]

## [1] 0.009610674

fitted(f1,"nu")[1]

## [1] 5.96016

fitted(f1, "tau")[1]

## [1] 1.417194
```

The second point is the interpretation of the t tests. They are asymptotic Wald tests, testing the hypotheses $H_0 : \eta_\mu = \mu = 0$, $H_0 : \eta_\sigma = 0$, $H_0 : \eta_\nu = 0$, and $H_0 : \eta_\tau = 0$. The last three hypotheses are equivalent to the hypotheses $H_o : \sigma = 1$, $H_o : \nu = 1$, and $H_o : \tau = 1$. While t tests are useful in normal distribution regression models (with constant variance σ^2) when by testing the significance of a linear coefficient in the mean model one checks whether the explanatory variable affects the parameter μ (the mean) of the response, their usefulness is dubious in the current situation. In general the generalized likelihood ratio test provides a more reliable test for parameter values, see Section 11.5.4. Similarly profile deviance confidence intervals are more reliable than Wald confidence intervals, see Section 11.5.2.

A histogram of the data with the superimposed fitted distribution is obtained using the function `histDist()`. Below we use the argument `nbins=30`, to provide a more detailed histogram:

```
fh<-histDist(Rdax, family=GT, nbins=30, line.col="black")
```

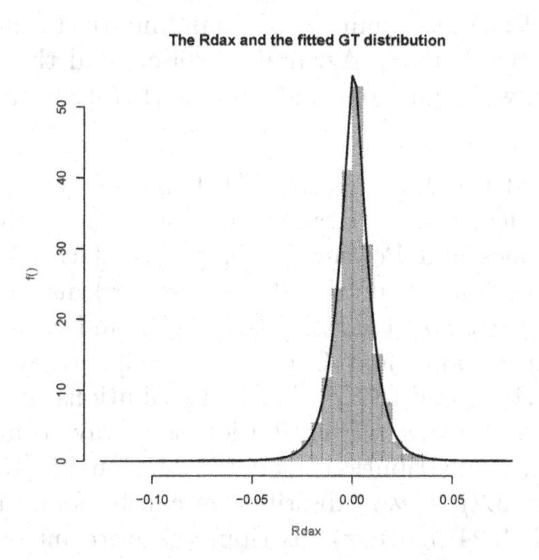

FIGURE 4.8: Histogram of DAX returns and the fitted $\texttt{GT}(\mu, \sigma, \nu, \tau)$ distribution.

4.7 Bibliographic notes

Most of the continuous distributions discussed in this and the next chapter can be found in Johnson et al. [1994] and Johnson et al. [1995], which are the classical reference books for continuous distributions. The normal (also known as the Gaussian) distribution, NO(μ, σ) and NO2(μ, σ), is the most famous distribution defined on $(-\infty, \infty)$. It was discovered independently by Gauss in 1809, by the mathematician Adrain in 1808, and also by Laplace in 1778 (in connection with the central limit theorem). The normal distribution, because of its association with least squares, was the main tool for distribution regression models until generalized linear models were introduced by Nelder and Wedderburn [1972].

The normal family, NOF(μ, σ, ν), was created to facilitate modeling of a power variance-to-mean relationship of the kind $\text{Var}(Y) = \sigma^2\mu^\nu$. The power exponential distribution, PE(μ, σ, ν), also known as the exponential power distribution, or the generalized error distribution [Box and Tiao, 1973] is a symmetric parametric family of distributions which contains the normal ($\nu = 2$) and Laplace ($\nu = 1$) distributions as special cases, and, as limiting cases when $\nu \to \infty$, includes all continuous uniform distributions on bounded intervals of the real line. The distribution can be used to model platy- ($\nu > 2$), meso- ($\nu = 2$), and leptokurtic ($\nu < 2$) response variables. The original t or 'Student' distribution was introduced by William Gosset for testing normal error means with common variance and further developed by R. A. Fisher. Its implementation in **gamlss** as t family, TF(μ, σ, ν), provides a way for modeling leptokurtosis especially when the values for the degrees of freedom parameter ν are small. For $\nu = 1$ we have the Cauchy, an extremely leptokurtic distribution, and as $\nu \to \infty$ the normal distribution. The exponential Gaussian, exGAUS(μ, σ, ν), distribution has been used in the literature for modeling reaction time, and the Gumbel and reverse Gumbel distributions are used to model the distribution of the maximum (or the minimum) of a number of samples of various distributions [Gumbel, 1958]. Azzalini [1986] created the skew normal type 1, SN1(μ, σ, ν), distribution, while Gibbons and Mylroie [1973] proposed the skew normal type 2, SN2(μ, σ, ν).

Johnson [1949] introduced the JSUo(μ, σ, ν, τ) distribution, while the authors of this book introduced the JSU(μ, σ, ν, τ) parameterization with μ the mean and σ the standard deviation. Jones and Pewsey [2009] proposed the SHASHo(μ, σ, ν, τ) and SHASHo2(μ, σ, ν, τ) distributions, while the SHASH(μ, σ, ν, τ) distribution was suggested by Jones [2005] in the discussion following the Rigby and Stasinopoulos [2005] paper. Azzalini [1986] and Azzalini and Capitanio [2003] created the SEP1(μ, σ, ν, τ), SEP2(μ, σ, ν, τ), ST1(μ, σ, ν, τ) and ST2(μ, σ, ν, τ) distributions. Rigby and Stasinopoulos [1994] proposed the NET(μ, σ, ν, τ) distribution as a way to fit robustly the mean and variance of a normal distribution model. McDonald [1991] introduced the GT(μ, σ, ν, τ), while the EGB2(μ, σ, ν, τ) distribution can be found in McDonald [1996]. The SEP3(μ, σ, ν, τ) and SEP4(μ, σ, ν, τ) distributions were introduced by Fernandez et al. [1995] and Jones [2005] respectively. The ST3(μ, σ, ν, τ) can be found in Fernandez and Steel [1998], while ST4(μ, σ, ν, τ) was created by the current authors. The reparameterization of ST3(μ, σ, ν, τ) to SST(μ, σ, ν, τ) can be found in Würtz et al. [2006]. ST5(μ, σ, ν, τ) was created by Jones and Faddy [2003].

4.8 Exercises

1. **Distributions for minima and maxima**

(a) Generate 1000 samples, each of length $n = 1000$, from a normal distribution with $\mu = 10$ and $\sigma = 2$. For each sample calculate its median, minimum, and maximum values.

```
Ymin <- Ymax <- Ymid <-rep(0,1000)
for (i in 1:1000)
{
  Sample <- rNO(1000, mu=10, sigma=2)
  Ymax[i] <- max(Sample)
  Ymin[i] <- min(Sample)
  Ymid[i] <- median(Sample)
}
```

(b) Plot the 1000 minima, maxima, and median values using `truehist()` and comment on the plots.

```
truehist(Ymax)
truehist(Ymin)
truehist(Ymid)
```

(c) According to theory, the appropriate asymptotic distribution for the minima values is the Gumbel GU(μ, σ), for the maxima the reverse Gumbel RG(μ, σ), and for the medians the normal NO(μ, σ). Fit those distributions to the 1000 generated values for the minima, maxima, and medians.

```
m1 <- histDist(Ymin,family=GU)
m2 <- histDist(Ymax,family=RG)
m3 <- histDist(Ymid,family=NO)
```

(d) Do the distributions fit well? Use worm plots to check :

```
wp(m1, ylim.all=1)
wp(m2, ylim.all=1)
wp(m3, ylim.all=1)
```

(e) Use `fitDist(..., type = "realline")` to fit all available distributions on the real line. How do the GU(μ, σ), RG(μ, σ) and NO(μ, σ) fitted in (c) compare?

```
m11<-fitDist(Ymin, type = "realline")
m11$fits
m22<-fitDist(Ymax, type = "realline")
m22$fits
m33<-fitDist(Ymid, type = "realline")
m33$fits
```

(f) Use worm plots to check the adequacy of the 'best' fitted distributions:

```
wp(m11,ylim.all=.5)
wp(m22,ylim.all=.5)
wp(m33,ylim.all=.5)
```

2. **Analyse the rest of the returns from the `EuStockMarkets` data.**

   (a) Get the UK FTSE returns

   ```
   ftse <- EuStockMarkets[,"FTSE"]
   Rftse<-diff(log(ftse))
   plot(Rftse)
   truehist(Rftse)
   ```

   (b) Find an appropriate distribution for the FTSE returns.

   ```
   m0 <- gamlssML(Rftse)
   m1 <-  chooseDist(m0, type="realline")
   getOrder(m1)
   ```

   (c) Repeat the above two steps to find the 'best' distributions for the Switzerland SMI and France CAC returns.

   (d) Check how well the distributions fit using residual diagnostics.

3. **Using prediction global deviance to select a distribution.** The idea here is to use the prediction deviance, rather than the GAIC, to select the fitted distribution. The function `fitDistPred()` provides the facility to do that when no explanatory variables exist in the data, while `chooseDistPred()` can be used when there are explanatory variables.

   (a) Read in the `DAX` returns and create a training data set (of around 60% of the original sample) and a test data set (around 40%).

   ```
   dax  <- EuStockMarkets[,"DAX"]
   Rdax <- diff(log(dax))
   set.seed(3210)
   rand <- sample(2, length(Rdax), replace=TRUE, prob=c(0.6,0.4))
   table(rand)/length(dax)
   traindata <- subset(Rdax,rand==1 )
   testdata  <- subset(Rdax,rand==2 )
   ```

   (b) Plot the training and test data sets and fit the 'best' distribution for the training set:

   ```
   truehist(traindata)
   truehist(testdata)
   m1 <- fitDist(traindata)
   m1$fits
   ```

   (c) Now use the test data set to find the best-fitting distribution using the validation/test deviance:

   ```
   m2 <- fitDistPred(traindata, newdata=testdata)
   m2$fits
   ```

(d) Note that by changing the seed, the results can be different, since the random split between the training and the test data will be different:

```
set.seed(1230)
rand <- sample(2, length(Rdax), replace=TRUE,
                  prob=c(0.6,0.4))
table(rand)/length(dax)
traindata <- subset(Rdax,rand==1 )
testdata  <- subset(Rdax,rand==2 )
m1.1 <- fitDist(traindata)
m2.1 <- fitDistPred(traindata, newdata=testdata)
m2.1$fits
```

(e) Comment on the results.

(f) This is an alternative way of fitting the above models in (d) using the functions `chooseDist()` and `chooseDistPred()`:

```
traindata <- data.frame(y =subset(Rdax,rand==1 ))
testdata <- data.frame(y =subset(Rdax,rand==2 ))
truehist(traindata$y)
truehist(testdata$y)
m0 <- gamlss(y~1, data=traindata)
T1 <- chooseDist(m0, type="realline")
getOrder(T1,1)
T2 <- chooseDistPred(m0, type="realline", newdata=testdata)
T2
getorder(T2,1)
```

5

Continuous distributions on $(0, \infty)$

CONTENTS

This chapter shows:

1. different types of continuous distributions defined on the $(0, \infty)$ within the GAMLSS family;
2. how these distributions model skewness and kurtosis;
3. transforming distributions from the real line $(-\infty, \infty)$ to the positive real line $(0, \infty)$; and
4. examples of using these distributions for real data.

5.1 Introduction

Table 5.1 gives the continuous distributions defined on the positive real line $(0, \infty)$ in the `gamlss.family`. This class contains one-, two-, three- and four-parameter distributions. Many of these distributions can be generated by one or more of the methods described in Chapter 13. Distributional properties can be found in Chapter 19.

TABLE 5.1: Continuous distributions on the positive real line, $\mathbb{R}_+ = (0, \infty)$, implemented within the **gamlss.dist** package.

Distribution	**gamlss** name	p	Parameter range μ	σ	ν	τ	skewness	kurtosis
Box-Cox Cole and Green	`BCCG`	3	$\mathbb{R}_+$	$\mathbb{R}_+$	$\mathbb{R}$	-	both	(both)
Box-Cox Cole and Green orig	`BCCGo`	3	$\mathbb{R}_+$	$\mathbb{R}_+$	$\mathbb{R}$		both	(both)
Box-Cox power exp	`BCPE`	4	$\mathbb{R}_+$	$\mathbb{R}_+$	$\mathbb{R}$	$\mathbb{R}_+$	both	both
Box-Cox power exp orig	`BCPEo`	4	$\mathbb{R}_+$	$\mathbb{R}_+$	$\mathbb{R}$	$\mathbb{R}_+$	both	both
Box-Cox t	`BCT`	4	$\mathbb{R}_+$	$\mathbb{R}_+$	$\mathbb{R}$	$\mathbb{R}_+$	both	lepto [a]
Box-Cox t orig	`BCTo`	4	$\mathbb{R}_+$	$\mathbb{R}_+$	$\mathbb{R}$	$\mathbb{R}_+$	both	lepto[a]
exponential	`EXP`	1	$\mathbb{R}_+$	-	-	-	(+ve)	(lepto)
gamma	`GA`	2	$\mathbb{R}_+$	$\mathbb{R}_+$	-	-	(+ve)	(lepto)
gamma family	`GAF`	3	$\mathbb{R}_+$	$\mathbb{R}_+$	$\mathbb{R}$	-	(+ve)	(lepto)
gen beta type 2	`GB2`	4	$\mathbb{R}_+$	$\mathbb{R}_+$	$\mathbb{R}_+$	$\mathbb{R}_+$	both[b]	both[b]
gen gamma	`GG`	3	$\mathbb{R}_+$	$\mathbb{R}_+$	$\mathbb{R}$	-	both	(both)
gen inverse Gaussian	`GIG`	3	$\mathbb{R}_+$	$\mathbb{R}_+$	$\mathbb{R}$	-	+ve[b]	(lepto)[b]
inverse gamma	`IGAMMA`	2	$\mathbb{R}_+$	$\mathbb{R}_+$	-	-	(+ve)	(lepto)
inverse Gaussian	`IG`	2	$\mathbb{R}_+$	$\mathbb{R}_+$	-	-	(+ve)	(lepto)
log normal	`LOGNO`	2	$\mathbb{R}$	$\mathbb{R}_+$	-	-	(+ve)	(lepto)
log normal 2	`LOGNO2`	2	$\mathbb{R}_+$	$\mathbb{R}_+$	-	-	(+ve)	(lepto)
log normal (Box-Cox)	`LNO`	2	$\mathbb{R}$	$\mathbb{R}_+$	fixed	-	both	(both)
Pareto 1 [c]	`PARETO1o`	2	fixed	$\mathbb{R}_+$	-	-	(+ve)	(lepto)
Pareto 2	`PARETO2`	2	$\mathbb{R}_+$	$\mathbb{R}_+$	-	-	(+ve)	(lepto)
Pareto 2 original	`PARETO2o`	2	$\mathbb{R}_+$	$\mathbb{R}_+$	-	-	(+ve)	(lepto)
Weibull	`WEI`	2	$\mathbb{R}_+$	$\mathbb{R}_+$	-	-	(both)	(both)
Weibull (PH)	`WEI2`	2	$\mathbb{R}_+$	$\mathbb{R}_+$	-	-	(both)	(both)
Weibull (μ the mean)	`WEI3`	2	$\mathbb{R}_+$	$\mathbb{R}_+$	-	-	(both)	(both)

Note that $\mathbb{R} = (-\infty, \infty)$.
[a] Usually.
[b] Conjecture, i.e. believed but not proven.
[c] `PARETO1o`(μ, σ) distribution has range (μ, ∞)

In Table 5.1, positive (+ve), negative (-ve), and both skewness, and leptokurtic (lepto), platykurtotic (platy), and both kurtosis, and brackets in either skewness or kurtosis columns are taken to have the same meanings as in Table 4.1.

Note that for distributions defined on $(0, \infty)$ an extreme outlier value is not only when $Y \to \infty$ but also when $Y \to 0$. Outlier values of Y close to zero often cannot be spotted easily in exploratory plots but, as for large outlier values of Y, may have unbounded influence on the MLE of one or more parameters of the distribution, see Chapter 12.

5.2 The scale family of distributions

All distributions in Table 5.1 are *scale* family distributions with scaling parameter μ, (for fixed σ, ν, and τ), except for `GAF`, `IG`, `LNO`, `LOGNO` and `WEI2`. A formal definition of the scale family of distributions is given in Section 19.1. Here we give an intuitive account.

If a random variable is distributed as

$$Y \sim \mathcal{D}(\mu, \sigma, \nu, \tau)$$

and

$$Z = (Y/\mu) \sim \mathcal{D}(1, \sigma, \nu, \tau)$$

then $\mathcal{D}(\mu, \sigma, \nu, \tau)$ is a scale family distribution with *scaling* parameter μ. The random variable $Y = \mu Z$ is a scaled version of Z and hence $Y_1 = aY \sim \mathcal{D}(a\mu, \sigma, \nu, \tau)$.

An advantage of $\mathcal{D}$ belonging to a scale family of distributions is that we can divide or multiply Y by any constant and the shape of its distribution will not change. So while the scaling parameter μ is important to determine the location and scale of the distribution, it does not affect its shape. A fitted GAMLSS model with a scale family response distribution is *invariant* to a scale change in the unit of measurement of the response variable, e.g. height from inches to centimeters, if either

- a log link is used for μ and the predictor model for μ includes a constant term, or
- an identity link is used for μ.

The model is invariant in the sense that the fitted model for Y_1 is the same as the fitted model for Y when it is converted back to the original unit of measurement.

A second advantage is that if we study the shape of a scale family distribution, there is no need to vary the value of μ, since the shape of the distribution for fixed σ, ν, and τ remains the same. This can be seen in Figure 5.1, where the shape of the distribution for each value of σ is the same when $\mu = 1$ (left panel) or $\mu = 2$ (right panel). Hence for most distributions displayed in this chapter and in Chapter 19 we use $\mu = 1$.

A consequence of a scale family distribution defined on $(0, \infty)$ is that they have a square variance-mean relationship. This can be shown from the fact that, since $Y = \mu Z$, $\text{E}(Y) = \mu \text{E}(Z)$ so $\mu = \text{E}(Y)/\text{E}(Z)$ and $\text{Var}(Y) = \mu^2 \text{Var}(Z) = [\text{E}(Y)]^2 \text{Var}(Z)/[\text{E}(Z)]^2$. Now $\text{E}(Z)$ and $\text{Var}(Z)$ do not depend on μ, but possibly on other distribution parameters. Therefore if those extra parameters are fixed then $\text{Var}(Y)$ is proportional to $[\text{E}(Y)]^2$. Hence the variance of Y is proportional to the square of the mean for scale family distributions with scaling parameter μ, for fixed parameters σ, ν, and τ.

The distributions `IG`(μ, σ), `LOGNO`(μ, σ), and `WEI2`(μ, σ) can be reparameterized to a scale family of distributions, but none of their current parameters is a scaling parameter.

5.3 One- and two-parameter distributions on $(0, \infty)$

The exponential distribution `EXP`(μ) is the only one-parameter distribution with range $(0, \infty)$ in **gamlss.dist**. It is only able to model the location of the distribution, since the scale, skewness, and kurtosis are determined by the location parameter μ. The exponential is a special case of the gamma distribution: `EXP`(μ) = `GA`$(\mu, 1)$.

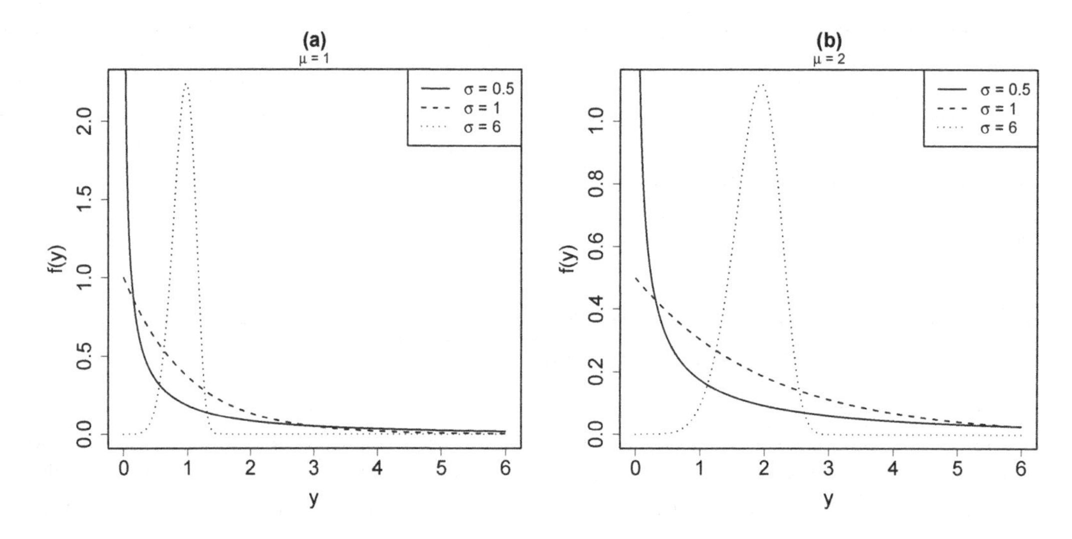

FIGURE 5.1: Plots of the Weibull distribution, WEI(μ,σ): (a) $\mu = 1$ and $\sigma = 0.5, 1, 6$ (b) $\mu = 2$ and $\sigma = 0.5, 1, 6$. The shape of the distribution in panels (a) and (b) is identical, since WEI(μ,σ) is a scale family distribution with scaling parameter μ.

The two-parameter distributions on $(0,\infty)$ are only able to model location and scale. The skewness and kurtosis of the distributions are determined by the location and scale values. The explicit two-parameter distributions on $(0,\infty)$ in **gamlss.dist** are:

gamma: GA(μ,σ). A very popular distribution and member of the exponential family of distributions;

inverse gamma: IGAMMA(μ,σ). A distribution much-used as a prior in Bayesian analysis;

inverse Gaussian: IG(μ,σ). A member of the exponential family. IG has a higher positive (moment) skewness and a heavier right tail than the gamma distribution GA, when they have the same mean and variance;

log normal: LOGNO(μ,σ) and LOGNO2(μ,σ). The log normal is associated with geometric Brownian motion in financial modeling. The difference between LOGNO(μ,σ) and LOGNO2(μ,σ) is that in the first μ can take values on $(-\infty,\infty)$ and $\log(\mu)$ is the median of the distribution, while in the second μ is the median with range $(0,\infty)$;

Pareto: PARETO1o(μ,σ), PARETO2(μ,σ) and PARETO2o(μ,σ). This is the well-known heavy-tailed distribution for modeling extreme events. Note $Y \sim$ PARETO1o(μ,σ) has a survival function, $S_Y(y)$, exactly (rather than asymptotically) proportional to $y^{-\sigma}$, and hence a log survival function, $\log S_Y(y)$, exactly linear in $\log y$. It is very important in the theory of heavy tails, see Chapter 17. PARETO2 and PARETO2o are different parameterizations of the same distribution, with PARETO2$(\mu,\sigma) =$ PARETO2o$(\mu, 1/\sigma)$.

Weibull: WEI(μ,σ), WEI2(μ,σ), and WEI3(μ,σ). The first parameterization is the standard parameterization of the Weibull with mean equal to $\mu(\log 2)^{1/\sigma}$; the second is the parameterization used for the proportional hazards model in survival analysis; and the third has μ as its mean.

The exponential, Weibull, and gamma distributions are widely used in survival and reliability analysis. In survival analysis the interest lies on how explanatory factors affect the survival time of individuals, and in reliability in how they affect the reliability of components of a system. Statistically, the important feature of the data in both cases is that some individuals may still be alive or some components may still be working at the end of the observation period. This gives rise to some cases having response variable values lying in an interval beyond the end of their final observed time, a phenomenon known as *censoring*. Censored (or interval) response variable values affect the way the likelihood function of the model is defined, see Chapter 11.

Figure 5.1 shows the pdf's for the Weibull (`WEI`) distribution, for $\mu = 1, 2$ and $\sigma = 0.5, 1, 6$. The moment skewness of this distribution is positive for $\sigma \leq 3.6023$ and negative otherwise. The survival and hazard functions, defined in Section 1.5, are important in survival and reliability analysis. These functions are shown in Figures 5.2(a) and 5.2(b), respectively, for the Weibull distribution. Hazard functions are not automatically given in **gamlss** but are generated using the function `gen.hazard()`.

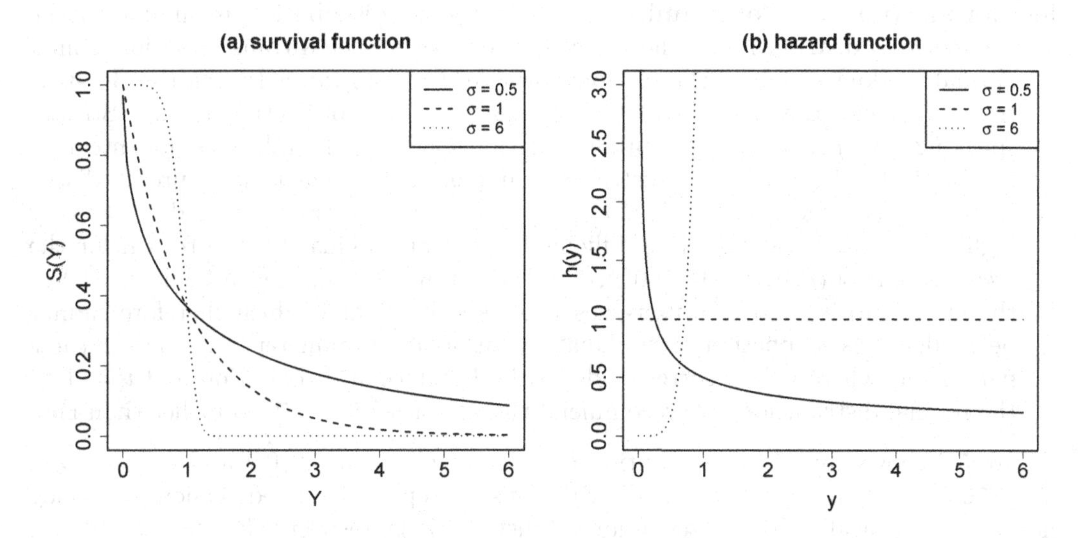

FIGURE 5.2: Weibull distribution, WEI(μ, σ) (a) survival function and (b) hazard function for $\mu = 1$ and $\sigma = 0.5, 1, 6$.

5.4 Three-parameter distributions on $(0, \infty)$

Three-parameter distributions on $(0, \infty)$ are able to model skewness in addition to location and scale. The explicit three-parameter distributions on $(0, \infty)$ in **gamlss.dist** are:

Box-Cox Cole and Green: The `BCCG`(μ, σ, ν) and `BCCGo`(μ, σ, ν) distributions are implementations of the distribution used in the LMS method in centile estimation [Cole and Green, 1992]. The only difference between BCCG and BCCGo is their default link function for the μ parameter in their **gamlss.dist** implementations: BCCG has the identity link as default, as the original LMS method; BCCGo has the log link, which is more appropriate in general since $\mu > 0$. The BCCG distribution is defined by assuming that a random variable Z has a truncated standard normal distribution and is transformed using equation (13.6) of Section 13.2.3. The parameter μ is approximately the median of the distribution, σ is often approximately the coefficient of variation, while ν is a skewness parameter.

generalized gamma: The GG(μ, σ, ν) distribution was introduced by Lopatatzidis and Green [2000] in an unpublished paper kindly provided to us by the first author. It can be derived by transformation by assuming that $Z = (Y/\mu)^\nu$ has a gamma GA$(1, \sigma\nu)$ distribution with mean 1 and variance $\sigma^2\nu^2$, see Section 13.2.3. The GG(μ, σ, ν) is a reparameterized version of the distribution given by Johnson et al. [1995, p. 401].

generalized inverse Gaussian: The GIG(μ, σ, ν) distribution is a parameterization of the generalized inverse Gaussian distribution of Jørgensen [1982]. The GG and GIG distributions may provide good fits to response variables that include values close to zero, see Figures 19.10 and 19.11.

log normal (i.e Box-Cox) family: The LNO(μ, σ, ν) distribution is similar to the BCCG(μ, σ, ν) distribution in the sense that it is derived by a transformation from a normal random variable. The random variable Z is transformed using the Box-Cox transformation [Box and Cox, 1964] of equation (13.7) of Section 13.2.3. (Strictly speaking, LNO(μ, σ, ν) is not a proper distribution since the pdf does not integrate to exactly 1. This is the only distribution in **gamlss.dist** that is not a proper distribution.) Also unfortunately in LNO(μ, σ, ν) the parameters μ, σ, and ν are very highly correlated and therefore difficult to estimate. (This was the reason for the creation of BCCG(μ, σ, ν).) In LNO(μ, σ, ν) the parameter ν (sometimes known as λ in the literature) is fixed to a user-chosen pre-specified value which therefore cannot be modeled as a function of explanatory variables. Parameter ν is the skewness parameter, where $\nu = 0$ corresponds to the log normal distribution and $\nu = 1$ to the normal distribution. We recommend the use of BCCG or BCCGo rather than LNO.

Figure 5.3 shows the BCCG distribution for $\mu = 1$, $\sigma = 0.15, 0.2, 0.5$ and $\nu = -2, 0, 4$. The BCCG$(\mu, \sigma, 0)$ and BCCG$(\mu, \sigma, 1)$ distributions are reparameterized versions of the log normal and truncated (below zero) normal distributions, respectively.

5.5 Four-parameter distributions on $(0, \infty)$

Four-parameter distributions on $(0, \infty)$ can model location, scale, and both skewness and kurtosis.

There are only three four-parameter distributions on $(0, \infty)$ implemented in **gamlss.dist**, but more can be generated using the techniques described in Section 5.6. The explicit four-parameter distributions in **gamlss.dist** are:

Box-Cox power exponential: There are two implementations of the Box-Cox power exponential, BCPE(μ, σ, ν, τ) and BCPEo(μ, σ, ν, τ). They differ only in the way the

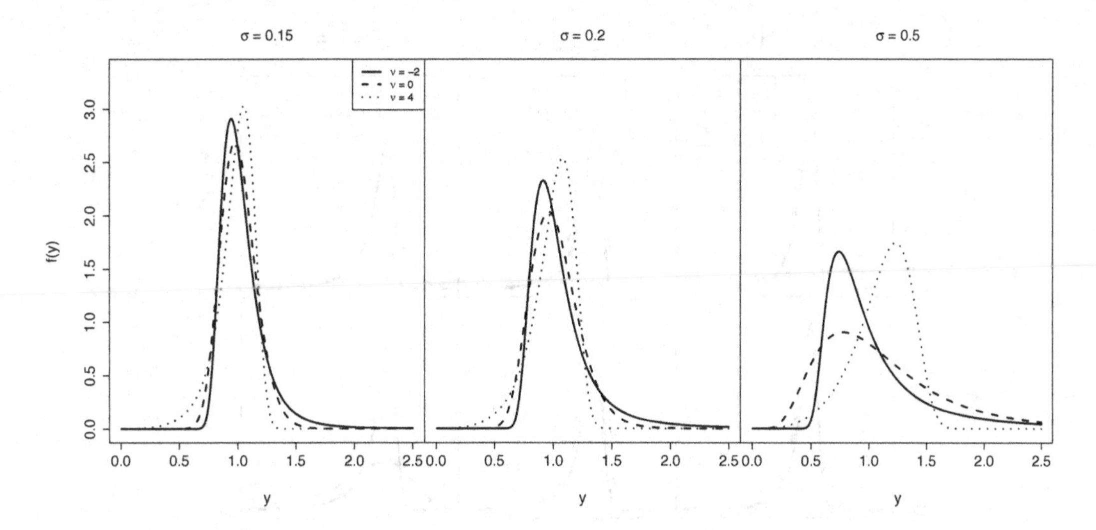

FIGURE 5.3: The $\texttt{BCCG}(\mu, \sigma, \nu)$ distribution, with $\mu = 1$, $\sigma = 0.15, 0.2, 0.5$, and $\nu = -2, 0, 4$.

default link for μ is implemented. $\texttt{BCPE}(\mu, \sigma, \nu, \tau)$ uses the identity link for μ, while $\texttt{BCPEo}(\mu, \sigma, \nu, \tau)$ uses the log link. The $\texttt{BCPE}(\mu, \sigma, \nu, \tau)$ is derived by assuming Z given by (13.6) in Section 13.2.3 has a truncated standard power exponential distribution, $\texttt{PE}(0, 1, \tau)$. $\texttt{BCPE}(\mu, \sigma, \nu, \tau)$ can model both positive and negative skewness and both platykurtosis and leptokurtosis;

Box-Cox *t*: $\texttt{BCT}(\mu, \sigma, \nu, \tau)$ and $\texttt{BCTo}(\mu, \sigma, \nu, \tau)$ are two implementations of the Box-Cox t distribution with default identity and log links for μ, respectively. The $\texttt{BCT}(\mu, \sigma, \nu, \tau)$ is derived by assuming Z given by (13.6) in Section 13.2.3 has a truncated t distribution, $\texttt{TF}(0, 1, \tau)$. It can model both positive and negative skewness, and leptokurtosis. In our experience both BCT and BCPE distributions are very flexible, reliable, and fit a variety of response variables well, especially when the values of the response variable are not close to zero. Figure 5.4 shows the $\texttt{BCT}(\mu, \sigma, \nu, \tau)$ distribution with $\mu = 1$ and different values of σ, ν, and τ;

generalized beta type 2: The $\texttt{GB2}(\mu, \sigma, \nu, \tau)$ is a flexible distribution which contains various known distributions as special cases. For example, setting $\sigma = 1$ we have a specific form of the Pearson type VI distribution. The Burr XII (or Singh-Maddala) distribution is given when $\nu = 1$. The Burr III (or Dagum) distribution is given when $\tau = 1$. For $\sigma = \nu = 1$ we have the Pareto distribution, $\texttt{PARETO2o}(\mu, \sigma)$, while $\nu = \tau = 1$ gives the log logistic distribution.

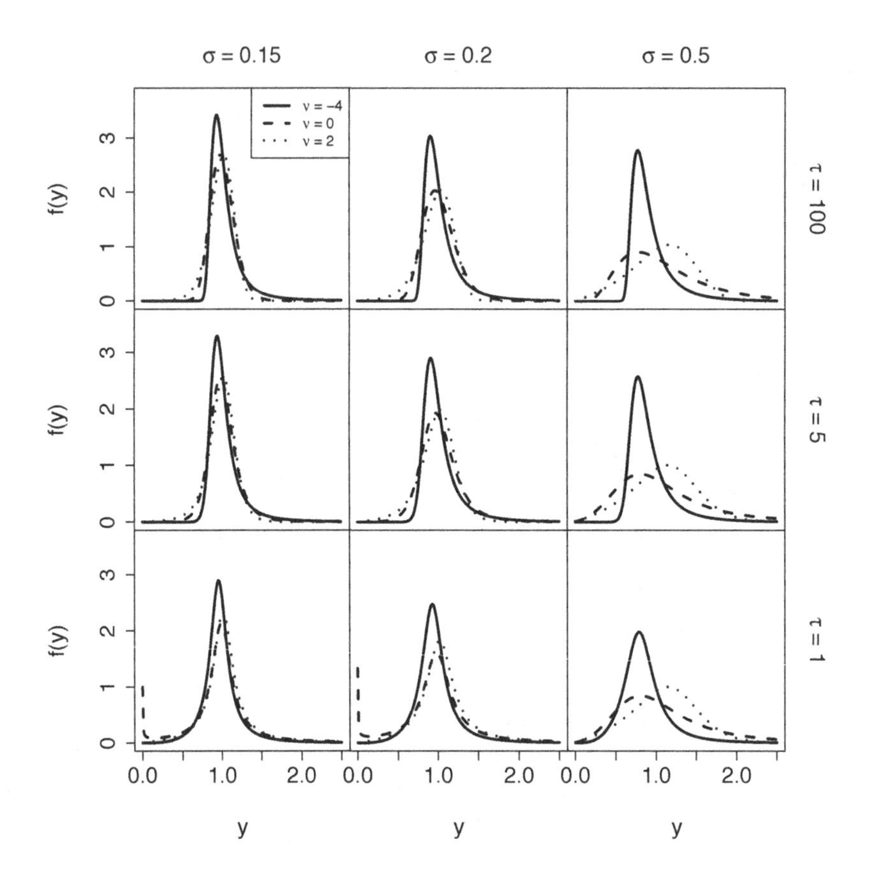

FIGURE 5.4: Box-Cox t distribution $\texttt{BCT}(\mu, \sigma, \nu, \tau)$. Parameter values $\mu = 1, \quad \sigma = 0.15, 0.2, 0.5, \quad \nu = -4, 0, 2, \quad \tau = 1, 5, 100$.

5.6 Creating distributions on $(0, \infty)$

Distributions defined on the positive real line $(0, \infty)$ may be created using transformations of a random variable defined on the real line $(-\infty, \infty)$. Section 5.6.1 deals with log transformations and Section 5.6.2 with truncated distributions.

5.6.1 Log transform distributions on $(0, \infty)$

Any continuous random variable Z defined on $(-\infty, \infty)$ can be transformed by the inverse log (i.e. exponential) transformation $Y = \exp(Z)$ to a random variable Y defined on $(0, \infty)$. The resulting distribution is called a log transform distribution. For example, if Z has the t family distribution, i.e. $Z \sim \texttt{TF}(\mu, \sigma, \nu)$, and the exponential transformation is applied, then $Y \sim \texttt{logTF}(\mu, \sigma, \nu)$, a log t family distribution on $(0, \infty)$. For more information about log transformations see Section 13.2.1.

The following is an example on how to take a `gamlss.family` distribution on $(-\infty, \infty)$ and create a corresponding log transform distribution. The function `gen.Family()` of the **gamlss.dist** package generates the d, p, q, and r functions of the log transform distribution, together with the function which is used for fitting within **gamlss**. We generate the log t family distribution logTF(μ, σ, ν), and plot the distribution for different values of μ, σ, and ν. Note that μ, σ and ν are defined on the original t family distribution ranges, i.e. $(-\infty, \infty)$ for μ, and $(0, \infty)$ for σ and ν. This implies that $\exp(\mu)$ is the median rather than the mean of the logTF(μ, σ, ν) distribution, since the t family distribution is symmetric with median μ. Also σ and ν are related to the scale and shape of the distribution.

```
gen.Family("TF", type="log")

## A  log  family of distributions from TF has been generated
##   and saved under the names:
##   dlogTF plogTF qlogTF rlogTF logTF
```

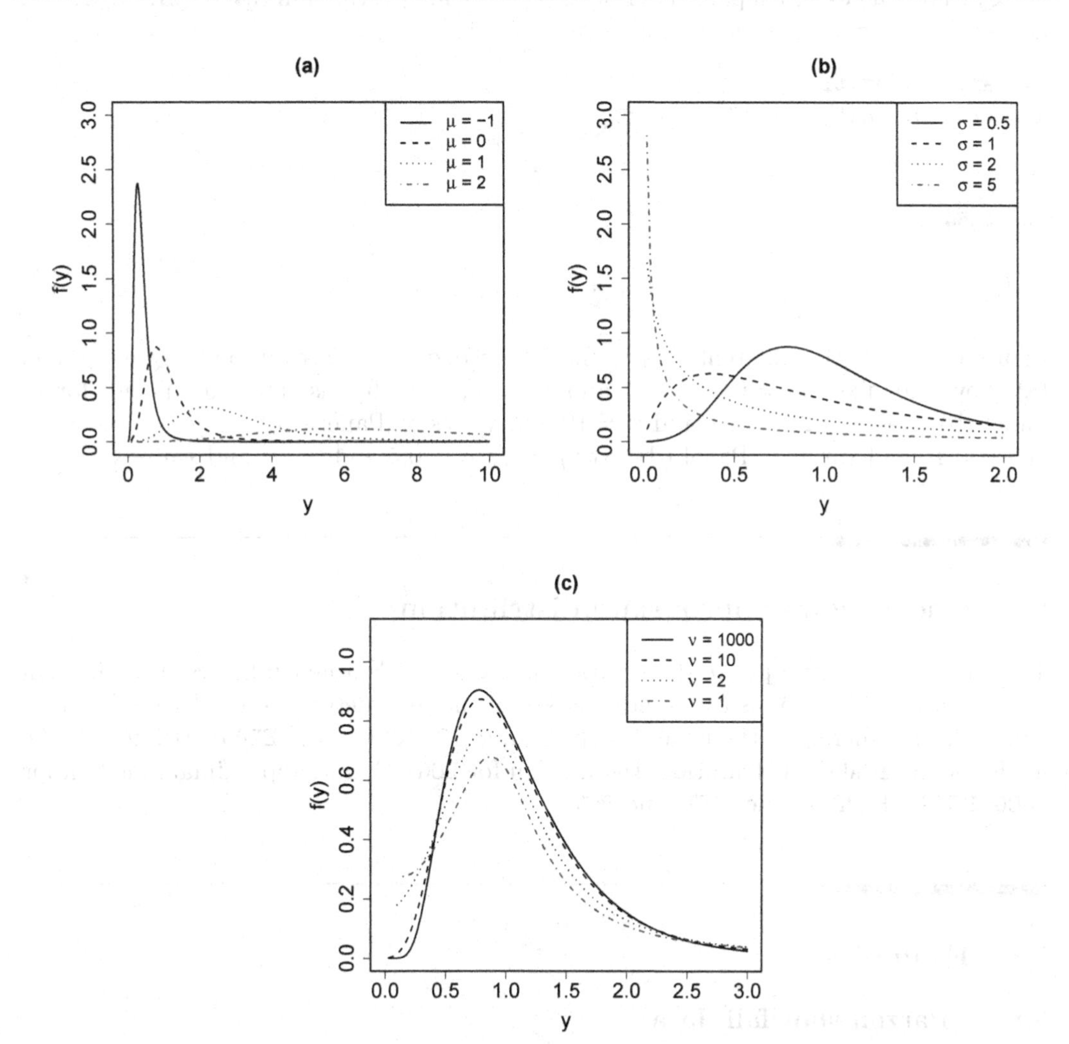

FIGURE 5.5: The log-t distribution: (a) $\mu = -1, 0, 1, 2,\ \ \sigma = 0.5$, and $\nu = 10$, (b) $\mu = 0,\ \ \sigma = 0.5, 1, 2, 5$, and $\nu = 10$, and (c) $\mu = 0,\ \ \sigma = 0.5$, and $\nu = 1000, 10, 2, 1$.

Figure 5.5 shows the different shapes of the logTF(μ, σ, ν) distribution. Panel (a) shows, for fixed $\sigma = 0.5$ and $\nu = 10$, how the distribution changes for different values of μ, panel (b) fixes $\mu = 0$ and $\nu = 10$ and varies σ, and finally panel (c) fixes $\mu = 0$ and $\sigma = 0.5$ and varies ν.

5.6.2 Truncated distributions on $(0, \infty)$

Any distribution defined on the real line $(-\infty, \infty)$ can be left-truncated at zero to give a truncated distribution on $(0, \infty)$ using the function `gen.trun()` from the **gamlss.tr** package.

In the example below we transform a skew t distribution, SST(μ, σ, ν, τ), by left-truncating it at zero to give a truncated SST distribution, SSTtr(μ, σ, ν, τ), defined on $(0, \infty)$. The range of each parameter of SSTtr(μ, σ, ν, τ) is the same as for SST(μ, σ, ν, τ), i.e. $-\infty < \mu < \infty$, $\sigma > 0$, $\nu > 0$ and $\tau > 2$.

```
library(gamlss.tr)
gen.trun(0,"SST",type="left")

## A truncated family of distributions from SST has been generated
##  and saved under the names:
##  dSSTtr pSSTtr qSSTtr rSSTtr SSTtr
## The type of truncation is left
##  and the truncation parameter is 0
```

Figure 6.3 shows the different shapes the SSTtr(μ, σ, ν, τ) distribution can take. Panel (a) shows, for fixed $\sigma = 0.5$, $\nu = 1$ and $\tau = 10$, how the distribution changes for μ. Panel (b) fixes $\mu = 2$, $\nu = 1$ and $\tau = 10$ and varies σ. Panel (c) fixes $\mu = 2$, $\sigma = 0.5$ and $\tau = 10$ and varies ν. Panel (d) fixes $\mu = 2$, $\sigma = 0.5$ and $\nu = 1$ and varies τ.

5.7 Choosing an appropriate distribution

The choice of an appropriate distribution was discussed in general in Section 4.5. An important consideration is the interpretation of the parameters, especially the location parameter μ, where μ is the mean for the GIG, WEI3, IG, GA, and EXP distributions, the mode for the IGAMMA distribution, the median for LOGNO2, and approximate median for BCCG, BCCGo, BCPE, BCPEo, BCT, and BCT.

5.8 Examples

5.8.1 Parzen snowfall data

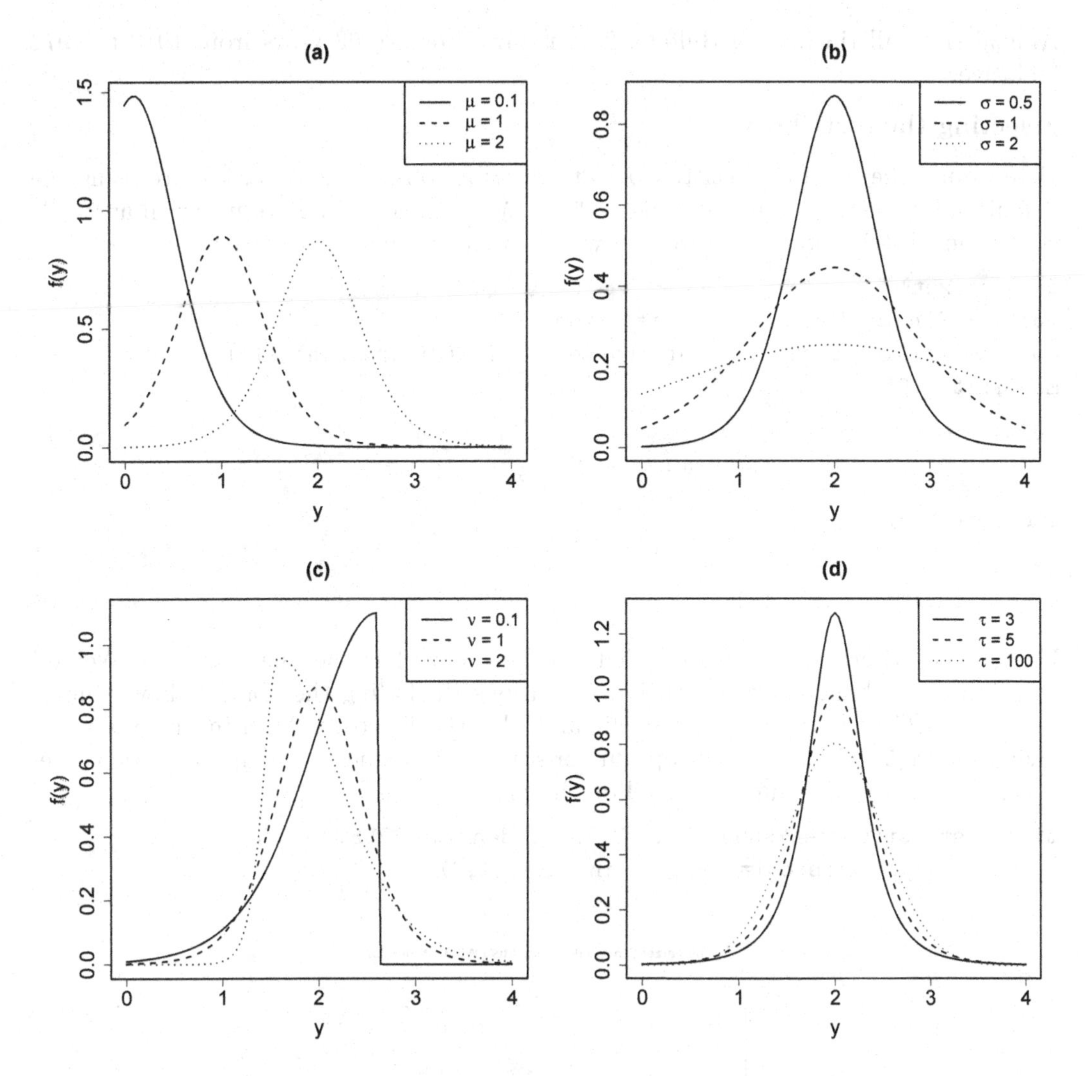

FIGURE 5.6: The truncated below zero SST distribution: (a) $\mu = 0.1, 1, 2$, $\sigma = 0.5$, $\nu = 1$, and $\tau = 10$, (b) $\mu = 2$, $\sigma = 0.5, 1, 2$, $\nu = 1$, and $\tau = 10$, (c) $\mu = 2$, $\sigma = 0.5$, $\nu = 0.1, 1, 2$, and $\tau = 10$, (d) $\mu = 2$, $\sigma = 0.5$, $\nu = 1$, and $\tau = 3, 5, 100$.

R data file: `parzen` in package **gamlss.data** of dimension 63×1
source: Hand et al. [1994]
variables
`snowfall` : the annual snowfall in Buffalo, NY (inches) from 1910 to 1972 inclusive.
purpose: to demonstrate the fitting of continuous distribution to a single variable.
conclusion: the Weibull distribution appears to fit best.

This data set is used by Parzen [1979] and is also in Hand et al. [1994, data set 278].

Annual snowfall (inches) in Buffalo, NY, is given for the 63 years from 1910 to 1972 inclusive.

Selecting the distribution

Here we use the function `fitDist()` to fit distributions to the data. We are using the default value for the argument `type = "realAll"`, meaning we are using all available continuous distributions. Also we try two different information criteria: AIC and SBC.

```
data(parzen)
mod1 <- fitDist(snowfall, data=parzen, k=2)
mod2 <- fitDist(snowfall, data=parzen, k=log(dim(parzen)[1]))
mod1$fit[1:6]

##      WEI     WEI2     WEI3       NO       PE      PE2
## 579.9043 579.9043 579.9043 580.7331 581.3779 581.3779

mod2$fit[1:6]

##      WEI     WEI2     WEI3       NO       PE      PE2
## 584.1905 584.1905 584.1905 585.0194 587.8073 587.8073
```

Using both criteria, it is obvious that the best model is the one using the Weibull distribution, although several other distributions (including the normal) have similar values of AIC and SBC. Next we refit and plot the fitted model using `histDist()`, giving Figure 5.7. Note that the option `density=TRUE` requests a nonparametric kernel density estimate to be superimposed on the plot.

```
m1 <-histDist(parzen$snowfall, "WEI3" , density=TRUE,
                  line.col=c(1,1), line.ty=c(1,2))
```

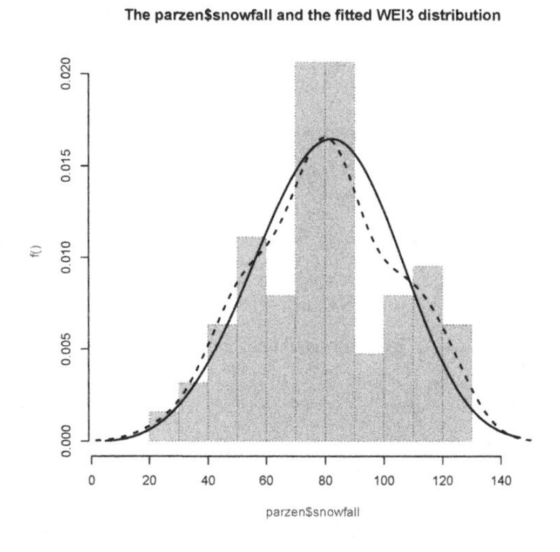

FIGURE 5.7: Histogram of Parzen's snowfall data together with the fitted Weibull distribution (solid) and a kernel density estimate (dashed).

The `WEI3`(μ, σ) distribution is the parameterization of the Weibull distribution with μ the mean.

Checking the model

Residuals are discussed in Section 4.5.3. A check of the normalized quantile residuals using a Q-Q and a worm plot (i.e. a detrended Q-Q plot) provides a guide to the adequacy of the fit. The **gamlss** package provides the functions `plot()` and `wp()` for this purpose.

Figure 5.8 shows the results of using `plot(m1)`, while Figure 5.9 shows the result of `wp(m1)`. Figure 5.8 plots the normalized quantile residuals against the fitted values and the case number (i.e. index number), together with their kernel density estimate and a normal Q-Q plot. Both the Q-Q plot (bottom right of Figure 5.8) and the worm plot in Figure 5.9 indicate that the model provides an adequate fit to the data. Adequacy of the `WEI3` distribution is indicated by over 95% of the deviations in the worm plot lying within the dashed (approximate 95%) confidence bands. Note that not all of the plots in Figure 5.8 are as useful as they will be in a regression type situation, especially the top left plot.

```
plot(m1)

## ******************************************************************
##        Summary of the Quantile Residuals
##                            mean   =  -0.002111081
##                        variance   =  1.014385
##                coef. of skewness   =  0.02768158
##                coef. of kurtosis   =  2.414167
## Filliben correlation coefficient   =  0.9939112
## ******************************************************************
```

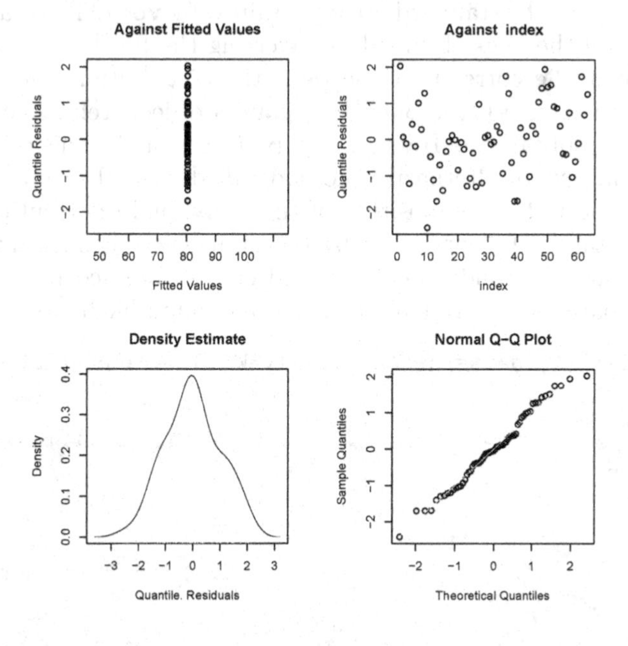

FIGURE 5.8: The residual plot from the fitted Weibull model to the snowfall data.

```
wp(m1)
```

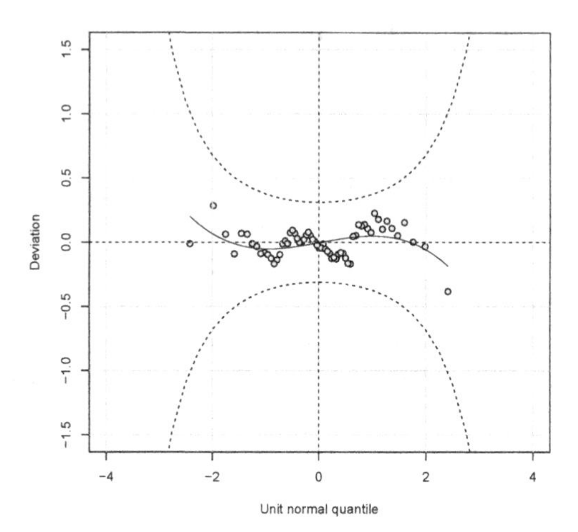

FIGURE 5.9: The worm plot from the fitted Weibull model to the snowfall data.

Confidence intervals for parameters in the model

There are several methods to check the reliability of the fitted parameters of the distribution. Standard errors for the fitted parameters are provided by two functions: (i) `summary()` and (ii) `vcov()`. In general the values obtained should be identical, since by default `summary()` gives the standard errors obtained by `vcov`. The standard errors obtained by `vcov()` are the ones obtained by inverting the full Hessian matrix and they do take into account the correlations between the distribution parameter estimates. Note that the function `vcov()`, applied to a **gamlss** object, refits the final model one more time in order to obtain the Hessian matrix. Occasionally this could fail, in which case `summary()` will use an alternative method called `qr` and give a *warning* that `qr` is used. This uses the QR decomposition of the individual distribution parameter estimation fits. The standard errors given by the `qr` method of `summary()` are not very reliable since they are the conditional standard errors obtained by assuming that the other distribution parameters are fixed at their maximum likelihood estimates.

```
m1<-gamlss(snowfall~1, data=parzen, family=WEI3, trace=FALSE)
summary(m1)

## *******************************************************************
## Family:  c("WEI3", "Weibull type 3")
##
## Call:
## gamlss(formula = snowfall ~ 1, family = WEI3, data = parzen,
##      trace = FALSE)
##
## Fitting method: RS()
##
## -------------------------------------------------------------------
```

```
## Mu link function:  log
## Mu Coefficients:
##             Estimate Std. Error t value Pr(>|t|)
## (Intercept)  4.38692    0.03676   119.3   <2e-16 ***
## ---
## Signif. codes:
## 0 '***' 0.001 '**' 0.01 '*' 0.05 '.' 0.1 ' ' 1
##
## -------------------------------------------------------------------
## Sigma link function:  log
## Sigma Coefficients:
##             Estimate Std. Error t value Pr(>|t|)
## (Intercept)   1.3439     0.0992   13.55   <2e-16 ***
## ---
## Signif. codes:
## 0 '***' 0.001 '**' 0.01 '*' 0.05 '.' 0.1 ' ' 1
##
## -------------------------------------------------------------------
## No. of observations in the fit:  63
## Degrees of Freedom for the fit:  2
##       Residual Deg. of Freedom:  61
##                       at cycle:  3
##
## Global Deviance:     575.9043
##             AIC:     579.9043
##             SBC:     584.1905
## *******************************************************************
vcov(m1, type="se")
## (Intercept) (Intercept)
##  0.03675988  0.09920447
```

The fitted Weibull distribution model is given by $Y_i \sim$ WEI3$(\hat{\mu}, \hat{\sigma})$ where $\log(\hat{\mu}) = 4.387$, $\hat{\mu} = \exp(4.387) = 80.399$; and $\log(\hat{\sigma}) = 1.344$, so $\hat{\sigma} = 3.834$. Note that $\hat{\mu}$ and $\hat{\sigma}$ are the maximum likelihood estimates of μ and σ.

The standard errors obtained are 0.0368 for $\log(\hat{\mu}) = \hat{\beta}_{01}$ and 0.0992 for $\log(\hat{\sigma}) = \hat{\beta}_{02}$ respectively, using either the `summary()` or `vcov()` functions. Note that since the Weibull fitting function `WEI3()` uses the log link for both μ and σ, the standard errors given are those for $\log(\hat{\mu}) = \hat{\beta}_{01}$ and for $\log(\hat{\sigma}) = \hat{\beta}_{02}$. For example, an approximate 95% confidence interval (CI) for $\log(\sigma) = \beta_{02}$, using the `vcov()` results, is

$$(1.344 - (1.96 \times 0.0992), 1.344 + (1.96 \times 0.0992)) = (1.150, 1.538) .$$

Hence an approximate 95% CI confidence interval for σ is given by

$$(\exp(1.150),\ \exp(1.538)) = (3.158, 4.655) .$$

This 95% CI for σ can be compared with the more reliable profile deviance 95% CI:

```
prof.dev(m1, "sigma", min=3, max=4.8, step=.01, type="l" )

## *****************************************************************
## The Maximum Likelihood estimator is  3.833493
## with a Global Deviance equal to  575.9043
## A  95 % Confidence interval is: ( 3.125857 , 4.61682 )
## *****************************************************************
```

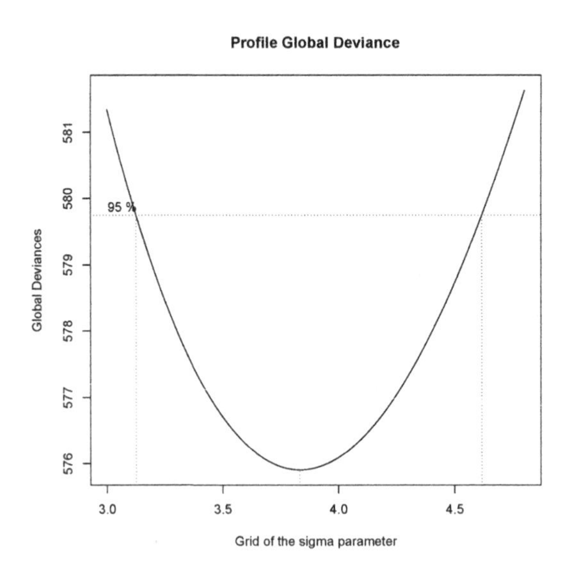

FIGURE 5.10: The profile deviance for the σ parameter for the Weibull (`WEI3`) model `m1`.

giving 95% CI (3.126, 4.617) and Figure 5.10. Note that `prof.dev()` works only with **gamlss** objects.

These CIs may also be compared with the bootstrap 95% CI for σ:

```
library(boot)
set.seed(1453)
mod1<-gamlss(snowfall~1, data=parzen, family=WEI3, trace=FALSE)
funB <- function(data, i)
     {
     d<-data.frame(snowfall=data[i,])
     coef(update(mod1, data=d),"sigma")
     }
(mod1.boot<-boot(parzen, funB, R=199, parallel="multicore",
                 ncpus = 4))

##
## ORDINARY NONPARAMETRIC BOOTSTRAP
##
##
## Call:
## boot(data = parzen, statistic = funB, R = 199, parallel = "multicore",
```

```
##      ncpus = 4)
##
##
## Bootstrap Statistics :
##      original    bias     std. error
## t1* 1.343915 0.0187897  0.08209921
```

```
boot.ci(mod1.boot, type=c("norm", "basic"))

## BOOTSTRAP CONFIDENCE INTERVAL CALCULATIONS
## Based on 199 bootstrap replicates
##
## CALL :
## boot.ci(boot.out = mod1.boot, type = c("norm", "basic"))
##
## Intervals :
## Level      Normal              Basic
## 95%   ( 1.164,  1.486 )   ( 1.163,  1.476 )
## Calculations and Intervals on Original Scale
## Some basic intervals may be unstable
```

We obtain two 95% bootstrap CIs intervals for σ, the 'normal'

$$(\exp(1.154), \exp(1.507)) = (3.171, 4.513),$$

and the 'basic"

$$(\exp(1.160), \exp(1.491)) = (3.190, 4.442).$$

More details about the `boot()` function can be found in Venables and Ripley [2000, p. 173].

5.8.2 Glass fiber strength data

R data file: `glass` in package **gamlss.data** of dimension 63×1
source: Smith and Naylor [1987]
variables
  `strength` : the strength of glass fibers (the unit of measurement is not given)
purpose: to demonstrate the fitting of a continuous parametric distribution to a single variable.
conclusion a `SEP4` distribution fits adequately

Data on the strength of glass fibers, measured at the National Physical Laboratory, England, is given by Smith and Naylor [1987]. Here we fit different distributions to the data and select the 'best' model using the AIC and SBC. We also demonstrate the use of the `gen.trun()` function, which creates a truncated distribution from a `gamlss.family` distribution. We generate a positive (i.e. left-truncated at zero) t distribution (`TFtr`) and fit it to the data together with other candidate distributions:

```
# truncated distributions package
library(gamlss.tr)
```

```
# create a truncated (at zero) t
gen.trun(par=0, family=TF)

## A truncated family of distributions from TF has been generated
##  and saved under the names:
##  dTFtr pTFtr qTFtr rTFtr TFtr
## The type of truncation is left
##  and the truncation parameter is 0
```

Next we fit all available continuous distributions using AIC and SBC information criteria. The argument `extra` is used to add the truncated t distribution to the default list `.realAll`.

```
data(glass)
m1<-fitDist(strength, data=glass, k=2, extra="TFtr") # AIC
m2<-fitDist(strength, data=glass, k=log(length(glass$strength)),
              extra="TFtr") # SBC
m1$fit[1:6]

##      SEP4     SEP3   SHASHo  SHASHo2     SEP1     SEP2
## 27.65361 27.97321 28.01800 28.01800 29.00072 29.05912

m2$fit[1:6]

##      SEP4     SEP3   SHASHo  SHASHo2     SEP1     SEP2
## 36.22615 36.54575 36.59054 36.59054 37.57326 37.63166
```

The best model for the glass fiber strength according to both AIC and SBC is the `SEP4` distribution. (Note however that `SEP4` has range $(-\infty, \infty)$, so a `SEP4` distribution left-truncated at zero (`SEPtr`) may be preferred, since the response variable `strength` is always positive.) Our truncated t distribution, `TFtr`, did not fit well. The fitted `SEP4` distribution is shown in Figure 5.11:

```
h1 <- histDist(glass$strength, family=SEP4, nbins=13,
         main="SEP4 distribution", line.col=gray(.2))
```

The fitted distribution has a left-side spike at its mode. Distributions which involve the power exponential distribution (e.g. all the `PE`s and `SEP`s) with values of the tail heaviness parameter(s), (i.e. τ for `PE`, `PE2`, `SEP1`, `SEP2`, `SEP3`, and both ν and τ for `SEP4`), less than or equal to 1 have a discontinuity in the gradient at the mode, leading to a spike at the mode. This often results in a multimodal likelihood function with respect to μ, and leads to inferential problems. In the `SEP4` distribution the parameters ν and τ adjust the left and right tail heaviness of the distribution, respectively. The estimates of these two parameters are $\hat{\nu} = \exp(-0.3338) = 0.7162$ and $\hat{\tau} = \exp(0.0602) = 1.0620$, (obtained from the output of `summary(m1)`), indicating a discontinuity (a left-side spike) at the mode, resulting in possible problems with the inferential procedures since $\nu \leq 1$. Note we can extract the fitted coefficients using either of the functions `coef()` or `fitted()`, e.g. `coef(msep4, "nu")` or `fitted(msep4, "nu")[1]`.

```
summary(m1)

## *******************************************************************
## Family:  c("SEP4", "skew exponential power type 4")
```

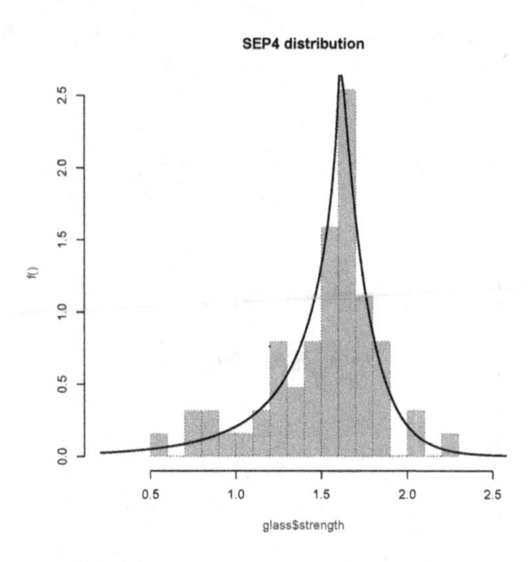

FIGURE 5.11: Histogram of the glass fiber strength data together with the fitted SEP4 distribution.

```
##
## Call:
## gamlssML(formula = y, family = DIST[i], data = sys.parent())
##
##
## Fitting method: "nlminb"
##
##
## Coefficient(s):
##              Estimate  Std. Error   t value   Pr(>|t|)
## eta.mu     1.61000000  0.00436385 368.94000 < 2.22e-16 ***
## eta.sigma -1.83908555  0.45814686  -4.01418 5.9652e-05 ***
## eta.nu    -0.33378926  0.20478263  -1.62997    0.10311
## eta.tau    0.06020469  0.33687862   0.17871    0.85816
## ---
## Signif. codes:
## 0 '***' 0.001 '**' 0.01 '*' 0.05 '.' 0.1 ' ' 1
##
##  Degrees of Freedom for the fit: 4 Residual Deg. of Freedom   59
## Global Deviance:     19.6536
##             AIC:     27.6536
##             SBC:     36.2262
```

A worm plot of the residuals of the fitted model in Figure 5.12 shows that all the points are close to the horizontal line and within the dashed confidence bands, indicating that the SEP4 distribution provides a good fit to the data.

```
wp(m1)
```

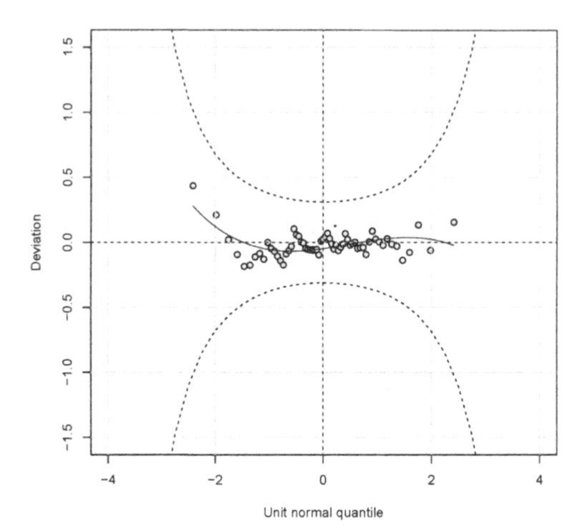

FIGURE 5.12: Worm plot of the residuals for model `m1` using the `SEP4` distribution fitted to the glass fiber strength data.

5.9 Centile estimation

Centile estimation of a response variable is widely used in medicine, nutrition, sport science and other disciplines, where an individual is checked to see whether they have an abnormally low or high value of the response variable, given their value(s) of the explanatory variable(s), and hence whether they are potentially at risk.

5.9.1 One explanatory variable

The most common centile estimation involves only one explanatory variable. Chapter 13 of Stasinopoulos et al. [2017] gives a detailed description of the appropriate methodology concerning centile estimation with one explanatory variable. Exercise 1 of this chapter is an example of centile estimation with one explanatory variable, in which we use height as a response variable against age. In the rest of this section we demonstrate how GAMLSS can be used to create centile curves with two explanatory variables.

5.9.2 Two explanatory variables: the Dutch boys data

The standard estimation of centile curves usually involves two continuous variables, the response variable and the explanatory variable. It is possible, however, to have more explanatory variables and in the following we give an example where the analysis involves two explanatory variables.

R data file: `dbhh` in package **gamlss.data** of dimension 6885×3
source: Fredriks et al. [2000a,b]
variables
`age` : age in years,
`ht` : height
`head` : head circumference
purpose: to create centiles of a continuous response variable
conclusion: the Box-Cox t, `BCTo`, distribution seems to fit best

The Fourth Dutch Growth Study [Fredriks et al., 2000a,b] is a cross-sectional study in which growth and development of the Dutch population between the ages of 0 and 22 years were measured. Amongst other variables, height, weight, head circumference, and age were recorded for 7482 males and 7018 females. Here we analyze 6885 observations for head circumference, height, and age of males, having removed observations with missing values for any of these three variables from the original data set. This analysis was also presented in Stasinopoulos et al. [2018].

The response variable is head circumference (`head`) and the explanatory variables are height (`ht`) and `age`. The objective here is to model the distribution of head circumference using height and age.

The data are shown in two dimensions in Figure 5.13 and in three dimensions in Figure 5.14, where we see that the response variable, head circumference, is defined only in a limited joint range of the age and height space. This has consequences in fitting a model to head circumference because prediction outside the data space of the explanatory variables will rely on extrapolation and therefore will be unreliable.

Figure 5.13 is explained in its caption.

```
require(ggplot2)
require(GGally)
pm <- ggpairs(
 dbhh[,c("age","ht","head")],
 upper = list(continuous = wrap("density", col="black")),
 lower = list(continuous = wrap("points", alpha = 0.05, size=0.01))
 )
pm
```

```
with(dbhh,cloud(head~ht*age,col="darkgray", pch=".",xlab="height",
                type=c("p"), screen=list(z=20, x=-70)))
```

In order to fit a GAMLSS model to the data we need:

1. to consider whether transformed versions of height and age will help the analysis,
2. a suitable distribution for head circumference, and
3. to determine how height and age affect the distribution of the response variable.

In order to choose a suitable transformation for age and height we plotted log head circumference (since we later use distributions with a default log link for μ), against each of six variables: (i) age and height, (ii) log age and log height, and (iii) the square roots

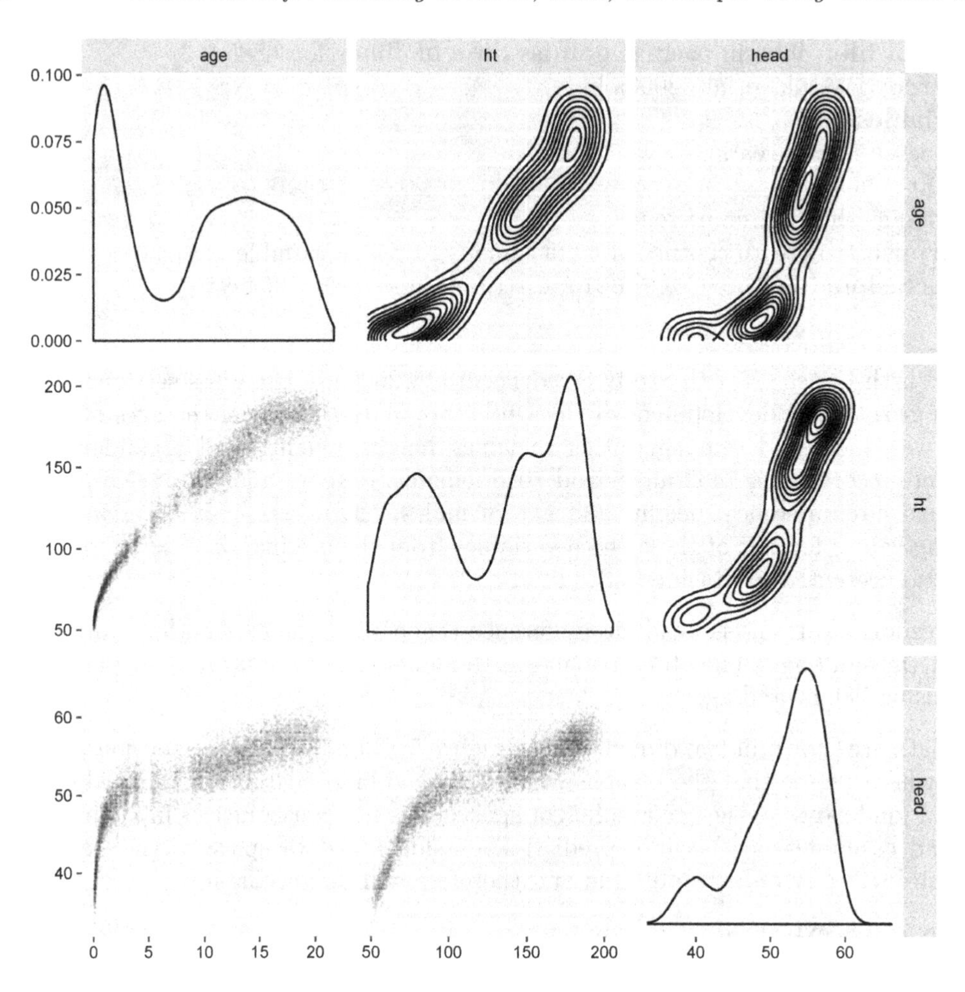

FIGURE 5.13: Scatterplot matrix of `head`, `age`, and `ht`. The lower diagonal plots are scatterplots; upper diagonals are contour plots of the bivariate density estimates; and the diagonal plots are univariate density estimates.

of age and height. The transformation that made the relationship between log(head) and each transformed explanatory variable closest to linear was the log transformation for both height and age.

To choose a suitable distribution, we used three distributions (`BCCGo`, `BCPEo`, and `BCTo`) defined for a positive response variable and initially fitted additive smooth functions (i.e. P-splines, `pb()`) in log age (`lage`) and log height (`lht`), for μ only.

```
dbhh<-transform(dbhh, lage=log(age),  lheight=log(ht))
mbccg<-gamlss(head~pb(lage)+pb(lheight), family=BCCGo,
              data=dbhh, c.crit=0.01)
mbct<-gamlss(head~pb(lage)+pb(lheight), family=BCTo,
             data=dbhh, c.crit=0.01)
mbcpe<-gamlss(head~pb(lage)+pb(lheight), family=BCPEo,
              data=dbhh, c.crit=0.01)
```

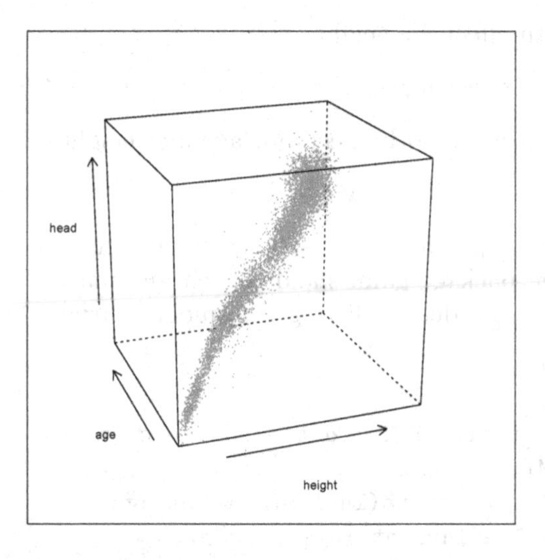

FIGURE 5.14: A three-dimensional plot of head circumference against age and height.

```
AIC(mbccg, mbct, mbcpe)

##               df      AIC
## mbct   22.95974 24988.82
## mbcpe  23.01714 25099.51
## mbccg  22.15571 25256.16

AIC(mbccg, mbct, mbcpe, k=log(6885))

##               df      AIC
## mbct   22.95974 25145.79
## mbcpe  23.01714 25256.88
## mbccg  22.15571 25407.64
```

The distribution that was 'best' using either the AIC or SBC was the `BCTo`. Note that instead of explicitly fitting the above three models we could have used the function `chooseDist()`, with argument `type="extra"` in order to just fit the three distributions:

```
mbno<-gamlss(head~pb(lage)+pb(lheight), sigma.fo=~1,family=NO,
             data=dbhh, n.cyc=100)
Tab <- chooseDist(mbno, type="extra",
                  extra=c("BCCGo", "BCTo", "BCPEo"))

## minimum GAIC(k= 2 ) family: BCTo
## minimum GAIC(k= 3.84 ) family: BCTo
## minimum GAIC(k= 8.84 ) family: BCTo
```

The `BCTo` model is preferred for the three GAICs computed.

In order to determine how the explanatory variables age and height affect the different distribution parameters of the `BCTo`(μ, σ, ν, τ), we use a selection technique which chooses between the following four candidate models for each distribution parameter:

- $s(u_h)$: main smooth effect for height
- $s(u_a)$: main smooth effect for age
- $s(u_a) + s(u_h)$: additive smooth effects for age and height
- $s(u_a, u_h)$: smooth interaction of age and height

where $u_h = \log(\text{height})$, $u_a = \log(\text{age})$, $s(\cdot)$ is a smooth function and $s(\cdot,\cdot)$ is a smooth surface. Note that the package **gamlss.add**, which automatically loads the **R** package **mgcv** of Wood [2017], is needed for fitting the smooth surface.

```
library(gamlss.add)
nC <- detectCores()
M1<-gamlss(head~1,  family=BCTo, data=dbhh, n.cyc=100)
M2<-stepGAICAll.A(M1,
scope=list(lower=~1, upper=~pb(lheight) + pb(lage) +
                 ga(~te(lheight,lage,k=10))),
                 k=4, parallel="multicore", ncpus = nC)
```

The function `stepGAICA.All()` is explained in Stasinopoulos et al. [2017], Section 11.5, and uses a stepping procedure to select the terms for each parameter μ, σ, ν, and τ of the `BCTo` distribution.

Note that we used $\kappa = 4$ as the penalty in the model selection criterion, GAIC(κ). The chosen model is

$$\begin{aligned}
\texttt{head} &\sim \texttt{BCTo}(\mu, \sigma, \nu, \tau) \\
\log(\mu) &= s_1(u_a, u_h) \\
\log(\sigma) &= s_2(u_h) \\
\nu &= s_3(u_a) \\
\log(\tau) &= s_4(u_h) \;.
\end{aligned} \tag{5.1}$$

The worm plot for the final chosen model (5.1), given in Figure 5.15(a), showed some extreme outliers in the tails (six in the upper tail and five in the lower tail, out of 6885 observations). These outliers distort the fitted centiles of head circumference as shown by the the distorted worm plot of Figure 5.15(a). When the eleven extreme outliers were removed, the worm plot improved greatly, as shown in Figure 5.15(b). This should result in improved fitted centiles. The case numbers of the the extreme residuals are found below:

```
which(resid(M2)>3.5)

## [1]  257 1484 1542 2834 5014 6325

which(resid(M2) < -3.5)

## [1]  631 1503 2523 5479 6530
```

```
dbhh1<-subset(dbhh,(resid(M2)>-3.5)&(resid(M2)<3.5))
M3<-gamlss(head~ga(~te(log(age), log(ht),k=10)),
           sigma.fo=~pb(log(ht)), nu.fo=~pb(log(age)),
           tau.fo=~pb(log(ht)), family=BCTo, data=dbhh1,
           n.cyc=100)
```

```
wp(M2, ylim.all=1.2); title("(a)")
wp(M3, ylim.all=1.2); title("(b)")
```

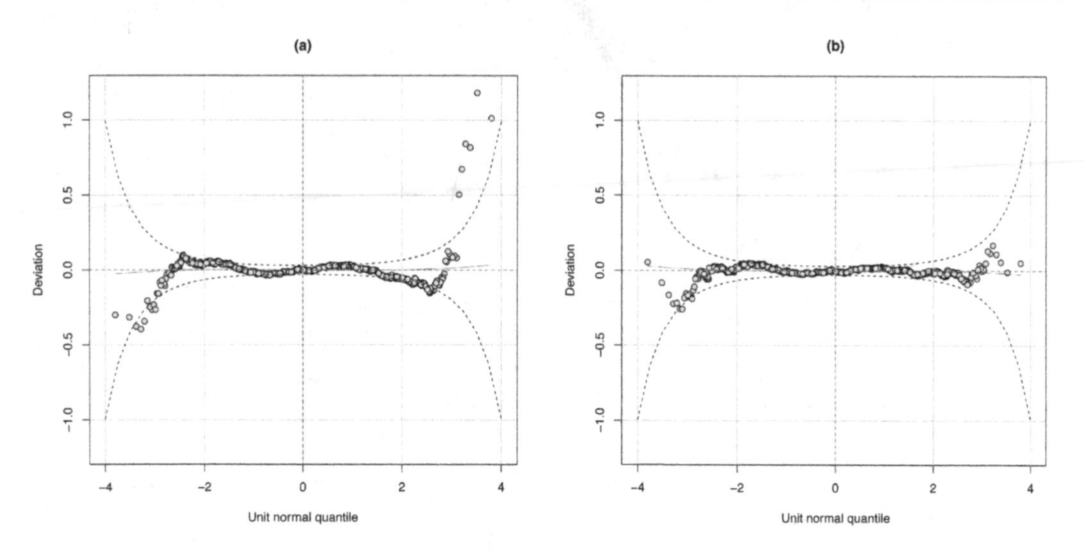

FIGURE 5.15: Worm plot of the residuals for the BCTo model (a) using all 6885 observations (M2) and (b) removing eleven observations with extreme residuals (M3).

Figure 5.16 gives term plots for the chosen model (5.1) fitted by M3. The top left panel is a contour plot of $s_1(u_h, u_a)$, while the top right and bottom two panels are plots of $s_2(u_h)$, $s_3(u_a)$, and $s_4(u_h)$, respectively. Note the functions are plotted using the original explanatory variables, but do not include the fitted constant intercept in each function.

```
term.plot(M3, what="mu", pages=1, main="(a)", col.term=1)
term.plot(M3, what="sigma", pages=1, main="(b)", col.term=1)
term.plot(M3, what="nu", pages=1, main="(c)", col.term=1)
term.plot(M3, what="tau", pages=1, main="(d)", col.term=1)
```

Model (5.1) can now be used to obtain the 5% and 95% centiles (i.e. the 0.05 and 0.95 quantiles) of head. We will use the function `centilesTwo()` of the package **gamlss.add** (which calls function `exclude.too.far()` of package **mgcv**) to obtain contour plots of a specified centile of head against height and age. Figures 5.17(a) and (b) show the 5% and 95% centiles for head. Figures 5.17(c) and (d) restrict the age range to 0 to 2 years to see more clearly the head contour values for that age group. From a practical viewpoint, given the height and age of the Dutch boy, an observed head circumference less than the 5% centile value indicates an unusually small head circumference, while a value greater than the 95% centile value indicates an unusually large head circumference. This can be used as a medical diagnostic tool.

```
newage <- seq(0.05,21,0.05)
newheight<-seq(30, 210, 1)
centilesTwo(M3, grid.x1=newage, grid.x2=newheight, age,ht,
        cent=.05, nlevels=20,xlab="age(years)", points=FALSE,
        ylab="height(cm)",  cex.axis=1.1, main="(a)",labcex =1)
```

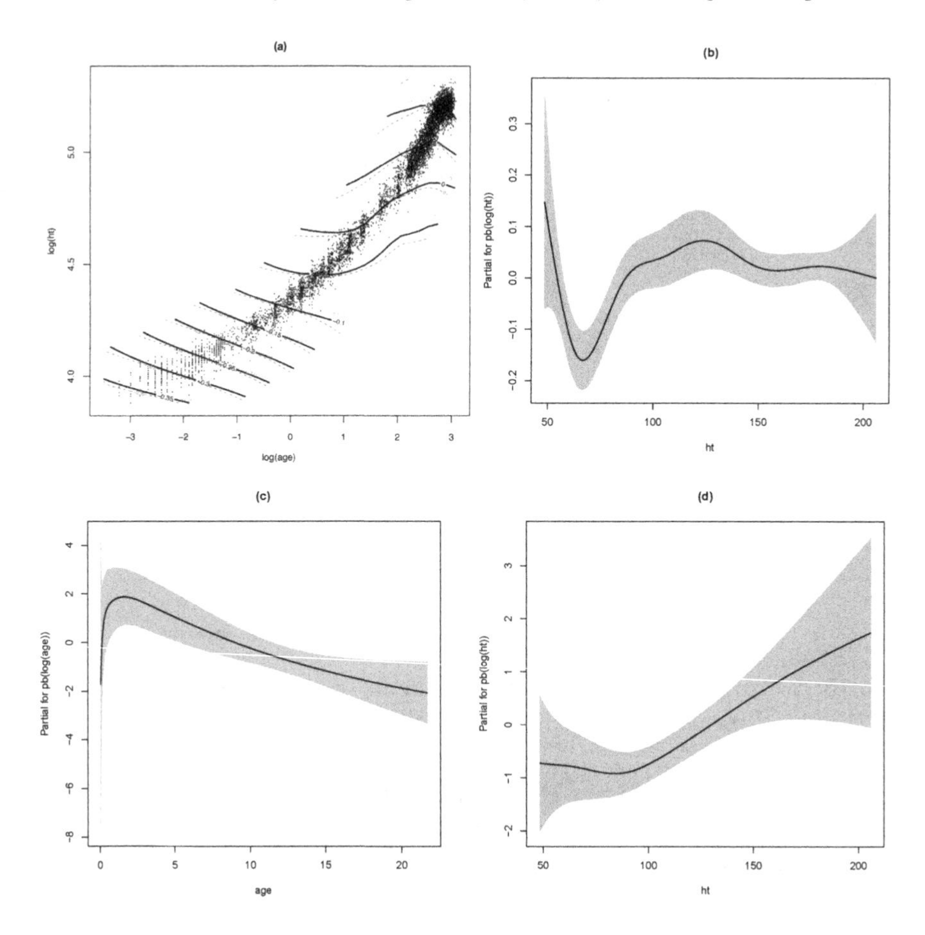

FIGURE 5.16: Term plots for the final `BCTo` model (5.1) fitted by `M3` for (a) $\log(\mu) = s_1(u_h, u_a)$, (b) $\log(\sigma) = s_2(u_h)$, (c) $\nu = s_3(u_a)$, and (d) $\log(\tau) = s_4(u_h)$.

```
centilesTwo(M3, grid.x1=newage, grid.x2=newheight, age,ht,
        cent=.95, nlevels=20,xlab="age(years)", points=FALSE,
        ylab="height(cm)",  cex.axis=1.1, main="(b)",labcex =1)
centilesTwo(M3, grid.x1=newage, grid.x2=newheight, age,ht,
        cent=.05, nlevels=20,xlab="age(years)",
        ylab="height(cm)",  cex.axis=1.1, xlim=c(0,2),
        ylim=c(45,100), main="(c)",labcex =1)
centilesTwo(M3, grid.x1=newage, grid.x2=newheight, age,ht,
        cent=.95, nlevels=20,xlab="age(years)",
        ylab="height(cm)",  cex.axis=1.1, xlim=c(0,2),
        ylim=c(45,100), main="(d)", labcex =1)
```

The function `getQuantile()` is designed to obtain conditional quantiles of the response variable against one explanatory variable, given that the rest of the explanatory variables are fixed at pre-specified values. Figure 5.18 shows the use of `getQuantile()`. It displays the 5%, 50%, and 95% conditional centiles of head circumference. Figure 5.18(a) shows the head circumference centiles against age (from 0 to 2 years) for height

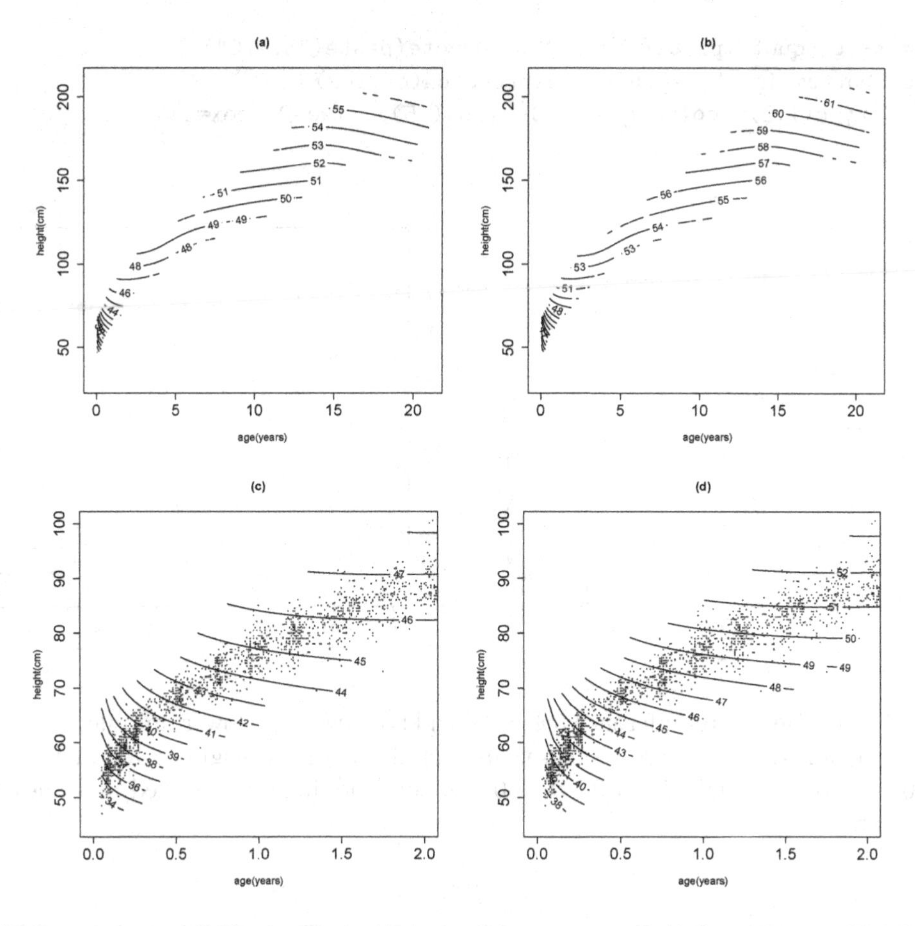

FIGURE 5.17: Contour plots for 5% and 95% centiles of head against height and age. Plots (a) and (c) are 5% centiles, plots (b) and (d) are 95% centiles. The top two plots have all ages, while the bottom two plots have age restricted from zero to two years.

of 60 and 70 centimetres. Figure 5.18(b) shows the head circumference centiles against height (from 90 to 150 cm) for age fixed at each of the values 4, 6, and 8 years. The code for plotting Figure 5.18(a) is shown below:

```
# age from 0 to 2 at heights 60 70
plot(head~age, data=dbhh1, type="n", xlim=c(0,2), ylim=c(35, 50),
     main="(a)")
for (i in c(.05, 0.5, 0.95))# height 60
{
  Qua <- getQuantile(M3, quantile=i, term="age", fixed.at = list(ht=60))
  curve(Qua, 0, 0.7,  lwd=1.5, lty=1, add=T, col=gray(.7))
}
for (i in c(.05, 0.5, 0.95))# height 70
{
  Qua <- getQuantile(M3, quantile=i,term="age", fixed.at = list(ht=70))
  curve(Qua, 0.3, 1.7,  lwd=1.5, lty=2, add=T, col=gray(.3))
}
```

```
ex.leg <- c(bquote(paste("ht=60")), bquote(paste("ht=70")))
legend("bottomright",legend=as.expression(ex.leg),
      lty=c(1,2), col=c(gray(.7), gray(.5)), lwd=2, cex=1)
```

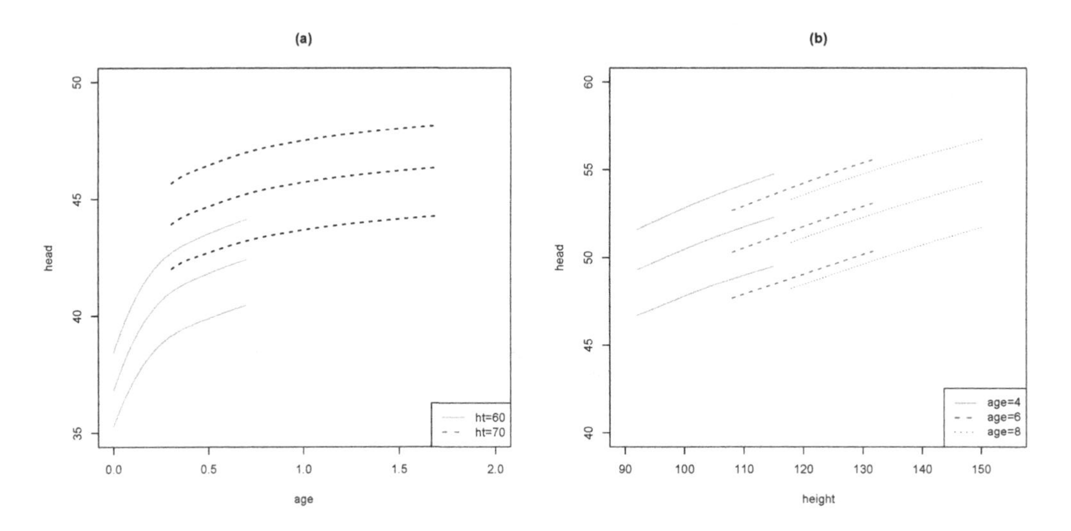

FIGURE 5.18: Head circumference 5%, 50% and 95% centiles (lower, middle, and upper curves): (a) for age from zero to two years, conditioning on heights fixed at values 60 and 70 cm, (b) for height from 90 to 150 cm, conditioning on age fixed at values 4, 6, and 8 years.

5.10 Bibliographic notes

We remind the reader that many of the continuous distributions discussed in this chapter can be found in Johnson et al. [1994] and Johnson et al. [1995], which are the classical reference books for continuous distributions.

The exponential, `EXP`(μ), gamma, `GA`(μ, σ), and inverse Gaussian, `IG`(μ, σ), are all members of the exponential family of distributions (see Section 13.8.3) and hence distributions belonging to a Generalized Linear Model [Nelder and Wedderburn, 1972]. As a consequence they have been studied extensively in the GLM literature, see for example Aitkin et al. [2009]. The `IG` has a heavier right tail than the `GA`, see Chapter 17. Chapters 15 and 17 of Johnson et al. [1994] discuss the inverse Gaussian and gamma distributions respectively. The log normal distribution (`LOGNO`, `LOGNO2`) has been very popular in finance and survival analysis. Chapter 4 of Johnson et al. [1994] is devoted to different versions of the log normal distribution. The inverse gamma distribution, which is derived as the reciprocal of a gamma distributed variable, has been widely used within the Bayesian framework as a prior distribution of precision parameters. Here we use it as a distribution in its own right. A general discussion about the different types of Pareto distribution can be found in Johnson et al. [1994, Chapter 20]. In GAMLSS we have included the Pareto type 1, `PARETO1o`(μ, σ) and two parameterizations of the Pareto type

2, PARETO2(μ, σ) and PARETO2o(μ, σ). The Weibull distribution is properly reviewed in Johnson et al. [1994, Chapter 21]. In GAMLSS there are three different parameterizations of the Weibull: (i) WEI(μ, σ) corresponding to equation (21.3) of Johnson et al. [1994], (ii) WEI2(μ, σ) which corresponds to the parameterization of the Weibull used for proportional hazards models, and (iii) WEI3(μ, σ), a parameterization created by the authors of this book in which μ is the mean of the distribution and therefore easier to interpret in regression models.

The gamma family, GAF(μ, σ, ν), was created by the authors of this book as a way of modeling variance-mean relationships that differ from square. The generalized gamma distribution is a three-parameter generalization of the gamma distribution, see Johnson et al. [1994, Section 8.7]. Our parameterization of the generalized gamma distribution, GG(μ, σ, ν), is based on the (unpublished) work of Lopatatzidis and Green [2000]. The generalized inverse Gaussian distribution, GIG(μ, σ, ν), was studied extensively by Jørgensen [1982].

The BCCG(μ, σ, ν) and BCCGo(μ, σ, ν) distributions are a way of ensuring that the LMS method of centile estimation of Cole and Green [1992] is based on a proper distribution, see Rigby and Stasinopoulos [2004]. The log normal family LNO(μ, σ, ν) is not a proper distribution but provides a way of implementing the well-known Box-Cox transformation to normality [Box and Cox, 1964]. It is the only distribution in **gamlss.dist** which is not a proper distribution. A reparameterized proper distribution is given by BCCG(μ, σ, ν). The Box-Cox power exponential, BCPE(μ, σ, ν, τ) and BCPEo(μ, σ, ν, τ), and Box-Cox t, BCT(μ, σ, ν, τ) and BCTo(μ, σ, ν, τ) distributions were created by Rigby and Stasinopoulos [2004] and Rigby and Stasinopoulos [2006], respectively, in order to add kurtosis modeling to the LMS method of centile estimation. The World Health Organization (WHO) used the BCPE distribution during the development of growth curves for children [WHO, 2006, 2007, 2009], although the final distribution used was BCCG. In our experience the Box-Cox t has proven to be one of the most reliable and flexible continuous distributions defined on $(0, \infty)$. Information about the generalized beta type 2 (GB2) distribution can be found in McDonald [1984].

5.11 Exercises

1. **Centile estimation with one explanatory variable**

   There are currently two major methodologies for creating centile (or quantile) curves: (i) the LMS method and its extensions [Cole and Green, 1992, Rigby and Stasinopoulos, 2004, 2006] and (ii) the quantile regression method [Koenker et al., 1994, He and Ng, 1999, Np and Maechler, 2007]. More details about both methodologies can be found in Stasinopoulos et al. [2017, Chapter 13]. Here we concentrate on the LMS methodology and its extensions, which are a subclass of GAMLSS and were also adopted by the World Health Organization for the construction of worldwide standard growth (centile) curves for children, see WHO [2006, 2007, 2009]. The model for the extended LMS methodology with one explanatory variable x can be written as:

$$
\begin{aligned}
Y &\sim \mathcal{D}(\mu, \sigma, \nu, \tau) \\
g_1(\mu) &= s_1(u) \\
g_2(\sigma) &= s_2(u) \\
g_3(\nu) &= s_3(u) \\
g_4(\tau) &= s_4(u) \\
u &= x^{\xi}
\end{aligned}
\tag{5.2}
$$

where $\mathcal{D}(\cdot)$, $g(\cdot)$, and $s(\cdot)$ have their usual meanings; and u is a power transformation function of the explanatory variable x with ξ the power transform parameter. The reason that a power transformation for x may be needed is to improve the estimation of the smoothing functions when spells of sharp growth in Y occur for low values of x. Model (5.2) has five parameters to be estimated: the four smoothing parameters for the functions $s_1(\cdot)$, $s_2(\cdot)$, $s_3(\cdot)$, and $s_4(\cdot)$ and the power transform parameter ξ.

The purpose of this exercise is to estimate the centiles curves for `height` against `age` using the data set `dbhh` first introduced in Section 5.9.2.

(a) To avoid estimating the power parameter ξ in model (5.2), we use an empirical method, which consists of plotting the log of the response variable height against age, log age, and the square root of age (corresponding to ξ effectively equal to 1, 0 or 0.5, respectively) and choose the one which looks most linear and therefore easiest to smooth. We use the log of the response since we later use a distribution with a default log link function for μ. The plotting commands are shown below:

```
plot(log(ht)~age, data=dbhh, col="darkgray", pch=".",
     cex.lab=1.5)
plot(log(ht)~log(age), data=dbhh, col="darkgray", pch=".",
     cex.lab=1.5)
plot(log(ht)~sqrt(age), data=dbhh, col="darkgray", pch=".",
     cex.lab=1.5)
```

Which plot looks most linear?

In the the analysis that follows use u = `age`$^{0.5}$ as the explanatory variable. (Note that estimation of the smoothing parameters in `pb()` is done automatically in **gamlss** using the methodology described in Rigby and Stasinopoulos [2013] and in Stasinopoulos et al. [2017, Chapter 3 and 9]. Note also that there are functions in **gamlss** for automatically choosing the power parameter ξ, see Stasinopoulos et al. [2017, Chapter 13].)

(b) Fit the `BCCGo`, `BCPEo` and `BCTo` distributions to the response variable height (`ht`) using P-spline smoothers (`pb()`) in `sqrt(age)` for each of μ, σ, ν, (and τ for `BCPEo` and `BCTo`):

```
# fit the BCT distribution
MBCT<-gamlss(ht~pb(sqrt(age)), sigma.fo=~pb(sqrt(age)),
             nu.fo=~pb(sqrt(age)), tau.fo=~pb(sqrt(age)),
             family=BCTo, data=dbhh)
```

```
# fit the BCPE distribution
MBCPE<-gamlss(ht~pb(sqrt(age)), sigma.fo=~pb(sqrt(age)),
              nu.fo=~pb(sqrt(age)), tau.fo=~pb(sqrt(age)),
              family=BCPEo, data=dbhh)
# fit the BCCG distribution
MBCCG<-gamlss(ht~pb(sqrt(age)), sigma.fo=~pb(sqrt(age)),
              nu.fo=~pb(sqrt(age)), tau.fo=~pb(sqrt(age)),
              family=BCCGo, data=dbhh)
```

(c) Choose between the BCCGo, BCPEo and BCTo distributions using the AIC and SBC.

```
GAIC(MBCT,MBCPE,MBCCG,k=2)
GAIC(MBCT,MBCPE,MBCCG,k=log(6885))
```

What is your conclusion?

(d) To help you decide between the two best models above use the centile curves and worm plots.

```
# centile curves for BCT model
centiles(MBCT, xvar=dbhh$age,
        cent=c(0.1,0.4,2,10,25,50,75,90,98, 99.6,99.9),
        ylab="ht", xlab="age", legend=FALSE)
# centile curves for BCPE and BCCG models
centiles.com(MBCPE, MBCCG, xvar=dbhh$age,
         cent=c(0.1,0.4,2,10,25,50,75,90,98,99.6,99.9),
         ylab="ht", xlab="age", no.data=TRUE, legend=FALSE)
wp(MBCT, ylim.all=.5)
wp(MBCPE, ylim.all=.5)
```

(e) Describe your final chosen model.

(f) An alternative procedure for choosing a suitable model for centile estimation is the **gamlss** function `lms()`, which automates the process of choosing between the LMS (BCCGo), LMSP (BCPEo), and LSMT (BCTo) methods. The function also includes an automated estimation procedure for both the power parameter ξ and the smoothing parameters for all the distribution parameters. The final selection of the model is based on the GAIC(k) with k chosen by the user. Below we use $k = 4$, a compromise between AIC and BIC.

```
MLMS<- lms(ht,age,families=c("BCCGo","BCPEo","BCTo"),
            data=dbhh, k=4, calibration=FALSE, trans.x=TRUE,
            cent=c(0.1,0.4,2,10,25,50,75,90,98,99.6,99.9))
MLMS$family ; MLMS$power
```

What is the distribution of the final chosen model using the `lms()` function? What is the estimate of the power parameter? How does the centiles plot of the current model differ from the one fitted previously?

2. **The `plasma` data**

R data file: `plasma` in package **gamlss.data** of dimension 315 × 14
source: Harrell [2002]
variables

- `age` : age (years)
- `sex` : factor sex (1=male, 2=female)
- `smokstat` : factor smoking status (1=never, 2=former, 3=current smoker)
- `bmi` : body mass index (weight/(height^2))
- `vituse` : factor vitamin use (1=yes, fairly often, 2=yes, not often, 3=no)
- `calories` : number of calories consumed per day
- `fat` : grams of fat consumed per day
- `fiber` : grams of fiber consumed per day
- `alcohol` : number of alcoholic drinks consumed per week
- `cholesterol` : cholesterol consumed (mg per day)
- `betadiet` : dietary beta-carotene consumed (mcg per day)
- `retdiet` : dietary retinol consumed (mcg per day)
- `betaplasma` : plasma beta-carotene (ng/ml)
- `retplasma` : plasma retinol (ng/ml)

purpose: to demonstrate modeling a continuous response variable

'Observational studies have suggested that low dietary intake or low plasma concentrations of retinol, beta-carotene, or other carotenoids might be associated with increased risk of developing certain types of cancer ... We designed a cross-sectional study to investigate the relationship between personal characteristics and dietary factors, and plasma concentrations of retinol, beta-carotene, and other carotenoids.' [Harrell, 2002]

(a) **Find an appropriate (marginal) distribution:**. Take the dietary beta-carotene (`betadiet`) to be your response variable (assuming that there are no explanatory variables).

  i. Determine which continuous distribution on $(0, \infty)$ best fits `betadiet`, and plot the fitted distribution.

  ii. Use a diagnostic plot to assess the fit of this distribution.

  iii. Obtain an approximate 95% confidence interval (CI) for the parameter μ, and compare it with the corresponding profile deviance and bootstrap CIs.

(b) **Regression model with two explanatory variables:** Consider a regression model with `betadiet` as response variable and `age` and `fiber` as predictors.

  i. Produce scatterplots of the three variables, to assess their relationships.

  ii. Use the function `chooseDist()` to find an appropriate distribution for response variable. You can start from the model:

```
m0 <- gamlss(betadiet ~ age + fiber, sigma.fo=~age + fiber,
             data=plasma, family=GA)
```

  iii. For the chosen distribution above, use the function `stepGAICAll.A()` to find appropriate models for the different parameters of the distribution.

Try the four alternative models discussed in Section 5.9.2, single main effects, additive main effects, and interaction. You will need the package **gamlss.add**. The code below could be helpful:

```
library(gamlss.add)
nC <- detectCores()
M1 <- gamlss(betadiet ~ 1, data=plasma, family=BCTo)
M2<-stepGAICAll.A(M1,scope=list(lower=~1, upper=~pb(age) +
                  pb(fiber) + ga(~te(age,fiber))),
                  k=4, parallel="multicore", ncpus = nC)
```

iv. Describe the chosen model.

v. Use residual diagnostics to check its adequacy.

vi. Use the function `centilesTwo()` to obtain contour plots for the 5% and 95% centiles of `betadiet` against age and fiber. Use the argument `dist=0.1` which controls how far the contour plot extends beyond the points (`age, fiber`).

vii. For age fixed at 30 and 50, find the 5% and 95% centiles curves for `betadiet` against fiber using the function `getQuantile()`. See for example the code below.

```
plot(betadiet~fiber, data=plasma)
 for (i in c(.05, 0.5, 0.95)) {
 Qua <- getQuantile(M2, quantile=i,term="fiber",
                    fixed.at = list(age=30))
   curve(Qua, 5, 35,  lwd=1, lty=1, add=T)
 }
 for (i in c(.05, 0.5, 0.95)) {
 Qua <- getQuantile(M2, quantile=i,term="fiber",
                    fixed.at = list(age=50))
   curve(Qua, 5, 35,  lwd=1, lty=1, col="red", add=T)
 }
```

(c) **Regression model with all the explanatory variables:** Consider a regression model with `betadiet` as response variable and `age`, `sex`, `smokstat`, `bmi`, `vituse`, `calories`, `fat`, `fiber`, `alcohol`, and `cholesterol` as the predictors.

i. Use the function `stepGAICAll.A()` to find an appropriate model for `betadiet`, using only linear terms in the explanatory variables (for μ, σ, ν, and τ in `BCTo`) to start with.

```
nC <- detectCores()
M1 <- gamlss(betadiet ~ 1, data=plasma, family=BCTo)
M3<-stepGAICAll.A(M1,scope=list(lower=~1,
             upper=~age+sex+smokstat+bmi+vituse+calories+fat+
             fiber+alcohol+cholesterol),
             k=4, parallel="multicore", ncpus = nC)
```

Comment on the final chosen model and use residual diagnostics to evaluate its adequacy.

ii. Try a model with two-way interactions in the explanatory variables:

```
M4<-stepGAICAll.A(M1,scope=list(lower=~1,
          upper=~(age+sex+smokstat+bmi+vituse+calories+fat+
          fiber+alcohol+cholesterol)^2),
          k=4, parallel="multicore", ncpus = nC)
```

Use residual diagnostics to evaluate its adequacy, and compare it to the model fitted in (b).

6

Continuous distributions on $(0, 1)$

CONTENTS

This chapter provides explanation for:

1. different types of continuous distributions defined on the interval $(0, 1)$;
2. how to create distributions on $(0, 1)$ using a logit transformation or truncation.

6.1 Introduction

This chapter examines situations where a continuous response variable has its range on the interval between zero and one, excluding the endpoints zero and one, denoted as $(0, 1)$. The situation where the endpoints are included, i.e. $[0, 1]$, is studied in Chapter 9 where mixed distributions are examined. Zero to one responses occur when we have proportions, percentages, or fractions. There are only five explicit distributions in **gamlss.dist** for modeling data on $(0, 1)$, see Table 6.1, but many more distributions can be created by transforming any continuous random variable defined on $(-\infty, \infty)$, using an inverse logit transformation, to a random variable with range $(0, 1)$. Also, any continuous distribution that contains the interval $(0, 1)$ in its range can be truncated to $(0, 1)$ giving a distribution suited to modeling responses in this range. Methods of choosing an appropriate distribution were discussed in Section 4.5.

For a random variable Y with range $(0, 1)$, an extreme outlier occurs when $Y \to 0$ or $Y \to 1$. These may have unbounded influence on the MLE of one or more specific parameters of the distribution, see Chapter 12.

TABLE 6.1: Continuous distributions on $(0,1)$ implemented within the **gamlss.dist** package. Note that $\mathbb{R}_+ = (0, \infty)$.

Distribution	**gamlss** name	p	Parameter range μ	σ	ν	τ	skewness
beta	`BE`	2	$(0,1)$	$(0,1)$	-	-	(both)
beta (orig)	`BEo`	2	$\mathbb{R}_+$	$\mathbb{R}_+$	-	-	(both)
gen beta type 1	`GB1`	4	$(0,1)$	$(0,1)$	$\mathbb{R}_+$	$\mathbb{R}_+$	both
logit normal	`LOGITNO`	2	$(0,1)$	$\mathbb{R}_+$	-	-	(both)
simplex	`SIMPLEX`	2	$(0,1)$	$\mathbb{R}_+$	-	-	(both)

6.2 Explicit distributions on $(0, 1)$

Table 6.1 and Chapter 21 give the explicit continuous distributions on $(0,1)$ in **gamlss.dist**:

beta distribution The beta distribution, $\texttt{BE}(\mu,\sigma)$ and $\texttt{BEo}(\mu,\sigma)$, is the most popular distribution on the range $(0,1)$ and has been used widely for regression type models, see Cribari-Neto and Zeileis [2010], Ferrari and Cribari-Neto [2004], and Kieschnick and McCullough [2003]. It has two parameters and therefore only the location and the scale of the distribution can be modeled. There are two parameterizations of the beta distribution in **gamlss**: (i) the original beta distribution (`BEo`) with mean $\mu/(\mu+\sigma)$ and (ii) a reparameterized version (`BE`) which has mean μ and therefore provides easier interpretation for regression models. Figure 6.1 shows the pdf of the `BE` distribution for different μ and σ combinations. The beta distribution can be uni-modal, U-shaped, or J-shaped. For $\mu = 0.5$ the distribution is symmetric; for $0 < \mu < 0.5$ it is positively skewed; and for $0.5 < \mu < 1$ it is negatively skewed.

generalized beta type 1 The generalized beta type 1, $\texttt{GB1}(\mu,\sigma,\nu,\tau)$, is a four-parameter distribution and therefore skewness and kurtosis may be modeled in addition to location and scale. Note that for $\nu = \tau = 1$ we have the beta distribution, i.e. $\texttt{GB1}(\mu,\sigma,1,1) \equiv \texttt{BE}(\mu,\sigma)$. Note also that if

$$Y \sim \texttt{GB1}(\mu,\sigma,\nu,\tau)$$

then

$$Z \sim \frac{Y^\tau}{\nu + (1-\nu)Y^\tau} \sim \texttt{BE}(\mu,\sigma) \; .$$

Hence $Y = \left[\nu Z/\left(1 - Z + \nu Z\right)\right]^{1/\tau}$, so Y is obtained by transformation from Z. No explicit moments are available for the `GB1` distribution.

logit normal The logit normal distribution, $\texttt{LOGITNO}(\mu,\sigma)$, is a two-parameter distribution derived by the inverse logit transformation, $Y = 1/\left[1+\exp(-Z)\right]$, of $Z \sim \texttt{NO}(\texttt{logit}(\mu),\sigma)$, see Section 6.3.1 and Section 21.1.2 for more details. Note that the median of the $\texttt{LOGITNO}(\mu,\sigma)$ distribution is μ, while σ is a shape parameter.

simplex The simplex distribution, $\texttt{SIMPLEX}(\mu,\sigma)$, is a two-parameter distribution on

$(0,1)$. It is a special case of the four-parameter generalized simplex distribution given by Barndorff-Nielsen and Jørgensen [1991], which itself is derived from the generalized inverse Gaussian distribution. The mean of the distribution is μ, and σ is a dispersion parameter. For more properties of the simplex distribution see Zhang et al. [2016] and references therein.

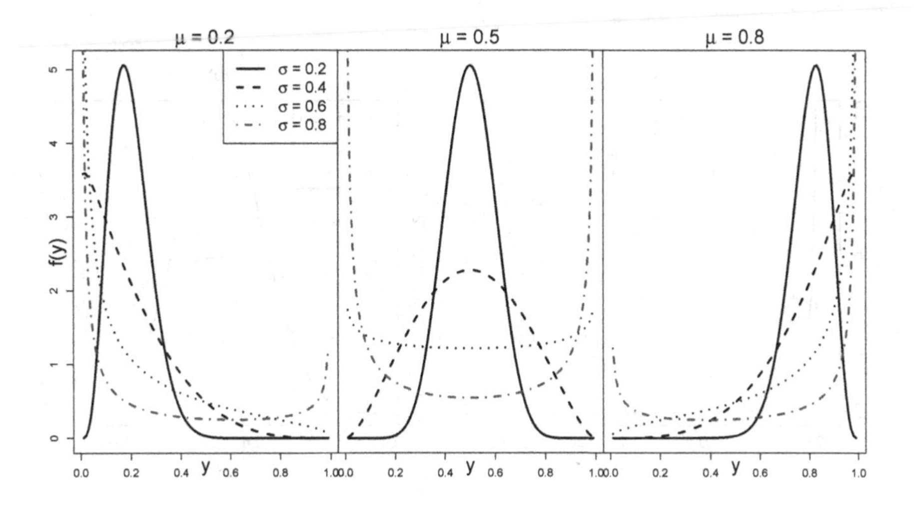

FIGURE 6.1: The beta distribution $\texttt{BE}(\mu,\sigma)$ for $\mu = (0.2, 0.5, 0.8)$, and $\sigma = (0.2, 0.4, 0.6, 0.8)$.

6.3 Creating distributions on $(0,1)$

6.3.1 Logit transform distributions on $(0,1)$

Any continuous random variable Z defined on the real line $(-\infty,\infty)$ can be transformed by the inverse logit transformation $Y = 1/[1+\exp(-Z)]$ to a random variable Y defined on $(0,1)$. For example, if Z is a t-family distributed variable, i.e. $Z \sim \texttt{TF}(\mu,\sigma,\nu)$, and the inverse logit transformation is applied, then $Y \sim \texttt{logitTF}(\mu,\sigma,\nu)$, a logit-$t$ family distribution on $(0,1)$.

The following is an example of how to take a `gamlss.family` distribution on $(-\infty,\infty)$ and create a corresponding logit distribution on $(0,1)$. The **gamlss** function `gen.Family()` generates the `d`, `p`, `q`, and `r` functions of the distribution, together with the function which is used for fitting within **gamlss**. Here we generate a logit t family distribution and plot the distribution for different values of μ, σ, and ν in Figure 6.2. Note that μ, σ, and ν are defined on the original t family distribution ranges, i.e. $-\infty < \mu < \infty$, $\sigma > 0$ and $\nu > 0$. The t-family distribution is symmetric about its median μ, and this implies that $1/[1+\exp(-\mu)]$ is the median of the logit t family distribution, `logitTF`, not the mean. Also σ and ν are related to the scale and shape of the distribution.

```
# generate the distribution
gen.Family("TF", type="logit")
```

```
## A  logit  family of distributions from TF has been generated
##  and saved under the names:
##  dlogitTF plogitTF qlogitTF rlogitTF logitTF
```

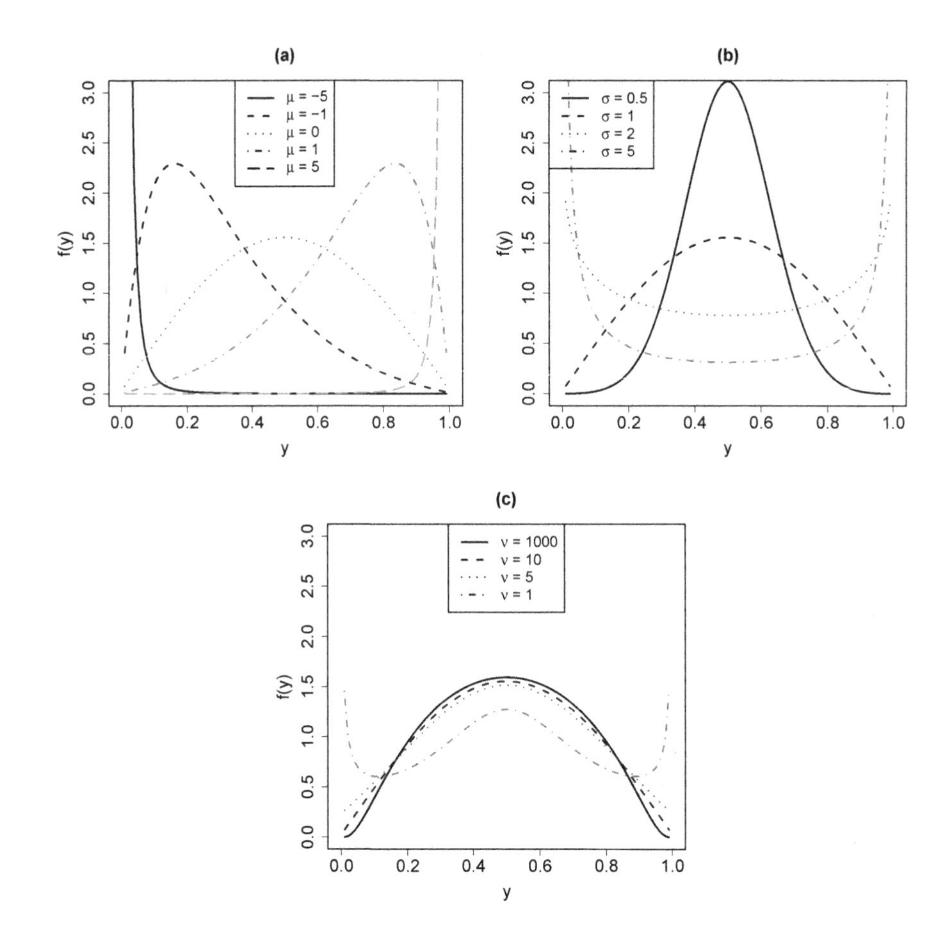

FIGURE 6.2: A logit t family distribution, `logitTF`(μ, σ, τ): (a) $\mu = -5, -1, 0, 1, 5$, $\sigma = 1$ and $\nu = 10$, (b) $\mu = 0$, $\sigma = 0.5, 1, 2, 5$, and $\nu = 10$, and (c) $\mu = 0$, $\sigma = 1$, and $\nu = 1, 5, 10, 1000$.

The plots in Figure 6.2 were generated with a series of calls to the `curve()` function such as the one below:

```
curve(dlogitTF(x, mu=-5, sigma=1, nu=10), 0,1, ylim=c(0,3), lwd=3,
      lty=2, col=2)
```

Figure 6.2 shows the different shapes the distribution can take. Panel (a) shows, for fixed $\sigma = 1$ and $\nu = 10$, how the distribution changes for different values of $\mu = -5, -1, 0, 1, 5$.

Panel (b), for fixed $\mu = 0$ and $\nu = 10$, varies $\sigma = 0.5, 1, 2, 5$. Finally panel (c), for fixed $\mu = 0$ and $\sigma = 1$, varies $\nu = 1, 5, 10, 1000$.

6.3.2 Truncated distributions on (0,1)

Any distribution defined on $(-\infty, \infty)$ can be left-truncated at zero and right-truncated at one to give a truncated distribution on (0,1), using the function `gen.trun()` in the **gamlss.tr** package. Note however that for distributions generated by truncating a continuous distribution, although the exact values zero and one are permitted, they have non-zero likelihood but zero probabilities. Hence they are not recommended if the response variable has values exactly zero or one. Instead mixed distributions studied in Chapter 9 are recommended. Note also that in the presence of exact zero or one values. the truncated distribution deviances cannot be compared to the deviances from mixed distributions in Chapter 9.

We demonstrate the truncation of a skew Student t distribution (SST(μ, σ, ν, τ)) below zero and above one, and plot it for different values of μ, σ, ν, and τ in Figure 6.3.

```
# generate the distribution
library(gamlss.tr)
gen.trun(c(0,1),"SST",type="both")
```

```
## A truncated family of distributions from SST has been generated
##  and saved under the names:
##  dSSTtr pSSTtr qSSTtr rSSTtr SSTtr
## The type of truncation is both
##  and the truncation parameter is 0 1
```

Similarly any distribution defined on $(0, \infty)$ can be right-truncated at one to give a truncated distribution on $(0, 1)$, using `gen.trun()`.

6.4 Examples

6.4.1 Tensile strength data

R data file: `tensile` in package **gamlss.data** of dimension 30 × 1
source: Hand et al. [1994]
variables
  `str` : the transformed values of strength of polyester fibers (the unit of measurement are not given)
purpose: to demonstrate the fitting of a parametric distribution to the data
conclusion: the data are use to demonstrate how different $(0, 1)$ distributions can be fitted in GAMLSS, the simplex distribution fits best according to AIC

These data come from Quesenberry and Hales [1980] and were reproduced in Hand et al. [1994, data set 180, p. 140]. They contain transformed measurements of tensile strength of polyester fibers and the authors were checking if the untransformed measurements

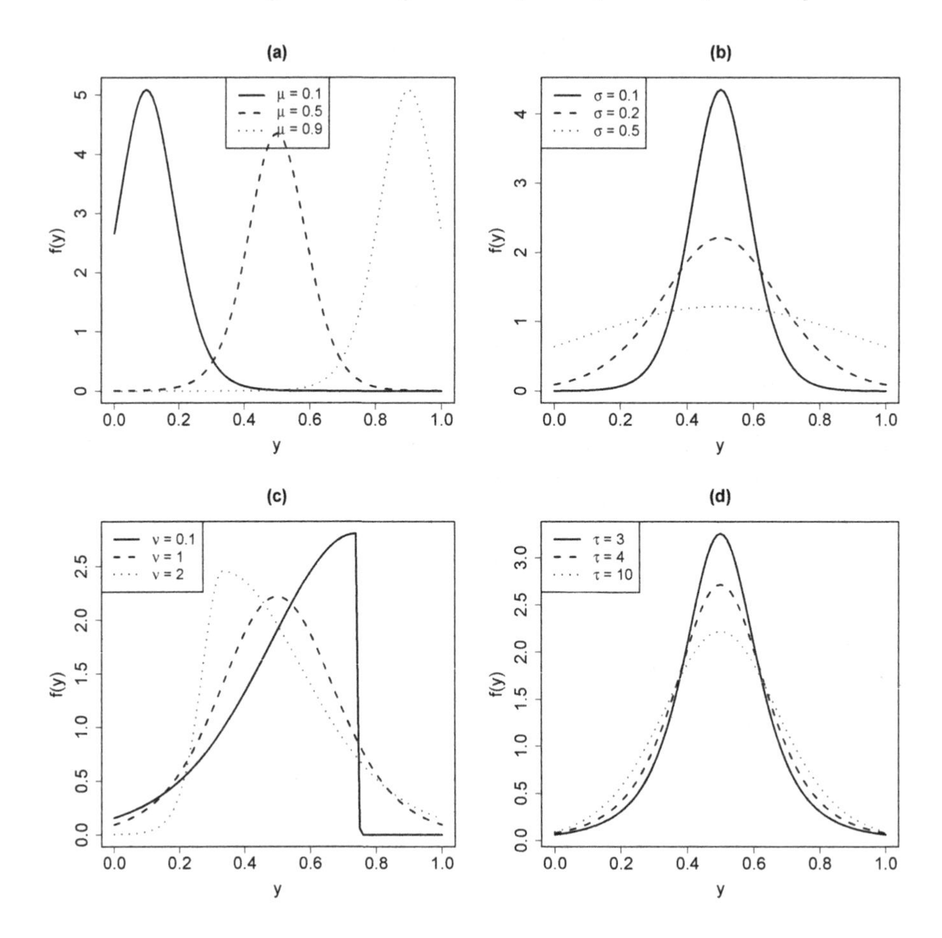

FIGURE 6.3: A truncated `SST` distribution, $\texttt{SSTtr}(\mu, \sigma, \nu, \tau)$: (a) $\mu = 0.1, 0.5, 0.9$, $\sigma = 0.1$, $\nu = 1$, and $\tau = 10$, (b) $\mu = 0.5$, $\sigma = 0.1, 0.2, 0.5$, $\nu = 1$, and $\tau = 10$, (c) $\mu = 0.5$, $\sigma = 0.2$, $\nu = 0.1, 1, 2$, and $\tau = 10$, (d) $\mu = 0.5$, $\sigma = 0.2$, $\nu = 1$, and $\tau = 3, 4, 10$.

were consistent with the log normal distribution. According to Hand et al. [1994] 'these data follow from a preliminary transformation. If the log normal hypothesis is correct, these data should have been uniformly distributed'.

Here we use the data as an example of a variable restricted to the range $(0, 1)$ and the fitting of alternative distributions. We create the logit t family, logit `SEP3` and logit `ST3` distributions using the `gen.Family()` function; and then the truncated normal, t family and `ST3` distributions with range $(0, 1)$ using `gen.trun()`. First we create the new distributions and then fit all distributions using `fitDist()` which selects the 'best' distribution using the AIC. The newly created distributions are included in the list of zero to one distributions using the argument `extra`.

```
data(tensile)
# creating new logit distributions
gen.Family("TF", "logit")
#Similarly for ST3 and SEP3
```

```
# creating new truncated distributions
library(gamlss.tr)
gen.trun(c(0,1),"TF",type="both")
#Similary for NO and ST3

# fitting the different distributions
mf1 <- fitDist(tensile$str, type="real0to1",
       extra=c("logitTF", "logitST3", "logitSEP3",
               "NOtr", "TFtr", "ST3tr"), k=2)
```

```
mf1$fits[1:6]

##    SIMPLEX   LOGITNO      NOtr        BE       BEo   logitTF
## -4.369151 -4.227934 -2.679559 -2.610127 -2.610127 -2.227934

mf1$failed

## list()
```

According to the AIC, (given by `k=2`), the simplex distribution gives the best fit. A worm plot of the residuals of the simplex distribution model (not shown here) confirms an adequate fit. For comparison of the shapes of the different fitted distribution, Figure 6.4 shows six fitted distributions. The plots were created using the function `histDist()`[a], for example:

```
msim <-histDist(tensile$str, "SIMPLEX" , nbins=14, xlim=c(0,1),
                ylim=c(0,3), main="(a) SIMPLEX")
```

6.4.2 PBSC study data

R data file: sdac in package **simplexreg** of dimension 239×5
source: Zhang et al. [2016]
variables
- `rcd` : the recovery rate of CD34+ cells (the response)
- `age` the age of the patient
- `ageadj` the adjusted age by setting `age` < 40 as the baseline age (0) and subtracting 40 from other ages
- `chemo` whether a patient receives chemotherapy on a one-day protocol (0) or on a 3-day protocol (1)
- `gender` the gender of the patient

purpose: to demonstrate the fitting of a parametric distribution on zero to one.
conclusion: A truncated t family distribution was found to fit best.

These data are from a study on peripheral blood stem cell (PBSC) transplant patients between 2003 and 2008, at the Cross Cancer Institute, Alberta Health Services [Zhang et al., 2016]. Zhang et al. [2016] fitted a simplex distribution regression model using the explanatory variables `ageadj` and `chemo` to model μ, and compared the resulting model to the heterogeneity model where in addition `age` is used to model σ. The two models fitted by Zhang et al. [2016] are reproduced below:

[a] The function `histDist()` is not working with mixed distributions.

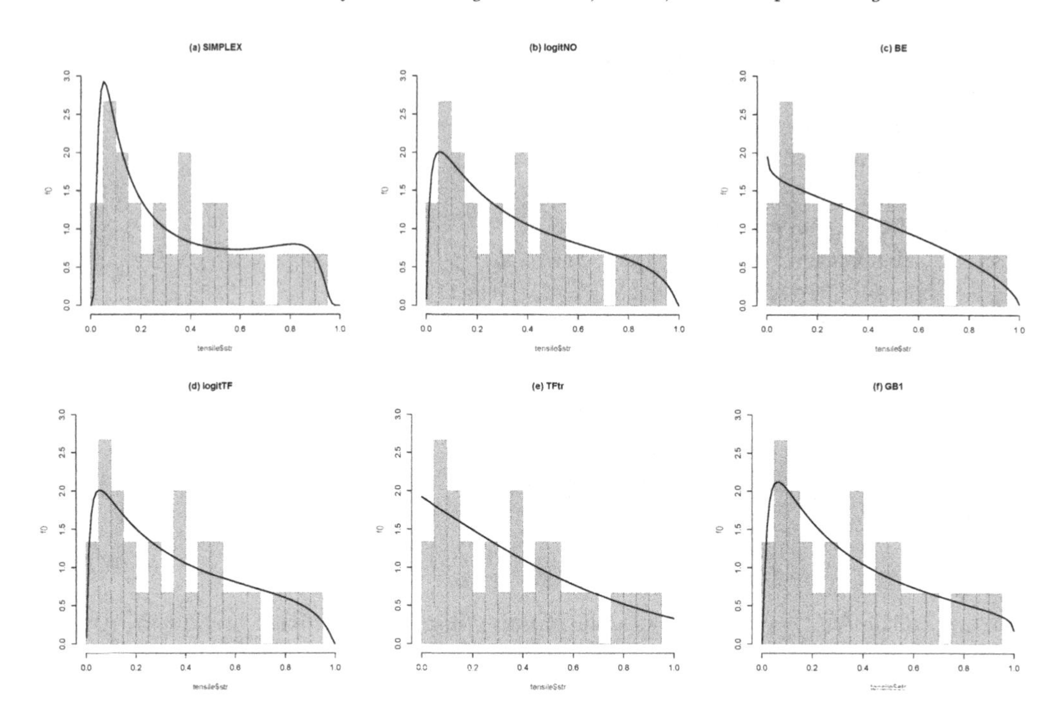

FIGURE 6.4: Tensile strength data with fitted (a) simplex, (b) logit normal, (c) beta, (d) logit t family, (e) truncated t family, and (f) the generalized beta type 1, (GB1), distributions.

```
library(simplexreg)
data(sdac)
# modeling only the mean mu
msi <- gamlss(rcd ~ ageadj + chemo, data = sdac, family=SIMPLEX,
              method=mixed(20,100), trace=FALSE)
# modeling heterogeneity
msi1 <- gamlss(rcd ~ ageadj + chemo, sigma.fo=~age, data = sdac,
               family=SIMPLEX,  method=mixed(20,100), trace=FALSE)
AIC(msi, msi1)
##       df       AIC
## msi1  5 -308.7456
## msi   4 -305.2478
```

The AIC results support the heterogeneity model `msi1`. Within the GAMLSS framework it is of interest to investigate whether any other distribution fits the data better than the simplex distribution. The function `chooseDist()` provides a table of GAICs for choosing between different distributions, given a specific model, but in our case we have two models of interest. So what we do next is to use `chooseDist()` twice, once for the simple model and once for the heterogeneity model. In addition to the explicit distributions on $(0, 1)$ we also try the `logitTF`, `logitSEP3`, and `logitSTT` distributions and the truncated normal, t family, and `ST3` distributions (generated as in Section 6.4.1

using the methods of Section 6.3). Then we merge the two tables and use the merged table to choose an appropriate model.

```
# find best distribution for the simple model
tm <- chooseDist(msi, type="real0to1", extra=c("logitTF",
          "logitSEP3", "logitSST","NOtr", "TFtr", "SSTtr"))

## minimum GAIC(k= 2 ) family: TFtr
## minimum GAIC(k= 3.84 ) family: TFtr
## minimum GAIC(k= 5.48 ) family: NOtr

# rename its rows for later
rownames(tm) <- paste(rownames(tm), "*",sep="")
# find best distribution for the heterogeneity model
tm1 <- chooseDist(msi1, type="real0to1",extra=c("logitTF",
          "logitSEP3", "logitSST","NOtr", "TFtr", "SSTtr"))

## minimum GAIC(k= 2 ) family: TFtr
## minimum GAIC(k= 3.84 ) family: TFtr
## minimum GAIC(k= 5.48 ) family: NOtr

# merge the two tables
tab <- rbind(tm, tm1)
#  best AIC
getOrder(tab, column = 1)[1:5]

## GAIG with k= 2
##       TFtr*       GB1* logitSEP3*       TFtr     SSTtr*
##  -389.8820  -389.8075  -389.2371  -388.6500  -387.8858

# best BIC
getOrder(tab, column = 3)[1:5]

## GAIG with k= 5.48
##      NOtr*     TFtr*      BEo*       BE*      GB1*
## -372.5215 -372.4820 -372.0576 -369.3842 -368.9275
```

Note that in the output '*' represents the fits for the simpler model. The best models using the AIC and BIC were the `TFtr` and `NOtr` distributions, respectively, both with no model for σ. It is worth noticing that the simplex distribution models come towards the end of the lists and therefore provide worse fits than most of the other `gamlss.family` distributions. The worm plots for the `TFtr` and `NOtr` distribution models shown in Figure 6.5 indicate that the first distribution `TFtr` provides a better fit.

```
mtftr1 <- gamlss(rcd ~ ageadj + chemo, data = sdac, family=TFtr,
               method=mixed(20,100), trace=FALSE)
mnotr <- gamlss(rcd ~ ageadj + chemo, data = sdac, family=NOtr,
               method=mixed(20,100), trace=FALSE)
wp(mtftr1); title("(a) truncated t family")
wp(mnotr); title("(b)  truncated normal")
```

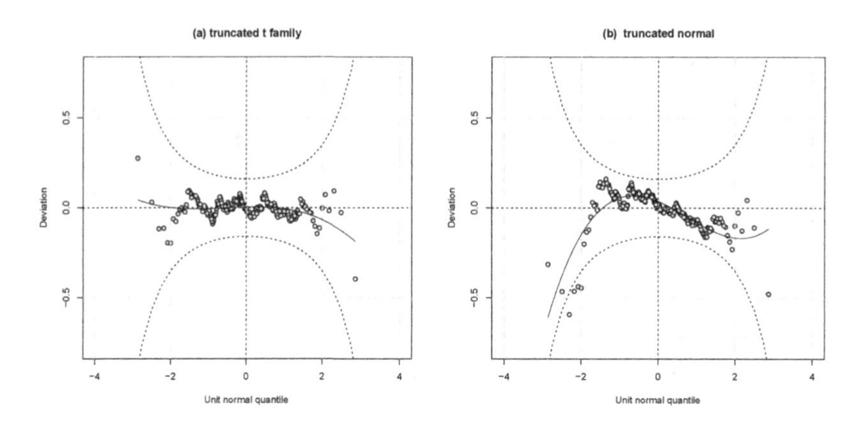

FIGURE 6.5: Worm plots for (a) truncated t family (b) truncated normal distributions.

6.5 Bibliographic notes

Distributions for bounded continuous random variables have a sparse literature, in comparison with those for unbounded variables. The beta distribution has dominated the analysis of bounded continuous response variables, being until fairly recently the only well-known candidate for this range. The beta distribution is also of use as a mixing distribution for bounded parameters in, for example the beta-binomial distribution (see Section 8.2.2). Beta regression is available in the **R** package **betareg** [Cribari-Neto and Zeileis, 2010], and simplex regression in the **simplexreg** package [Zhang et al., 2016]. Kotz and van Dorp [2004] present a monograph on alternatives to the beta distribution, being mainly the triangular distribution and the two-sided power distribution family.

6.6 Exercises

1. The `cable` data set concerns the penetration of cable television in $n = 283$ market areas in the USA. The data were collected in a mailed survey questionnaire in 1992 [Kieschnick and McCullough, 2003]. The aim of the study was to explain cable television uptake (the proportion `pen5`) as a function of area demographics.

R data file: `cable` in package **gamlss.data** of dimension 283×6
source: Kieschnick and McCullough [2003]
variables

- `pen5` : proportion of households having cable TV in market area
- `lin` : log median income
- `child` : percentage of households with children
- `ltv` : number of local TV stations
- `dis` : consumer satisfaction index
- `agehe` : age of cable TV headend

purpose: to demonstrate the fitting of a parametric distribution to the response variable `pen5` with range $(0,1)$.
conclusion: For the marginal distribution, the truncated normal fits best; for the regression model, the truncated skew t (`SSTtr`) is best.

(a) Examine a histogram of `pen5`. What is the range of `pen5`?

```
truehist(cable$pen5, nbins=20, xlim=c(0,1))
```

(b) Using `fitDist()`, find the distribution that best fits `pen5`. Use generated and truncated distributions, as well as the explicit `gamlss.family` distributions. Display the fit of the 'best' distribution (using the default, AIC).

```
library(gamlss.tr)
# logit transformations
gen.Family("TF", "logit")
gen.Family("ST3", "logit")
gen.Family("SEP3", "logit")
# truncated distributions
gen.trun(c(0,1),"TF",type="both")
gen.trun(c(0,1),"NO",type="both")
gen.trun(c(0,1),"SST",type="both")

a <- fitDist(cable$pen5, type="real0to1", extra=c("logitTF",
          "logitST3", "logitSEP3","TFtr", "NOtr", "SSTtr"))
a$fits
histDist(cable$pen5, family=NOtr)
```

(c) Now select a distribution for the regression model for `pen5`, with μ predictor `pb(lin)+ltv+agehe`.

```
m0 <- gamlss(pen5~pb(lin)+ltv+agehe, family=BE, data=cable,
             n.cyc=50)
c1 <- chooseDist(m0, type="real0to1", extra=c("logitTF",
          "logitST3", "logitSEP3","TFtr", "NOtr", "SSTtr"))
m1 <- gamlss(pen5~pb(lin)+ltv+agehe, family=SSTtr, data=cable,
             n.cyc=50)
plot(m1)
wp(m1)
```

(d) Investigate whether model `m1` can be improved on by the addition or deletion of covariates in the model for μ, and the addition of covariates in the model for σ.

2. Kieschnick and McCullough [2003] analyse US election data, at the state level, in the 2000 Presidential Election. The response variable is the proportion of the state that voted for George Bush; and the predictors are state demographic indicators.

R data file: `bush2000` in package **gamlss.data** of dimension 51×10
source: Kieschnick and McCullough [2003]
variables

- `state` : name of state
- `bush` : proportion of state's vote for George Bush
- `male` : percentage of population male
- `pop` : population
- `rural` : percentage of population living in rural areas
- `bpovl` : percentage of population with income below the poverty level
- `clfu` : unemployment rate (%)
- `mgt18` : percentage of male population older than 18 years
- `pgt65` : percentage of population older than 65 years
- `numgt75` : percentage of population with income > \$75K

purpose: to demonstrate the fitting of a parametric distribution to the response variable `bush` with range $(0, 1)$.
conclusion: For the marginal distribution, the truncated t family fits best; for the regression model, the truncated normal is best.

3. Examine a histogram of `bush`. What type of distribution is suggested?

```
truehist(bush2000$bush, nbins=20, xlim=c(0,1))
```

4. Using `fitDist()`, find the distribution that best fits `bush`. Use generated and truncated distributions, as well as the explicit `gamlss.family` distributions as in question 1. Display the fit of the 'best' distribution.

5. Now select a distribution for the regression model for `bush`, with μ predictor `male+log(pop)+rural+bpovl+clfu`.

```
m0 <- gamlss(bush~male+log(pop)+rural+bpovl+clfu,
             family=BE, data=bush2000, n.cyc=100)
c1 <- chooseDist(m0, type="real0to1", extra=c("logitTF",
         "logitST3", "logitSEP3","TFtr", "NOtr", "SSTtr"))
m1 <- gamlss(bush~male+log(pop)+rural+bpovl+clfu,
             family=NOtr, data=bush2000, n.cyc=100)
plot(m1)
wp(m1)
```

6. Investigate whether model `m1` can be improved on by the addition or deletion of covariates in the model for μ, and the addition of covariates in the model for σ.

7

Discrete count distributions

CONTENTS

This chapter provides explanation for:

1. different types of discrete count data distributions with range $\{0, 1, 2, \ldots\}$ within the GAMLSS family, and
2. how these distributions model overdispersion, underdispersion, and zero-inflation.

7.1 Introduction

This chapter deals with discrete distributions for an unlimited count variable Y having range $R_Y = \{0, 1, 2, \ldots\}$. [There are also two discrete count distributions with range $R_Y = \{1, 2, 3, \ldots\}$, the logarithmic (`LG`) (Section 22.1.2), and the Zipf (`ZIPF`) (Section 22.1.5) distributions, but these are not discussed in this chapter.] Table 7.1 lists all the explicit discrete distributions in **gamlss.dist** with their default link functions. Discrete distributions with limited range $R_Y = \{0, 1, 2, \ldots, n\}$, for known upper limit n, are dealt with in Chapter 8.

Sections 7.1.1 and 7.1.2 consider the Poisson distribution and its limitations, respectively. Section 7.2 considers explicit `gamlss.family` discrete count distributions on $\{0, 1, 2, \ldots\}$. Section 7.3 considers overdispersed and underdispersed discrete count distributions relative to the Poisson distribution, including mixture distributions and discretized continuous distributions. Section 7.4 considers zero-inflated and zero-adjusted distributions for modeling an excess or shortage of zero values in a response variable, i.e. an increased or decreased probability that $Y = 0$. It also considers K-inflated distribution where the inflation occurs at $Y = K$. Section 7.5 compares the skewness and kurtosis of the distributions. Section 7.6 considers the negative binomial family for modeling the variance-mean relationship. Section 7.7 explains how to model count data using **gamlss** packages, and presents three examples.

TABLE 7.1: Discrete distributions implemented within **gamlss.dist**, with default link functions. (inf=inflated, adj=adjusted, inv=inverse)

Distribution	**gamlss** name	Parameter link function			
		μ	σ	ν	τ
$R_Y = \{0,1,2,\ldots\}$					
beta neg binomial	`BNB`	log	log	log	-
Delaporte	`DEL`	log	log	logit	-
discrete Burr XII	`DBURR12`	log	log	log	-
double Poisson	`DPO`	log	log	-	-
generalized Poisson	`GPO`	log	log	-	-
geometric	`GEOM`	log	-	-	-
geometric (original)	`GEOMo`	logit	-	-	-
negative binomial type I	`NBI`	log	log	-	-
negative binomial type II	`NBII`	log	log	-	-
neg binomial family	`NBF`	log	log	log	-
Poisson	`PO`	log	-	-	-
Poisson-inv Gaussian	`PIG`	log	log	-	-
Poisson-inv Gaussian 2[a]	`PIG2`	log	log	-	-
Poisson shifted GIG[a]	`PSGIG`	log	log	ident	logit
Sichel	`SI`	log	log	ident	-
Sichel (μ the mean)	`SICHEL`	log	log	ident	-
Waring (μ the mean)	`WARING`	log	log	-	-
Yule (μ the mean)	`YULE`	log	-	-	-
zero-adj beta neg binom	`ZABNB`	log	log	log	logit
zero-adj logarithmic	`ZALG`	logit	logit	-	-

zero-adj neg binomial	ZANBI	log	log	logit	-
zero-adj PIG	ZAPIG	log	log	logit	-
zero-adj Sichel	ZASICHEL	log	log	ident	logit
zero-adj Poisson	ZAP	log	logit	-	-
zero-adj Zipf	ZAZIPF	log	logit	-	-
zero-inf beta neg binom	ZIBNB	log	log	log	logit
zero-inf neg binomial	ZINBI	log	log	logit	-
zero-inf neg binom fam[a]	ZINBF	log	log	log	logit
zero-inf Poisson	ZIP	log	logit	-	-
zero-inf Poisson (μ the mean)	ZIP2	log	logit	-	-
zero-inf PIG	ZIPIG	log	log	logit	-
zero-inf Sichel	ZISICHEL	log	log	ident	logit
$R_Y = \{0, 1, \ldots, n\}$					
binomial	BI	logit	-	-	-
beta binomial	BB	logit	log	-	-
double binomial	DBI	logit	log	-	-
zero-adj beta binomial	ZABB	logit	log	logit	-
zero-adj binomial	ZABI	logit	logit	-	-
zero-inf beta binomial	ZIBB	logit	log	logit	-
zero-inf binomial	ZIBI	logit	logit	-	-
$R_Y = \{1, 2, \ldots\}$					
logarithmic	LG	logit	-	-	-
Zipf	ZIPF	log	-	-	-

7.1.1 Poisson distribution

For a count variable Y with range $R_Y = \{0, 1, 2, \ldots\}$, the classical model is the Poisson distribution, first published by Poisson [1837], which has had a major influence on discrete distribution modeling for almost two centuries. The Poisson distribution, $\texttt{PO}(\mu)$, was introduced in Chapter 1. Its probability function is given by

$$P(Y = y \mid \mu) = \frac{e^{-\mu}\mu^y}{y!} \qquad \text{for } y = 0, 1, 2, \ldots \tag{7.1}$$

where $\mu > 0$. However the Poisson distribution has only a single parameter μ, which equals the mean $E(Y)$. The variance $\text{Var}(Y) = \mu$, moment skewness $\gamma_1 = \mu^{-\frac{1}{2}}$, and moment excess kurtosis $\gamma_2 = \mu^{-1}$ all depend on μ, and so cannot be modeled independently of μ. The moment skewness and moment excess kurtosis both tend to zero as μ increases to infinity.

7.1.2 Limitations of the Poisson distribution

There are four major problems often encountered when modeling count data using the Poisson distribution:

- overdispersion (and underdispersion), discussed in Section 7.3,

[a] Currently under development

- heavy right tail (i.e. high positive skewness), discussed in Sections 7.3, 7.5, and also in Section 17.4,
- excess (or shortage) of zero values, discussed in Section 7.4. Excess at a value other than zero, i.e. K, is also discussed in Section 7.4.
- variance-mean relationship, discussed in Section 7.6.

Overdispersion is usually defined as the extra variation in a count response variable which is not explained by the Poisson distribution. The problem occurs because the population variance of a Poisson random variable is equal to its mean, i.e. $\text{Var}(Y) = \text{E}(Y) = \mu$. Unfortunately very often for a count response variable $\text{Var}(Y) > \text{E}(Y)$, i.e. the distribution is overdispersed relative to the Poisson distribution. A less common problem occurs when $\text{Var}(Y) < \text{E}(Y)$, i.e. the distribution is underdispersed relative to the Poisson.

The second problem is with the *right tail* of the count data distribution. There are many situations where high values of the response variable Y, i.e. values in the right tail of the distribution, happen more often than the Poisson distribution would suggest. Because the Poisson is a one-parameter distribution, its skewness and kurtosis are fixed given the mean μ and consequently events happening in the right tail of the distribution may not be modeled properly.

The third problem of *excess (or shortage) of zero values* occurs when the response variable has a higher (or lower) probability of a zero value than that given by the Poisson distribution. This again is a phenomenon that occurs frequently in practice.

The fourth problem is the variance-mean relationship. For the Poisson distribution the variance equals (and hence increases with) the mean. However in practice the variance may increase with a power of the mean, i.e. Taylor's power law, see Enki et al. [2017].

7.2 Explicit count distributions

Here we consider explicit discrete distributions in **gamlss.dist** for a count variable Y having range $R_Y = \{0, 1, 2, \ldots\}$. The details of the explicit discrete distributions are given in Chapter 22.

7.2.1 One-parameter distributions

There are four one-parameter discrete count distributions in **gamlss.dist**. One-parameter distributions are only able to model the location of the distribution.

Geometric `GEOM`(μ) and `GEOMo`(μ) are alternative parameterizations of the geometric distribution, which has its mode at $Y = 0$ and declining probabilities as Y increases. In the `GEOM`(μ) parameterization μ is the mean, while in the (original) `GEOMo`(μ) parameterization μ is the probability $P(Y = 0)$. `GEOMo`(μ) is an appropriate model when Y counts the number of 'misses' before the first 'hit', in an independent sequence of trials having equal probabilities μ of a 'hit'.

Poisson `PO`(μ) is the classic model for a discrete count variable, see Section 7.1.1.

Yule YULE(μ) is a distribution with mean μ and a heavy right tail (especially for large μ), with a mode at $Y = 0$ and declining probabilities as Y increases.

7.2.2 Two-parameter distributions

(a) There are several two-parameter discrete count distributions in **gamlss.dist** with the capability for modeling location and scale:

Negative binomial: NBI(μ, σ) and NBII(μ, σ). NBI(μ, σ) is the classic distribution for modeling overdispersed count data. It has mean μ and variance $\mu + \sigma\mu^2$. NBI is a gamma mixture of Poisson distributions, see Section 7.3.3. NBII(μ, σ) is a reparameterization of NBI(μ, σ) with mean μ and variance $\mu + \sigma\mu$.

Poisson-inverse Gaussian: PIG(μ, σ) and PIG2(μ, σ). PIG(μ, σ) has mean μ and variance $\mu + \sigma\mu^2$ and is similar to NBI(μ, σ), but has a heavier right tail and a higher positive moment skewness than a NBI(μ, σ) with the same mean and variance. It is an inverse Gaussian mixture of Poisson distributions, see Section 7.3.3. PIG2(μ, σ) is a reparameterization of PIG(μ, σ) where the parameters μ and σ are information orthogonal [Heller et al., 2018].

Waring: WARING(μ, σ) has mean μ and a heavy right tail (especially for large σ), with a mode at $Y = 0$ and declining probabilities as Y increases. This is an overdispersed geometric distribution.

Double Poisson: DPO(μ, σ) is a special case of the double exponential family of Efron [1986], with approximate mean μ and approximate variance $\sigma\mu$. Hence (approximately) it is overdispersed Poisson if $\sigma > 1$ and underdispersed Poisson if $\sigma < 1$. It is the only explicit two-parameter distribution in **gamlss.dist** which can model underdispersed (as well as overdispersed) Poisson data.

Generalized Poisson: GPO(μ, σ) has mean μ and variance $\mu(1 + \sigma\mu)^2$.

Note that the PO(μ) distribution is a limiting case of NBI(μ, σ), NBII(μ, σ), PIG(μ, σ), and GPO(μ, σ) as $\sigma \to 0$; and of PIG2(μ, σ) as $\sigma \to \infty$. It is also a special case of DPO(μ, σ) when $\sigma = 1$.

Note also that the GEOM(μ) distribution is a limiting case of WARING(μ, σ) as $\sigma \to 0$; a special case of NBI(μ, σ) when $\sigma = 1$; and a special case of NBII(μ, σ) when $\sigma = \mu$.

(b) There are five two-parameter discrete count distributions in **gamlss.dist** for modeling location and probability at value zero, i.e. $P(Y = 0)$. They are either zero-inflated or zero-adjusted distributions and are discussed in Section 7.4. The zero-inflated distributions allow the probability of a zero value to be inflated, but not deflated. The zero-adjusted distributions allow the probability of a zero value to be inflated or deflated.

Zero-inflated Poisson: ZIP(μ, σ) and ZIP2(μ, σ). This allows the probability of a zero value to be inflated, but not deflated, relative to the PO(μ) distribution.

Zero-adjusted Poisson: ZAP(μ, σ). This allows the probability of a zero value to be deflated or inflated, relative to the PO(μ) distribution.

Zero-adjusted logarithmic: ZALG(μ, σ) has declining probabilities after $Y = 1$.

Zero-adjusted Zipf: ZAZIPF(μ, σ) has a very heavy right tail (especially for μ close to zero), and has declining probabilities after $Y = 1$.

7.2.3 Three-parameter distributions

(a) There are four three-parameter discrete count distributions in **gamlss.dist** for modeling location, scale, and skewness.

Beta negative binomial: BNB(μ, σ, ν) has mean μ and a heavy right tail (especially for large σ). This is an overdispersed negative binomial distribution.

Discrete Burr XII: DBURR12(μ, σ, ν) is a discretized continuous distribution. It is the only explicit discrete three-parameter distribution in **gamlss.dist** which can model underdispersed (as well overdispersed) Poisson data. It has a heavy right tail.

Delaporte: DEL(μ, σ, ν) has mean μ and a heavier right tail and higher skewness and kurtosis than an NBI(μ, σ) distribution with the same mean and variance. It can also have lower probabilities at or close to zero. It is a shifted gamma mixture of Poisson distributions, see Section 7.3.3.

Sichel: SICHEL(μ, σ, ν) and SI(μ, σ, ν). SICHEL(μ, σ, ν) is a parameterization of the Sichel distribution with mean μ. It is a generalized inverse Gaussian mixture of Poisson distributions, see Section 7.3.3, and so is also known as the generalized inverse Gaussian Poisson (GIGP) distribution. SI(μ, σ, ν) is an alternative parameterization in which μ is not the mean.

(b) There are four three-parameter discrete count distributions in **gamlss.dist** for modeling location, scale, and the probability at value zero, i.e. $P(Y = 0)$:

Zero-adjusted negative binomial type I: ZANBI(μ, σ, ν)

Zero-inflated negative binomial type I: ZINBI(μ, σ, ν)

Zero-adjusted Poisson-inverse Gaussian: ZAPIG(μ, σ, ν)

Zero-inflated Poisson-inverse Gaussian: ZIPIG(μ, σ, ν)

See Section 7.4 for an explanation and properties of zero-adjusted and zero-inflated distributions.

(c) There is one three-parameter discrete count distribution in **gamlss.dist** for modeling location, scale, and the variance-mean relationship:

Negative binomial family: NBF(μ, σ, ν) is a family of parameterizations of the negative binomial distribution with mean μ and variance $\mu + \sigma\mu^{\nu}$.

7.2.4 Four-parameter distributions

(a) There is one four-parameter discrete count distribution (under development in **gamlss.dist**) for modeling the location, scale, skewness, and kurtosis of the distribution:

Poisson shifted generalized inverse Gaussian: PSGIG(μ, σ, ν, τ) is a shifted generalized inverse Gaussian mixture of Poisson distributions. It has mean μ.

(b) There are four four-parameter discrete count distributions in **gamlss.dist** for modeling the location, scale, skewness, and $P(Y = 0)$:

Zero-adjusted beta negative binomial: `ZABNB`(μ, σ, ν, τ)

Zero-inflated beta negative binomial: `ZIBNB`(μ, σ, ν, τ)

Zero-adjusted Sichel: `ZASICHEL`(μ, σ, ν, τ)

Zero-inflated Sichel: `ZISICHEL`(μ, σ, ν, τ)

7.3 Overdispersion and underdispersion

7.3.1 Introduction

Overdispersion and underdispersion (relative to the Poisson distribution) has been recognized for a long time as a potential problem within the distributions literature [Haight, 1967] and the literature of generalized linear models [Nelder and Wedderburn, 1972]. Over the years several solutions to the problem of overdispersion have been suggested, see e.g. Consul [1989] and Dossou-Gbété and Mizère [2006].

Here we consider three approaches to modeling overdispersion and underdispersion:

(a) mixture distributions (Sections 7.3.2 and 7.3.3),

(b) discretized continuous distributions (Section 7.3.4), and

(c) *ad-hoc* solutions (Section 7.3.5).

It should be noted that mixture distributions can deal with overdispersion but not underdispersion, while discretized continuous distributions and *ad-hoc* solutions can potentially deal with both overdispersion and underdispersion. There are currently only two explicit discrete count distributions that model both overdispersion and underdispersion: the double Poisson, `DPO`(μ, σ), and the discrete Burr XII, `DBURR12`(μ, σ, ν). However other discretized continuous distributions, which model overdispersion and underdispersion, can also be fitted as explained in Section 7.3.4.

7.3.2 Mixture distributions

Mixture distributions account for overdispersion by assuming that the distribution of Y depends on a random variable γ whose distributional form is known. This can also solve the problems of excess of zero values and heavy tails in the response variable. The methodology works like this.

Assume that the distribution of the response variable Y has a discrete probability function $P(Y = y \mid \gamma)$, conditional on a continuous random variable γ with pdf $f_\gamma(\gamma)$.

Then the marginal (or unconditional) probability function of Y is given by

$$\mathrm{P}(Y=y)=\int_{R_\gamma}\mathrm{P}(Y=y\,|\,\gamma)f_\gamma(\gamma)\,d\gamma \tag{7.2}$$

where R_γ is the range of γ. The resulting distribution of Y is called a continuous mixture of discrete distributions.

When γ is a discrete random variable with probability function $\mathrm{P}(\gamma=\gamma_j)$ then

$$\mathrm{P}(Y=y)=\sum_{\gamma_j\in R_\gamma}\mathrm{P}(Y=y\,|\,\gamma=\gamma_j)\mathrm{P}(\gamma=\gamma_j) \tag{7.3}$$

where R_γ is the discrete range of γ. The resulting distribution of Y is called a discrete mixture of discrete distributions. When γ takes only a finite number of possible values, then the resulting distribution is called a finite mixture of discrete distributions.

Within the mixture distributions, category (a) in Section 7.3.1, we distinguish three main types:

(a) (i) when $\mathrm{P}(Y=y)$ given by (7.2) exists explicitly. This is dealt with in Section 7.3.3,

(a) (ii) when $\mathrm{P}(Y=y)$ given by (7.2) is not explicit, but is approximated by integrating out γ using an approximation, e.g. Gaussian quadrature. This is discussed in Section 7.7,

(a) (iii) when $\mathrm{P}(Y=y)$ is the probability function of a finite mixture of discrete distributions given by (7.3), where γ takes only a finite number of possible values. This is discussed in Section 7.7.

7.3.3 Explicit continuous mixtures of Poisson distributions

These are discrete distributions included in category (a)(i) above. Suppose that, given a fixed value of γ, Y has a Poisson distribution with mean $\mu\gamma$, i.e. $Y\,|\,\gamma\sim \texttt{PO}(\mu\gamma)$. Suppose further that γ is a continuous random variable with pdf $f_\gamma(\gamma)$ defined on $(0,\infty)$. Then the marginal distribution of Y is a continuously mixed Poisson distribution. Provided γ has mean 1, then Y has mean μ. (The model can be considered as a multiplicative Poisson random effect model, provided the distribution of γ does not depend on μ.)

For example suppose $Y\,|\,\gamma\sim\texttt{PO}(\mu\gamma)$ where $\gamma\sim\texttt{GA}(1,\sigma^{1/2})$. The marginal distribution of Y is $Y\sim\texttt{NBI}(\mu,\sigma)$.

To show this result

$$\mathrm{P}(Y=y\,|\,\gamma)=\frac{e^{-\mu\gamma}(\mu\gamma)^y}{y!}\,,$$

where

$$f_\gamma(\gamma)=\frac{\gamma^{1/\sigma-1}\exp\left(-\gamma/\sigma\right)}{\sigma^{(1/\sigma)}\Gamma(1/\sigma)}\qquad \text{for } \gamma>0\ .$$

TABLE 7.2: Explicit continuously mixed Poisson distributions implemented in **gamlss.dist**. Note $h(\sigma,\nu) = 2\sigma(\nu+1)/b + 1/b^2 - 1$, $b = K_{\nu+1}(1/\sigma)/K_{\nu}(1/\sigma)$, and $K_{\lambda}(t)$ is the modified Bessel function of the second kind. If $\gamma \sim \texttt{SG}(1,\sigma^{1/2},\nu)$, a shifted gamma distribution, then $(\gamma-\nu)/(1-\nu) \sim \texttt{GA}(1,\sigma^{1/2})$.

Distribution	**gamlss** name	Mixing distribution	E(Y)	Var(Y)
Delaporte	$\texttt{DEL}(\mu,\sigma,\nu)$	$\texttt{SG}(1,\sigma^{1/2},\nu)$	μ	$\mu+\sigma(1-\nu)^2\mu^2$
NB type I	$\texttt{NBI}(\mu,\sigma)$	$\texttt{GA}(1,\sigma^{1/2})$	μ	$\mu+\sigma\mu^2$
NB type II	$\texttt{NBII}(\mu,\sigma)$	$\texttt{GA}(1,\sigma^{1/2}\mu^{-1/2})$	μ	$\mu+\sigma\mu$
NB family	$\texttt{NBF}(\mu,\sigma,\nu)$	$\texttt{GA}(1,\sigma^{1/2}\mu^{\nu/2-1})$	μ	$\mu+\sigma\mu^{\nu}$
PIG	$\texttt{PIG}(\mu,\sigma)$	$\texttt{IG}(1,\sigma^{1/2})$	μ	$\mu+\sigma\mu^2$
Sichel	$\texttt{SICHEL}(\mu,\sigma,\nu)$	$\texttt{GIG}(1,\sigma^{1/2},\nu)$	μ	$\mu+h(\sigma,\nu)\mu^2$
ZI Poisson	$\texttt{ZIP}(\mu,\sigma)$	$\texttt{BI}(1,1-\sigma)$	$(1-\sigma)\mu$	$(1-\sigma)(\mu+\sigma\mu^2)$
ZI Poisson 2	$\texttt{ZIP2}(\mu,\sigma)$	$(1-\sigma)^{-1}\texttt{BI}(1,1-\sigma)$	μ	$\mu+\frac{\sigma}{(1-\sigma)}\mu^2$
ZI NB	$\texttt{ZINBI}(\mu,\sigma,\nu)$	$\texttt{ZAGA}(1,\sigma^{1/2},\nu)$	$(1-\nu)\mu$	$(1-\nu)\left[\mu+(\sigma+\nu)\mu^2\right]$
ZI PIG	$\texttt{ZIPIG}(\mu,\sigma,\nu)$	$\texttt{ZAIG}(1,\sigma^{1/2},\nu)$	$(1-\nu)\mu$	$(1-\nu)\left[\mu+(\sigma+\nu)\mu^2\right]$

Then equation (7.2) gives:

$$\mathrm{P}(Y=y) = \int_0^{\infty} \frac{e^{-\mu\gamma}(\mu\gamma)^y}{y!} \cdot \frac{\gamma^{1/\sigma-1}\exp(-\gamma/\sigma)}{\sigma^{(1/\sigma)}\Gamma(1/\sigma)}\, d\gamma$$

$$= \frac{\Gamma(y+\frac{1}{\sigma})}{\Gamma(\frac{1}{\sigma})\Gamma(y+1)} \left(\frac{\sigma\mu}{1+\sigma\mu}\right)^y \left(\frac{1}{1+\sigma\mu}\right)^{1/\sigma} \quad \text{for } y=0,1,2,\ldots \tag{7.4}$$

where $\mu > 0$ and $\sigma > 0$. This is the probability function of the negative binomial type I, $\texttt{NBI}(\mu,\sigma)$, distribution. This is a genesis of the negative binomial distribution which is quite distinct from its classical development as the distribution of the number of failures until the kth success in independent Bernoulli trials, and was originally developed by Greenwood and Yule [1920].

By using different distributions for γ, we can generate a variety of continuously mixed Poisson count distributions. Table 7.2 shows the explicit continuously mixed Poisson distributions currently available in **gamlss.dist**, together with their **gamlss.dist** name, corresponding mixing distributions for γ, where $Y \mid \gamma \sim \texttt{PO}(\mu\gamma)$, and their mean and variance.

The probability (density) functions for all the distributions in Table 7.2 are given in Part III of the book with the exception of the shifted gamma (SG) distribution which is defined in the caption of Table 7.2.

Many previous parameterizations of continuously mixed Poisson distributions (e.g. the Sichel and Delaporte distributions [Johnson et al., 2005, Wimmer and Altmann, 1999]) have been defined such that none of the parameters of the distribution is the mean of Y, and indeed the mean of Y is often a complex function of the distribution parameters. This feature makes it difficult to interpret regression models. Fortunately most of the continuously mixed Poisson distributions given in Table 7.2 have $\mathrm{E}(Y) = \mu$. This allows easier interpretation of models for μ.

7.3.4 Discretized continuous distributions

Discretized continuous distributions provide a method of obtaining a discrete distribution from a continuous distribution. This allows the creation of very flexible discrete distributions, and is underexplored, undervalued, and underused in the statistical literature and applied statistics.

Let W be a continuous random variable defined on $(0, \infty)$, with pdf $f_W(w)$, cdf $F_W(w)$, and survivor function $S_W(w) = 1 - F_W(w)$. Then the corresponding discretized random variable Y has probability function given by

$$P(Y = y) = P(y < W < y + 1) = F_W(y + 1) - F_W(y) = S_W(y) - S_W(y + 1),$$

and cdf $F_Y(y) = F_W(y + 1)$, for $y = 0, 1, 2, 3, \ldots$.

If W has an explicit cdf $F_W(y)$, then Y has both an explicit pdf and an explicit cdf.

Note that discrete distributions created by the discretized method can be very flexible and can often cope with underdispersion as well as overdispersion, in a count response variable, as well as heavy and light right tails (see Sections 17.4 and 17.8). Discretized continuous distributions that have been developed include the discrete Weibull distribution [Nakagawa and Osaki, 1975] and the discrete Burr XII [Para and Jan, 2016] (`DBURR12`), discussed below.

In **gamlss**, a discretized version of any `gamlss.family` continuous distribution on $(0, \infty)$ can be fitted easily using the **gamlss.cens** package, as discussed later in this section.

One potential criticism of the above methods of generating discrete distributions is the fact that if the parameter μ_W is the mean of the continuous random variable W, then the mean of the discrete random variable Y will not in general be exactly μ_W.

Discrete Burr XII distribution, DBURR12(μ, σ, ν, τ)

The *continuous* Burr XII distribution, BURR12(μ, σ, ν), is obtained by setting $Y = 1$ and then $\tau = \nu$ in the GB2(μ, σ, ν, τ) distribution, i.e. BURR12(μ, σ, ν) $\equiv$ GB2($\mu, \sigma, 1, \nu$). Let $W \sim$ BURR12(μ, σ, ν), then W has pdf

$$f_W(w) = \frac{\sigma \nu w^{\sigma - 1}}{\mu^\sigma} \left[1 + \left(\frac{w}{\mu}\right)^\sigma\right]^{-\nu - 1} \qquad \text{for } 0 < w < \infty\ ,$$

where $\mu > 0$, $\sigma > 0$ and $\nu > 0$. It has cdf

$$F_W(w) = 1 - \left[1 + \left(\frac{w}{\mu}\right)^\sigma\right]^{-\nu} .$$

Hence the *discretized* (or *discrete*) Burr XII distribution, DBURR12(μ, σ, ν), has probability function

$$P(Y = y) = \left[1 + \left(\frac{y}{\mu}\right)^\sigma\right]^{-\nu} - \left[1 + \left(\frac{y + 1}{\mu}\right)^\sigma\right]^{-\nu} \qquad \text{for } y = 0, 1, 2, \ldots\ ,$$

TABLE 7.3: Counts of kidney cysts.

y	0	1	2	3	4	5	6	7	8	9	10	11
frequency	65	14	10	6	4	2	2	2	1	1	1	2

where $\mu > 0$, $\sigma > 0$ and $\nu > 0$, and cdf

$$F_Y(y) = 1 - \left[1 + \left(\frac{y+1}{\mu}\right)^{\sigma}\right]^{-\nu} .$$

The DBURR12(μ, σ, ν) distribution can be underdispersed or overdispersed relative to the Poisson distribution, and can also have a very heavy right tail when $\sigma\nu$ is close to zero. The DBURR12(μ, σ, ν) is an explicit distribution in **gamlss.dist**, see Section 22.3.2.

Fitting a general discretized continuous distribution using the gamlss.cens package

A discretized distribution, derived from any continuous distribution on $(0, \infty)$ in **gamlss.dist**, can be fitted easily using the **gamlss.cens** package.

Let W be a continuous random variable on $(0, \infty)$ and Y the corresponding discretized random variable on $\{0, 1, 2, \ldots\}$. Since

$$\mathrm{P}(Y = y) = \mathrm{P}(y < W < y + 1) \ ,$$

Y is obtained by interval censoring of W to lie in the interval $(y, y + 1)$.

Example: Counts of kidney cysts

Para and Jan [2016] analyze a univariate sample of 110 counts (y) of kidney cysts in mice fetuses. The data are given in Table 7.3.

R data file: cysts in package **gamlss.data** of dimension 12×2
source: [Para and Jan, 2016]
variables used in our analysis
y : number of cysts
f : frequency
purpose: to demonstrate the fitting of discretized continuous distributions

Here we fit the explicit discrete Burr XII (DBURR12) distribution, and also use the **gamlss.cens** package to fit the discretized gamma (GA), Weibull (WEI3), generalized gamma (GG), generalized inverse Gaussian (GIG), and generalized beta type II (GB2) distributions. We compare them using the SBC.

```
data(cysts)
library(gamlss.cens)
m1 <- gamlss (y ~ 1, weights = f, data = cysts,
```

```
              family = DBURR12,
              method=mixed(20,100), trace=FALSE)

cysts$y[1]<-0.000000001 #to avoid problems with zero values
gen.cens("GA", type="interval")

## A censored family of distributions from GA has been generated
##  and saved under the names:
##  dGAic pGAic qGAic GAic
## The type of censoring is interval
```

Similarly `gen.cens()` is used for `WEI3`, `GG`, `GIG` and `GB2` distributions.

```
m2 <- gamlss(Surv(y, y+1, type = "interval2")~1, weights=f,
             data=cysts, family=GAic, trace=FALSE)
m3 <- gamlss(Surv(y, y+1, type = "interval2")~1, weights=f,
             family=WEI3ic, data=cysts, trace=FALSE)
m4 <- gamlss(Surv(y, y+1, type = "interval2")~1, weights=f,
             family=GGic, data=cysts, n.cyc=50,trace=FALSE)
m5 <- gamlss(Surv(y, y+1, type = "interval2")~1, weights=f,
             family=GIGic, data=cysts, trace=FALSE )
m6 <- gamlss(Surv(y, y+1, type = "interval2")~1, weights=f,
             family=GB2ic, data=cysts,n.cyc=1000,trace=FALSE)
```

```
GAIC(m1,m2,m3,m4,m5,m6, k=log(110))

##    df      AIC
## m2  2 343.9951
## m3  2 345.3786
## m4  3 348.0283
## m1  3 350.5299
## m6  4 352.9512
## m5  3 445.0624
```

The best model according to SBC is `m2`, the discretized gamma.

Note that `Surv()` in model `m2` creates an interval $(y, y+1)$ response variable, while `gen.cens()` creates a fitting function `GAic` for an interval censored gamma distribution; similarly for `m3` to `m6`.

Note also that the deviance and fitted distribution for model `m1` is very close to that given in Para and Jan [2016], however the parameter estimates look different because they use a different parameterization (see Section 22.3.2) and also two of the distribution parameters (μ and ν of `DBURR12`(μ, σ, ν)) are informationally highly correlated.

7.3.5 Ad-hoc methods

We refer to *ad-hoc* solutions, i.e. category (c) in Section 7.3.1, as those that have been implemented in the past, mainly for their computational convenience, and some also for good asymptotic properties for the estimation of the mean regression function, but which do *not* assume an explicit proper distribution for the response variable. The quasi-likelihood approach proposed by Wedderburn [1974], for example, requires assumptions on the first two moments of the response variable. As this approach is incapable of modeling the second moment parameter, the dispersion, as a function of explanatory variables, the extended quasi-likelihood (EQL) was proposed by Nelder and Pregibon [1987]. Alternative approaches are the pseudo-likelihood (PL) method introduced by Carroll and Ruppert [1982] and Efron's double exponential (EDE) family [Efron, 1986]. The PL method effectively approximates the probability function by a normal distribution with a chosen variance-mean relationship, but does not properly maximize the resulting likelihood. See Davidian and Carroll [1988] and Nelder and Lee [1992] for a comparison of the EQL and the PL. The problem with all these methods is that, while they work well with moderate under- or overdispersion, they have difficultly modeling heavy tails in the response distribution. They also suffer from the fact, that, for a given data set, the adequacy of the fit of those methods cannot be compared using a properly maximized log likelihood function $\hat{\ell}$ and criteria based on $\hat{\ell}$, such as the AIC. The problem is that they do not fit a proper discrete distribution. For the EQL and EDE methods the distribution probabilities do not sum to one, see for example Stasinopoulos [2006].

Note that with increasing computing power the constant of summation, missing from the EQL and EDE methods, can be calculated so that they represent proper distributions, resulting in a true likelihood function that can be maximized. However these models are still computationally slow to fit to large data sets; the true probability function cannot be expressed explicitly (except by including an infinite sum for the constant of summation); and their flexibility is limited by usually having at most two parameters. See Lindsey [1999] for a similar criticism of the *ad-hoc* methods.

The double Poisson distribution, a member of the EDE family, is implemented in **gamlss.dist** as `DPO`(μ, σ), a proper distribution summing to one, with probability function given in Section 22.2.1. This distribution can model underdispersion ($\sigma < 1$) as well as overdispersion ($\sigma > 1$) relative to a Poisson distribution ($\sigma = 1$).

7.4 Excess or shortage of zero values

A solution to excess zero values in a discrete distribution is a zero-inflated discrete distribution, dealt with in Section 7.4.1. Occasionally excess values in count data occur not at zero but at some other value (K) of the distribution. The K-inflated discrete distributions cater for those cases and are discussed in Section 7.4.2. A solution to both a shortage or excess of zero values in a discrete distribution is a zero-adjusted (or zero-altered) discrete distribution, discussed in Section 7.4.3.

7.4.1 Zero-inflated discrete distributions

A random variable Y has a *zero-inflated* distribution, $Y \sim$ `ZID`, if it takes value zero with probability p, and has the discrete distribution $\mathcal{D}$ with probability $1 - p$.

Hence, if $Y \sim$ `ZID`, then $Y = 0$ with probability p, and $Y = Y_1$ with probability $1 - p$, where $Y_1 \sim \mathcal{D}$, for $0 < p < 1$. Hence if $Y \sim$ ZID, then

$$\mathrm{P}(Y = y) = \begin{cases} p + (1-p)\mathrm{P}(Y_1 = 0) & \text{if } y = 0 \\ (1-p)\mathrm{P}(Y_1 = y) & \text{if } y = 1, 2, 3, \ldots \end{cases} \tag{7.5}$$

where $Y_1 \sim \mathcal{D}$. (Note that ZIP2 is an exception.) The probability that $Y = 0$ has two components: p, and $(1-p)\mathrm{P}(Y_1 = 0)$ from $Y_1 \sim \mathcal{D}$. Hence $\mathrm{P}(Y = 0) > \mathrm{P}(Y_1 = 0)$ and the distribution is called a 'zero-inflated' distribution.

Let

$$\begin{aligned} {\mu'}_{rY} &= \mathrm{E}(Y^r) \\ \mu_{rY} &= \mathrm{E}\{[Y - \mathrm{E}(Y)]^r\} \\ {\mu'}_r &= \mathrm{E}(Y_1^r) \\ \mu_r &= \mathrm{E}\{[Y_1 - \mathrm{E}(Y_1)]^r\} \,. \end{aligned}$$

Hence the mean, variance, and third and fourth central moments of Y are denoted as ${\mu'}_{1Y}$, μ_{2Y}, μ_{3Y} and μ_{4Y}; and the corresponding quantities for Y_1 as ${\mu'}_1$, μ_2, μ_3 and μ_4, respectively.

Note that $\mathrm{E}(Y^r) = (1-p)\mathrm{E}(Y_1^r)$, i.e. ${\mu'}_{rY} = (1-p){\mu_r}'$. Hence using equations (2.1) and (2.3), the mean, variance, and third and fourth central moments of $Y \sim ZID$ are given by

$$\begin{aligned} {\mu'}_{1Y} &= \mathrm{E}(Y) = (1-p)\mathrm{E}(Y_1) = (1-p){\mu'}_1 \\ \mu_{2Y} &= \mathrm{Var}(Y) = (1-p)\mathrm{Var}(Y_1) + p(1-p)\left[\mathrm{E}(Y_1)\right]^2 \\ \mu_{3Y} &= (1-p)\left(\mu_3 + 3p\mu_2{\mu'}_1 + p(2p-1){\mu'}_1^3\right) \\ \mu_{4Y} &= (1-p)\left(\mu_4 + 4p\mu_3{\mu'}_1 + 6p^2\mu_2{\mu'}_1^2 + p(1-3p+3p^2){\mu'}_1^4\right) \,. \end{aligned} \tag{7.6}$$

The cdf of $Y \sim$ ZID is given by

$$\mathrm{P}(Y \le y) = p + (1-p)\mathrm{P}(Y_1 \le y) \tag{7.7}$$

for $y = 0, 1, 2, \ldots$.

The probability generating function (pgf) of $Y \sim$ ZID is given by

$$G_Y(t) = p + (1-p)G_{Y_1}(t) \tag{7.8}$$

where $G_{Y_1}(t)$ is the pgf of Y_1.

Below are two examples of zero-inflated distributions: ZIP and ZINBI.

Zero-inflated Poisson distribution The zero-inflated Poisson distribution, denoted ZIP, is a discrete mixture with two components: zero with probability σ, and a PO(μ) distribution with probability $1-\sigma$.

Hence, if $Y \sim$ ZIP(μ, σ), then $Y = 0$ with probability σ, and $Y = Y_1$ with probability

$(1-\sigma)$, where $Y_1 \sim \texttt{PO}(\mu)$ and $0 < \sigma < 1$. Hence from (7.5), the pf of $Y \sim \texttt{ZIP}(\mu, \sigma)$ is given by

$$P(Y = y \mid \mu, \sigma) = \begin{cases} \sigma + (1-\sigma)e^{-\mu} & \text{if } y = 0 \\ (1-\sigma)\frac{\mu^y}{y!}e^{-\mu} & \text{if } y = 1, 2, 3, \ldots \end{cases} \tag{7.9}$$

where $\mu > 0$ and $0 < \sigma < 1$. The mean and variance are

$$\begin{aligned} E(Y) &= (1-\sigma)\mu \\ \text{Var}(Y) &= (1-\sigma)\mu + \sigma(1-\sigma)\mu^2 \\ &= E(Y) + \frac{\sigma}{1-\sigma}\{E(Y)\}^2 . \end{aligned}$$

Zero-inflated negative binomial type I distribution The zero-inflated negative binomial type I distribution, denoted `ZINBI`, is a discrete mixture with two components: value zero with probability ν, and a $\texttt{NBI}(\mu, \sigma)$ distribution with probability $1-\nu$.

Hence, if $Y \sim \texttt{ZINBI}(\mu, \sigma, \nu)$, then $Y = 0$ with probability ν, and $Y = Y_1$ with probability $(1-\nu)$, where $Y_1 \sim \texttt{NBI}(\mu, \sigma)$, and $0 < \nu < 1$. Hence, from (7.5), the pf of $Y \sim \texttt{ZINBI}(\mu, \sigma, \nu)$ is given by

$$P(Y = y \mid \mu, \sigma, \nu) = \begin{cases} \nu + (1-\nu)P(Y_1 = 0 \mid \mu, \sigma) & \text{if } y = 0 \\ (1-\nu)P(Y_1 = y \mid \mu, \sigma) & \text{if } y = 1, 2, 3, \ldots \end{cases}$$

where $\mu > 0$, $\sigma > 0$ and $0 < \nu < 1$ and $Y_1 \sim \texttt{NBI}(\mu, \sigma)$. The mean and variance are

$$\begin{aligned} E(Y) &= (1-\nu)\mu \\ \text{Var}(Y) &= (1-\nu)\mu + (1-\nu)(\sigma+\nu)\mu^2 . \end{aligned}$$

Further zero-inflated distributions
Other zero-inflated distributions available in **gamlss.dist** are:

$\texttt{ZIP2}(\mu, \sigma)$: zero-inflated Poisson type II,

$\texttt{ZIPIG}(\mu, \sigma, \nu)$: zero-inflated `PIG`,

$\texttt{ZIBNB}(\mu, \sigma, \nu, \tau)$: zero-inflated `BNB`,

$\texttt{ZISICHEL}(\mu, \sigma, \nu, \tau)$: zero-inflated `SICHEL`.

7.4.2 K-inflated discrete distributions

Assume Y is a discrete count response variable, and that $\mathcal{D}$ is a discrete distribution with range $R = \{0, 1, 2, \ldots\}$. Suppose that $Y \sim \mathcal{D}$, with the exception that Y exhibits a greater probability of a value K than that specified by $\mathcal{D}$.

A *K-inflated* distribution, denoted `KID`, is a discrete mixture with two components: value K with probability p, and the discrete distribution $\mathcal{D}$ with probability $1-p$.

Hence, if $Y \sim$ KID, then $Y = K$ with probability p, and $Y = Y_1$ with probability $(1-p)$, where $Y_1 \sim \mathcal{D}$. Hence if $Y \sim$ KID, then

$$P(Y=y) = \begin{cases} p + (1-p)P(Y_1 = K) & \text{if } y = K \\ (1-p)P(Y_1 = y) & \text{if } y = 0, 1, 2, \ldots \text{ but not } K \end{cases} \tag{7.10}$$

K-inflated distributions have been found useful in actuarial statistics, where the number of claims are sometimes greater than the theoretical count distribution suggests at a specific value K. Payandeh and Mohammadpour [2018] provide a nice actuarial example in which the K-inflated negative binomial distribution is used.

K-inflated distributions can be fitted in GAMLSS using the package **gamlss.countKinf** [Stasinopoulos and Mohammadpour, 2018]. In order to fit any count data distribution inflated at a specified value K, the distribution is generated using the function `gen.Kinf()` which takes arguments `family` and `kinf`. In the following example we generate a one-inflated negative binomial distribution, $Y \sim$ inf1NBI(μ, σ, ν), i.e. $Y_1 \sim$ NBI(μ, σ), $p = \nu$ and $K = 1$ in (7.10).

```
library(gamlss.countKinf)
gen.Kinf(family=NBI, kinf=1)

## inflated at 1 distribution is generated from NBI and saved under the
## names:
##  dinf1NBI pinf1NBI qinf1NBI rinf1NBI inf1NBI
```

Generate a sample of 1000 observations from the one-inflated negative binomial distribution with $\mu = 5$, $\sigma = 0.5$, and excess probability $\nu = 0.2$ at value 1, and fit the distribution to the data:

```
Y<-rinf1NBI(1000, mu=5, sigma=0.5 , nu=0.2)
m1<-histDist(Y, family=inf1NBI, main="inf1NBI")
```

7.4.3 Zero-adjusted (or zero-altered) discrete distributions

Consider a discrete count response variable Y which can exhibit either a greater or lesser probability of a zero value than that of a discrete count distribution $\mathcal{D}$ having range $R = \{0, 1, 2, \ldots\}$. This can be modeled by a 'zero-adjusted' (or 'zero-altered') distribution, denoted by ZAD, which is a discrete mixture with two components: value zero with probability p, and a distribution $\mathcal{D}$tr (the distribution $\mathcal{D}$ truncated at zero), with probability $1-p$.

Hence, if $Y \sim$ ZAD, then $Y = 0$ with probability p, and $Y = Y_1$ with probability $1-p$, where $Y_1 \sim \mathcal{D}$tr and $0 < p < 1$. Hence if $Y \sim$ ZAD, then

$$P(Y=y) = \begin{cases} p & \text{if } y = 0 \\ (1-p)P(Y_1 = y) & \text{if } y = 1, 2, 3, \ldots \end{cases}$$

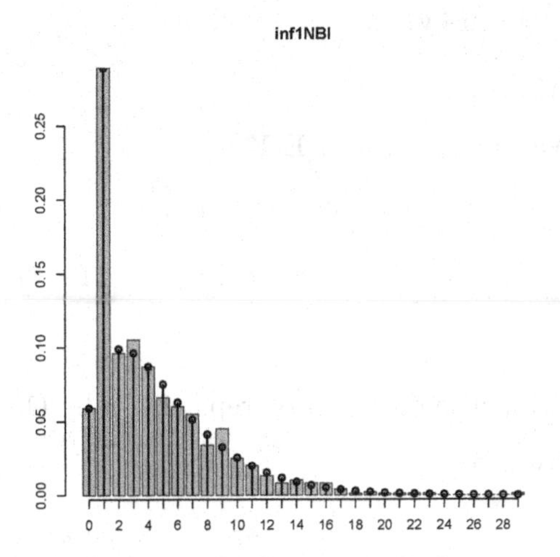

FIGURE 7.1: Observed and fitted one-inflated negative binomial distribution.

where $Y_1 \sim \mathcal{D}\texttt{tr}$, and so

$$P(Y = y) = \begin{cases} p & \text{if } y = 0 \\ \dfrac{(1-p)P(Y_2 = y)}{1 - P(Y_2 = 0)} & \text{if } y = 1, 2, 3, \ldots \end{cases} \tag{7.11}$$

where $Y_2 \sim \mathcal{D}$.

Note that $P(Y = 0) = p$, which can be greater than or less than $P(Y_2 = 0)$ and hence the distribution is called a 'zero-adjusted' (or 'zero-altered') distribution. If $p = P(Y_2 = 0)$ then $Y = Y_2 \sim$ `D` and `ZAD` becomes distribution $\mathcal{D}$. If $p > P(Y_2 = 0)$ then the zero-adjusted distribution `ZAD` has an inflated probability at zero relative to $\mathcal{D}$ and is a reparameterization of the zero-inflated distribution `ZID` in Section 7.4.1. (Because of the reparameterization, these models are, in general, different when the parameters are modeled by explanatory variables.) However if $p < P(Y_2 = 0)$ then `ZAD` has a deflated probability at zero relative to $\mathcal{D}$ and is different from (i.e. is not a reparameterization of) `ZID`. Hence `ZID` is a reparameterized submodel of `ZAD`.

Let

$$\begin{aligned} \mu'_{rY} &= E(Y^r) \\ \mu_{rY} &= E\{[Y - E(Y)]^r\} \\ \mu'_r &= E(Y_2^r) \\ \mu_r &= E\{[Y_2 - E(Y_2)]^r\} \,. \end{aligned}$$

Hence the mean, variance, and third and fourth central moments of Y and Y_2 are denoted by μ'_{1Y}, μ_{2Y}, μ_{3Y}, μ_{4Y} and μ'_1, μ_2, μ_3, μ_4 respectively.

Note that $E(Y^r) = cE(Y_2^r)$, i.e. $\mu'_{rY} = c\mu_r'$, where $c = (1-p)/[1 - P(Y_2 = 0)]$. Hence using equations (2.1) and (2.3), the mean, variance, and third and fourth central moments of $Y \sim$ `ZAD` are given by

$$\begin{aligned}
\mu'_{1Y} &= \mathrm{E}(Y) = c\mathrm{E}(Y_2) = c\mu_1' \\
\mu_{2Y} &= \mathrm{Var}(Y) = c\mathrm{Var}(Y_2) + c(1-c)\left[\mathrm{E}(Y_2)\right]^2 \\
\mu_{3Y} &= c\left[\mu_3 + 3(1-c)\mu_2\mu_1' + (1-3c+2c^2)\mu'3_1\right] \\
\mu_{4Y} &= c\left[\mu_4 + 4(1-c)\mu_3\mu_1' + 6(1-2c+c^2)\mu_2\mu_1'^2 + (1-4c+6c^2-3c^3)\mu_1'^4\right] .
\end{aligned} \tag{7.12}$$

Note that equation (7.12) is also obtained by replacing p by $(1-c)$ in (7.6).

The cdf of $Y \sim$ `ZAD` is given by

$$\mathrm{P}(Y \le y) = \begin{cases} p & \text{if } y = 0 \\ p + c\left[\mathrm{P}(Y_2 \le y) - \mathrm{P}(Y_2 = 0)\right] & \text{if } y = 1, 2, \ldots \end{cases} \tag{7.13}$$

and its probability generating function (pgf) by

$$G_Y(t) = (1-c) + c\,G_{Y_1}(t) \tag{7.14}$$

where $G_{Y_1}(t)$ is the pgf of Y_1.

Below are two examples of zero adjusted distributions: `ZAP` and `ZANBI`.

Zero-adjusted Poisson distribution A zero-adjusted Poisson distribution, denoted `ZAP`, is a discrete mixture with two components: zero with probability σ, and the distribution $\texttt{POtr}(\mu)$, (i.e. $\texttt{PO}(\mu)$ truncated at zero), with probability $1-\sigma$.

Hence, if $Y \sim \texttt{ZAP}(\mu, \sigma)$, then $Y = 0$ with probability σ, and $Y = Y_1$ with probability $(1-\sigma)$, where $Y_1 \sim \texttt{POtr}(\mu)$ and $0 < \sigma < 1$.

Hence in (7.11) $Y_2 \sim \texttt{PO}(\mu)$ and the pf of $Y \sim \texttt{ZAP}(\mu, \sigma)$ is given by

$$P(Y = y \mid \mu, \sigma) = \begin{cases} \sigma & \text{if } y = 0 \\ \dfrac{(1-\sigma)e^{-\mu}\mu^y}{y!\,(1-e^{-\mu})} & \text{if } y = 1, 2, \ldots \end{cases}$$

where $\mu > 0$ and $0 < \sigma < 1$. The mean and variance are

$$\begin{aligned}
\mathrm{E}(Y) &= (1-\sigma)\,\mu/\left(1-e^{-\mu}\right) \\
\mathrm{Var}(Y) &= (1+\mu)\mathrm{E}(Y) - \left[\mathrm{E}(Y)\right]^2 .
\end{aligned}$$

Zero-adjusted negative binomial type I distribution A zero-adjusted negative binomial type I distribution, denoted `ZANBI`, is a discrete mixture of two components: zero with probability ν, and the distribution $\texttt{NBItr}(\mu, \sigma)$ (i.e. $\texttt{NBI}(\mu, \sigma)$ truncated at zero), with probability $1-\nu$.

Hence, if $Y \sim \texttt{ZANBI}(\mu, \sigma, \nu)$, then $Y = 0$ with probability ν, and $Y = Y_1$ with probability $(1-\nu)$, where $Y_1 \sim \texttt{NBItr}(\mu, \sigma)$ and $0 < \nu < 1$.

Hence in (7.11) $Y_2 \sim$ NBI(μ, σ) and the pf of $Y \sim$ ZANBI(μ, σ, ν) is given by

$$P(Y = y \mid \mu, \sigma, \nu) = \begin{cases} \nu & \text{if } y = 0 \\ \dfrac{(1-\nu)\,\mathrm{P}(Y_2 = y \mid \mu, \sigma)}{1 - \mathrm{P}(Y_2 = 0 \mid \mu, \sigma)} & \text{if } y = 1, 2, \ldots \end{cases}$$

for $\mu > 0$, $\sigma > 0$ and $0 < \nu < 1$ where $Y_2 \sim$ NBI(μ, σ). The mean and variance of Y are given by

$$\mathrm{E}(Y) = (1-\nu)\,\mu \left[1 - (1+\mu\sigma)^{-1/\sigma}\right]^{-1}$$
$$\mathrm{Var}(Y) = [1 + (\sigma+1)\mu]\,\mathrm{E}(Y) - [\mathrm{E}(Y)]^2 \ .$$

Further zero-adjusted distributions
Other zero-adjusted distributions available in **gamlss.dist** are:

ZALG(μ, σ): zero-adjusted LG,

ZAZIPF(μ, σ): zero-adjusted ZIPF,

ZAPIG(μ, σ, ν): zero-adjusted PIG,

ZABNB(μ, σ, ν, τ): zero-adjusted BNB,

ZASICHEL(μ, σ, ν, τ): zero-adjusted SICHEL.

7.5 Comparison of the count distributions

Count distributions can be compared using a diagram of their moment kurtosis against moment skewness. Figure 7.2 displays the moment skewness-kurtosis combinations for different count distributions with fixed mean 5 and variance 30. Similar figures are obtained for other combinations of fixed mean and variance.

The zero-inflated Poisson (ZIP), negative binomial (NBI), negative binomial truncated at zero (NBItr), and Poisson-inverse Gaussian (PIG) distributions each have two parameters, so fixing the mean and variance of Y results in a single combination of moment skewness-kurtosis, displayed as circles in Figure 7.2.

The zero-inflated negative binomial (ZINBI) and zero-adjusted negative binomial (ZANBI) distributions each have three parameters, so their possible moment skewness-kurtosis combinations are represented by curves. The ZINBI curve goes from the moment skewness-kurtosis of the ZIP to that of the NBI.

The ZANBI curve goes from the skewness-kurtosis of the ZIP to that of the NBItr. The ZANBI can lead to a zero-inflated or zero-deflated probability relative to the NBI distribution. In the zero-inflated case, the ZANBI is a reparameterization of the ZINBI distribution, and hence has the same moment skewness-kurtosis curve (between ZIP and NBI). In the zero-deflated case, the ZANBI is different from (i.e. not a reparameterization of) the ZINBI, and its moment skewness-kurtosis curve lies between NBI and NBItr,

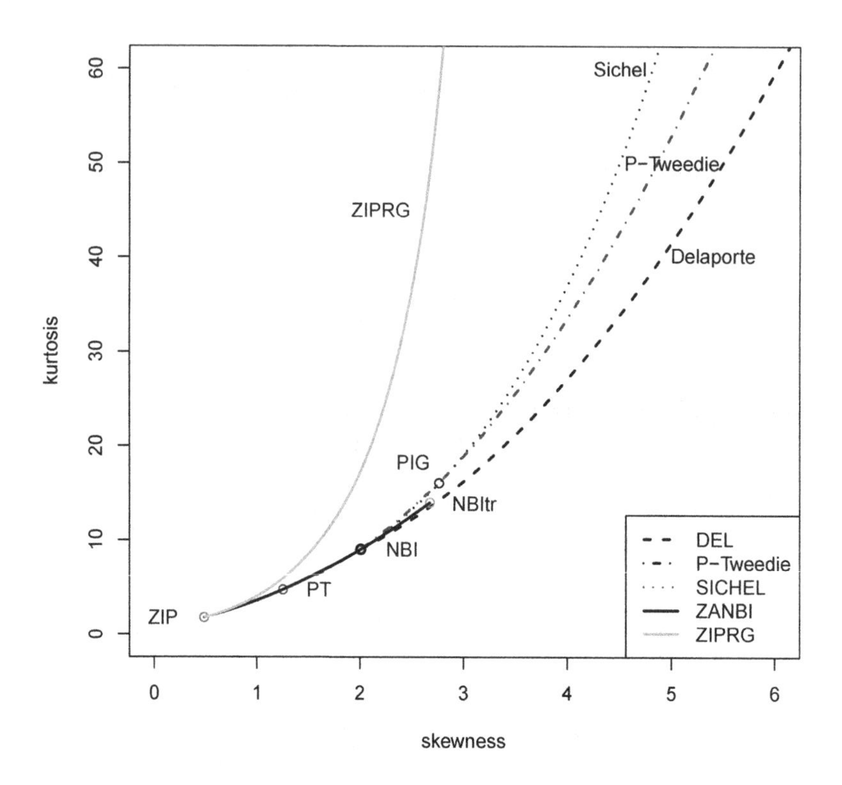

FIGURE 7.2: Moment skewness-kurtosis combinations for different count distributions (with fixed mean 5 and variance 30).

where the exact probability ν at $Y = 0$ decreases from 0.1667 (equal to that of the `NBI`(5,1) distribution) to zero.

The Sichel (`SICHEL`), Poisson-Tweedie[a] (P-Tweedie) and Delaporte (`DEL`) distributions also have three parameters, so their moment skewness-kurtosis combinations are represented by curves. The three curves meet at the skewness-kurtosis point of the (`NBI`) which is a limiting case of the `DEL` and `SICHEL`, and an internal special case of the Poisson-Tweedie. The Poisson-Tweedie curve alone continues (as its power parameter decreases from two to one) and stops at the circle marked PT between `ZIP` and `NBI`. (Note also that the `PIG` is a special case of both the `SICHEL` and the Poisson-Tweedie distributions.) The zero-inflated Poisson reciprocal gamma distribution (`ZIPRG`) has three parameters and its moment skewness-kurtosis curve has the highest kurtosis for a given skewness. The Poisson-shifted generalized inverse Gaussian (`PSGIG`) is a four-parameter distribution and has moment skewness-kurtosis combinations covering the region between the `SICHEL` and `DEL` curves, while the zero-inflated Sichel (`ZISICHEL`) covers the region between the `ZIPRG` and `SICHEL` curves. Note however that the `ZIPRG` and `PSGIG` distributions are not, at the time of writing, implemented in **gamlss.dist**.

[a]The Poisson-Tweedie distribution is not implemented in the **gamlss.dist** package.

7.6 Families modeling the variance-mean relationship

Consider the mixed Poisson model where $Y \mid \gamma \sim \texttt{PO}(\mu\gamma)$ and $\gamma \sim \texttt{D}(\sigma, \nu, \tau)$ for some distribution D, where $\text{E}(\gamma) = 1$ and $\text{Var}(\gamma) = v(\sigma, \nu, \tau)$ for some function v of the parameters of the mixing distribution D. This leads to the mean of Y given by

$$\text{E}(Y) = \text{E}_\gamma\left[\text{E}(Y \mid \gamma)\right] = \text{E}_\gamma\left(\mu\gamma\right) = \mu$$

and a variance-mean relationship for Y given by $\text{Var}(Y) = \mu + \mu^2\, v(\sigma, \nu, \tau)$, since

$$\text{Var}(Y) = \text{E}_\gamma\left[\text{Var}(Y \mid \gamma)\right] + \text{Var}_\gamma\left[\text{E}(Y \mid \gamma)\right] = \text{E}_\gamma(\mu\gamma) + \text{Var}_\gamma(\mu\gamma) = \mu + \mu^2 v(\sigma, \nu, \tau).$$

Hence in particular the negative binomial type I (NBI), Poisson-inverse Gaussian (PIG), Sichel (SICHEL), Delaporte (DEL), and Poisson-shifted generalized inverse Gaussian (PSGIG) distributions, which are all mixed Poisson models, all have this quadratic variance-mean relationship.

Alternative variance-mean relationships can be obtained by reparameterization.

For example the negative binomial type I (NBI) distribution has $\text{E}(Y) = \mu$ and $\text{Var}\,(Y) = \mu + \sigma\mu^2$. If σ is reparameterized to σ/μ then $\text{Var}(Y) = (1 + \sigma)\mu$, giving the negative binomial type II (NBII) distribution. If σ is reparameterized to $\sigma\mu$ then $\text{Var}(Y) = \mu + \sigma\mu^3$.

More generally, a family of reparameterizations of the negative binomial type I distribution can be obtained by reparameterizing σ to $\sigma\mu^{\nu-2}$, giving $\text{Var}(Y) = \mu + \sigma\mu^\nu$. This gives a three-parameter distribution model with parameters μ, σ, and ν, called the negative binomial family distribution, denoted $\texttt{NBF}(\mu, \sigma, \nu)$, (see Section 22.3.4). Note that for constant parameters μ, σ, and ν, this distribution model is not identifiable. However if the parameters depend on explanatory variables then, in general, the model is identifiable.

A family of reparameterizations can be applied similarly to other mixed Poisson distributions. In particular the Poisson-inverse Gaussian and Delaporte can be easily extended to reparameterization families using an extra parameter in a similar way to the negative binomial type I above.

7.7 Modeling a discrete count response variable using the **gamlss** packages

Explicit discrete count distributions available in **gamlss** are given in Section 7.2. They include explicit mixed Poisson distributions in category (a)(i) in Section 7.3.2; zero-inflated and zero-adjusted discrete count distributions in Section 7.4; and the double Poisson (DPO) distribution (category (c) in Section 7.3.1).

Models in category (a)(ii) in Section 7.3.2, in which the continuous mixture of discrete distributions is approximated by integrating out a normally distributed random variable using Gaussian quadrature, can be fitted using the package **gamlss.mx**, as can a finite mixture of discrete distributions in category (a)(iii). Package **gamlss.mx** fits type (a)(ii) and (a)(iii) models, allowing a general discrete conditional distribution in equation (7.3). For example, a negative binomial distribution may be used, resulting in a negative binomial-normal mixture model and a negative binomial nonparametric finite mixture model, respectively (see Stasinopoulos et al. [2017], Chapter 7 and Section 14.2).

Discretized continuous distributions (category (b) in Section 7.3.1) can be fitted using the add-on package **gamlss.cens**. An example of this is given in Section 7.3.4.

All of the discrete distributions above allow for fitting overdispersion (relative to a Poisson distribution). In addition the double Poisson (`DPO`), discrete Burr XII (`DBURR12`) and many other discretized continuous distributions fitted using **gamlss.cens** accommodate underdispersion as well as overdispersion.

Note that the Waring and the beta negative binomial (`BNB`) distributions are overdispersed geometric and negative binomial distributions, respectively, while the Yule distribution is a special case of the Waring. All three are explicit distributions included in **gamlss.dist**. They all have heavy right tails.

7.7.1 Lice data

R data file: `lice` in package **gamlss.data** of dimension 71×2
source: Williams [1944]
variables
  `head` : the number of head lice
  `freq` : the frequency of prisoners with the number of head lice
purpose: to demonstrate the fitting of a parametric discrete distribution.
conclusion a `BNB`(μ, σ, ν) distribution fits best

```
plot(lice$head, lice$freq, type="h", lwd=3,
     ylab="frequency", xlab="number of lice")
```

The `lice` data set, shown in Figure 7.3, gives the frequencies (`freq`) of prisoners with number of head lice (`head`), for Hindu male prisoners in Cannamore, South India, in the period 1937-39 [Williams, 1944]. We firstly fit a Poisson distribution to `head` using `freq` as prior weights, and then use `chooseDist()` for comparison of alternative distributions. Note that because we are using prior weights, the total number of observations, which is the sum of the frequencies, has to be entered explicitly in the calculation of the SBC in `chooseDist()`.

```
mPO <- gamlssML(head ~ 1, data = lice, family = PO,
             weights = freq, trace = FALSE)
N <- sum(lice$freq)
table1 <- chooseDist(mPO, type="counts", k=c(2, 3.84, log(N)))

## Error in solve.default(oout$hessian) :
##   Lapack routine dgesv: system is exactly singular: U[1,1] = 0
## minimum GAIC(k= 2 ) family: BNB
## minimum GAIC(k= 3.84 ) family: BNB
## minimum GAIC(k= 6.98749 ) family: ZALG
```

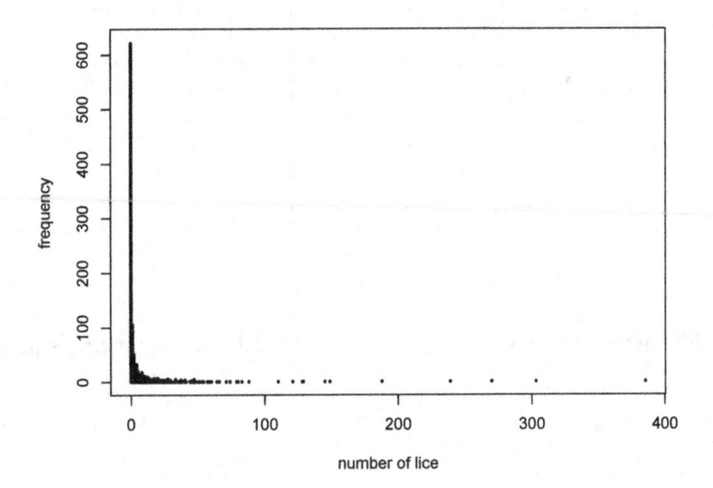

FIGURE 7.3: The lice data.

```
getOrder(table1)[1:5]
## GAIG with k= 2
##       BNB    ZABNB    ZIBNB     ZALG   SICHEL
## 4641.563 4642.769 4643.563 4644.220 4646.199
```

The three-parameter beta negative binomial, BNB(μ, σ, ν), and the two-parameter zero-adjusted logarithmic, ZALG(μ, σ), distributions fit best for AIC and SBC, respectively.

The worm plot provides a good diagnostic check of the fit of the discrete distribution, especially of the *right* tail. An alternative diagnostic check, particularly of the fit of the *left* tail of the discrete distribution, is given by the hanging rootogram [Kleiber and Zeileis, 2016]. This plot compares the square root of observed (O_v) and expected (E_v) frequencies for $v = 0, 1, 2, \ldots, V$, where v is the number of head lice. Hence $O_v = \sum_{i=1}^{n} \mathrm{I}(y_i = v)$ is the observed number of prisoners having v head lice, while $\mathrm{E}_v = \sum_{i=1}^{n} \mathrm{P}(Y_i = v \mid \hat{\mu}_i, \hat{\sigma}_i, \hat{\nu}_i, \hat{\tau}_i)$ is the expected number of prisoners having v head lice obtained from the fitted model. Note $\mathrm{I}(\cdot)$ is the indicator function, i.e. $\mathrm{I}(y_i = v) = 1$ if $y_i = v$, and 0 otherwise.

The two selected models ZALG and BNB are refitted below and their fitted distribution and worm plots are plotted in Figure 7.4. Their hanging rootograms are plotted in Figure 7.5.

```
m1 <- histDist(lice$head, "ZALG", freq = lice$freq, xlim = c(0, 20),
        main = "(a) ZALG", trace = FALSE)
m2 <- histDist(lice$head, "BNB", freq = lice$freq, xlim = c(0, 20),
        main = "(b) BNB", trace = FALSE)
rqres.plot(m1,howmany=40,plot="all", ylim.all=.7); title("(c)")
rqres.plot(m2,howmany=40,plot="all", ylim.all=.7); title("(d)")
```

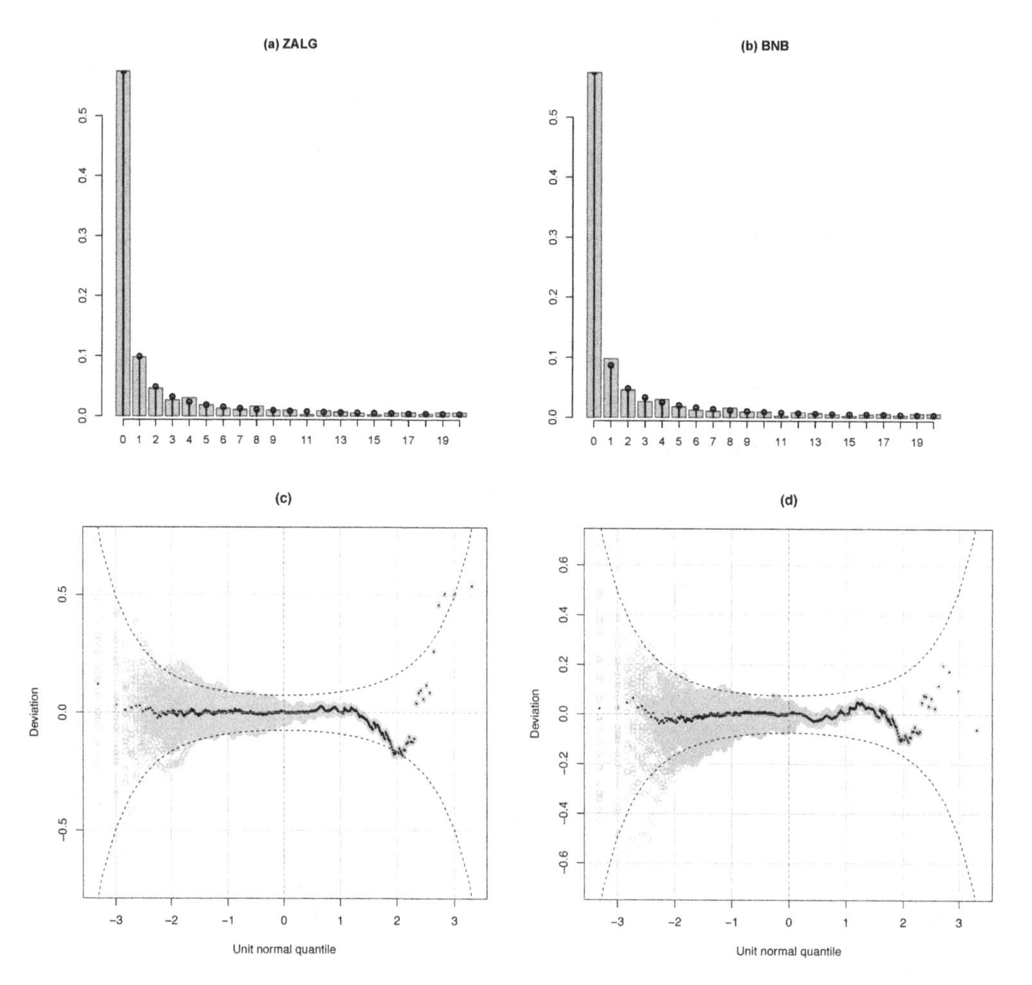

FIGURE 7.4: The `lice` data. The fitted distributions: (a) `ZALG` and (b) `BNB`. The worm plots (c) `ZALG` and (d) `BNB`. Gray points show 40 realizations of the (normalized) randomized quantile residuals, while black points show their median.

The fitted distributions for `ZALG` and `BNB` both give good fits to the `lice` data, as shown by Figure 7.4(a) and (b), respectively. (Note however that the fit to the right tail cannot be assessed in Figure 7.4(a) and (b) since the barplots end at `lice`=20, missing the right tail.) However the worm plot for `BNB` in Figure 7.4(d) indicates a better fit to the `lice` data in the right tail than the `ZALG` in Figure 7.4(c).

```
m11 <- gamlss(head~1, family="ZALG", weights = lice$freq,
        n.cyc=200, data=lice, trace=FALSE)
m22 <- gamlss(head~1, family="BNB", weights = lice$freq,
        n.cyc=1000, data=lice, trace=FALSE)
rootogram(m11, cex.lab=1.5, cex.axis=1.5, main="(a) ZALG",
        style="hanging", max=40, col=1)
abline(h=c(-1,1), col=gray(.2))
rootogram(m22, cex.lab=1.5, cex.axis=1.5, main="(b) BNB",
        style="hanging", max=40, col=1)
```

```
abline(h=c(-1,1), col=gray(.2))
```

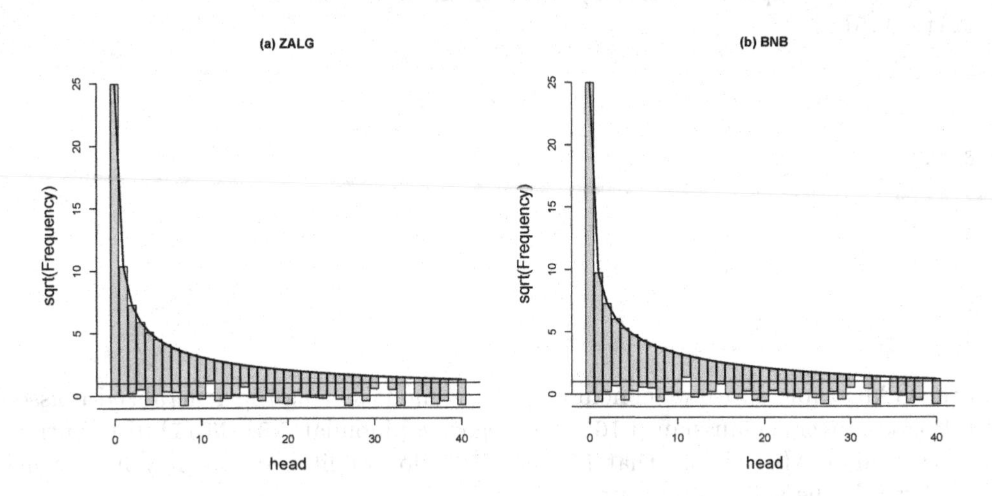

FIGURE 7.5: The `lice` data: hanging rootograms for fitted distributions (a) `ZALG` and (b) `BNB`.

Figure 7.5 shows the hanging rootograms for the chosen `ZALG` and `BNB` models `m11` and `m22`, respectively. The curve shows the values of $\sqrt{E_v}$, while the vertical shaded bars are drawn from $\sqrt{E_v}$ down to $\sqrt{E_v} - \sqrt{O_v}$. Hence the heights of the shaded bars are $\sqrt{O_v}$, for $v = 0, 1, 2, \ldots, 40$. For a 'perfect' fitted model the bottom of the shaded bars, i.e $\sqrt{E_v} - \sqrt{O_v}$, would be aligned along the horizontal axis at zero. The plots also show the "warning limits" of Tukey [1972, p. 314], set at ± 1, which are very rough 95% limits using the approximate normal distribution with mean $\sqrt{E_v}$ and standard deviation 0.5, for $\sqrt{O_v}$. (Tukey also suggests 'control' limits set at ± 1.5, which are very rough 99.8% limits.) For both the `ZALG` and `BNB` models, the hanging rootogram shows four violations of the warning limits.

7.7.2 Computer failure data

R data file: `computer` in package **gamlss.data** of dimension 128 × 1
source: Hand et al. [1994]
variables
  `failure` : the number of computers that broke down
purpose: to demonstrate the fitting of a parametric discrete distribution
conclusion the $\texttt{GPO}(\mu, \sigma)$ distribution fits best

The `computer` data relate to DEC-20 computers which operated at the Open University in the 1980s. They give the number of computers that broke down in each of 128 consecutive weeks of operation, starting in late 1983, see Hand et al. [1994, p. 109, data set 141] . Using `fitDist()`, we fit count data distributions and compare them using the AIC:

```
mPO <- gamlssML(computer$failure, "PO", main = "PO", trace = FALSE)
mFin <- fitDist(computer$failure, type="counts", k=2)
mFin$fits[1:5]

##      GPO      PIG     NBII      NBI   SICHEL
## 636.1393 636.4159 636.8405 636.8405 638.0528

mFin$failed

## [[1]]
## [1] "LG"
##
## [[2]]
## [1] "ZIPF"
```

From the GAIC table above we conclude that the models using the generalized Poisson (GPO), Poisson-inverse Gaussian (PIG), and negative binomial (NBI, NBII) distributions have very similar AICs. [Note that LG and ZIPF do not fit because they have range $\{1, 2, \ldots\}$, while the variable `failure` has zero values.]

The fitted GPO$(\hat{\mu}, \hat{\sigma})$ and its corresponding worm plot are shown in Figure 7.6.

```
mGPO <- histDist(computer$failure, "GPO", main = "(a)",
                 trace = FALSE)
wp(mGPO)
title("(b)")
```

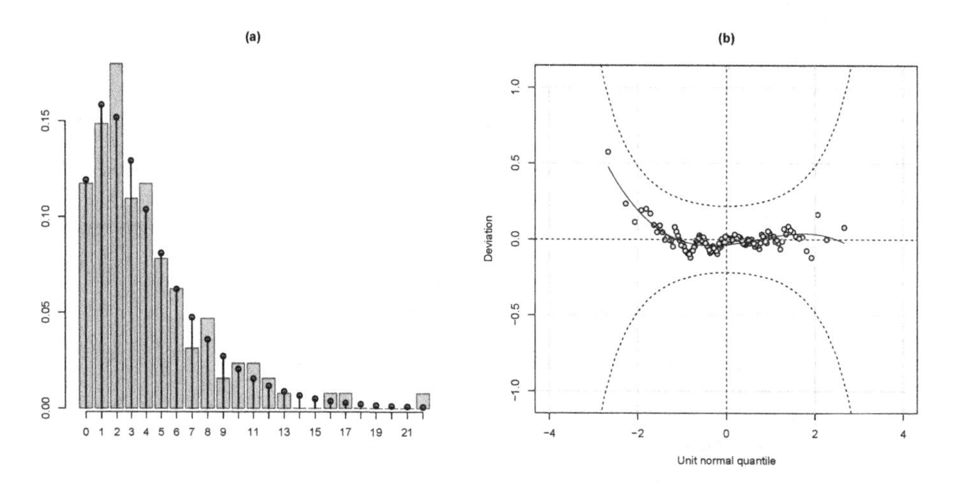

FIGURE 7.6: (a) Barplot and fitted generalized Poisson, GPO(μ, σ), distribution for the computer failure data, and (b) the corresponding worm plot of the residuals.

Now we refit the GPO distribution model and display its summary.

```
mod1 <- gamlss(failure ~ 1, data = computer, family = GPO,
               trace = FALSE)
summary(mod1)
```

```
## ******************************************************************
## Family:  c("GPO", "Generalised Poisson")
##
## Call:
## gamlss(formula = failure ~ 1, family = GPO, data = computer,
##     trace = FALSE)
##
## Fitting method: RS()
##
## ------------------------------------------------------------------
## Mu link function:  log
## Mu Coefficients:
##             Estimate Std. Error t value Pr(>|t|)
## (Intercept)  1.39019    0.08323    16.7   <2e-16 ***
## ---
## Signif. codes:
## 0 '***' 0.001 '**' 0.01 '*' 0.05 '.' 0.1 ' ' 1
##
## ------------------------------------------------------------------
## Sigma link function:  log
## Sigma Coefficients:
##             Estimate Std. Error t value Pr(>|t|)
## (Intercept)  -1.5103     0.1589  -9.504   <2e-16 ***
## ---
## Signif. codes:
## 0 '***' 0.001 '**' 0.01 '*' 0.05 '.' 0.1 ' ' 1
##
## ------------------------------------------------------------------
## No. of observations in the fit:  128
## Degrees of Freedom for the fit:  2
##       Residual Deg. of Freedom:  126
##                      at cycle:  2
##
## Global Deviance:     632.1393
##             AIC:     636.1393
##             SBC:     641.8434
## ******************************************************************
```

Hence the fitted distribution model for the computer failure data is given by $Y \sim$ $\texttt{GPO}(\hat{\mu}, \hat{\sigma})$ where $\hat{\mu} = \exp(1.390) = 4.016$ and $\hat{\sigma} = \exp(-1.510) = 0.221$, (since both μ and σ have log link functions), with fitted mean $\hat{\mathrm{E}}(Y) = \hat{\mu} = 4.016$ and fitted variance $\widehat{\mathrm{Var}}(Y) = \hat{\mu}\,(1 + \hat{\sigma}\hat{\mu})^2 = 14.296$.

7.7.3 Demand for medical care

R data file: `NMES1988` in package **AER** of dimension 4406×19
source: Deb and Trivedi [1997]
variables used in our analysis

- `visits` : number of physician office visits
- `hospital` : number of hospital stays,
- `health` : health status: a factor indicating whether self-perceived health is poor, average (reference category) or excellent,
- `chronic` : number of chronic conditions,
- `gender` : a factor indicating gender,
- `school` : number of years of education,
- `insurance` : a factor indicating whether the individual is covered by private insurance.

purpose: to demonstrate the fitting of a discrete count response variable
conclusion: the zero-inflated beta negative binomial distribution fits best

This cross-sectional data set originates from the United States National Medical Expenditure Survey (NMES) conducted in 1987 and 1988. The response variable is `visits` (number of physician office visits), a count variable with range $0, 1, 2, \ldots$. The data frame `NMES1988` has 4406 observations on 19 variables, but we consider only seven variables. This analysis was presented in Stasinopoulos et al. [2018].

```
library(AER)        #for the data set
data(NMES1988)
# select variables for analysis
nmes <- NMES1988[,c("visits","hospital","health","chronic","gender",
              "school","insurance")]
hist(nmes$visits, breaks = seq(-0.5,90.5,1), main = "",
     xlab = "visits")
```

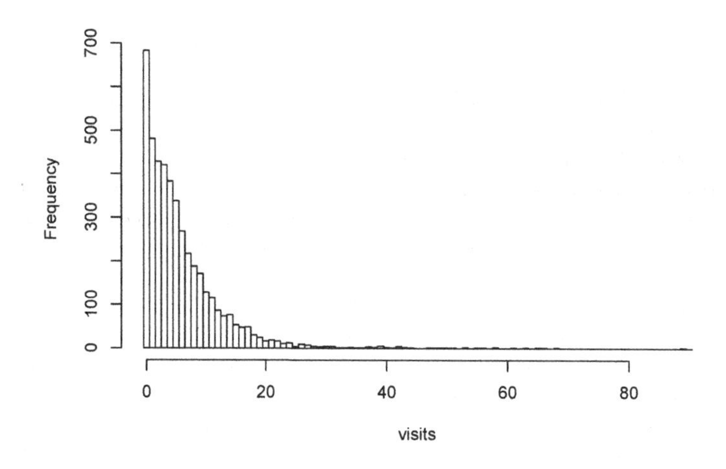

FIGURE 7.7: Histogram of `visits`.

The histogram of `visits` is shown in Figure 7.7 and shows strong positive skewness and possible zero-inflation. The plots of `visits` against each explanatory variable are given in Figure 7.8. These suggest possible relationships between mean `visits` and each of the quantitative variables `hospital`, `chronic`, and `school`. Boxplots of `visits` are given for each of the categorical covariates showing the following. The median number of visits decreases as health status improves from poor to average and then to excellent. The median number of visits is similar for male and female, with a few higher values for male. The median number of visits is slightly higher for a person covered by private insurance than for a person not covered. The variation (as measured by the interquartile range) varies with health status. The longer upper than lower tails in the boxplots indicate positive skewness.

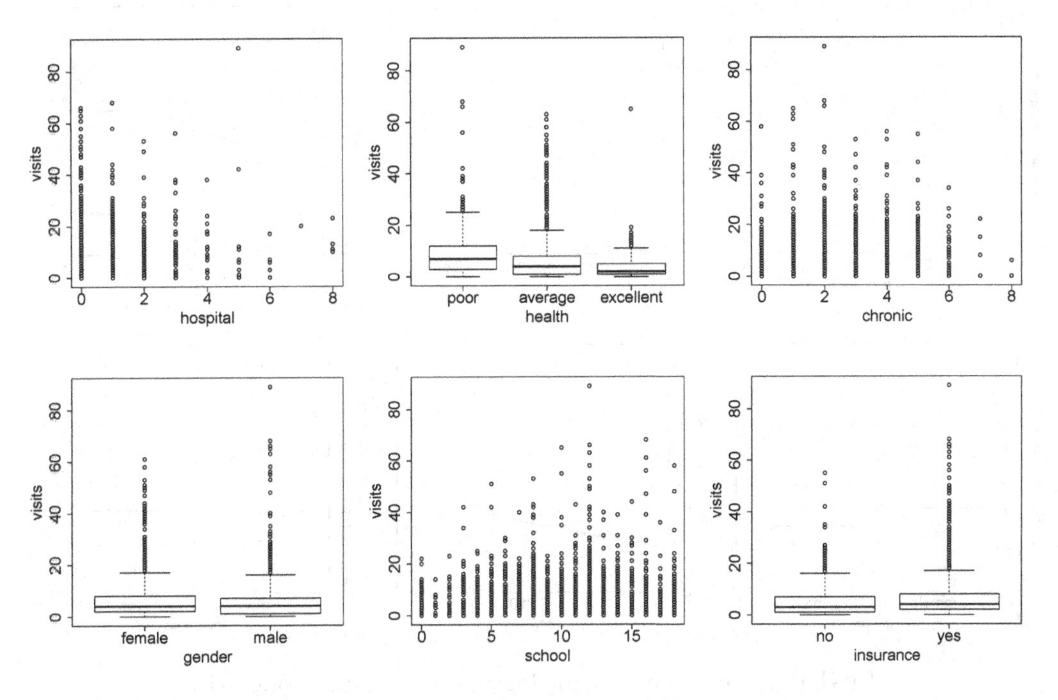

FIGURE 7.8: Plot of `visits` against explanatory variables `hospital`, `health`, `chronic`, `gender`, `school` and `insurance`.

Therefore, any statistical model used for the analysis of this data should be able to accommodate overdispersion, high positive skewness, and also the possibility of an excess of zeros. The mean, variance, skewness, and excess zeros of the response variable `visits` may depend on explanatory variables. One way to deal with the complexity in these data is to fit different distributions and model each of the parameters of the distribution as additive functions of the explanatory variables. The two-parameter negative binomial (`NBI`) distribution accommodates overdispersion and is a good initial candidate, with the capability of modeling the response mean and dispersion. Initially we use an `NBI`(μ, σ) model with μ and σ modeled as additive functions of the categorical variables and smoothing terms (i.e. `pb`) for quantitative variables:

```
m1 <- gamlss(visits ~ pb(hospital) + health + pb(chronic) +
             gender + pb(school) + insurance,
```

```
          sigma.fo=~ pb(hospital) + health +
          pb(chronic) + gender + pb(school) + insurance,
          family=NBI, data=nmes)
```

The term plot for the predictor for μ, i.e. $\log(\mu)$, is given in Figure 7.9, and suggests that by using square root transformed variables for `hospital` and `chronic` we could avoid smoothing for those two variables.

```
term.plot(m1, cex.axis=2, cex.lab=1.8, col.term=1)
nmes <-transform(nmes, sh=sqrt(hospital), sc=sqrt(chronic))
```

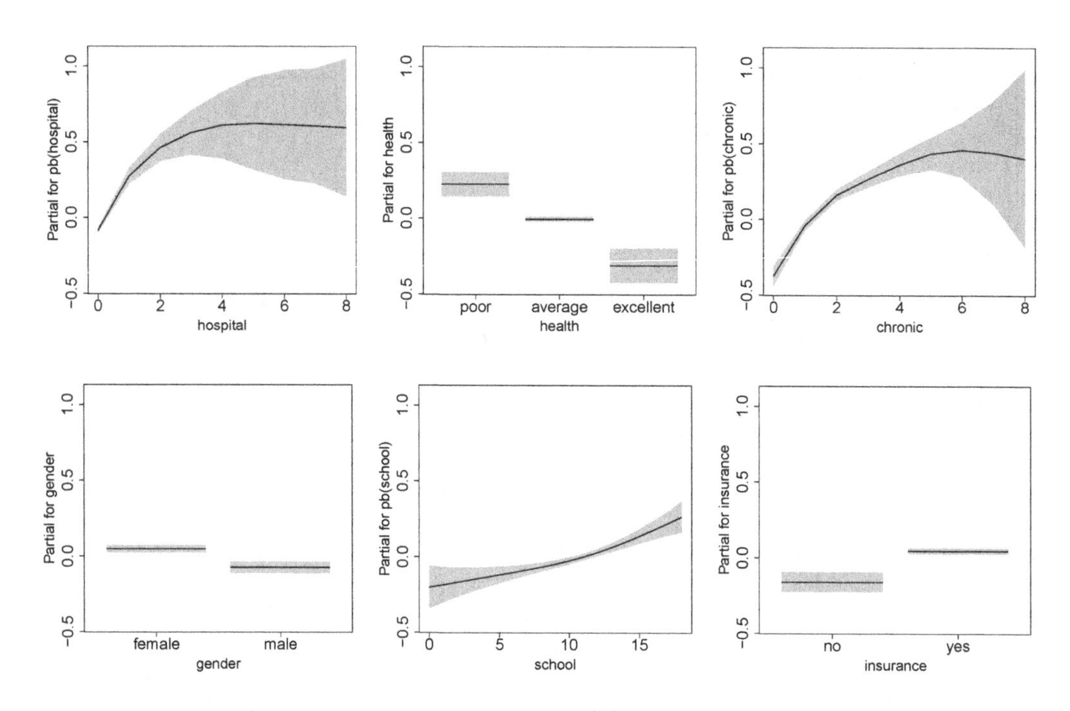

FIGURE 7.9: Term plot for $\log(\mu)$ for the `NBI` model `m1`.

Next we proceed using a stepwise model selection procedure for both μ and σ using `NBI`(μ, σ) and only linear terms:

```
m2<-gamlss(visits~1, family=NBI, data=nmes)
m2b<-stepGAICAll.A(m2, scope=list(lower=~ 1, upper=~ sh +
            health + sc + gender + school + insurance), k=4)
```

For an explanation of the `stepGAICAll.A()` function, see Stasinopoulos et al. [2017] p 397-399. The chosen `NBI` model for visits using GAIC($\kappa = 4$) is

```
m2c<-gamlss(visits~sh + health + sc + gender + school + insurance,
            sigma.fo=~sh + health + sc + gender + insurance,
            family=NBI, data=nmes, trace=FALSE)
```

A problem with the `NBI`(μ, σ) model is that, since it has only two parameters, it cannot model the skewness or kurtosis in the response variable independently of the mean and variance. A solution to modeling the skewness is to consider a discrete distribution with more than two parameters, i.e. with a parameter that can model the skewness of the distribution. Three-parameter candidates are the `SICHEL` and the beta negative binomial (`BNB`). Due to the possible excess of zero values, the zero-inflated or zero-adjusted versions of the discrete distributions should also be considered. We have evaluated the zero-inflated and zero-adjusted versions of the negative binomial (`ZINBI`, `ZANBI`), Sichel (`ZISICHEL`, `ZASICHEL`), negative binomial family (`ZINBF`) and beta negative binomial (`ZIBNB`, `ZABNB`) distributions. For example, the code for model selection using the `ZINBI` distribution is

```
m3<-gamlss(visits~1, family=ZINBI, data=nmes)
m3b<-stepGAICAll.A(m3, scope = list(lower=~ 1, upper=~ sh +
                health + sc + gender + school + insurance), k=4)
```

There were two clear candidates for 'best' distribution, the `ZINBI` and the `ZIBNB` distributions. The chosen model for `ZINBI` was:

```
m3c<-gamlss(visits ~ sc + sh + school + health + insurance + gender,
             sigma.fo=~sc + insurance +  gender + sh + health,
             nu.fo=~sh + insurance, trace=FALSE,
             family=ZINBI,data=nmes)
```

The chosen model for `ZIBNB` was:

```
m4c<-gamlss(visits~sc + sh + insurance + gender +school + health,
              sigma.fo=~sc + health + gender,
              nu.fo=~sh,
              tau.fo=~sc + insurance, trace=FALSE,
              family=ZIBNB, data=nmes, n.cyc=100, nu.start=1)
```

The `ZIBNB` model was a lot better than both `NBI` and `ZINBI` using GAIC with $k = 4$.

```
GAIC(m2c,m3c, m4c, k=4)

##     df      AIC
## m4c 18 24030.52
## m3c 18 24135.92
## m2c 15 24150.62
```

To make the term plots (given later) clearer, the selected `ZIBNB` model was refitted using the original variables:

```
m4d<-gamlss(visits~sqrt(chronic)+ sqrt(hospital) + insurance
             + gender +school+ health ,
             sigma.fo=~health + sqrt(chronic)  + gender,
             nu.fo=~sqrt(hospital),
             tau.fo=~sqrt(chronic)  + insurance, trace=FALSE,
             family=ZIBNB,data=nmes,n.cyc=100, nu.start=1)
```

The fitted model parameters are given by

```
summary(m4d)
```

For brevity we omit the output. The fitted model using stepwise selection is given by

$$
\begin{aligned}
Y &\sim \texttt{ZIBNB}(\hat{\mu}, \hat{\sigma}, \hat{\nu}, \hat{\tau}) \\
\log(\hat{\mu}) &= 0.980 + 0.332\sqrt{\text{chronic}} + 0.382\sqrt{\text{hospital}} \\
&\quad + 0.025\text{school} + 0.255(\text{if health=poor}) \\
&\quad - 0.313(\text{if health=excellent}) - 0.112(\text{if gender=male}) \\
&\quad + 0.123(\text{if insurance=yes}) \\
\log(\hat{\sigma}) &= -1.7026 - 0.208\sqrt{\text{chronic}} + 0.394(\text{if health=poor}) \\
&\quad - 0.345(\text{if health=excellent}) + 0.197(\text{if gender=male}) \\
\log(\hat{\nu}) &= -2.679 + 0.966\sqrt{\text{hospital}} \\
\log[\hat{\tau}/(1-\hat{\tau})] &= -1.077 - 0.744\sqrt{\text{chronic}} - 1.546(\text{if insurance=yes}),
\end{aligned}
\tag{7.15}
$$

where Y = number of visits.

In order to interpret the parameters of the above model, $\texttt{ZIBNB}(\mu, \sigma, \nu, \tau)$ is a mixture with two components: zero with probability τ and a $\texttt{BNB}(\mu, \sigma, \nu)$ distribution with probability $(1-\tau)$.

- For the $\texttt{BNB}(\mu, \sigma, \nu)$ component:
  - μ is the mean,
  - σ is a right tail heaviness parameter (and increases the variance for $\sigma < 1$, while the variance is infinite for $\sigma \geq 1$),
  - ν increases the variance (for $\nu^2 > \sigma/\mu$ and $\sigma < 1$), and
- τ is the probability of excess of zeros.
- The mean of $Y \sim \texttt{ZIBNB}(\mu, \sigma, \nu, \tau)$ is $\text{E}(Y) = (1-\tau)\mu$.

```
term.plot(m4d)
term.plot(m4d,what='sigma')
term.plot(m4d,what='nu')
term.plot(m4d,what='tau')
```

Figure 7.10 displays the fitted parametric terms in the final chosen model (7.15). The top two rows in Figure 7.10 display the terms in $\log(\hat{\mu})$: their effects are additive for $\log(\hat{\mu})$ and hence multiplicative for the fitted mean visits $(1-\hat{\tau})\hat{\mu}$. Assuming all other explanatory variables are fixed, then, due to $\hat{\mu}$, the fitted mean visits increases with the number of chronic conditions (`chronic`), the number of hospital stays (`hospital`), and the number of years schooling (`school`). A poor self-perceived health status results in a 29.0%[b] *increase* in fitted mean visits (relative to average health status) and an excellent health status results in a 26.9% *decrease* in the fitted mean visits (relative to average health). Being male results in a 10.6% decrease in fitted mean visits, compared with being female. Being covered by private insurance increases the fitted mean visits

[b]calculated from the parameter estimate in (7.15) as $(e^{0.255} - 1) \times 100$

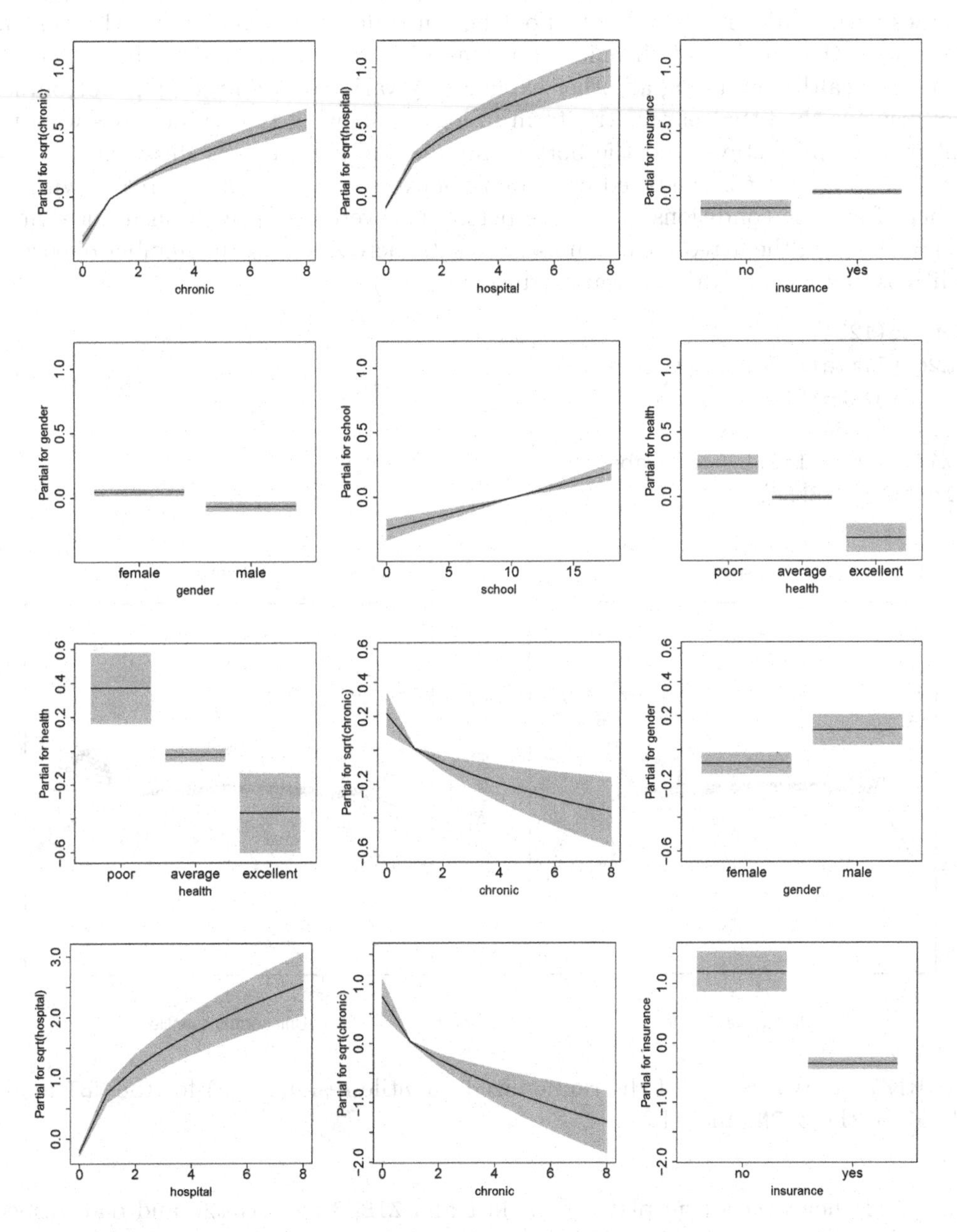

FIGURE 7.10: Term plots for the `ZIBNB` model `m4d`: Top two rows: $\log(\mu)$; third row: $\log(\sigma)$; bottom row left: $\log(\nu)$; bottom row middle and right: $\log[\tau/(1-\tau)]$.

by 13.1%, due to $\hat{\mu}$, but also results in an additional increase due to $(1-\hat{\tau})$. Also increasing `chronic` results in an additional increase in fitted mean visits due to $(1-\hat{\tau})$.

The third row in Figure 7.10 displays the fitted parametric terms in $\log(\hat{\sigma})$. Since $\hat{\sigma}$ controls the heaviness of the right tail of the distribution of visits, this heaviness increases with gender male and with poor health (relative to average health), but decreases with the number of chronic conditions and if the health is excellent (relative to average health), assuming all other explanatory variables are fixed. The bottom left panel displays the fitted parametric term in $\log(\hat{\nu})$, showing that $\hat{\nu}$ increases with the number of hospital stays; and the bottom middle and right panels show the terms in $\log[\hat{\tau}/(1-\hat{\tau})]$. Since $\hat{\tau}$ is the fitted probability of excess of zeros, this decreases with the number of chronic conditions and if the person is covered by private insurance. Since $\mathrm{E}(Y)=(1-\tau)\mu$, the fitted mean number of visits increases with the number of chronic conditions and with private insurance, due to $(1-\hat{\tau})$.

```
set.seed(1234)
wp(m2c,ylim.all=2.1, cex.lab=1.5)
title("(a) NBI")
set.seed(1234)
wp(m4d,ylim.all=2.1, cex.lab=1.5)
title("(b) ZIBNB")
```

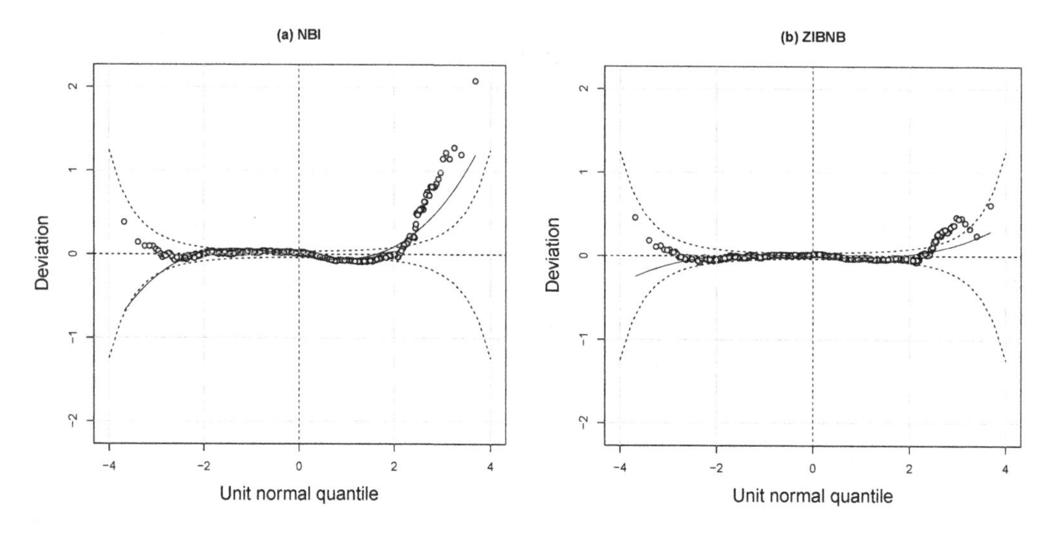

FIGURE 7.11: Worm plot of the randomized quantile residuals (a) for the `NBI` model and (b) for the `ZIBNB` model.

Figure 7.11 shows the worm plots of the `NBI` and `ZIBNB` models `m2c` and `m4d`, respectively. The `NBI` worm plot (left) indicates that the `NBI` does not provide a good fit to the data, since many points lie well outside the elliptical approximate 95% pointwise interval bands, in the right tail. The `ZIBNB` worm plot (right) shows that the `ZIBNB` provides a better fit, but is still inadequate in the right tail.

There is an outlier not shown in the right tail of the `NBI` plot in Figure 7.11(a), because its vertical axis deviation value is much greater than 2. The outlier observation is case

1522 and has `visits` = 65 and `health` = `excellent`. Omitting case 1522 makes a considerable difference to the fitted parameter estimates for `health` = `excellent` in μ and σ for both the `NBI` and `ZIBNB` models, so it may be reasonable to omit this case.

Figure 7.12 shows the hanging rootograms for the chosen `NBI` and `ZIBNB` models. The `NBI` hanging rootogram in Figure 7.12(a) shows seven violations of the warning limits (at ± 1), including two potentially important ones at `visits=1` and `visits=>40`, suggesting that the `NBI` model may be inadequate. For the important violation of the control limits for `visits` > 40, we have $E_{>40} = 10.20$, $O_{>40} = 25$ and $\sqrt{E_{>40}} - \sqrt{O_{>40}} = -1.81$, indicating that the right tail of the `NBI` model is inadequate. In contrast, Figure 7.12(b), the `ZIBNB` hanging rootogram, shows only two violations of the warning limits (which is to be expected for 41 values of v), suggesting that the `ZIBNB` model is an improved, and potentially adequate, model according to this diagnostic. In the `ZIBNB` model there is no violation of the warning or control limits for `visits` > 40, since $E_{>40} = 17.88$ and $O_{>40} = 25$ and $\sqrt{E_{>40}} - \sqrt{O_{>40}} = -0.77$.

```
# NBI
rootogram(m2c, main="(a) NBI", col=1, max=40)

## ******************************************************************
##                 Summary of the rootogram:
##                 Tail observed Frequency = 25
##                 Tail expected Frequency = 10.20455
## Tail [sqrt(expected)-sqrt(observed)] = -1.805544
##          Number Tukey warning violations = 7
##          Number Tukey control violations = 2
## ******************************************************************

abline(h=c(-1,1), col=gray(.2))
# ZIBNB
rootogram(m4d, main="(b) ZIBNB", col=1, max=40)

## ******************************************************************
##                 Summary of the rootogram:
##                 Tail observed Frequency = 25
##                 Tail expected Frequency = 17.88081
## Tail [sqrt(expected)-sqrt(observed)] = -0.7714294
##          Number Tukey warning violations = 2
##          Number Tukey control violations = 1
## ******************************************************************

abline(h=c(-1,1), col=gray(.2))
```

The adequacy of the fitted models can be further investigated by multiple worm plots (i.e. for different ranges of an explanatory variable), see van Buuren and Fredriks [2001] or Stasinopoulos et al. [2017, p. 428-433], (and analogously multiple rootograms, not yet developed).

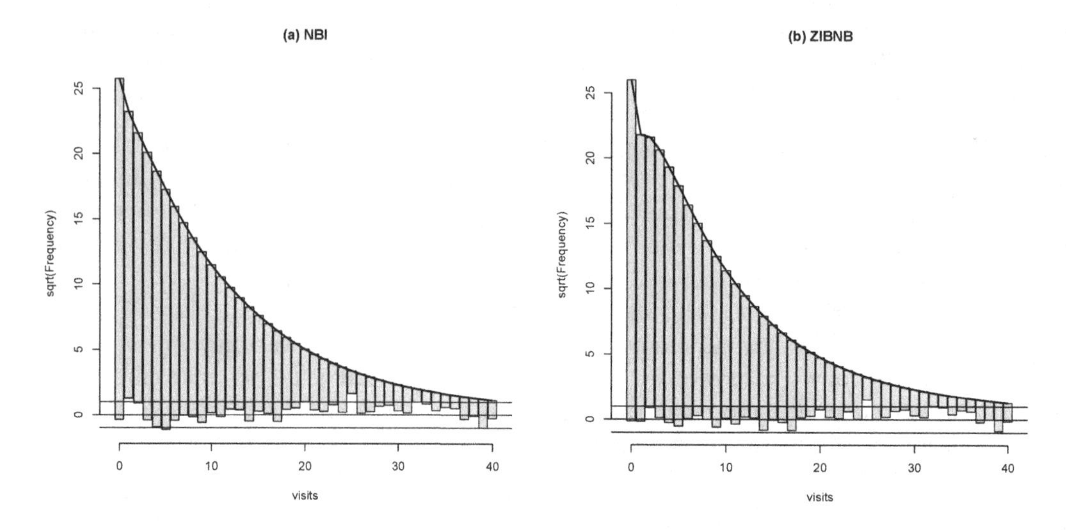

FIGURE 7.12: Hanging rootograms for the chosen models: (a) NBI and (b) ZIBNB.

7.8 Appendix: skewness and kurtosis for a mixture of Poisson distributions

Let $Y \mid \gamma \sim \texttt{PO}(\mu\gamma)$ and let γ have a distribution with cumulant generating function $\mathcal{K}_\gamma(t)$. Then the cumulant generating function of the marginal distribution of Y, $\mathcal{K}_Y(t)$, is given by

$$\mathcal{K}_Y(t) = \mathcal{K}_\gamma\left[\mu\left(e^t - 1\right)\right] \ .$$

Hence, assuming that $\mathrm{E}(\gamma) = 1$, the moments and cumulants of Y and γ are related by

$$\begin{aligned}
\mathrm{E}\,(Y) &= \mu \\
\mathrm{Var}\,(Y) &= \mu + \mu^2 \mathrm{Var}(\gamma) \\
\kappa_{3Y} &= \mu + 3\mu^2 \mathrm{Var}(\gamma) + \mu^3 \kappa_{3\gamma} \\
\kappa_{4Y} &= \mu + 7\mu^2 \mathrm{Var}\,(\gamma) + 6\mu^3 \kappa_{3\gamma} + \mu^4 \kappa_{4\gamma} \ ,
\end{aligned} \tag{7.16}$$

where κ_{3Y} and κ_{4Y} are the third and fourth cumulants of Y; and $\kappa_{3\gamma}$ and $\kappa_{4\gamma}$ are the third and fourth cumulants of γ.

The moment skewness and moment kurtosis of Y are

$$\begin{aligned}
\sqrt{\beta_1} &= \frac{\kappa_{3Y}}{[\mathrm{Var}(Y)]^{1.5}} \\
\beta_2 &= 3 + \frac{\kappa_{4Y}}{[\mathrm{Var}(Y)]^2}
\end{aligned}$$

respectively.

Example: Sichel distribution If $Y \sim \texttt{SICHEL}(\mu, \sigma, \nu)$ then the mixing distribution is $\gamma \sim \texttt{GIG}(1, \sigma^{1/2}, \nu)$, with pdf defined by (19.18), and

$$\begin{aligned} \mathrm{E}(\gamma) &= 1 \\ \mathrm{Var}\,(\gamma) &= g_1 \\ \kappa_{3\gamma} &= g_2 - 3g_1 \\ \kappa_{4\gamma} &= g_3 - 4g_2 + 6g_1 - 3g_1^2 \,, \end{aligned}$$

where

$$\begin{aligned} g_1 &= 1/b^2 + 2\sigma(\nu+1)/b - 1 \\ g_2 &= 2\sigma(\nu+2)/b^3 + \left[4\sigma^2(\nu+1)(\nu+2)+1\right]/b^2 - 1 \\ g_3 &= \left[1+4\sigma^2(\nu+2)(\nu+3)\right]/b^4 + \left[8\sigma^3(\nu+1)(\nu+2)(\nu+3) + 4\sigma(\nu+2)\right]/b^3 - 1 \,, \end{aligned}$$

and where $b = K_{\nu+1}(1/\sigma^2)[K_\nu(1/\sigma^2)]^{-1}$, $g_1 = \mathrm{E}(\gamma^2) - 1 = \mathrm{Var}(\gamma)$, $g_2 = \mathrm{E}(\gamma^3) - 1$ and $g_3 = \mathrm{E}(\gamma^4) - 1$. and $K(\cdot)$ is a modified Bessel function of the second kind [Abramowitz and Stegun, 1965].

The cumulants for the Sichel distribution are given by (7.16), from which its mean, variance, skewness and kurtosis are obtained.

7.9 Bibliographic notes

In the literature of discrete distributions, Johnson et al. [2005] is a well-known comprehensive reference; and Wimmer and Altmann [1999] give a very extensive technical documentation of the properties of approximately 750 discrete distributions. A number of texts give comprehensive overviews of discrete distributions, and count data modeling in particular. Winkelmann [2008] gives a detailed account of count data models and their genesis. The coverage of the material is not limited to econometric applications, as the title suggests, but is quite generic. Claim counts are of particular importance in the actuarial world and Denuit et al. [2007] give a comprehensive treatment of the analysis of claims. Cameron and Trivedi [2013] discuss regression models for count data.

Zero-inflated and zero-adjusted Poisson distributions are discussed by Johnson et al. [2005, p351-356] as 'zero-modified' distributions defined by (7.9), where zero-deflation is accommodated by allowing σ to be negative, under the condition that

$$\sigma + (1-\sigma)e^{-\mu} \geq 0 \,.$$

Hurdle models for count data [Mullahy, 1986] assume that different statistical processes govern observations above and below a hurdle. Zero-adjusted models may be regarded as hurdle models with the hurdle at zero. Lambert [1992] and Mullahy [1986] first discussed regression modeling for the zero-inflated Poisson and Poisson hurdle models, respectively.

Accommodating Poisson overdispersion by continuous mixtures of Poisson distributions dates back to Greenwood and Yule [1920] and their use of the gamma as mixing distribution to derive the negative binomial. Other mixing distributions used are the generalized inverse Gaussian, yielding the Sichel (`SICHEL`) distribution [Sichel, 1971]; its two-parameter special case, the Poisson-inverse Gaussian (`PIG`), [Stein et al., 1987, Dean et al., 1989]; and the normal, giving the Poisson-normal [Hinde, 1982].

Flexible regression modeling for count data is implemented by several **R** packages besides **gamlss**, including **pscl** [Zeileis et al., 2008, Jackman, 2017] and **VGAM** Yee [2019], (in particular zero-inflated and zero-adjusted (hurdle) models); and **mgcv** [Wood, 2017].

7.10 Exercises

1. Gupta et al. [1996] present the following data giving the number of lamb foetal movements `y` recorded by ultrasound over 240 consecutive five-second intervals:

y	0	1	2	3	4	5	6	7
frequency	182	41	12	2	2	0	0	1

   (a) Use the function `chooseDist()` to find an appropriate count distribution for the data. You should specify the correct penalties, i.e. `k=c(2, 3.84, 5.48)`, corresponding to AIC, $\chi^2_{1,0.05}$ and SBC, respectively. Note the total number of observations is the sum of the frequencies (i.e. 240) rather the length of y which is the default value in the `chooseDist()` function, so SBC uses $k = \log(240) = 5.48$.

   (b) Display and check the adequacy of the fitted distribution.

2. The USA National AIDS Behavioral Study recorded `y`, the number of times individuals engaged in risky sexual behavior during the previous six months, together with two explanatory factors: sex of individual (male or female) and whether they have a risky partner (no or yes), giving the following frequency distribution:

y	0	1	2	3	4	5	6	7	10	12	15	20	30	37	50
male, no	541	19	17	16	3	6	5	2	6	1	0	3	1	0	0
male, yes	102	5	8	2	1	4	1	0	0	0	1	0	0	1	0
female, no	238	8	0	2	1	1	1	1	0	0	1	0	0	0	0
female, yes	103	6	4	2	0	1	0	0	0	0	0	0	0	0	1

   The data were previously analyzed by Heilbron [1994].

   (a) Read the above frequencies (corresponding to the male no, male yes, female no, female yes rows of the above table) into a single variable `f`. Read the corresponding count values into `y` by using:

```
y <-rep(c(0:7,10,12,15,20,30,37,50),4)
```

   Generate a single factor `dtype` for type of individual with four levels (corresponding to male no, male yes, female no, female yes) by

```
dtype <- gl(4,15)
```

(b) Find an appropriate count distribution for the data, (using the `chooseDist()` function in combination with `getOrder()`), using factor `dtype` for the mean model and a constant scale (and shape).

(c) Check whether your chosen distribution model needs the factor `dtype` for the rest of the parameters of the distribution. Recheck whether `dtype` is needed for the mean model.

(d) Use diagnostics to check the model.

(e) Output and interpret the parameter estimates for your chosen model. Note that `tapply(fitted(model), dtype, mean)` will get you the mean fitted values for different types.

3. **The tidal data**: The data set `tidal` [McArdle and Anderson, 2004] gives counts of the organism 'intertidal bivalve *A. stutchburyi*' in three tidal areas in the Bay of Plenty, New Zealand. Each observation is the count of the number of these organisms in a 0.25m^2 quadrat, as well as the vertical tidal height of the quadrat. The vertical heights have been classified into three tidal areas: upper (vertical height >0.66m), middle (0.33 - 0.66 m), and lower (<0.33 m). Ecologists are interested in the effect of tidal height (either raw or classified) on the number of organisms.

R data file: `tidal` in package **gamlss.data** of dimension 90 × 3
source: McArdle and Anderson [2004]
variables used in our analysis
`number` : count of *A. stutchburyi* organisms
`vertht` : vertical tidal height (m)
`ht` : factor, tidal area
1: `vertht`<0.33m (lower)
2: 0.33 ≤ `vertht` ≤0.66m (middle)
3: `vertht`>0.66m (upper)
purpose: to demonstrate the fitting of a discrete count response variable

(a) Construct a scatterplot of `number` against `vertht`. What is your impression of the relationship between the two variables?

```
data(tidal)
plot(number ~ vertht, data=tidal)
```

(b) Construct boxplots of `number` against the factor `ht`. What do you conclude about the variability of `number`?

```
plot(number ~ ht, data=tidal)
```

(c) The following three models explore different ways of modeling the response variable `number` as a function of vertical heights.

i. `number` ∼ `vertht`, (linearly);

ii. `number` ∼ `pb(vertht)`, (smoothly);

iii. `number` $\sim$ `ht`, (as a factor);

In order to choose systematically between the three different models above and between different distributions, fit, for each model, all count distributions using the function `chooseDist()`. (Note in the following code the renaming of the row names of the saved tables.)

```
m1 <- gamlss(number~vertht, data=tidal, family="PO")
T1 <- chooseDist(m1, type="count", trace=T)
rownames(T1) <- paste(rownames(T1), "1",sep="_")
m2 <- gamlss(number~pb(vertht), data=tidal, family="PO")
T2 <- chooseDist(m2, type="count", trace=T)
rownames(T2) <- paste(rownames(T2), "2",sep="_")
m3 <- gamlss(number~ht, data=tidal, family="PO")
T3 <- chooseDist(m3, type="count", trace=T)
rownames(T3) <- paste(rownames(T3), "3",sep="_")
TT <- rbind(T1,T2,T3)
getOrder(TT, column=1)[1:5]
```

What is your best fitted model according to AIC (i.e. column 1)?

(d) Refit the 'best' three models according to AIC using the function `update()`.

```
M1 <- update(m2, family="WARING")
M2 <- update(m2, family="ZINBI")
M3 <- update(m2, family="NBI")
```

(e) Critically assess the suitability of the three models. Note the convergence problems of the Waring distribution model and the fact that a lot of the residuals of the model are NA's. (This highlights the problem of any automatic procedure. What happens here is that the flexibility of smoothing the μ model using `pb()` allows the σ to become zero, creating instability in the estimation procedure. One way to avoid this is to use a shifted log link for σ function which does not allow this to happen.)

```
M11 <- gamlss(number~pb(vertht), data=tidal,
              family=WARING(sigma.link="Slog"))
```

The resulting fitted `WARING` model now has a poor AIC and so `WARING` is rejected.

(f) Hence use the `ZINBI` model `M2` and check for a suitable model for `sigma` and `nu` using the `stepGAICAll.A()` function.

```
M21 <-stepGAICAll.A(M2,scope=list(lower=~1,
                     upper=~vertht+pb(vertht)+ht) )
```

(g) Construct term plots for all of the parameters of your chosen model. Describe the relationship between the distribution of the number of organisms and the vertical height.

8

Binomial type distributions

CONTENTS

This chapter provides explanation for:

1. different distributions for a binomial type response variable within the GAMLSS family; and
2. how these distributions model overdispersion, underdispersion, and an excess or shortage of zeros.

8.1 Introduction

This chapter is concerned with the binomial distribution and distributions with the same range, which we refer to as binomial type distributions. The discrete random variable Y has finite range $R_Y = \{0, 1, 2, \ldots, n\}$, where the positive integer n is assumed to be known, and is called the *binomial denominator.*

The underlying notion of a binomial distribution is that of a 'Bernoulli trial', which is any action that results in an event of interest either occurring or not occurring. The probability of event occurrence is assumed to be μ. Y is defined as the number of times that the event occurred in n independent trials, and has the binomial distribution, $\texttt{BI}(n, \mu)$, with probability function

$$P(Y = y \mid n, \mu) = \binom{n}{y} \mu^y (1 - \mu)^{n-y} \qquad \text{for } y = 0, 1, \ldots, n \;, \tag{8.1}$$

where $0 < \mu < 1$ and $\binom{n}{y} = \frac{n!}{y!(n-y)!}$. The special case $n = 1$ is called the Bernoulli distribution.

For example, in an analysis of motor vehicle accident fatalities, a trial would be an accident, and the event of interest would be the occurrence of a fatality in the accident. Assuming that the probability of a fatality in an accident is μ, and that we observe n independent accidents, then the total number of accidents which had a fatality (Y) has the binomial distribution BI(n, μ), i.e. $Y \sim$ BI(n, μ).

The special characteristic of any binomial type distribution is the fact that the binomial denominator n appears in all probability functions. A probability function of a binomial type distribution has the form $P(Y = y \mid n, \boldsymbol{\theta})$, where n is assumed to be known and $\boldsymbol{\theta}$ is the vector of (usually unknown) parameters of the distribution. For any `gamlss.family` binomial type distribution, the argument `bd` defines the binomial denominator n. The response variable in a formula of a binomial type response model is in matrix form, where the first column of the matrix is y and the second is $n - y$, i.e. `cbind(y, bd-y)`.

8.2 Explicit distributions

There are seven distributions explicitly available in **gamlss.dist** for modeling a binomial type response variable, i.e. with range $\mathbf{R}_Y = \{0, 1, \ldots, n\}$: BI(n, μ) is a one-parameter distribution; ZIBI(n, μ, σ), ZABI(n, μ, σ), DBI(n, μ, σ), and BB(n, μ, σ) have two parameters; while ZIBB(n, μ, σ, ν) and ZABB(n, μ, σ, ν) have three parameters. Table 8.1 lists the binomial type distributions in **gamlss.dist** with their default link functions. See Chapter 23 for details and plots for these distributions.

TABLE 8.1: Discrete binomial type distributions implemented within **gamlss.dist**, with default link functions. (inf=inflated, adj=adjusted)

Distribution	**gamlss** name	Parameter link function μ	σ	ν	τ
	$R_Y = \{0, 1, \ldots, n\}$				
binomial	BI	logit	-	-	-
beta binomial	BB	logit	log	-	-
double binomial	DBI	logit	log	-	-
zero-adj beta binomial	ZABB	logit	log	logit	-
zero-adj binomial	ZABI	logit	logit	-	-
zero-inf beta binomial	ZIBB	logit	log	logit	-
zero-inf binomial	ZIBI	logit	logit	-	-

8.2.1 Binomial distribution

The binomial, BI(n, μ) is the most common finite range discrete distribution. It has a single parameter μ, which lies in the range $0 < \mu < 1$. The binomial distribution has probability function given by (8.1), mean $E(Y) = n\mu$ and variance $\text{Var}(Y) = n\mu(1 - \mu)$. The BI$(n, \mu)$ distribution may be inadequate if the response variable has underdispersion or overdispersion, and/or excess or lack of zeros, relative to BI(n, μ).

8.2.2 Overdispersion and underdispersion

Beta binomial $\texttt{BB}(n, \mu, \sigma)$. Overdispersion relative to the Poisson distribution was discussed in Section 7.3. Overdispersion relative to the binomial distribution may be accommodated similarly via a continuous mixture, in particular the beta binomial (BB) distribution. Consider that, conditional on probability π, Y has the binomial distribution: $Y \mid \pi \sim \texttt{BI}(n, \pi)$. Consider further that π is a random variable with a beta distribution: $\pi \sim \texttt{BEo}(\frac{\mu}{\sigma}, \frac{1-\mu}{\sigma})$, with mean $\text{E}(\pi) = \mu$ (see Section 21.1.1). Then the unconditional (or marginal) distribution of Y is beta binomial: $Y \sim \texttt{BB}(n, \mu, \sigma)$ where $0 < \mu < 1$, $\sigma > 0$ and n is a known positive integer. It has mean $\text{E}(Y) = n\mu$ and variance

$$\text{Var}(Y) = n\mu(1-\mu)\left[1 + \frac{\sigma(n-1)}{(1+\sigma)}\right] > n\mu(1-\mu)$$

and hence provides a model for an overdispersed binomial response variable.

Double binomial $\texttt{DBI}(n, \mu, \sigma)$ is a special case of the double exponential family of Efron [1986] with approximate mean $n\mu$ and approximate variance $\sigma n\mu(1-\mu)$. Hence (approximately) it is overdispersed binomial if $\sigma > 1$ and underdispersed binomial if $\sigma < 1$. It is the only explicit two-parameter distribution in **gamlss.dist** which can model an underdispersed (as well as overdispersed) binomial response variable.

8.2.3 Excess or shortage of zero values

The excess or shortage of zero values relative to the Poisson distribution was discussed in Section 7.4. We use the same methods to define zero-inflated and zero-adjusted versions of the binomial and beta binomial distributions.

Zero-inflated binomial $\texttt{ZIBI}(n, \mu, \sigma)$. This allows the probability of a zero value to be inflated (but not deflated) relative to a binomial $\texttt{BI}(n, \mu)$ distribution. The $\texttt{ZIBI}(n, \mu, \sigma)$ is a mixture with two components: value 0 with probability σ, and a $\texttt{BI}(n, \mu)$ distribution with probability $(1-\sigma)$.

Zero-adjusted binomial $\texttt{ZABI}(n, \mu, \sigma)$. This distribution is a mixture with two components: value 0 with probability σ, and a zero-truncated binomial distribution with probability $(1-\sigma)$. The $\texttt{ZABI}(n, \mu, \sigma)$ can cope with both inflated and deflated probability of a zero value relative to $\texttt{BI}(n, \mu)$.

Zero-inflated beta binomial $\texttt{ZIBB}(n, \mu, \sigma, \nu)$. This allows the probability of a zero value to be inflated (but not deflated) relative to a beta binomial $\texttt{BB}(n, \mu, \sigma)$ distribution. The $\texttt{ZIBB}(n, \mu, \sigma, \nu)$ is a mixture with two components: value 0 with probability ν, and a $\texttt{BB}(n, \mu, \sigma)$ distribution with probability $(1-\nu)$.

Zero-adjusted beta binomial $\texttt{ZABB}(n, \mu, \sigma, \nu)$. This distribution is a mixture with two components: value 0 with probability ν, and a zero-truncated beta binomial distribution with probability $(1-\nu)$. The $\texttt{ZABB}(n, \mu, \sigma, \nu)$ can cope with both inflated and deflated probability of a zero value relative to $\texttt{BB}(n, \mu, \sigma)$.

8.3 Examples

8.3.1 Alveolar-bronchiolar adenomas

R data file: `alveolar` in package **gamlss.data** of dimension 23×2
source: Tamura and Young [1987], Hand et al. [1994]
variables
`r` : number of mice having alveolar-bronchiolar adenomas
`n` : total number of mice
purpose: to demonstrate the fitting of a binomial type distribution
conclusion the binomial distribution is adequate

Here we consider the alveolar-bronchiolar adenomas data used by Tamura and Young [1987] and reproduced in Hand et al. [1994]. The data are the number of mice having alveolar-bronchiolar adenomas out of the total numbers of mice in 23 independent groups. Below we create the response variable[a] (which should be a matrix with columns `r` and `n-r`), initially fit a binomial distribution and then select an appropriate distribution using `chooseDist()`.

```
data(alveolar)
alveolar$y <- with(alveolar, cbind(r, n - r))
m1 <- gamlss(y ~ 1, data = alveolar, family = BI, trace = F)
mf <- chooseDist(m1, type="binom")

## minimum GAIC(k= 2 ) family: BI
## minimum GAIC(k= 3.84 ) family: BI
## minimum GAIC(k= 3.14 ) family: BI
```

For a binomial response variable with no explanatory variables, `histDist()` can be used to display the data and the fit but with a limited scope. The plot works well if the binomial denominator is *constant* for all observations, where `histDist()` plots a histogram of the number of events `r` and superimposes the fitted binomial distribution. When the binomial denominator is not constant for all observations, as in the `alveolar` data, then `histDist()` plots a histogram of the proportions (which may be of some interest) and indicates where the fitted proportion lies:

```
m1 <- histDist(y, data=alveolar, family=BI,
         xlim=c(0,.25),  border.hist= 1, line.col=1)
wp(m1)
```

The worm plot indicates that the binomial, $\texttt{BI}(n, \mu)$, distribution with constant μ provides an adequate model for r, the number of mice having alveolar-bronchiolar adenomas, out of total numbers n of mice, i.e. $r \sim \texttt{BI}(n, \mu)$.

[a] Binomial type data need as a response a matrix with the number of 'successes' and 'failures' as columns.

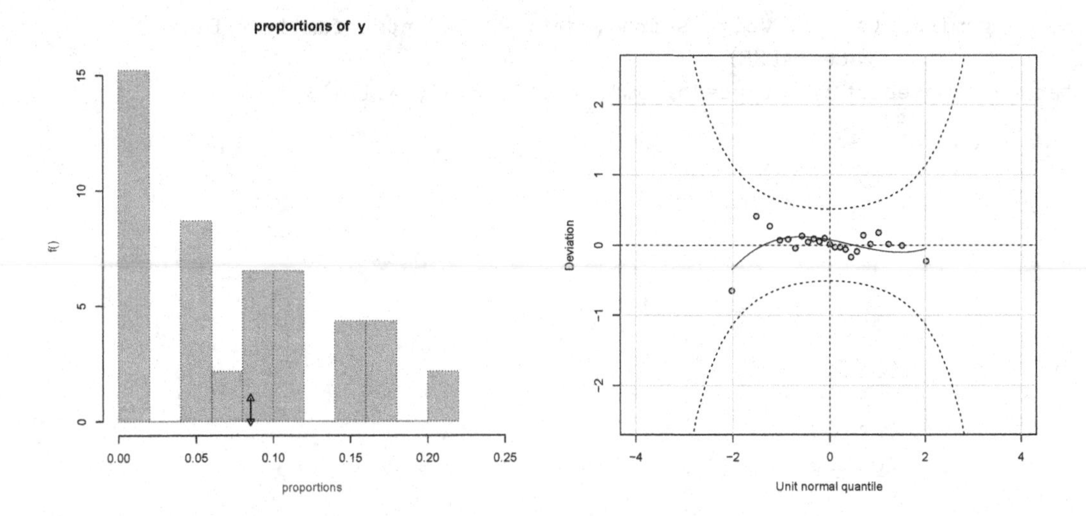

FIGURE 8.1: Left: proportion of alveolar-bronchiolar adenomas (with arrow indicating the fitted proportion); right: worm plot from the fitted BI(n, μ) model.

8.3.2 First-year student examination results

Here we demonstrate the fitting of a binomial type data response variable given that the binomial denominator n is constant. The data shown in Table 8.2 [Karlis and Xekalaki, 2009] refer to the numbers of course examinations passed, `r`, and their frequency, `freq`, from a class of 65 first-year students. All students enrolled for eight courses during the year. Hence, the variable n-`r` in Table 8.2 is defined as $8 - r$.

TABLE 8.2: The first-year student examination results data where the binomial denominator n is constant at 8.

r	n-r	freq
0	8	1
1	7	4
2	6	4
3	5	8
4	4	9
5	3	6
6	2	8
7	1	12
8	0	13

Now we create the data, initially fit a binomial distribution, and then use `chooseDist()` to find an appropriate distribution model using different GAICs.

```
r <- 0:8
freq <- c(1,4,4,8,9,6,8,12,13)
y <- cbind(r,8-r)
colnames(y) <- c("r", "n-r")
students <- data.frame(y,freq)
```

```
m1 <- gamlssML(y ~ 1, weights=freq, data = students, family = BI,
               trace=FALSE)
tab <- chooseDist(m1, type="binom", k=c(2, 3.84, 4.174))

## minimum GAIC(k= 2 ) family: BB
## minimum GAIC(k= 3.84 ) family: BB
## minimum GAIC(k= 4.174 ) family: BB

tab

##              2     3.84    4.174
## BI    339.6467 341.4867 341.8207
## BB    273.4987 277.1787 277.8467
## DBI   274.8705 278.5505 279.2185
## ZIBI  334.9268 338.6068 339.2748
## ZABI  334.9268 338.6068 339.2748
## ZIBB  275.4987 281.0187 282.0207
## ZABB  274.7575 280.2775 281.2795
```

The beta binomial is the best-fitting model. Next we refit the model using `histDist()`. Note that, since `weights=freq` is used, the response variable y needs to be expanded to individual cases in Y, in order to obtain the correct worm plot (using `rqres.plot`). Figure 8.2(a) shows the data and their fitted probabilities and Figure 8.2(b) shows a worm plot of the residuals. The fitted probabilities are increasing almost linearly with the number of examinations passed, while the worm plot indicates an adequate fit.

```
Y <-with(students, cbind(rep(y[,1],freq), rep(y[,2],freq)))
m1<-histDist(Y, family=BB, ylim=c(0,0.25), line.col=1,
          xlab ="number of exams passed",
          ylab ="probability", main="(a)")
rqres.plot(m1, howmany=40, plot="all", ylim.all=.7)
title("(b)")
```

8.4 Bibliographic notes

The binomial distribution dates back to Bernoulli [1713], while its continuous mixture the beta binomial, was developed by Skellam [1948]. Zero-inflated binomial models were introduced by Hall [2000]. Zero-adjusted models may be regarded as special cases of hurdle models in which the hurdle is zero [Winkelmann, 2008].

8.5 Exercises

1. **Infant mortality**: The following data set is not a real data set but is created for the purpose of demonstrating a binomial type response variable. The data set is based on some real data obtained from the Parana State in Brazil in 2010. There are 399

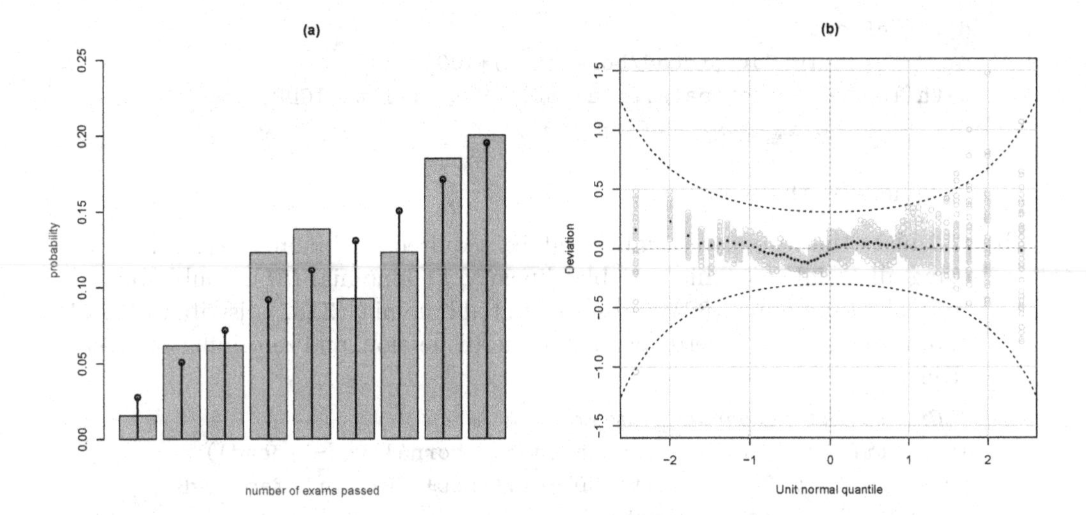

FIGURE 8.2: (a) The sample distribution of `r` with the fitted `BB`(n, μ, σ) distribution, (b) worm plot of the residuals from the fitted model, with gray points showing 40 different realizations of the (normalized) randomized quantile residuals, and black points showing their median.

observations, and the response variable is the numbers of infant deaths out of the total number of live births. There are several social-economic variables which can be used as explanatory variables. There are also longitude and latitude coordinates, but this geographical information is not used here.

R data file: `InfMort` in package **gamlss.data** of dimension 399 × 11
source: simulated data
variables

- `x` the x-coordinate
- `y` the y-coordinate
- `dead` the number of dead infants
- `bornaliv` the number of infants born alive
- `IFDM` : FIRJAN index of city development
- `illit` : the illiteracy index
- `lGNP` : the logarithm of the gross national product
- `cli` : the proportion of children living in a household with half the basic salary.
- `lpop` : the logarithm of the number of people living in each city
- `PSF` : the proportion covered by the family health program
- `Poor` the proportion of individuals low household income per capita.

purpose: to demonstrate the fitting of a binomial type distribution

(a) Read the data into **R** and plot the percentage of infant deaths against the explanatory variables.

```
data(InfMort)
mr <- with(InfMort,(dead/bornalive)*100)
with(InfMort, plot(data.frame(mr, IFDM, illit, lGDP,
        cli, lpop, PSF, poor)))
```

Comment on the plot.

(b) Calculate the response variable matrix, required for binomial type data, and, using all social-economic variables linearly, fit binomial (`BI`), double-binomial (`DBI`), beta-binomial (`BB`), and zero inflated binomial (`ZIBI`) distributions. This allows you to detect overdispersion, underdispersion, and zero-inflation, respectively.

```
# Create the response  matrix
InfMort$Y<-with(InfMort, cbind(dead, bornalive -  dead))
m0 <- gamlss(Y~IFDM+illit+lGDP+cli+lpop+PSF+poor, family=BI,
             data=InfMort)
# underdispersed
m1 <- gamlss(Y~IFDM+illit+lGDP+cli+lpop+PSF+poor, family=DBI,
             data=InfMort)
# overdispersed
m2 <- gamlss(Y~IFDM+illit+lGDP+cli+lpop+PSF+poor, family=BB,
             data=InfMort)
# zero-inflation
m3 <- gamlss(Y~IFDM+illit+lGDP+cli+lpop+PSF+poor, family=ZIBI,
             data=InfMort, n.cyc=500)
AIC(m0,m1,m2,m3)
```

What is your conclusion? (You can find the estimated value for σ for each model by using `fitted(model,"sigma")[1]`).

(c) For the two best fitted models try to add a full (all socio-economic variables) σ model:

```
m11 <- gamlss(Y~IFDM+illit+lGDP+cli+lpop+PSF+poor,
      sigma.fo=~IFDM+illit+lGDP+cli+lpop+PSF+poor,family=DBI,
      data=InfMort)
m21 <- gamlss(Y~IFDM+illit+lGDP+cli+lpop+PSF+poor,
      sigma.fo=~IFDM+illit+lGDP+cli+lpop+PSF+poor, family=BB,
      data=InfMort)
AIC(m0,m1,m2,m3, m11, m21)
```

State your conclusions and support your arguments using residual diagnostics such as `wp()` or `rqres.plot( , plot="all")`, (which generates several randomized quantile residuals).

(d) Let us concentrate on the `BB`(μ, σ) model and use the following code to select using AIC the 'best' linear terms for the models for μ and σ.

```
m20 <- gamlss(Y~1, family=BB, data=InfMort)# null model
fm2 <- stepGAICAll.A(m20, scope=list(lower=~1,
      upper=~IFDM+illit+lGDP+cli+lpop+PSF+poor) )
```

```
fm2# final model
```

Comment on the model.

(e) Let us try to use smooth terms rather than linear.

```
m20 <- gamlss(Y~1, family=BB, data=InfMort)# null model
sm2 <- stepGAICAll.A(m20, scope=list(lower=~1,
      upper=~pb(IFDM)+pb(illit)+pb(lGDP)+pb(cli)+pb(lpop)+pb(PSF)
      +pb(poor)) )
sm2# final model
```

Compare model `sm2` with model `fm2`. Check your final chosen model using residual diagnostics.

(f) Check whether a smooth surface, taking into account the geographical nature of the data, improves the fit. The following code uses the interface of `gamlss()` with the **mgcv** package:

```
library(gamlss.add)
m5 <- gamlss( Y ~ IFDM + lpop+ga(~s(x,y)), sigma.formula = ~IFDM,
               family = BB, data = InfMort)
```

9

Mixed distributions

CONTENTS

This chapter provides explanation for:

1. a general introduction to mixed distributions and in particular,
2. explicit and generated zero-adjusted distributions defined on the interval $[0, \infty)$,
3. explicit and generated inflated distributions defined on $[0, 1]$, and
4. generalized Tobit model providing mixed distributions defined on any interval, including $[0, \infty)$ and $[0, 1]$.

9.1 Introduction

A mixed distribution, first defined in Section 1.2.3, is a mixture of two components: a continuous distribution and a discrete distribution. It is a continuous distribution where the range of Y also includes specific discrete values, usually on the boundary of the range of Y, with non-zero probabilities. (A mixed distribution is a special case of a finite mixture distribution, described in Stasinopoulos et al. [2017, Chapter 7].)

Here we treat mixed distributions as a special class of distributions because of their general applicability and practical usefulness. Table 9.1 shows the explicit mixed distributions in **gamlss.dist**, others can be generated as explained at the end of Section 9.1.

There are two types of mixed distributions in GAMLSS:

- *Zero-adjusted* distributions on $[0, \infty)$, which are continuous distributions on $(0, \infty)$ with their range extended to include the exact discrete value zero. Hence their range is the interval $[0, \infty)$, where the square bracket indicates that the endpoint zero is included. (Min and Agresti [2002] refer to such distributions as *semicontinuous.*) These are appropriate for a response variable which can either be exactly zero or a continuous positive quantity, and are commonly encountered. Some examples are: daily rainfall, individual weekly alcohol consumption, and total claim amounts on insurance policies over a year. Zero-adjusted distributions are discussed further in Section 9.2.

- *Inflated* distributions on $[0, 1]$, which are continuous distributions, on $(0, 1)$, with their range extended to include the exact discrete value zero and/or one. Hence their range is $[0, 1)$, $(0, 1]$, or $[0, 1]$. The following are some examples of response variables requiring an inflated distribution:

  (i) *Loss given default* (LGD) is the proportion of a credit exposure that is lost if the debtor defaults on a loan. LGD can contain values between zero and one including both endpoints, where zero means that the balance is fully recovered, while one means total loss of exposure.

  (ii) The *visual analogue scale* is used to measure intangible quantities such as pain and quality of life. For the measurement of pain, for example, patients are required to make a mark on a 100 mm scale reflecting their perception of their pain, where the left endpoint has the descriptor "no pain" and the right endpoint "worst pain imaginable". The distance from the left endpoint to the patient's mark is measured and the measurements are scaled to the interval $[0, 1]$. Clearly observations at both endpoints are possible.

  (iii) *Spirometric lung function* is the ratio of forced expiratory volume in one second (FEV_1) to forced vital capacity (FVC), which is an established index for diagnosing airway obstruction. This variable often lies in the range $(0, 1]$.

  Inflated distributions are discussed further in Section 9.3. Note a random variable Y with any fixed range $[a, b]$ where a and b are known, can be transformed to $Z = (Y - a)/(b - a)$ with range [0,1].

Table 9.1 shows the explicit mixed `gamlss.family` distributions. The first five are inflated distributions on $[0, 1]$ (see Section 21.3 for more details), while the last two are zero-adjusted distributions on $[0, \infty)$ (see Chapter 20 for more details). More inflated and zero-adjusted distributions are easily generated and fitted using the **gamlss.inf** functions `gamlssInf0to1()` and `gamlssZadj()`, respectively. Also the generalized Tobit model, which can be used to provide mixed distributions on any interval with point probabilities at the endpoints, including $[0, \infty)$ and $[0, 1]$, can be created using the **gamlss.cens** function `gen.cens()`. These are all described in the next two sections.

TABLE 9.1: Mixed distributions implemented within the **gamlss.dist** package, with default link functions.

Distribution	gamlss name	Range R_Y	Parameter link functions μ	σ	ν	τ
zero-inflated beta	`BEINF0`	$[0, 1)$	logit	logit	log	-
zero-inflated beta	`BEOI`	$[0, 1)$	logit	log	logit	-
one-inflated beta	`BEINF1`	$(0, 1]$	logit	logit	log	-
one-inflated beta	`BEZI`	$(0, 1]$	logit	log	logit	-
zero- and one-inflated beta	`BEINF`	$[0, 1]$	logit	logit	log	log
zero-adjusted gamma	`ZAGA`	$[0, \infty)$	log	log	logit	-
zero-adjusted inv Gaussian	`ZAIG`	$[0, \infty)$	log	log	logit	-

9.2 Zero-adjusted distributions on $[0,\infty)$

This section defines a zero-adjusted distribution on $[0, \infty)$ and describes several ways of fitting a zero-adjusted response variable within the GAMLSS framework. A zero-adjusted distribution with range $[0, \infty)$, denoted here by $Y \sim$ `ZAW`, takes value $Y = 0$ with probability p, and has a continuous distribution `W` with range $(0, \infty)$ with probability $(1-p)$. Hence, for $Y \sim$ `ZAW`, $Y = 0$ with probability p, and $Y = Y_1$ with probability $(1 - p)$, where $Y_1 \sim$ `W`, and $0 < p < 1$. Informally, the mixed probability function of Y is given by

$$f_Y(y) = \begin{cases} p & \text{if } y = 0 \\ (1-p) f_{Y_1}(y) & \text{if } y > 0\ . \end{cases} \tag{9.1}$$

Note that $\mathrm{E}(Y^r) = (1 - p)\mathrm{E}(Y_1^r)$, i.e. $\mu'_{rY} = (1 - p)\mu_r'$, where $\mu'_{rY} = \mathrm{E}(Y^r)$ and $\mu_r' = \mathrm{E}(Y_1^r)$. Hence using equations (2.1) and (2.3), the mean, variance, and third and fourth central moments of $Y \sim$ `ZAW` are given by

$$\begin{aligned} \mu'_{1Y} &= \mathrm{E}(Y) = (1-p)\mathrm{E}(Y_1) = (1-p)\mu'_1 \\ \mu_{2Y} &= \mathrm{Var}(Y) = (1-p)\mathrm{Var}(Y_1) + p(1-p)\left[\mathrm{E}(Y_1)\right]^2 \\ \mu_{3Y} &= (1-p)\left(\mu_3 + 3p\mu_2\mu'_1 + p(2p-1)\mu_1'^3\right) \\ \mu_{4Y} &= (1-p)\left(\mu_4 + 4p\mu_3\mu'_1 + 6p^2\mu_2\mu_1'^2 + p(1-3p+3p^2)\mu_1'^4\right)\ . \end{aligned} \tag{9.2}$$

where $Y_1 \sim W$, $\mu_1' = \mathrm{E}(Y_1)$, and $\mu_{rY} = \mathrm{E}\left\{[Y - \mathrm{E}(Y)]^r\right\}$ and $\mu_r = \mathrm{E}\left\{[Y_1 - \mathrm{E}(Y_1)]^r\right\}$ for $r = 1, 2, 3, 4$.

The cumulative distribution function (cdf) of $Y \sim$ `ZAW` is given by

$$\mathrm{P}(Y \le y) = p + (1-p)\mathrm{P}(Y_1 \le y) \tag{9.3}$$

for $y \ge 0$, where $Y_1 \sim$ `W`.

The moment generating function (mgf) of $Y \sim$ `ZAW` is given by

$$M_Y(t) = p + (1-p)M_{Y_1}(t), \tag{9.4}$$

where $M_{Y_1}(t)$ is the mgf of $Y_1 \sim$ `W`.

9.2.1 Explicit zero-adjusted distributions

There are two explicitly defined zero-adjusted distributions on $[0, \infty)$ in **gamlss.dist**: (i) the zero-adjusted gamma, $\texttt{ZAGA}(\mu, \sigma, \nu)$, and (ii) the zero-adjusted inverse Gaussian, $\texttt{ZAIG}(\mu, \sigma, \nu)$. These two distributions can be used to fit an explicit GAMLSS model. Unfortunately, we have found that for many data sets, both distributions fail to capture some feature of the response distribution adequately. That is understandable since both the gamma and inverse Gaussian distributions have only two parameters and therefore lack sufficient flexibility in some cases. In such cases a generated zero-adjusted distribution, explained in Section 9.2.2, may be more appropriate.

The zero-adjusted gamma, $\texttt{ZAGA}(\mu, \sigma, \nu)$, distribution is a mixture with two components: a value zero with probability ν, and a gamma, $\texttt{GA}(\mu, \sigma)$, distribution on $(0, \infty)$ with probability $(1 - \nu)$.

Hence $Y \sim \texttt{ZAGA}(\mu, \sigma, \nu)$ has mixed probability function given by

$$f_Y(y \mid \mu, \sigma, \nu) = \begin{cases} \nu & \text{if } y = 0 \\ (1-\nu) f_{Y_1}(y \mid \mu, \sigma) & \text{if } y > 0 \end{cases} \tag{9.5}$$

for $y \geq 0$, where $\mu > 0$, $\sigma > 0$ and $0 < \nu < 1$, and where $Y_1 \sim \texttt{GA}(\mu, \sigma)$. See Section 20.1 for details and plots of $\texttt{ZAGA}(\mu, \sigma, \nu)$.

The zero-adjusted inverse Gaussian, $\texttt{ZAIG}(\mu, \sigma, \nu)$, distribution is obtained by replacing the pdf of the gamma distribution, $\texttt{GA}(\mu, \sigma)$, for Y_1 above by the inverse Gaussian, $\texttt{IG}(\mu, \sigma)$, pdf. See Section 20.2 for details and plots of $\texttt{ZAIG}(\mu, \sigma, \nu)$.

The default link functions relating the parameters (μ, σ, ν) of ZAGA or ZAIG to the predictors (η_1, η_2, η_3), which may depend on explanatory variables, are

$$\begin{aligned} \log \mu &= \eta_1 \\ \log \sigma &= \eta_2 \\ \log\left(\frac{\nu}{1-\nu}\right) &= \eta_3 \ . \end{aligned}$$

The model (9.5) is equivalent to a gamma distribution, $\texttt{GA}(\mu, \sigma)$, model for $Y > 0$, together with a binary (or binomial $\texttt{BI}(1, \nu)$) distribution for the recoded variable Y_0 given by

$$Y_0 = \begin{cases} 0 & \text{if } Y > 0 \\ 1 & \text{if } Y = 0 \end{cases} \tag{9.6}$$

i.e.

$$\mathrm{P}(Y_0 = y_0) = \begin{cases} 1-\nu & \text{if } y_0 = 0 \\ \nu & \text{if } y_0 = 1 \ . \end{cases} \tag{9.7}$$

It can be shown easily that the log-likelihood function for the $\texttt{ZAGA}(\mu, \sigma, \nu)$ model (9.5) is equal to the sum of the log-likelihood functions of the gamma model $\texttt{GA}(\mu, \sigma)$ for $Y > 0$ and the binary model (9.7) for Y_0. The ZAGA model can be fitted explicitly in **gamlss** and provides the deviance and randomized quantile residuals for comparing models. Alternatively the gamma model can be fitted after deleting all cases with $y = 0$, and the binary model fitted to the recoded variable Y_0. This methodology is used in the package **gamlss.inf** to fit the generated zero-adjusted distributions described in the next section.

Examples using the `ZAGA` and `ZAIG` distributions on substantive data sets are given in Tong et al. [2013] and Heller et al. [2006], respectively.

9.2.2 Generating a zero-adjusted distribution

Any continuous distribution defined on $(0, \infty)$ can be extended to a zero-adjusted distribution on $[0, \infty)$ by adding a point probability at zero. The resulting mixed probability function takes the form:

$$f_Y(y \mid \boldsymbol{\theta}, \xi_0) = \begin{cases} \xi_0 & \text{if } y = 0 \\ (1 - \xi_0) f_W(y \mid \boldsymbol{\theta}) & \text{if } y > 0 \end{cases} \tag{9.8}$$

where $f_W(\cdot \mid \boldsymbol{\theta})$ is the pdf of any continuous distribution on $(0, \infty)$ with parameters $\boldsymbol{\theta}^\top = (\theta_1, \theta_2, \ldots, \theta_p)$, and $0 < \xi_0 < 1$, where $\xi_0 = \mathrm{P}(Y = 0)$ is the point probability at zero. Note that the point probability ξ_0 does not depend on the parameters $\boldsymbol{\theta}$ and can be modeled independently of $\boldsymbol{\theta}$, potentially giving a very flexible model for $\mathrm{P}(Y = 0)$.

Note that, since a `gamlss.family` distribution in the current implementation can have up to four parameters denoted by $\boldsymbol{\theta}^\top = (\mu, \sigma, \nu, \tau)$, the total number of parameters of a zero-adjusted distribution can be up to five, denoted as $(\mu, \sigma, \nu, \tau, \xi_0)$. With the extra ξ_0 parameter, the GAMLSS regression model takes the form:

$$\begin{aligned}
Y &\overset{\text{ind}}{\sim} \mathcal{D}(\boldsymbol{\mu}, \boldsymbol{\sigma}, \boldsymbol{\nu}, \boldsymbol{\tau}, \boldsymbol{\xi}_0) \\
\boldsymbol{\eta}_1 &= g_1(\boldsymbol{\mu}) = \mathbf{X}_1\boldsymbol{\beta}_1 + s_{11}(\mathbf{x}_{11}) + \ldots + s_{1J_1}(\mathbf{x}_{1J_1}) \\
\boldsymbol{\eta}_2 &= g_2(\boldsymbol{\sigma}) = \mathbf{X}_2\boldsymbol{\beta}_2 + s_{21}(\mathbf{x}_{21}) + \ldots + s_{2J_2}(\mathbf{x}_{2J_2}) \\
\boldsymbol{\eta}_3 &= g_3(\boldsymbol{\nu}) = \mathbf{X}_3\boldsymbol{\beta}_3 + s_{31}(\mathbf{x}_{31}) + \ldots + s_{3J_3}(\mathbf{x}_{3J_3}) \\
\boldsymbol{\eta}_4 &= g_4(\boldsymbol{\tau}) = \mathbf{X}_4\boldsymbol{\beta}_4 + s_{41}(\mathbf{x}_{41}) + \ldots + s_{4J_4}(\mathbf{x}_{4J_4}) \\
\boldsymbol{\eta}_5 &= g_5(\boldsymbol{\xi}_0) = \mathbf{X}_5\boldsymbol{\beta}_5 + s_{51}(\mathbf{x}_{51}) + \ldots + s_{5J_5}(\mathbf{x}_{5J_5})
\end{aligned} \tag{9.9}$$

where $\mathcal{D}(\boldsymbol{\mu}, \boldsymbol{\sigma}, \boldsymbol{\nu}, \boldsymbol{\tau}, \boldsymbol{\xi}_0)$ is a zero-adjusted distribution, and $\mathbf{X}_k$, $\boldsymbol{\beta}_k$, $s_{kj}(\mathbf{x}_{kj})$ and $\boldsymbol{\eta}_j$ have their usual meanings, see Section 3.1. Since $\xi_0 = \mathrm{P}(Y = 0)$ has range $0 < \xi_0 < 1$, the link function $g_5(\cdot)$ should be a logit or similar function.

The package **gamlss.inf** implements model (9.9). The function `gen.Zadj()` is used to generate a zero-adjusted distribution by combining any `gamlss.family` distribution defined on $(0, \infty)$ and a point probability at zero. A generated zero-adjusted response variable model is fitted using the function `gamlssZadj()`. Because of the orthogonality of the parameter ξ_0 with the rest of the distribution parameters, the model can be fitted by maximizing the log-likelihood separately for ξ_0 and for (μ, σ, ν, τ). Maximizing the log-likelihood for (μ, σ, ν, τ) is achieved by fitting a GAMLSS model after deleting all cases with $y = 0$, and maximizing for ξ_0 by fitting a binary model to variable Y_0 defined in equation (9.6). The deviances of the two models are then combined to produce the deviance of the zero-adjusted model, and the residuals recalculated to produce the randomized quantile residuals of the zero-adjusted model. More details can be found in the **gamlss.inf** package vignettes (available on the GAMLSS web site `www.gamlss.com`).

Note that we have argued above that any distribution defined on $(0, \infty)$ can be extended to a zero-adjusted version. This includes distributions on $(0, \infty)$ which have been generated using the log transformation (Section 5.6.1) or by truncation (Section 5.6.2). The example in Section 9.2.4 demonstrates this.

9.2.3 The generalized Tobit model

The 'Tobit' model was first proposed by Tobin [1958], who used it to model household expenditure on durable goods using household income. The original Tobit model is based on the idea that the response variable Y is related to the random variable Z as

$$Y = \begin{cases} Z & \text{if } Z > 0 \\ 0 & \text{otherwise} \end{cases} \tag{9.10}$$

where $Z \sim \texttt{NO}(\mu, \sigma)$. That is, Y is equal to Z if Z is positive and equal to zero if not. Figure 9.1(a) illustrates the original Tobit model. The probability that Y is equal to zero is given by the probability that Z is less or equal to zero, i.e. $\text{P}(Y = 0) = \text{P}(Z \leq 0)$, which is shown in Figure 9.1(a) as the left shaded area for $Z \leq 0$ and also as the point probability at zero for Y. The model can be fitted by treating any zero value of the response variable Y as a left-censored observation.

There are several variations of the Tobit model in the literature depending on the range of the values for the response variable and the assumed distribution for Z. We will refer to any model in which the assumed distribution is any (explicit or generated) continuous `gamlss.family` distribution and where any left or right restriction is imposed on the range of the response variable, as the *generalized Tobit model.* Figure 9.1(b) shows an example of a generalized Tobit model. The distribution for Z is the skew t type 3 distribution, i.e. $Z \sim \texttt{SST}(\mu.\sigma, \nu, \tau)$. The range of Y is restricted to $[0, 1]$ and the point probabilities for Y are $P(Y = 0) = \text{P}(Z \leq 0)$ and $\text{P}(Y = 1) = \text{P}(Z \geq 1)$. These probabilities are shown as the left and right shaded areas below the `SST` pdf curve for Z and also as point probabilities at zero and one for Y, respectively. Note that the generalized Tobit model can be used as a general method for fitting a distribution to any bounded continuous response variable. Demonstrations of the fitting of generalized Tobit models are given in Sections 9.2.4 and 9.3.4.

A very important property of the generalized Tobit model is that the point probabilities (at the endpoints of the range of Y) depend on the parameters of the continuous distribution for Z. This clearly restricts their flexibility.

9.2.4 Example of fitting zero-adjusted distributions

The data we analyze here are motor vehicle insurance data, i.e. claim amounts on motor vehicle insurance policies over a one-year period. The data frame `mvi` is a sample of 2000 observations from the data frame `mviBig` which has 67,143 observations.

Policies which did not have a claim during the observation period have `numclaims`=0 and `claimcst0`=0; policies which had at least one claim have `numclaims`$\geq$ 1 and

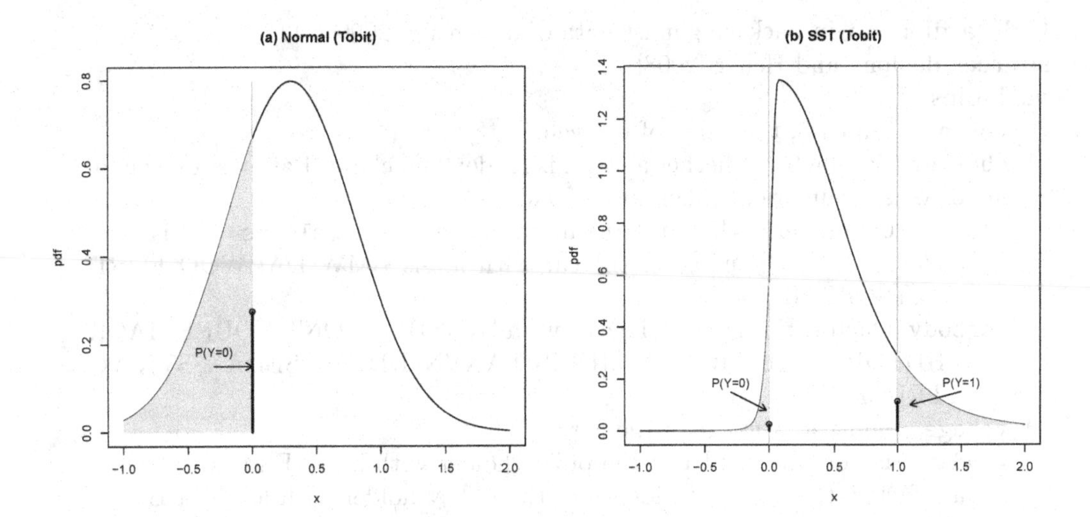

FIGURE 9.1: Mixed probability functions for Y resulting from (a) the original Tobit model and (b) a generalized Tobit model. The original Tobit model in (a) assumes that $Z \sim$ NO$(\mu = 0.3,\ \sigma = 0.5)$, and that Y values equal to zero are treated as left-censored at zero Z observations. The generalized Tobit model (b) assumes that $Z \sim$ SST$(\mu = 0.5,\ \sigma = 0.45,\ \nu = 3,\ \tau = 5)$, and that Y values at zero are treated as left-censored at zero Z observations, and Y values at one as right-censored at one Z observations.

`claimcst0`>0. Note that any substantive analysis of the data should take into account the variable `exposure`, which is the proportion of the observation period for which the policy was active. However here we ignore `exposure` and use the data as an illustration of fitting a zero-adjusted distribution for the response variable total amount of claim, `claimcst0` with range $[0, \infty)$. Variables `retval` and `agecat` are not used in the analysis below.

R data file: `mvi` in package **gamlss.data** of dimension 2000×11
source: de Jong and Heller [2008]
variables

`retval` : numeric, the value of the vehicle
`whetherclm` : factor, whether a claim is made: 0 no claim, 1 at least one claim
`numclaims` : numeric, the number of claims
`claimcst0` : numeric, the total amount of claim. For `numclaims`=0 it is zero.
`vehmake` : factor, the make of the car with levels BMW DAEWOO FORD MITSUBISHI
`vehbody` : factor, the type of the car, with levels BUS CONT COUPE HACK HDTOP HRSE MCARA MIBUS PANVN RDSTR SEDAN STNWG TRUCK UTE
`vehage` : numeric, the age of the car
`gender` : factor, the gender of the policy holder with levels F M
`area` : factor, the area of residence of the policy holder with levels A B C D E F
`agecat` : factor, the age band of the policy holder with levels 1 2 3 4 5 6 (one is youngest)
`exposure` : numeric, the time of exposure with values from zero to one

purpose: to demonstrate the use of inflated distributions in GAMLSS.

Fitting explicit zero-adjusted distributions

We start by fitting the two explicit distributions `ZAGA`(μ, σ, ν) and `ZAIG`(μ, σ, ν). Note that in the analysis below, models for σ were not considered; we model μ (the mean of the non-zero claims) and ν (the probability of a zero claim).

```
data(mvi)
# zero-adjusted GA
m1 <- gamlss(claimcst0~vehmake+vehbody+vehage+gender+area,
             nu.fo=~vehmake+vehbody+vehage+gender+area,
             family=ZAGA, data=mvi, trace=FALSE )
# zero-adjusted IG
m2 <- gamlss(claimcst0~vehmake+vehbody+vehage+gender+area,
             nu.fo=~vehmake+vehbody+vehage+gender+area,
             family=ZAIG, data=mvi, trace=FALSE )
```

```
AIC(m1,m2)

##    df      AIC
## m2 47 3319.317
## m1 47 3323.011
```

The `ZAIG`(μ, σ, ν) appears to provide a better fit, according to AIC.

Fitting generated zero-adjusted distributions

We now use the `gamlssZadj()` function of the package **gamlss.inf** to fit a zero-adjusted generalized inverse Gaussian, which we denote as `GIGZadj`$(\mu, \sigma, \nu, \xi_0)$, and a zero-adjusted `BCTo`, `BCToZadj`$(\mu, \sigma, \nu, , \tau, \xi_0)$, distribution. Some features of the `gamlssZadj()` function should be noted here. The first is that the response variable and the μ formula are each declared separately, not as part of the μ formula as in `gamlss()`. The second is that the `family` argument takes the original `gamlss.family` distribution, not the zero-adjusted one. (In addition note that the number of cycles (`n.cyc`) in the zero-adjusted `GIG` model below is increased to 100.)

```
library(gamlss.inf)
# zero-adjusted GIG
m3 <- gamlssZadj(claimcst0,
            mu.fo= ~vehmake+vehbody+vehage+gender+area,
            xi0.fo= ~vehmake+vehbody+vehage+gender+area,
            family=GIG, data=mvi, trace=FALSE, n.cyc=100)
# zero-adjusted BCTo
m4 <- gamlssZadj(claimcst0,
            mu.fo=~vehmake+vehbody+vehage+gender+area,
            xi0.fo=~vehmake+vehbody+vehage+gender+area,
            family=BCTo, data=mvi, trace=FALSE)
```

```
AIC(m1,m2,m3,m4)

##    df      AIC
## m3 48 3299.820
## m4 49 3301.129
## m2 47 3319.317
## m1 47 3323.011
```

The `GIGZadj` distribution seems to be better than the `ZAIG` and slightly better than the `BCToZadj`.

Fitting generated zero-adjusted truncated distributions

The zero-adjusted truncated `TF`(μ, σ) and `SST`(μ, σ, ν, τ) distributions, denoted by `TFtrZadj`$(\mu, \sigma, \nu, \xi_0,)$ and `SSTtrZadj`$(\mu, \sigma, \nu, \tau, \xi_0)$, respectively, are fitted below.

```
library(gamlss.tr)
gen.trun(0, "TF", type="left")

## A truncated family of distributions from TF has been generated
##  and saved under the names:
##  dTFtr pTFtr qTFtr rTFtr TFtr
## The type of truncation is left
##  and the truncation parameter is 0
```

```
m5 <- gamlssZadj(claimcst0,
            mu.fo=~vehmake+vehbody+vehage+gender+area,
            xi0.fo=~vehmake+vehbody+vehage+gender+area,
            family=TFtr, data=mvi, trace=FALSE, n.cyc=100)
```

```
gen.trun(0, "SST", type="left")

## A truncated family of distributions from SST has been generated
##  and saved under the names:
##  dSSTtr pSSTtr qSSTtr rSSTtr SSTtr
## The type of truncation is left
##  and the truncation parameter is 0
```

```
m6 <- gamlssZadj(claimcst0,
            mu.fo=~vehmake+vehbody+vehage+gender+area,
            xi0.fo=~vehmake+vehbody+vehage+gender+area,
            family=SSTtr, data=mvi, trace=FALSE )
```

The zero-adjusted truncated SST improves the AIC.

Fitting a generalized Tobit model

Two generalized Tobit models are fitted here, using: (a) the t family distribution TF(μ, σ, ν), and (b) the skew t type 3, SST(μ, σ, ν, τ). Note that the function `Surv()` below generates a response variable which is left-censored at zero.

```
# create the survival response variable
library(survival)
mvi$claimcst0surv<- with(mvi, Surv(claimcst0, claimcst0!=0, type="left"))
# create the distributions for fitting generalized Tobit models
library(gamlss.cens)
# TF distribution
gen.cens("TF", type="left")

## A censored family of distributions from TF has been generated
##  and saved under the names:
##  dTFlc pTFlc qTFlc TFlc
## The type of censoring is left

# SST distribution
gen.cens("SST", type="left")

## A censored family of distributions from SST has been generated
##  and saved under the names:
##  dSSTlc pSSTlc qSSTlc SSTlc
## The type of censoring is left
```

```
# fit the  models
# gereralized Tobit TF
m7 <- gamlss(claimcst0surv~vehmake+vehbody+vehage+gender+area,
               family=TFlc, data=mvi, trace=FALSE, n.cyc=400)
# generalized Tobit SST
m8 <- gamlss( claimcst0surv~vehmake+vehbody+vehage+gender+area,
               family=SSTlc, data=mvi, n.cyc=500, trace=FALSE,
               start.from=m7)
```

```
GAIC(m1, m2 ,m3 ,m4, m5, m6, m7, m8)

##    df      AIC
## m7 25 3274.383
## m6 49 3274.871
## m8 26 3277.191
## m3 48 3299.820
## m4 49 3301.129
## m5 48 3303.840
```

```
## m2 47 3319.317
## m1 47 3323.011
```

Based on the AIC the generalized Tobit `TF` model `m7` is best, with the zero-adjusted truncated `SST` model `m6` second. The generalized Tobit models use fewer degrees of freedom than the rest, since no model for the probability at zero is fitted. It is a feature of any generalized Tobit model that there is no way of explicitly modeling the zero probability as a function of covariates. The zero probability is given in general by $F(0 \mid \hat{\mu}, \hat{\sigma}, \hat{\nu}, \hat{\tau})$, that is, the fitted cdf evaluated at zero. Therefore, the probability at zero is a function of the parameters which themselves can be functions of explanatory variables, but the relationship is implicit rather than explicit. Depending on the data and the interpretation that is required, that can be an advantage, since it results in a simpler model. On the other hand it can also be a disadvantage, since no simple formula for the probability of zero results and hence no simple mechanism for explaining how the zeros are generated is provided. For the current data modeling, the probability of no claims is an important part of the model and therefore models `m3` and `m6` are easier to interpret. Note, however, that we have not made any attempt here to select appropriate terms for all the parameters of the distribution. (This is left as an exercise at the end of the chapter.)

Figure 9.2 provides an idea on how well the best model, using each of the four different approaches described above, fits. It shows worm plots for models:

(a) `m2`, the zero-adjusted inverse Gaussian distribution `ZAIG`(μ, σ, ν),

(b) `m3`, the zero-adjusted generalized inverse Gaussian distribution `GIGZadj`$(\mu, \sigma, \nu, \xi_0)$,

(c) `m6`, zero-adjusted truncated `SST` distribution `SSTtrZajd`$(\mu, \sigma, \nu, \xi_0)$ and

(d) `m7`, generalized Tobit `TF` distribution `TF1c`(μ, σ, ν).

All models, apart from possibly model `m2`, show reasonably adequate fits. The adequacy of the models can be further explored using multiple worm plots and Q statistics, see Chapter 12 of Stasinopoulos et al. [2017].

```
wp(m2, ylim.all=1); title('(a) m2')
wp(m3, ylim.all=1); title('(b) m3')
wp(m6, ylim.all=1); title('(c) m6')
wp(m7, ylim.all=1); title('(d) m7')
```

9.3 Inflated distributions on [0,1]

9.3.1 Introduction

There are three types of inflated distribution on $[0, 1]$ in GAMLSS. These are mixed distributions which are continuous in $(0, 1)$ with additional point probabilities at zero, one, or both.

case 1: inflated at zero, Figure 9.3(a),

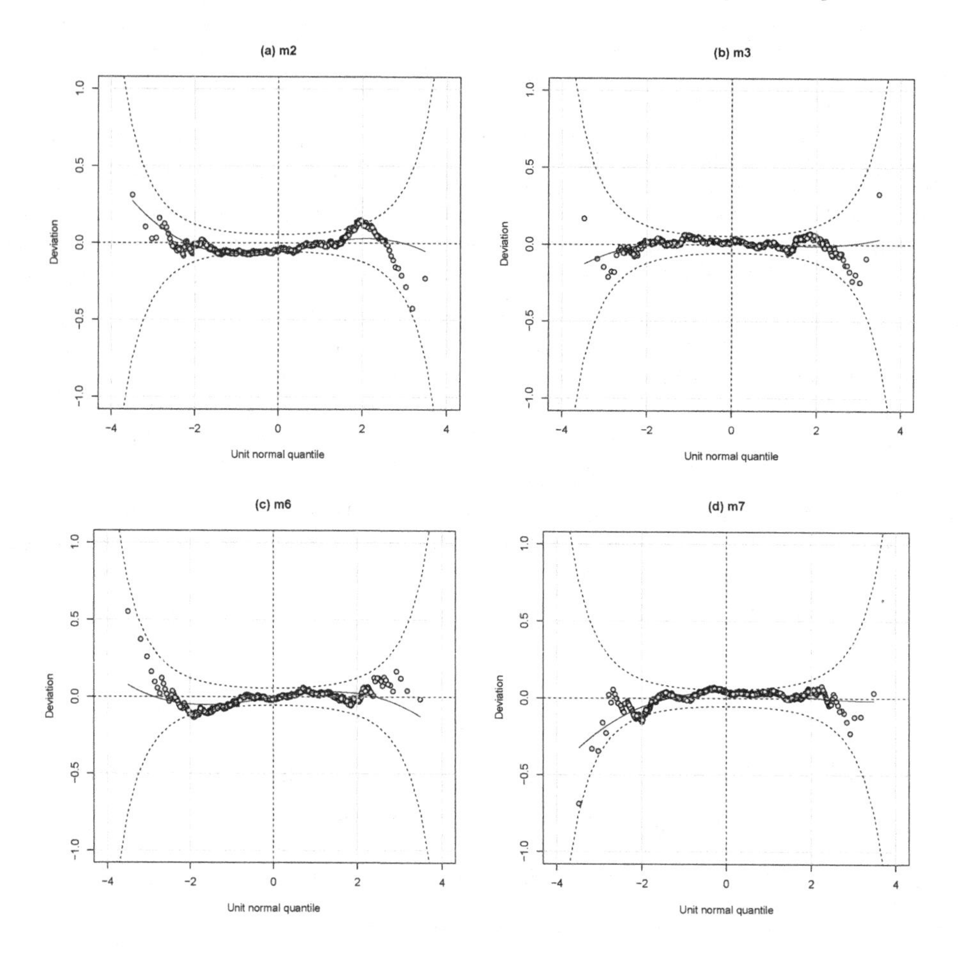

FIGURE 9.2: The worm plots of the residuals from model (a) m2, the zero-adjusted inverse Gaussian distribution, (b) m3, the zero-adjusted generalized inverse Gaussian distribution, (c) m6, the zero-adjusted truncated SST distribution, and (d) m7, the generalized Tobit TF distribution.

case 2: inflated at one, Figure 9.3(b), and

case 3: inflated at both zero and one, Figure 9.3(c).

Let $f_W(y \mid \boldsymbol{\theta})$ be any continuous distribution defined on the range $(0,1)$.

For case 1, the mixed probability function for Y, with range $[0,1)$, is given by:

$$f_Y(y \mid \boldsymbol{\theta}, \xi_0) = \begin{cases} \xi_0 & \text{if } y = 0 \\ (1-\xi_0) f_W(y \mid \boldsymbol{\theta}) & \text{if } 0 < y < 1 \end{cases} \tag{9.11}$$

where the probability at zero is $\mathrm{P}(Y=0) = \xi_0$ and $0 < \xi_0 < 1$.

For case 2, the mixed probability function for Y, with range $(0,1]$, is given by:

$$f_Y(y \mid \boldsymbol{\theta}, \xi_1) = \begin{cases} (1-\xi_1) f_W(y \mid \boldsymbol{\theta}) & \text{if } 0 < y < 1 \\ \xi_1 & \text{if } y = 1 \end{cases} \tag{9.12}$$

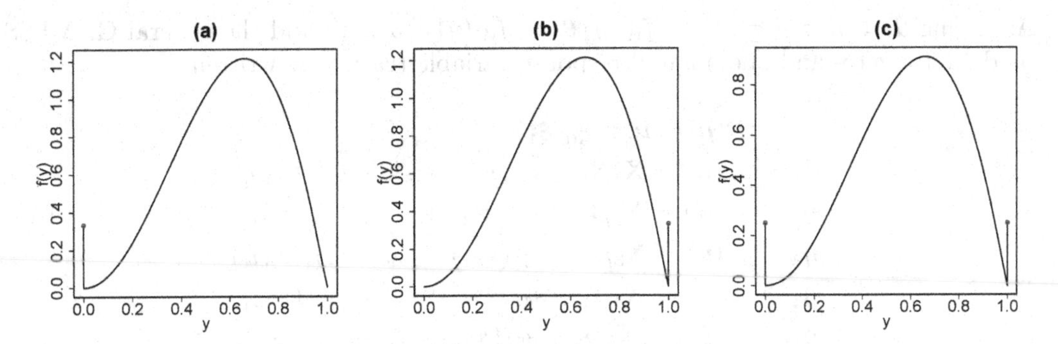

FIGURE 9.3: The three different types of inflated models: (a) zero-inflated, (b) one-inflated and (c) zero- and one-inflated.

where $P(Y = 1) = \xi_1$ and $0 < \xi_1 < 1$.

For case 3, inflated at both zero and one, the mixed probability function for Y, with range $[0, 1]$, is given by:

$$f_Y(y \mid \boldsymbol{\theta}, p_0, p_1) = \begin{cases} p_0 & \text{if } y = 0 \\ (1 - p_0 - p_1) f_W(y \mid \boldsymbol{\theta}) & \text{if } 0 < y < 1 \\ p_1 & \text{if } y = 1 \end{cases} \tag{9.13}$$

where $P(Y = 0) = p_0$, $P(Y = 1) = p_1$ and therefore $0 < p_0 < 1$, $0 < p_1 < 1$ and $0 < p_0 + p_1 < 1$. This formulation presents some difficulties, in that (i) the parameters p_0 and p_1 are not orthogonal, and their MLEs are strongly negatively correlated; and (ii) the constraint $p_0 + p_1 < 1$ is awkward to implement in the regression framework, where models for p_0 and p_1 may be specified. We have reparameterized (9.13) to the following model:

$$f_Y(y \mid \boldsymbol{\theta}, \xi_0, \xi_1) = \begin{cases} \frac{\xi_0}{1+\xi_0+\xi_1} & \text{if } y = 0 \\ \frac{1}{1+\xi_0+\xi_1} f_W(y \mid \boldsymbol{\theta}) & \text{if } 0 < y < 1 \\ \frac{\xi_1}{1+\xi_0+\xi_1} & \text{if } y = 1 \end{cases} \tag{9.14}$$

where $\xi_0 > 0$, $\xi_1 > 0$ and

$$p_0 = P(Y = 0) = \frac{\xi_0}{1 + \xi_0 + \xi_1},$$
$$p_1 = P(Y = 1) = \frac{\xi_1}{1 + \xi_0 + \xi_1},$$

hence $\xi_0 = p_0/p_2$ and $\xi_1 = p_1/p_2$, where $p_2 = 1 - p_0 - p_1 = P(0 < Y < 1)$. The parameters ξ_0 and ξ_1 do not present the difficulties associated with p_0 and p_1 in model (9.13). While in cases 1 and 2, ξ_0 and ξ_1 are probabilities, that is not true for case 3 and the interpretation for these parameters is not as straightforward.

Assuming $\boldsymbol{\theta} = (\mu, \sigma, \nu, \tau)$, then $f_W(y \mid \boldsymbol{\theta}) = f_W(y \mid \mu, \sigma, \nu, \tau)$ and the general GAMLSS model for a zero- and one-inflated response variable (case 3) is written as:

$$
\begin{aligned}
Y &\stackrel{\text{ind}}{\sim} \mathcal{D}(\boldsymbol{\mu}, \boldsymbol{\sigma}, \boldsymbol{\nu}, \boldsymbol{\tau}, \boldsymbol{\xi}_0, \boldsymbol{\xi}_1) \\
\boldsymbol{\eta}_1 &= g_1(\boldsymbol{\mu}) = \mathbf{X}_1\boldsymbol{\beta}_1 + s_{11}(\mathbf{x}_{11}) + \ldots + s_{1J_1}(\mathbf{x}_{1J_1}) \\
\boldsymbol{\eta}_2 &= g_2(\boldsymbol{\sigma}) = \mathbf{X}_2\boldsymbol{\beta}_2 + s_{21}(\mathbf{x}_{21}) + \ldots + s_{2J_2}(\mathbf{x}_{2J_2}) \\
\boldsymbol{\eta}_3 &= g_3(\boldsymbol{\nu}) = \mathbf{X}_3\boldsymbol{\beta}_3 + s_{31}(\mathbf{x}_{31}) + \ldots + s_{3J_3}(\mathbf{x}_{3J_3}) \\
\boldsymbol{\eta}_4 &= g_4(\boldsymbol{\tau}) = \mathbf{X}_4\boldsymbol{\beta}_4 + s_{41}(\mathbf{x}_{41}) + \ldots + s_{4J_4}(\mathbf{x}_{4J_4}) \\
\boldsymbol{\eta}_5 &= g_5(\boldsymbol{\xi}_0) = \mathbf{X}_5\boldsymbol{\beta}_5 + s_{51}(\mathbf{x}_{51}) + \ldots + s_{5J_5}(\mathbf{x}_{5J_5}) \\
\boldsymbol{\eta}_6 &= g_6(\boldsymbol{\xi}_1) = \mathbf{X}_6\boldsymbol{\beta}_6 + s_{61}(\mathbf{x}_{61}) + \ldots + s_{6J_6}(\mathbf{x}_{6J_6})
\end{aligned} \tag{9.15}
$$

where $\mathcal{D}(\boldsymbol{\mu}, \boldsymbol{\sigma}, \boldsymbol{\nu}, \boldsymbol{\tau}, \boldsymbol{\xi}_0, \boldsymbol{\xi}_1)$ is a zero- and one-inflated distribution and $\mathbf{X}_k$, $\boldsymbol{\beta}_k$, $s_{kj}(\mathbf{x}_{kj})$ and $\boldsymbol{\eta}_k$ have their usual meanings, see Section 3.1. The default link functions for ξ_0 and ξ_1 in (9.15) are $\eta_5 = \log(\xi_0)$ and $\eta_6 = \log(\xi_1)$, since both parameters are constrained to be positive. Models of the type in equations (9.11 and (9.12) are discussed in Hossain et al. [2015, 2016a] and Hossain [2017], while those in equation (9.13) are discussed in Hossain et al. [2016b] and Hossain [2017].

Case 1, the zero-inflated model, can be fitted similarly to the zero-adjusted models in Section 9.2. A binary model (`BI`) is fitted with response variable taking value 1 if $y = 0$ and zero otherwise, and a continuous $(0, 1)$ distribution is fitted after deleting all cases with $y = 0$. Equivalently, case 2, one-inflated models, can be fitted with a binary response variable with value 1 if $y = 1$ and zero otherwise, and a $(0, 1)$ distribution after deleting all cases with $y = 1$. For case 3, when the inflation appears in both zero and one, a different approach is needed. Instead of a binary model, a multinomial model with three levels, `MN3`, with response variable taking values 0 if $y = 0$, 1 if $y = 1$, and 2 if $0 < y < 1$ can be used, see Section 23.4. Then a $(0, 1)$ continuous distribution is fitted after deleting all cases having $y = 0$ or $y = 1$. The above fits are possible because the likelihood function of an inflated response factorizes into two components. One component is a binomial (or multinomial) likelihood and the other is a likelihood for a continuous random variable with range $(0, 1)$. This methodology is used in the package **gamlss.inf** to fit the generated inflated distribution models described in Section 9.3.3.

9.3.2 Explicit inflated distributions

All explicitly defined inflated distributions in **gamlss** are related to the beta distribution `BE`(μ, σ). The explicit distributions, listed by their cases, are:

case 1: `BEINF0`(μ, σ, ν) and `BEZI`(μ, σ, ν). `BEZI`(μ, σ, ν) was developed by Ospina and Ferrari [2012] and is a reparameterized version of `BEINF0`(μ, σ, ν).

case 2: `BEINF1`(μ, σ, ν) and `BEOI`(μ, σ, ν). `BEOI`(μ, σ, ν) was developed by Ospina and Ferrari [2012] and is a reparameterized version of `BEINF1`(μ, σ, ν).

case 3: `BEINF`(μ, σ, ν, τ)

Figure 9.3 shows instances of the above distributions.

- For $Y \sim$ `BEINF`(μ, σ, ν, τ) the mixed probability function is given by model (9.14), with $\xi_0 = \nu$, $\xi_1 = \tau$ and $W \sim$ `BE`(μ, σ) with pdf $f_W(\cdot \mid \mu, \sigma)$ as given by equation (21.1):

$$f_Y(y \mid \mu, \sigma, \nu, \tau) = \begin{cases} \frac{\nu}{1+\nu+\tau} & \text{if } y = 0 \\ \\ \frac{1}{1+\nu+\tau} f_W(y \mid \mu, \sigma) & \text{if } 0 < y < 1 \\ \\ \frac{\tau}{1+\nu+\tau} & \text{if } y = 1 \end{cases} \tag{9.16}$$

where $0 < \mu < 1$, $0 < \sigma < 1$, $\nu > 0$ and $\tau > 0$. The correspondence with the point probabilities in (9.13) is $p_0 = \nu/(1 + \nu + \tau)$ and $p_1 = \tau/(1 + \nu + \tau)$. Hence $\nu = p_0/(1 - p_0 - p_1)$ and $\tau = p_1/(1 - p_0 - p_1)$. The default link functions are logit for μ and σ, and log for ν and τ. See Section 21.3.1 for more details about `BEINF`.

- For $Y \sim$ `BEINF0`(μ, σ, ν), set $\tau = 0$ (and hence $p_1 = 0$) in (9.16). Therefore $\nu = p_0/(1 - p_0)$ and ν is interpreted as the odds of a zero value. See Section 21.3.2 for more details about `BEINF0`.
- For $Y \sim$ `BEINF1`(μ, σ, ν) set $\nu = 0$ (and hence $p_0 = 0$) and then set $\tau = \nu$ in (9.16). In this case $\nu = p_1/(1 - p_1)$ is the odds of a one. See Section 21.3.3 for more details about `BEINF1`.

The `BEZI`(μ, σ, ν) and `BEOI`(μ, σ, ν) distributions are based on the work of Ospina and Ferrari [2012], who use a parameterization for the beta part of the distribution that is different to `BE`(μ, σ). Their mixed probability functions are, as in models (9.11) and (9.12), respectively.

For $Y \sim$ `BEZI`(μ, σ, ν),

$$f_Y(y \mid \mu, \sigma, \nu) = \begin{cases} \nu & \text{if } y = 0 \\ (1 - \nu) f_W(y \mid \mu, \sigma) & \text{if } 0 < y < 1 \end{cases} \tag{9.17}$$

and for $Y \sim$ `BEOI`(μ, σ, ν),

$$f_Y(y \mid \mu, \sigma, \nu) = \begin{cases} (1 - \nu) f_W(y \mid \mu, \sigma) & \text{if } 0 < y < 1 \\ \nu & \text{if } y = 1 \end{cases} \tag{9.18}$$

where $0 < \mu < 1$, $\sigma > 0$ and $0 < \nu < 1$ and $f_W(\cdot \mid \mu, \sigma)$ is the beta distribution pdf as given in Ospina and Ferrari [2012]. In this case ν is the probability at zero and one, in (9.17), and (9.18) respectively, and the default link for ν is logit. The default link functions for μ and σ are logit and log, respectively.

Note that the log-likelihood function for the `BEINF` model (9.16) is equal to the sum of the log-likelihood functions of a beta (`BE`) model and a multinomial (`MN3`) model. That is, the `BEINF` model can be fitted in two parts: fit an `MN3` model to a response variable taking value 0 if $y = 0$, 1 if $y = 1$ and 2 if $0 < y < 1$, and a `BE` model to cases $0 < y < 1$. While this technique is used to fit the generated inflated distributions described in Section 9.3.3, it is not used for the explicit distributions described in this section, where the standard **gamlss** maximization procedure is used.

9.3.3 Generating an inflated distribution

The idea here is to generate an inflated distribution from any `gamlss.family` distribution by taking a continuous distribution with range $(0, 1)$ for W and adding point probabilities at zero and/or one. This is achieved by the function `gen.Inf0to1()` of the package **gamlss.inf** which has two arguments, `family` and `type.of.Inflation`. The first specifies a distribution family on $(0, 1)$, for which any explicit, logit, or truncated distribution defined on $(0, 1)$ can be used (see Sections 3.4.2 and 3.4.3). The second specifies the type of inflation, for which the options are `"Zero"`, `"One"`, and `"Zero&One"`.

In the example below first we take the skew t-family distribution, `SST` (defined on $(-\infty, \infty)$) and use `gen.Family()` to generate the distribution `logitSST` defined on $(0, 1)$. Then, by using `gen.Inf0to1()` on the generated `logitSST` distribution, a zero- and one-inflated `logitSST` distribution, denoted `logitSSTInf0to1`, is created.

```
library(gamlss.inf)
gen.Family(family="SST", type="logit")

## A  logit  family of distributions from SST has been generated
##   and saved under the names:
##   dlogitSST plogitSST qlogitSST rlogitSST logitSST

gen.Inf0to1(family="logitSST",  type.of.Inflation="Zero&One")

## A  0to1 inflated logitSST distribution has been generated
##   and saved under the names:
##   dlogitSSTInf0to1 plogitSSTInf0to1 qlogitSSTInf0to1 rlogitSSTInf0to1
##   plotlogitSSTInf0to1
```

The `d`, `p`, `q` and `r` functions for the `logitSSTInf0to1` distribution are created, as well as the plotting function. Figure 9.4 shows several realizations of the `logitSSTInf0to1`$(\mu, \sigma, \nu, \tau, \xi_0, \xi_1)$ distribution, which were obtained using the function `plotlogitSSTInf0to1()` as, for example

```
plotlogitSSTInf0to1(mu= 1, sigma=1, nu=1, tau=10, xi0=.1, xi1=.2)
```

To fit a generated inflated distribution regression model, the function `gamlssInf0to1()` is used. A demonstration of this is given in Section 9.3.4.

9.3.4 Example of fitting an inflated distribution

In this section we model 3164 observations of male lung function related to data previously analyzed by Stanojevic et al. [2008]. The response variable `slf` $= FEV_1/FVC$ is the ratio of forced expiratory volume in one second (FEV_1) to forced vital capacity (FVC). Spirometric lung function is an established index for diagnosing airway obstruction (see, for example Quanjer et al. [2010]) and the explanatory variable is height. The data are plotted in Figure 9.5. The purpose here is to create centile curves of `slf` against height; it is clear from Figure 9.5 that the response variable `slf` is bounded above at one and includes many cases with `slf=1`. More details on the analysis of the FEV_1/FVC data using **gamlss** packages can be found in Hossain et al. [2016a] and Hossain [2017].

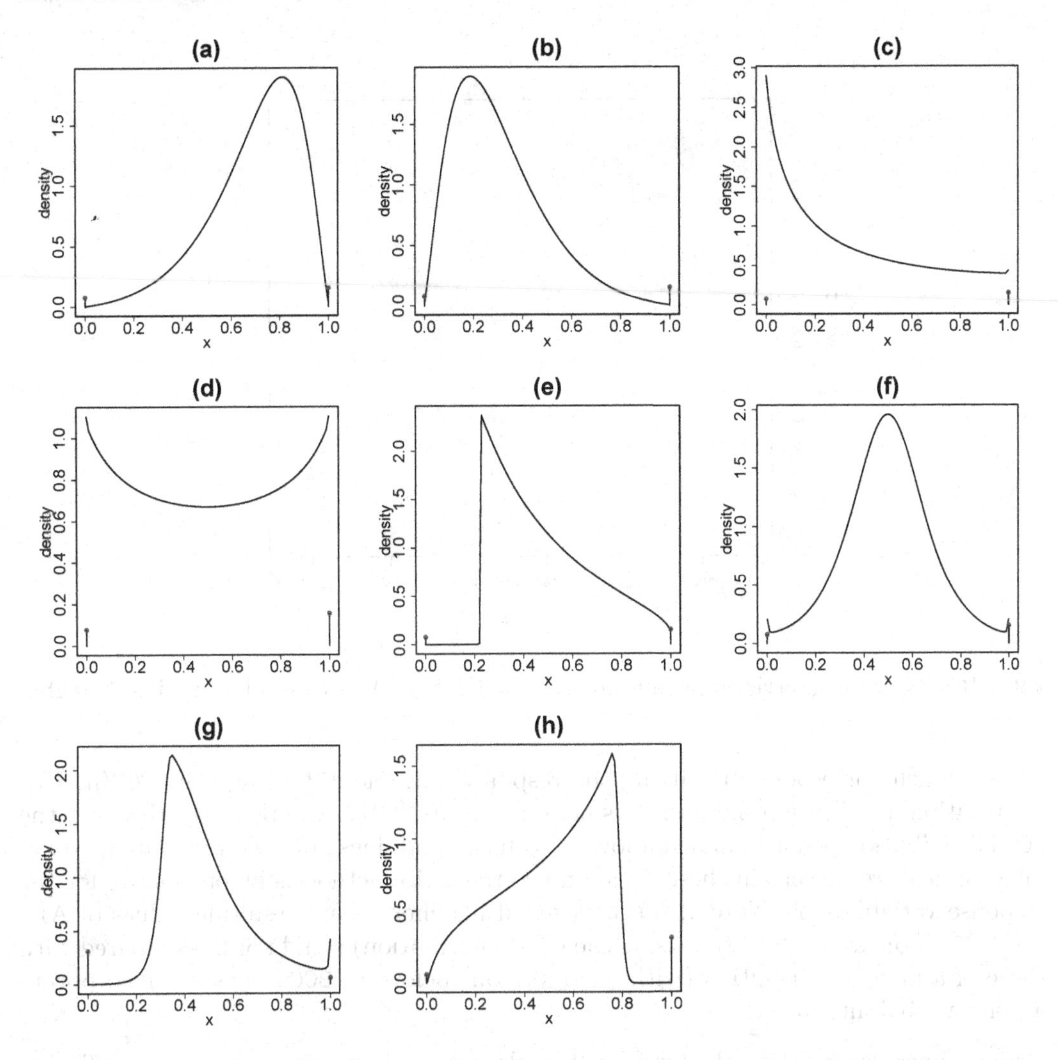

FIGURE 9.4: The zero- and one-inflated logit-SST distribution with $(\mu, \sigma, \nu, \tau, \xi_0, \xi_1) =$ (a) $(1, 1, 1, 10, 0.1, 0.2)$, (b) $(-1, 1, 1, 10, 0.1, 0.2)$, (c) $(-1, 2, 1, 10, 0.1, 0.2)$, (d) $(0, 2, 1, 10, 0.1, 0.2)$, (e) $(0, 1, 10, 10, 0.1, 0.2)$, (f) $(0, 1, 1, 3, 0.1, 0.2)$, (g) $(0, 1, 2, 3, 0.5, 0.1)$, (h) $(0, 1, 0.3, 100, 0.1, 0.5)$.

R data file: `lungFunction` in package **gamlss.data** of dimension 3164×3
source: Stanojevic et al. [2008]
var `slf` : spirometric lung function FEV_1/FVC
`height` : height in centimeters
`age` : subject's age
purpose: to demonstrate the use of inflated distributions in GAMLSS.

```
plot(slf~height, data=lungFunction, pch=20, col="darkgrey",
     ylab="FEV1/FVC")
```

The standard method for centile estimation is the LMS method of Cole and Green

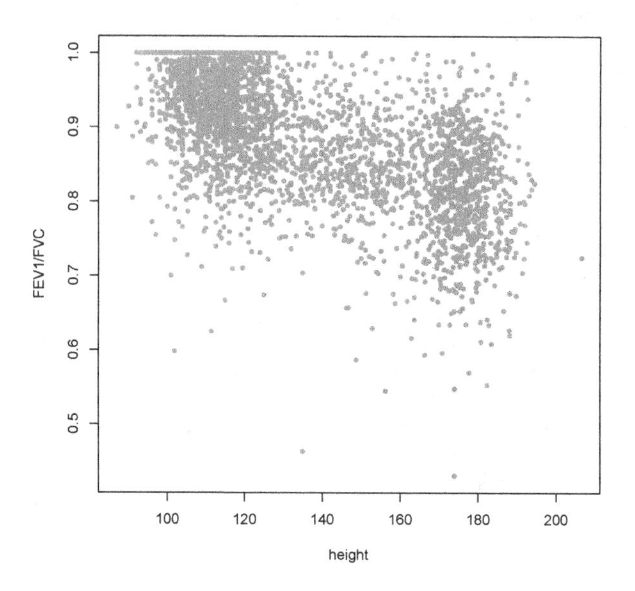

FIGURE 9.5: Spirometric lung function `slf` $= FEV_1/FVC$ for males, against `height`.

[1992], which corresponds to fitting the response variable `slf` using the BCCG(μ, σ, ν) distribution, see Section 5.9 and Stasinopoulos et al. [2017, Chapter 13]. However the BCCG distribution is not bounded above by value 1, and does not accommodate the one-inflation feature present in these data, and so the LMS method is inappropriate for the response variable `slf`. [Note also that a fitted likelihood (and resulting values of AIC and SBC) from an LMS fit (i.e. using the BCCG distribution) could not be compared with those of a mixed distribution (with one-inflation) below, as BCCG does not incorporate a point probability at one.]

Here we investigate fitting the lung function data using four methods with one-inflation discussed above, and produce centile curves based on these. After some initial analysis a log transformation for height proved to be beneficial since it improved the fit. Also the default method for choosing the smoothing parameter in the P-spline function `pb()` produced rather wiggly centiles. We therefore decided to choose the smoothing parameter in `pb()` using a local GAIC with penalty $k = 6$, which is a compromise between the AIC with $k = 2$ and the SBC with $k = \log(n) = 8.06$.

Fitting the one-inflated beta distribution Since the one-inflated beta, BEINF1(μ, σ, ν), is an explicit distribution, it is straightforward to fit it using the `gamlss()` function. We use a P-spline (i.e. smooth function, `pb()`) for each distribution parameter, μ, σ, and ν.

```
lungFunction <- transform(lungFunction, logHeight=log(height))
mbeinf1 <- gamlss(slf ~ pb(logHeight, method="GAIC", k=6),
                sigma.formula=~pb(logHeight, method="GAIC", k=6),
                nu.formula=~pb(logHeight, method="GAIC", k=6),
                data=lungFunction, family=BEINF1, trace=FALSE)
```

The centiles for the fitted model, shown in Figure 9.6(a), are obtained using

```
centiles(mbeinf1, xvar=lungFunction$height, legend=FALSE,
        main="(a) One-inflated beta BEINF1")
```

This gives the default centile curves, i.e. $(0.4, 2, 10, 25, 50, 75, 90, 98, 99.6)\%$, (which can be amended by the `cent` argument), see Figure 9.6(a).

Fitting a generated one-inflated distribution To fit a generated one-inflated distribution, the function `gamlssInf0to1()` is used. Below we first generate a logit SST(μ, σ, ν, τ) distribution, and then fit its one-inflated version to the response variable `slf`:

```
gen.Family("SST", "logit")
library(gamlss.inf)
mlogitSST  <- gamlssInf0to1( y=slf,
                 mu.formula=~ pb(logHeight, method="GAIC", k=6),
                 sigma.formula=~pb(logHeight, method="GAIC", k=6),
                 nu.formula=~pb(logHeight, method="GAIC", k=6),
                 tau.formula=~pb(logHeight, method="GAIC", k=6),
                 xi1.formula=~pb(logHeight, method="GAIC", k=6),
                 data=lungFunction, family=logitSST,trace = FALSE,
                 setseed=351, gd.tol=100)
```

Note that the function `gamlssInf0to1()` checks whether the response variable (`slf`) has any value exactly zero, or exactly one, or both, and then fits a 'zero', or 'one', or 'zero and one' inflated distribution, respectively.

The centiles for the fitted model shown in Figure 9.6(b), are obtained using

```
centiles.Inf0to1(mlogitSST, xvar=lungFunction$height, legend =
                  FALSE, main="(b) One-inflated logitSST")
```

Fitting a one-inflated truncated distribution Next we create an SST(μ, σ, ν, τ) distribution truncated below zero and above one. This distribution is then fitted with one-inflation.

```
library(gamlss.tr)
gen.trun(c(0,1), type="both",family="SST")
mSSTtr  <- gamlssInf0to1( y=slf,
                 mu.formula=~ pb(logHeight, method="GAIC", k=6),
                 sigma.formula=~pb(logHeight, method="GAIC", k=6),
                 nu.formula=~pb(logHeight, method="GAIC", k=6),
                 tau.formula=~pb(logHeight, method="GAIC", k=6),
                 xi1.formula=~pb(logHeight, method="GAIC", k=6),
                 data=lungFunction, family=SSTtr,trace = FALSE,
                 setseed=351, gd.tol=100)
```

The centiles for the fitted model, shown in Figure 9.6(c), are obtained using

```
centiles.Inf0to1(mSSTtr, xvar=lungFunction$height, legend = FALSE,
                 main="(c) One-inflated truncated SST")
```

Fitting a generalized Tobit model Next we fit two generalized Tobit models. We first create a right censored at one response variable using the function `Surv()` of the **survival** package. Then we generate a normal (for the standard Tobit) and a BCCGo(μ, σ, ν) survival fitting function. These are needed because they are used within `gamlss()` for calculating the appropriate log-likelihood function.

```
library(survival)
# creating the response variable
lungFunction<- transform(lungFunction,
          slfS=Surv(d2m$f, d2m$f!=1, type="right"))
# creating the distribution
library(gamlss.cens)
gen.cens("NO", type="right")
gen.cens("BCCGo", type="right")
mnorc <- gamlss(slfS ~ pb(logHeight, method="GAIC", k=6),
      sigma.formula= ~pb(logHeight, method="GAIC", k=6),
      data=lungFunction, family=NOrc)
mbccgorc <- gamlss(slfS ~ pb(logHeight, method="GAIC", k=6),
      sigma.formula=~pb(logHeight, method="GAIC", k=6),
      nu.formula=~pb(logHeight, method="GAIC", k=6),
      data=lungFunction, family=BCCGorc)
```

The fitted centiles for the model `mbccgorc` shown in Figure 9.6(d) are obtained using the `centiles.T()` function, which produces centiles for a generalized Tobit model on $(0, 1]$:

```
centiles.T(mbccgorc, xvar=lungFunction$height, legend = FALSE,
           ylim=c(.4,1), main="(d) gen. Tobit BCCGo")
```

Comparing the models

Figure 9.6 shows the fitted centile curves using the above methods for the `slf` $= FEV_1/FVC$ response variable. The curves look similar with the exception of (b), which shows a lower 0.4% centile curve.

For model selection, the AIC and SBC values of the fitted models are shown below:

```
AIC(mbeinf1, mlogitSST, mSSTtr, mnorc, mbccgorc)

##                  df       AIC
## mlogitSST 14.387924 -6362.628
## mbccgorc   8.972146 -6344.114
## mSSTtr    13.462850 -6343.027
## mnorc      6.803020 -6257.222
## mbeinf1    6.008017 -6168.898

AIC(mbeinf1, mlogitSST, mSSTtr, mnorc, mbccgorc, k=log(3164))

##                  df       AIC
## mbccgorc   8.972146 -6289.747
## mlogitSST 14.387924 -6275.443
## mSSTtr    13.462850 -6261.447
## mnorc      6.803020 -6215.998
```

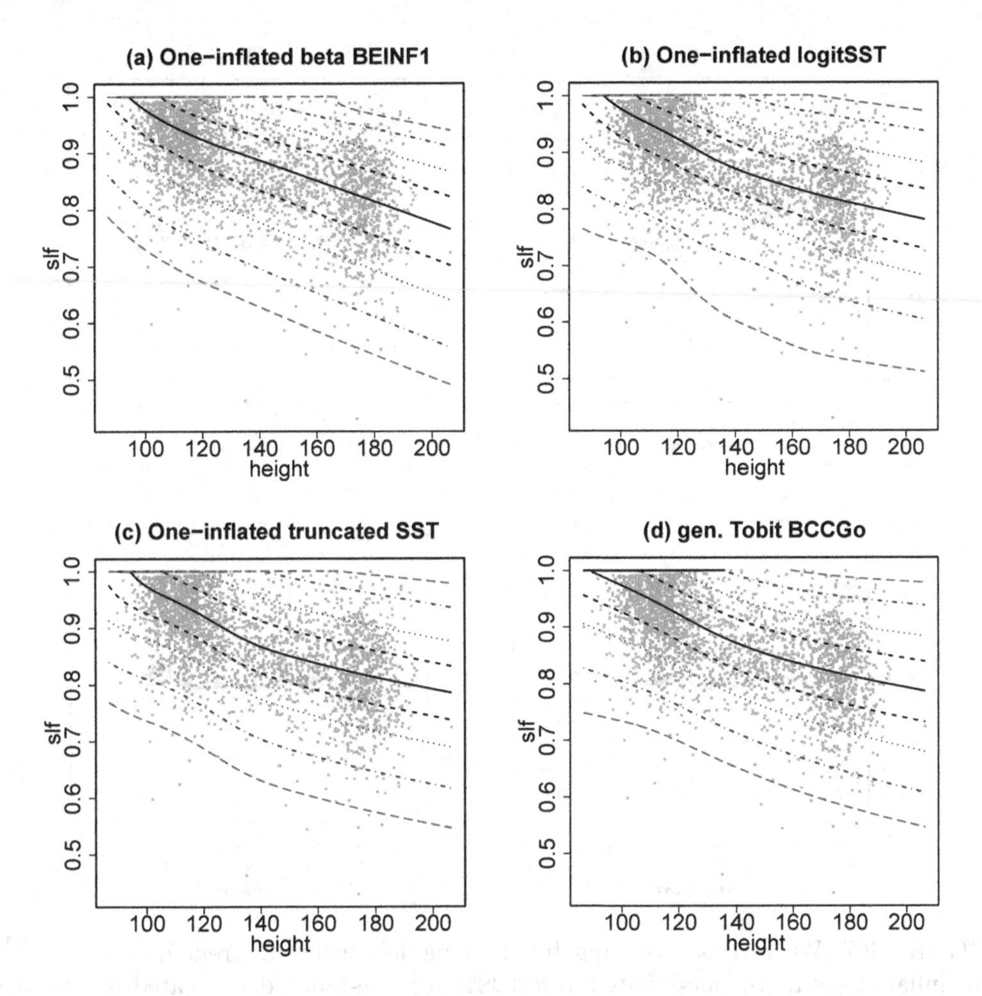

FIGURE 9.6: Fitted centiles for slf=FEV_1/FVC for males, against `height` using different distributions: (a) one-inflated beta, (b) one-inflated logit `SST`, (c) one inflated truncated `SST`, and (d) generalized Tobit `BCCGo` model.

```
## mbeinf1     6.008017 -6132.491
```

The AIC selects the one-inflated `logitSST` model, while the SBC prefers the generalized Tobit `BCCGo` model, which uses fewer degrees of freedom. The one-inflated beta (`BEINF1`) model fares worst for both criteria. The adequacy of the model fits can be investigated via worm plots of the models, shown in Figure 9.7. The one-inflated `logitSST` distribution, Figure 9.7(b), seems to have the best worm plot. The adequacy of the models (within non-overlaping intervals of height) can be explored using multiple worm plots and Q statistics, see Chapter 12 of Stasinopoulos et al. [2017].

Note that for the generalized Tobit `BCCGo` model, the probability that `slf` equals one is not modeled explicitly as a function of `height`, in contrast to the inflated models. Nevertheless, it can be calculated as $\hat{p}_1 = \text{P}(\texttt{slf} = 1) = 1 - F(1, \hat{\mu}, \hat{\sigma}, \hat{\nu})$ and plotted against `height`. Figure 9.8 shows the fitted probabilities at one, i.e. $P(\texttt{slf} = 1)$, for the

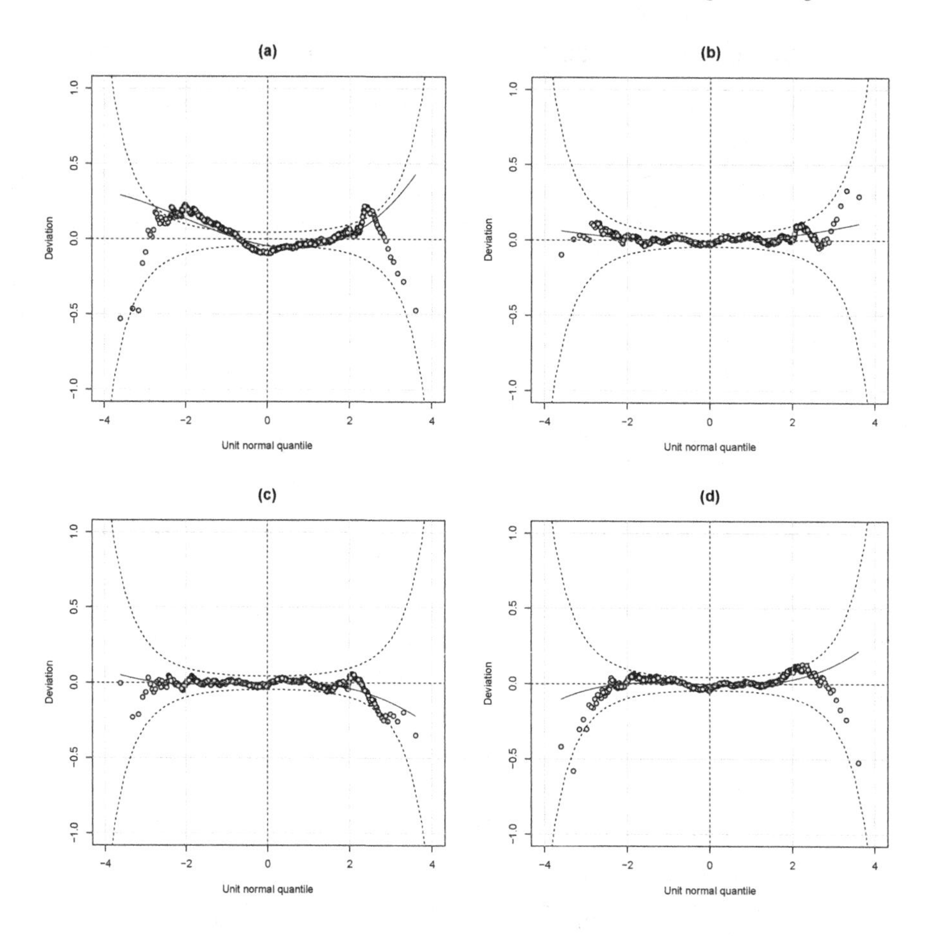

FIGURE 9.7: Worm plots for lung function models using different distributions: (a) one-inflated beta, (b) one-inflated `logitSST`, (c) one-inflated truncated `SST`, and (d) generalized Tobit `BCCGo` model.

models `mbeinf1`, `mlogitSST`, `mSSTtr` and `mbccgorc`, plotted against `height`. Note that for the one-inflated beta model $\hat{p}_1 = \hat{\nu}/(1+\hat{\nu})$:

```
fitted(mbeinf1, "nu")/(1+fitted(mbeinf1, "nu"))
```

while for the `mlogitSST` and `mSSTtr`, $\hat{p}_1 = \hat{\xi}_1$:

```
fitted(mlogitSST, "xi1")
```

For the Tobit model we have fitted probabilities:

```
1-pBCCGo(1, mu=fitted(mbccgorc), sigma=fitted(mbccgorc,"sigma"),
         nu=fitted(mbccgorc,"nu"))
```

The fitted probabilities at one (shown in Figure 9.8) for the three inflated models are the same, and are very similar to the estimated probabilities of the generalized Tobit model apart from at small heights (where there are not many observations).

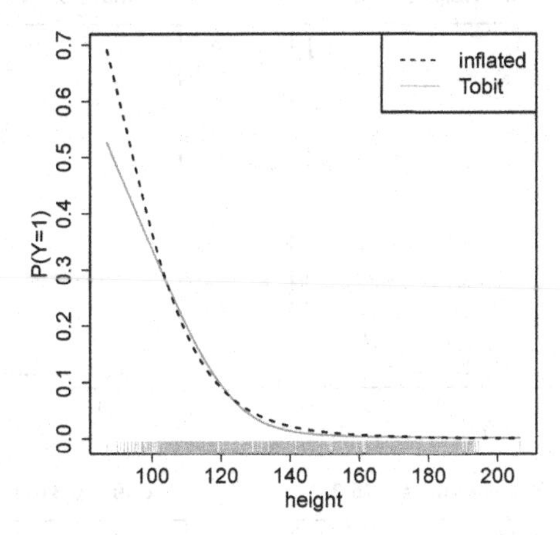

FIGURE 9.8: The fitted probability $\hat{\mathrm{P}}(\texttt{slf} = 1)$ against height for the inflated models `mbeinf1`, `mlogitSST`, and `mSSTtr` and the generalized Tobit model `mbccgorc`.

```
par(cex.axis=1.6,lwd=2,cex.lab=1.6,mgp=c(2.5,1,0),
par(mar=c(5, 6, 4, 1)+0.1))
plotmlogitSST1 <- function(i)
{
 curve(dlogitSSTInf0to1(x, mu=fitted(mlogitSST)[i],
                        sigma=fitted(mlogitSST, "sigma")[i],
                        nu=fitted(mlogitSST, "nu")[i],
                        tau=fitted(mlogitSST, "tau")[i]), 0,1,
       ylab="f(y)", xlab="y", add=FALSE)
points(1,fitted(mlogitSST, "xi1")[i])
lines(1,fitted(mlogitSST, "xi1")[i])
title(paste("obs. no", i, "  height = ",lungFunction$height[i]),cex.main=2)
}
 plotmlogitSST1(100)
 plotmlogitSST1(1630)
 plotmlogitSST1(1991)
 plotmlogitSST1(3164)
```

Figure 9.9 shows the fitted mixed probability function for selected cases for model `mlogitSST`. As the height increases, the fitted probability $\hat{\mathrm{P}}(\texttt{slf} = 1)$ decreases and the location (e.g. median) of the continuous part of the distribution of `y=slf` also decreases.

9.4 Bibliographic notes

The Tobit model [Tobin, 1958] was the earliest approach to the modeling of a zero-adjusted response variable; however, it suffers from the disadvantage of not enabling

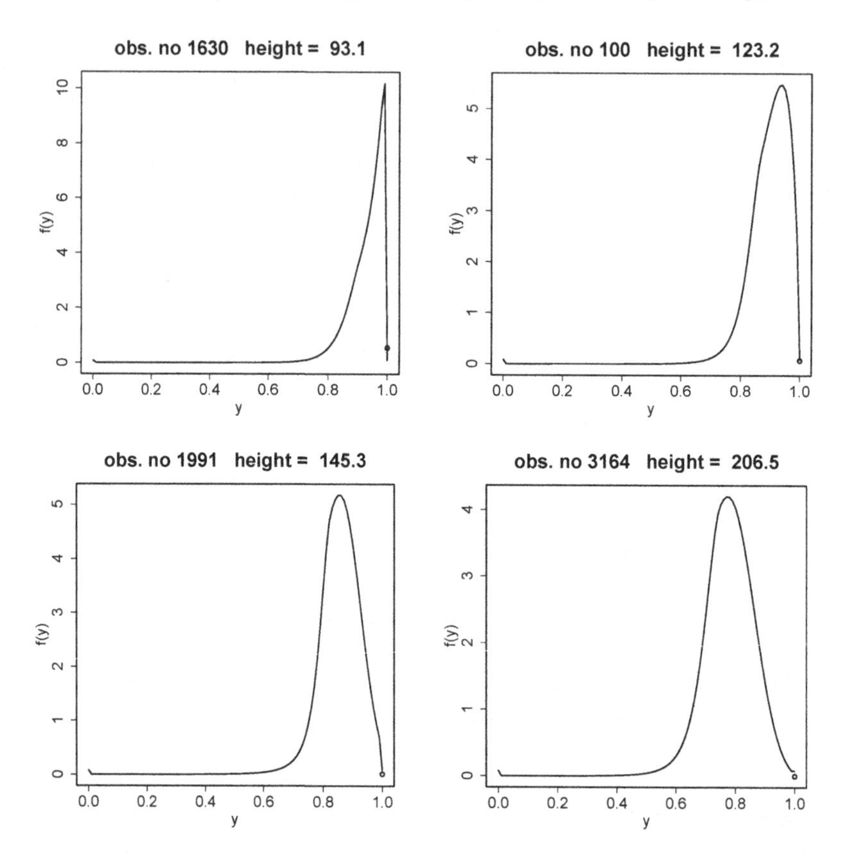

FIGURE 9.9: Fitted mixed probability functions for selected observations, one-inflated `logitSST` model.

the modeling of the zero probability separately from the distribution of the positive response. Duan et al. [1983] compared alternative methods for the modeling of medical expenses, a zero-adjusted response. They proposed a two-part regression model for such responses, the first part for the occurrence of a positive response and the second for the size of the response, conditional on a positive response. In particular, they specified a probit model for the occurrence of positive expenses, and a normal model for the logarithm of positive expenses, where the parameters in the two parts are distinct. Heller et al. [2006] formulated the two-part regression model in the GAMLSS framework. Their response variable was the size of insurance claims; a logistic model was specified for the occurrence of a claim, and the inverse Gaussian distribution for the size (severity) of a positive claim.

The Tweedie distributions are a family of distributions belonging to the exponential dispersion family, with $\text{Var}(Y) = \mu^p$ [Jørgensen, 1997]. When $1 < p < 2$, the Tweedie is a mixed distribution, with a probability mass at zero and a positive component which is a random (Poisson) sum of gamma random variables. It could be used for modeling zero-adjusted response variables, but is not attractive as it suffers from the same drawback as the Tobit model, in that there are not distinct parameters for the zero and positive components of the model; in addition the positive component does not provide the flexibility of other modeling approaches presented in the chapter as it only has two parameters and its distribution is usually multi-modal. Saei et al. [1996] suggest grouping the positive response values into ordered categories and analyzing them as an ordinal response, where zero is the lowest ordinal category. Min and Agresti [2002] give an overview of models for the analysis of zero-adjusted and zero-inflated response variables.

The literature on inflated bounded response variables is more sparse. While beta regression has been studied by, for example Kieschnick and McCullough [2003] and Ferrari and Cribari-Neto [2004], it is only more recently that boundary inflation has been considered for these models [Ospina and Ferrari, 2010, 2012, Hossain et al., 2015, 2016a,b, Hossain, 2017] and incorporated in **gamlss**. Unified computational methods for zero-adjusted and zero-inflated data are considered by Yang and Simpson [2010]. **R** packages besides **gamlss** that implement zero-adjusted and zero-inflated models include **censReg** for Tobit regression [Henningsen, 2017], **ZOIP** [Diaz Zapata, 2017], and **Tweedie** [Dunn, 2017].

9.5 Exercises

1. **The VAS data:**

> **R data file:** `vas5` in package **gamlss.data** of dimension 364 × 3.
> **variables**
> `patient` : factor indicating the patient
> `treat` : the treatment factor with levels 1 2 3 4 5 6 7
> `vas` : the response variable with values 0 to 100
> **purpose:** to demonstrate a 0 to 1 response variable.

The `vas5` data set has 364 observations (one from each of 364 patients), with a response variable `vas` and one explanatory factor with seven levels, `treat` (where level 1 is placebo). Variable `vas` takes values from 0 to 100 so by dividing by 100 we can obtain a response `VAS` with range $[0, 1]$ and therefore we can model it using one of the distributions discussed in this chapter.

(a) Plot the response variable `VAS` against treatment (`treat`), check for 0 or 1 values, and comment.

```
library(gamlss)
data(vas5)
vas5$VAS <- with(vas5, vas/100)
plot(VAS~treat, data=vas5)
sum(vas5$vas==0)
sum(vas5$vas==1)
```

(b) First try the explicit `BEINF` distribution. Given that there are two possible models for each parameter, i) ~ 1 and ii) $\sim$ treat, and 4 different parameters, there are 4^2 possible different models which can be fitted using the `BEINF` distribution. Fit all those models and choose between them using GAIC with $k = 2$ and $k = \log(n)$.

(c) Use diagnostics to check the adequacy of the two chosen models (i.e. with $\kappa = 2$ and $\kappa = \log(n)$. Do you think that a different distribution is worth pursuing? State your preferred model.

(d) For your chosen model, plot the different fitted distributions of VAS for each treatment. You can use the following code (by replacing 'model' with the name of your chosen model). Note that ind gives the case number of observations having each of the seven levels of factor treat. Comment on the plots.

```
op <- par(mfrow = c(4, 2))
lev <- c(1, 2, 3, 4, 5, 6, 7)
ind <- c(3, 2, 9, 12, 1, 7, 4)
j <- 0
for (i in ind) {
     j = 1 + j
     xlab <- paste("treatment = ", eval(substitute(lev[j])))
     ylab <- paste("p(y)")
 plotBEINF( mu = fitted(model)[i],
         sigma = fitted(model, "sigma")[i],
            nu = fitted(model, "nu")[i],
           tau = fitted(mode;, "tau")[i],
from = 0, to = 1, n = 101, xlab = xlab, ylab = ylab)
}
par(op)
```

```
op <- par(mfrow = c(4, 2))
lev <- c(1, 2, 3, 4, 5, 6, 7)
ind <- c(3, 2, 9, 12, 1, 7, 4)
j <- 0
for (i in ind) {
     j = 1 + j
     xlab <- paste("treatment = ", eval(substitute(lev[j])))
     ylab <- paste("p(y)")
 plotBEINF( mu = fitted(mod22)[i],
         sigma = fitted(mod22, "sigma")[i],
            nu = fitted(mod22, "nu")[i],
           tau = fitted(mod22;, "tau")[i],
from = 0, to = 1, n = 101, xlab = xlab, ylab = ylab)
}
par(op)
```

2. **The brown fat data**: Brown fat (or brown adipose tissue) is found in hibernating mammals, its function being to increase tolerance to the cold. It is also present in newborn humans. In adult humans it is more rare and is known to vary considerably with ambient temperature. Routhier-Labadie et al. [2011] analyzed data on 4,842 subjects over the period 2007-2008, of whom 328 (6.8%) were found to have brown fat. Brown fat mass and other demographic and clinical variables were recorded. The purpose of the study was to investigate the factors associated with brown fat occurrence and mass in humans.

R data file: `brownfat` in package **gamlss.data** of dimension 4842×6
source: Routhier-Labadie et al. [2011], Statistical Society of Canada [2011]
variables

- `sex` : factor, 1=female, 2=male
- `diabetes` : factor, 0=no, 1=yes
- `age` : age in years
- `day` : day of observation (1=1 January, ..., 365=31 December)
- `exttemp` : external temperature (degrees Centigrade)
- `season` : factor, 1=Spring, 2=Summer, 3=Autumn, 4=Winter
- `weight` : weight in kg
- `height` : height in cm
- `BMI` : body mass index
- `glycemy` : glycemia (mmol/L)
- `LBW` : lean body weight
- `cancerstatus` : factor, 0=no, 1=yes, 99=missing
- `brownfat` : factor, presence of brown fat (0=no, 1=yes)
- `bfmass` : brown fat mass (g) (zero if `brownfat=0`)

purpose: to demonstrate the use of zero-adjusted distributions in GAMLSS.

(a) The response variable is `bfmass`, which is either zero or a continuous positive quantity. Construct a histogram of `bfmass`. Is this useful? Construct also a histogram of positive `bfmass`. Find the number of cases with `bfmass=0`. What features will we need in the continuous distribution component?

```
truehist(brownfat$bfmass)
bfpos <- subset(brownfat$bfmass,brownfat$bfmass>0)
truehist(bfpos, nbins=40)
sum(brownfat$bfmass==0)
```

(b) Use `fitDist()` to select a distribution for the continuous component.

```
a <- fitDist(bfpos, type="realplus")
a$fits
```

(c) Fit zero-inflated regression models for `bfmass`:

- use the following distributions for the continuous component of the distribution of `bfmass`: gamma (`GA`), inverse Gaussian (`IG`), generalized inverse Gaussian (`GIG`), truncated t family (`TFtr`) and the generalized Tobit model with the t family (`TFtr`) distribution,
- μ predictor: `age + sex + exttemp`,
- ξ_0 (or ν for explicit distributions `ZAGA` and `ZAIG`) predictor: `sex`+ `diabetes` + `pb(BMI)` + `pb(LBW)` + `exttemp`.

```
library(gamlss.inf)
library(gamlss.tr)

m1 <- gamlss(bfmass~ age+sex+exttemp,
             nu.fo=~ sex+diabetes+pb(BMI)
             +pb(LBW)+exttemp,
```

```
                  family=ZAGA, data=brownfat)

m2 <- gamlss(bfmass~ age+sex+exttemp,
                  nu.fo=~ sex+diabetes+pb(BMI)
                  +pb(LBW)+exttemp,
                  family=ZAIG, data=brownfat)

m3 <- gamlssZadj(bfmass,
                  mu.fo=~ age+sex+exttemp,
                  xi0.fo=~ sex+diabetes+pb(BMI)
                  +pb(LBW)+exttemp,
                  family=GIG, data=brownfat)

gen.trun(0, "TF", type="left")
m4 <- gamlssZadj(bfmass,
                  mu.fo=~ age+sex+exttemp,
                  xi0.fo=~ sex+diabetes+pb(BMI)
                  +pb(LBW)+exttemp,
                  family=TFtr, data=brownfat, n.cyc=50)

library(survival)
brownfat$bfmass.surv<- with(brownfat,
          Surv(bfmass, bfmass!=0, type="left"))
library(gamlss.cens)
gen.cens("TF", type="left")
m5 <- gamlss(bfmass.surv ~ age+sex+exttemp,
                 family=TFlc, data=brownfat, trace=FALSE)
```

Compare the models using the AIC, and construct the worm plot of the best model. Is the best model the one you expected?

```
AIC(m1, m2, m3, m4, m5)
wp(m3)
```

(d) Investigate the inclusion of other covariates not in the model. You should consider the inclusion of covariates for σ (and possibly ν for m3, m4, and m5).

(e) Interpret the effect of the covariates in the models for μ and ξ_0 (or ν for ZAGA and ZAIG).

3. **The sleep data**: This question concerns a study conducted on 133 patients thought to have the condition Obstructive Sleep Apnea (OSA). These patients have undergone a sleep study at a Canadian sleep clinic [Ahmadi et al., 2008]. While the focus on the study was the relationship between the Berlin Questionnaire for sleep apnea to polysomnographic measurements of respiratory disturbance, in particular the arousal index, we will analyze the proportion of sleep time that is REM sleep (REM). This variable is in the interval [0,1), so necessitates the use of zero-inflated models. We have removed patients with missing values, giving $n = 106$ observations.

R data file: `sleep` in package **gamlss.data** of dimension 106 × 9
source: Ahmadi et al. [2008], Statistical Society of Canada [2006]
variables

`age` : age in years
`gender` : factor, 1=female, 0=male
`BMI` : body mass index
`necksize` : neck circumference (cm)
`sbp` : systolic blood pressure (mmHg)
`alcohol` : factor, alcohol usage (1=yes, 0=no)
`caffeine` : factor, caffeine usage (1=yes, 0=no)
`REM` : proportion of rapid eye movement (REM) sleep time
`AI` : arousal index (number of arousals from sleep per hour of sleep)

purpose: to demonstrate the use of inflated distributions in GAMLSS.

(a) Construct a histogram of `REM`. Count the number of cases with `REM=0` and check if any cases have `REM=1`. Is zero-inflation a prominent feature?

```
truehist(sleep$REM, nbins=30, xlim=c(0,1))
sum(sleep$REM==0)
sum(sleep$REM==1)
```

Note there are only 3 zero values for `REM`. However these must be dealt with. One question is whether the zero values are errors in the data, or genuine true exact zero values, or rounded zero values. If the zeros are errors, then their cases should be ommited. If the zeros are genuine exact zero values, then a mixed distribution (as discussed in this chapter and used in (b) below) is needed. If the zeros are rounded values, e.g. if values of `REM` in the interval $(0.000, 0.0005)$ were recoded as 0, then `REM` could be analyzed using an interval censored response, or values of 0 could be recoded to 0.00025, (but note that the fitted model might be sensitive to the recoded value chosen, see Li et al. [2014] for more details).

(b) Using `age` as the predictor for μ, and no predictors for σ and ξ_0 (or ν for `BEINF0`), compare the fits of the following models for `REM`: zero-inflated beta, zero-inflated simplex, and zero-inflated logit skew t.

```
library(gamlss.inf)
mbeta <- gamlss(REM~age, family=BEINF0, data=sleep)
msimplex <- gamlssInf0to1(REM, mu.fo=~age, family=SIMPLEX,
                 data=sleep)
gen.Family("SST", "logit")
mlogitSST <- gamlssInf0to1(REM, mu.fo=~age, family=logitSST,
                 data=sleep)
AIC(mbeta, msimplex, mlogitSST)
wp(mlogitSST)
```

(c) Create centile curves for your best fitted model.

```
centiles.Inf0to1(mlogitSST, xvar=sleep$age)
```

(d) Based on your findings above, what is the relationship between the proportion of REM sleep time and age?

(e) Investigate the use of other zero-inflated generated and truncated response distributions.

(f) Investigate the inclusion of other covariates into the models for σ and ξ_0 (or ν for `BEINF0`). (Do not include `AI` as this is an outcome rather than a predictor.)

Part II

Advanced topics

10

Statistical inference

CONTENTS

This chapter provides the theoretical background for fitting a distribution to data. In particular it explains:

1. basic concepts of statistical inference;
2. the likelihood function; and
3. the classical frequentist and Bayesian approaches to inference.

10.1 Introduction: population, sample, and model

Statistical inference is the process in which a *sample* is used to make inference about the *population* distribution. This is done by the use of a *statistical model*. The model involves a set of assumptions on how the population is behaving and how the sample is generated from the population. A common assumption is that the model has a specific parametric theoretical distribution (as described in Part I). In this case we have a *parametric distribution* model. If the assumptions do not explicitly assume a theoretical distribution, the model is usually called *nonparametric*. GAMLSS are regression models (see Section 3.1) in which it is assumed that the population distribution of the response variable can be approximated by a parametric theoretical distribution, therefore GAMLSS are parametric distribution models. GAMLSS also allows nonparametric smoothing terms in the predictors of the parameters of the distribution as in equation (3.1) and for this reason can be called *semi-parametric* models. If no smoother appears in any of the models for the distribution parameters then we have a purely parametric model as in equation (3.4). Inference for parametric distribution regression models is well established, while for semi-parametric distribution regression models (such as the semi-parametric GAMLSS model (3.1)) it is less so. Stasinopoulos et al. [2017, Section 5.1.2] identifies several aspects where inference is needed within a GAMLSS model. This includes inference for:

1. constant distribution parameters, i.e. one of μ, σ, ν, or τ,
2. coefficients of linear parametric terms in a predictor model,
3. smoothing curves, e.g. $s(\mathbf{x}_{13})$,
4. smoothing parameters, e.g. $\lambda_{1,3}$,
5. prediction of values of a distribution parameter,
6. prediction of future values of the response, and
7. model selection.

This chapter deals primarily with case 1. Section 10.2 introduces the likelihood function, a concept which is important for parameter estimation. Section 10.3 considers the two main approaches, classical frequentist inference and Bayesian inference. Chapter 11 focuses on classical frequentist inference.

The basic concepts of *population*, *sample*, and *model* as used within statistical inference are introduced next.

The population

The *population* is the set of the particular 'subjects' we would like to study. The interest usually lies in some features of the population which manifest themselves as a set of response variables. Here we focus on a single response variable Y. Populations can be *real*, e.g. the height of adults living in the UK, or can be *conceptual*, e.g. in a clinical trial we have only a limited number of people taking the actual treatment, but our interest lies in all possible people who could have taken the same treatment. It is rather difficult to try to define precisely what we mean by 'population'. Problems arise with trying to determine (i) whether the number of elements in the population is finite or not, and (ii) the true range of possible values for the population. For the sake of simplicity and since those arguments hardly add to a better understanding of statistical inference, we will only assume that the 'true' *population distribution* for Y, denoted as $f_P(y)$, exists but is *unknown*. Statistical inference argues that we can say something about the population distribution on the basis of a sample taken from the population. Note that the *population cumulative distribution function* is defined as $F_P(y) = \mathrm{P}(Y \leq y)$, which is the proportion of the population with value $Y \leq y$, for any value y.

The sample

A *sample* is a subset of the population. The sample is called a *simple random sample* if each element of the population has an equal chance of being selected. The sample is denoted here as a vector $\mathbf{y} = (y_1, y_2, \ldots, y_n)^\top$ of length n. Let d be the number of distinct values of Y in the sample $\mathbf{y}$, and n_I the number of times that the distinct value y_I occurs in the sample for $I = 1, 2, \ldots, d$. The total sample size is $n = \sum_{I=1}^{d} n_I$. Let

$p_I = n_I/n$, the proportion of the sample with value y_I, then the *empirical probability function* (epf) of $\mathbf{y}$ is given by

$$f_E(y) = \hat{\mathrm{P}}(Y = y) = \begin{cases} p_I & \text{if } y = y_I \quad \text{for } I = 1, 2, \ldots, d \\ 0 & \text{otherwise.} \end{cases} \tag{10.1}$$

The epf can be plotted as a bar plot for discrete variables or a needle plot for continuous variables, see Figure 10.1(a). The needle plot is not a very informative graph; instead histograms or a smooth version of the needle plot called a nonparametric density estimator is preferred, see Härdle [1990], Silverman [1986], and Wand and Jones [1999].

The *empirical cumulative distribution function* (ecdf) $F_E(y) = \hat{\mathrm{P}}(Y \leq y)$, is the sum of p_I over all I for which $y_I \leq y$. It is a step function increasing each time a distinct value y_I appears in the sample. Figure 10.1 shows a typical (a) epf and (b) ecdf, for a continuous variable. The data are the **parzen** data set first introduced in Section 5.8.1. A nonparametric density estimator function is superimposed on the epf plot of Figure 10.1(a), representing approximately the sample distribution of snowfall in a more informative way.

```
data(parzen)
tp <- with(parzen, table(snowfall))
tp <- tp/length(parzen$snowfall)    # to create proportions
Ecdf <- with(parzen,ecdf(snowfall)) # the empirical cdf
plot(tp, xlab="y (snowfall)", ylab="fE(y)", main="(a) epdf",lwd=0.5)
dpar<-density(parzen$snowfall)
lines(dpar, col="black",lwd=2)
plot(Ecdf, xlab="y (snowfall)", ylab="FE(y)", main="(b) ecdf")
```

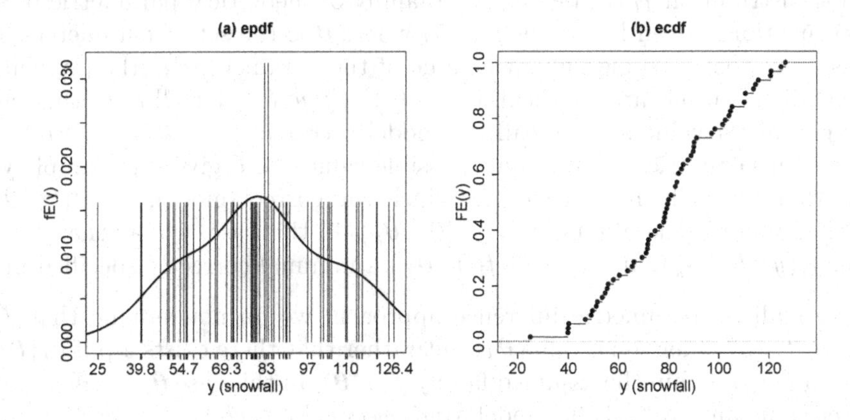

FIGURE 10.1: (a) The empirical probability function (with a noparametric density estimate) and (b) the empirical cumulative distribution function, of the Parzen snowfall data.

The ecdf plays a very important role in statistical inference, both parametric and nonparametric, since for a random sample of size n drawn from the true population, the ecdf $F_E(y)$ is a consistent estimator, as $n \to \infty$, of the 'true' population cdf $F_P(y)$. This is obtained by assuming either random sampling from a population of infinite size, or random sampling with replacement from a finite population. The ecdf also provides, through the *plug-in principle* [Efron and Tibshirani, 1993, p. 35], a simple method of nonparametric estimation. For example, let us assume that we are interested in a specific characteristic of the population distribution, say $\vartheta = t(F_P)$. This characteristic, ϑ, is a function $t(\cdot)$ of the 'true' population cdf $F_P(Y)$, (e.g. the mean). The plug-in estimate of ϑ is given by replacing the true cdf F_P in $t(\cdot)$ by the ecdf F_E, i.e. $\hat{\vartheta} = t(f_E)$. Therefore according to this principle, the sample mean $\bar{y}$ is the plug-in estimate of the population mean. (Of course this principle can fail in situations where the true population characteristic does not exist. For example, the mean of the theoretical Cauchy distribution does not exist, and therefore the plug-in-principle will fail.)

The model

A *statistical model* is a set of assumptions about the population and how the sample is generated from the population. The model is called *parametric* if a specified parametric theoretical distribution is assumed for the response variable. A *nonparametric* model requires less restrictive assumptions. For example, assuming that the distribution for the response is symmetric is a less restrictive assumption than assuming that the distribution is normal. As a general rule, the more assumptions one makes, the more information can be extracted from the data, if the assumptions are correct, but the more things can go wrong, if they are not. For GAMLSS models this means that selection of the appropriate distribution is particularly important.

'Standard' (or 'naïve') classical parametric statistical inference assumes that the true population distribution $f_P(y)$ belongs to a family of theoretical parametric probability (density) functions, given by model $f(y \,|\, \boldsymbol{\theta})$ where $\boldsymbol{\theta}$ is the set of parameters. In addition it assumes that at some unknown value of the parameter $\boldsymbol{\theta}_T$, the population and theoretical distributions are identical, i.e. $f_P(y) = f(y \,|\, \boldsymbol{\theta}_T)$ for all y. In this case $\boldsymbol{\theta}_T$ is the *true* parameter value and we call the model a *correctly specified model.* Statistical inference in this case tries to determine possible values for $\boldsymbol{\theta}$ given the sample $\mathbf{y}$. [Note, however, that we could have a case in which the correct model, say $f_1(y \,|\, \boldsymbol{\theta}_T)$, is a submodel of a more general family $f_2(y \,|\, (\boldsymbol{\theta}_T, \boldsymbol{\theta}_k)$. In this case, we say that $f_1(y \,|\, \boldsymbol{\theta}_T)$ is nested in $f_2(y \,|\, (\boldsymbol{\theta}_T, \boldsymbol{\theta}_k)$, and we call $f_1(y \,|\, \boldsymbol{\theta}_T)$ a minimal correctly specified model.]

In a more realistic parametric inference approach, we do not assume that $f_P(Y)$ is identical to $f(y \,|\, \boldsymbol{\theta}_T)$ for some value $\boldsymbol{\theta}_T$, but rather that there exists a pdf $(y \,|\, \boldsymbol{\theta}_c)$ which is *closest* to $f_P(y)$ within the assumed family $f(y \,|\, \boldsymbol{\theta})$. In this case $\boldsymbol{\theta}_c$ is called the *closest* parameter value and we call the model a *misspecified model.*

Figure 10.2 shows schematic plots of the true population distribution $f_P(y)$, the empirical sample distribution $f_E(y)$, and the model $f(y \,|\, \boldsymbol{\theta})$ under (a) a correctly specified model and (b) a misspecified model. A schematic plot is a very informal representation in which we show the basic concepts without concern about the correct mathematical representation, as for example the plot of Hastie et al. [2009, p. 225]. Both population $f_P(y)$ and empirical $f_E(y)$ distributions are denoted as points, while the model $f(y \,|\, \boldsymbol{\theta})$ is represented as a line. The different elliptical regions around the true population distribution $f_P(y)$ represent loosely the variation of the empirical sample distribution $f_E(y)$ around the population distribution $f_P(y)$. A different sample will result in a different $f_E(y)$. Different values of the parameter $\boldsymbol{\theta}$ represent different points on the model line $f(y \mid \boldsymbol{\theta})$. The point $f(y \mid \boldsymbol{\theta}_c)$ in Figure 10.2 (b) represents the model *closest* to the true distribution $f_P(y)$, based on the Kullback-Leibler distance (11.19), and therefore is the 'best' model we can achieve under the assumption of model $f(y \mid \boldsymbol{\theta})$. Under the standard or naïve classical parametric statistical inference, $f(y \mid \boldsymbol{\theta})$ passes through the population distribution $f_P(y)$ and therefore $f_P(y) = f(y \mid \boldsymbol{\theta}_c) = f(y \mid \boldsymbol{\theta}_T)$ as shown in Figure 10.2(a). Note that the model line represents only one possible theoretical distribution. A different theoretical distribution, say $f_1(y \mid \boldsymbol{\theta}_1)$, will be represented by a completely different line. If this passes closer to $f_P(y)$ than the model $f(y \mid \boldsymbol{\theta})$, we will do better modeling the true population under the assumption of $f_1(y \mid \boldsymbol{\theta}_1)$ than $f(y \mid \boldsymbol{\theta})$. This highlights the problem with the standard classical parametric statistical inference approach where $f_P(y)$ and $f(y \mid \boldsymbol{\theta}_T)$ are assumed to coincide, which therefore excludes the possibility that the model $f(y \mid \boldsymbol{\theta})$ can be wrong.

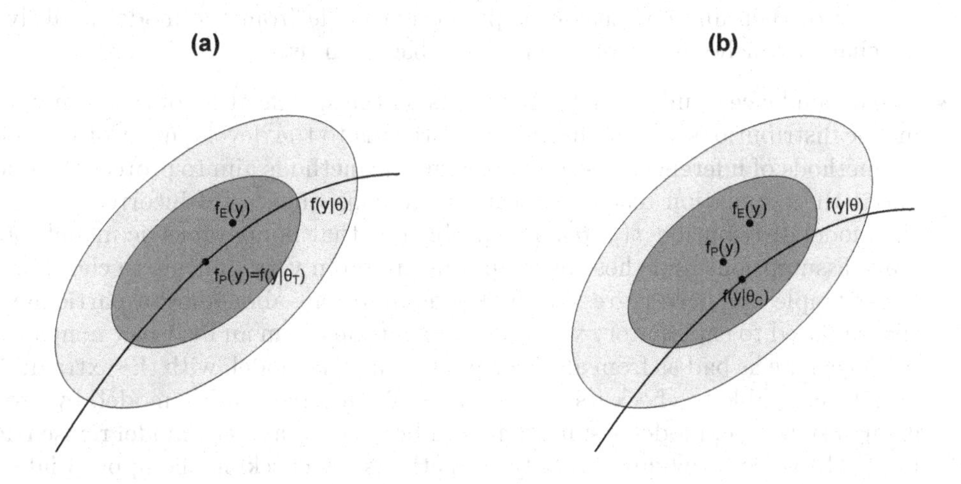

FIGURE 10.2: Schematic plot of true population distribution, $f_P(y)$, the empirical distribution of the sample, $f_E(y)$, and a model $f(y \mid \boldsymbol{\theta})$ (shown as a line) representing different values of the parameter $\boldsymbol{\theta}$. **(a)** The model at $\boldsymbol{\theta}_T$ represents the true population distribution, i.e. $f_P(y) = f(y \mid \boldsymbol{\theta}_T)$, for a correctly specified model. **(b)** The model at $\boldsymbol{\theta}_c$, i.e.$f(y \mid \boldsymbol{\theta}_c)$, represents the closest to, or 'best' approximation of, the population distribution $f_P(y)$ by the model $f(y \mid \boldsymbol{\theta})$, for a misspecified model.

The following comments are also appropriate here:

- $\boldsymbol{\theta}$ is used as generic notation for a parameter vector, but any one-to-one transformation of $\boldsymbol{\theta}$ to $\boldsymbol{\phi}$ (a reparameterization) will also perfectly define the theoretical distribution $f(y \mid \boldsymbol{\theta})$. If no explanatory variables are involved then $f(y \mid \boldsymbol{\theta})$ and $f(y \mid \boldsymbol{\phi})$ are equivalent models if $\boldsymbol{\phi} = g(\boldsymbol{\theta})$ and the function $g(\cdot)$ is a one-to-one transformation from the parameter space of $\boldsymbol{\theta}$ to the parameter space of $\boldsymbol{\phi}$. This is not the case if the parameters are modeled as functions of explanatory variables. In this case, different parameterizations of the model distribution can result in completely different fitted models.

- The model pdf $f(y \mid \boldsymbol{\theta})$ is, in general, not correctly specified ('All models are wrong but some are useful' [Box, 1979]), but it helps us to understand the behaviour of the unknown population distribution $f_P(Y)$. Given that $f(y \mid \boldsymbol{\theta}_c)$ is '*close*' to $f_P(Y)$ and given that we estimate $\boldsymbol{\theta}$ carefully, we should have a good approximation of $f_P(Y)$. Within this more general setup there are two unknown quantities:

  (a) the theoretical pdf $f(\cdot)$, which can be selected from a large number of appropriate distributions, and

  (b) the parameter(s) $\boldsymbol{\theta}$.

  Standard (or naïve) classical statistical inference puts more emphasis on (b), i.e. on how to choose $\boldsymbol{\theta}$, assuming a particular theoretical model $f(y \mid \boldsymbol{\theta})$, rather than finding the most appropriate family of distributions for the response variable in the first place. Much time has been spent on the debate between the Bayesian or the classical approaches for inference about $\boldsymbol{\theta}$. The problem with this is that if the model $f(y \mid \boldsymbol{\theta})$ is not a good approximation to the true population distribution $f_P(y)$ for any $\boldsymbol{\theta}$, then any conclusion or prediction made from the model is likely to be unreliable, whichever inference approach has been used.

- The dependence of inferential conclusions on the specification of the family of parametric distributions $f(y \mid \boldsymbol{\theta})$ has led statisticians to the development of *nonparametric* methods of inference. Those nonparametric methods aim to protect the inference about the population distribution from the consequences of incorrectly specifying the model distribution $f(y \mid \boldsymbol{\theta})$. The problem is that nonparametric models often *do* make assumptions and those assumptions are often very difficult to check or verify. For example, quantile regression makes assumptions about how a particular quantile is related to explanatory variables. Conclusions from an incorrect nonparametric model can be as bad as from an incorrect parametric model, with the extra difficultly of not being able to check its assumptions. With a parametric model, by providing enough structure, model assumptions can be checked and the model revised accordingly. GAMLSS provides a variety of methods for checking the appropriateness of the fitted distribution and the model for its parameters, see Stasinopoulos et al. [2017, Chapter 12].

10.2 The likelihood function

The concept of the likelihood function is of vital importance in parametric statistical inference. It is based on the reasoning that 'parameter values that make the observed data appear relatively probable are more likely to be correct than parameter values that make the observed data appear relatively improbable' [Wood, 2015]. The major statistical schools of inference, *classical frequentist* and *Bayesian*, use the likelihood function as their main inferential tool. Bayesian inference uses the likelihood function as the source of information given by the data, which is combined with a prior distribution for the parameters, to form the posterior distribution of the parameters (see Section 10.3.2 and Gelman et al. [2013]). Classical inference treats parameters as unknown constants, assumes the data are a realization of potentially repeatable sampling from the assumed model, and makes inference about the parameters using the likelihood (see Section 10.3.1, Chapter 11 and Cox and Hinkley [1979]).

The *likelihood function*, $L(\boldsymbol{\theta})$, is the probability of observing the sample, viewed not as a function of the sample $\mathbf{y}$ but as a function of the parameter(s) $\boldsymbol{\theta}$.

10.2.1 Discrete response variable

Let $\mathbf{y} = (y_1, y_2, \ldots, y_n)$ be an observed random sample from a discrete response variable, where we assume the model distribution is $f(y \mid \boldsymbol{\theta})$, with unknown parameter(s) $\boldsymbol{\theta}$. The likelihood function, $L(\boldsymbol{\theta})$, is the probability of observing the sample assuming that the model is correctly specified (and that the observations are independent) is:

$$L(\boldsymbol{\theta}) = \prod_{i=1}^{n} f(y_i \mid \boldsymbol{\theta}) \,.$$

Because of the independence assumption, the joint probability of the sample is the product of the individual probabilities. Note the change of emphasis in the argument of the likelihood function from y in $f(y \mid \boldsymbol{\theta})$ to $\boldsymbol{\theta}$ in $L(\boldsymbol{\theta})$.[a] Given the sample values $\mathbf{y}$, the likelihood is *not* a function of the sample $\mathbf{y}$, since this has been observed and is considered fixed, but a function of the parameter(s) $\boldsymbol{\theta}$. In practice it is more convenient to work with the logarithm of the likelihood. Assuming independent observations, the log-likelihood, $\ell(\boldsymbol{\theta})$, is defined as

$$\ell(\boldsymbol{\theta}) = \sum_{i=1}^{n} \log f(y_i \mid \boldsymbol{\theta}) \,. \tag{10.2}$$

10.2.2 Continuous response variable

For a continuous variable Y, $\mathrm{P}(Y = y) = 0$ for all values $y \in R_Y$. Therefore a slightly different definition of the likelihood is required for a continuous variable. A specific value y_i is observed to a certain level of accuracy, say $y_i \pm \Delta_i$. (For example, if y_i is rounded to the nearest first decimal place then $\Delta_i = 0.05$ and an observed value $y_i = 5.7$ corresponds to $5.65 < y < 5.75$.) Hence the likelihood (i.e. the probability of observing the data $\mathbf{y}$ given the model), assuming independent observations, is defined as:

$$\begin{aligned} L(\boldsymbol{\theta}) &= \prod_{i=1}^{n} \mathrm{P}\left(y_i - \Delta_i < Y < y_i + \Delta_i \mid \boldsymbol{\theta}\right) \\ &= \prod_{i=1}^{n} \left[F\left(y_i + \Delta_i \mid \boldsymbol{\theta}\right) - F\left(y_i - \Delta_i \mid \boldsymbol{\theta}\right)\right] \end{aligned} \tag{10.3}$$

[a] As the likelihood and log-likelihood are conditioned on the observed sample, we should denote them as $L(\boldsymbol{\theta}|\mathbf{y})$ and $\ell(\boldsymbol{\theta}|\mathbf{y})$, respectively. For simplicity of notation, we omit the conditioning on $\mathbf{y}$.

where $F(\cdot)$ is the cdf of Y. Assuming the Δ_i's are sufficiently small, we have:

$$L(\boldsymbol{\theta}) \approx 2\prod_{i=1}^{n} f(y_i \,|\, \boldsymbol{\theta})\Delta_i = 2\left(\prod_{i=1}^{n} f(y_i \,|\, \boldsymbol{\theta})\right)\left(\prod_{i=1}^{n} \Delta_i\right)$$

and hence the log-likelihood is given approximately by:

$$\ell(\boldsymbol{\theta}) \approx \sum_{i=1}^{n} \log f(y_i \,|\, \boldsymbol{\theta}) + \sum_{i=1}^{n} \log \Delta_i + \log 2 \; . \tag{10.4}$$

The most popular method of fitting a theoretical distribution family model to an observed (continuous or discrete) random sample is the method of maximum likelihood estimation, consisting of finding the value of $\boldsymbol{\theta}$ which maximizes the likelihood or log-likelihood. Since the second and third terms in (10.4) do not depend on $\boldsymbol{\theta}$, maximizing $\ell(\boldsymbol{\theta})$ over $\boldsymbol{\theta}$ involves only the first term. The fact that equations (10.2) and (10.4) produce the same maximum has led to the use of equation (10.2) as the definition of the log-likelihood, independent of whether the response variable is discrete or continuous. In fact by default `gamlss()` uses equation (10.2) for all types of response variables, including mixed. [Note, however, (10.3) can be used in `gamlss()` by first generating an interval censored version of a particular distribution using function `gen.cens()` of the package **gamlss.cens**, see Stasinopoulos et al. [2017] p164-166.]

Occasionally for continuous variables the definition in (10.2) creates problems. To demonstrate the point consider a single observation y_i from a normal distribution, i.e. $Y_i \sim \texttt{NO}(\mu, \sigma)$. The likelihood for this observation is maximized as $\mu \to y_i$, $\sigma \to 0$ and the likelihood (and log-likelihood) goes to infinity. Within a flexible regression model like GAMLSS, maximizing the likelihood occasionally leads to σ going to zero. (To avoid the problem σ can be bounded not to go below a fixed small value by using a log-shifted link function, see link `Slog` in Table 3.1.) Note that the likelihood for continuous variables defined in (10.3) is bounded above by one so it cannot go to infinity. Nevertheless numerical problems do arise in regions where the cdf $F(\cdot)$ is flat, resulting in differences in (10.3) which are close to or almost zero.

10.2.3 Interval response variable

The likelihood definition (10.3) is useful for censored or interval response variables. In this case the response variable can be:

left censored: $(-\infty, y_{i2}]$ i.e. at or below y_{i2},

right censored: (y_{i1}, ∞) i.e. above y_{i1},

interval response: $(y_{i1}, y_{i2}]$ i.e. above y_{i1} and at or below y_{i2}.

In all three cases the likelihood takes the form

$$L(\boldsymbol{\theta}) = \prod_{i=1}^{n} \left[F_Y(y_{i2} \,|\, \boldsymbol{\theta}) - F_Y(y_{i1} \,|\, \boldsymbol{\theta})\right] \; . \tag{10.5}$$

The package **gamlss.cens** allows users to use censored or interval response variables, see Stasinopoulos et al. [2017] p164-166.

10.2.4 Likelihood with time dependence

Note that the likelihood function becomes more complicated if we do not assume independence between the observations. For example, consider a time series data set, where we assume that observation y_t at time t depends on all the previous observations. The likelihood in this case is:

$$L(\boldsymbol{\theta}) = f(y_1 \,|\, \boldsymbol{\theta}) f(y_2 \,|\, y_1, \boldsymbol{\theta}) f(y_3 \,|\, y_1, y_2, \boldsymbol{\theta}) \ldots f(y_n \,|\, y_1, \ldots, y_{n-1}, \boldsymbol{\theta}) \;,$$

which could be very complicated. A further Markov assumption that y_t depends only on the observation y_{t-1} simplifies the likelihood to:

$$L(\boldsymbol{\theta}) = f(y_1 \,|\, \boldsymbol{\theta}) f(y_2 \,|\, y_1, \boldsymbol{\theta}) f(y_3 \,|\, y_2, \boldsymbol{\theta}) \ldots f(y_n \,|\, y_{n-1}, \boldsymbol{\theta}) \;.$$

10.2.5 Air conditioning example: likelihood function

For demonstrating the basic ideas of the likelihood function we use the air conditioning data `aircond` first used in Section 1.2.4. The data consist of 24 observations of intervals, in service-hours, between failures of the air-conditioning equipment. Here we investigate the likelihood function under two different scenarios assuming (a) an exponential distribution, and (b) a gamma distribution.

Model 1: exponential distribution model, EXP(μ). First assume that the data are independent identically distributed observations. An exponential distribution model has only one parameter, see equation (19.1), and hence $\theta = \mu$. Under this model the likelihood function is

$$L(\mu) = \prod_{i=1}^{n} \frac{1}{\mu} e^{-y_i/\mu} \tag{10.6}$$

with log-likelihood function given by

$$\ell(\mu) = \sum_{i=1}^{n} \left\{ -\frac{y_i}{\mu} - \log(\mu) \right\} = -\frac{1}{\mu} \sum_{i=1}^{n} y_i - n \log(\mu) \;. \tag{10.7}$$

Note that the likelihood function depends on the data only through the function $S = \sum_{i=1}^{n} y_i$. Functions of the data only are generally called *statistics.* Functions of the data appearing in the likelihood, such as S, are called *sufficient statistics.* Sufficient statistics are important within the classical methodology of inference because they provide a way of finding good estimators. Sufficient statistics exist if the assumed distribution belongs to the exponential family (see Section 13.8.3).

The likelihood and log-likelihood functions for the parameter μ for the air conditioning data are shown in Figure 10.3.

```
data(aircond)
llike <- function(x) sum(dEXP(aircond, mu=x, log=TRUE))
like <- function(x) prod(dEXP(aircond, mu=x))
logL <- Vectorize(llike)
L <- Vectorize(like)
```

```
curve(L, 40, 112, xlab=expression(mu), ylab=expression(L(mu)))
title("(a) Likelihood")
curve(logL, 40, 112, xlab=expression(mu), ylab=expression(logL(mu)))
title("(b) Log-likelihood")
```

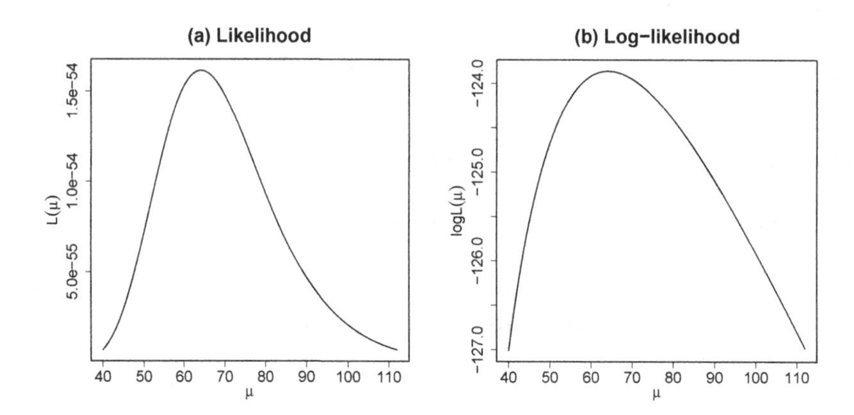

FIGURE 10.3: (a) Likelihood and (b) log-likelihood function for parameter μ from an exponential distribution model, EXP(μ), for the air conditioning data.

Model 2: gamma distribution model, GA(μ, σ). Under the second model, assume that the data are independent identically distributed observations and assume a gamma distribution model, i.e. the two-parameter distribution defined in equation (19.2), with mean $\text{E}(Y) = \mu$ and variance $\text{Var}(Y) = \sigma^2\mu^2$. Here $\boldsymbol{\theta} = (\mu, \sigma)^\top$. The likelihood function is

$$L(\mu, \sigma) = \prod_{i=1}^{n} \frac{1}{(\sigma^2\mu)^{1/\sigma^2}} \frac{y_i^{1/\sigma^2-1} \, e^{-y_i/(\sigma^2\mu)}}{\Gamma(1/\sigma^2)} \tag{10.8}$$

with log-likelihood

$$\begin{aligned} \ell(\mu, \sigma) &= \sum_{i=1}^{n} \left[-\frac{1}{\sigma^2}\left(\log\sigma^2 + \log\mu\right) + \left(\sigma^{-2} - 1\right)\log y_i - \frac{y_i}{\sigma^2\mu} - \log\Gamma\left(\sigma^{-2}\right) \right] \\ &= -\frac{n}{\sigma^2}\left(\log\sigma^2 + \log\mu\right) - n\log\Gamma\left(\sigma^{-2}\right) \\ &+ \left(\sigma^{-2} - 1\right)\sum_{i=1}^{n}\log y_i - \frac{\sum_{i=1}^{n} y_i}{\sigma^2\mu} \,. \end{aligned} \tag{10.9}$$

There are two sufficient statistics in the gamma case: $S = \sum_{i=1}^{n} y_i$ and $T = \sum_{i=1}^{n} \log y_i$. (The gamma distribution belongs to the exponential family.) Figure 10.4 shows (a) a contour and (c) a surface plot of the two-dimensional log-likelihood function $\ell(\mu, \sigma)$ for a gamma distribution model for the air conditioning data. Panels (b) and (d) show the corresponding contour and surface plots of the likelihood function $L(\mu, \sigma)$, respectively.

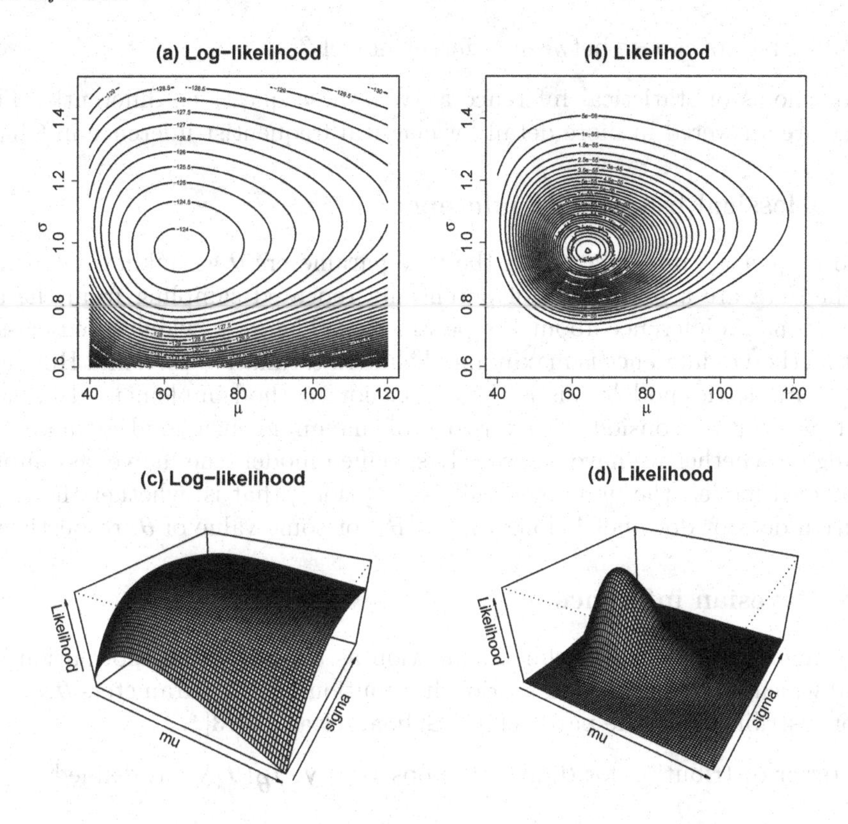

FIGURE 10.4: Contour plot of (a) the log-likelihood and (b) likelihood, and surface plot of (c) the log-likelihood and (d) likelihood, for parameters μ and σ from a gamma distribution model, GA(μ, σ), for the air conditioning data.

10.3 Using the likelihood function for statistical inference

In the previous section we defined the likelihood and the log-likelihood functions. The next question is how those functions are used for statistical inference about the parameters. The Bayesian and classical frequentist schools of statistics use the likelihood function differently. Bayesian inference uses the likelihood as the only source of information about the parameters coming from the data, which is then combined with a prior distribution for the parameters to give the posterior distribution of the parameters. Classical frequentist inference uses the likelihood for inference about the parameters, assuming repeatable sampling of the data.

Sections 10.3.1 and 10.3.2 briefly discuss how the likelihood function is used by classical frequentist and the Bayesian schools of inference, respectively. Particular attention is given to the following questions:

- how is the likelihood function used for inference about the parameter(s) $\boldsymbol{\theta}$?
- how are nuisance parameters eliminated?

- how do analysts choose between different models?

The two schools of statistical inference answer these questions differently. The three questions are answered in more detail for classical frequentist inference in Chapter 11.

10.3.1 Classical frequentist inference

Classical frequentist inference treats the true parameters $\boldsymbol{\theta}$ as unknown constants, assumes the data are a realization of potentially repeated sampling from the assumed model, and makes inference about the parameters. The most widely-used classical frequentist method of inference is maximum likelihood estimation [Cox and Hinkley, 1979]. For fully parametric models this is the estimation method implemented in **gamlss**. In the next chapter we consider two versions of maximum likelihood estimation theory depending on whether we have a correctly specified model (the 'naïve' assumption), or a misspecified model (the 'more general' assumption); that is, whether the population distribution does or does not belong to $f(y \mid \boldsymbol{\theta})$ for some value of $\boldsymbol{\theta}$, respectively.

10.3.2 Bayesian inference

Bayesian inference uses the likelihood function as the source of information given by the data, which is combined with a prior distribution for the parameters $\boldsymbol{\theta}$, to form the posterior distribution of the parameters [Gelman et al., 2013].

The posterior distribution for $\boldsymbol{\theta}$ given the observed $\mathbf{y}$, $f_{\boldsymbol{\theta}}(\boldsymbol{\theta} \mid \mathbf{y})$, is defined as:

$$\begin{aligned} f_{\boldsymbol{\theta}}(\boldsymbol{\theta} \mid \mathbf{y}) &= \frac{L(\boldsymbol{\theta})\,\pi(\boldsymbol{\theta})}{\int_{\boldsymbol{\theta}'} L(\boldsymbol{\theta}')\pi(\boldsymbol{\theta}')d\boldsymbol{\theta}'} \\ &\propto L(\boldsymbol{\theta})\,\pi(\boldsymbol{\theta}) \\ &\propto \text{likelihood} \times \text{prior} \end{aligned} \tag{10.10}$$

where $\pi(\boldsymbol{\theta})$ is a prior distribution for $\boldsymbol{\theta}$ and $\left[\int_{\boldsymbol{\theta}'} L(\boldsymbol{\theta}')\pi(\boldsymbol{\theta}')d\boldsymbol{\theta}'\right]^{-1}$ is a normalizing constant. It is obvious from (10.10) that the only information about $\boldsymbol{\theta}$ coming from the data is contained in the likelihood function $L(\boldsymbol{\theta})$.

The following comments are related to Bayesian inference:

- By having a posterior distribution for all the parameters in $\boldsymbol{\theta}$, probabilistic conclusions can be drawn about them. For example the mean, mode (*maximum a posteriori* or MAP), variance, or quantiles of the posterior distribution of $\boldsymbol{\theta}$ can be obtained. The parameters in $\boldsymbol{\theta}$ are random variables as far as Bayesian inference is concerned.
- The Bayesian school fully conditions its inference on the given data (and the appropriateness of the assumed model). It is not concerned with what could have happened, but only on what did happen.
- The derivation of $f_{\boldsymbol{\theta}}(\boldsymbol{\theta} \mid \mathbf{y})$ involves integration, possibly in high dimensional spaces, something that had held up the spread of Bayesian techniques for a long time. Nowadays the use of computer simulation techniques such as Markov Chain Monte Carlo (MCMC) has changed this, see Gilks et al. [1996] and Gill [2014].
- A problem with the Bayesian school of inference has to do with the priors $\pi(\boldsymbol{\theta})$. Bayesian theory requires any prior information about $\boldsymbol{\theta}$ to be expressed as a prior

pdf. These can be *informative* or *non-informative* priors. Informative priors are based on prior information or relative beliefs, while non-informative priors are uninformative relative to the information contained in the data and therefore have minimal influence on the inference. Note that $\pi(\boldsymbol{\theta})$ does not have to be a proper distribution, as long as the posterior $f_{\boldsymbol{\theta}}(\boldsymbol{\theta} \mid \mathbf{y})$ is. For more information about priors see Barnett [1999], Gelman et al. [2013], or Aitkin [2010, Chapter 2].

- If we are interested in only one of the parameters in $\boldsymbol{\theta}$, say θ_1, we can eliminate the rest of the parameters (sometimes called the *nuisance parameters*) by integrating them out:

$$f_{\theta_1}(\theta_1) = \int_{\theta_2} \dots \int_{\theta_k} f_{\boldsymbol{\theta}}(\boldsymbol{\theta} \mid \mathbf{y}) \, d\theta_k \dots d\theta_2 \, .$$

  That is, Bayesian inference has a simple way of eliminating nuisance parameters.

- From the statistical modeling point of view, note that the likelihood $L(\boldsymbol{\theta})$ refers to only one possible model. In practice, for a given data set, we are often faced with the situation of choosing between several different models. *Bayes factors* and the *deviance information criterion* are used in those circumstances to choose between models [Raftery, 1999, Gelman et al., 2013].

- The adequacy of a particular model should be checked using residual diagnostics.

- Bayesian inference for a particular model deals well with parameter uncertainty, but is vulnerable to model misspecification.

10.4 Bibliographic notes

A thorough treatment of the important topic of statistical inference is beyond the scope of this book. Excellent texts on the subject include Cox and Hinkley [1979], Lindsey [1996], Barnett [1999], and Held and Sabanés Bové [2014], which cover classical frequentist and Bayesian approaches, and Gelman et al. [2013] and Aitkin [2010] on the Bayesian approach.

10.5 Exercises

1. Construct plots of the likelihood and log-likelihood functions for the Poisson parameter μ, for the number of deaths by horse kicks in the Prussian army, introduced in Section 1.2.5 (data set `prussian` in package **pscl**).

2. The zero-inflated Poisson, `ZIP`, distribution may be suitable for the horse kicks data. (Confirm this by examining a histogram.) Construct contour and surface plots for μ and σ of the `ZIP` distribution, for these data. (You will need to find reasonable ranges for μ and σ, by trial and error.)

11

Maximum likelihood estimation

CONTENTS

This chapter provides the theoretical background for fitting a distribution to a random sample of observations. In particular it explains:

1. maximum likelihood estimators (MLEs) of the parameters;
2. properties of MLEs under a correctly specified model and how to obtain standard errors, confidence intervals, and tests; and
3. properties of MLEs under a misspecified model.

11.1 Introduction

In Section 11.2, maximum likelihood estimation of the parameters $\boldsymbol{\theta}$ of a distribution model $f(y|\boldsymbol{\theta})$ is considered. In Section 11.3, the statistical properties of the maximum likelihood estimator of $\boldsymbol{\theta}$ are given, assuming a random sample of observations from a correctly specified population distribution model. Approximation of the expected information matrix and log-likelihood function are discussed in Section 11.4. Standard error based approximate confidence intervals and Wald tests for a parameter θ are given in Section 11.5. More accurate profile confidence intervals and generalized likelihood ratio tests for a parameter θ are also given in 11.5, and model selection using the generalized Akaike information criterion is investigated. In Section 11.6, the statistical properties of the MLE are given assuming the population distribution model is misspecified. Robust standard error based approximate confidence intervals and tests under model misspecification are given in Section 11.7.

11.2 Maximum likelihood estimators (MLE) of the parameters

Assume $\mathbf{y} = (y_1, y_2, \ldots, y_n)^\top$ is a random sample of observations from a population, and a model distribution is $f(y \mid \boldsymbol{\theta})$, which may or may not be correctly specified, with unknown parameter(s) $\boldsymbol{\theta}$. The *maximum likelihood estimate* (MLe) is the value of $\boldsymbol{\theta}$, say $\hat{\boldsymbol{\theta}}$, which maximizes the likelihood function $L(\boldsymbol{\theta})$ or equivalently maximizes the log-likelihood $\ell(\boldsymbol{\theta})$. [It is common in statistics to use a circumflex ('hat') above a parameter symbol (e.g. $\hat{\boldsymbol{\theta}}$) to indicate an estimate.] The MLe is called a *point estimate* in classical frequentist statistical inference. Point estimates are the solutions to the following inferential problem: if I have to guess the value of the parameters $\boldsymbol{\theta}$ what should I use? Point estimators are rather naïve in the sense that even if we believe that there is a 'true' parameter value our guess would inevitably be wrong.

The MLe of $\boldsymbol{\theta}$ is a function of the random sample of observations $\mathbf{y}$, while the *maximum likelihood estimator* (MLE) of $\boldsymbol{\theta}$ is the same function of the random sample of variables[a] $\mathbf{Y} = (Y_1, Y_2, \ldots, Y_n)^\top$. Hence the MLe of $\boldsymbol{\theta}$ is a vector of constant numerical values calculated from the observed $\mathbf{y}$, while the MLE of $\boldsymbol{\theta}$ is a vector of random variables which are functions of $\mathbf{Y}$.

MLe's can be derived analytically or numerically. Analytical solutions are rather rare, so numerical computation of MLe's is frequently needed.

Air conditioning example continued: MLE

Model 1: exponential distribution model

This demonstrates how to find the MLe of $\theta = \mu$ analytically, using Model 1 in Section 10.2.5 where an exponential distribution model, EXP(μ), was used. Differentiating $\ell(\mu)$ of equation (10.7) with respect to μ gives:

$$\frac{d\ell(\mu)}{d\mu} = \frac{\sum_{i=1}^{n} y_i}{\mu^2} - \frac{n}{\mu}. \tag{11.1}$$

By setting the result to zero and solving for μ gives

$$\frac{d\ell(\mu)}{d\mu} = 0 \quad \rightarrow \quad \hat{\mu} = \frac{\sum_{i=1}^{n} y_i}{n} = \bar{y} \tag{11.2}$$

[a] Note Y_i for $i = 1, 2, \ldots, n$ are independent and identically distributed random variables from a population distribution $f_P(y)$.

where $\hat{\mu}$ is the MLe of μ. (Since $\frac{d^2\ell}{d\mu^2} < 0$ at $\hat{\mu} = \bar{y}$, it is a maximum point.) In this case the MLe $\hat{\mu}$ is the sample mean $\bar{y}$, while the corresponding random variable $\bar{Y}$ is the MLE of μ. For the example in Section 10.2.5 on the time intervals, in hours, between failures of the air-conditioning equipment, the MLe of μ is $\hat{\mu} = \bar{y} = 64.125$ hours, for an exponential, EXP(μ), distribution model for Y.

Model 2: gamma distribution model

For Model 2 in Section 10.2.5 the gamma distribution model, GA(μ, σ) was used, where $\boldsymbol{\theta} = (\mu, \sigma)$. Differentiating $\ell(\mu, \sigma)$ in (10.9) with respect to μ and σ gives:

$$\frac{\partial \ell(\mu,\sigma)}{\partial \mu} = \frac{\sum_{i=1}^n y_i - n\mu}{\sigma^2 \mu^2} \tag{11.3}$$

$$\frac{\partial \ell(\mu,\sigma)}{\partial \sigma} = \frac{2}{\sigma^3}\left[\left(\frac{\sum_{i=1}^n y_i}{\mu}\right) - \left(\sum_{i=1}^n \log y_i\right) + n\log(\mu) + n\log(\sigma^2) - n + n\Psi\left(\frac{1}{\sigma^2}\right)\right] \tag{11.4}$$

where $\Psi(x) = \frac{d}{dx}\log\Gamma(x)$ is the psi or digamma function. Setting (11.3) and (11.4) to zero and solving gives the MLe's for μ and σ. Clearly $\hat{\mu} = \bar{y}$, but $\hat{\sigma}$ cannot be obtained in closed form. The next section shows how $\hat{\mu}$ and $\hat{\sigma}$ are obtained numerically.

Numerical maximization

For Model 1, in order to find the MLe of μ numerically in **R**, there are several options. We can use the general optimization function `optim()`, or more easily the **gamlss** functions `gamlssML()` or `gamlss()`.

To use `optim()` first we define the log-likelihood function, using the definition of log-likelihood in equation (10.2) or more precisely minus the log-likelihood, since by default `optim()` minimizes rather than maximizes functions. We then call the function. For Model 1 we have :

```
data(aircond)
logl <- function(mu) -sum(dEXP(aircond, mu=mu, log=TRUE))
o1 <- optim(45, logl, method="Brent", lower=0.01, upper=1000)
o1$par
## [1] 64.125
o1$value
## [1] 123.86
```

The MLe is $\hat{\mu} = \bar{y} = 64.125$, while the fitted log-likelihood is $\hat{\ell} = -123.860$. Note that we use the value `45` as starting value for μ, while `0.01` and `1000` are used as the lower and upper limits for the search, respectively. Note that in the definition of the function `logl` the `gamlss.family` function `dEXP()` with argument `log=TRUE` is used to get the log-likelihood of the exponential distribution. In `optim()`, the option `method="Brent"` is used, which is recommended for one-parameter minimization. For multidimensional $\boldsymbol{\theta}$ (as below in Model 2) the option `method="L-BFGS-B"` should be used.

Note that in the air conditioning data, the values are rounded to integers, so for example value $y = 50$ corresponds to $49.5 < y < 50.5$. Hence using the definition of the likelihood for a continuous variable given in (10.3) and taking logs we obtain a similar MLe of μ:

```
logl1 <- function(mu) -sum(log(pEXP(aircond+.5, mu=mu)-
                        pEXP(aircond-.5, mu=mu)))
optim(45, logl1, method="Brent", lower=0.01, upper=1000)$par
## [1] 64.1237
```

For Model 2 (ignoring the rounding):

```
loglgamma <- function(p) -sum(dGA(aircond, mu=p[1], sigma=p[2], log=TRUE))
o2 <- optim(c(45,1), loglgamma, method="L-BFGS-B",
            lower=c(0.01, 0.01), upper=c(Inf, Inf))
o2$par
## [1] 64.122747  0.972409
o2$value
## [1] 123.8364
```

The MLe are given as $\hat{\mu} = 64.123$ (slightly different from $\bar{y} = 64.125$, probably due to numerical approximation) and $\hat{\sigma} = 0.972$ while the fitted log-likelihood is $\hat{\ell} = -123.836$. The methods `optim()` and `mle()` allow restrictions on the values of the parameter space, which is important in this case since μ in the exponential distribution and μ and σ in the gamma distribution are constrained to be positive.

The function `gamlssML()` maximizes the log-likelihood with respsect to the coeficients $\beta_1 = \log(\mu)$ and $\beta_2 = \log(\sigma)$ and therefore the restrictions on μ and σ to be positive are satisfied automatically.

```
# Model 1: exponential
m1 <- gamlssML(aircond, family=EXP)
coef(m1)
## [1] 4.160834
fitted(m1,"mu")[1]
## [1] 64.125
deviance(m1)
## [1] 247.72
# Model 2: gamma
m2 <- gamlssML(aircond, family=GA)
coef(m2)
## [1] 4.160834
fitted(m2,"mu")[1]
## [1] 64.12501
coef(m2,"sigma")
## [1] -0.02796259
fitted(m2,"sigma")[1]
## [1] 0.9724247
deviance(m2)
## [1] 247.6728
```

For example, for Model 2 we have $\exp(\hat{\beta}_1) = \hat{\mu} = 64.125$ and $\exp(\hat{\beta}_2) = \hat{\sigma} = 0.972$. The deviance is defined as $\text{GDEV} = -2\ell$ therefore:

```
deviance(m1)/2
## [1] 123.86
o1$value
```

```
## [1] 123.86
```

are identical. Deviances are better for model comparison and we can see that there is only a very small difference between the deviances of models `m1` and `m2`.

Based on a given sample, a point estimate $\hat{\theta}$ is a 'best' guess of the value of parameter θ. However $\hat{\theta}$ does not give information on its precision. For this we need some theoretical properties of the maximum likelihood estimators. Those properties are examined in Section 11.3, in which the case of a correctly specified model is considered, i.e. the naïve inferential assumption that $f_P(y) = f(y \mid \boldsymbol{\theta}_T)$, where $\boldsymbol{\theta}_T$ is the true value of $\boldsymbol{\theta}$. In Section 11.6 the more general case of a misspecified model where $f(y \mid \boldsymbol{\theta}_c)$ is the closest approximation of $f_P(y)$, is discussed.

11.3 Properties of the MLE under a correctly specified model

Assume $\mathbf{Y} = (Y_1, Y_2, ..., Y_n)$ is a random sample (i.e. independently identically distributed, iid, random variables) from population distribution $f_P(y)$. A parametric model family of probability (density) functions is given by $f(y \mid \boldsymbol{\theta})$ for a range of values of $\boldsymbol{\theta}$, where $f(y \mid \boldsymbol{\theta})$ is a known function of y, except for parameters $\boldsymbol{\theta}$. The naïve statistical inference assumption is that $f_P(y) = f(y \mid \boldsymbol{\theta}_T)$ for all y, where $\boldsymbol{\theta}_T$ is the true value of $\boldsymbol{\theta}$. In this case the model pdf $f(y \mid \boldsymbol{\theta})$ is correctly specified. Let $\hat{\boldsymbol{\theta}}$ be the MLE of $\boldsymbol{\theta}$ from the correctly specified model pdf $f(y \mid \boldsymbol{\theta})$ given the random sample $\mathbf{Y}$. There are three basic properties of $\hat{\boldsymbol{\theta}}$, assuming certain conditions hold: *invariance*, *consistency* and *asymptotic normality*. Let $\boldsymbol{\theta} = (\theta_1, \theta_2, \ldots, \theta_K)^\top$.

Invariance

Invariance means that if $\boldsymbol{\phi} = g(\boldsymbol{\theta})$ is a one-to-one transformation of $\boldsymbol{\theta}$, then the MLEs $\hat{\boldsymbol{\phi}}$ and $\hat{\boldsymbol{\theta}}$ are related by $\hat{\boldsymbol{\phi}} = g(\hat{\boldsymbol{\theta}})$. It therefore does not matter which parameterization is used, at the point of maximum the likelihood will be the same. Also the transformation from one parameterization to another does not affect the estimates. Note though that, while the MLEs are invariant, their estimated standard errors are not, since they depend on the curvature of the log likelihood function at the point of maximum.

Consistency

The MLE $\hat{\boldsymbol{\theta}}$ is, under certain conditions, a (weakly) consistent estimator of the true parameter $\boldsymbol{\theta}_T$, i.e. $\hat{\boldsymbol{\theta}}$ converges in probability to $\boldsymbol{\theta}_T$ as $n \to \infty$. This means that for all $\varepsilon > 0$,

$$\lim_{n \to \infty} \mathrm{P}(|\hat{\boldsymbol{\theta}} - \boldsymbol{\theta}_T| > \varepsilon) = 0 \,. \tag{11.5}$$

Sufficient conditions for weak consistency (11.5) to hold are given by Newey and McFadden [1994], Theorem 2.5. A derivation of and sufficient conditions for strong consistency, which implies weak consistency (11.5), is given by Wald [1949], and with less restrictive conditions by White [1982].

Asymptotic normality

First we define convergence in distribution. Let $\{U_n; n = 1, 2, \ldots\}$ be a sequence of random variables and U another random variable, with cumulative distribution functions $F_{U_n}(u)$ for $n = 1, 2, \ldots$ and $F_U(u)$, respectively. Then U_n converges in distribution to U as $n \to \infty$, written $U_n \overset{d}{\to} U$, means $\lim_{n\to\infty} F_{U_n}(u) = F_U(u)$ for all continuity points of $F_U(u)$. Under certain conditions, $\sqrt{n}(\hat{\boldsymbol{\theta}} - \boldsymbol{\theta}_T)$ converges in distribution to a K-dimensional normal distribution as $n \to \infty$:

$$\sqrt{n}(\hat{\boldsymbol{\theta}} - \boldsymbol{\theta}_T) \overset{d}{\to} \mathcal{N}_K(\mathbf{0}, \boldsymbol{J}(\boldsymbol{\theta}_T)^{-1}) \,, \tag{11.6}$$

where $\boldsymbol{J}(\boldsymbol{\theta}_T)$ is the (Fisher) expected information matrix for a single observation Y_i, evaluated at $\boldsymbol{\theta}_T$:

$$\boldsymbol{J}(\boldsymbol{\theta}_T) = -\mathrm{E}_{f_P}\left[\frac{\partial^2 \ell_i(\boldsymbol{\theta})}{\partial\boldsymbol{\theta}\partial\boldsymbol{\theta}^\top}\right]_{\boldsymbol{\theta}_T},$$

where $\ell_i(\boldsymbol{\theta}) = \log f_Y(Y_i \mid \boldsymbol{\theta})$ and the expectation is over the population distribution of Y_i, i.e. $f_P(y) = f_{Y_i}(y \mid \boldsymbol{\theta}) = f(y \mid \boldsymbol{\theta})$. An outline of the derivation of (11.6) is given in Appendix 11.8, Ripley [1996, page 32] and Claeskens and Hjort [2008, page 26-27]. A more rigorous derivation of (11.6) with sufficient conditions is given by Cramér [1946] and with less restrictive conditions by White [1982]. Sufficient conditions for (11.6) are also given by Newey and McFadden [1994, Theorem 3.3].

Note also (11.6) should be interpreted in terms of the limit of probabilities associated with $\sqrt{n}(\hat{\boldsymbol{\theta}} - \boldsymbol{\theta}_T)$, and not in terms of the limit of moments of $\sqrt{n}(\hat{\boldsymbol{\theta}} - \boldsymbol{\theta}_T)$. For example the mean (or variance) of $\sqrt{n}(\hat{\boldsymbol{\theta}} - \boldsymbol{\theta}_T)$ does not necessarily converge to the mean (or variance) of the asymptotic distribution. Informally[b], asymptotically we have

$$\hat{\boldsymbol{\theta}} \sim \mathcal{N}_K(\boldsymbol{\theta}_T, n^{-1}\boldsymbol{J}(\boldsymbol{\theta}_T)^{-1}) = \mathcal{N}_K(\boldsymbol{\theta}_T, \boldsymbol{i}(\boldsymbol{\theta}_T)^{-1}) \tag{11.7}$$

where

$$\begin{aligned} \boldsymbol{i}(\boldsymbol{\theta}_T) &= -\mathrm{E}_{\mathbf{Y}}\left[\frac{\partial^2 \ell(\boldsymbol{\theta})}{\partial\boldsymbol{\theta}\partial\boldsymbol{\theta}^\top}\right]_{\boldsymbol{\theta}_T} = -\sum_{i=1}^{n} \mathrm{E}_{f_P}\left[\frac{\partial^2 \ell_i(\boldsymbol{\theta})}{\partial\boldsymbol{\theta}\partial\boldsymbol{\theta}^\top}\right]_{\boldsymbol{\theta}_T} \\ &= -n\mathrm{E}_{f_P}\left[\frac{\partial^2 \ell_i(\boldsymbol{\theta})}{\partial\boldsymbol{\theta}\partial\boldsymbol{\theta}^\top}\right]_{\boldsymbol{\theta}_T} = n\boldsymbol{J}(\boldsymbol{\theta}_T) \end{aligned} \tag{11.8}$$

is the *expected information matrix* of the n iid random variables $\mathbf{Y}$, evaluated at $\boldsymbol{\theta}_T$, and $\ell(\boldsymbol{\theta})$ in (11.8) is the log-likelihood for the sample, as defined in (10.2), but with y_i replaced by Y_i. Note strictly (11.8) uses $\ell(\boldsymbol{\theta})$ as a function of $\mathbf{Y}$ and $\ell_i(\boldsymbol{\theta})$ as a function of Y_i.

Asymptotic efficiency

For a single parameter θ, the maximum likelihood estimator $\hat{\theta}$ of θ_T is asymptotically a more efficient estimator of θ_T than a wide class of alternative estimators. This means that for their asymptotic distributions, the ratio of the mean square error of the alternative estimator of θ_T to that of the MLE is greater than or equal to 1.

[b]Note the formal asymptotic distribution is given by (11.6) while (11.7) is informal because $n^{-1}\mathbf{J}(\boldsymbol{\theta}_T)^{-1} \to 0$ as $n \to \infty$.

11.4 Approximation

11.4.1 Observed information matrix

As $\boldsymbol{i}(\boldsymbol{\theta}_T)$ is not always easy to derive analytically, *the observed information matrix* $\boldsymbol{I}(\boldsymbol{\theta}_T)$ is often used instead. It is defined as

$$\boldsymbol{I}(\boldsymbol{\theta}_T) = -\left[\frac{\partial^2 \ell(\boldsymbol{\theta})}{\partial \boldsymbol{\theta} \partial \boldsymbol{\theta}^\top}\right]_{\boldsymbol{\theta}_T} = -\sum_{i=1}^{n} \left[\frac{\partial^2 \ell_i(\boldsymbol{\theta})}{\partial \boldsymbol{\theta} \partial \boldsymbol{\theta}^\top}\right]_{\boldsymbol{\theta}_T} . \tag{11.9}$$

Note (11.9) uses $\ell(\boldsymbol{\theta})$ as a function of $\mathbf{y}$ and $\ell_i(\boldsymbol{\theta})$ as a function of y_i.

Note that $\boldsymbol{I}(\boldsymbol{\theta}_T)$ is equal to the negative of the Hessian matrix of the log-likelihood function at $\boldsymbol{\theta}_T$. The variance of the asymptotic distribution of $\hat{\boldsymbol{\theta}}$ is then approximated by $\boldsymbol{I}(\boldsymbol{\theta}_T)^{-1}$ instead of $\boldsymbol{i}(\boldsymbol{\theta}_T)^{-1}$. Of course $\boldsymbol{\theta}_T$ is unknown, so it is estimated by $\hat{\boldsymbol{\theta}}$ in both the expected and observed information matrices, giving $\boldsymbol{i}(\hat{\boldsymbol{\theta}})$ and $\boldsymbol{I}(\hat{\boldsymbol{\theta}})$.

Tests using the `summary()` function on a **gamlss** object use the following approximate distribution for $\hat{\boldsymbol{\theta}}$, for large n:

$$\hat{\boldsymbol{\theta}} \sim \mathcal{N}_K(\boldsymbol{\theta}_T, \boldsymbol{I}(\hat{\boldsymbol{\theta}})^{-1}) . \tag{11.10}$$

Let $\boldsymbol{\theta}_T = (\theta_{T1}, \theta_{T2}, .., \theta_{TK})^\top$ and $\hat{\boldsymbol{\theta}} = (\hat{\theta}_1, \hat{\theta}_2, .., \hat{\theta}_K)^\top$. Hence, for $k = 1, 2, .., K$, the estimated standard error of $\hat{\theta}_k$:

$$\widehat{\text{se}}(\hat{\theta}_k) = \left[\widehat{\text{Var}}(\hat{\theta}_k)\right]^{1/2} \tag{11.11}$$

is calculated as the square root of the kth diagonal element of $\boldsymbol{I}(\hat{\boldsymbol{\theta}})^{-1}$.

11.4.2 Log-likelihood function

The log-likelihood function of $\boldsymbol{\theta}$ can be approximated around the point $\hat{\boldsymbol{\theta}}$ by a second-order Taylor series expansion, which is a quadratic function:

$$\begin{aligned} \ell(\boldsymbol{\theta}) &\approx \ell(\hat{\boldsymbol{\theta}}) + \left[\frac{\partial \ell}{\partial \boldsymbol{\theta}}\right]_{\hat{\boldsymbol{\theta}}} (\boldsymbol{\theta} - \hat{\boldsymbol{\theta}}) + \frac{1}{2}(\boldsymbol{\theta} - \hat{\boldsymbol{\theta}})^\top \left[\frac{\partial^2 \ell}{\partial \boldsymbol{\theta} \partial \boldsymbol{\theta}^\top}\right]_{\hat{\boldsymbol{\theta}}} (\boldsymbol{\theta} - \hat{\boldsymbol{\theta}}) \\ &= \ell(\hat{\boldsymbol{\theta}}) + \frac{1}{2}(\boldsymbol{\theta} - \hat{\boldsymbol{\theta}})^\top \left[\frac{\partial^2 \ell}{\partial \boldsymbol{\theta} \partial \boldsymbol{\theta}^\top}\right]_{\hat{\boldsymbol{\theta}}} (\boldsymbol{\theta} - \hat{\boldsymbol{\theta}}) . \end{aligned} \tag{11.12}$$

The second term disappears since $\ell(\boldsymbol{\theta})$ has its maximum at $\boldsymbol{\theta} = \hat{\boldsymbol{\theta}}$ and hence $\left[\frac{\partial \ell}{\partial \boldsymbol{\theta}}\right]_{\hat{\boldsymbol{\theta}}} = 0$. (Note the matrix $\left[\frac{\partial \ell^2}{\partial \boldsymbol{\theta} \partial \boldsymbol{\theta}^\top}\right]_{\hat{\boldsymbol{\theta}}}$ can be obtained in **R** using the function `optimHess()`.) To demonstrate this, the one-dimensional log-likelihood for the `aircond` data for the exponential distribution model and its quadratic approximation are plotted in Figure 11.1. This shows that the quadratic approximation is very good close to the MLe $\hat{\mu}$, but because of the asymmetry of the log-likelihood, it becomes less accurate for points away from the maximum. By the time the two curves reach the horizontal line of Figure 11.1

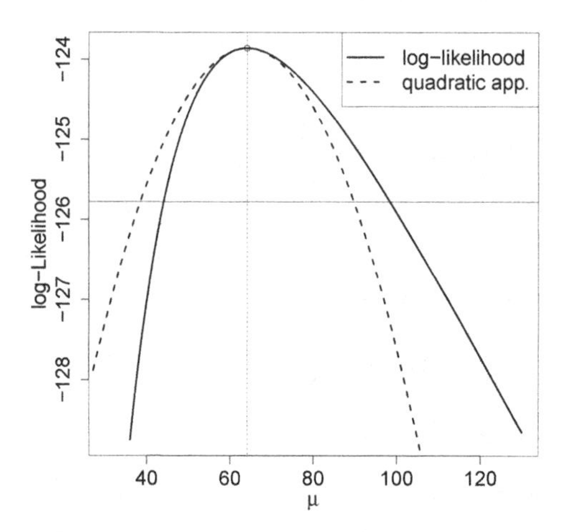

FIGURE 11.1: The log-likelihood of the `aircond` data together with the quadratic approximation of the log-likelihood, against $\theta = \mu$.

they are considerably different. The horizontal line is drawn at a distance 1.92 below $\ell(\hat{\theta})$. The significance of the value 1.92 is that it is half of $\chi^2_{1,0.05} = 3.84$ and is related to an asymptotic result in the construction of a 95% confidence interval for μ.

Confidence intervals based on standard errors rely on the quadratic approximation of the log-likelihood (see Section 11.5.1). Alternative likelihood-based confidence intervals (including profile confidence intervals) are discussed in Section 11.5.2. The two types of confidence intervals can be very different if the quadratic approximation of the likelihood is not accurate as in Figure 11.1. There are two other important points to be made here. The first is that a quadratic shaped log-likelihood is associated with the normal distribution. That is, the log-likelihood for μ of a normal distribution is quadratic in μ, so in this case the log-likelihood and its quadratic approximation are identical. The second point is that the log-likelihood becomes closer to a quadratic shape as the sample size increases, corresponding to the normal asymptotic distribution of the MLE. The shape of the log-likelihood for a finite sample size mainly depends on how the probability (density) function is parameterized. As a general rule the closer the log-likelihood is to a quadratic shape the better, firstly because the search for the maximum is easier, but also because the estimated standard errors of the parameter estimates will be more accurate.

11.5 Confidence intervals, tests, and model selection

A $100(1-\alpha)\%$ *confidence interval* (CI) for a single parameter θ, having true value θ_T, is a range of values which we are $100(1-\alpha)\%$ confident contains θ_T. It is common to use $\alpha = 0.05$ and therefore a 95% confidence interval. A narrow confidence interval indicates an estimate with high precision, and conversely a wide confidence interval indicates an estimate with low precision. A $100(1-\alpha)\%$ confidence interval for θ is also often interpreted as the set of values θ_0 for θ for which when testing, at significance level $100\,\alpha\%$, we would not reject the hypothesis $H_0 : \theta = \theta_0$.

11.5.1 Standard error based confidence intervals and tests

Consider a model distribution with a single parameter θ, having true value θ_T and MLE $\hat{\theta}$. From the asymptotic normality of MLEs, we have the informal asymptotic distribution of $\hat{\theta}$ as $n \to \infty$:

$$\hat{\theta} \sim \mathcal{N}(\theta_T, \left[\mathrm{se}(\hat{\theta})\right]^2),$$

where, from (11.7), with $k = 1$, $\mathrm{se}(\hat{\theta}) = \left[\mathrm{Var}(\hat{\theta})\right]^{1/2} = [i(\theta_T)]^{-1/2}$.

Esimating $\mathrm{se}(\hat{\theta})$ by $\hat{\mathrm{se}}(\hat{\theta}) = \left[\hat{\mathrm{Var}}(\hat{\theta})\right]^{1/2} = \left[i(\hat{\theta})\right]^{-1/2}$ or $\left[I(\hat{\theta})\right]^{-1/2}$ gives

$$\mathrm{P}\left(\hat{\theta} - z_{\alpha/2}\,\hat{\mathrm{se}}(\hat{\theta}) < \theta_T < \hat{\theta} + z_{\alpha/2}\,\hat{\mathrm{se}}(\hat{\theta})\right) \approx 1 - \alpha$$

where $z_{\alpha/2}$ is the value of a standard normal distribution corresponding to upper tail probability $\alpha/2$. Hence a standard error based approximate $100(1-\alpha)\%$ confidence interval for θ_T is then given by $\left(\hat{\theta} \pm z_{\alpha/2}\,\hat{\mathrm{se}}(\hat{\theta})\right)$. A standard error based confidence interval is usually much less accurate than the profile confidence interval given in Section 11.5.2. [The horizontal line in Figure 11.1 gives an example of the difference.]

A test of $H_0 : \theta = \theta_0$ against $H_1 : \theta \neq \theta_0$ can be based on the Wald test statistic

$$Z = \frac{\hat{\theta} - \theta_0}{\hat{\mathrm{se}}(\hat{\theta})} \sim \mathcal{N}(0, 1) \qquad (11.13)$$

asymptotically as $n \to \infty$, if H_0 is true. Hence in a $100\alpha\%$ *significance level test*, we reject H_0 if the observed test statistic z satisfies:

$$|\,z\,| = \left|\frac{\hat{\theta} - \theta_0}{\hat{\mathrm{se}}(\hat{\theta})}\right| > z_{\alpha/2}\,. \qquad (11.14)$$

The Wald test is usually much less accurate than the generalized likelihood ratio test given in Section 11.5.4.

For a vector parameter $\boldsymbol{\theta}$, having true value $\boldsymbol{\theta}_T$ and MLE $\hat{\boldsymbol{\theta}}$, then, from (11.7), a confidence interval for one of the the parameters in $\boldsymbol{\theta}$, say θ_k, is obtained approximately from

$$\mathrm{P}\left(\hat{\theta}_k - z_{\alpha/2}\,\hat{\mathrm{se}}(\hat{\theta}_k) < \theta_{Tk} < \hat{\theta}_k + z_{\alpha/2}\,\hat{\mathrm{se}}(\hat{\theta}_k)\right) \approx 1 - \alpha$$

where $\hat{\mathrm{se}}(\hat{\theta}_k)$ is the square root of the kth diagonal element of $\left[\boldsymbol{i}(\hat{\boldsymbol{\theta}})\right]^{-1}$ or $\left[\boldsymbol{I}(\hat{\boldsymbol{\theta}})\right]^{-1}$.

The following points are important here:

- Confidence intervals based directly on standard errors are symmetrical about the fitted parameter $\hat{\theta}$. This may not be a good idea, if the log-likelihood for the parameter is far from a quadratic shape (as, for example, in Figure 11.1), because the resulting confidence intervals are not reliable, i.e. their coverage may be far from the nominal coverage $100(1-\alpha)\%$. (*Coverage* is the percentage of confidence intervals that capture the true parameter value, over repeated drawings of random samples.)

- Profile confidence intervals are generally more reliable than standard error-based confidence intervals, and provide a way of eliminating nuisance parameters, see Section 11.5.2.

- If there is very high correlation between the parameters $\boldsymbol{\theta}$, that is, if the values of the off-diagonal elements of the information matrix are relatively large, individual standard errors and CI (even profile CI) may be misleading. Bayesian elimination of nuisance parameters may work better in those cases.

Air conditioning example continued: se based CI

Model 1 Here we obtain a standard error based CI for parameter μ in the exponential distribution model, EXP(μ). For the `aircond` example using the exponential distribution model, using (11.1) the observed information at $\mu = \mu_T$ is

$$\boldsymbol{I}(\mu_T) = -\left[\frac{d^2\ell(\mu)}{d\mu^2}\right]_{\mu_T} = \frac{2\sum_{i=1}^n y_i}{\mu_T^3} - \frac{n}{\mu_T^2} .$$

The expected information matrix evaluated at $\mu = \mu_T$ is

$$\begin{aligned} \boldsymbol{i}(\mu_T) = -\mathrm{E}\left[\frac{d^2\ell(\mu)}{d\mu^2}\right]_{\mu_T} &= \frac{2\sum_{i=1}^n \mathrm{E}\left(Y_i\right)}{\mu_T^3} - \frac{n}{\mu_T^2} \\ &= \frac{2n\mu_T}{\mu_T^3} - \frac{n}{\mu_T^2} \\ &= \frac{n}{\mu_T^2} , \end{aligned} \tag{11.15}$$

where, in (11.15), $\ell(\mu)$ is treated as a function of the random sample of iid random variables $\mathbf{Y}$, rather than the observed sample $\mathbf{y}$.

Hence $\boldsymbol{J}(\mu_T) = \boldsymbol{i}(\mu_T)/n = 1/\mu_T^2$ and as $n \to \infty$

$$\sqrt{n}(\hat{\mu} - \mu_T) \xrightarrow{d} \mathcal{N}\left(0, \boldsymbol{J}(\mu_T)^{-1}\right) = \mathcal{N}\left(0, \mu_T^2\right) \approx \mathcal{N}\left(0, \hat{\mu}^2\right)$$

and, using (11.7) with $k = 1$, informally, asymptotically as $n \to \infty$

$$\hat{\mu} \sim \mathcal{N}\left(\mu_T, \boldsymbol{i}(\mu_T)^{-1}\right) = \mathcal{N}\left(\mu_T, \frac{\mu_T^2}{n}\right) \approx \mathcal{N}\left(\mu_T, \frac{\hat{\mu}^2}{n}\right) .$$

Using **R** we obtain:

```
# estimated observed information (calculated explicitly)
2*sum(aircond)/mean(aircond)^3-(length(aircond)/mean(aircond)^2)

## [1] 0.005836554

# estimated observed information (calculated numerically)
-optimHess(o1$par, logl)
```

```
##              [,1]
## [1,] 0.005836554

# estimated expected information matrix
(length(aircond)/mean(aircond)^2)

## [1] 0.005836554
```

In this example (a special case) the estimated observed information and the estimated expected information are identical. Note how the numerical Hessian of the negative of the log-likelihood is calculated using the standard **R** function `optimHess()`. The estimated standard error for $\hat{\mu}$ is given in this case by $\hat{\text{se}}(\hat{\mu}) = [\boldsymbol{i}(\hat{\mu})]^{-1/2} = \hat{\mu}/\sqrt{n} = \sqrt(1/0.005836554) = 13.0895$.

An approximate (se based) 95% confidence interval for μ_T is

$$(\hat{\mu} \pm 1.96 \times \hat{\text{se}}(\hat{\mu})) = (64.12 \pm 1.96 \times 13.0895) = (38.47, 89.78)\ . \tag{11.16}$$

It is, however, much easier to obtain (potentially more accurate) se based confidence intervals using the `gamlss()` function.

The following **R** code shows how the `gamlss()` function can be used to obtain se based confidence intervals for μ. Note however that the default link for μ in the exponential `gamlss.family` distribution, `EXP`, is log. What this means is that the parameter fitted in the predictor model is not μ but $\log(\mu)$ and therefore the corresponding confidence interval is for $\log \mu$. By exponentiating the endpoints of the confidence interval for $\log \mu$ we obtain a confidence interval for μ.

For parameters defined on $(0, \infty)$, modeling the log of the parameter, constructing a confidence interval on the log scale and then transforming it (by exponentiating) generally produces more reliable confidence intervals than those constructed directly on the (untransformed) original scale.

```
m1 <- gamlss(aircond~1, family=EXP, trace=FALSE)
summary(m1)

## ******************************************************************
## Family:  c("EXP", "Exponential")
##
## Call:
## gamlss(formula = aircond ~ 1, family = EXP, trace = FALSE)
##
##
## Fitting method: RS()
##
## ------------------------------------------------------------------
## Mu link function:  log
## Mu Coefficients:
##             Estimate Std. Error t value Pr(>|t|)
## (Intercept)   4.1608     0.2041   20.38 3.19e-16 ***
## ---
## Signif. codes:
## 0 '***' 0.001 '**' 0.01 '*' 0.05 '.' 0.1 ' ' 1
```

```
##
## ------------------------------------------------------------------
## No. of observations in the fit:  24
## Degrees of Freedom for the fit:  1
##       Residual Deg. of Freedom:  23
##                       at cycle:  2
##
## Global Deviance:     247.72
##             AIC:     249.72
##             SBC:     250.8981
## ******************************************************************
```

```
confint(m1)          # 95% CI for log mu
```

```
##                2.5 %  97.5 %
## (Intercept) 3.760758 4.56091
```

```
exp(confint(m1))     # 95% CI for mu
```

```
##                2.5 %   97.5 %
## (Intercept) 42.98101 95.67052
```

The approximate 95% confidence interval for $\beta_T = \log \mu_T$ is computed by `confint(m1)` as $\left(\hat{\beta} \pm 1.96 \times \text{se}(\hat{\beta})\right) = (4.16 \pm 1.96 \times 0.20) = (3.76, 4.56)$, which is symmetric about the estimate $\hat{\beta} = 4.16$. However the transformed 95% confidence interval for μ is $(\exp(3.76), \exp(4.56)) = (42.99, 95.68)$, which is not symmetric about $\hat{\mu} = \exp(\hat{\beta}) = 64.12$. Note also that this 95% confidence interval for μ is very different from and probably more reliable than that of (38.47, 89.78) obtained in equation (11.16).

Model 2

Here we obtain se based confidence intervals for parameters μ and σ in the gamma distribution model for the `aircond` example.

```
m2 <- gamlss(aircond~1, family=GA, trace=FALSE)
confint(m2)                               # 95% CI for log mu
```

```
##               2.5 %   97.5 %
## (Intercept) 3.77179 4.549878
```

```
confint(m2, parameter="sigma")           # 95% CI for log sigma
```

```
##                  2.5 %    97.5 %
## (Intercept) -0.4170062 0.3610815
```

```
exp(confint(m2))                          # 95% CI for mu
```

```
##                2.5 %   97.5 %
## (Intercept) 43.45781 94.62087
```

```
exp(confint(m2, parameter="sigma")) # 95% CI for sigma
```

```
##                 2.5 %  97.5 %
## (Intercept) 0.6590169 1.43488
```

Note that the 95% CI for σ is $(0.659, 1.434)$, which contains $\sigma = 1$, corresponding to the exponential distribution. Therefore the exponential distribution null hypothesis $H_0 : \sigma = 1$ cannot be rejected using a 5% significance level test.

11.5.2 Profile log-likelihood function and profile confidence intervals

The profile log-likelihood function generally provides more reliable confidence intervals than standard error based confidence intervals. It provides a way of eliminating nuisance parameters. Let $\boldsymbol{\theta} = (\boldsymbol{\theta}_1, \boldsymbol{\theta}_2)$ be the model parameters. Let us assume that we are interested in parameters $\boldsymbol{\theta}_1$, since for example they answer the scientific question we are looking at. In this case $\boldsymbol{\theta}_1$ are the *parameters of interest*, and we call $\boldsymbol{\theta}_2$ *nuisance parameters*. The question is how we can eliminate $\boldsymbol{\theta}_2$ and concentrate only on inference about $\boldsymbol{\theta}_1$. One answer to the question is to use the *profile log-likelihood* for inference about $\boldsymbol{\theta}_1$.

Let $\ell(\boldsymbol{\theta}) = \ell(\boldsymbol{\theta}_1, \boldsymbol{\theta}_2)$ be the log-likelihood for parameters $\boldsymbol{\theta}$. The profile log-likelihood $p\ell(\boldsymbol{\theta}_1)$ for $\boldsymbol{\theta}_1$ is given by maximizing $\ell(\boldsymbol{\theta}_1, \boldsymbol{\theta}_2)$ over $\boldsymbol{\theta}_2$ for each value of $\boldsymbol{\theta}_1$, i.e.

$$p\ell(\boldsymbol{\theta}_1) = \max_{\boldsymbol{\theta}_2} \ell(\boldsymbol{\theta}_1, \boldsymbol{\theta}_2) \ .$$

For a single parameter $\theta = \theta_1$, $p\ell(\theta) = \max_{\boldsymbol{\theta}_2} \ell(\theta, \boldsymbol{\theta}_2)$. By definition of the MLE, $\hat{\theta}$ is the first element of

$$\hat{\boldsymbol{\theta}} = \max_{\boldsymbol{\theta}} \ell(\boldsymbol{\theta})$$

and clearly we also have

$$\hat{\theta} = \max_{\theta} p\ell(\theta) \ .$$

In order to define a confidence interval based on the profile likelihood, we need to introduce the *generalized likelihood ratio test*. For the hypothesis on the single parameter θ:

$$H_0 : \theta = \theta_0 \quad \text{against } H_1 : \theta \neq \theta_0 \ , \tag{11.17}$$

the test statistic is based on the ratio of profile likelihoods

$$\mathrm{LR} = \frac{pL(\theta_0)}{pL(\hat{\theta})} \ ,$$

where the profile likelihood $pL(\theta)$ is defined similarly to the profile log-likelihood. If H_0 is true, we have the asymptotic result as $n \to \infty$,

$$\Lambda = -2 \log \mathrm{LR} = 2 \left(p\ell(\hat{\theta}) - p\ell(\theta_0) \right) \sim \chi_1^2 \ .$$

This provides a test for (11.17) by comparing the observed generalized likelihood ratio test statistic Λ with a χ_1^2 distribution. Reject H_0 at the $100\alpha\%$ significance level if $\Lambda > \chi^2_{1,1-\alpha}$.

A $100(1-\alpha)\%$ *profile likelihood confidence interval* for a single parameter θ includes all values θ_0 for which we would not reject the hypothesis $H_0 : \theta = \theta_0$ in (11.17) at a $100\alpha\%$ significance level, i.e.

$$\theta_0 : 2 \left(p\ell(\hat{\theta}) - p\ell(\theta_0) \right) < \chi^2_{1,\alpha} \ . \tag{11.18}$$

The *profile global deviance* is defined as $p\text{GDEV}(\theta) = -2\,p\ell(\theta)$. Equation (11.18) may be expressed in terms of pGDEV as

$$\theta_0 : p\text{GDEV}(\theta_0) < p\text{GDEV}(\hat{\theta}) + \chi^2_{1,\alpha} \; .$$

See also Section 11.5.4 and Venzon and Moolgavkar [1988].

Air conditioning example continued: profile confidence intervals

Model 1

In the air conditioning example, assume the data are a random sample from an exponential or a gamma distribution. For the exponential distribution we have only one parameter so there is no need for profile likelihood. The code below shows the use of the function `prof.dev()` for obtaining a likelihood-based confidence interval for the single parameter μ. Note that the function uses the global deviance GDEV, first defined in Section 4.5 as

$$\text{GDEV} = -2 \log L(\hat{\boldsymbol{\theta}}) = -2\,\ell(\hat{\boldsymbol{\theta}}) \; ,$$

rather than the log-likelihood.

```
m1 <- gamlss(aircond~1, family=EXP, trace=FALSE)
m1A <-prof.dev(m1,"mu", min=30, max=120, length=50, col=1)
```

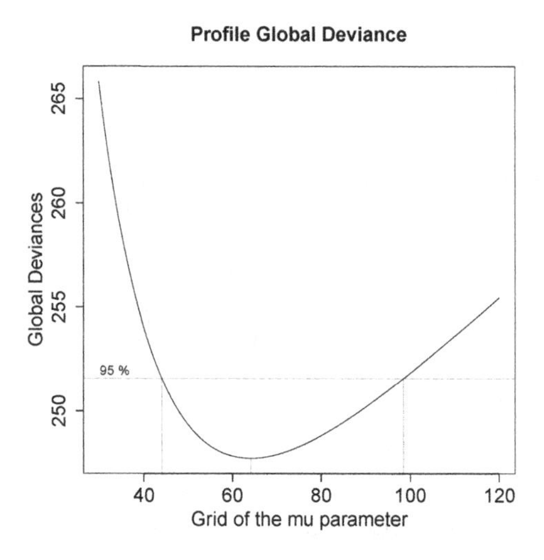

FIGURE 11.2: Profile global deviance plot for the parameter μ under the exponential distribution assumption.

The output for `m1A` (not shown here) gives a likelihood-based 95% confidence interval for μ as $(44.1, 98.4)$. This is different from the standard error-based intervals and is very likely to be more accurate.

Model 2

Under the gamma distribution assumption the profile log-likelihoods for μ and σ are given by $p\ell(\mu) = \max_\sigma \ell(\mu,\sigma)$ and $p\ell(\sigma) = \max_\mu \ell(\mu,\sigma)$, respectively. The left and right panels of Figure 11.3 show the profile global deviance for μ and σ, respectively. The profile global deviance for μ is $p\text{GDEV}(\mu) = -2\,p\ell(\mu)$. The profile global deviance plots of $p\text{GDEV}(\mu)$ and $p\text{GDEV}(\sigma)$ are obtained by:

```
m2 <- gamlss(aircond~1, family=GA, trace=FALSE)
m2A <- prof.dev(m2,"mu", min=30, max=120, length=50, col=1)
m2B <- prof.dev(m2,"sigma", min=0.7, max=1.5, length=50, col=1)
```

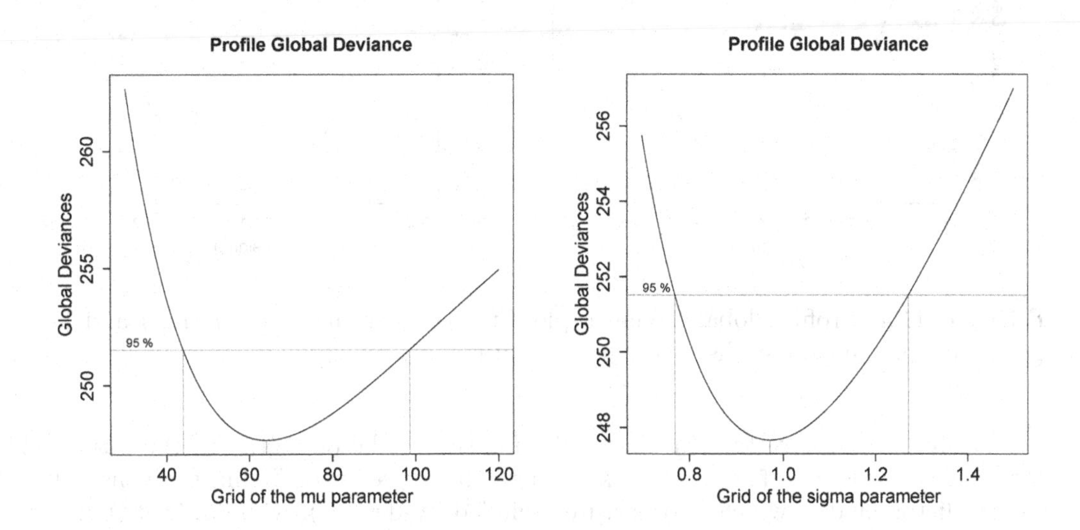

FIGURE 11.3: Profile global deviance plots for parameters μ (left) and σ (right) under the gamma distribution assumption.

Using the output for `m2A` (not shown here), the MLe $\hat{\mu} = 64.125$ corresponds to the minimum global deviance, the vertical dotted lines in the left panel of Figure 11.3 mark the 95% profile confidence interval for μ, i.e. (44.01, 98.60), given by all μ for which

$$p\text{GDEV}(\mu) < p\text{GDEV}(\hat{\mu}) + 3.84 \ ,$$

since $\chi^2_{1,0.05} = 3.84$. Similarly for the profile global deviance plot of $p\text{GDEV}(\sigma)$ in the right panel of Figure 11.3, the MLe $\hat{\sigma} = 0.972$ and the resulting 95% profile confidence interval for σ is $(0.77, 1.27)$. The profile confidence intervals for μ and σ are both different from the corresponding standard error based intervals, and are very likely to be more accurate.

Alternatively the profile deviance function and profile confidence interval for the predictors $\log \mu$ and $\log \sigma$ can be obtained using the `prof.term()` function, which results in the same confidence intervals for μ and σ as above:

```
x<-rep(1,length(aircond))
m2C <- quote(gamlss(aircond~-1+offset(this*x), family=GA))
prof.term(m2C,min=3.6,max=4.8, xlab="mu", col=1)
m2D <- quote(gamlss(aircond~1, sigma.fo=~-1+offset(this*x),family=GA))
prof.term(m2D,min=-0.4,max=0.4,xlab="sigma", col=1)
```

The resulting 95% profile confidence intervals are $(3.78, 4.59)$ for $\log \mu$ and $(-0.26, 0.24)$ for $\log \sigma$, and hence (44.0, 98.6) for μ and (0.77, 1.27) for σ, the same intervals as

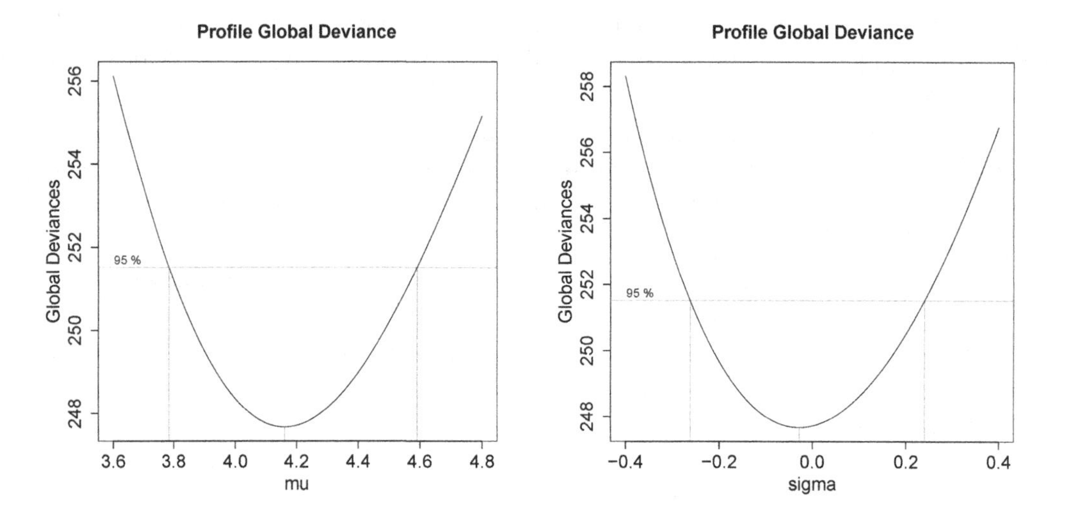

FIGURE 11.4: Profile global deviance plots for the parameter $\log \mu$ (left) and $\log \sigma$ (right) under the gamma distribution assumption.

previously obtained. Note that the model for the predictor $\log(\mu)$ in the `gamlss()` function above is `-1+offset(this*x)`, where the `-1` removes fitting the constant in the predictor model (which is otherwise included by default), while `offset(this*x)` offsets constant `this` in the predictor model, i.e. fits a predictor model with a constant exactly equal to `this`. The profile deviance function is plotted against parameter `this`, i.e. $\beta = \log(\mu)$. The profile deviance function and profile confidence interval for $\log \sigma$ are obtained similarly. It is worth noticing that the profile deviances of the predictors $\log(\mu)$ and $\log(\sigma)$ in Figure 11.4 are a lot closer to a quadratic form than the deviances of the original μ and σ parameters in Figure 11.3.

It is important to note that the method above can also be used to obtain the profile deviance function and profile confidence interval for the coefficient β of an explanatory variable in regression models using the **gamlss** function `prof.term()`, see Stasinopoulos et al. [2017, p130-134].

11.5.3 Model selection

Choosing between models is important because the GAMLSS models are flexible and therefore allow many different possible models for a response variable in a data set to be tried. We should be able to choose between those models in a consistent way. Note that the global deviance plays an important role in testing between different models.

11.5.4 Nested models: the generalized likelihood ratio (GLR) test

Let $\mathcal{M}_0$ and $\mathcal{M}_1$ be two different models. Model $\mathcal{M}_0$ is *nested* within $\mathcal{M}_1$ if it is a special case of $\mathcal{M}_1$. In this case $\mathcal{M}_0$ is the simpler model and $\mathcal{M}_1$ the more complicated one. Consider the hypotheses for the parametric GAMLSS models $\mathcal{M}_0$ and $\mathcal{M}_1$:

$$H_0 : \mathcal{M}_0 \quad \text{and} \quad H_1 : \mathcal{M}_1 \ ,$$

where $\mathcal{M}_0$ and $\mathcal{M}_1$ have fitted likelihoods $\hat{L}_0$ and $\hat{L}_1$, fitted log-likelihoods $\hat{\ell}_0$ and $\hat{\ell}_1$, and fitted global deviances GDEV_0 and GDEV_1 with error degrees of freedom df_{e0} and df_{e1}, respectively. The error degrees of freedom for each model are defined by $\text{df}_e = n - p$, where p is the number of model parameters. The likelihood ratio is $\text{LR} = \hat{L}_0/\hat{L}_1$, and the models are compared using the generalized likelihood ratio (GLR) test statistic:

$$\Lambda = -2\log \text{LR} = 2\left(\hat{\ell}_1 - \hat{\ell}_0\right) = \text{GDEV}_0 - \text{GDEV}_1 \; .$$

Under $\mathcal{M}_0$, assuming that certain conditions are satisfied, we have the asymptotic distribution $\Lambda \sim \chi^2_d$, where $d = \text{df}_{e0} - \text{df}_{e1}$. Hence, at the $100\alpha\%$ significance level, H_0 is rejected if $\Lambda \geq \chi^2_{d,\alpha}$, i.e. $\text{GDEV}_0 - \text{GDEV}_1 \geq \chi^2_{d,\alpha}$.

Air conditioning example continued: GLR test

We consider the `aircond` data again, and here we are interested in which of the exponential or gamma distributions is appropriate for the data. Note that the exponential, $\texttt{EXP}(\mu)$ distribution is a special case of the gamma, $\texttt{GA}(\mu, \sigma)$, where $\sigma = 1$, so the exponential model is a submodel of the gamma model. Hence in the gamma, $\texttt{GA}(\mu, \sigma)$, distribution model, we are testing the null hypothesis $H_0 : \sigma = 1$ against the alternative $H_1 : \sigma \neq 1$. We fit the exponential (H_0) and gamma (H_1) distributions, and check the difference in global deviance against a chi-square distribution with one degree of freedom.

```
m1 <- gamlss(aircond~1, family=EXP, trace=FALSE)
m2 <- gamlss(aircond~1, family=GA, trace=FALSE)
LR.test(m1,m2)
## Likelihood Ratio Test for nested GAMLSS models.
## (No check whether the models are nested is performed).
##
##        Null model: deviance= 247.72 with  1 deg. of freedom
##  Altenative model: deviance= 247.6728 with  2 deg. of freedom
##
## LRT = 0.04720985 with 1 deg. of freedom and p-value= 0.8279915
```

Observed $\Lambda = \text{GDEV}_0 - \text{GDEV}_1 = 247.72 - 247.6728 = 0.0472$, where $\Lambda \dot{\sim} \chi^2_1$ if H_0 is true. Since $\chi^2_{1,0.05} = 3.84$, we reject H_0 if $\Lambda > 3.84$ and hence here we accept H_0, at the 5% significance level. Alternatively, the p-value is $p = \text{P}(\chi^2_1 > 0.0472) = 0.828 > 0.05$, indicating acceptance of H_0 at the 5% significance level.

11.5.5 The generalized Akaike information criterion

For comparing non-nested GAMLSS models, the generalized Akaike information criterion (GAIC) [Akaike, 1983] is appropriate. The GAIC was defined in Section 4.5.1 and is obtained by adding to the fitted global deviance a fixed penalty κ for each effective degree of freedom used in a model. The penalty term avoids overfitting, and the model with the smallest value of $\text{GAIC}(\kappa)$ is selected. The Akaike information criterion (AIC) [Akaike, 1974] and the Schwarz Bayesian criterion (SBC)[c] [Schwarz, 1978] are special

[c]The SBC is also commonly referred to as the Bayesian information criterion, or BIC.

cases of GAIC(κ) corresponding to $\kappa = 2$ and $\kappa = \log n$, respectively. Justification for the use of SBC comes also as a crude approximation to Bayes factors [Raftery, 1996, 1999]. In practice it is usually found that, while the AIC is very generous in model selection, the SBC is too restrictive. Consequently the model chosen by AIC is usually too complicated (leading to overfitting and unstable parameter estimates), while the model chosen by SBC is usually too simple (leading to underfitting and biased parameter estimates). One possibility is to use $\kappa = (2+\log n)/2$ which is a compromise between AIC and SBC (as suggested to us by Ramires et al. [2019]) Our experience is that a value of κ in the range $2.5 \leq \kappa \leq 4$ often works well. Using GAIC(κ) allows different penalties κ to be tried for different modeling purposes. The sensitivity or robustness of the selected model to the choice of k can also be investigated by using a selection of different values of κ, e.g. $\kappa = 2, 2.5, 3, 3.5, 4$. The function `GAIC.table()` facilitates this comparison. Claeskens and Hjort [2003] consider a focused information criterion (FIC) in which the criterion for model selection depends on the objective of the study, in particular on the specific parameter of interest. Claeskens and Hjort [2008] is a good reference for model selection based on information criteria. Chapter 11 in Stasinopoulos et al. [2017] explains in more detail the functions available for model selection in the **gamlss** software.

11.6 Properties of the MLE under a misspecified model

Assume $\mathbf{Y} = (Y_1, Y_2, \ldots, Y_n)$ is a random sample of iid random variables with population distribution $f_P(y)$. A parametric model family of probability (density) functions is given by $f_Y(y \mid \boldsymbol{\theta})$ for a range of values of $\boldsymbol{\theta}$, where $f_Y(y \mid \boldsymbol{\theta})$ is a known function of y, except for parameters $\boldsymbol{\theta}$. Assume the model is incorrect, i.e. assume that $f_P(y)$ does not belong to the model family $f_Y(y \mid \boldsymbol{\theta})$ for any value of $\boldsymbol{\theta}$, but that $\boldsymbol{\theta}_c$ is the value of $\boldsymbol{\theta}$ for which $f_Y(y \mid \boldsymbol{\theta}_c)$ is 'closest' to $f_P(y)$. Let $\hat{\boldsymbol{\theta}}$ be the MLE of $\boldsymbol{\theta}$ in model $f_Y(y \mid \boldsymbol{\theta})$ given the random sample $\mathbf{Y}$.

Figure 11.5 shows a schematic plot similar to Figure 10.2 (b). The population, $f_P(y)$ and the sample, $f_E(y)$, distributions are represented as points, while the model, $f_Y(y \mid \boldsymbol{\theta})$, is represented as a line for different values of $\boldsymbol{\theta}$. In addition two "directed" lines are shown, one from the population and the other from the sample, to the model line. The directed lines represent a form of minimal distance from the points to the line. The point $\boldsymbol{\theta}_c$ on the model line represents the value of $\boldsymbol{\theta}$ which is closest to the population distribution as measured by the distance function $d\left(f_P(y), f_Y(y \mid \boldsymbol{\theta})\right)$:

$$\boldsymbol{\theta}_c = \min_{\boldsymbol{\theta}} d\left(f_P(y), f_Y(y \mid \boldsymbol{\theta})\right) \ ,$$

where $d(\cdot, \cdot)$ is the Kullback-Leibler distance:

$$d\left(f_P(y), f_Y(y \mid \boldsymbol{\theta})\right) = \int \left(\log f_P(y) - \log f_Y(y \mid \boldsymbol{\theta})\right) f_P(y)\, dy \ . \tag{11.19}$$

The point $\hat{\boldsymbol{\theta}}$ represents the point on the model (line) which is at minimum distance from the sample. Distance is measured here similarly to equation (11.19) but where the population distribution $f_P(y)$ is replaced by the empirical distribution $f_E(y)$. In fact

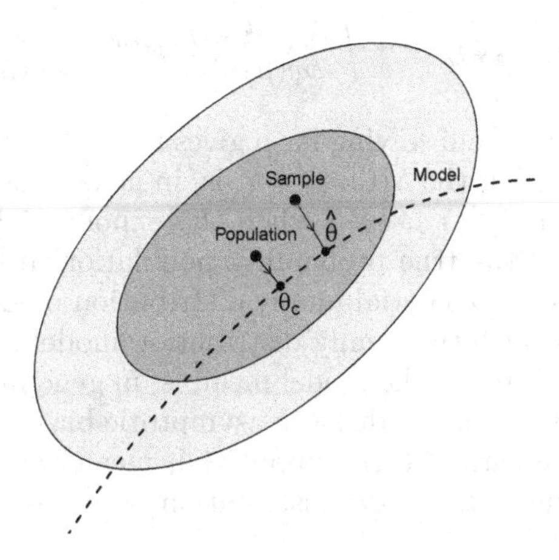

FIGURE 11.5: A schematic presentation of $\boldsymbol{\theta}_c$, the 'closest' or 'best' value for $\boldsymbol{\theta}$ under the model $f_Y(y \,|\, \boldsymbol{\theta})$, and $\hat{\boldsymbol{\theta}}$ its MLE. $\boldsymbol{\theta}_c$ is the value of $\boldsymbol{\theta}$ 'closest' to the population using the Kullback-Leibler distance (or risk function), while $\hat{\boldsymbol{\theta}}$ is the value of $\boldsymbol{\theta}$ closest to the sample using the empirical risk function.

the point $\hat{\boldsymbol{\theta}}$ is the MLE of $\boldsymbol{\theta}$. The model represents an (incorrect) assumption made about the population. A different model assumption will generate a different line and its equivalent point 'closest' to the true population. We will refer to $\boldsymbol{\theta}_c$ as the 'closest' value for $\boldsymbol{\theta}$ under the model $f_Y(y \,|\, \boldsymbol{\theta})$, to emphasize that $f_Y(y \,|\, \boldsymbol{\theta})$ is just one model among other possible models, i.e. $\boldsymbol{\theta}_c$ is the value of $\boldsymbol{\theta}$ for which $f_Y(y \,|\, \boldsymbol{\theta}$ is closest to $f_P(y)$ as judged by the Kullback-Leibler distance. Note, however, that $f_Y(y \,|\, \boldsymbol{\theta}_c)$, although closest, may still not be a good approximation to $f_P(y)$.

The three basic classical properties of the MLE of $\boldsymbol{\theta}$, invariance, consistency, and asymptotic normality, under a correctly specified model were discussed in Section 11.3. Properties of the MLE of $\boldsymbol{\theta}$ under a misspecified model are considered below. White [1982] gives the derivations of sufficient conditions for the properties to hold under model misspecification.

Invariance

The invariance property of $\hat{\boldsymbol{\theta}}$ given in section 11.3 still holds, except the true value $\boldsymbol{\theta}_T$ is replaced by the 'closest' value $\boldsymbol{\theta}_c$.

Consistency

The consistency property of $\hat{\boldsymbol{\theta}}$ given in Section 11.3 still holds, except the true value $\boldsymbol{\theta}_T$ is replaced by the 'closest' value $\boldsymbol{\theta}_c$. There are situations in which $\boldsymbol{\theta}_c$ still represents a true population distribution measure. In particular, for a misspecified exponential family distribution model $\mathcal{E}(\mu, \phi)$ with mean parameter μ and dispersion parameter ϕ, μ_c is always equal to the true population mean. This follows from (11.19), by substituting

$\log f(y \mid \boldsymbol{\theta}) = \log f(y \mid \mu, \phi)$ for an exponential family distribution and differentiating with respect to μ:

$$\frac{\partial}{\partial \mu} d\left(f_P(y), f(y \mid \mu, \phi)\right) = -\int \frac{y-\mu}{\phi v(\mu)} f_P(y)\, dy = \frac{1}{\phi v(\mu)} \left(\mathrm{E}_{f_P}(Y) - \mu\right) . \tag{11.20}$$

Setting this equal to zero and solving for μ gives $\mu_c = \mathrm{E}_{f_P}(Y)$. Hence for an exponential family distribution model, the MLE $\hat{\mu} = \bar{Y}$ is, in general, a consistent estimator of the true population mean $\mathrm{E}_{f_P}(Y)$, even when the exponential family distribution model is a misspecification of the true probability population distribution. This does not, in general, hold for a non-exponential family distribution model [Gourieroux et al., 1984]. Hence when a non-exponential family distribution model misspecifies the true population distribution, the MLE of the model mean is, in general, not a consistent estimator of the true population mean, resulting in asymptotic bias. This is important in practice because if we want to estimate the population mean using a non-exponential family distribution model, then the model distribution needs to be a good approximation of the population distribution.

Asymptotic normality

Under certain conditions, $\sqrt{n}(\hat{\boldsymbol{\theta}} - \boldsymbol{\theta}_c)$ converges in distribution to

$$\mathcal{N}_K\left(\mathbf{0}, \boldsymbol{J}(\boldsymbol{\theta}_c)^{-1} \boldsymbol{K}(\boldsymbol{\theta}_c) \boldsymbol{J}(\boldsymbol{\theta}_c)^{-1}\right)$$

i.e.

$$\sqrt{n}(\hat{\boldsymbol{\theta}} - \boldsymbol{\theta}_c) \xrightarrow{d} \mathcal{N}_K(\mathbf{0}, \boldsymbol{J}(\boldsymbol{\theta}_c)^{-1} \boldsymbol{K}(\boldsymbol{\theta}_c) \boldsymbol{J}(\boldsymbol{\theta}_c)^{-1}) , \tag{11.21}$$

where $\boldsymbol{J}(\boldsymbol{\theta}_c)$ is the expected information matrix for a single observation Y_i evaluated at $\boldsymbol{\theta}_c$:

$$\boldsymbol{J}(\boldsymbol{\theta}_c) = -\mathrm{E}_{f_P}\left[\frac{\partial^2 \ell_i(\boldsymbol{\theta})}{\partial \boldsymbol{\theta} \partial \boldsymbol{\theta}^\top}\right]_{\boldsymbol{\theta}_c} \tag{11.22}$$

and $\boldsymbol{K}(\boldsymbol{\theta}_c)$ is the variance of the first derivative of the log-likelihood for a single observation Y_i, evaluated at $\boldsymbol{\theta}_c$:

$$\boldsymbol{K}(\boldsymbol{\theta}_c) = \mathrm{Var}_{f_P}\left[\frac{\partial \ell_i(\boldsymbol{\theta})}{\partial \boldsymbol{\theta}}\right]_{\boldsymbol{\theta}_c} . \tag{11.23}$$

Note that the expectation and variance in equations (11.22) and (11.23), respectively, are taken over the true population distribution $f_P(y)$.

An outline of the derivation of (11.8) is given in Appendix 11.8, Ripley [1996, p32] and Claeskens and Hjort [2008, p26-27]. A more rigorous derivation of (11.8) with sufficient conditions is given by White [1982]. Equation (11.8) shows that the asymptotic variance-covariance matrix of the MLE is a function of both the expected information matrix and the variance-covariance matrix of the first derivative of the log-likelihood function, both for a single observation Y_i and both evaluated at $\boldsymbol{\theta}_c$. Hence, informally, asymptotically as $n \to \infty$,

$$\hat{\boldsymbol{\theta}} \sim \mathcal{N}_K(\boldsymbol{\theta}_c, n^{-1} \boldsymbol{J}(\boldsymbol{\theta}_c)^{-1} \boldsymbol{K}(\boldsymbol{\theta}_c) \boldsymbol{J}(\boldsymbol{\theta}_c)^{-1}) . \tag{11.24}$$

[Note that if the population distribution belongs to the model parametric family of distributions, then $\boldsymbol{\theta}_c = \boldsymbol{\theta}_T$, $\boldsymbol{K}(\boldsymbol{\theta}_T) = \boldsymbol{J}(\boldsymbol{\theta}_T)$ and the asymptotic variance-covariance matrix of $\hat{\boldsymbol{\theta}}$ is $\boldsymbol{i}(\boldsymbol{\theta}_T)^{-1} = n^{-1}\boldsymbol{J}(\boldsymbol{\theta}_T)^{-1}$. We then have $\sqrt{n}(\hat{\boldsymbol{\theta}} - \boldsymbol{\theta}_T) \xrightarrow{d} \mathcal{N}_K(\mathbf{0}, \boldsymbol{J}(\boldsymbol{\theta}_T)^{-1})$, and informally, asymptotically as $n \to \infty$, $\hat{\boldsymbol{\theta}} \sim \mathcal{N}_k(\boldsymbol{\theta}, \boldsymbol{i}(\boldsymbol{\theta}_T)^{-1}$, as in Section 11.3.]

Approximating the asymptotic variance-covariance matrix

From (11.24), informally, the asymptotic variance-covariance matrix of $\hat{\boldsymbol{\theta}}$ is given by $n^{-1}\boldsymbol{J}(\boldsymbol{\theta}_c)^{-1}\boldsymbol{K}(\boldsymbol{\theta}_c)\boldsymbol{J}(\boldsymbol{\theta}_c)^{-1}$. $\boldsymbol{J}(\boldsymbol{\theta})$ and $\boldsymbol{K}(\boldsymbol{\theta})$ are unknown expected information and variance-covariance matrices, respectively, which can be approximated by the corresponding sample estimates, i.e.

$$\hat{\boldsymbol{J}}(\boldsymbol{\theta}_c) = -\frac{1}{n}\sum_{i=1}^{n}\left[\frac{\partial^2 \ell_i(\boldsymbol{\theta})}{\partial\boldsymbol{\theta}\partial\boldsymbol{\theta}^\top}\right]_{\boldsymbol{\theta}_c} \quad \text{and} \quad \hat{\boldsymbol{K}}(\boldsymbol{\theta}_c) = \frac{1}{n-K}\sum_{i=1}^{n}\left[\frac{\partial \ell_i(\boldsymbol{\theta})}{\partial\boldsymbol{\theta}}\frac{\partial \ell_i(\boldsymbol{\theta})}{\partial\boldsymbol{\theta}^T}\right]_{\boldsymbol{\theta}_c} .$$

Of course $\boldsymbol{\theta}_c$ is unknown, so is estimated by $\hat{\boldsymbol{\theta}}$, giving $\hat{\boldsymbol{J}}(\hat{\boldsymbol{\theta}})$ and $\hat{\boldsymbol{K}}(\hat{\boldsymbol{\theta}})$.

The output from the **gamlss** `summary()` command with option `robust=TRUE` uses the following approximate distribution for $\hat{\boldsymbol{\theta}}$, for large n:

$$\hat{\boldsymbol{\theta}} \sim \mathcal{N}_K(\boldsymbol{\theta}_c, n^{-1}\hat{\boldsymbol{J}}(\hat{\boldsymbol{\theta}})^{-1}\hat{\boldsymbol{K}}(\hat{\boldsymbol{\theta}})\hat{\boldsymbol{J}}(\hat{\boldsymbol{\theta}})^{-1}) .$$

The estimated variance-covariance matrix of $\hat{\boldsymbol{\theta}}$, given by $n^{-1}\hat{\boldsymbol{J}}(\hat{\boldsymbol{\theta}})^{-1}\hat{\boldsymbol{K}}(\hat{\boldsymbol{\theta}})\hat{\boldsymbol{J}}(\hat{\boldsymbol{\theta}})^{-1}$, is called a 'sandwich' estimate [Huber, 1967].

11.7 Robust confidence intervals and tests

Let $\boldsymbol{\theta}_c = (\theta_{c1}, \theta_{c2}, .., \theta_{cK})^\top$ and $\hat{\boldsymbol{\theta}} = (\hat{\theta}_1, \hat{\theta}_2, .., \hat{\theta}_K)^\top$, where $\boldsymbol{\theta}_c$ is the value of parameter $\boldsymbol{\theta}$ which makes the model distribution $f(y \mid \boldsymbol{\theta})$ 'closest' to the true population distribution $f_P(y)$, (as judged by the Kullback-Leibler distance).

First note that the usual standard error based confidence intervals and Wald tests for an individual parameter θ_{ck} are, in general, not valid when the model is misspecified, because they use an incorrect estimated standard error for $\hat{\theta}_k$. Profile likelihood confidence intervals and generalized likelihood ratio tests are also not valid when the model is misspecified. *Robust* standard error based confidence intervals and robust Wald tests should be used when there is model misspecification, as described below.

For $k = 1, 2, .., K$, the robust estimated standard error, $\hat{\text{se}}(\hat{\theta}_k)$, of $\hat{\theta}_k$ is calculated from the square root of the kth diagonal element of $n^{-1}\hat{\boldsymbol{J}}(\hat{\boldsymbol{\theta}})^{-1}\hat{\boldsymbol{K}}(\hat{\boldsymbol{\theta}})\hat{\boldsymbol{J}}(\hat{\boldsymbol{\theta}})^{-1}$. A robust standard error based approximate $100(1-\alpha)\%$ confidence interval for a single parameter say θ, e.g. θ_{ck}, is given by $(\hat{\theta} \pm z_{\alpha/2}\, \hat{\text{se}}(\hat{\theta}))$, where $\hat{\text{se}}(\hat{\theta})$ is the robust estimated standard error. A robust $100\alpha\%$ significance level Wald test of $H_0 : \theta = \theta_0$ against $H_1 : \theta \neq \theta_0$ is based on the robust Wald test statistic, computed as in equation (11.13), with $\hat{\text{se}}(\hat{\theta})$ the robust estimated standard error as above. The corresponding rejection region is given by (11.14).

Air conditioning example continued: robust CI and robust Wald test

Model 1

Here we consider again the air conditioning example under model 1 where we assume an exponential distribution. If the population distribution does not belong to the exponential distribution, then robust confidence intervals and tests should be used. Assume a misspecified exponential distribution model. Then $\frac{d\ell}{d\mu}$ is still given by (11.1). Hence $\boldsymbol{J}(\mu_c) = \frac{1}{n}\boldsymbol{i}(\mu_c) = \mu_c^{-2}$ from Section 11.5.1, and

$$\boldsymbol{K}(\mu_c) = \text{Var}_{f_P}\left[\frac{d\ell_i}{d\mu}\right]_{\mu_c} = \text{Var}_{f_P}\left[\frac{Y_i}{\mu^2} - \frac{1}{\mu}\right]_{\mu_c} = \frac{\text{Var}_{f_P}(Y)}{\mu_c^4}$$

and hence $\boldsymbol{J}(\mu_c)^{-1}\boldsymbol{K}(\mu_c)\boldsymbol{J}(\mu_c)^{-1} = \text{Var}_{f_P}(Y)$ giving as $n \to \infty$,

$$\sqrt{n}(\hat{\mu} - \mu_c) \xrightarrow{d} \mathcal{N}(0, \text{Var}_{f_P}(Y)) \approx \mathcal{N}\left(0, s_y^2\right) \ ,$$

where s_y^2 is the sample variance of Y. Hence the approximate distribution of $\hat{\mu}$, for large n, is

$$\hat{\mu} \sim \mathcal{N}\left(\mu_c, s_y^2/n\right).$$

Note that, since the exponential distribution is a member of the exponential family, μ_c equals the population mean, i.e. $\mu_c = \text{E}_{f_P}(Y)$, see Section 11.6. We demonstrate below the use of the `gamlss()` function to obtain robust confidence intervals for μ_c. Note however that, as in the computation of the standard error based confidence interval for μ in Section 11.5.1, the confidence interval given is for $\log\mu$; the corresponding confidence interval for μ is obtained by exponentiating its endpoints. This generally produces more reliable confidence intervals than those obtained directly from the parameter itself.

```
m1 <- gamlss(aircond~1, family=EXP, trace=FALSE)
summary(m1, robust=T)

## ******************************************************************
## Family:  c("EXP", "Exponential")
##
## Call:
## gamlss(formula = aircond ~ 1, family = EXP, trace = FALSE)
##
##
## Fitting method: RS()
##
## ------------------------------------------------------------------
## Mu link function:  log
## Mu Coefficients:
##             Estimate Std. Error t value Pr(>|t|)
## (Intercept)   4.1608     0.1994   20.86   <2e-16 ***
## ---
## Signif. codes:
## 0 '***' 0.001 '**' 0.01 '*' 0.05 '.' 0.1 ' ' 1
##
## ------------------------------------------------------------------
## No. of observations in the fit:  24
```

```
## Degrees of Freedom for the fit:  1
##       Residual Deg. of Freedom:  23
##                       at cycle:  2
##
## Global Deviance:      247.72
##              AIC:      249.72
##              SBC:      250.8981
## ******************************************************************

vcov(m1, "se", robust=T)  # the standard error for log mu

## (Intercept)
##   0.1994367

confint(m1, robust=T)          # robust 95% CI for log mu

##                  2.5 %   97.5 %
## (Intercept) 3.769946 4.551723

exp(confint(m1, robust=T)) # robust 95% CI for mu

##                  2.5 %  97.5 %
## (Intercept) 43.37771 94.7956
```

Note that using the `robust=T` option above does not affect the parameter estimates, it only affects the estimated standard errors of the parameter estimates. In this example the distribution of the times to breakdown are close to an exponential distribution, and using robust standard errors makes only a small difference. The robust 95% confidence interval for $\log \mu_c = \log \mathrm{E}_{f_P}(Y)$ is calculated as $(4.16 \pm 1.96 \times 0.1994) = (4.16 \pm 0.39) = (3.77, 4.55)$. Hence the robust 95% confidence interval for $\mu_c = \mathrm{E}_{f_P}(Y)$ is $(\exp(3.77), \exp(4.55)) = (43.4, 94.8)$.

Model 2

The robust confidence intervals for the `aircond` example using the gamma distribution are obtained as follows:

```
m2 <- gamlss(aircond~1, family=GA, trace=FALSE)
confint(m2, robust=T)                              # 95% CI for log mu

##                 2.5 %   97.5 %
## (Intercept) 3.76116 4.560508

confint(m2, parameter="sigma", robust=T)         # 95% CI for log sigma

##                    2.5 %     97.5 %
## (Intercept) -0.4276362 0.3717115

exp(confint(m2, robust=T))                         # 95% CI for mu

##                 2.5 %   97.5 %
## (Intercept) 42.9983 95.63206

exp(confint(m2, parameter="sigma", robust=T))  # 95% CI for sigma

##                   2.5 %    97.5 %
```

```
## (Intercept) 0.6520486 1.450215
```

11.8 Appendix: asymptotic normality of MLE under model misspecification

The asymptotic normality property of the maximum likelihood estimator $\hat{\boldsymbol{\theta}}$ under model misspecification was given by equation (11.8) in Section 11.6.

Outline proof

The summation of the first derivatives of the log-likelihood functions (called the score functions) is

$$\mathbf{U}_n(\boldsymbol{\theta}) = \sum_{i=1}^{n} \frac{\partial \ell_i(\boldsymbol{\theta})}{\partial \boldsymbol{\theta}} ,$$

where $\ell_i(\boldsymbol{\theta}) = \log f_Y(Y_i \,|\, \boldsymbol{\theta})$. The MLE $\hat{\boldsymbol{\theta}}$ solves $\mathbf{U}_n(\hat{\boldsymbol{\theta}}) = 0$. A first-order Taylor series expansion gives

$$\mathbf{U}_n(\hat{\boldsymbol{\theta}}) \simeq \mathbf{U}_n(\boldsymbol{\theta}_c) + \boldsymbol{I}_n(\tilde{\boldsymbol{\theta}})(\hat{\boldsymbol{\theta}} - \boldsymbol{\theta}_c) = 0 \tag{11.25}$$

where $\tilde{\boldsymbol{\theta}}$ is a value of $\boldsymbol{\theta}$ which lies between $\hat{\boldsymbol{\theta}}$ and $\boldsymbol{\theta}_c$ and

$$\boldsymbol{I}_n(\boldsymbol{\theta}) = \sum_{i=i}^{n} \frac{\partial^2 \ell_i(\boldsymbol{\theta})}{\partial \boldsymbol{\theta} \partial \boldsymbol{\theta}^\top} .$$

Hence from (11.25),

$$\sqrt{n}(\hat{\boldsymbol{\theta}} - \boldsymbol{\theta}_c) = \left[-\frac{1}{n} \boldsymbol{I}_n(\tilde{\boldsymbol{\theta}}) \right]^{-1} n^{-\frac{1}{2}} \mathbf{U}_n(\boldsymbol{\theta}_c) . \tag{11.26}$$

By the law of large numbers,

$$-\frac{1}{n}\boldsymbol{I}_n(\tilde{\boldsymbol{\theta}}) = -\frac{1}{n} \sum_{i=i}^{n} \left[\frac{\partial^2 \ell_i(\boldsymbol{\theta})}{\partial \boldsymbol{\theta} \partial \boldsymbol{\theta}^\top} \right]_{\tilde{\boldsymbol{\theta}}} \xrightarrow{p} -\mathrm{E}_{f_P} \left[\frac{\partial^2 \ell_i(\boldsymbol{\theta})}{\partial \boldsymbol{\theta} \partial \boldsymbol{\theta}^\top} \right]_{\boldsymbol{\theta}_c} = \boldsymbol{J}(\boldsymbol{\theta}_c) . \tag{11.27}$$

Note that $\mathbf{U}_n(\boldsymbol{\theta}_c) = \sum_{i=1}^{n} \mathbf{U}_i(\boldsymbol{\theta}_c)$ where $\mathbf{U}_i(\boldsymbol{\theta}_c) = \left[\frac{\partial \ell_i(\boldsymbol{\theta})}{\partial \boldsymbol{\theta}} \right]_{\boldsymbol{\theta}_c}$. Also,

$$\mathrm{E}_{f_P}\left(\mathbf{U}_i(\boldsymbol{\theta}_c)\right) = \mathbf{0}$$

since

$$\mathrm{E}_{f_P}\left(\mathbf{U}_i(\boldsymbol{\theta}_c)\right) = \int \frac{\partial}{\partial \boldsymbol{\theta}} \left[\log f_Y(y_i \,|\, \boldsymbol{\theta})\right]_{\boldsymbol{\theta}_c} f_P(y_i)\, dy_i = \mathbf{0}$$

because $\boldsymbol{\theta}_c$ minimizes the Kullback-Leibler distance

$$d\left(f_P(y), f_y(y \,|\, \boldsymbol{\theta})\right) = \int \left(\log f_P(y) - \log f_y(y \,|\, \boldsymbol{\theta})\right) f_P(y)\, dy$$

with respect to $\boldsymbol{\theta}$. Also $V_{f_P}\left(\mathbf{U}_i(\boldsymbol{\theta}_c)\right) = \boldsymbol{K}(\boldsymbol{\theta}_c)$. Hence $E_{f_P}\left(\mathbf{U}_n(\boldsymbol{\theta}_c)\right) = 0$ and $V_{f_P}\left(\mathbf{U}_n(\boldsymbol{\theta}_c)\right) = n\boldsymbol{K}(\boldsymbol{\theta}_c)$. Therefore by the central limit theorem,

$$n^{-\frac{1}{2}}\mathbf{U}_n(\boldsymbol{\theta}_c) \xrightarrow{d} \mathcal{N}_K(\mathbf{0}, \boldsymbol{K}(\boldsymbol{\theta}_c)). \tag{11.28}$$

Hence applying Slutsky's theorem to (11.26) using (11.27) and (11.28) gives

$$\sqrt{n}(\hat{\boldsymbol{\theta}} - \boldsymbol{\theta}_c) \xrightarrow{d} \boldsymbol{J}(\boldsymbol{\theta}_c)^{-1}\mathcal{N}_K(\mathbf{0}, \boldsymbol{K}(\boldsymbol{\theta}_c)) = \mathcal{N}_K\left(\mathbf{0}, \boldsymbol{J}(\boldsymbol{\theta}_c)^{-1}\boldsymbol{K}(\boldsymbol{\theta}_c)\boldsymbol{J}(\boldsymbol{\theta}_c)^{-1}\right). \tag{11.29}$$

Note that if the true population distribution $f_P(y)$ belongs to the parametric family of distributions $f_Y(y \mid \boldsymbol{\theta})$, then we have $\boldsymbol{\theta}_c = \boldsymbol{\theta}_T$ and $\boldsymbol{K}(\boldsymbol{\theta}_T) = \boldsymbol{J}(\boldsymbol{\theta}_T)$, so $\sqrt{n}(\hat{\boldsymbol{\theta}} - \boldsymbol{\theta}_T) \xrightarrow{d} \mathcal{N}_K\left(\mathbf{0}, \boldsymbol{J}(\boldsymbol{\theta}_T)^{-1}\right)$.

11.9 Bibliographic notes

Millar [2011] is an accessible reference for both the practical and theoretical aspects of maximum likelihood estimation. Other excellent references are Silvey [1975] and Held and Sabanés Bové [2014].

11.10 Exercises

1. Show that the empirical distribution function can be thought of as a nonparametric maximum likelihood estimator.

2. Let Y_i be a random variable having a binomial distribution, $\texttt{BI}(n_i, \theta)$, with probability function

$$f(y_i \mid \theta) = \frac{n_i!}{(n_i - y_i)!\, y_i!}\theta^{y_i}(1-\theta)^{(n_i - y_i)} \qquad \text{for } y_i = 0, 1, \ldots, n_i\ ,$$

where $0 < \theta < 1$. Let $\mathbf{y} = (y_1, y_2, \ldots, y_n)^\top$ be a random sample of observations from the above distribution, and $\mathbf{Y} = (Y_1, Y_2, \ldots, Y_n)^\top$ the corresponding random variables.

(a) Write down the likelihood function $L(\theta)$ and show that the log-likelihood for θ is given by:

$$\ell(\theta) = \sum_{i=1}^{n} Y_i \log(\theta) + \sum_{i=1}^{n}(n_i - Y_i)\log(1-\theta)\ .$$

(b) Find the MLE for θ.

(c) Show that the expected information for θ is

$$i(\theta) = \frac{\sum_{i=1}^{n} n_i}{\theta} + \frac{\sum_{i=1}^{n} n_i}{(1-\theta)} \ .$$

Note that $E(Y_i) = n_i\theta$.

3. A manufacturing process produce fibers of varying lengths. It is assumed that the length of a fiber is a continuous random variable with pdf

$$f(y \mid \theta) = \theta^{-2} y e^{-y/\theta} \qquad \text{for } y > 0 \ ,$$

where $\theta > 0$ is an unknown parameter. Suppose that n randomly selected fibers have lengths $(y_1, y_2, \ldots, y_n)$ with corresponding random variables $(Y_1, Y_2, \ldots, Y_n)$. Find an expression for the MLE for θ.

12

Robustness of parameter estimation to outlier contamination

CONTENTS

12.1 The influence function for investigating robustness of the MLE to outlier contamination

12.1.1 Introduction

Robustness is discussed in detail in Huber [1981] and Hampel et al. [1986]. Let $Y_1, Y_2, \ldots, Y_n$ be a random sample from a continuous distribution with cdf $F_Y(y \mid \boldsymbol{\theta})$ where $\boldsymbol{\theta}^\top = (\theta_1, \theta_2, \ldots, \theta_K)$ is a vector of distribution parameters. Here the robustness of the maximum likelihood estimator (MLE) of $\boldsymbol{\theta}$ to outlier contamination is considered. An outlier is defined as a value of y lying in the left or right tail of the distribution of Y, i.e. a value of y for which $F_Y(y \mid \boldsymbol{\theta})$ is close to 0 or 1. Note that robustness to outlier contamination should not be confused with robustness to model distribution misspecification, which is considered in Sections 11.6 and 11.7.

A fundamental concept in robustness to outliers is the influence function (or curve), introduced by Hampel [1968, 1974]. See also Huber [1981, p13] and Hampel et al.

[1986, p81-87, p100-103]. "The importance of the influence function lies in its heuristic interpretation: it describes the effect of an infinitesimal contamination at the point x on the estimate, standardized by the mass of the contamination" [Hampel et al., 1986, p84]. For a definition of the influence function see Appendix A (Section 12.3).

For a sufficiently large sample of size $n-1$, the change in the parameter estimate caused by an additional observation at y is approximately $1/n$ times the influence function at y, see Hampel et al. [1986, p84]. For a robust estimator of parameter $\boldsymbol{\theta}$ we want the influence functions, for parameters θ_k for $k = 1, 2, \ldots, K$, to be bounded as y moves into the left or right tail of the distribution of Y, i.e. as y moves towards one of the ends of the range of Y, resulting in a bounded influence on the estimator of $\boldsymbol{\theta}$.

12.1.2 The influence function for the MLE

Lemma 12.1. *Let random variable Y have pdf $f_Y(y \mid \boldsymbol{\theta})$ and cdf $F_Y(y \mid \boldsymbol{\theta})$. Let $\mathrm{IC}(y, \hat{\theta}_k)$ be the influence function (or curve) for the maximum likelihood estimator $\hat{\theta}_k$ of parameter θ_k for $k = 1, 2, \ldots, K$, assuming a random sample from $f_Y(y \mid \theta)$. Let $\mathrm{IC}(y, \hat{\boldsymbol{\theta}})$ be the corresponding vector of influence functions, i.e. $\mathrm{IC}(y, \hat{\boldsymbol{\theta}}) = [\mathrm{IC}(y, \hat{\theta}_1), \mathrm{IC}(y, \hat{\theta}_2), \ldots, \mathrm{IC}(y, \hat{\theta}_K)]^\top$. Then*

$$\mathrm{IC}(y, \hat{\boldsymbol{\theta}}) = \mathbf{A}^{-1} \frac{\partial \ell}{\partial \boldsymbol{\theta}} \tag{12.1}$$

where $\mathbf{A} = -E_Y \left[\frac{\partial^2 \ell}{\partial \boldsymbol{\theta} \partial \boldsymbol{\theta}^\top} \right]$ is the expected information matrix and $\frac{\partial \ell}{\partial \boldsymbol{\theta}}$ is the vector of first derivatives of the log density function, i.e. $\ell = \log f_Y(y \mid \boldsymbol{\theta})$, so $\frac{\partial \ell}{\partial \boldsymbol{\theta}} = \left(\frac{\partial \ell}{\partial \theta_1}, \frac{\partial \ell}{\partial \theta_2}, \ldots, \frac{\partial \ell}{\partial \theta_K} \right)^\top$.

[Note strictly ℓ is a function of Y rather than y.]

The above lemma is given by applying Hampel et al. [1986, p230] to the special case of the MLE of $\boldsymbol{\theta}$. Note that $\hat{\theta}_k$ has an unconditionally bounded influence function if $\mathrm{IC}(y, \hat{\theta}_k)$ is bounded.

Lemma 12.2. *If, in Lemma 12.1, θ_k is unknown, but the true values of all other parameters (denoted $\boldsymbol{\theta}^*$) are known, then the conditional influence function for $\hat{\theta}_k$ (given known $\boldsymbol{\theta}^*$) is*

$$\mathrm{IC}(y, \hat{\theta}_k \mid \boldsymbol{\theta}^*) = A_{kk}^{-1} \frac{\partial \ell}{\partial \theta_k} \propto \frac{\partial \ell}{\partial \theta_k}$$

where $A_{kk} = -E_Y \left[\frac{\partial^2 \ell}{\partial \theta_k^2} \right]$ and the proportionality is with respect to y.

Hence $\hat{\theta}_k$ has a conditionally (given known $\boldsymbol{\theta}^*$) bounded influence function if $\frac{\partial \ell}{\partial \theta_k}$ is bounded.

12.1.3 Examples

Here we find the influence functions for the MLEs of all parameters of each of four example distributions for Y: the normal, beta, gamma, and t family distributions, i.e. $\texttt{NO}(\mu,\sigma)$, $\texttt{BEo}(\mu,\sigma)$, $\texttt{GA}(\mu,\sigma)$, and $\texttt{TF}(\mu,\sigma,\nu)$, respectively. The resulting influence functions are derived later in this section.

For the MLE of each parameter of each distribution, we check whether the unconditional and conditional (given the other parameters of the distribution are known) influence functions (as defined in Lemmas 12.1 and 12.2, respectively) are bounded (✓✓, ✓, respectivelly), or both unbounded (x), as outlier value y moves towards one of the ends of the range of Y. Table 12.1 gives a summary of these results. For example for $Y \sim \texttt{TF}(\mu,\sigma,\nu)$ the unconditional influence function for the MLE $\hat{\mu}$ for μ is bounded, for $\hat{\sigma}$ the conditional influence function is bounded and for $\hat{\nu}$ they are both unbounded, as $y \to -\infty$ (left tail) and as $y \to \infty$ (right tail).

TABLE 12.1: Bounded or unbounded influence functions for MLE of parameters of distributions

	Left tail			Right tail		
	μ	σ	ν	μ	σ	ν
$\texttt{NO}(\mu,\sigma)$	x	x		x	x	
$\texttt{BEo}(\mu,\sigma)$	x	✓		✓	x	
$\texttt{GA}(\mu,\sigma)$	✓✓	x		x	x	
$\texttt{TF}(\mu,\sigma,\nu)$	✓✓	✓	x	✓✓	✓	x

x=unbounded, ✓= conditional bounded, ✓✓= unconditional bounded

Example 1: Normal distribution, $\texttt{NO}(\mu,\sigma)$

Let $Y \sim \texttt{NO}(\mu,\sigma)$ then

$$f_Y(y|\mu,\sigma) = \frac{1}{\sqrt{2\pi}\sigma} \exp\left[-\frac{(y-\mu)^2}{2\sigma^2}\right]$$

for $-\infty < y < \infty$, so

$$\ell = \log f_Y(y \mid \mu,\sigma) = -\frac{1}{2}\log(2\pi) - \log\sigma - \frac{(y-\mu)^2}{2\sigma^2} \, ,$$

$$\frac{\partial \ell}{\partial \mu} = \frac{(y-\mu)}{\sigma^2} \quad \text{and} \quad \frac{\partial \ell}{\partial \sigma} = \frac{1}{\sigma}\left[\frac{(y-\mu)^2}{\sigma^2} - 1\right].$$

Hence the conditional influence functions for the MLE $\hat{\mu}$ (given σ) and $\hat{\sigma}$ (given μ) are both unbounded as $y \to \infty$ and as $y \to -\infty$ (since $\frac{\partial \ell}{\partial \mu}$ and $\frac{\partial \ell}{\partial \sigma}$ are unbounded).

The vector of unconditional influence functions for the MLE $\hat{\boldsymbol{\theta}} = (\hat{\mu},\hat{\sigma})^\top$ of $\boldsymbol{\theta} = (\mu,\sigma)^\top$

is given by equation (12.1) where $\frac{\partial \ell}{\partial \boldsymbol{\theta}} = \left(\frac{\partial \ell}{\partial \mu}, \frac{\partial \ell}{\partial \sigma}\right)^\top$ and

$$A = -E_Y\left(\frac{\partial^2 \ell}{\partial \boldsymbol{\theta} \partial \boldsymbol{\theta}^\top}\right) = -E_Y\begin{pmatrix} \frac{\partial^2 \ell}{\partial \mu^2} & \frac{\partial^2 \ell}{\partial \mu \partial \sigma} \\ \frac{\partial^2 \ell}{\partial \mu \partial \sigma} & \frac{\partial^2 \ell}{\partial \sigma^2} \end{pmatrix} = \begin{pmatrix} \frac{1}{\sigma^2} & 0 \\ 0 & \frac{2}{\sigma^2} \end{pmatrix}.$$

Hence

$$\begin{aligned} \text{IC}(y, \hat{\mu}) &= y - \mu \\ \text{IC}(y, \hat{\sigma}) &= \frac{\sigma}{2}\left[\frac{(y-\mu)^2}{\sigma^2} - 1\right]. \end{aligned}$$

Since A is a diagonal matrix, μ and σ are expected information orthogonal and the unconditional influence functions for each of $\hat{\mu}$ and $\hat{\sigma}$ are the same as the conditional influence functions for $\hat{\mu}$ (given σ) and for $\hat{\sigma}$ (given μ).

Note both the influence functions for $\hat{\mu}$ and $\hat{\sigma}$ are unbounded as $y \to \infty$ and as $y \to -\infty$. A single observation y can have an unbounded effect on the MLEs of μ and σ. This is seen from the MLEs of μ and σ, i.e. $\overline{Y} = \frac{1}{n}\sum Y_i$ and $S^2 = \frac{1}{n}\sum(Y_i - \overline{Y})^2$, which are both unbounded as a single value Y_i tends to $+\infty$ or $-\infty$.

Example 2: Beta distribution, BEo(μ, σ)

Let $Y \sim$ BEo(μ, σ) then

$$f_Y(y \mid \mu, \sigma) = \frac{1}{B(\mu, \sigma)} y^{\mu-1}(1-y)^{\sigma-1} \qquad \text{for } 0 < y < 1,$$

$$\begin{aligned} \ell = \log f_Y(y \mid \mu, \sigma) &= -\log\Gamma(\mu) - \log\Gamma(\sigma) + \log\Gamma(\mu+\sigma) + (\mu-1)\log y \\ &\quad + (\sigma-1)\log(1-y) \\ \frac{\partial \ell}{\partial \mu} &= -\Psi(\mu) + \Psi(\mu+\sigma) + \log y \qquad (12.2) \\ \frac{\partial \ell}{\partial \sigma} &= -\Psi(\sigma) + \Psi(\mu+\sigma) + \log(1-y), \qquad (12.3) \end{aligned}$$

where $\Psi(x) = \frac{\partial}{\partial x}[\log\Gamma(x)]$ is the psi or digamma function.

Hence, from $\partial\ell/\partial\mu$, the conditional influence function for $\hat{\mu}$ (given σ) is unbounded as $y \to 0$, but bounded as $y \to 1$, while, from $\partial\ell/\partial\sigma$, the conditional influence function for $\hat{\sigma}$ (given μ) is unbounded as $y \to 1$, but bounded as $y \to 0$.

The vector of unconditional influence functions is given by equation (12.1) where $\boldsymbol{\theta} = (\mu, \sigma)^\top$, $\frac{\partial \ell}{\partial \boldsymbol{\theta}} = \left(\frac{\partial \ell}{\partial \mu}, \frac{\partial \ell}{\partial \sigma}\right)^\top$ and $\mathbf{A} = \begin{pmatrix} A_{11} & A_{12} \\ A_{21} & A_{22} \end{pmatrix}$ with matrix elements $A_{11} = \Psi^{(1)}(\mu) - \Psi^{(1)}(\mu+\sigma)$, $A_{22} = \Psi^{(1)}(\sigma) - \Psi^{(1)}(\mu+\sigma)$ and $A_{12} = A_{21} = -\Psi^{(1)}(\mu+\sigma)$, where $\Psi^{(1)}(x) = \frac{d}{dx}\Psi(x)$ is the trigamma function.

Hence

$$\begin{aligned}\text{IC}(y,\hat{\mu}) =& \frac{1}{|\mathbf{A}|}\left[A_{22}\frac{\partial \ell}{\partial \mu} - A_{12}\frac{\partial \ell}{\partial \sigma}\right]\\ \sim & \begin{cases} \dfrac{A_{22}}{|\mathbf{A}|}\log y & \text{as } y \to 0\\ -\dfrac{A_{12}}{|\mathbf{A}|}\log(1-y) & \text{as } y \to 1\end{cases}\end{aligned}$$

$$\begin{aligned}\text{IC}(y,\hat{\sigma}) =& \frac{1}{|\mathbf{A}|}\left[A_{11}\frac{\partial \ell}{\partial \sigma} - A_{12}\frac{\partial \ell}{\partial \mu}\right]\\ \sim & \begin{cases} -\dfrac{A_{12}}{|\mathbf{A}|}\log y & \text{as } y \to 0\\ \dfrac{A_{11}}{|\mathbf{A}|}\log(1-y) & \text{as } y \to 1.\end{cases}\end{aligned}$$

Hence both $\text{IC}(y,\hat{\mu})$ and $\text{IC}(y,\hat{\sigma})$ are unbounded as $y \to 0$ and as $y \to 1$. Hence the unconditional influence functions for $\hat{\mu}$ and $\hat{\sigma}$ are both unbounded as $y \to 0$ and as $y \to 1$.

Example 3: Gamma distribution, $\texttt{GA}(\mu,\sigma)$

Let $Y \sim \texttt{GA}(\mu,\sigma)$ then

$$f_Y(y\,|\,\mu,\sigma) = \frac{y^{1/\sigma^2-1}\exp\left[-y/(\sigma^2\mu)\right]}{(\sigma^2\mu)^{1/\sigma^2}\Gamma(1/\sigma^2)} \qquad \text{for } y > 0,$$

$$\begin{aligned}\ell = \log f_Y(y\,|\,\mu,\sigma) &= \left(\frac{1}{\sigma^2}-1\right)\log y - \frac{y}{\sigma^2\mu} - \frac{1}{\sigma^2}\log\sigma^2 - \frac{1}{\sigma^2}\log\mu - \log\Gamma(1/\sigma^2)\\ \frac{\partial \ell}{\partial \mu} &= \frac{1}{\sigma^2\mu^2}(y-\mu)\\ \frac{\partial \ell}{\partial \sigma} &= \frac{2}{\sigma^3}\left[\frac{y}{\mu} + \Psi(1/\sigma^2) - \log\left(\frac{y}{\sigma^2\mu}\right) - 1\right].\end{aligned}$$

Hence the conditional influence function for $\hat{\mu}$ (given σ) is unbounded as $y \to \infty$ (but bounded as $y \to 0$). However the conditional influence function for $\hat{\sigma}$ (given μ) is unbounded as $y \to 0$ or as $y \to \infty$.

The vector of unconditional influence functions is given by (12.1) where $\boldsymbol{\theta} = (\mu,\sigma)^\top$, $\dfrac{\partial \ell}{\partial \boldsymbol{\theta}} = \left(\dfrac{\partial \ell}{\partial \mu}, \dfrac{\partial \ell}{\partial \sigma}\right)^\top$ and $\mathbf{A} = \begin{pmatrix} A_{11} & A_{12}\\ A_{21} & A_{22}\end{pmatrix}$ with matrix elements $A_{11} = \dfrac{1}{\sigma^2\mu^2}$, $A_{22} = \dfrac{4}{\sigma^6}\Psi^{(1)}(1/\sigma^2) - \dfrac{4}{\sigma^4}$, $A_{12} = A_{21} = 0$. Hence

$$\begin{aligned}\text{IC}(y,\hat{\mu}) =& A_{11}^{-1}\frac{\partial \ell}{\partial \mu} = y - \mu\\ \text{IC}(y,\hat{\sigma}) =& A_{22}^{-1}\frac{\partial \ell}{\partial \sigma} = \left[\frac{2}{\sigma^3}\Psi^{(1)}(1/\sigma^2) - \frac{2}{\sigma}\right]^{-1}\left[\frac{y}{\mu} + \Psi(1/\sigma^2) - \log\left(\frac{y}{\sigma^2\mu}\right) - 1\right],\end{aligned}$$

where $\hat{\mu}$ and $\hat{\sigma}$ are the MLEs of μ and σ, respectively.

Hence the unconditional influence function for $\hat{\mu}$, $\text{IC}(y, \hat{\mu})$, is unbounded as $y \to \infty$ but bounded as $y \to 0$. However the unconditional influence function for $\hat{\sigma}$, $\text{IC}(y, \hat{\sigma})$, is unbounded as $y \to 0$ and as $y \to \infty$.

Example 4: *t* family distribution, $\texttt{TF}(\mu, \sigma, \nu)$

Let $Y \sim \texttt{TF}(\mu, \sigma, \nu)$ then

$$f_Y(y \mid \mu, \sigma, \nu) = \frac{1}{\sigma B(1/2, \nu/2)\nu^{1/2}} \left[1 + \frac{z^2}{\nu}\right]^{-(\nu+1)/2}$$

where $z = (y - \mu)/\sigma$, so

$$\begin{aligned}
\ell = \log f_Y(y \mid \mu, \sigma, \nu) &= -\log \sigma - \frac{(\nu+1)}{2} \log \left[1 + \frac{z^2}{\nu}\right] + \log \Gamma\left[(\nu+1)/2\right] \\
&\quad - \log \Gamma(\nu/2) - \log \Gamma(\tfrac{1}{2}) - \tfrac{1}{2} \log \nu \\
\frac{\partial \ell}{\partial \mu} &= \frac{\omega}{\sigma^2}(y - \mu) \qquad \text{where } \omega = \frac{(\nu+1)}{(\nu + z^2)} \\
\frac{\partial \ell}{\partial \sigma} &= \frac{1}{\sigma}(\omega z^2 - 1) \\
\frac{\partial \ell}{\partial \nu} &= -\tfrac{1}{2} \log \left[1 + \frac{z^2}{\nu}\right] + \frac{\omega z^2}{2\nu} + \tfrac{1}{2}\Psi[(\nu+1)/2] - \tfrac{1}{2}\Psi(\nu/2) - \frac{1}{2\nu}\,.
\end{aligned}$$

Hence as $y \to \infty$ or $y \to -\infty$, the conditional influence function for $\hat{\mu}$ (given σ and ν) is bounded, the conditional influence function for $\hat{\sigma}$ (given μ and ν) is bounded, but the conditional influence function for $\hat{\nu}$ (given μ and σ) is unbounded.

The vector of unconditional influence functions is given by (12.1), where $\boldsymbol{\theta} = (\mu, \sigma, \nu)^\top$ and

$$\mathbf{A} = \begin{bmatrix} A_{11} & 0 & 0 \\ 0 & A_{22} & A_{23} \\ 0 & A_{32} & A_{33} \end{bmatrix}$$

with non-zero matrix elements

$$\begin{aligned}
A_{11} &= \frac{(\nu+1)}{(\nu+3)\sigma^2} \\
A_{22} &= \frac{2\nu}{(\nu+3)\sigma^2} \\
A_{33} &= -\frac{(\nu+5)}{2\nu(\nu+1)(\nu+3)} - \frac{1}{4}\Psi^{(1)}[(\nu+1)/2] + \frac{1}{4}\Psi^{(1)}(\nu/2) \\
A_{23} = A_{32} &= -\frac{2}{(\nu+1)(\nu+3)\sigma}\,.
\end{aligned}$$

Hence

$$\text{IC}(y, \hat{\mu}) = A_{11}^{-1} \frac{\partial \ell}{\partial \mu} = \frac{(\nu+3)\omega(y-\mu)}{(\nu+1)} = \frac{(\nu+3)}{(\nu+z^2)}(y - \mu) \to 0 \text{ as } y \to -\infty \text{ or } y \to \infty,$$

$$\mathrm{IC}(y,\hat{\sigma}) = \frac{1}{d}\left[A_{33}\frac{\partial \ell}{\partial \sigma} - A_{23}\frac{\partial \ell}{\partial \nu}\right] \sim \frac{A_{23}}{2d}\log\left[1+\frac{z^2}{\nu}\right] \text{ as } y \to -\infty \text{ or } y \to \infty,$$

where $d = \begin{vmatrix} A_{22} & A_{23} \\ A_{32} & A_{33} \end{vmatrix} = A_{22}A_{33} - A_{23}^2$.

$$\mathrm{IC}(y,\hat{\nu}) = \frac{1}{d}\left[A_{22}\frac{\partial \ell}{\partial \nu} - A_{23}\frac{\partial \ell}{\partial \sigma}\right] \sim -\frac{A_{22}}{2d}\log\left[1+\frac{z^2}{\nu}\right] \text{ as } y \to -\infty \text{ or } y \to \infty.$$

Hence the unconditional influence function for $\hat{\mu}$, $\mathrm{IC}(y,\hat{\mu})$, is bounded as $y \to -\infty$ or $y \to \infty$, but the unconditional influence functions for σ and ν, $\mathrm{IC}(y,\hat{\sigma})$ and $\mathrm{IC}(y,\hat{\nu})$, respectively, are unbounded as $y \to -\infty$ or $y \to \infty$.

Note however that the conditional influence function of $\hat{\sigma}$ (given that ν is known) is given by $A_{22}^{-1}\frac{\partial \ell}{\partial \sigma}$ and is bounded as $y \to -\infty$ or $y \to \infty$.

Some authors have suggested fixing ν to be a low value, e.g. $\nu = 5$, when fitting the t family distribution. Although this results in bounded influence functions for the resulting MLEs $\hat{\mu}$ and $\hat{\sigma}$, clearly the value of ν is in general not correct, and the resulting MLE of σ is asymptotically biased. However the MLE of μ is asymptotically unbiased.

Note that it is important to distinguish between the true value of ν being known, and the value of ν being unknown but fixed at an (almost certainly) wrong value, leading to MLEs of μ and σ from a wrong model distribution.

12.2 Robust fitting of a GAMLSS model

12.2.1 Discussion of robust fitting

In general, maximum likelihood estimation leads to parameter estimators with unbounded influence functions for some or all parameters. See Section 12.1.3 for examples. Hence MLEs are generally vulnerable to the unbounded influence of even a single outlier y. An outlier value y of a response variable Y relative to the model is a value of y for which $F_Y(y \mid \boldsymbol{\theta})$ is very close to 0 or 1. Outlier values may be gross errors due to 'contamination', in which case they need to be identified and removed. A robust fitted model is a good way to identify gross errors. Alternatively outlier values may be genuine values, in which case the model may be wrong, since extreme outliers are unlikely to occur. We could try changing the model, e.g. change the model distribution of Y or the predictors for the distribution parameters, in order to accommodate the outliers. A distribution with heavier tails may be more appropriate. Assuming this has been tried and genuine outlier values still remain, then we may wish to robustly fit the model so that the outliers do not distort the fit to the rest of the observations.

We suggest the following practical robustness procedure:

Iterate between:

1. (Re)fit the model robustly;

2. Identify and remove (or weight out) gross outliers;

For references on robustness see the bibliographic notes in Section 12.5. Below we apply robust fitting to a GAMLSS model for a continuous random variable.

12.2.2 Robust fitting of a GAMLSS model for a continuous response variable (with bias correction)

We robustly fit a GAMLSS model for a continuous response variable Y by obtaining parameter estimators with bounded influence functions. Our approach is based on an adapted generalization of the algorithm given in the Appendix of Alimadad and Salibian-Barrera [2011], and is outlined in our Appendix B (Section 12.4). The resulting algorithm bounds the first derivatives of the log-likelihood function ℓ with respect to each of the distribution parameters μ, σ, ν, and τ.

For example for μ, $\psi\left(\frac{\partial \ell_i}{\partial \mu_i}\right)$ is a bounded $\frac{\partial \ell_i}{\partial \mu_i}$ obtained by bounding y_i by setting

$$y_i^* = \begin{cases} y_i & \text{if } F_Y^{-1}(\alpha_1 \,|\, \boldsymbol{\theta}_i) \le y_i \le F_Y^{-1}(1-\alpha_1 \,|\, \boldsymbol{\theta}_i) \\ F_Y^{-1}(\alpha_1 \,|\, \boldsymbol{\theta}_i) & \text{if } y_i < F_Y^{-1}(\alpha_1 \,|\, \boldsymbol{\theta}_i) \\ F_Y^{-1}(1-\alpha_1 \,|\, \boldsymbol{\theta}_i) & \text{if } y_i > F_Y^{-1}(1-\alpha_1 \,|\, \boldsymbol{\theta}_i) \end{cases} \tag{12.4}$$

i.e.

$$y_i^* = \begin{cases} y_i & \text{if } \Phi^{-1}(\alpha_1) \le r_i \le \Phi^{-1}(1-\alpha_1) \\ F_Y^{-1}(\alpha_1 \,|\, \boldsymbol{\theta}_i) & \text{if } r_i < \Phi^{-1}(\alpha_1) \\ F_Y^{-1}(1-\alpha_1 \,|\, \boldsymbol{\theta}_i) & \text{if } r_i > \Phi^{-1}(1-\alpha_1) = -\Phi^{-1}(\alpha_1) \end{cases} \tag{12.5}$$

where α_1 is a small probability (e.g. $\alpha_1 = 0.01$), $r_i = \Phi^{-1}\left[F_Y(y_i \,|\, \boldsymbol{\theta}_i)\right]$ is the normalized quantile residual and where, in this section only, $\boldsymbol{\theta}_i = (\mu_i, \sigma_i, \nu_i, \tau_i)^\top$. We then set

$$\psi\left(\frac{\partial \ell_i}{\partial \mu_i}\right) = \psi\left(\frac{\partial \ell(y_i)}{\partial \mu_i}\right) = \frac{\partial \ell(y_i^*)}{\partial \mu_i} \,.$$

Then in the **gamlss** fitting algorithm [Rigby and Stasinopoulos, 2005, Appendix A], [Stasinopoulos et al., 2017, p66], replace $\frac{\partial \ell_i}{\partial \mu_i}$ by

$$h_i = \psi\left(\frac{\partial \ell_i}{\partial \mu_i}\right) - \mathrm{E}\left[\psi\left(\frac{\partial \ell_i}{\partial \mu_i}\right)\right]$$

where the expectation term is a correction to ensure unbiased estimating equations and hence Fisher consistency of the estimator, i.e. a 'bias correction'. Similarly for $\psi\left(\frac{\partial \ell_i}{\partial \sigma_i}\right)$, $\psi\left(\frac{\partial \ell_i}{\partial \nu_i}\right)$ and $\psi\left(\frac{\partial \ell_i}{\partial \tau_i}\right)$.

[Note within the fitting algorithm, $\boldsymbol{\theta}_i$ is automatically replaced by $\hat{\boldsymbol{\theta}}_i$, the current estimate of $\boldsymbol{\theta}_i$.]

Example: Beta distribution, BEo(μ, σ)

Here we fit the beta, BEo(μ, σ), distribution robustly, including the bias correction. The first derivatives of the log density function, $\ell = \log(f_Y(y \mid \mu, \sigma)$, with respect to μ and σ are given by equations (12.2) and (12.3), respectively. First $\psi\left(\frac{\partial \ell_i}{\partial \mu_i}\right)$ and $\psi\left(\frac{\partial \ell_i}{\partial \sigma_i}\right)$ are bounded versions of $\frac{\partial \ell_i}{\partial \mu_i}$ and $\frac{\partial \ell_i}{\partial \sigma_i}$ obtained by setting

$$y_i^* = \begin{cases} y_i & \text{if } a_i \leq y_i \leq b_i \\ a_i & \text{if } y_i < a_i \\ b_i & \text{if } y_i > b_i, \end{cases}$$

where $a_i = F_Y^{-1}(\alpha_1 | \mu_i, \sigma_i)$ and $b_i = F_Y^{-1}(1 - \alpha_1 | \mu_i, \sigma_i)$ and α_1 is a small probability. We then set

$$\begin{aligned} \psi\left(\frac{\partial \ell_i}{\partial \mu_i}\right) &= \psi\left(\frac{\partial \ell(y_i)}{\partial \mu_i}\right) = \frac{\partial \ell(y_i^*)}{\partial \mu_i} \quad \text{and} \\ \psi\left(\frac{\partial \ell_i}{\partial \sigma_i}\right) &= \psi\left(\frac{\partial \ell(y_i)}{\partial \sigma_i}\right) = \frac{\partial \ell(y_i^*)}{\partial \sigma_i}. \end{aligned}$$

Within the **gamlss** algorithm, $\frac{\partial \ell_i}{\partial \mu_i}$ and $\frac{\partial \ell_i}{\partial \sigma_i}$ are replaced by

$$\begin{aligned} h_{1i} &= \psi\left(\frac{\partial \ell_i}{\partial \mu_i}\right) - \mathrm{E}\left[\psi\left(\frac{\partial \ell_i}{\partial \mu_i}\right)\right] \quad \text{and} \\ h_{2i} &= \psi\left(\frac{\partial \ell_i}{\partial \sigma_i}\right) - \mathrm{E}\left[\psi\left(\frac{\partial \ell_i}{\partial \sigma_i}\right)\right] \end{aligned}$$

respectively, where the expected values provide the bias corrections, and for example

$$\mathrm{E}\left[\psi\left(\frac{\partial \ell_i}{\partial \mu_i}\right)\right] = \int_{a_i}^{b_i} \frac{\partial \ell_i}{\partial \mu_i} f_Y(y|\mu_i, \sigma_i)\, dy + \alpha_1 \left[\left.\frac{\partial \ell_i}{\partial \mu_i}\right|_{a_i} + \left.\frac{\partial \ell_i}{\partial \mu_i}\right|_{b_i}\right],$$

and $\left.\frac{\partial \ell_i}{\partial \mu_i}\right|_{a_i}$ is $\frac{\partial \ell_i}{\partial \mu_i}$ evaluated at $y_i = a_i$.

The integral is not tractable and so is evaluated by numerical integration. The robust fitting of the BEo distribution (with bias correction), is implemented in **gamlss** as BEor(μ, σ), but currently only for constant μ and σ. To demonstrate its use we simulated data, presented in Section 12.2.4.

12.2.3 Automated robust fitting procedure (without bias correction)

A function `gamlssRobust()` is available to automate a robust procedure. It first fits the model and then repeats step 2 below until convergence to provide an initial robust fit. It then iterates between steps 1 and 2 below until convergence.

1. Weight out gross outliers with $|r_i| >$ `CD.bound` using the current fitted model values of r_i, for $i = 1, 2, \ldots, n$;

2. Bound y_i to y_i^* using (12.5) with `bound` for $|r_i|$ equal to $-\Phi^{-1}(\alpha_1)$ using the current fitted model values of r_i for $i = 1, 2, \ldots, n$, and refit the model (*but unfortunately currently without bias correction*).

In order to reduce the bias, the high default values, `bound` $= -\Phi^{-1}(1/n)$ and `CD.bound` $= -\Phi^{-1}(1/(3n))$, are recommended. The values of `bound` and `CD.bound` may need to be adjusted for a particular data set. Lower values lead to increased robustness, but also increased bias.

12.2.4 Robust fitting of a beta distribution using simulated data

Here we simulate a random sample of size 490 from a `BEo`$(5, 5)$ distribution and contaminate it with random samples of size 5 from each of two uniform distributions, `unif`$(0, 0.1)$ and `unif`$(0.9, 1)$.

We firstly fit a `BEo` distribution to the contaminated data of size 500. It is clear from the worm plot of the residuals from the contaminated data fit, shown in Figure 12.1, that the contamination results in a distorted fit to the data.

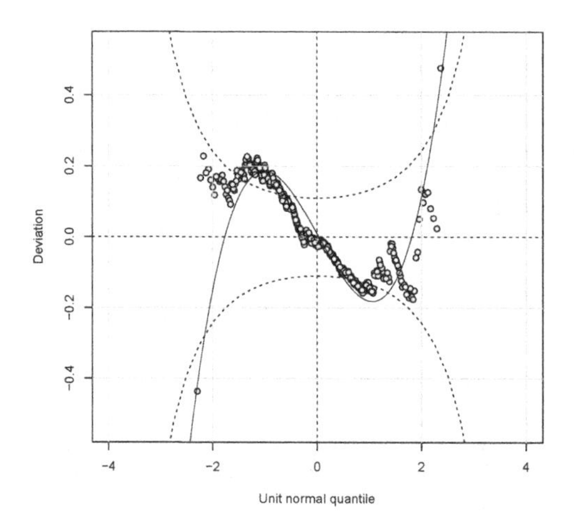

FIGURE 12.1: Worm plot for the `BEo`(μ, σ) fit to the contaminated data.

Robust fitting (with bias correction)

We alternate between two steps:

1. (Re)fit `BEo`(μ, σ) robustly (with bias correction) by using `BEor`(μ, σ) to the case weighted data set, using e.g. $\alpha_1 = 0.01$, specified by bounding the normalized quantile residual $|r_i|$, i.e. use `bound=` $-\ \Phi^{-1}(0.01) = 2.33$. The code looks like:

```
b1 <- gamlss(y~1,family=BEor, n.cyc=100, bound=2.33)
```

2. Weight out cases with $|r_i| >$ `CD.bound` i.e. greater than a specified value, e.g. `CD.bound=`$\Phi^{-1}(1/(3n))$ where $n = 500$ is the total number of cases.

Robust fitting (without bias correction)

We now use the automated function `gamlssRobust()` to robustly fit the beta `BEo`(μ, σ) distribution without bias correction, using `bound=`$-\Phi^{-1}(1/n) = 2.88$ and `CD.bound=`$-\Phi^{-1}(1/(3n)) = 3.21$, chosen to reduce the bias.

```
b2 <- gamlss(y~1, family=BEo)
b3 <- gamlssRobust(b2, bound=2.88, CD.bound=3.21)
```

Further research is needed on the robust fitting of a GAMLSS model.

Comparison of the fits to the contaminated beta data

Figure 12.2 displays a histogram of the contaminated data, the true pdf for `BEo`$(5, 5)$, the (non-robust) beta fit model `b2`, and the robust beta fit (with bias correction).

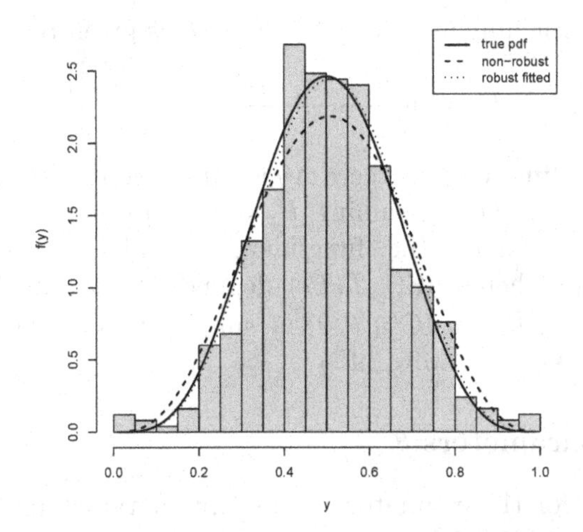

FIGURE 12.2: Histogram of the contaminated beta data, with the true `BEo` pdf (solid line), the non-robust fitted `BEo` pdf (dashed line), and the robust (with bias correction) fitted `BEo` pdf (dotted line).

12.3 Appendix A: influence function

12.3.1 Single parameter θ

The influence function (IF) or influence curve (IC) for the estimator of a single parameter θ is formally defined by Hampel et al. [1986, p81-88].

Let $\mathbf{y}^\top = (y_1, y_2, \ldots, y_n)$ be a random sample from a distribution with cdf F. Let $F_\theta = f_Y(y \mid \theta)$ be the model distribution of Y for some value of θ. Let real valued statistic $T_n(\mathbf{y})$ be an estimator of parameter θ which depends on the sample only through the empirical distribution function F_n, where

$$F_n(y) = \frac{1}{n}\sum_{i=1}^{n} \mathrm{I}(y_i \le y)$$

and $\mathrm{I}(y_i \le y) = \begin{cases} 1 & \text{if } y_i \le y \\ 0 & \text{if } y_i > y \ . \end{cases}$

Hence the estimator $T_n(\mathbf{y})$ is given by $T_n(\mathbf{y}) = T(F_n)$, for all n and F_n, for some functional T which transforms a cdf function (e.g. F_n) to a real value constant (e.g. $T(F_n)$).

Assume $T_n(\mathbf{y}) = T(F_n) \to T(F)$ as $n \to \infty$ in probability, where F is the cdf of Y and $T(F)$ is called the asymptotic value of $\{T_n : n \ge 1\}$ at F. We would like the functional to be Fisher consistent, i.e. $T(F_\theta) = \theta$ for all θ, which means that at the model the estimator $\{T_n : n \ge 1\}$ asymptotically measures the right quantity.

The influence function (or influence curve) of T at F is given by

$$\mathrm{IF}(y, T, F) = \lim_{t \downarrow 0} \frac{T((1-t)F + t\Delta y) - T(F)}{t} ,$$

for those y for which the limit exists, where Δy is a $0-1$ step function at y corresponding to a point mass 1 at y. Hence, assuming $F = F_\theta$ for some θ, $\mathrm{IF}(y, T, F_\theta)$ depends on y, the form of the estimator (i.e. functional T) and the true distribution cdf of the observations F_θ, and hence $\mathrm{IF}(y, T, F_\theta)$ depends on y and θ. Hence we replace $\mathrm{IF}(y, T, F_\theta)$ by the lazy notation $\mathrm{IC}(y, \hat{\theta}, \theta)$ or simply $\mathrm{IC}(y, \hat{\theta})$. (We prefer 'IC' to 'IF' to avoid any confusion with the word 'if'.)

12.3.2 Multiple parameters θ

The influence function for the estimators of multiple parameters $\boldsymbol{\theta}$ is formally defined by Hampel et al. [1986, p225-231].

Let $\mathbf{y}^\top = (y_1, y_2, \ldots, y_n)$ be a random sample from a distribution with cdf F. Let $F_Y(y \,|\, \boldsymbol{\theta})$ be the model distribution of Y for some value of $\boldsymbol{\theta}$, where $\boldsymbol{\theta}^\top = (\theta_1, \theta_2, \ldots, \theta_K)$. Let a vector of real value statistics $\mathbf{T}_n(\mathbf{y})$ be an estimator of $\boldsymbol{\theta}$ which depends on the sample only through the empirical distribution function F_n. Hence $\mathbf{T}_n(\mathbf{y}) = \mathbf{T}(F_n)$, for all n and F_n, for functionals $\mathbf{T}$, which transform a cdf function (e.g. F_n) to a vector (e.g. $\mathbf{T}(F_n)$). Assume $\mathbf{T}_n(\mathbf{y}) = \mathbf{T}(F_n) \to \mathbf{T}(F)$ as $n \to \infty$ in probability.

The K-dimensional vector of influence functions of functionals $\mathbf{T}$ at F is given by

$$\mathrm{IF}(y, \mathbf{T}, F) = \lim_{t \downarrow 0} \frac{\mathbf{T}\left((1-t)F + t\Delta y\right) - \mathbf{T}(F)}{t}$$

for those y for which the limit exists. Hence, assuming $F = F_{\boldsymbol{\theta}}$ for some $\boldsymbol{\theta}$, we denote $\mathrm{IF}(y, \mathbf{T}, F_{\boldsymbol{\theta}})$ by $\mathrm{IC}(y, \hat{\boldsymbol{\theta}}, \boldsymbol{\theta})$ or simply $\mathrm{IC}(y, \hat{\boldsymbol{\theta}})$ and it depends only on y and $\boldsymbol{\theta}$, although the way it depends on y and $\boldsymbol{\theta}$ is determined by $\mathbf{T}$ and $F_{\boldsymbol{\theta}}$.

Note $\mathrm{IC}(y, \hat{\boldsymbol{\theta}}, \boldsymbol{\theta}) = \left(\mathrm{IC}(y, \hat{\theta}_1, \boldsymbol{\theta}), IC(y, \hat{\theta}_2, \boldsymbol{\theta}), \ldots, IC(y, \hat{\theta}_K, \boldsymbol{\theta})\right)^\top$ or equivalently in simple notation $\mathrm{IC}(y, \hat{\boldsymbol{\theta}}) = \left(\mathrm{IC}(y, \hat{\theta}_1), \mathrm{IC}(y, \hat{\theta}_2), \ldots, \mathrm{IC}(y, \hat{\theta}_K)\right)^\top$.

12.4 Appendix B: algorithm for robustly fitting a GAMLSS model

In this Appendix we outline the justification of the algorithm for robustly fitting a **gamlss** model. The algorithm is based on an adapted generalization of the algorithm in the Appendix of Alimadad and Salibian-Barrera [2011]. For simplicity we focus on robustly fitting a linear model to the predictor of a single distribution parameter (μ below), assuming the other parameters are fixed.

In the Appendix of Alimadad and Salibian-Barrera [2011] we replace their $\psi_c(r_i)\frac{1}{\sqrt{\nu_i}}$ by $\psi\left(\frac{\partial \ell_i}{\partial \mu_i}\right)$, defined in Section 12.2.2. Hence we wish to solve $f(\boldsymbol{\beta}) = 0$ where

$$f(\boldsymbol{\beta}) = \sum_{i=1}^{n} \psi\left(\frac{\partial \ell_i}{\partial \mu_i}\right) \frac{\partial \mu_i}{\partial \eta_i} \mathbf{x}_i - a(\boldsymbol{\beta}) \ ,$$

where $a(\boldsymbol{\beta}) = \sum_{i=1}^{n} E\left[\psi\left(\frac{\partial \ell_i}{\partial \mu_i}\right)\right] \frac{\partial \mu_i}{\partial \eta_i} \mathbf{x}_i$. Hence we solve

$$f(\boldsymbol{\beta}) = \sum_{i=1}^{n} h_i \frac{\partial \mu_i}{\partial \eta_i} \mathbf{x}_i = 0,$$

where $h_i = \psi\left(\frac{\partial \ell_i}{\partial \mu_i}\right) - E\left[\psi\left(\frac{\partial \ell_i}{\partial \mu_i}\right)\right]$.

A Newton-Raphson algorithm gives

$$-E\left[\triangledown f(\boldsymbol{\beta}^{(j)})\right]\left(\boldsymbol{\beta}^{(j+1)} - \boldsymbol{\beta}^{(j)}\right) = f(\boldsymbol{\beta}^{(j)})$$

where $\triangledown f(\boldsymbol{\beta}^{(j)})$ denotes the gradient of f. Unlike Alimadad and Salibian-Barrera [2011] we approximate $-E\left[\triangledown f(\boldsymbol{\beta}^{(j)})\right]$ by

$$-E\left[\frac{\partial^2 \ell}{\partial \boldsymbol{\beta} \partial \boldsymbol{\beta}^\top}\right] = \sum_{i=1}^{n} \omega_i \mathbf{x}_i \mathbf{x}_i^\top ,$$

where $\omega_i = -\left[E\left(\frac{\partial^2 \ell_i}{\partial \mu_i}\right)\right]\left(\frac{\partial \mu_i}{\partial \eta_i}\right)^2$.

Hence

$$\left[\sum_{i=1}^{n} \omega_i \mathbf{x}_i \mathbf{x}_i^\top\right]\left(\boldsymbol{\beta}^{(j+1)} - \boldsymbol{\beta}^{(j)}\right) = \sum_{i=1}^{n} h_i \frac{\partial \mu_i}{\partial \eta_i} \mathbf{x}_i \ ,$$

then

$$\left[\sum_{i=1}^{n} \omega_i \mathbf{x}_i \mathbf{x}_i^\top\right] \boldsymbol{\beta}^{(j+1)} = \sum_{i=1}^{n} \omega_i \mathbf{x}_i z_i,$$

where $z_i = \eta_i + \frac{h_i}{\omega_i}\left(\frac{\partial \mu_i}{\partial \eta_i}\right)$.

This corresponds to replacing $\dfrac{\partial \ell_i}{\partial \mu_i}$ by h_i in the `gamlss()` fitting algorithm. (If no approximation to $-E\left[\nabla f(\boldsymbol{\beta}^{(j)})\right]$ is made, then this corresponds to replacing $\dfrac{\partial \ell_i}{\partial \mu_i}$ by h_i and $\dfrac{\partial^2 \ell_i}{\partial \mu_i^2}$ by $\mathrm{E}\left[\dfrac{\partial}{\partial \mu_i}\left\{\psi\left(\dfrac{\partial \ell_i}{\partial \mu_i}\right) - \mathrm{E}\left[\psi\left(\dfrac{\partial \ell_i}{\partial \mu_i}\right)\right]\right\}\right]$ in the `gamlss()` fitting algorithm.)

12.5 Bibliographic notes

Robustness is discussed in detail by Huber [1981] and Hampel et al. [1986]. Robust fitting is applied to generalized linear models by Cantoni and Ronchetti [2001], to generalized additive models by Alimadad and Salibian-Barrera [2011], and to estimation of the mean and dispersion functions in extended generalized additive models by Croux et al. [2012].

The vector of influence functions, and variance-covariance matrix, of robust estimators of the vector of parameters is given by Hampel et al. [1986], p 230-231 and Cantoni and Ronchetti [2001], p 1023.

12.6 Exercises

Let $\hat{\mu}$, $\hat{\sigma}$, $\hat{\nu}$, and $\hat{\tau}$ be the MLEs of parameters μ, σ, ν, and τ in the exercises below, assuming a random sample of observations from the distribution.

1. Let $Y \sim$ `PARETO2`(μ, σ).
   (a) Show that the conditional influence function for $\hat{\mu}$ is bounded as $y \to 0$ and as $y \to \infty$.
   (b) Show that the conditional influence function for $\hat{\sigma}$ is bounded as $y \to 0$, but unbounded as $y \to \infty$.
   (c) Hence show that the (unconditional) influence functions for $\hat{\mu}$ and $\hat{\sigma}$ are each bounded as $y \to 0$.
   (d) Show that the (unconditional) influence functions for $\hat{\mu}$ and $\hat{\sigma}$ are each unbounded as $y \to \infty$. (Hint: show that $\mathrm{E}\left[\dfrac{\partial^2 \ell}{\partial \mu \partial \sigma}\right] \neq 0$.)
2. Let $Y \sim$ `IGAMMA`(μ, σ).
   (a) Show that the conditional influence function for $\hat{\mu}$ is bounded as $y \to \infty$, but unbounded as $y \to 0$.
   (b) Show that the conditional influence function for $\hat{\sigma}$ is unbounded as $y \to 0$ or as $y \to \infty$.
3. Let $Y \sim$ `ST3`(μ, σ, ν, τ).

(a) Show that the conditional influence functions for $\hat{\mu}$, $\hat{\sigma}$, and $\hat{\nu}$ are each bounded as $y \to -\infty$ or as $y \to \infty$.

(b) Show that the conditional influence function for $\hat{\tau}$ is unbounded as $y \to -\infty$ or as $y \to \infty$.

(c) Are the (unconditional) influence functions for each of $\hat{\mu}$, $\hat{\sigma}$, and $\hat{\nu}$ bounded as $y \to -\infty$ or as $y \to \infty$? (The authors have not yet proved whether this is true.)

4. Let $Y \sim \texttt{GT}(\mu, \sigma, \nu, \tau)$.

   (a) Show that the conditional influence functions for $\hat{\mu}$ and $\hat{\sigma}$ are each bounded as $y \to -\infty$ or as $y \to \infty$.

   (b) Show that the conditional influence functions for $\hat{\nu}$ and $\hat{\tau}$ are each unbounded as $y \to -\infty$ or as $y \to \infty$.

5. Let $Y \sim \texttt{BCT}(\mu, \sigma, \nu, \tau)$.

   (a) Show that the conditional influence functions for $\hat{\mu}$ and $\hat{\sigma}$ are each bounded as $y \to -\infty$ or as $y \to \infty$.

   (b) Show that the conditional influence functions for $\hat{\nu}$ and $\hat{\tau}$ are each unbounded as $y \to -\infty$ or as $y \to \infty$.

6. Let $Y \sim \text{logit}\texttt{TF}(\mu, \sigma, \nu)$ (see Section 6.3.1).

   (a) Show that the (unconditional) influence function for $\hat{\mu}$ is bounded as $y \to 0$ or as $y \to 1$.

   (b) Show that the conditional influence function for $\hat{\sigma}$ is bounded as $y \to 0$ or as $y \to 1$.

   (c) Show that the conditional influence function for $\hat{\nu}$ is unbounded as $y \to 0$ or as $y \to 1$.

   Hint: Note that logitTF inherits the influence function properties of TF.

7. **The Dutch boys head circumference data:** The purpose of this exercise is to estimate centiles robustly. We will use the data set `dbhh`, first introduced in Section 5.9.2, where the `head` circumference was modeled as a function of two explanatory variables `age` and `ht` (height). Here, similar to exercise 1 in Chapter 5 where we modeled `ht` against `age`, we are interested in creating 'robust' centiles curves for `head` against `age`.

   (a) Input the data into **R** and plot log `head` against `age`, log `age`, and $\sqrt{\texttt{age}}$ to determine a suitable transformation for age for the the analysis. Which transformation do you think is suitable and why?

   (b) Find which of the `BCCGo`, `BCPEo`, and `BCTo` distributions fits the response variable `head` best. For example, for fitting `BCTo` use:

```
m1 <- gamlss(head~pb(log(age)), sigma.fo=~pb(log(age)),
             nu.fo=~pb(log(age)),
             tau.fo=~pb(log(age)), family=BCTo, data=dbhh)
```

State which distribution you think is appropriate for head and why.

(c) Check the residuals of the fitted model using a worm plot and comment. You can use the command:

```
which(abs(resid(m1))>3)
which(abs(resid(m1))>3.5)
```

to identify observations which have quantile residuals greater in absolute value than 3 and 3.5, respectively.

(d) Plot the fitted centile curves.

```
wp(m1, ylim.all=1)
centiles(m1, xvar=dbhh$age)
```

(e) Use the function gamlssRobust() to obtain a robustified fit of the m1 model.

```
m2 <- gamlssRobust(m1, trace=T)
```

Note that the above model used the default values for bound = abs(qNO(1/n)) and CD.bound = abs(qNO(1/(3*n))), the actual values of which can be found using m2$bound and m2$CD.bound. Experiment by refitting the model with different values for bound and CD.bound, e.g. 3.0 and 3.5, respectively. (Try to find values of bound and CD.bound which give a fitted model which fits most of the data well, e.g. a flat worm plot for most of the data. Note cases outside CD.bound are weighted out and so do not have residuals.)

(f) Plot the centile curves of your chosen robust fit.

(g) Use the function centiles.comp() to compare the centiles of the original fit to the centiles of the robust fit and comment.

```
centiles.com(m1,m2, xvar=dbhh$age, color=gray(.4))
```

(h) m2$obs.weighted and m2$obs.bounded will give you the observation numbers of the case deletion (weighted out) observations and the ones in which the response variable was bounded, respectively. Compare those observations with the ones obtained with high residuals from the original model in (c) above.

13

Methods of generating distributions

CONTENTS

This chapter provides explanation for:

1. how new continuous distributions can be generated,
2. how some of the distributions are interconnected,
3. families of distributions.

13.1 Introduction

Here we examine how several of the **gamlss.dist** distributions in Table 3.2 can be generated. Many of these distributions can be generated by one or more of the following methods:

1. transformation from a single random variable
2. transformation from two or more random variables
3. Azzalini type methods
4. splicing distributions
5. mixtures of distributions
6. truncation distributions
7. systems of distributions.

These methods are discussed in Sections 13.2 to 13.8, respectively.

This chapter focuses primarily on generating **gamlss** distributions. There are many other methods of generating distributions not covered in this chapter, some of which are discussed in Section 13.9.

13.2 Transformation from a single random variable

The following is the general rule applied when transforming from one random variable to another. Let Z be a continuous random variable with known pdf defined on range R_Z. Let the new random variable be $Y = g(Z)$, where the function $g(\cdot)$ is a one-to-one transformation that maps the set $Z \in R_Z$ onto the set $Y \in R_Y$, where R_Y is the range of Y. Let the inverse function of $g(\cdot)$ be $z = g^{-1}(y) = h(y)$ with continuous and non-zero first derivative $\frac{dz}{dy} = h'(y)$ for all $y \in R_Y$. Then the pdf of Y is given by

$$f_Y(y) = f_Z(h(y)) \, |h'(y)| = f_Z(z) \left| \frac{dz}{dy} \right| , \tag{13.1}$$

and $|\cdot|$ is the absolute value.

The cdf of Y is $F_Y(y) = F_Z((h(y))$, provided that $g(\cdot)$ is a monotonically increasing function, since

$$F_Y(y) = \text{P}(Y \le y) = \text{P}(h(Y) \le h(y)) = \text{P}(Z \le h(y)) = F_Z(h(y)),$$

and, in this case, if the location parameter for Z, say μ_Z, is the median of the distribution of Z then $\mu_Y = g(\mu_Z)$ is the median of the distribution of Y, since

$$F_Y(\mu_Y) = \text{P}(Y \le g(\mu_Z)) = \text{P}(g^{-1}(Y) \le \mu_Z) = \text{P}(Z \le \mu_Z) = 0.5.$$

This can provide a useful parameterization, see Section 13.2.1.

The **gamlss** function `gen.Family()` can take any `gamlss.family` distribution defined on $(-\infty, \infty)$ and transform it to $(0, \infty)$ or $(0, 1)$. These two families of transformed distributions are discussed next.

13.2.1 The log family of distributions

Consider the case of Z defined on $(-\infty, \infty)$ and let $Y = \exp(Z)$, defined on $(0, \infty)$. Then $Z = \log(Y)$ and the pdf of Y is

$$f_Y(y) = f_Z(\log(y)) \cdot \frac{1}{y} , \qquad \text{for } y > 0 , \tag{13.2}$$

since $\frac{dz}{dy} = \frac{1}{y}$.

For example, let Z have a normal, `NO`(μ, σ), distribution. Then $Y = \exp(Z)$ has the log normal distribution, `LOGNO`(μ, σ), with pdf given by

$$f_Y(y \,|\, \mu, \sigma) = \frac{1}{\sqrt{2\pi\sigma^2}} \frac{1}{y} \exp\left[-\frac{(\log(y) - \mu)^2}{2\sigma^2} \right], \qquad \text{for } y > 0 . \tag{13.3}$$

Since μ is the median of Z, $\exp(\mu)$ is the median of Y in (13.3). A different parameterization is obtained by replacing μ by $\log(\mu)$ in (13.3) as implemented in `LOGNO2`(μ, σ), so that μ is then the median of Y.

The `LOGNO`(μ, σ) can also be obtained using the `gen.Family()` function:

```
gen.Family("NO", type="log")

## A  log  family of distributions from NO has been generated
##  and saved under the names:
##  dlogNO plogNO qlogNO rlogNO logNO
```

The newly created distribution `logNO`(μ, σ) is equivalent to `LOGNO`(μ, σ). There is no facility at the moment to automatically create the reparameterized distribution `LOGNO2`(μ, σ). The advantage of course of the function `gen.Family()` is that it can be applied to any **gamlss.dist** distribution with range $(-\infty, \infty)$, including distributions with three or four parameters, for example to create `logSST`(μ, σ, ν, τ) on $(0, \infty)$, from `SST`(μ, σ, ν, τ) on $(-\infty, \infty)$.

13.2.2 The logit family of distributions

Consider the case of $Z \in (-\infty, \infty)$ and $Y = \dfrac{1}{1 + \exp(-Z)}$. Then $Y \in (0, 1)$ with pdf:

$$f_Y(y) = f_Z(\text{logit}(y)) \cdot \frac{1}{y(1-y)}, \qquad \text{for } 0 < y < 1 \;, \tag{13.4}$$

since $z = \log\left(\frac{y}{1-y}\right) = \text{logit}(y)$ and $\frac{dz}{dy} = \frac{1}{y(1-y)}$.

For example, let Z have a normal, `NO`(μ, σ), distribution. Then Y has a logit normal, `LOGITNO`(μ, σ) distribution, with pdf

$$f_Y(y \mid \mu, \sigma) = \frac{1}{\sqrt{2\pi\sigma^2}\, y\,(1-y)} \exp\left[-\frac{(\text{logit}(y) - \mu)^2}{2\sigma^2}\right], \qquad \text{for } 0 < y < 1 \;. \tag{13.5}$$

A different parameterization is obtained by replacing μ by $\text{logit}(\mu)$ in (13.5), so that μ is then the median of Y. This parameterization is not currently implemented in **gamlss.dist**.

The following code creates the distribution `logitNO`(μ, σ), which is equivalent to `LOGITNO`(μ, σ):

```
gen.Family("NO", type="logit")

## A  logit  family of distributions from NO has been generated
##  and saved under the names:
##  dlogitNO plogitNO qlogitNO rlogitNO logitNO
```

Note that `gen.Family()` can be applied to any **gamlss.dist** distribution with range $(-\infty, \infty)$ to create a corresponding logit distribution with range $(0, 1)$.

13.2.3 More general transformations

Many three- and four-parameter continuous distributions can be defined by assuming that a transformed variable Z has a simple well-known distribution. The parameters of the distribution of Y may come from parameters of the univariate transformation or from parameters of the distribution of Z or both. Below we consider distributions available in **gamlss.dist** which can be obtained by a univariate transformation. Table 13.1 provides a summary of these distributions.

13.2.3.1 Box-Cox Cole and Green: `BCCG`

The *Box-Cox Cole and Green* distribution for $Y > 0$ denoted by $\texttt{BCCG}(\mu, \sigma, \nu)$, assumes that Z has a truncated standard normal distribution, i.e. a truncated $\texttt{NO}(0, 1)$, where

$$Z = \begin{cases} \frac{1}{\sigma\nu}\left[\left(\frac{Y}{\mu}\right)^{\nu} - 1\right] & \text{if } \nu \neq 0 \\ \frac{1}{\sigma}\log\left(\frac{Y}{\mu}\right) & \text{if } \nu = 0. \end{cases} \tag{13.6}$$

See Section 19.4.1 for further details about `BCCG`. Cole and Green [1992] were the first to use this distribution to model all three parameters as nonparametric smooth functions of a single explanatory variable.

Note that the parameterization above is different from, and more orthogonal than, the one used originally by Box and Cox [1964]. Rigby and Stasinopoulos [2000] and Stasinopoulos et al. [2000] used the original Box-Cox parameterization:

$$Z = \begin{cases} (Y^{\nu} - 1)/\nu & \text{if } \nu \neq 0 \\ \log(Y) & \text{if } \nu = 0\,, \end{cases} \tag{13.7}$$

where $Z \sim \texttt{NO}(\mu, \sigma)$, to model μ and σ^2 as functions of explanatory variables for constant power parameter ν. They obtained the maximum likelihood estimate of ν from its profile likelihood. This model (and equivalently the distribution) is denoted as $Y \sim \texttt{LNO}(\mu, \sigma, \nu)$ in the **gamlss.dist** package, where ν is fixed by the user. See Section 19.4.5.

13.2.3.2 Box-Cox power exponential: `BCPE`

The *Box-Cox power exponential* distribution for $Y > 0$, denoted by $\texttt{BCPE}(\mu, \sigma, \nu, \tau)$, is defined by assuming Z given by (13.6) has a truncated standard power exponential distribution, i.e. a truncated $\texttt{PE}(0, 1, \tau)$, see Rigby and Stasinopoulos [2004]. This distribution is useful for modeling positive or negative skewness combined with (lepto or platy) kurtosis in a continuous response variable on $(0, \infty)$. See Section 19.5.2.

13.2.3.3 Box-Cox *t*: `BCT`

The *Box-Cox t* distribution for $Y > 0$, denoted by $\texttt{BCT}(\mu, \sigma, \nu, \tau)$, is defined by assuming Z given by (13.6) has a truncated standard t distribution with τ degrees of freedom, i.e. a truncated $\texttt{TF}(0, 1, \tau)$, [Rigby and Stasinopoulos, 2006]. See Section 19.5.1.

13.2.3.4 Exponential generalized beta type 2: `EGB2`

The *exponential generalized beta type 2* distribution for $-\infty < Y < \infty$, denoted by $\texttt{EGB2}(\mu, \sigma, \nu, \tau)$, assumes that $\exp(Y)$ has a generalized beta type 2 distribution. This

TABLE 13.1: Distributions generated by univariate transformation.

Distribution of Y	Distribution of Z	Transformation to Z	References
BCCG	truncated NO$(0,1)$	(13.6)	Cole and Green [1992]
BCPE	truncated PE$(0,1,\tau)$	(13.6)	Rigby and Stasinopoulos [2004]
BCT	truncated TF$(0,1,\tau)$	(13.6)	Rigby and Stasinopoulos [2006]
EGB2	F$_{2\nu,2\tau}$	(13.8)	Johnson et al. [1995] p.142
GB1	BE(μ,σ)	(13.9)	see Section 21.2
GB2	F$_{2\nu,2\tau}$	$(\tau/\nu)\,(Y/\mu)^{\sigma}$	McDonald [1984] p 648
GG	GA$(1,\sigma\nu)$	$(Y/\mu)^{\nu}$	Lopatatzidis and Green [2000]
JSUo	NO$(0,1)$	(13.10)	Johnson [1949]
PE	GA$(1,\nu^{1/2})$	$\nu\left\lvert\frac{Y-\mu}{c\sigma}\right\rvert^{\nu}$	Nelson [1991]
SHASHo	NO$(0,1)$	(13.11)	Jones and Pewsey [2009]
SHASH	NO$(0,1)$	(13.12)	Jones [2005]
ST3	BEo(a,b)	(13.13)	Jones and Faddy [2003]

distribution was called the exponential generalized beta of the second kind by McDonald [1991] and was investigated by McDonald and Xu [1995]. The distribution may also be defined by assuming that $Z \sim F_{2\nu,2\tau}$, defined by (13.36), where, from Johnson et al. [1995, p142],

$$Z = (\tau/\nu)\exp\left[(Y-\mu)/\sigma\right] \ . \tag{13.8}$$

See Section 18.4.1. The distribution has been called a generalized logistic distribution type IV, see Johnson et al. [1995, p142], who report its long history starting with Perks [1932]. Note also that $R = \exp\left[(Y-\mu)/\sigma\right]$ has a beta distribution of the second kind (BE2(ν,τ)), [Johnson et al., 1995, p248, 325]; and $B = R/(1+R)$ has an original beta distribution (BEo(ν,τ)).

13.2.3.5 Generalized beta type 1: GB1

The *generalized beta type 1* distribution for $0 < Y < 1$, denoted by GB1(μ,σ,ν,τ), is defined by assuming $Z \sim$ BE(μ,σ), where

$$Z = \frac{Y^{\tau}}{\nu + (1-\nu)\,Y^{\tau}} \tag{13.9}$$

where $0 < \mu < 1$, $0 < \sigma < 1$, $\nu > 0$ and $\tau > 0$. Note that GB1(μ,σ,ν,τ) always has range $0 < Y < 1$ and so is different from the generalized beta of the first kind [McDonald and Xu, 1995], whose range depends on the parameters. See Section 21.2.

13.2.3.6 Generalized beta type 2: GB2

The *generalized beta type 2* distribution (or generalized beta distribution of the second kind) for $Y > 0$ [McDonald, 1984, p648], denoted by GB2(μ,σ,ν,τ), can be defined by assuming $Z \sim F_{2\nu,2\tau}$, defined by (13.36), where $Z = (\tau/\nu)(Y/\mu)^{\sigma}$. Note also that $R = (Y/\mu)^{\sigma}$ has a beta distribution of the second kind (BE2(ν,τ)), [Johnson et al., 1995, p248, 325], and $B = R/(1+R)$ has an original beta (BEo(ν,τ)) distribution. See Section 19.5.3.

13.2.3.7 Generalized gamma: GG

The *generalized gamma* distribution for $Y > 0$, parameterized by Lopatatzidis and Green [2000] and denoted by GG(μ,σ,ν), assumes that Z has a gamma, GA$(1,\sigma\nu)$, distribution with mean 1 and variance $\sigma^2\nu^2$, where $Z = (Y/\mu)^{\nu}$. See Section 19.4.3.

A reparameterization of GG(μ,σ,ν), [Johnson et al., 1995, p401], given by setting $\mu = \alpha_2\alpha_3^{1/\alpha_1}$, $\sigma = \left(\alpha_1^2\alpha_3\right)^{-1/2}$ and $\nu = \alpha_1$, is denoted GG2$(\alpha_1,\alpha_2,\alpha_3)$, (not implemented in **gamlss.dist**). See Section 19.4.3.

13.2.3.8 Johnson's Su: JSUo, JSU

The original *Johnson Su* distribution for $-\infty < Y < \infty$, denoted by JSUo(μ,σ,ν,τ), Johnson [1949], is defined by assuming

$$Z = \nu + \tau \sinh^{-1}[(Y-\mu)/\sigma] \tag{13.10}$$

has a standard normal distribution. See Section 18.4.3. The *reparameterized Johnson Su* distribution, for $-\infty < Y < \infty$, denoted by JSU(μ,σ,ν,τ), has mean μ and standard deviation σ for all values of ν and τ. See Section 18.4.3.

13.2.3.9 Power exponential: PE, PE2

The *power exponential* distribution for $-\infty < Y < \infty$, denoted by PE(μ,σ,ν), is a symmetric distribution defined by assuming that $Z = \nu\left|\frac{Y-\mu}{c\sigma}\right|^{\nu}$ has a GA$(1,\nu^{1/2})$ distribution, where $c^2 = \Gamma(1/\nu)\,[\Gamma(3/\nu)]^{-1}$. See Section 18.3.3.

This parameterization, used by Nelson [1991], ensures that μ and σ are the mean and standard deviation of Y, respectively, for all $v > 0$. A reparameterization of PE(μ,σ,ν) used by Nandi and Mämpel [1995], denoted by PE2(μ,σ,ν), is given by replacing σ by σ/c in PE(μ,σ,ν). See Section 18.3.3.

The Subbotin distribution (Subbotin [1923], Johnson et al. [1995, p195]), which uses as parameters (θ,ϕ,δ), is also a reparameterization of PE2(μ,σ,ν) given by setting $\mu = \theta$, $\sigma = \phi 2^{\delta/2}$ and $\nu = 2/\delta$. Box and Tiao [1973, p157] equations (3.2.3) and (2.2.5) are reparameterizations of the Subbotin parameterization and PE(μ,σ,ν), in which $\delta = 1+\beta$ and $\nu = 2/(1+\beta)$, respectively. The distribution is also called the exponential power distribution or the Box-Tiao distribution.

13.2.3.10 Sinh-arcsinh: SHASHo, SHASH, SHASHo2

The original *sinh-arcsinh* distribution for $-\infty < Y < \infty$ [Jones and Pewsey, 2009], denoted by SHASHo(μ,σ,ν,τ), is defined by assuming that Z has a standard normal distribution where

$$Z = \sinh\left\{\tau \sinh^{-1}[(Y-\mu)/\sigma] - \nu\right\} . \tag{13.11}$$

See Section 18.4.6. Jones and Pewsey [2009] suggest a more stable reparameterization SHASHo2(μ,σ,ν,τ) by replacing σ by $\sigma\tau$ in (13.11). See Section 18.4.6.

The *sinh-arcsinh* distribution for $-\infty < Y < \infty$ [Jones, 2005], denoted by $\texttt{SHASH}(\mu, \sigma, \nu, \tau)$, is defined by assuming that $Z \sim \texttt{NO}(0, 1)$, where

$$Z = \frac{1}{2}\left[\exp\left\{\tau \sinh^{-1}[(Y-\mu)/\sigma]\right\} - \exp\left\{-\nu \sinh^{-1}[(Y-\mu)/\sigma]\right\}\right]. \quad (13.12)$$

See Section 18.4.5.

13.2.3.11 Skew t type 5: ST5

The *skew t type 5* distribution for $-\infty < Y < \infty$ [Jones and Faddy, 2003], denoted by $\texttt{ST5}(\mu, \sigma, \nu, \tau)$, assumes that $Z \sim \texttt{BEo}(a, b)$ where

$$Z = \frac{1}{2}\left[1 + R/(a+b+R^2)^{1/2}\right], \quad (13.13)$$

where $R = (Y-\mu)/\sigma$, $a = \tau^{-1}\left[1+\nu(2\tau+\nu^2)^{-1/2}\right]$, and $b = \tau^{-1}\left[1-\nu(2\tau+\nu^2)^{-1/2}\right]$. See Section 18.4.16.

13.3 Transformation from two or more random variables

Distributions can be generated from a function of two (or more) random variables.

13.3.1 Distributions generated by transformation

13.3.1.1 Student t family: TF

The *Student t* family for $-\infty < Y < \infty$ [Lange et al., 1989], denoted by $\texttt{TF}(\mu, \sigma, \nu)$, is defined by assuming that $Y = \mu + \sigma T$ where $T \sim t_\nu$ has a standard t distribution with ν degrees of freedom, defined itself by $T = Z(W/\nu)^{-1/2}$ where $Z \sim \texttt{NO}(0, 1)$ and $W \sim \chi^2_\nu \equiv \texttt{GA}(\nu, [2/\nu]^{1/2})$. The chi-square degrees of freedom ν is treated as a continuous parameter; and Z and W are independent random variables. See Section 18.3.6.

13.3.1.2 Skew t type 2: ST2

The skew t type 2 distribution for $-\infty < Y < \infty$ [Azzalini and Capitanio, 2003], denoted $\texttt{ST2}(\mu, \sigma, \nu, \tau)$, is defined by assuming that $Y = \mu + \sigma T$ where $T = Z\,(W/\tau)^{-1/2}$ and $Z \sim \texttt{SN1}(0, 1, \nu)$ (skew normal type 1 distribution, see Section 13.4), and $W \sim \chi^2_\tau$, and where Z and W are independent random variables. See Section 18.4.13.

The $\texttt{ST2}(\mu, \sigma, \nu, \tau)$ distribution is the one-dimensional special case of the multivariate skew t used in the **R** package **Sn** [Azzalini, 2018] .

13.3.2 Distributions formed by convolution

An important special case of a function of two independent random variables is their sum, i.e. $Y = Z_1 + Z_2$. The pdf of Y is obtained by convolution, i.e.

$$f_Y(y) = \int_{-\infty}^{y} f_{Z_1}(z) f_{Z_2}(y-z)\, dz\ . \quad (13.14)$$

The following are two examples.

13.3.2.1 Exponential Gaussian: exGAUS

If $Z_1 \sim$ NO(μ, σ) and $Z_2 \sim$ EXP(ν) in (13.14), then $Y = Z_1 + Z_2$ has an exponential Gaussian distribution, denoted by exGAUS(μ, σ, ν), for $-\infty < Y < \infty$. The distribution has been also called a lagged normal distribution [Johnson et al., 1994, p 172]. See Section 18.3.1.

13.3.2.2 Generalized Erlangian

As pointed out by Johnson et al. [1994, p172], the convolution of two or more exponential pdfs with different mean parameters gives the generalized Erlangian distribution, while the convolution of a normal with a generalized Erlangian pdf gives a generalized lagged normal distribution, see Davis and Kutner [1976].

13.4 Azzalini type methods

There are two Azzalini type methods: the first was proposed in 1985 and the second in 2003. These methods are described in Sections 13.4.1 and 13.4.2, respectively.

Note that, as we have mentioned in Section 4.5.4, an important disadvantage of distributions generated by Azzalini type methods is that their cdf is not explicitly available, but requires numerical integration. Their inverse cdf requires a numerical search and many integrations. Consequently both functions can be slow, particularly for large data sets. Centiles and centile-based measures (e.g. the median) are not explicitly available. Moment-based measures are usually complicated, if available. However the distributions can be very flexible in modeling skewness and kurtosis.

13.4.1 Azzalini (1985) method

Lemma 1 of Azzalini [1985] proposed the following method of introducing skewness into a symmetric pdf. Let $f_{Z_1}(z)$ be a pdf symmetric about zero and let $F_{Z_2}(z)$ be an absolutely continuous cdf such that $dF_{Z_2}(z)/dz$ is symmetric about zero. Then, for any $-\infty < \nu < \infty$, $f_Z(z)$ is a proper pdf where

$$f_Z(z) = 2 f_{Z_1}(z) F_{Z_2}(\nu z) \ . \tag{13.15}$$

Let $Y = \mu + \sigma Z$, where $-\infty < \mu < \infty$ and $0 < \sigma < \infty$, then

$$f_Y(y) = \frac{2}{\sigma} f_{Z_1}(z) F_{Z_2}(\nu z) \tag{13.16}$$

where $z = (y - \mu)/\sigma$. This allows the generation of families of skew distributions including the skew normal type 1, SN1(μ, σ, ν), skew exponential power type 1, SEP1(μ, σ, ν, τ), and skew t type 1, ST1(μ, σ, ν, τ), given below. Note switching from ν to $-\nu$ reflects $f_Y(y)$ about $y = \mu$.

13.4.1.1 Skew normal type 1: SN1

The *skew normal type 1* distribution for $-\infty < Y < \infty$ [Azzalini, 1985], denoted by $\texttt{SN1}(\mu, \sigma, \nu)$, is defined by assuming Z_1 and Z_2 have $\texttt{NO}(0, 1)$ distributions in (13.16). Consider $Y \sim \texttt{SN1}(\mu, \sigma, \nu)$. First note that $Z = (Y - \mu)/\sigma \sim \texttt{SN1}(0, 1, \nu)$ has pdf given by (13.15) where Z_1 and Z_2 have $\texttt{NO}(0, 1)$ distributions.

Figure 13.1 shows how $\texttt{SN1}(0, 1, \nu)$ is obtained using (13.15): (a) $f_{Z_1}(z)$ of $Z_1 \sim \texttt{NO}(0, 1)$; (b) $2\,F_{Z_2}(\nu z)$ of $Z_2 \sim \texttt{NO}(0, 1)$, for $\nu = 0, 1, 2, 1000$; and (c) $f_Z(z)$ of $Z \sim \texttt{SN1}(0, 1, \nu)$, for $\nu = 0, 1, 2, 1000$.

Clearly from equation (13.15), the pdf $f_Z(z)$ is pdf $f_{Z_1}(z)$ weighted by $2\,F_{Z_2}(\nu z)$ for each value for z for $-\infty < z < \infty$. When $\nu = 0$ then $2\,F_{Z_2}(\nu z) = 1$ for all z, i.e. a constant weight 1, hence $f_Z(z) = f_{Z_1}(z)$, a standard normal pdf. For $\nu > 0$, $2\,F_{Z_2}(\nu z)$ provides heavier weights for $z > 0$ than for $z < 0$, and results in a positively moment skew $f_Z(z)$.

As $\nu \to \infty$, $2\,F_{Z_2}(\nu z)$ tends to a 0-1 step function resulting in a half normal distribution $f_Z(z)$, the most positively moment skew distribution in the SN1 distribution. Switching from ν to $-\nu$ reflects $F_Z(z)$ about $z = 0$, leading to negatively skew distributions. Finally $Y = \mu + \sigma Z$, so the distribution of $Y \sim \texttt{SN1}(\mu, \sigma, \nu)$ is a scaled and shifted version of the distribution of $Z \sim \texttt{SN1}(0, 1, \nu)$.

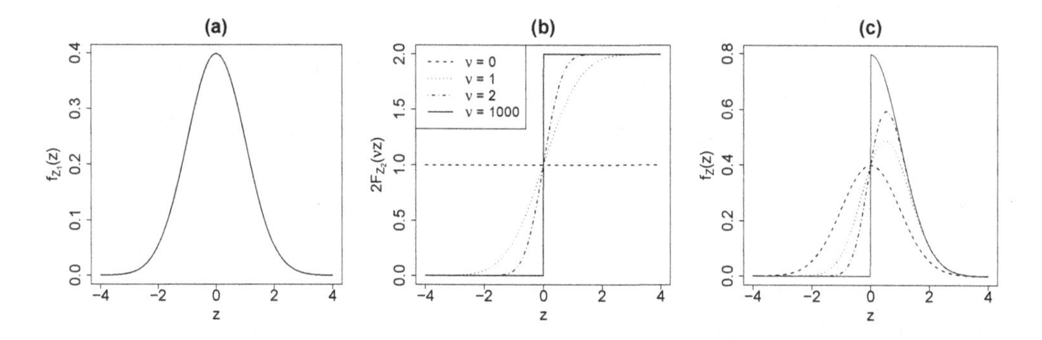

FIGURE 13.1: Azzalini's method (a) pdf: $f_{Z_1}(z)$ of $Z_1 \sim \texttt{NO}(0,1)$, (b) 2 cdf: $2\,F_{Z_2}(\nu z)$ of $Z_2 \sim \texttt{NO}(0,1)$ for ν=0, 1, 2, 1000 (note that the solid line $\nu = 1000$ is effectively a step function), (c) pdf: $\texttt{SN1}(0, 1, \nu)$ for ν=0, 1, 2, 1000.

See Section 18.3.4 for further information about $\texttt{SN1}(\mu, \sigma, \nu)$.

13.4.1.2 Skew exponential power type 1: SEP1

The *skew exponential power type 1* distribution for $-\infty < Y < \infty$ [Azzalini, 1986], denoted by $\texttt{SEP1}(\mu, \sigma, \nu, \tau)$, is defined by assuming Z_1 and Z_2 have power exponential type 2, $\texttt{PE2}\left(0, \tau^{1/\tau}, \tau\right)$, distributions in (13.16). Azzalini [1986] called this distribution type I. See Section 18.4.7. The skew normal type 1, $\texttt{SN1}\,(\mu, \sigma, \nu)$, is a special case of $\texttt{SEP1}(\mu, \sigma, \nu, \tau)$ obtained by setting $\tau = 2$.

TABLE 13.2: Distributions generated by Azzalini type methods using (13.18).

Y distribution	Z_1 distribution	Z_2 distribution	$w(z)$
$\texttt{SN1}(\mu,\sigma,\nu)$	$\texttt{NO}(0,1)$	$\texttt{NO}(0,1)$	νz
$\texttt{SEP1}(\mu,\sigma,\nu,\tau)$	$\texttt{PE2}\left(0,\tau^{1/\tau},\tau\right)$	$\texttt{PE2}\left(0,\tau^{1/\tau},\tau\right)$	νz
$\texttt{SEP2}(\mu,\sigma,\nu,\tau)$	$\texttt{PE2}\left(0,\tau^{1/\tau},\tau\right)$	$\texttt{NO}(0,1)$	$\nu\,(2/\tau)^{1/2}\,\text{sign}(z)\|z\|^{\tau/2}$
$\texttt{ST1}(\mu,\sigma,\nu,\tau)$	$\texttt{TF}(0,1,\tau)$	$\texttt{TF}(0,1,\tau)$	νz
$\texttt{ST2}(\mu,\sigma,\nu,\tau)$	$\texttt{TF}(0,1,\tau)$	$\texttt{TF}(0,1,\tau+1)$	$\nu\lambda^{1/2}z$

13.4.1.3 Skew t type 1: ST1

The *skew t type 1* distribution for $-\infty < Y < \infty$ [Azzalini, 1986], denoted by $\texttt{ST1}(\mu,\sigma,\nu,\tau)$, is defined by assuming Z_1 and Z_2 have Student t distributions with $\tau > 0$ degrees of freedom, i.e. $t_\tau = \texttt{TF}(0,1,\tau)$, in (13.16). See Section 18.4.12.

13.4.2 Azzalini and Capitanio method

Equation (13.15) was generalized, in Azzalini and Capitanio [2003] Proposition 1, to

$$f_Z(z) = 2f_{Z_1}(z)F_{Z_2}(w(z)) \tag{13.17}$$

where $w(z)$ is any odd function of z, i.e. $w(-z) = -w(z)$. Hence if $Y = \mu + \sigma Z$, where $-\infty < \mu < \infty$ and $0 < \sigma < \infty$, then

$$f_Y(y) = \frac{2}{\sigma} f_{Z_1}(z)F_{Z_2}(w(z)) \tag{13.18}$$

where $z = (y-\mu)/\sigma$. This allows a wider generation of families of distributions than the Azzalini [1985] method, including the skew exponential power type 2, $\texttt{SEP2}(\mu,\sigma,\nu,\tau)$, and skew t type 2, $\texttt{ST2}(\mu,\sigma,\nu,\tau)$ distributions below. A summary of distributions generated by (13.18) is given in Table 13.2.

13.4.2.1 Skew exponential power type 2: SEP2

The *skew exponential power type 2* distribution for $-\infty < Y < \infty$, denoted by $\texttt{SEP2}(\mu,\sigma,\nu,\tau)$ [Azzalini, 1986, DiCiccio and Monti, 2004], is expressed in the form (13.18) by letting $Z_1 \sim \texttt{PE2}(0,\tau^{1/\tau},\tau)$, $Z_2 \sim \texttt{NO}(0,1)$ and $w(z) = \nu(2/\tau)^{1/2}\text{sign}(z)|z|^{\tau/2}$. Azzalini [1986] developed a reparameterization of this distribution given by setting $\nu = \text{sign}(\lambda)|\lambda|^{\tau/2}$ and called it type II. See Section 18.4.8.

The skew normal type 1, $\texttt{SN1}(\mu,\sigma,\nu)$, distribution is a special case of $\texttt{SEP2}(\mu,\sigma,\nu,\tau)$ obtained by setting $\tau = 2$.

13.4.2.2 Skew t type 2: ST2

The *skew t type 2* distribution, denoted by $\texttt{ST2}(\mu,\sigma,\nu,\tau)$, is expressed in the form (13.18) by letting $Z_1 \sim \texttt{TF}(0,1,\tau)$, $Z_2 \sim \texttt{TF}(0,1,\tau+1)$, and $w(z) = \nu\lambda^{1/2}z$ where $\lambda = (\tau+1)/(\tau+z^2)$ [Azzalini and Capitanio, 2003]. An alternative derivation of $\texttt{ST2}(\mu,\sigma,\nu,\tau)$ is given in Section 13.3.1.2. See also Section 18.4.13 for further information about $\texttt{ST2}(\mu,\sigma,\nu,\tau)$.

TABLE 13.3: Distributions generated by splicing.

Distr. of Y	Distr. of Y_1	Distr. of Y_2	References
SN2(μ, σ, ν)	NO$(\mu, \sigma/\nu)$	NO$(\mu, \sigma\nu)$	Gibbons and Mylroie [1973]
SEP3(μ, σ, ν, τ)	PE2 $(\mu, \sigma 2^{1/\tau}/\nu, \tau)$	PE2 $(\mu, \sigma\nu 2^{1/\tau}, \tau)$	Fernandez et al. [1995]
SEP4(μ, σ, ν, τ)	PE2(μ, σ, ν)	PE2(μ, σ, τ)	Jones [2005]
ST3(μ, σ, ν, τ)	TF$(\mu, \sigma/\nu, \tau)$	TF$(\mu, \sigma\nu, \tau)$	Fernandez and Steel [1998]
ST4(μ, σ, ν, τ)	TF(μ, σ, ν)	TF(μ, σ, τ)	Section 18.4.15

13.5 Distributions generated by splicing

13.5.1 Splicing using two components

Splicing has been used to introduce skewness into a symmetric distribution family. Let Y_1 and Y_2 have pdfs that are symmetric about μ. A spliced distribution for Y may be defined by

$$f_Y(y) = \pi_1 f_{Y_1}(y)\mathrm{I}(y < \mu) + \pi_2 f_{Y_2}(y)\mathrm{I}(y \geq \mu) \tag{13.19}$$

where $\mathrm{I}(\cdot)$ is an indicator function taking value 1 if the condition in $(\cdot)$ is true and 0 otherwise. Ensuring that $f_Y(y)$ is a proper pdf requires $(\pi_1 + \pi_2)/2 = 1$. Ensuring continuity at $y = \mu$ requires $\pi_1 f_{Y_1}(\mu) = \pi_2 f_{Y_2}(\mu)$. Hence $\pi_1 = 2/(1+k)$ and $\pi_2 = 2k/(1+k)$ where $k = f_{Y_1}(\mu)/f_{Y_2}(\mu)$ and

$$f_Y(y) = \frac{2}{(1+k)}\left\{f_{Y_1}(y)\mathrm{I}(y < \mu) + k f_{Y_2}(y)\mathrm{I}(y \geq \mu)\right\}. \tag{13.20}$$

A summary of distributions generated by (13.20) is given in Table 13.3.

13.5.2 Splicing using two components with different scale parameters

A 'scale-spliced' distribution for Y may be defined by assuming that $f_Z(z)$ is symmetric about zero and that $Y_1 = \mu + \sigma Z/\nu$ and $Y_2 = \mu + \sigma\nu Z$ in (13.20), where $-\infty < \mu < \infty$, $0 < \sigma < \infty$, and $0 < \nu < \infty$, i.e. Y_1 and Y_2 have different scale parameters. Hence from (13.20),

$$f_Y(y) = \frac{2}{(1+k)}\left\{\frac{\nu}{\sigma} f_Z(\nu z)\,\mathrm{I}(y < \mu) + \frac{k}{\nu\sigma} f_Z(z/\nu)\,\mathrm{I}(y \geq \mu)\right\} \tag{13.21}$$

for $z = (y-\mu)/\sigma$ and $k = f_{Y_1}(\mu)/f_{Y_2}(\mu) = \nu^2$. Hence

$$f_Y(y) = \frac{2\nu}{\sigma(1+\nu^2)}\left\{f_Z(\nu z)\,\mathrm{I}(y < \mu) + f_Z(z/\nu)\,\mathrm{I}(y \geq \mu)\right\}. \tag{13.22}$$

The formulation (13.22) was used by Fernandez et al. [1995] and Fernandez and Steel [1998]. This allows the generation of 'scale-spliced' families of distributions including the skew normal type 2, skew exponential power type 3, and skew t type 3 described below. The distribution of Y is symmetric for $\nu = 1$, and appears to be positively moment skew for $\nu > 1$ and negatively moment skew for $0 < \nu < 1$, assuming Z has its mode at 0. Switching from ν to $1/\nu$ reflects $f_Y(y)$ about $y = \mu$.

13.5.2.1 Skew normal type 2: SN2

A *skew normal type 2* distribution (or two-piece normal distribution) for $-\infty < Y < \infty$, denoted by $\texttt{SN2}(\mu, \sigma, \nu)$, is defined by assuming $Z \sim \texttt{NO}(0, 1)$ in (13.22) or equivalently $Y_1 \sim \texttt{NO}(\mu, \sigma/\nu)$ and $Y_2 \sim \texttt{NO}(\mu, \sigma\nu)$ in (13.20), giving

$$f_Y(y) = \frac{2\nu}{\sqrt{2\pi}\sigma\,(1+\nu^2)} \left\{ \exp\left[-\frac{1}{2}(\nu z)^2\right] \mathrm{I}(y < \mu) + \exp\left[-\frac{1}{2}\left(\frac{z}{\nu}\right)^2\right] \mathrm{I}(y \geq \mu) \right\} \tag{13.23}$$

where $z = (y - \mu)/\sigma$. References to this distribution are given in Johnson et al. [1995, p173] and Jones and Faddy [2003]. The earliest reference appears to be Gibbons and Mylroie [1973]. See also Section 18.3.5.

For example consider $\mu = 0$ and $\sigma = 1$ in (13.23), then $Y \sim \texttt{SN2}(0, 1, \nu)$, where $Y_1 \sim \texttt{NO}(0, 1/\nu)$ and $Y_2 \sim \texttt{NO}(0, \nu)$, i.e. a spliced two-piece normal distribution. Figure 13.2 plots $Y \sim \texttt{SN2}(0,1,\nu)$ for ν=1, 2, 3, 5. Switching from ν to $1/\nu$ reflects $f_Y(y)$ about $y = 0$.

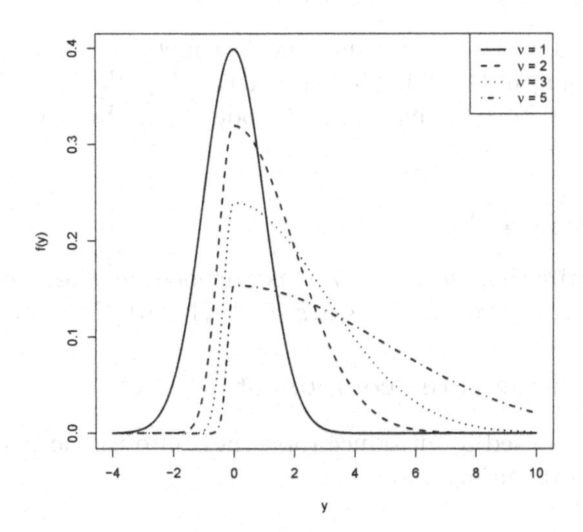

FIGURE 13.2: Splicing method $Y \sim \texttt{SN2}(0,1,\nu)$ for ν=1, 2, 3, 5.

13.5.2.2 Skew exponential power type 3: SEP3

A *skew exponential power type 3* distribution for $-\infty < Y < \infty$ [Fernandez et al., 1995], denoted by $\texttt{SEP3}(\mu, \sigma, \nu, \tau)$, is defined by assuming $Z \sim \texttt{PE2}(0, 2^{1/\tau}, \tau)$ in (13.22) or equivalently, $Y_1 \sim \texttt{PE2}(\mu, \sigma 2^{1/\tau}/\nu, \tau)$ and $Y_2 \sim \texttt{PE2}(\mu, \sigma\nu 2^{1/\tau}, \tau)$ in (13.20). See Section 18.4.9. Note that the $\texttt{SN2}(\mu, \sigma, \nu)$ is the special case $\texttt{SEP3}(\mu, \sigma, \nu, \tau)$ obtained by setting $\tau = 2$, i.e. $\texttt{SN2}(\mu, \sigma, \nu) \equiv \texttt{SEP3}(\mu, \sigma, \nu, 2)$.

13.5.2.3 Skew *t* type 3: ST3

A *skew t type 3* distribution for $-\infty < Y < \infty$ [Fernandez and Steel, 1998], denoted by $\texttt{ST3}(\mu, \sigma, \nu, \tau)$, is defined by assuming $Z \sim \texttt{TF}(0, 1, \tau) = t_\tau$ in (13.22), or equivalently $Y_1 \sim \texttt{TF}(\mu, \sigma/\nu, \tau)$ and $Y_2 \sim \texttt{TF}(\mu, \sigma\nu, \tau)$ in (13.20). See Section 18.4.14.

Two alternative reparameterizations of ST3(μ, σ, ν, τ) in which μ and σ are the mean and the standard deviation of Y are given by SST(μ, σ, ν, τ) and by Hansen [1994]. Theodossiou [1998] extended the Hansen reparameterization to a five-parameter skew generalized t distribution.

13.5.3 Splicing using two components with different shape parameters

A 'shape-spliced' distribution for Y may be defined by assuming Y_1 and Y_2 in (13.20) have pdfs that are symmetric about μ, with the same scale parameter σ, but with different shape parameters, ν and τ, respectively. This allows the generation of 'shape-spliced' families of distributions, including skew exponential power type 4, SEP4(μ, σ, ν, τ), and skew t type 4, ST4(μ, σ, ν, τ), given below.

13.5.3.1 Skew exponential power type 4: SEP4

A *skew exponential power type 4* distribution for $-\infty < Y < \infty$ [Jones, 2005], denoted by SEP4(μ, σ, ν, τ), is defined by assuming $Y_1 \sim$ PE2(μ, σ, ν) and $Y_2 \sim$ PE2(μ, σ, τ) in (13.20). See Section 18.4.10. Note that μ is the mode of Y.

A similar distribution was used by Nandi and Mämpel [1995] who set $Y_1 \sim$ PE2(μ, σ, ν) and $Y_2 \sim$ PE2$(\mu, \sigma/q, \tau)$ in (13.20), where $q = \Gamma\left[1 + (1/\tau)\right] / \Gamma\left[1 + (1/\nu)\right]$. However this distribution constrains both the median and mode of Y to be μ, which is perhaps rather restrictive.

13.5.3.2 Skew t type 4: ST4

A *skew t type 4* distribution for $-\infty < Y < \infty$, denoted by ST4(μ, σ, ν, τ), is defined by assuming $Y_1 \sim$ TF(μ, σ, ν) and $Y_2 \sim$ TF(μ, σ, τ) in (13.20). See Section 18.4.15.

13.5.4 Splicing using three components

Splicing has also been used to introduce robustness into the normal distribution, as in the NET(μ, σ, ν, τ) distribution below.

13.5.4.1 Normal-exponential-t: NET

The *normal-exponential-t* distribution for $-\infty < Y < \infty$, denoted by NET(μ, σ, ν, τ) [Rigby and Stasinopoulos, 1994], is defined by $Y = \mu + \sigma Z$, where Z is standard normal for $|Z| < \nu$, exponential with mean $1/\nu$ for $\nu \leq |Z| < \tau$, and has Student t with $(\nu\tau - 1)$ degrees of freedom type tails for $|Z| \geq \tau$. See Section 18.4.4 for details. In **gamlss.dist** the parameters ν and τ are treated as constants with default values $\nu = 1.5$ and $\tau = 2$.

13.6 Distributions generated by a mixture of distributions

A distribution for Y can be generated by a mixture of distributions. Assume that, given γ, Y has conditional probability (density) function $f(y \mid \gamma)$ and marginally γ has probability (density) function $f(\gamma)$. Then the marginal distribution of Y is given by

$$f_Y(y) = \begin{cases} \int f(y \mid \gamma)\, f(\gamma)\, d\gamma & \text{if } \gamma \text{ is continuous,} \\ \sum_i f(y \mid \gamma_i) \mathrm{P}(\gamma = \gamma_i) & \text{if } \gamma \text{ is discrete.} \end{cases} \tag{13.24}$$

The marginal distribution of Y is called a continuous mixture distribution if γ is continuous and a discrete mixture distribution if γ is discrete. If γ is discrete with a finite range of values, then the marginal distribution of Y is a finite mixture distribution. General finite mixture distributions are considered in detail in Stasinopoulos et al. [2017, Chapter 7].

Continuous mixture probability (density) functions for Y are explicitly defined if the integral in (13.24) is tractable, and this is dealt with in this section for the case where the conditional distribution of Y is continuous, and in Section 7.3 if the conditional distribution of Y is discrete. However the integral is often intractable and consequently $f_Y(y)$ is not explicitly defined, but may be approximated, using e.g. Gaussian quadrature points, and this is dealt with in Chapter 9 of Stasinopoulos et al. [2017], where the model is viewed as a random effect model at the observational level.

13.6.1 Explicitly defined continuous mixture distributions

The marginal distribution of Y will, in general, be continuous if the conditional distribution of Y is continuous. A summary of explicit continuous mixture distributions for Y generated by (13.24) is given in Table 13.4.

13.6.1.1 Student *t* family: TF

The *(Student) t* family distribution for $-\infty < Y < \infty$, denoted as $\texttt{TF}(\mu, \sigma, \nu)$, may be generated from a continuous mixture by assuming $Y|\gamma \sim \texttt{NO}(\mu, \gamma)$; and $\gamma \sim \sqrt{\nu}\sigma\chi_\nu^{-1} = \texttt{GG}(\sigma, [2\nu]^{-1/2}, -2)$, a scale inverted Chi distribution, which is a special case of the generalized gamma distribution, [Box and Tiao, 1973].

13.6.1.2 Generalized *t*: GT

The *generalized t* distribution for $-\infty < Y < \infty$, denoted as $\texttt{GT}(\mu, \sigma, \nu, \tau)$, may be generated by assuming $Y|\gamma \sim \texttt{PE2}(\mu, \gamma, \tau)$ and $\gamma \sim \texttt{GG2}(-\tau, \sigma\nu^{1/\tau}, \nu)$, [McDonald, 1991].

13.6.1.3 Generalized beta type 2: GB2

The *generalized beta type 2* distribution for $Y > 0$, denoted $\texttt{GB2}(\mu, \sigma, \nu, \tau)$, may be generated by assuming $Y|\gamma \sim \texttt{GG2}(\sigma, \gamma, \nu)$ and $\gamma \sim \texttt{GG2}(-\sigma, \mu, \tau)$, [McDonald, 1996].

13.6.1.4 Exponential generalized beta type 2: EGB2

The *exponential generalized beta type 2* distribution for $-\infty < Y < \infty$, denoted $\texttt{EGB2}(\mu, \sigma, \nu, \tau)$, may be generated by assuming $Y|\gamma \sim \texttt{EGG2}(1/\sigma, \gamma, \nu)$ and $\gamma \sim \texttt{GG2}(-1/\sigma, e^{\mu}, \tau)$, [McDonald, 1996]. (Note that the exponential generalized gamma type 2 distribution is defined by: if $Z \sim \texttt{EGG2}(\mu, \sigma, \nu)$ then $e^Z \sim \texttt{GG2}(\mu, \sigma, \nu)$.)

TABLE 13.4: Distributions generated by continuous mixtures.

Distr. of Y	Distr. of $Y\|\gamma$	Distr. of γ	Reference
$\texttt{TF}(\mu, \sigma, \nu)$	$\texttt{NO}(\mu, \gamma)$	$\texttt{GG}(\sigma, [2\nu]^{-1/2}, -2)$	Box and Tiao [1973]
$\texttt{GT}(\mu, \sigma, \nu, \tau)$	$\texttt{PE2}\ (\mu, \gamma, \tau)$	$\texttt{GG2}\ \left(-\tau, \sigma\nu^{1/\tau}, \nu\right)$	McDonald [1991]
$\texttt{GB2}(\mu, \sigma, \nu, \tau)$	$\texttt{GG2}(\sigma, \gamma, \nu)$	$\texttt{GG2}(-\sigma, \mu, \tau)$	McDonald [1996]
$\texttt{EGB2}(\mu, \sigma, \nu, \tau)$	$\texttt{EGG2}(1/\sigma, \gamma, \nu)$	$\texttt{GG2}(-1/\sigma, e^{\mu}, \tau)$	McDonald [1996]

13.7 Truncated distributions

Let Z be a random variable with any continuous or discrete distribution, and R_Z be the range of all possible values of Z. A random variable Y with a truncated distribution can be created from Z by truncating Z, i.e. by restricting the range of Y to a truncated subset of R_Z. There are three types of truncation, depending on which side the truncation is performed. Let a and b be constants defined within R_Z, with $a < b$. Then the resulting distribution is called:

1. *left* truncated if $Y > a$,
2. *right* truncated if $Y < b$, and
3. truncated on *both* sides if $a < Y < b$.

Note that for truncated continuous distributions, replacing $<$ by $\leq$ above does not matter, but it is vital in the definition of truncated discrete distributions. In the implementation of truncated distributions in the package **gamlss.tr**, we have adopted the convention that in left truncation the value a is not included in the range, and in right truncation the value of b is not included in the range. Hence in both sides truncation neither a nor b is included in the range.

In general the following results are relevant to left, right, and both sides truncation.

13.7.1 Left truncation

Let Y denote the random variable Z left truncated at a, so $Y > a$; and let Z denote the original random variable (before truncation) with probability (density) function $f_Z(z)$ and cdf $F_Z(z)$. Then the probability (density) function of Y is

$$f_Y(y) = \frac{f_Z(y)}{1 - F_Z(a)}\ , \qquad \text{for } y > a \tag{13.25}$$

with cdf

$$F_Y(y) = \frac{F_Z(y) - F_Z(a)}{1 - F_Z(a)}, \qquad \text{for } y > a \tag{13.26}$$

and inverse cdf

$$y_p = F_Y^{-1}(p) = F_Z^{-1}\left\{F_Z(a) + p\left[1 - F_Z(a)\right]\right\} \tag{13.27}$$

for $0 < p < 1$. The median of Y is given by $y_{0.5}$, i.e., set $p = 0.5$ in (13.27).

13.7.2 Right truncation

Let Y denote the random variable Z right truncated at b, so $Y < b$. As before, we denote the original untruncated random variable as Z. Then the probability (density) function of Y is

$$f_Y(y) = \begin{cases} \dfrac{f_Z(y)}{F_Z(b-1)} & \text{if } Z \text{ is discrete} \\ \dfrac{f_Z(y)}{F_Z(b)} & \text{if } Z \text{ is continuous} \end{cases} \tag{13.28}$$

for $y < b$, with cdf

$$F_Y(y) = \begin{cases} \dfrac{F_Z(y)}{F_Z(b-1)} & \text{if } Z \text{ is discrete} \\ \dfrac{F_Z(y)}{F_Z(b)} & \text{if } Z \text{ is continuous} \end{cases} \tag{13.29}$$

for $y < b$, and inverse cdf

$$y_p = F_Y^{-1}(p) = \begin{cases} F_Z^{-1}\left[pF_Z(b-1)\right] & \text{if } Z \text{ is discrete} \\ F_Z^{-1}\left[pF_Z(b)\right] & \text{if } Z \text{ is continuous} \end{cases} \tag{13.30}$$

for $0 < p < 1$. The median of Y is given by $y_{0.5}$ from (13.30) .

13.7.3 Both sides truncation

Let Y denote the random variable Z left truncated at a and right truncated at b, so $a < Y < b$; and Z is the original untruncated random variable. Then the probability (density) function of Y is

$$f_Y(y) = \begin{cases} \dfrac{f_Z(y)}{F_Z(b-1) - F_Z(a)} & \text{if } Z \text{ is discrete} \\ \dfrac{f_Z(y)}{F_Z(b) - F_Z(a)} & \text{if } Z \text{ is continuous} \end{cases} \tag{13.31}$$

for $a < y < b$, with cdf

$$F_Y(y) = \begin{cases} \dfrac{F_Z(y) - F_Z(a)}{F_Z(b-1) - F_Z(a)} & \text{if } Z \text{ is discrete} \\ \dfrac{F_Z(y) - F_Z(a)}{F_Z(b) - F_Z(a)} & \text{if } Z \text{ is continuous} \end{cases} \tag{13.32}$$

for $a < y < b$, and inverse cdf

$$y_p = \begin{cases} F_Z^{-1}\left[pF_Z(b-1) + (1-p)F_Z(a)\right] & \text{if } Z \text{ is discrete} \\ F_Z^{-1}\left[pF_Z(b) + (1-p)F_Z(a)\right] & \text{if } Z \text{ is continuous} \end{cases} \tag{13.33}$$

for $0 < p < 1$. The median of Y is given by $y_{0.5}$ from (13.33).

Examples of creating truncated distributions are given in Sections 3.4.3, 5.6.2, and 6.3.2.

13.8 Systems of distributions

13.8.1 Pearson system

The Pearson system of probability density functions $f_Y(y \mid \boldsymbol{\theta})$, where $\boldsymbol{\theta}^\top = (\theta_1, \theta_2, \theta_3, \theta_4)$, is defined by solutions of the equation:

$$\frac{\frac{d}{dy} f_Y(y \mid \boldsymbol{\theta})}{f_Y(y \mid \boldsymbol{\theta})} = -\frac{\theta_1 + y}{\theta_2 + \theta_3 y + \theta_4 y^2}\,, \tag{13.34}$$

see Johnson et al. [1994, Section 4.1, p15-25]. The solutions of (13.34) fall into one of seven families of distributions called Type I to Type VII. Types I, IV, and VI cover disjoint regions of the moment skewness-kurtosis $(\sqrt{\beta_1}, \beta_2)$ space, while the other four types are boundary types, see Johnson et al. [1994, p23, Figure 12.2]. Type I is a shifted and scaled beta distribution, with the resulting arbitrary range defined by two extra parameters. Type II is a symmetrical form of type I. Type III is a shifted gamma distribution. Types IV and V are not well-known distributions, probably because the constants of integration are intractable. Type VI is a shifted and scaled F distribution. Type VII is a scaled t distribution, i.e. $\texttt{TF}(0, \sigma, \nu)$.

13.8.2 Stable distribution system

Stable distributions are defined through their characteristic function. In general their pdf cannot be obtained explicitly except by integration [Nolan, 2007], or by using complicated infinite summations [Johnson et al., 1994, p58-60]. McDonald [1996] and Lambert and Lindsey [1999] discuss the application of stable distributions to modeling stock returns.

The stable distribution parameterization used in the **gamlss.dist** distribution $\texttt{SB}(\mu, \sigma, \nu, \tau)$ is the default "S0" parameterization used in the **R** package **stabledist** [Wuertz and Maechler, 2016], where our parameters (μ, σ, ν, τ) correspond to their $(\delta, \gamma, \beta, \alpha)$.

In this "S0" parameterization, $\texttt{SB}(\mu, \sigma, \nu, \tau)$ is a location-scale family of distributions with location shift parameter μ and scaling parameter σ. Parameter ν is a skewness parameter and τ is a kurtosis parameter. The 'S0' parameterization is given by Nolan [2007] and is defined as follows. If $Y \sim \texttt{SB}(\mu, \sigma, \nu, \tau)$, then the standardized variable $Z = (Y - \mu)/\sigma \sim \texttt{SB}(0, 1, \nu, \tau)$, and the distribution of Z is defined through its characteristic function given by Nolan [2007, equation (2)]:

$$\mathrm{E}[\exp(itZ)] = \begin{cases} \exp\left\{-|t|^\tau \left[1 + i\nu(\text{sign } t)(\tan\frac{\pi\tau}{2})(|t|^{1-\tau} - 1)\right]\right\} & \text{if } \tau \neq 1 \\ \exp\left\{-|t| \left[1 + i\nu(\text{sign } t)(\frac{2}{\pi}) \log|t|\right]\right\} & \text{if } \tau = 1, \end{cases}$$

where $-\infty < Y < \infty$, $-\infty < \mu < \infty$, $\sigma > 0$, $-1 < \nu < 1$ and $0 < \tau \leq 2$. Nolan [2007] gives integrals for calculating the pdf and cdf of Z. These are numerically evaluated in **stabledist**, which is used by $\texttt{SB}(\mu, \sigma, \nu, \tau)$ in **gamlss.dist**. Note that $\texttt{SB}(\mu, \sigma, \nu, 2) \equiv \texttt{NO}(\mu, \sqrt{2}\sigma)$, irrespective of the value of ν. Note also that $\texttt{SB}(\mu, \sigma, \nu, \tau)$ is a symmetric distribution if $\nu = 0$.

13.8.3 Exponential family

The *exponential family* of distributions $\mathcal{E}(\mu, \phi)$ is defined by the probability (density) function $f_Y(y \mid \mu, \phi)$ of Y having the form:

$$f_Y(y \mid \mu, \phi) = \exp\left\{\frac{y\theta - b(\theta)}{\phi} + c(y, \phi)\right\} \tag{13.35}$$

where $\text{E}(Y) = \mu = b'(\theta)$, $\text{Var}(Y) = \phi V(\mu)$ and $V(\mu)$ is the *variance function* $V(\mu) = b''(\theta(\mu))$ and ϕ is called the dispersion parameter. The form of (13.35) includes many important distributions including the normal, Poisson, gamma, inverse Gaussian, and Tweedie distributions having variance functions $V(\mu) = 1, \mu, \mu^2, \mu^3$ and μ^p for $p \leq 0$ or $p > 1$, respectively, and also the binomial distribution with variance function $V(\mu) = \mu(1-\mu)/n$.

The exponential family with mean μ and variance having the form $\phi\mu^p$ is called the Tweedie family [McCullagh and Nelder, 1989, Tweedie, 1984]. The probability (density) function exists only for $p \leq 0$ or $p > 1$ and suffers from being intractable (except using complicated series approximations) except for the specific values $p = 0, 2, 3$. Furthermore, in general for $1 < p < 2$, the distribution is a combination of a probability mass at $Y = 0$ and a continuous distribution for $Y > 0$, which cannot be modeled independently and is inappropriate for a continuous dependent variable Y, see Gilchrist [2000] and the discussion in Section 9.4. This distribution is not currently available in **gamlss.dist**.

13.9 Bibliographic notes

Systems of distributions are discussed in Johnson et al. [1994, Section 12.4, p15-63]. Azzalini [1985, 1986] and Azzalini and Capitanio [2003] presented a skewed family of distributions. Fernandez et al. [1995] and Fernandez and Steel [1998] presented an alternative skewed family of distributions generated by splicing. Mudholkar and Srivastava [1993] proposed a method to introduce an extra parameter to a two-parameter Weibull distribution. See for example Gupta et al. [1998] for several other exponentiated distributions. The Kumaraswamy generalized family is presented in Cordeiro and de Castro [2011]. The geometric exponential-Poisson family is presented in Nadarajah et al. [2013]. Aljarrah et al. [2014] used quantile functions to generate the T-X family of distributions. Cordeiro et al. [2017] introduced general mathematical properties of a new generator of continuous distributions with one extra parameter, called the generalized odd half-Cauchy family.

Lee et al. [2013] presented a review of methods for generating families of univariate continuous distributions in the recent decades. Jones [2015] reviewed and compared some of the main general techniques for providing families of typically unimodal distributions with one or more shape parameters. Tahir and Nadarajah [2015] discuss the exponential, Marshal-Olkin extended, beta generated, McDonald generalized, exponentiated generalized and Kumaraswamy generalized families of distributions and provide additional literature in chronological order about these families of distributions. Nadarajah and Rocha [2016] present an **R** package **Newdistns** for generating new distributions.

13.10 Exercises

1. **Transformations**

   (a) If $Y \sim$ BCCG(μ, σ, ν, τ) and Z is given by (13.6), show that Z has a truncated normal distribution, with $-1/(\sigma\nu) < Z < \infty$ if $\nu > 0$, and $-\infty < Z < -1/(\sigma\nu)$ if $\nu < 0$. [Hint: $f_Z(z) = f_Y(y)\left|\frac{dy}{dz}\right|$ and $f_Y(y)$ is given by (19.14).]

   (b) If $Y \sim$ EGB2(μ, σ, ν, τ) and $Z = (\tau/\nu)\exp[(Y - \mu/\sigma)]$, show that Z has an $F_{2\nu,2\tau}$ distribution with pdf given by

$$f_Z(z) = \left(\frac{\nu}{\tau}\right)^{\nu} \frac{z^{\nu-1}}{B(\nu,\tau)\left(1 + \frac{\nu z}{\tau}\right)^{\nu+\tau}} \qquad \text{for } z > 0 \,. \tag{13.36}$$

   (c) If $Y \sim$ GB1(μ, σ, ν, τ) and Z is given by (13.9), show that Z has a BE(μ, σ) distribution.

   (d) If $Y \sim$ GB2(μ, σ, ν, τ) and $Z = (\tau/\nu)(Y/\mu)^{\sigma}$, show that Z has a $F_{2\nu,2\tau}$ distribution.

   (e) If $Y \sim$ GG(μ, σ, ν) and $Z = (Y/\mu)^{\nu}$, show that Z has a GA$(1, \sigma\nu)$ distribution.

   (f) If $Y \sim$ JSUo(μ, σ, ν, τ) and $Z = \nu + \tau \sinh^{-1}[(Y - \mu)/\sigma]$, show that Z has a NO$(0, 1)$ distribution.

   (g) If $Y \sim$ PE(μ, σ, ν) and $Z = \nu\left|\frac{Y - \mu}{c\sigma}\right|^{\nu}$, where $c = [\Gamma(1/\nu)/\Gamma(3/\nu)]^{1/2}$, show that Z has a GA$(1, \nu^{1/2})$ distribution.

2. Find the pdf of a skew generalized t distribution for Y, with parameters $(\mu, \sigma, \nu, \tau, \xi)$, obtained by scale-splicing from $Z \sim$ GT$(0, 1, \nu, \tau)$, using equation (13.22) with skewness parameter ν replaced by ξ in (13.22).

3. Find the pdf of a Box-Cox generalized t distribution for Y with parameters $(\mu, \sigma, \nu, \tau, \xi)$, obtained by transformation from Z, assuming Z has a truncated GT$(0, 1, \nu, \tau)$ distribution with $-1/(\sigma\nu) < Z < \infty$ if $\nu > 0$ and $-\infty < Z < -1/(\sigma\nu)$ if $\nu < 0$, where Z is related to Y by (13.6) with skewness parameter ν replaced by ξ in (13.6).

4. Find the pdf of Johnson's logistic Su distribution for Y, obtained by transformation from Z, assuming Z has a standard logistic, LO$(0, 1)$, distribution, where Z is related to Y by $Z = \nu + \tau \sinh^{-1}[(Y - \mu)/\sigma]$, [Tadikamalla and Johnson, 1982].

5. **Mixtures**

   (a) If $Y/\gamma \sim$ NO(μ, γ) and $\gamma \sim$ GG$(\sigma, [2\nu]^{-1/2}, -2)$, show that the marginal distribution of Y is $Y \sim$ TF(μ, σ, ν).

   (b) If $Y/\gamma \sim$ PE2(μ, γ, τ) and $\gamma \sim$ GG2$(-\tau, \sigma\nu^{1/\tau}, \nu)$, show that the marginal distribution of Y is $Y \sim$ GT(μ, σ, ν, τ).

(c) If $Y/\gamma \sim$ GG2(μ, γ, ν) and $\gamma \sim$ GG2$(-\sigma, \mu, \tau)$, show that the marginal distribution of Y is $Y \sim$ GB2(μ, σ, ν, τ).

6. Implement in **gamlss** the beta exponential G distribution due to Alzaatreh et al. [2013], presented in the paper of Nadarajah and Rocha [2016].

14

Discussion of skewness

CONTENTS

14.1 Introduction

The concept of skewness can be thought of informally as a lack of symmetry. However while symmetry is unambiguous, a rigorous definition of skewness is neither obvious nor unique.

A skewness ordering compares two distributions to define which is more 'skew to the right', or equivalently less 'skew to the left'. A distribution of a random variable Y is defined to be right skewed (i.e. 'positively skewed') if Y is more skew to the right than $-Y$, according to a particular skewness ordering. In this chapter three skewness orderings are considered based on: moment skewness, centile skewness, and van Zwet skewness [van Zwet, 1964a,b]. These will be used to provide three alternative definitions of positive skewness, and they will then be applied to compare the skewness of two distributions.

For a thorough investigation of the concept of skewness see MacGillivray [1986], where a wide variety of skewness measures and orderings are given.

14.2 Positive skewness

The following are three criteria for defining positive skewness, or skewness to the right. The criteria are not equivalent.

(1) Moment positive skewness

Moment skewness of the distribution of Y was defined in Section 2.2.4 as

$$\gamma_1 = \sqrt{\beta_1} = \mu_3/(\mu_2)^{1.5} \tag{14.1}$$

where μ_k is the kth central moment. It is also known as Pearson's moment coefficient of skewness. Using this measure, 'moment positive skewness' of the distribution of Y is defined by $\gamma_1 > 0$. Problems with using γ_1 are that it may not be defined (e.g. for $Y \sim \texttt{TF}(\mu, \sigma, \nu)$ when $\nu \leq 3$), and is very sensitive to the heaviness of the tail(s) of the distribution, see example 2 in Section 14.3.

The condition $\gamma_{1Y} > 0$ is equivalent to $\gamma_{1Y} > \gamma_{1W}$ where $W = -Y$, since $\gamma_{1W} = -\gamma_{1Y}$, where γ_{1Y} and γ_{1W} are the moment skewness of Y and W, respectively.

Note that the distribution of Y is moment negative skew $\Longleftrightarrow \gamma_1 < 0$.

(2) Centile positive skewness

The centile skewness function of the distribution of Y was defined in Section 2.3.3 as

$$s_p = \frac{(y_p + y_{1-p})/2 - y_{0.5}}{(y_{1-p} - y_p)/2} \tag{14.2}$$

for $0 < p < 0.5$, where $y_p = F_Y^{-1}(p)$ and $F_Y^{-1}(\cdot)$ is the inverse cdf of Y. Using this measure, 'centile positive skewness' of the distribution of Y is defined by $s_p \geq 0$ for all $0 < p < 0.5$, with $s_p > 0$ for some p.

This condition is equivalent to $s_{pY} \geq s_{pW}$ for all $0 < p < 0.5$, with $s_{pY} > s_{pW}$ for some p, where $W = -Y$, since $s_{pW} = -s_{pY}$, where s_{pY} and s_{pW} are the centile skewness functions of Y and W, respectively.

Note that the distribution of Y is centile negatively skew $\Longleftrightarrow s_p \leq 0$ for all $0 < p < 0.5$ with $s_p < 0$ for some p.

(3) van Zwet positive skewness

A continuous distribution can be defined as van Zwet skewed to the right (i.e. positively van Zwet skew) using van Zwet [1964a,b] and MacGillivray [1986, p997]. Let Y be a continuous random variable with pdf $f_Y(y)$ and cdf $F_Y(y)$. Then $W = -Y$ has pdf $f_W(w) = f_Y(-w)$ and cdf $F_W(w) = 1 - F_Y(-w)$. Let $F_Y(y)$ be continuously differentiable and $dF_Y(y)/dy = f_Y(y) > 0$ on the continuous range of Y. We also assume that $f_Y(y)$ is continuously differentiable on the range of Y.

The distribution of Y is van Zwet skewed to the right (i.e. positively van Zwet skew):

$\Longleftrightarrow$ Y is more van Zwet skewed to the right than $W = -Y$, denoted by $F_W <_2 F_Y$ (see Section 14.4)

$\Longleftrightarrow$ the monotonic increasing transformation $Y = F_Y^{-1}[F_W(W)]$, i.e. $Y = F_Y^{-1}[1 - F_Y(-W)]$, from W to Y is convex but not linear. (The definition of a convex function is given below.)

$\Longleftrightarrow$ a Q-Q plot of Y against $W = -Y$ is convex but not linear, i.e. a plot of y_p against $w_p = -y_{1-p}$ for $0 < p < 1$ is convex but not linear, where $w_p = F_W^{-1}(p) = -F_Y^{-1}(1-p) = -y_{1-p}$.

$\Longleftrightarrow$ $\dfrac{f_W(w_p)}{f_Y(y_p)}$, i.e. $\dfrac{f_Y(y_{1-p})}{f_Y(y_p)}$, is non-decreasing with p for $0 < p < 1$, and increases for some p.

Note that the distribution of Y is van Zwet negative skew $\Longleftrightarrow F_Y <_2 F_W$.

Notes

- The notation '$\Longleftrightarrow$' means 'if and only if'.
- Non-negative van Zwet skewness implies non-negative moment skewness, provided γ_1 exists. However the reverse is not true in general.
- Non-negative van Zwet skewness implies non-negative centile skewness. However the reverse is not true in general.

Definition 14.1. ***Definition of a convex function***

$G(x)$ is convex in x

$\Longleftrightarrow$ $G[\lambda x_1 + (1-\lambda)x_2] \leq \lambda G(x_1) + (1-\lambda)G(x_2)$ *for* $0 \leq \lambda \leq 1$.

$\Longleftrightarrow$ *in a plot of* $G(x)$ *against* x, $(G(x), x)$ *for* $x_1 < x < x_2$ *lies at or below a line from* $(G(x_1), x_1)$ *to* $(G(x_2), x_2)$.

$\Longleftrightarrow$ $\frac{dG(x)}{dx}$ *is constant or increasing (i.e. non-decreasing) with* x, *provided* $\frac{dG(x)}{dx}$ *exists.*

$\Longleftrightarrow$ $\frac{d^2G(x)}{dx^2} \geq 0$, *provided* $\frac{d^2G(x)}{dx^2}$ *exists.*

14.3 Examples

Example 1

The Weibull, $\texttt{WEI}(\mu, \sigma)$, distribution skewness has been considered by MacGillivray [1986, p1008]. Since μ is a scaling parameter and does not affect the moment, centile, or van Zwet skewness, let $\mu = 1$ without loss of generality. Therefore, let $Y \sim \texttt{WEI}(1, \sigma)$ with pdf $f_Y(y) = \sigma y^{\sigma-1} \exp(-y^\sigma)$, cdf $F_Y(y) = 1 - \exp(-y^\sigma)$ and inverse cdf $y_p = F_Y^{-1}(p) = [-\log(1-p)]^{1/\sigma}$.

In Figure 14.1(a), $f_Y(y)$ is plotted for a range of values of σ, and in Figure 14.1(b) the moment skewness γ_1 is plotted against σ, where $\gamma_1 > 0$ for $\sigma \leq 3.60$ and $\gamma_1 < 0$ for $\sigma > 3.60$. Hence, according to γ_1, Y is positively moment skew if $\sigma \leq 3.60$ and negatively moment skew if $\sigma > 3.60$.

Figure 14.2(a) shows the centile skewness function s_p plotted against p for a range of values of σ. Since the range of Y is $(0, \infty)$, $s_p \to 1$ as $p \to 0$ for all values of σ. For $\sigma \leq (1 - \log 2)^{-1} = 3.26$, $s_p > 0$ for all $0 < p < 0.5$, so s_p is always positive, and so Y is positively centile skew. However for $\sigma > (1 - \log 2)^{-1} = 3.26$, s_p switches from positive to negative as p increases, i.e. s_p is positive for low values of p and negative for large values of p. Hence if $\sigma > (1 - \log 2)^{-1}$ the distribution of Y is positively centile skew in the tail, and negatively centile skew in the center.

Figure 14.2(b) provides a Q-Q plot of Y against $W = -Y$, i.e. y_p against $-y_{1-p}$ for a

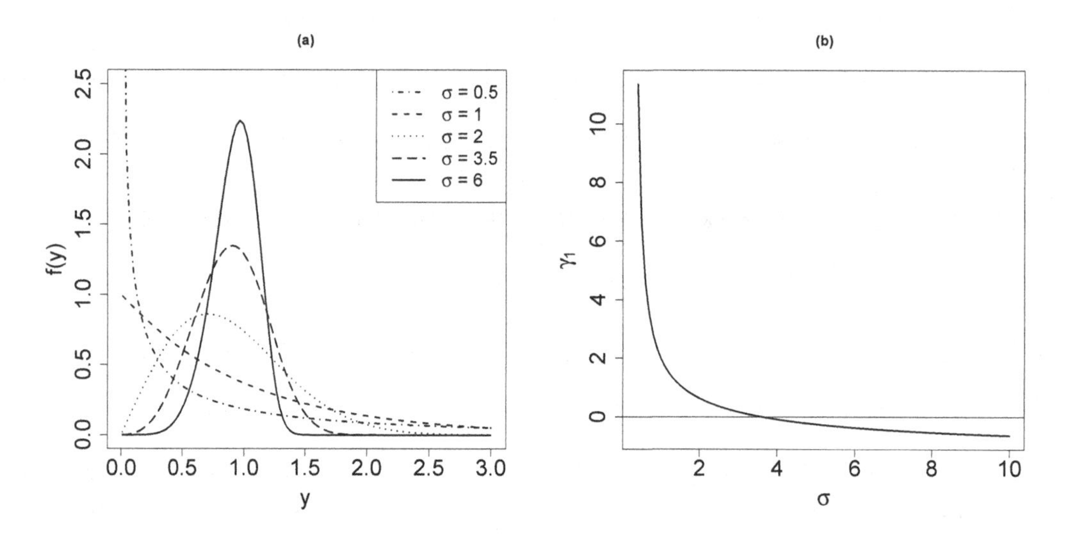

FIGURE 14.1: (a) pdf of $\texttt{WEI}(1, \sigma)$, (b) Moment skewness of $\texttt{WEI}(1, \sigma)$ against σ.

range of values of σ. For $\sigma \leq (1 - \log 2)^{-1} = 3.26$, the Q-Q plot of Y against $W = -Y$ is convex, indicating that the distribution of Y is positively van Zwet skewed. However for $\sigma > (1 - \log 2)^{-1}$, the Q-Q plot of Y against $W = -Y$ is not convex (or concave), but is concave at the center and convex in the tails, showing that the distribution of Y is negatively van Zwet skew in the center, and positively van Zwet skew in the tail.

The monotonic increasing transformation from $W = -Y$ to Y is given by

$$Y = F_Y^{-1}[F_W(W)] = F_Y^{-1}[1 - F_Y(-W)] = [-\log\{1 - \exp[-(-W)^\sigma]\}]^{1/\sigma}.$$

Plotting Y against W for $W < 0$ gives the same plot as the right panel in Figure 14.2. Figure 14.3 gives an alternative plot to check for positive van Zwet skewness by plotting the ratio of the densities of $W = -Y$ and Y at their quantiles, i.e.

$$\frac{f_W(w_p)}{f_Y(y_p)} = \frac{f_Y(y_{1-p})}{f_Y(y_p)}$$

against p for a range of values for σ. A horizontal line at a ratio of 1 is also shown.

For each $\sigma \leq (1 - \log 2)^{-1} = 3.26$, $f_Y(y_{1-p})/f_Y(y_p)$ increases with p, indicating positive van Zwet skewness. However for $\sigma > 3.26$, $f_Y(y_{1-p})/f_Y(y_p)$ increases, then decreases, then increases again with increasing p, indicating positive van Zwet skewness in the tail, but negative van Zwet skewness in the center.

In conclusion, if $Y \sim \texttt{WEI}(1, \sigma)$ then for $\sigma \leq 3.26$ the distribution of Y is positively skew using any of the three criteria, while for $\sigma > 3.26$ the centile and van Zwet criteria indicate negative skewness (i.e. skewness to the left) in the center, and positive skewness (i.e. skewness to the right) in the tail. For $\sigma \leq 3.60$ the moment skewness γ_1 is positive, while for $\sigma > 3.60$ it is negative.

Example 2

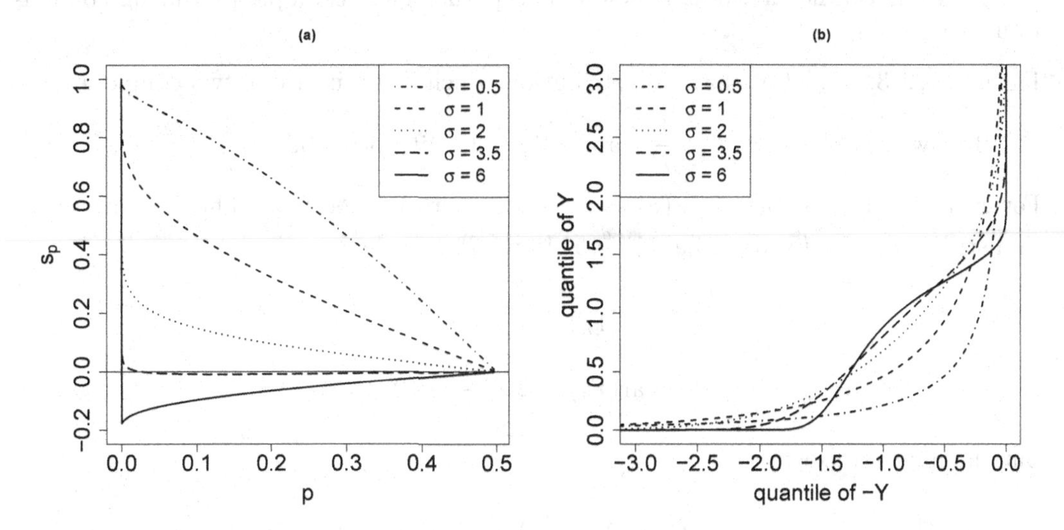

FIGURE 14.2: (a) Centile skewness function s_p of $\texttt{WEI}(1, \sigma)$ against p, (b) Q-Q plot of Y against $W = -Y$, i.e., y_p against $-y_{1-p}$.

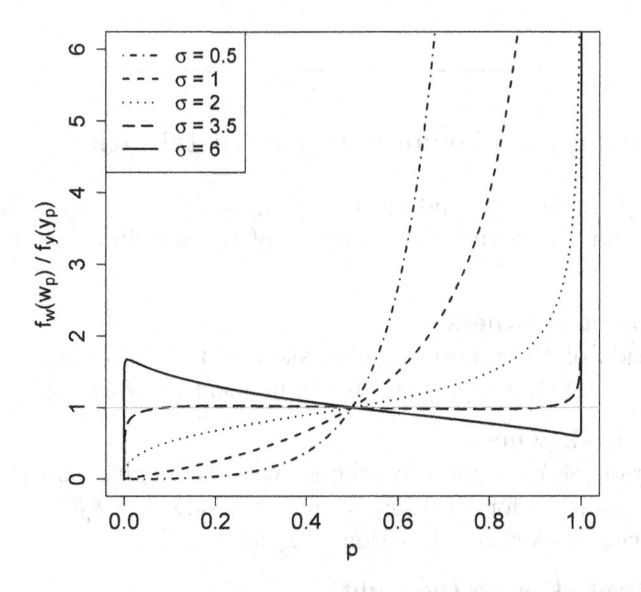

FIGURE 14.3: For Weibull distribution, plot of $\dfrac{f_W(w_p)}{f_Y(y_p)} = \dfrac{f_Y(y_{1-p})}{f_Y(y_p)}$ against p, where $W = -Y$.

This example shows that moment skewness is very sensitive to the tail(s) of the distribution, while centile skewness is not. It is a pathological example producing counter-intuitive results.

For $n = 1, 2, 3, \ldots$, let Y_n have a distribution which is a mixture of two components:

$$\texttt{NO}(0,1) \text{ with probability } 1 - \frac{1}{n^2}\,, \text{ and } \quad \texttt{NO}(n,1) \text{ with probability } \frac{1}{n^2}\,.$$

For every fixed y, $f_{Y_n}(y) \to f_Y(y)$ as $n \to \infty$, where $Y \sim \texttt{NO}(0,1)$. This is convergence in distribution (for fixed y), i.e. $Y_n \xrightarrow{d} Y$.However as $n \to \infty$,

$$\begin{aligned} \mathrm{E}(Y_n) &= \frac{1}{n} \to 0 \\ \mathrm{Var}(Y_n) &= 2 - \frac{1}{n^2} \to 2 \end{aligned}$$

but the moment skewness,

$$\gamma_1(Y_n) = \left(n - \frac{3}{n} + \frac{2}{n^3}\right)\left(2 - \frac{1}{n^2}\right)^{-3/2} \to \infty\,.$$

Furthermore, since $Y_n \xrightarrow{d} Y$, $F_{Y_n}(y) \to F_Y(y)$ as $n \to \infty$, for any fixed y. Similarly the inverse cdf converges: $F_{Y_n}^{-1}(p) \to F_Y^{-1}(p)$ for any fixed $p \in (0,1)$. Hence the centile skewness of Y_n converges to zero, i.e. $s_p(Y_n) \to 0$, as $n \to \infty$, for any fixed $p \in (0,1)$.

The centile skewness is robust to the right tail, however the moment skewness is sensitive to the right tail.

14.4 Comparing the skewness of two distributions

Let Y_i have pdf $f_i(y)$, cdf $F_i(y)$ and inverse cdf $y_{ip} = F_i^{-1}(p)$, for $i = 1, 2$. The following are three criteria for comparing the skewness of the distributions of Y_1 and Y_2. The criteria are not equivalent.

(1) **Higher moment skewness**
The distribution of Y_2 is more moment skew to the right than the distribution of $Y_1 \Leftrightarrow \gamma_1(Y_2) > \gamma_1(Y_1)$, where $\gamma_i(Y_i)$ is the moment skewness of Y_i for $i = 1, 2$.

(2) **Higher centile skewness**
The distribution of Y_2 is more centile skew to the right than the distribution of $Y_2 \Leftrightarrow s_p(Y_2) \geq s_p(Y_1)$ for all $0 < p < 0.5$ with $s_p(Y_2) > s_p(Y_1)$ for some p, where $s_p(Y_i)$ is the centile skewness function of Y_i for $i = 1, 2$.

(3) **More van Zwet skew to the right**
MacGillivray [1986, p997] provided a definition of more van Zwet skew to the right for continuous distributions, using van Zwet [1964a,b]. Assume that $F_i(y)$ is continuously differentiable and $dF_i(y)/dy = f_i(y) > 0$ on the continuous range of Y_i, for $i = 1, 2$. We also assume that $f_i(y)$ is continuously differentiable on the range of Y_i, for $i = 1, 2$.

The distribution of Y_2 is said to be more van Zwet skew to the right than the distribution of Y_1, denoted $F_1 <_2 F_2$

$\iff$ the monotonic increasing transformation $Y_2 = F_2^{-1}[F_1(Y_1)]$ from Y_1 to Y_2 is convex (but not linear), see Lemma 1.

$\iff$ a Q-Q plot of Y_2 against Y_1 is convex (but not linear) i.e. a plot of y_{2p} against y_{1p} for $p \in (0,1)$ is convex (but not linear), see Lemma 2.

$\iff$ $\dfrac{f_1(y_{1p})}{f_2(y_{2p})}$ is non-decreasing with p for $p \in (0,1)$ and increases for some p, see Lemma 3.

When the transformation from Y_1 to Y_2 is convex (including linear), Y_1 is said to be not more van Zwet skew to the right than Y_2, denoted $F_1 \le_2 F_2$, [MacGillivray, 1986].

Note:

(a) Y_1 not more van Zwet skew to the right than Y_2, implies Y_1 is not more moment skew to the right, since $F_1 \le_2 F_2 \Rightarrow \gamma_1(Y_1) \le \gamma_1(Y_2)$, provided $\gamma_1(Y_1)$ and $\gamma_1(Y_2)$ exist, see van Zwet [1964b] for the proof. However the reverse is not true in general.

(b) Y_1 not more van Zwet skew to the right than Y_2, implies Y_1 is not more centile skew to the right, since $F_1 \le_2 F_2 \Rightarrow s_p(Y_1) \le s_p(Y_2)$ for $p \in (0, 0.5)$, see Groeneveld and Meeden [1984] for the proof. However the reverse is not true in general.

Lemma 14.1. *Let $Y = F_2^{-1}[F_1(Y_1)]$ be a transformation from Y_1 to Y, where Y_1 has cdf F_1 and F_2 is a cdf. Then Y has cdf F_2.*

Proof.

$$\mathrm{P}(Y \le y) = \mathrm{P}(F_2^{-1}[F_1(Y_1)] \le y) = \mathrm{P}(F_1(Y_1) \le F_2(y)) = \mathrm{P}(U \le F_2(y)),$$

where $U = F_1(Y_1) \sim U(0,1)$ a uniform distribution over the interval $(0,1)$, since if $U = F_1(Y_1)$, and if f_U and f_1 are the pdf of U and Y_1, respectively, then

$$f_U(u) = f_1(y_1)\frac{dy_1}{du} = \frac{f_1(y_1)}{f_1(y_1)} = 1\ ,$$

so $U \sim U(0,1)$. Hence $\mathrm{P}(Y \le y) = \mathrm{P}(U \le F_2(y)) = F_2(y)$, so Y has cdf F_2. □

Lemma 14.2. *The transformation $Y_2 = F_2^{-1}[F_1(Y_1)]$ is a convex function of $Y_1 \Leftrightarrow$ a Q-Q plot of y_{2p} against y_{1p} is convex.*

Proof. Let $Y_2 = F_2^{-1}[F_1(Y_1)]$ be a convex function of Y_1. Let $p = F_1(Y_1)$ then $Y_1 = F_1^{-1}(p) = y_{1p}$ and $Y_2 = F_2^{-1}(p) = y_{2p}$. Hence a Q-Q plot of y_{2p} against y_{1p} is the same as the plot of Y_2 against Y_1, and so is convex. □

Lemma 14.3. *The transformation* $Y_2 = F_2^{-1}[F_1(Y_1)]$ *is a convex function of* $Y_1 \Leftrightarrow \dfrac{f_1(y_{1p})}{f_2(y_{2p})}$ *is non-decreasing with* p *for* $p \in (0,1)$.

Proof. $F_2^{-1}[F_1(y_{1p})]$ is convex in y_{1p}

$\Longleftrightarrow \dfrac{d}{dy_{1p}}\{F_2^{-1}[F_1(y_{1p})]\}$ is non-decreasing with y_{1p} for $p \in (0,1)$

$\Longleftrightarrow \dfrac{f_1(y_{1p})}{f_2(y_{2p})}$ is non-decreasing with y_{1p} and hence non-decreasing with p for $p \in (0,1)$,

since y_{1p} increases with p and

$$\begin{aligned}
\frac{d}{dy_{1p}}\{F_2^{-1}[F_1(y_{1p})]\} &= \frac{d}{dy_{1p}}[F_2^{-1}(p)] && \text{where } p = F_1(y_{1p}) \\
&= \frac{dy_{2p}}{dp}\frac{dp}{dy_{1p}} && \text{where } y_{2p} = F_2^{-1}(p) \\
&= \frac{f_1(y_{1p})}{f_2(y_{2p})}\,.
\end{aligned}$$

since $dp/dy_{ip} = f_i(y_{ip})$ for $i = 1, 2$. □

14.5 True skewness parameter

GAMLSS distributions can have up to four distributional parameters: μ, σ, ν, and τ. [Usually, for example, in continuous distributions with range $(-\infty, \infty)$, μ and σ are location and scale parameters respectively, while ν and τ are shape parameters which control features such as skewness or kurtosis.]

In order to distinguish a loosely-defined skewness parameter from the explicitly defined skewness in this chapter, we define below a 'true' skewness parameter as a parameter which changes monotonically with one of the moment, centile, or van Zwet skewness orderings.

In the tables in Part III of this book, a parameter is loosely referred to as a 'skewness parameter' if it affects the skewness of the distribution, or explicitly referred to as a 'primary true skewness parameter' or a 'true skewness parameter', if it has been proven to satisfy Definition 14.2 or 14.3 below, respectively.

Definition 14.2. *Let continuous random variable* Y *have a cdf* $F_Y(y \,|\, \theta, \theta^*)$ *which depends on a parameter* θ *and potentially other parameters* θ^*. *Let* θ_1 *and* θ_2 *be any two values of* θ *and let random variable* Y_i *have cdf* $F_Y(y \,|\, \theta_i, \theta_i^*)$ *for* $i = 1, 2$. *Then*

(a) θ *is a primary true moment skewness parameter if*

$$\theta_1 < \theta_2 \Rightarrow \gamma_1(Y_1) < \gamma_1(Y_2) \qquad \textit{for any } (\theta_1, \theta_2)$$

or alternatively if

$$\theta_1 > \theta_2 \Rightarrow \gamma_1(Y_1) < \gamma_1(Y_2) \qquad \text{for any } (\theta_1, \theta_2) \text{ ,}$$

where $\gamma_1(Y_i)$ *is the moment skewness of* Y_i *for* $i = 1, 2$.

(b) θ *is a primary true centile skewness parameter if*

$$\theta_1 < \theta_2 \Rightarrow Y_1 \text{ is more centile skew to the right than } Y_2 \text{ for any } (\theta_1, \theta_2)$$

or alternatively if

$$\theta_1 > \theta_2 \Rightarrow Y_1 \text{ is more centile skew to the right than } Y_2 \text{ for any } (\theta_1, \theta_2) \text{ .}$$

(c) θ *is a primary true van Zwet skewness parameter if*

$$\theta_1 < \theta_2 \Rightarrow F_1 <_2 F_2 \text{ for any } (\theta_1, \theta_2) \text{ ,}$$

or alternatively if

$$\theta_1 > \theta_2 \Rightarrow F_1 <_2 F_2 \text{ for any } (\theta_1, \theta_2)$$

where $F_i = F_Y(y \mid \theta_i, \theta_i^*)$ *for* $i = 1, 2$ *and* $F_1 <_2 F_2$ *is defined in Section 14.4.*

Definition 14.3. *If* $\theta_1^* = \theta_2^*$ *in the above conditions in Definition 14.2, we say that* θ *is a true (moment, centile, or van Zwet) skewness parameter, albeit not a primary true skewness parameter.*

Note that a (primary) true van Zwet skewness parameter is therefore also a (primary) true centile skewness parameter and a (primary) true moment skewness parameter.

Definition 14.4. *A continous random variable* Y*, defined on* $(0, \infty)$*, is said to have a scale-van Zwet skewness family of distributions with scaling parameter* θ_1 *and a true van Zwet skewness parameter* θ_2 *if* $Z = Y/\theta_1$ *has a cdf that does not depend on* θ_1*, and* θ_2 *is a true van Zwet skewness parameter.*

Definition 14.5. *A continuous random variable* Y*, defined on* $(-\infty, \infty)$*, is said to have a location-scale-van Zwet skewness family of distributions with location shift parameter* θ_1*, scaling parameter* θ_2 *and a true van Zwet skewness parameter* θ_3 *if* $Z = (Y - \theta_1)/\theta_2$ *has a cdf that does not depend on* θ_1 *or* θ_2*, and* θ_3 *is a true van Zwet skewness parameter.*

Example 1

Let $Y \sim$ `WEI`(μ, σ) for $\mu > 0$ and $\sigma > 0$, then Y has a scale-van Zwet skewness family of distributions with scaling parameter μ and true van Zwet skewness parameter σ.

Proof. Let $Z = Y/\mu$ then $Z \sim$ `WEI`$(1, \sigma)$, which does not depend on μ, so μ is a scaling parameter. Since van Zwet skewness ordering is unaffected by scaling, let $Y_1 \sim$ `WEI`$(1, \sigma_1)$ and $Y_2 \sim$ `WEI`$(1, \sigma_2)$. Then

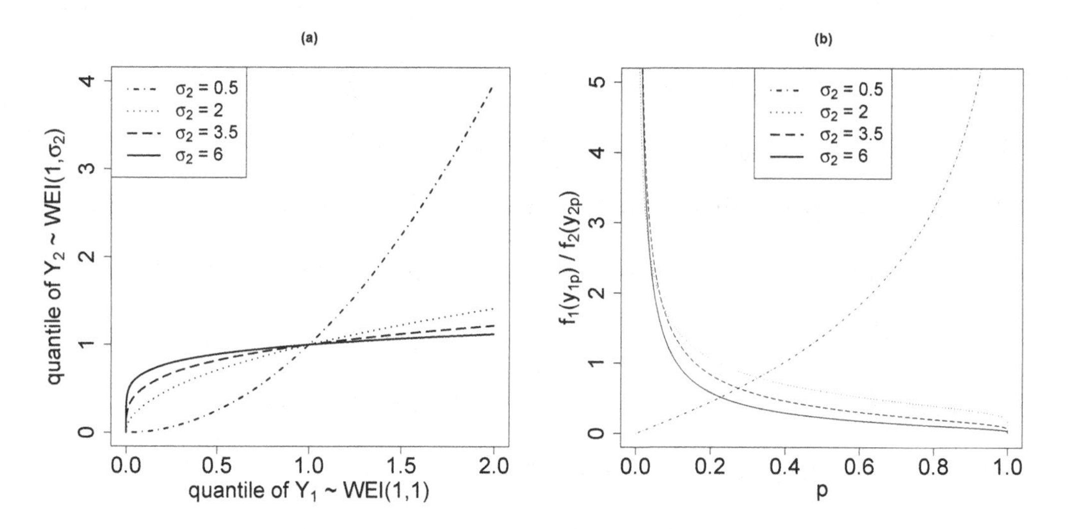

FIGURE 14.4: (a) Q-Q plot of of y_{2p} against y_{1p}, where $y_{ip} = F_i^{-1}(p)$ for $i = 1, 2$, where $Y_1 \sim$ `WEI`$(1, 1)$, and $Y_2 \sim$ `WEI`$(1, \sigma_2)$ for $\sigma_2 = 0.5, 2, 3.5, 6$, (b) $f_1(y_{1p})/f_2(y_{2p})$ against p, where $Y_1 \sim$ `WEI`$(1, 1)$, and $Y_2 \sim$ `WEI`$(1, \sigma_2)$ for $\sigma_2 = 0.5, 2, 3.5, 6$.

- $F_{Y_2}^{-1}[F_{Y_1}(y)] = [-\log(1 - p_1)]^{1/\sigma_2}$, where $p_1 = F_{Y_1}(y) = 1 - \exp(-y^{\sigma_1})$.
- Hence $F_{Y_2}^{-1}[F_{Y_1}(y)] = y^{\sigma_1/\sigma_2}$.
- Hence $F_{Y_2}^{-1}[F_{Y_1}(y)]$ is a convex function of y if $\sigma_2 < \sigma_1$, since $\frac{d^2}{dy^2}\left(y^{\sigma_1/\sigma_2}\right) > 0$ if $\sigma_2 < \sigma_1$.
- Hence $F_1 <_2 F_2$, i.e. Y_2 is more van Zwet skew to the right than Y_1, if $\sigma_2 < \sigma_1$.
- Hence σ is a true van Zwet skewness parameter for $Y \sim$ `WEI`(μ, σ), [and hence is also a true moment skewness parameter and a true centile skewness parameter].
- Hence $Y \sim$ `WEI`(μ, σ) has a scale-van Zwet skewness family of distributions.

□

For example let $Y_1 \sim$ `WEI`$(1, 1)$, i.e. $\sigma_1 = 1$, and $Y_2 \sim$ `WEI`$(1, \sigma_2)$ for $\sigma_2 = 0.5, 2, 3.5, 6$. See Figure 14.1(a) for a plot of their pdfs. Figure 14.4(a) shows a Q-Q plot of y_{2p} against y_{1p}, where $y_{ip} = F_i^{-1}(p)$ for $i = 1, 2$, for each of the four values of σ_2. When $\sigma_2 = 0.5$, then $\sigma_2 < \sigma_1 \Rightarrow F_1 <_2 F_2$, i.e. F_2 is more van Zwet skew to the right than F_1, resulting in a Q-Q curve which is convex. When $\sigma_2 = 2, 3.5$, or 6 then $\sigma_2 > \sigma_1 \Rightarrow F_1 >_2 F_2$, i.e. F_2 is less van Zwet skew to the right than F_1, resulting in a Q-Q curve which is concave. Figure 14.4(b) shows a plot of the ratio of the densities of Y_1 and Y_2 at their quantiles, i.e. $f_1(y_{1p})/f_2(y_{2p})$, against p for $0 < p < 1$. When $\sigma_2 = 0.5$, $F_1 <_2 F_2$, and the curve increases with p. When $\sigma_2 = 2, 3.5$ or 6, then $F_1 >_2 F_2$ and the curve decreases with p.

Example 2

Let $Y \sim$ `SHASHo`(μ, σ, ν, τ), then ν is a true van Zwet skewness parameter. (This was originally proven by Jones and Pewsey [2009].)

Proof. Let $Z = (Y - \mu)/\sigma$ then $Z \sim$ `SHASHo`$(0, 1, \nu, \tau)$ which does not depend on μ or σ, so μ is a location shift parameter and σ is a scaling parameter. Since van Zwet skewness ordering is unaffected by location shift or scaling, let $Y_j \sim$ `SHASHo`$(0, 1, \nu_j, \tau)$ for $j = 1, 2$, where $\nu_1 < \nu_2$.

From Table 18.19,

$$F_{Y_j}(y) = \Phi\left(\sinh[\tau \sinh^{-1}(y) - \nu_j]\right)$$

and

$$F_{Y_j}^{-1}(p) = \sinh\left\{\frac{\nu_j}{\tau} + \frac{1}{\tau}\sinh^{-1}[\Phi^{-1}(p)]\right\}.$$

for $j = 1, 2$.

Hence $F_{Y_2}^{-1}[F_{Y_1}(y)] = \sinh\{d + \sinh^{-1}(y)\}$ where $d = (\nu_2 - \nu_1)/\tau > 0$. Let $g(y) = F_{Y_2}^{-1}[F_{Y_1}(y)]$. Now $\sinh(x) = \frac{1}{2}\left(e^x - e^{-x}\right)$ and $\sinh^{-1}(x) = \log\left[(x^2+1)^{1/2} + x\right]$. Hence $g(y) = F_{Y_2}^{-1}[F_{Y_1}(y)] = \frac{1}{2}\left\{e^d[(y^2+1)^{1/2} + y] - e^{-d}[(y^2+1)^{1/2} - y]\right\}$.

Hence $\dfrac{d^2 g(y)}{dy^2} = \dfrac{\sinh(d)}{(y^2+1)^{1/2}}$, which is positive for all y, since $\sinh(d) > 0$ for $d > 0$. Hence $F_{Y_2}^{-1}[F_{Y_1}(y)]$ is a convex function of y for $\nu_1 < \nu_2$. Hence parameter ν in the `SHASHo`(μ, σ, ν, τ) distribution is a true van Zwet skewness parameter and $\nu_1 < \nu_2 \Rightarrow F_1 <_2 F_2$. □

Note that ν in `SHASHo`(μ, σ, ν, τ) is therefore also a true centile skewness parameter and a true moment skewness parameter.

14.6 Bibliographic notes

A thorough investigation of the concept of skewness is given by MacGillivray [1986], where a wide variety of skewness measures and orderings are given, including a definition of the centile skewness function. An original skewness ordering (called 'van Zwet skewness' throughout this chapter) was given by van Zwet [1964a,b].

14.7 Exercises

1. Let $Y \sim$ `LOGNO2`(μ, σ). Show that σ is a primary true van Zwet skewness parameter. (Note that σ is therefore also a primary true centile skewness parameter and a primary true moment skewness parameter.)

   [Hint: Show that $\sigma_1 < \sigma_2 \Rightarrow F_1 <_2 F_2$ where F_j is the cdf of $Y_j \sim$ `LOGNO2`(μ_j, σ_j) for $j = 1, 2$.]

2. Let $Y \sim$ `PARETO2`(μ, σ). Show that σ is a primary true van Zwet skewness parameter.

3. Let $Y \sim$ `JSUo`(μ, σ, ν, τ). Show that ν is a true van Zwet skewness parameter.

4. Let $Y \sim$ `GA`(μ, σ). Show that σ is a primary true moment skewness parameter.

5. Let $Y \sim$ `IGAMMA`(μ, σ). Show that σ is a primary true moment skewness parameter when the moment skewness is finite, i.e. for $\sigma^2 < 1/3$.

6. Let $Y \sim$ `SN1`(μ, σ, ν). Show that ν is a primary true moment skewness parameter.

7. Check whether any of the parameters labelled just 'skewness parameter' in the distribution summary tables in Chapters 18 and 19 satisfy the conditions to be a true, or primary true, moment, centile, or van Zwet skewness parameter, as defined in Section 14.5. (The authors have so far tried and failed to prove if any of these conditions hold!)

8. The gamma, `GA`(μ, σ), and inverse Gaussian, `IG`(μ, σ_1), distribution, where $\sigma_1 = \sigma/\sqrt{\mu}$ have the same μ and the same variance $\sigma^2\mu^2$. Which has the higher moment skewness?

9. `NBI`(μ, σ) and `PIG`(μ, σ) have the same mean and the same variance. Which has the higher moment skewness?

10. `NBI`(μ, σ) and `DEL`(μ, σ_1, ν) where $\sigma_1 = \sigma/(1 - \nu)^2$ have the same mean and the same variance. Which has the higher moment skewness?

15

Discussion of kurtosis

CONTENTS

Informally, a distribution which is 'more kurtotic' has heavier tails, is more peaked at its center (e.g. the mode or median), and has lighter shoulders (e.g. lighter density around the quartiles). "An increase in kurtosis is achieved through the location- and scale-free movement of probability mass from the 'shoulders' of a distribution into its center and tails" [Balanda and MacGillivray, 1990, p 17]. A kurtosis ordering compares two distributions to define which (if either) is more kurtotic. A distribution is defined to be leptokurtic (platykurtic) if it is more (less) kurtotic than the normal distribution, according to a particular kurtosis ordering. The problem is that 'more kurtotic' is not well defined as there are many different kurtosis orderings, which are not equivalent.

In this chapter three kurtosis orderings are considered based on: moment kurtosis, centile kurtosis, and Balanda-MacGillivray kurtosis. These will be defined and then used to compare the kurtosis of different distributions, initially comparing with the normal distribution to see if it is leptokurtic or platykurtic, using each of the three orderings.

For a thorough investigation of the concept of kurtosis see MacGillivray and Balanda [1988] and Balanda and MacGillivray [1990], where a wide variety of kurtosis measures and orderings are given.

15.1 Comparing the kurtosis of two distributions

Let Y_i be a random variable with pdf $f_i(y)$, cdf $F_i(y)$ and inverse cdf $y_{ip} = F_i^{-1}(p)$, for $p \in (0,1)$, for $i = 1, 2$. Assume $f_i(y) > 0$ on a continuous range R_i for $i = 1, 2$.

The following are three criteria for comparing the kurtosis of the distributions of Y_1 and Y_2. The criteria are not equivalent.

(1) **Higher moment kurtosis**

Moment excess kurtosis was defined in Section 2.2.4 as

$$\gamma_2 = \beta_2 - 3 = \mu_4/(\mu_2)^2 - 3, \tag{15.1}$$

where μ_k is the kth central moment.

The distribution of Y_2 is 'more moment kurtotic' than the distribution of Y_1
$\iff \gamma_2(Y_2) > \gamma_2(Y_1)$,
where $\gamma_2(Y_i)$ is the moment excess kurtosis of Y_i, for $i = 1, 2$.

(2) **Higher centile kurtosis**
The centile kurtosis function of a random variable Y was defined in Section 2.3.3 as

$$k_p(Y) = \frac{y_{1-p} - y_p}{y_{0.75} - y_{0.25}} \qquad \text{for } 0 < p < 0.5, \tag{15.2}$$

where $y_p = F_Y^{-1}(p)$. The distribution of Y_2 is 'more centile kurtotic' than the distribution of Y_1

$\iff$

$$k_p(Y_2) \geq k_p(Y_1), \text{ i.e. } \frac{k_p(Y_2)}{k_p(Y_1)} \geq 1, \text{ for all } 0 < p < 0.25, \tag{15.3}$$

and

$$k_p(Y_2) \leq k_p(Y_1), \text{ i.e. } \frac{k_p(Y_2)}{k_p(Y_1)} \leq 1, \text{ for all } 0.25 < p < 0.5 \tag{15.4}$$

with $k_p(Y_2) \neq k_p(Y_1)$ for some p.

Note condition (15.3) is one definition of Y_2 having heavier tails than Y_1, while condition (15.4) is one definition of Y_2 being more peaked than Y_1 around their medians.

Note also that, for $p \in (0, 0.5)$, $k_p(Y_2)/k_p(Y_1)$ equals the ratio of the interquantile ranges (with probability $1-2p$) of Y_2 and Y_1, when both Y_1 and Y_2 have been scaled to have the same interquartile range.

(3) **More Balanda-MacGillivray kurtotic**
The distribution of Y_2 is more Balanda-MacGillivray kurtotic than the distribution of Y_1, denoted here $F_1 <_S F_2$,

$\iff$ a plot of $k_p(Y_2)$ against $k_p(Y_1)$ for $0 < p < 0.5$ is convex (but not linear).

See Balanda and MacGillivray [1990]. Alternative equivalent conditions are given in Section 15.3.

Note that more Balanda-MacGillivray kurtotic implies more centile kurtotic, because a plot of $k_p(Y_2)$ against $k_p(Y_1)$ for $0 < p < 0.5$ is convex and passes through points (0,0) and (1,1) when $p = 0$ and $p = 0.25$, respectively. But the reverse does *not* hold in general.

Note the above three criteria can be used to compare the kurtosis of any two distributions (symmetric or skew), although sometimes neither distribution is more kurtotic than the other.

15.2 Leptokurtic and platykurtic distributions

A distribution of a random variable Y is defined to be leptokurtic (platykurtic) if it is more (less) kurtotic than the normal distribution, according to a particular kurtosis ordering. Here we apply each of the above three criteria. Since all three criteria are unaffected by shifting or scaling the random variables, let $Z \sim \texttt{NO}(0,1)$.

(1) **Positive moment excess kurtosis**
The distribution of Y is 'moment leptokurtotic' $\Longleftrightarrow \gamma_{2Y} > 0$.

This is equivalent to the moment excess kurtosis γ_{2Y} of the distribution of Y being greater than that of a normal distribution which has zero moment excess kurtosis, i.e. $\gamma_{2Y} > \gamma_{2Z}$, since $\gamma_{2Z} = 0$.

Note that the distrubution of Y is 'moment platykurtic' $\Longleftrightarrow \gamma_{2Y} < 0$.

(2) **Centile kurtosis greater than the normal distribution**
The distribution of Y is 'centile leptokurtic'

$$\Longleftrightarrow k_p(Y) \geq k_p(Z), \text{ i.e. } \frac{k_p(Y)}{k_p(Z)} \geq 1, \text{ for all } 0 < p < 0.25 \text{ and}$$
$$k_p(Y) \leq k_p(Z), \text{ i.e. } \frac{k_p(Y)}{k_p(Z)} \leq 1, \text{ for all } 0.25 < p < 0.5$$

with $k_p(Y) \neq k_p(Z)$ for some p.

Note $k_p(Y)/k_p(Z)$ is called the standardized centile kurtosis frunction, $sk_p(Y)$, in equation (2.10).

Note also that the distribution of Y is 'centile platykurtic' if $\geq$ and $\leq$ and interchanged above.

(3) **More Balanda-MacGillivray kurtotic than the normal distribution**
The distribution of Y is Balanda-MacGillivray leptokurtic

$\Longleftrightarrow F_Z <_S F_Y$ (using the notation for $<_S$ in Section 15.1)

$\Longleftrightarrow$ a plot of $k_p(Y)$ against $k_p(Z)$ is convex for $0 < p < 0.5$, but not linear.

Alternative equivalent conditions are given in Section 15.3. Note Balanda-MacGillivray leptokurtic implies centile leptokurtic.

Note that the distribution of Y is 'Balanda-MacGillivray platykurtic' $\Longleftrightarrow F_Y <_S F_Z$.

Example 1

Consider the power exponential distribution $\texttt{PE}(\mu, \sigma, \nu)$. Since μ is a location shift parameter and σ is a scaling parameter, both μ and σ do not affect the moment, centile and Balanda-MacGillivray kurtosis orderings. We therefore let $\mu = 0$ and $\sigma = 1$ without loss of generality, i.e. let $Y \sim \texttt{PE}(0, 1, \nu)$. (See Sections 4.3 and 18.3.3 for more information on the power exponential distribution.) Let $Z \sim \texttt{NO}(0,1)$.

Figure 15.1(a) plots the pdf of Y for $\nu = 0.5, 1, 2, 10$. Figure 15.1(b) plots the moment excess kurtosis γ_2 of Y against ν, where $\gamma_2 > 0$ for $\nu < 2$ and $\gamma_2 < 0$ for $\nu > 2$, i.e. the distribution of Y is moment leptokurtic if $\nu < 2$ and moment platykurtic if $\nu > 2$. If $\nu = 2$, the distribution of Y is $\texttt{NO}(0, 1)$, with $\gamma_2 = 0$.

Figure 15.1(c) plots the standardized centile kurtosis function $k_p(Y)/k_p(Z)$ against p for $0 < p < 0.5$ and $\nu = 0.5, 1, 2, 10$. For $\nu < 2$, the distribution of Y appears to be centile leptokurtic, since $k_p(Y)/k_p(Z) > 1$ for $0 < p < 0.25$ and $k_p(Y)/k_p(Z) < 1$ for $0.25 < p < 0.5$, while for $\nu > 2$, it appears to be centile platykurtic, since $k_p(Y)/k_p(Z) < 1$ for $0 < p < 0.25$ and $k_p(Y)/k_p(Z) > 1$ for $0.25 < p < 0.5$.

Figure 15.1(d) plots $k_p(Y)$ against $k_p(Z)$ for $0 < p < 0.5$, for $\nu = 0.5, 1, 2, 10$. For $\nu < 2$, the distribution of Y appears to be Balanda-MacGillivray leptokurtic since the plot is convex, while for $\nu > 2$ it appears to be platykurtic since the plot is concave.

In the above example the three criteria appear to be in agreement, since it appears that the distribution of Y is moment, centile, and Balanda-MacGillivray leptokurtic if $\nu < 2$ and platykurtic if $\nu > 2$.

15.3 Conditions for more Balanda-MacGillivray kurtotic

Assume that $F_i(y)$ is twice continuously differentiable on the continuous range of Y_i, for $i = 1, 2$.

The distribution of Y_2 is more Balanda-MacGillivray kurtotic than the distribution of Y_1 [Balanda and MacGillivray, 1990], denoted here $F_1 <_S F_2$,

$\iff$ a plot of $k_2(p) = k_p(Y_2)$ against $k_1(p) = k_p(Y_1)$ for $0 < p < 0.5$ is convex (but not linear), where $k_p(Y_i)$ is the kurtosis function of Y_i defined in (15.2), i.e. $k_2[k_1^{-1}(t)]$ for $t > 0$ is a convex function of t (but not linear).

$\iff$ a plot of $S_2(p)$ against $S_1(p)$ for $0 < p < 0.5$ is convex (but not linear), where $S_i(p) = y_{i,1-p} - y_{i,p}$ is the *spread function*, i.e. $S_2[S_1^{-1}(t)]$ for $0 < t < \infty$ is a convex function of t (but not linear).

$\iff$

$$\frac{[f_2(y_{2,p})]^{-1} + [f_2(y_{2,1-p})]^{-1}}{[f_1(y_{1,p})]^{-1} + [f_1(y_{1,1-p})]^{-1}}$$

is non-decreasing with p for $0.5 < p < 1$ and increases for some p, see Lemma 15.1.

The following are additional equivalent conditions specifically for symmetric distributions only:

$F_1 <_S F_2$

$\iff$ a Q-Q plot of Y_2 against Y_1 is convex (but not linear) for $Y_1 > m_1$, where $m_1 = \text{median}(Y_1)$
i.e. a plot of $y_{2,p}$ against $y_{1,p}$ for $0.5 < p < 1$ is convex (but not linear),
i.e. the transformation $Y_2 = F_2^{-1}[F_1(Y_1)]$ for $Y_1 > m_1$ is convex (but not linear)

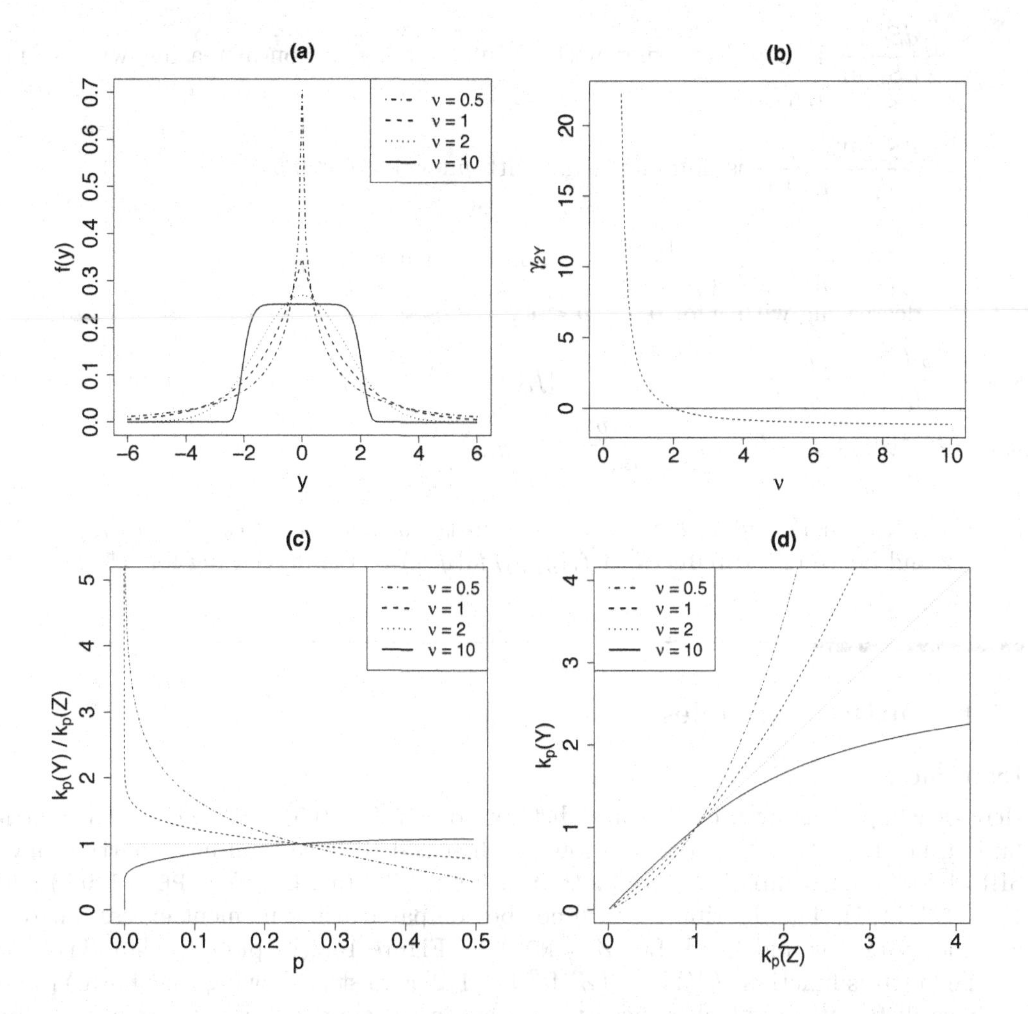

FIGURE 15.1: (a) pdf of PE$(0, 1, \nu)$ for $\nu = 0.5, 1, 2, 10$, (b) γ_{2Y} against ν, (c) $k_p(Y)/k_p(Z)$ against p for $\nu = 0.5, 1, 2, 10$, and (d) $k_p(Y)$ against $k_p(Z)$ for $\nu = 0.5, 1, 2, 10$.

$\iff \dfrac{f_1(y_{1,p})}{f_2(y_{2,p})}$ is non-decreasing with p for $0.5 < p < 1$ and increases for some p.

Lemma 15.1. *A plot of* $S_2(p)$ *against* $S_1(p)$ *for* $0 < p < 0.5$ *is convex*

$\iff$

$$\frac{[f_2(y_{2,p})]^{-1} + [f_2(y_{2,1-p})]^{-1}}{[f_1(y_{1,p})]^{-1} + [f_1(y_{1,1-p})]^{-1}} \tag{15.5}$$

is non-decreasing for $0.5 < p < 1$*, where* $S_i(p) = y_{i,1-p} - y_{i,p}$ *is the spread function of* Y_i *and* $y_{i,p} = F_i^{-1}(p)$ *for* $i = 1, 2$.

Proof. A plot of $S_2(p)$ against $S_1(p)$ for $0 < p < 0.5$ is convex

$\iff \dfrac{dS_2(p)}{dS_1(p)}$ is non-decreasing with $S_1(p)$, and hence non-increasing with p, for $0 < p < 0.5$.

$\iff \dfrac{dS_2(p)}{dp}\dfrac{dp}{dS_1(p)}$ is non-increasing with p for $0 < p < 0.5$.

$\iff \dfrac{[f_2(y_{2,p})]^{-1} + [f_2(y_{2,1-p})]^{-1}}{[f_1(y_{1,p})]^{-1} + [f_1(y_{1,1-p})]^{-1}}$ is non-increasing with p for $0 < p < 0.5$, i.e. non-decreasing with p for $0.5 < p < 1$,

since $\dfrac{dS_i(p)}{dp} = \dfrac{d}{dp}[y_{i,1-p} - y_{i,p}] = -[f_i(y_{i,1-p})]^{-1} - [f_i(y_{i,p})]^{-1}$ for $i = 1, 2$, since $y_{i,p} = F_i^{-1}(p) \Rightarrow p = F_i(y_{i,p}) \Rightarrow \dfrac{dp}{dy_{i,p}} = f_i(y_{i,p})$. □

Note that if both Y_1 and Y_2 have symmetric distributions then $f_i(y_{i,p}) = f_i(y_{i,1-p})$ for $i = 1, 2$ and hence (15.5) reduces to $f_1(y_{1,p})/f_2(y_{2,p})$ is non-decreasing for $0.5 < p < 1$.

15.4 Further examples

Example 2

Here we compare a power exponential distribution, PE$(0, 1, 0.5)$, with the Cauchy distribution, i.e. TF$(0, 1, 1)$. The distributions, each scaled to have semi-interquartile range SIR=1 for easier comparison, are plotted in Figure 15.2(a). Let $Y_1 \sim$ PE$(0, 1, 0.5)$ and $Y_2 \sim$ TF$(0, 1, 1)$. The distributions cannot be compared using moment excess kurtosis γ_2, since $\gamma_2(Y_2)$ is undefined. Let $Z \sim$ NO$(0, 1)$. Figure 15.2(b) plots the standardized centile kurtosis functions $k_p(Y_i)/k_p(Z)$ for $i = 1, 2$ against p, showing that both Y_1 and Y_2 have distributions which appear to be centile leptokurtic. Note Y_2 appears to be more centile kurtotic than Y_1 in the tail, since $k_p(Y_2) > k_p(Y_1)$ for very small p, while Y_1 appears to be more centile kurtotic than Y_2 in the center (i.e. more 'peaked'), since $k_p(Y_1) < k_p(Y_2)$ for p close to 0.5.

Figure 15.2(c) plots $k_p(Y_2)$ against $k_p(Y_1)$ for $0 < p < 0.5$ showing that neither Y_1 nor Y_2 is overall more Balanda-MacGillivray kurtotic since the curve is neither convex nor concave. Figure 15.2(d) plots $f_1(y_{1,p})/f_2(y_{2,p})$ against p for $0.5 < p < 1$, again showing that neither Y_1 nor Y_2 is overall more Balanda-MacGillivray kurtotic since the curve neither increases nor decreases throughout $0.5 < p < 1$.

Example 3

We consider the generalized t distribution, GT(μ, σ, ν, τ), which is symmetric with location shift and scaling parameters μ and σ which do not affect moment, centile, or Balanda-MacGillivray kurtosis. Hence let $\mu = 0$ and $\sigma = 1$ without loss of generality. The GT distribution has two kurtosis parameters $\nu > 0$ and $\tau > 0$, which allow partial decoupling of the peakedness (at the center) and tail heaviness of the distribution. The parameter τ affects the peakedness, with decreasing τ increasing the peakedness and $\tau \leq 1$ resulting in a spiky peak. For fixed τ, decreasing ν increases the tail heaviness. In

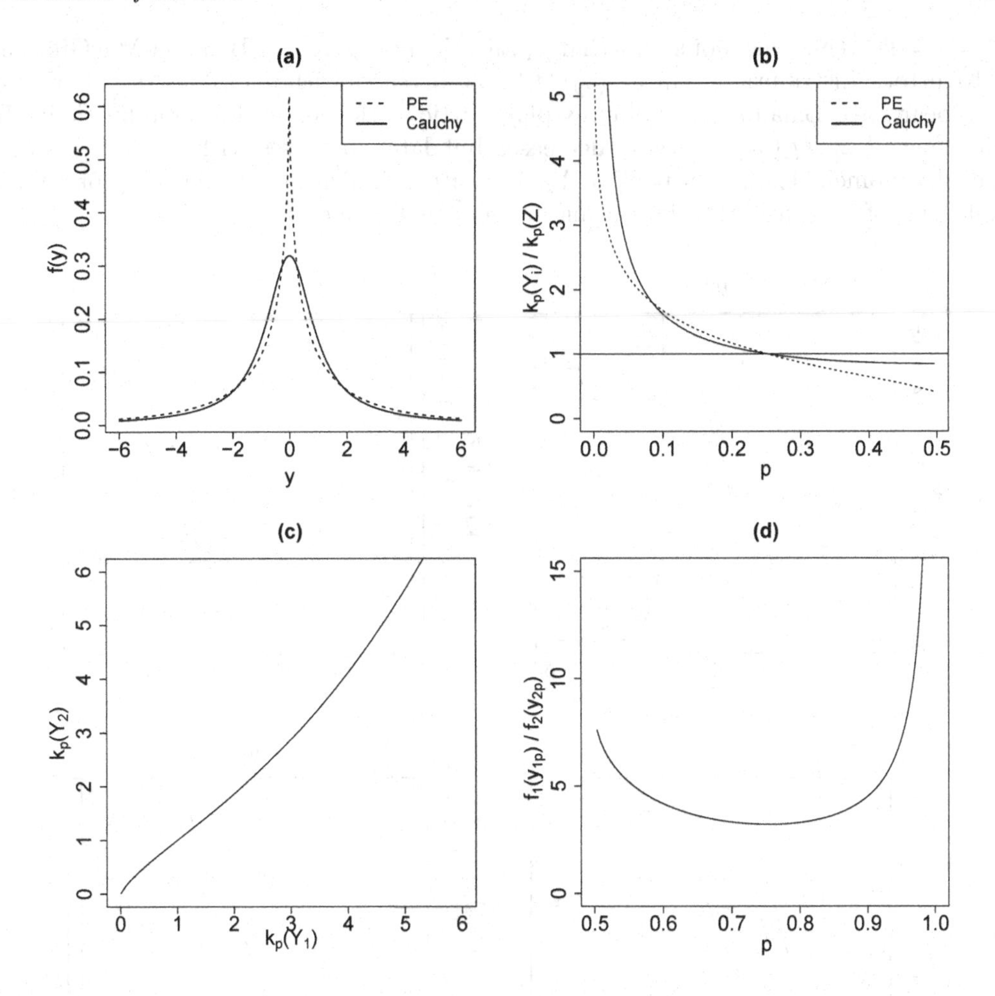

FIGURE 15.2: (a) pdf of $Y_1 \sim \texttt{PE}(0,1,0.5)$ and $Y_2 \sim \texttt{TF}(0,1,1)$, (both scaled in the plot to have SIR=1), (b) the standardized centile kurtosis functions $k_p(Y_i)/k_p(Z)$ for $i = 1, 2$ against p, where $Z \sim \texttt{NO}(0,1)$ (c) $k_p(Y_2)$ against $k_p(Y_1)$ for $0 < p < 0.5$, and (d) $f_1(y_{1,p})/f_2(y_{2,p})$ against p, for $0.5 < p < 1$.

the tail the pdf has order $\mathcal{O}(|y|^{-\nu\tau-1})$ as $|y| \to \infty$, the same order as a $t_{\nu\tau}$ distribution. Hence GT, ultimately as $|y| \to \infty$, always has heavier tails than the normal distribution.

Let $Y_i \sim \texttt{GT}(0,1,\nu_i,\tau_i)$ where (ν_i, τ_i) takes values $(1,1), (5,1), (1,5)$, and $(5,5)$, respectively, for $i = 1,2,3,4$. Figure 15.3(a) plots the four distributions, each scaled to have semi-interquartile range SIR=1 for easier comparison. The distributions have moment excess kurtosis $\gamma_2(Y_i)$, which is undefined for $i = 1$, and equal to $33.0, 1.29$ and -0.85 for $i = 2, 3$, and 4, respectively, so only Y_4 is moment platykurtic. Figure 15.3(b) plots the standardized centile kurtosis functions $k_p(Y_i)/k_p(Z)$ for $i = 1,2,3,4$ against p, for $0 < p < 0.5$. This shows that Y_1 and Y_2 appear to have centile leptokurtic distributions, since $k_p(Y_i)/K_p(Z) > 1$ for $0 < p < 0.25$ and < 1 for $0.25 < p < 0.5$. Y_3 and Y_4 appear to be centile platykurtic at the center, but centile leptokurtic in the extreme tails.

Figure 15.3(c) plots $k_p(Y_i)$ against $k_p(Z)$ for $0 < p < 0.5$ and $i = 1,2,3,4$. Figure 15.3(d) plots $f(z_p)/f_i(y_{i,p})$ against p for $0.5 < p < 1$ and $i = 1,2,3,4$ where $z_p =$

$F_Z^{-1}(p) = \Phi^{-1}(p)$. This plot shows that Y_1 and Y_2 appear to have Balanda-MacGillivray leptokurtotic distributions, since $f(z_p)/f_i(y_{i,p})$ increases with p for $0.5 < p < 1$. Y_3 and Y_4 appear to be Balanda-MacGillivray platykurtic at the center, but leptokurtic in the tails, since $f(z_p)/f_i(y_{i,p})$ initially decreases but later increases with p. Note however that the moment excess kurtosis of Y_3 is positive, indicating moment leptokurtosis, while that of Y_4 is negative, indicating moment platykurtosis.

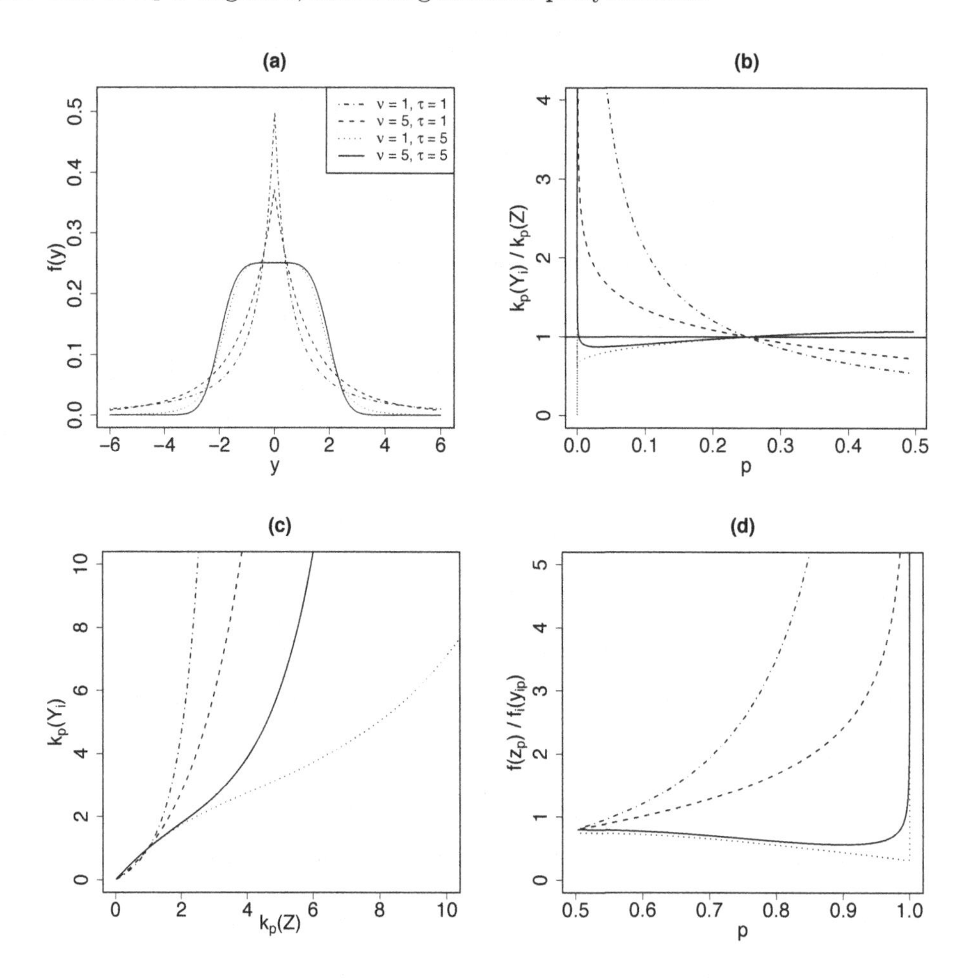

FIGURE 15.3: (a) pdf of $Y_i \sim$ GT$(0, 1, \nu_i, \tau_i)$ where (ν_i, τ_i) takes values $(1,1), (5,1), (1,5)$ and $(5,5)$, for $i = 1,2,3,4$ (scaled in the plot to have SIR=1), (b) the standardized centile kurtosis functions $k_p(Y_i)/k_p(Z)$ against p, for $i = 1,2,3,4$, (c) $k_p(Y_i)$ against $k_p(Z)$ for $0 < p < 0.5$, and (d) $f(z_p)/f_i(y_{i,p})$ against p, for $0.5 < p < 1$.

Example 4

The kurtosis of a skew distribution can be investigated. Here we continue Example 1 of Section 14.3: let $Y_i \sim$ `WEI`$(1, \sigma_i)$ where σ_i takes values $0.5, 1, 2, 3.5, 6$, respectively for $i = 1, \ldots, 5$. Figure 15.4(a) plots the five distributions. Figure 15.4(b) plots the moment excess kurtosis of `WEI`$(1, \sigma)$ against σ. Figure 15.4(c) plots the standardized centile kurtosis function $k_p(Y_i)/k_p(Z)$ against p for $i = 1, \ldots, 5$. The curves for $\sigma = 2$ and $\sigma = 3.5$ are very close for most of the range $0 < p < 0.5$. Surprisingly, the centile kurtosis function of $Y_5 \sim$ WEI$(1, 6)$ is very close to that of $Z \sim$ NO$(0, 1)$, except in the extreme tails, i.e. p very close to 0. Figure 15.4(d) plots the centile kurtosis function $k_p(Y_i)$ against $k_p(Z)$ for $0 < p < 0.5$, for $i = 1, \ldots, 5$.

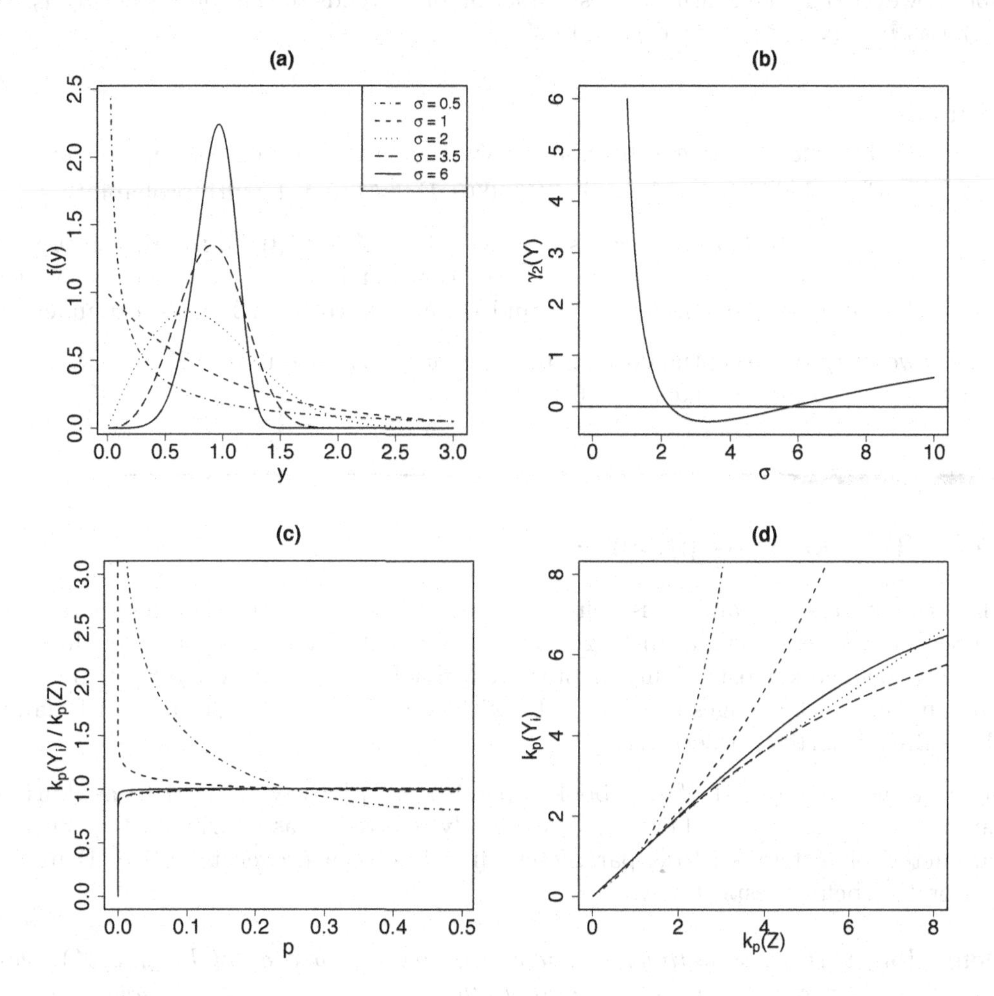

FIGURE 15.4: (a) pdf of WEI$(1, \sigma_i)$ where σ_i takes values $0.5, 1, 2, 3.5, 6$, for $i = 1, \ldots, 5$, respectively, (b) the moment excess kurtosis of WEI$(1, \sigma)$ against σ, (c) the standardized centile kurtosis functions $k_p(Y_i)/k_p(Z)$ against p, and (d) the centile kurtosis functions $k_p(Y_i)$ against $k_p(Z)$ for $0 < p < 0.5$.

Example 5

This example demonstrates that moment excess kurtosis is very sensitive to the tails of the distribution, while centile kurtosis is not. Balanda and MacGillivray [1990, p112] report the following example from Ali [1974]. For $n = 2, 3, \ldots$, let Y_n have a distribution which is a mixture of two components:

$$\texttt{NO}(0,1) \text{ with probability } 1 - \frac{1}{n^2-1} \quad \text{and} \quad \texttt{NO}(0,n) \text{ with probability } \frac{1}{n^2-1}\,.$$

For every fixed y, $f_{Y_n}(y) \rightarrow f_Y(y)$ as $n \rightarrow \infty$, where $Y \sim$ NO$(0, 1)$, i.e. Y_n converges in distribution (uniformly in y) to Y as $n \to \infty$. However $\mathrm{E}(Y_n) = 0$, $\mathrm{Var}(Y_n) = 2$, and the moment excess kurtosis is given by $\gamma_2 = 3(n^2 - 2)/4 \to \infty$ as $n \to \infty$.

Note however that the centile kurtosis function of Y_n tends to that of Y since $k_p(Y_n) \to k_p(Y)$ as $n \to \infty$ for any fixed $p \in (0, 0.5)$.

Example 6

For $n = 2, 3, \ldots$, let Y_n have a distribution which is a mixture of two components: $\texttt{NO}(0,1)$ with probability $1 - \frac{1}{n}$ and $\texttt{CAUCHY}(0,1) = \texttt{TF}(0,1,1)$ with probability $\frac{1}{n}$.

For every fixed y, $f_{Y_n}(y) \to f_Y(y)$ as $n \to \infty$, where $Y \sim \texttt{NO}(0,1)$, i.e. Y_n converges in distribution (uniformly in y) to Y as $n \to \infty$. However $\mathrm{E}(Y_n)$ is undefined for every n, and so the variance, moment skewness, and moment kurtosis of Y_n are also undefined.

Note however that the centile kurtosis function of Y_n tends to that of Y since $k_p(Y_n) \to k_p(Y)$ as $n \to \infty$ for any fixed $p \in (0, 0.5)$.

15.5 True kurtosis parameter

The 'true' kurtosis parameter is defined analogously to the 'true' skewness parameter, see Section 14.5. In order to distinguish a loosely-defined kurtosis parameter from the explicitly defined kurtosis in this chapter, we define below a 'true' kurtosis parameter as a parameter which changes monotonically with one of the moment, centile, or Balanda-MacGillivray kurtosis orderings.

In the tables in Part III of this book a parameter is loosely referred to as 'kurtosis parameter' if it affects the kurtosis, or explicitly referred to as a 'primary true kurtosis parameter' or a 'true kurtosis parameter', if it has been proven to satisfy Definition 15.1 or 15.2 below, respectively.

Definition 15.1. *Let continuous random variable Y have a cdf $F_Y(y \mid \theta, \theta^*)$ which depends on a parameter θ and potentially other parameters θ^*. Let θ_1 and θ_2 be any two values of θ, and Y_i have cdf $F_Y(y \mid \theta_i, \theta_i^*)$ for $i = 1, 2$. Then*

(a) θ is a primary true moment kurtosis parameter if

$$\theta_1 < \theta_2 \Rightarrow \gamma_2(Y_1) < \gamma_2(Y_2) \text{ for any } (\theta_1, \theta_2)$$

or alternatively if

$$\theta_1 > \theta_2 \Rightarrow \gamma_2(Y_1) < \gamma_2(Y_2) \text{ for any } (\theta_1, \theta_2) \; .$$

(b) θ is a primary true centile kurtosis parameter if

$$\theta_1 < \theta_2 \Rightarrow Y_1 \text{ is more centile kurtotic than } Y_2 \text{ for any } (\theta_1, \theta_2),$$

or alternatively if

$$\theta_1 > \theta_2 \Rightarrow Y_1 \text{ is more centile kurtotic than } Y_2 \text{ for any } (\theta_1, \theta_2).$$

(c) θ is a primary true Balanda-MacGillivray kurtosis parameter if

$$\theta_1 < \theta_2 \Rightarrow F_1 <_S F_2 \text{ for any } (\theta_1, \theta_2),$$

or alternatively if

$$\theta_1 > \theta_2 \Rightarrow F_1 <_S F_2 \text{ for any } (\theta_1, \theta_2),$$

where $F_i = F_Y(y|\theta_i, \theta_i^*)$ for $i = 1, 2$ and $F_1 <_S F_2$ is defined in Section 15.3.

Definition 15.2. *If $\theta_1^* = \theta_2^*$ in the above conditions in Definition 15.1, we say that θ is a true (moment, centile, or Balanda-MacGillivray) kurtosis parameter, albeit not a primary true kurtosis parameter.*

Note that a (primary) true Balanda-MacGillivray kurtosis parameter is therefore also a (primary) true centile kurtosis parameter.

Example 7

This example was originally derived by Balanda and MacGillivray [1990, p21]. Let $Y \sim \texttt{JSUo}(\mu, \sigma, \nu, \tau)$, then τ is a primary true Balanda-MacGillivray kurtosis parameter.

Proof. Let $Y_i \sim \texttt{JSUo}(\mu_i, \sigma_i, \nu_i, \tau_i)$ for $i = 1, 2$. Then $y_{i,p} = \mu_i + \sigma_i \sinh[(z_p - \nu_i)/\tau_i]$ where $z_p = \Phi^{-1}(p)$ and the centile kurtosis function of Y_i, for $i = 1, 2$, is given by

$$k_i(p) = k_p(Y_i) = \frac{y_{i,1-p} - y_{i,p}}{y_{i,0.75} - y_{i,0.25}} = \frac{\sinh(z_p/\tau_i)}{\sinh(z_{0.25}/\tau_i)} \ .$$

Note that $k_i(p)$ does not depend on μ_i, σ_i or ν_i.

$F_1 <_S F_2$

$\Longleftrightarrow$ a plot of $k_2(p)$ against $k_1(p)$ is convex (but not linear) for $0 < p < 0.5$

$\Longleftrightarrow$ $\dfrac{dk_2(p)}{dk_1(p)}$ is non-decreasing with p for $0 < p < 0.5$ and increases for some p.

But

$$\frac{dk_2(p)}{dk_1(p)} = \frac{dk_2(p)}{dp} \frac{dp}{dk_1(p)} \propto_p \frac{\cosh(z_p/\tau_2)}{\cosh(z_p/\tau_1)} \ ,$$

which increases with p for $0 < p < 0.5$ if $\tau_1 < \tau_2$, since

$$\frac{d}{dp}\{\log[\cosh(z_p/\tau_2)]\} - \frac{d}{dp}\{\log[\cosh(z_p/\tau_1)]\} \propto \tau_2^{-1}\tanh(z_p/\tau_2) - \tau_1^{-1}\tanh(z_p/\tau_1) > 0$$

for $0 < p < 0.5$, as $z_p < 0$ for $0 < p < 0.5$. Hence $\tau_1 < \tau_2 \Rightarrow F_1 <_S F_2$ and hence τ is a primary true Balanda-MacGillivray kurtosis parameter. □

15.6 Bibliographic notes

A thorough investigation of the concept of kurtosis is given by MacGillivray and Balanda [1988] and Balanda and MacGillivray [1990], where a wide variety of kurtosis measures and ordering are given. The centile kurtosis function was defined by Andrews et al. [1972]. An original kurtosis ordering, called 'Balanda-MacGillivray kurtosis' throughout this chapter, was given by Balanda and MacGillivray [1990].

15.7 Exercises

1. Let $Y \sim$ `SHASHo`(μ, σ, ν, τ). Then from Table 18.19 the inverse cdf is

$$y_p = F_Y^{-1}(p) = \mu + \sigma \sinh\left\{\frac{\nu}{\tau} + \frac{1}{\tau}\sinh^{-1}[\Phi^{-1}(p)]\right\}.$$

   (a) Show that the centile kurtosis function k_p is given by

$$k_p = \frac{\sinh\left[\frac{1}{\tau}\sinh^{-1}(z_p)\right]}{\sinh\left[\frac{1}{\tau}\sinh^{-1}(z_{0.25})\right]}$$

   where $z_p = \Phi^{-1}(p)$. Hence k_p depends only on distribution parameter τ.

   (b) Hence show that τ is a primary true Balanda-MacGillivray kurtosis parameter. (τ is therefore also a primary true centile kurtosis parameter.)

2. Let $Y \sim$ `PE`(μ, σ, ν). Then from Table 18.7 the moment kurtosis of Y is given by $\beta_2 = \Gamma(5\nu^{-1})\Gamma(\nu^{-1})/[\Gamma(3\nu^{-1})]^2$. Show that ν is a primary true moment kurtosis parameter, and β_2 decreases as ν increases.

   [Hint: $\Psi(x) = \frac{d}{dx}\log\Gamma(x) = -C + (x-1)\sum_{j=0}^{\infty}\frac{1}{(j+1)(x+j)}$ where $C = 0.57722$ is Euler's constant, from Johnson et al. [2005, p9].]

3. Let $Y \sim$ `GT`(μ, σ, ν, τ). Then from Table 18.14 the moment kurtosis of Y is given by $\beta_2 = \dfrac{B(5\tau^{-1}, \nu - 4\tau^{-1})B(\tau^{-1}, \nu)}{[B(3\tau^{-1}, \nu - 2\tau^{-1})]^2}$ for $\nu\tau > 4$.

   (a) Show that ν is a true moment kurtosis parameter for $\nu\tau > 4$, and β_2 decreases as ν increases.

   (b) Show that τ is a true moment kurtosis parameter for $\nu\tau > 4$, and β_2 decreases as τ increases.

4. Check whether any of the parameters labelled just 'kurtosis parameter' in the distribution summary tables in Chapters 18 and 19, satisfy the conditions to be a true or primary true, moment, centile or Balanda-MacGillivray kurtosis parameter, as defined in Section 15.5. (The authors have so far tried and failed to prove if any of these conditions hold!)

16

Skewness and kurtosis comparisons of continuous distributions

CONTENTS

This chapter compares the skewness and kurtosis of continuous distributions. In particular it:

1. compares their moment skewness and kurtosis;
2. compares their centile skewness and kurtosis;
3. shows the flexibility in skewness and kurtosis of different distributions which may be informative in the selection of an appropriate response variable distribution; and
4. shows how changing the distribution parameters ν and τ affects the skewness and kurtosis.

This chapter provides important information in terms of selecting appropriately flexible continuous distributions in terms of skewness and kurtosis.

16.1 Moment skewness and kurtosis comparisons

16.1.1 Introduction

In this section we compare the moment skewness and kurtosis of five continuous distributions. For each distribution considered (JSU, SHASHo, SEP3, ST3, and EGB2) we plot the region (or domain) of possible combinations of moment skewness and moment kurtosis. (See Chapter 18 for details about these distributions.) As these distributions are location-scale families with location shift parameter μ and scaling parameter σ, μ and σ do not affect the moment skewness and kurtosis, which therefore only depend on ν and τ. Hence, for each distribution, we plot moment skewness-kurtosis contours for constant ν and varying τ, and for constant τ and varying ν, to evaluate the effect of ν and τ on the moment skewness and kurtosis.

In Section 2.2.4, the moment skewness and moment excess kurtosis were defined as

$$\gamma_1 = \sqrt{\beta_1} = \mu_3/(\mu_2)^{1.5} \ , \quad \gamma_2 = \beta_2 - 3 = \mu_4/(\mu_2^2) - 3 \ ,$$

where μ_k is the kth central moment of Y. To show the range of (positive) moment skewness and moment excess kurtosis we apply the transformation of Jones and Pewsey [2009, p765] in their Figure 2, effectively:

$$\gamma_{jt} = \frac{\gamma_j}{1+|\gamma_j|} \quad \text{for } j = 1, 2 \ .$$

Note that the region of possible combinations of transformed moment skewness, γ_{1t} and transformed moment excess kurtosis, γ_{2t} is the rectangle within the points $(-1, 0)$, $(1, 0)$, $(1, 1)$, and $(-1, 1)$. Figure 16.1, shows only the positive side of γ_{1t} since the negative transformed moment skewness γ_{1t} is the reflection about the vertical axis at the origin.

16.1.2 Transformed moment kurtosis against transformed moment skewness

Here we investigate the regions of possible combinations of transformed moment kurtosis, γ_{2t}, against transformed (positive) moment skewness, γ_{1t}. For each of five distributions (JSU, SHASHo, SEP3, ST3, and EGB2) the approximate boundary of the region is plotted in Figure 16.1. (To understand why a plotted boundary may be approximate, see, for example, the discussion of Figure 16.2(b).) The boundary for all possible distributions is also given as the continuous line below all the rest. A vertical line at transformed moment skewness $\gamma_{1t} = 0$, and a horizontal line at transformed moment kurtosis $\gamma_{2t} = 1$, provide outer boundaries for the regions. The approximate 'active' region of possible combinations of $(\gamma_{2t}, \gamma_{1t})$ is above each curve for JSU, SHASHo, SEP3, and ST3 and between the two curves for EGB2.

The corresponding plot for negative moment skewness is a reflection (or mirror image) of Figure 16.1 about the vertical axis at $\gamma_{1t} = 0$.

Note that the normal distribution is at the point $(0, 0)$ in Figure 16.1. Transformed moment kurtosis $\gamma_{2t} < 0$ corresponds to moment excess kurtosis $\gamma_2 < 0$ and hence moment

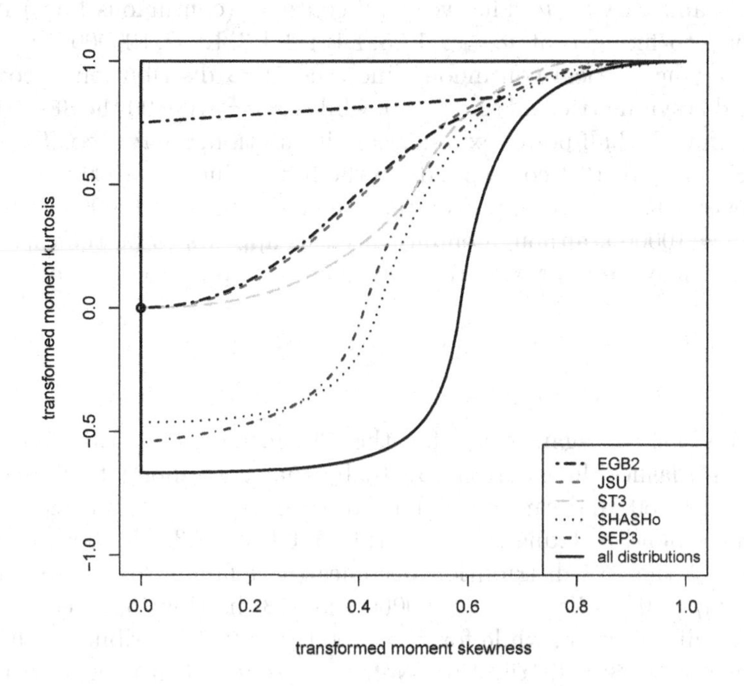

FIGURE 16.1: The regions of possible combinations of transformed moment kurtosis, γ_{2t}, and transformed (positive) moment skewness γ_{1t} for five distributions: EGB2, JSU, ST3, SHASHo, and SEP3 together with the boundary for all distributions.

platykurtosis, while $\gamma_{2t} > 0$ corresponds to moment leptokurtosis. The JSU, ST3, and EGB2 distributions appear to only allow moment leptokurtic distributions, while SEP3 and SHASHo allow both moment platykurtic and moment leptokurtic distributions. For very low kurtosis, only SEP3 is available, allowing a range of skewness, while for higher kurtosis, SHASHo allows the largest range of skewness.

SHASHo

Figure 16.2(a) plots γ_{2t} against γ_{1t}, showing the region of transformed moment kurtosis and transformed (positive) moment skewness, for the SHASHo(μ, σ, ν, τ) distribution. The 'horizontal' contours (dashed lines) from top to bottom correspond to fixing τ at $\tau = 0.1, 0.5, 0.75, 1, 1.25$ and varying ν. The 'vertical' contours (continuous lines) from left to right correspond to fixing ν at $\nu = 0, 0.05, 0.1, 0.2, 0.4, 1, 100$, and varying τ. Note the SHASHo$(\mu, \sigma, 0, \tau)$ distribution is symmetric, so $\gamma_{1t} = 0$, when $\nu = 0$. The normal distribution is at the point (0,0) in Figure 16.2. A high value of γ_{1t} is only achievable for a very high value of γ_{2t}, while a high value of γ_{2t} is achievable for any value of γ_{1t}. Hence a high value of moment skewness is only achievable for a (relatively) high value of τ.

SEP3

Similarly, Figure 16.2(b) plots γ_{2t} against γ_{1t} for the SEP3(μ, σ, ν, τ) distribution. The 'horizontal' contours (dashed lines) from top to bottom correspond to fixing τ at $\tau = 0.1, 1, 1.5, 2, 2.5$, and varying ν. The 'vertical' contours (continuous lines) from left to right correspond to fixing ν at $\nu = 1, 1.05, 1.1, 1.2, 1.3, 1.5, 2, 100000$, and varying τ. Note for $\nu = 1$ (the vertical continuous line) the `SEP3` distribution becomes the `PE` distribution and is symmetric and so $\gamma_{1t} = 0$, while for $\nu = 100000$ the `SEP3` distribution effectively becomes the half power exponential distribution, i.e. $\nu = \infty$. The 'boundary' curve for `SEP3` in Figure 16.1 corresponds to the high value $\nu = 100000$ (i.e. effectively $\nu = \infty$). However it has not been proven that all curves for $\nu < 100000$ lie to the left of the curve for $\nu = 100000$, although empirically this appears to be the case. Hence the 'boundary' curve may be approximate. This may also apply to the 'boundary' curves for the other distributions.

`ST3`

Figure 16.2(c) plots γ_{2t} against γ_{1t} for the `ST3`(μ, σ, ν, τ) distribution. The 'horizontal' contours (dashed lines) from top to bottom correspond to fixing τ at $\tau = 4.0001, 6, 10, 20, 100000$, and varying ν. The 'vertical' contours (continuous lines) from left to right correspond to fixing ν at $\nu = 1, 1.05, 1.1, 1.2, 1.3, 1.5, 100000$, and varying τ. Note for $\nu = 1$ the `ST3` distribution becomes the t family (`TF`) distributon and is symmetric, so $\gamma_{1t} = 0$, while for $\nu = 100000$ the `ST3` distribution effectively becomes the half t family distribution, while for $\tau = 100000$ the `ST3` distribution effectively becomes the skew normal `SN2` distribution. Note also τ must be greater than 4 for a finite moment kurtosis. [Note as τ changes from 4 to 3, γ_{1t} increases to 1, while the moent kurtosis is infinite, so $\gamma_{2t} = 1$. This is not shown in Figure 16.2(c).]

`JSU`

Figure 16.2(d) plots γ_{2t} against γ_{1t} for the `JSU`(μ, σ, ν, τ) distribution. The 'horizontal' contours (dashed lines) from top to bottom correspond to fixing τ at $\tau = 0.2, 1, 1.5, 2, 3, 5$, and varying ν. The 'vertical' contours (continuous lines) from left to right correspond to fixing ν at $\nu = 0, 0.1, 0.2, 0.5, 1, 100$, and varying τ. Note for $\nu = 0$ the `JSU` distribution is symmetric, so $\gamma_{1t} = 0$, while for $\nu = 100$ the `JSU` distribution effectively becomes the log normal distribution.

`EGB2`

Figure 16.2(e) plots γ_{2t} against γ_{1t} for the `EGB2`(μ, σ, ν, τ) distribution. The dashed line contours from top to bottom correspond to fixing τ at $\tau = 0.00001, 0.5, 1, 2, 4, 10$, and varying ν (above τ). The continuous line contours from top to bottom correspond to fixing ν at $\nu = 0.1, 0.5, 1, 2, 4, 1000$, and varying τ (below ν). The vertical line at $\gamma_{1t} = 0$ corresponds to setting $\nu = \tau$ (and varying ν) giving a symmetric distribution. Note the `EGB2`(μ, σ, ν, τ) distribution is positively skew for $\nu > \tau$ (and negatively skew for $\nu < \tau$).

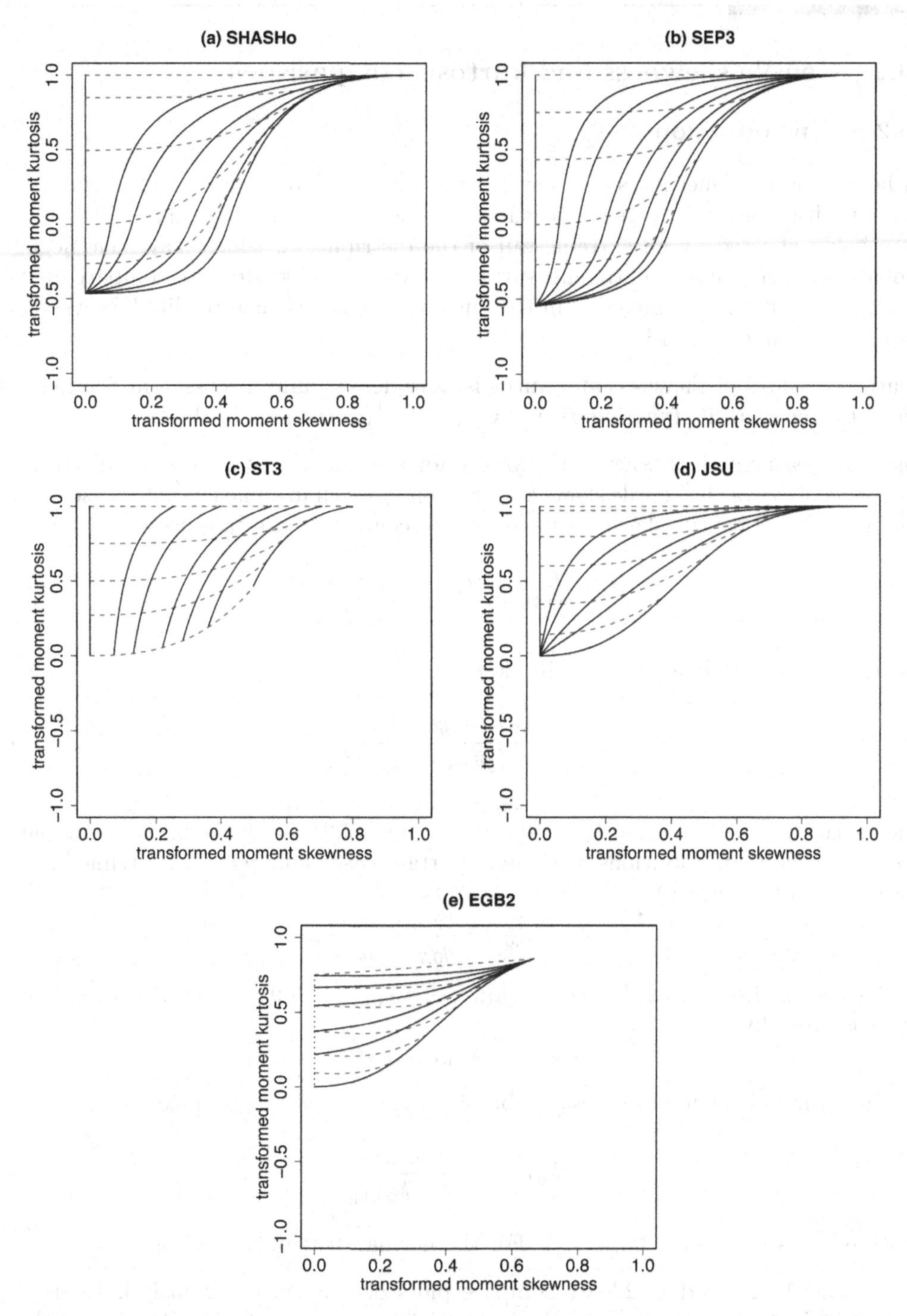

FIGURE 16.2: Transformed moment kurtosis against transformed (positive) moment skewness for the (a) SHASHo, (b) SEP3, (c) ST3, (d) JSU, and (e) EGB2 distributions, showing contours fixing τ and varying ν (dashed lines), and contours fixing ν and varying τ (continuous lines).

16.2 Centile skewness and kurtosis comparisons

16.2.1 Introduction

Although moment measures of skewness and kurtosis have traditionally been used to compare distributions, moment measures may not exist or may be unreliable, suffering from being affected by an extreme tail of the distribution, which may have negligible probability. Here we use centile measures of skewness and kurtosis to compare distributions. Distributions not included in the current comparison can readily be investigated using the methods given here.

Centile measures of skewness and kurtosis were defined and discussed in Section 2.3.3. They use the quantile function given by $y_p = F_Y^{-1}(p)$ for $p \in (0,1)$.

The two measures of skewness that we use for the purpose of comparing distributions are special cases of the centile skewness function s_p given in equation (2.6): case $p = 0.25$ (giving Galton's centile skewness measure) for central centile skewness:

$$\gamma = s_{0.25} = \frac{(Q_1 + Q_3)/2 - m}{(Q_3 - Q_1)/2} \tag{16.1}$$

and the case $p = 0.01$ for tail centile skewness:

$$s_{0.01} = \frac{(y_{0.01} + y_{0.99})/2 - y_{0.5}}{(y_{0.99} - y_{0.01})/2} . \tag{16.2}$$

The centile kurtosis function k_p given by equation (2.8) forms the basis of comparison of kurtosis across distributions, with an important case being $p = 0.01$ (giving Andrew's centile kurtosis measure):

$$\delta = k_{0.01} = \frac{y_{0.99} - y_{0.01}}{y_{0.75} - y_{0.25}} .$$

As the normal distribution has centile kurtosis $k_{0.01} = 3.449$, the centile excess kurtosis $ek_{0.01}$ is given by

$$ek_{0.01} = k_{0.01} - 3.449 .$$

To allow the full range of kurtosis to be plotted, $ek_{0.01}$ is transformed to

$$tk_{0.01} = \frac{ek_{0.01}}{1 + |ek_{0.01}|} , \tag{16.3}$$

which ensures that $-1 < tk_{0.01} < 1$. For the normal distribution, $tk_{0.01} = 0$.

In Sections 16.2.2 and 16.2.3 we compare plots of transformed centile kurtosis $tk_{0.01}$ against each of the central centile skewness functions $s_{0.25}$ (equation (16.1)) and the tail centile skewness $s_{0.01}$ (equation (16.2)), for important distributions on $(-\infty, \infty)$. The following distributions are considered: exponential generalized beta type 2 (`EGB2`), Johnson's SU (`JSU`), sinh-arcsinh original (`SHASHo`), skew exponential power type 3 (`SEP3`), skew t type 3 (`ST3`), and stable (`SB`) distributions. See Section 13.8.2 for information about `SB`.

16.2.2 Transformed centile kurtosis against central centile skewness

Here we investigate the relationship between transformed centile kurtosis $tk_{0.01}$ given by (16.3) and the (positive) central centile skewness $s_{0.25} \in (0, 1)$ given by (16.1). For each of the six distributions, the approximate boundary of (positive) central centile skewness $s_{0.25}$ is plotted against the transformed centile kurtosis $tk_{0.01}$, in Figure 16.3. (To understand why a plotted boundary can be approximate, see, for example, the discussion of Figures 16.4(a) and 16.5(a).) The vertical line at zero and the horizontal line at one form outer boundaries of each of the six regions of the distributions.

The approximate 'active' region of possible combinations of transformed centile kurtosis and (positive) central centile skewness, i.e. $(tk_{0.01}, s_{0.25})$, is above each curve for JSU, SHASHo, SEP3, ST3, and SB, and between the two curves for EGB2. The corresponding plot for negative central skewness is a mirror image around the vertical origin axis.

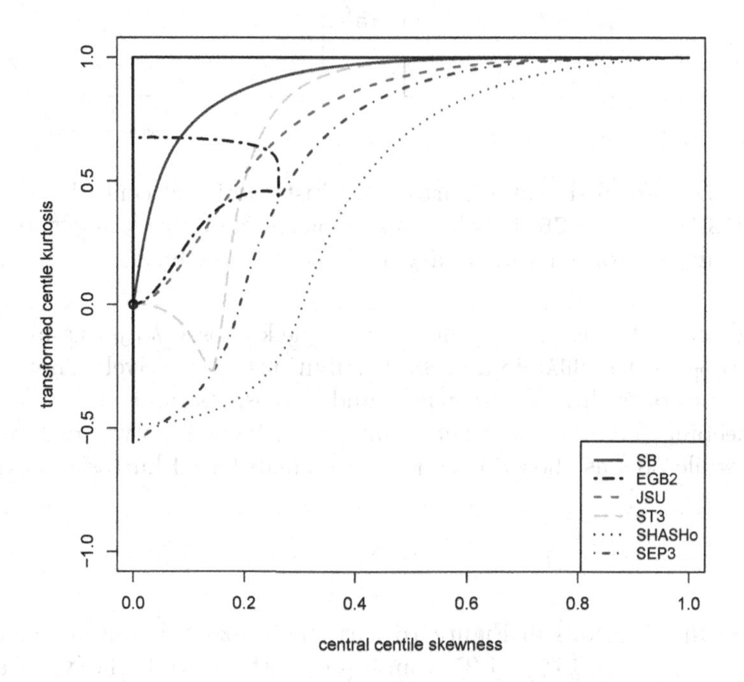

FIGURE 16.3: The boundary of central centile skewness against the transformed centile kurtosis for six distributions SB, EGB2, JSU, ST3, SHASHo, and SEP3.

The normal distribution is at the point (0, 0) in Figure 16.3. Transformed centile kurtosis below zero can be considered as 'centile platykurtic' (for $p = 0.01$), while above zero can be considered 'centile leptokurtic' (for $p = 0.01$). The SB, EGB2, and JSU distributions do not appear to allow 'centile platykurtic' distributions (for $p = 0.01$).

For very low transformed centile kurtosis only SEP3 is available, allowing a range of central skewness, while for higher kurtosis SHASHo allows the largest range of skewness. The SHASHo is generally the most versatile. The SB, JSU and ST3 distributions have more restricted central centile skewness for a given transformed centile kurtosis. The EGB2 distribution is more restricted in central centile skewness and transformed centile kurtosis, with the transformed kurtosis at or moderately above that of the normal distribution. The stable distribution SB is very restricted in central centile skewness for a given transformed centile kurtosis, with the transformed centile kurtosis is generally much higher than the normal distribution. The range of possible central centile skewness increases with the transformed centile kurtosis for all distributions, with the exception of EGB2.

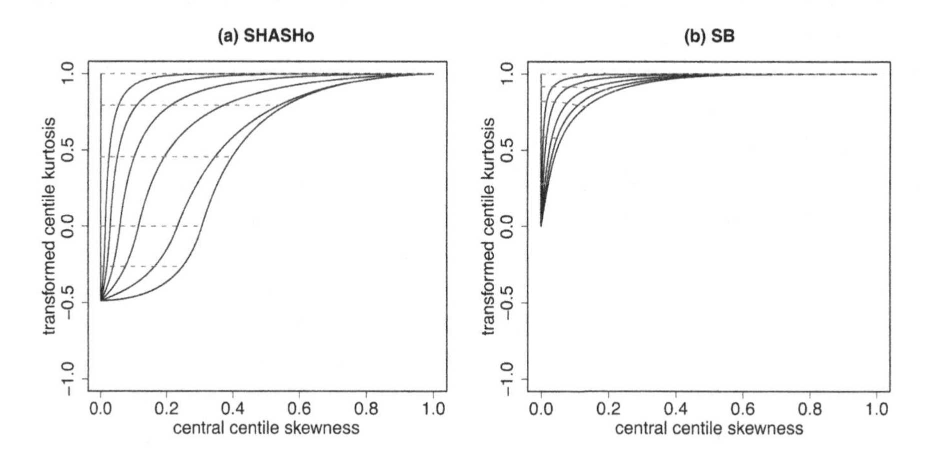

FIGURE 16.4: Transformed centile kurtosis against (positive) central centile skewness for the (a) SHASHo and (b) SB distributions, showing contours fixing τ and varying ν (dashed lines), and contours fixing ν varying τ (continuous lines).

Figure 16.4 (a) and (b) show the transformed centile kurtosis $tk_{0.01}$ against central centile skewness $s_{0.25}$ for the SHASHo and SB distributions, respectively, showing contours for different values of each of the skewness and kurtosis parameters, ν and τ, respectively, while keeping the other parameter constant. The SHASHo was chosen because of its flexibility, while SB was chosen because its moment-based kurtosis-skewness plot is not possible.

SHASHo

For the SHASHo distributions in Figure 16.4(a) the horizontal contours (dashed lines) correspond to $\tau = 0.1, 0.5, 0.75, 1, 1.25$ from top to bottom, while the 'vertical' contours (continuous lines) correspond to $\nu = 0, 0.05, 0.1, 0.2, 0.4, 1, 100$ from left to right. Note that $\tau = 0.1$ and $\nu = 100$ effectively correspond to the limits $\tau = 0$ and $\nu = \infty$ as no change in the contours was observed as τ was decreased below 0.1 and ν increased above 100, respectively.

Note that for the SHASHo(μ, σ, ν, τ) distribution, parameter τ is a primary true Balanda-MacGillivray kurtosis parameter (see Section 15.5). Consequently, τ is a primary true centile kurtosis parameter. Hence the centile kurtosis depends only on parameter τ, and in Figure 16.4(a) the (dashed) contours, for fixed τ values are exactly horizontal. So for SHASHo with a fixed τ, ν affects the central centile skewness only.

Additionally, for the SHASHo(μ, σ, ν, τ) distribution, parameter ν is a true van Zwet skewness parameter (see Section 14.5) and hence a true centile skewness parameter.

Hence for each fixed τ, the centile skewness increases with ν. Consequently, in Figure 16.4(a), the continuous contours, for different fixed ν values, cannot cross each other.

Hence the boundary is given by $\nu = \infty$. The boundary curve plotted in Figure 16.3 for the SHASHo distribution corresponds to $\nu = 100$ (effectively $\nu = \infty$ since the curve showed negligible change above $\nu = 100$), and hence should be an accurate boundary.

The above comments also apply to the JSU distribution.

SB

For the stable (SB) distribution in Figure 16.4(b) the 'horizontal' contours (dashed lines) correspond to $\tau = 0.1, 1.25, 1.5, 1.75, 1.9$ from top to bottom, while the 'vertical' contours (continuous lines) correspond to $\nu = 0, 0.1, 0.25, 0.5, 0.75, 1$ from left to right. Note that $\tau = 0.1$ effectively corresponds to the limit $\tau = 0$.

SEP3

Similarly, Figure 16.5(a) plots $tk_{0.01}$ against $s_{0.25}$ for the SEP3(μ, σ, ν, τ) distribution. The 'horizontal' contours (dashed lines) from top to bottom correspond to fixing τ at $\tau = 0.01, 1, 1.5, 2, 2.5$ and varying ν. The 'vertical' contours (continuous lines) from left to right correspond to fixing ν at $\nu = 1, 1.05, 1.1, 1.2, 1.3, 1.5, 2, 100000$, and varying τ.

Note that in Figure 16.5(a) the continuous line contours for $\nu = 2$ and $\nu = 100000$ cross at high values of central centile skewness and transformed centile kurtosis. Hence, since the 'boundary' curve plotted in Figure 16.3 for the SEP3(μ, σ, ν, τ) distribution corresponds to the high value $\nu = 10000$ (i.e. effectively $\nu = \infty$), this 'boundary' curve is only approximate. This may also apply to the 'boundary' curves for the other distributions, except for SHASHo and JSU, whose boundary curves should be accurate.

ST3

Figure 16.5(b) plots $tk_{0.01}$ against $s_{0.25}$ for the ST3(μ, σ, ν, τ) distribution, for $\tau = 0.01, 2, 4, 6, 10, 20, 100000$, giving 'horizontal' contours (dashed lines) from top to bottom, and for $\nu = 1, 1.05, 1.1, 1.2, 1.3, 1.5, 2, 100000$ giving 'vertical' contours (continuous lines) from left to right.

JSU

Figure 16.5(c) plots $tk_{0.01}$ against $s_{0.25}$ for the JSU(μ, σ, ν, τ) distribution, for $\tau = 0.2, 1, 1.5, 2, 3, 5$ giving 'horizontal' contours (dashed lines) from top to bottom, and for $\nu = 0, 0.1, 0.2, 0.5, 1, 100$ giving 'vertical' contours (continuous lines) from left to right.

EGB2

Figure 16.5(d) plots $tk_{0.01}$ against $\gamma = s_{0.25}$ for the EGB2(μ, σ, ν, τ) distribution, for fixed $\tau = 0.2, 0.5, 1, 2, 4, 10$ (dashed lines) from top to bottom and varying ν (above τ), and for $\nu = 0.1, 0.5, 1, 2, 4, 1000$ (continuous lines) top to bottom and varying τ (below ν).

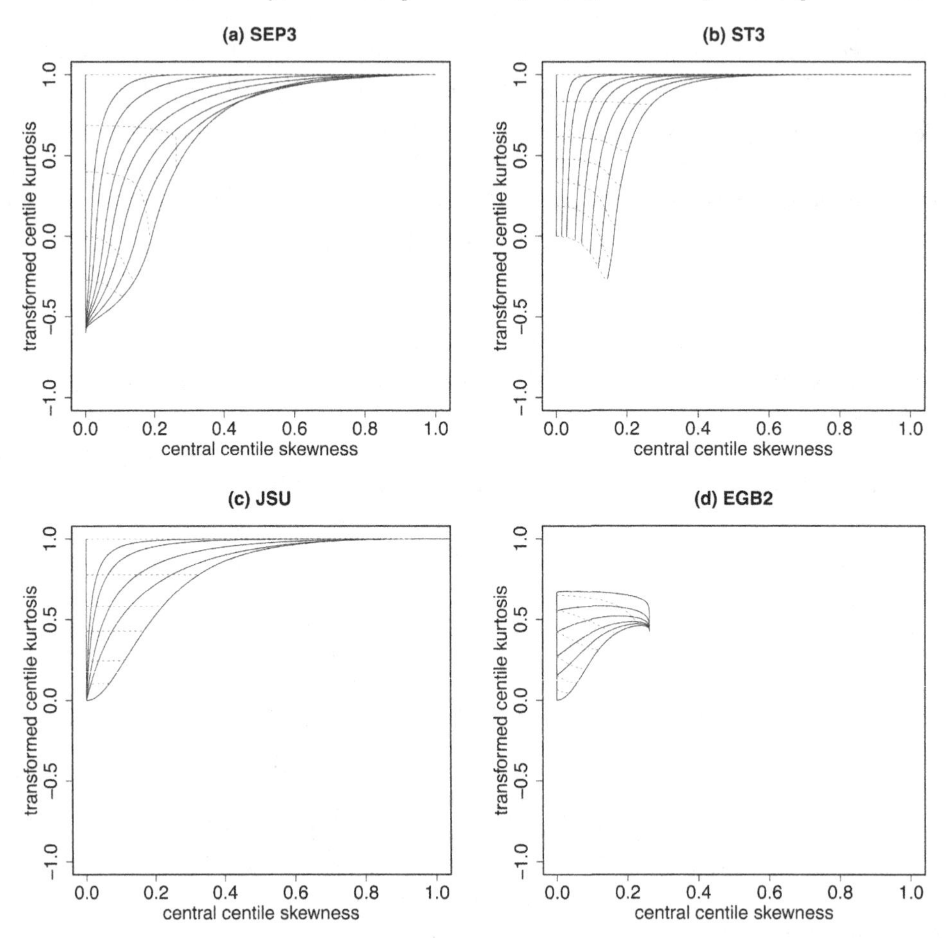

FIGURE 16.5: Transformed centile kurtosis against (positive) central centile skewness for the (a) SEP3, (b) ST3, (c) JSU, and (d) EGB2 distributions, showing contours fixing τ and varying ν (dashed lines), and contours fixing ν varying τ (continuous lines).

16.2.3 Transformed centile kurtosis against tail centile skewness

Section 16.2.2 is amended to replace the central centile skewness $s_{0.25}$ with the tail centile skewness $s_{0.01}$ (equation (16.2)). Figures 16.6, 16.7, and 16.8 correspond to Figures 16.3, 16.4, and 16.5, respectively. The contour values of ν and τ in Figures 16.7 and 16.8 are the same as used in Figures 16.4 and 16.5. The general comments about the kurtosis-skewness relationship for the six distributions still apply. However there is a much wider spread of values of tail centile skewness than central centile skewness, for any particular value of centile kurtosis.

16.2.4 Conclusions

The boundary of central and tail centile skewness against the transformed centile kurtosis is given for six important four-parameter distributions with range $(-\infty, \infty)$. Overall the sinh-arcsinh (SHASHo) is the most flexible distribution in modeling the skewness and kurtosis. However its tails are not as heavy as the stable (SB) or skew t type 3 (ST3). Hence the SHASHo and SEP3 are flexible enough to model a response variable which can

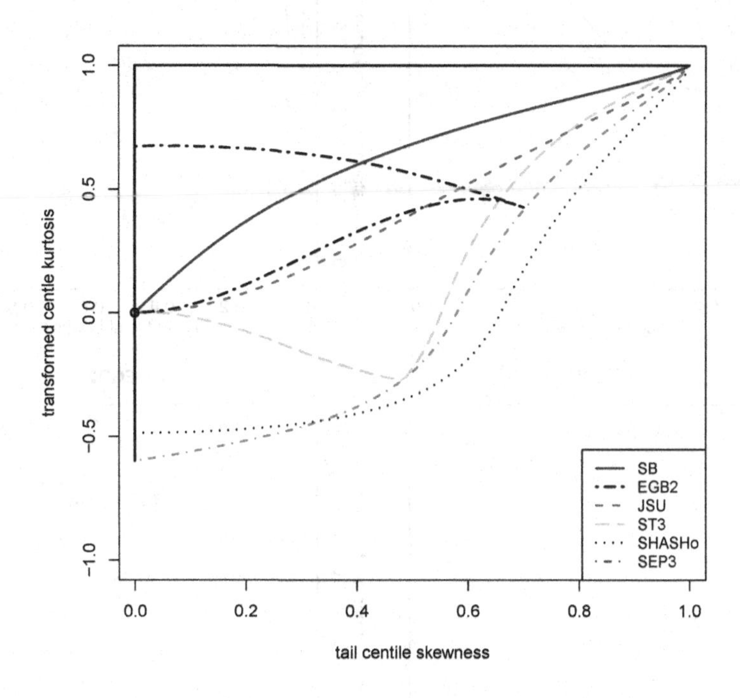

FIGURE 16.6: The boundary of tail centile skewness against the transformed centile kurtosis for six distributions SB, EGB2, JSU, ST3, SHASHo, and SEP3.

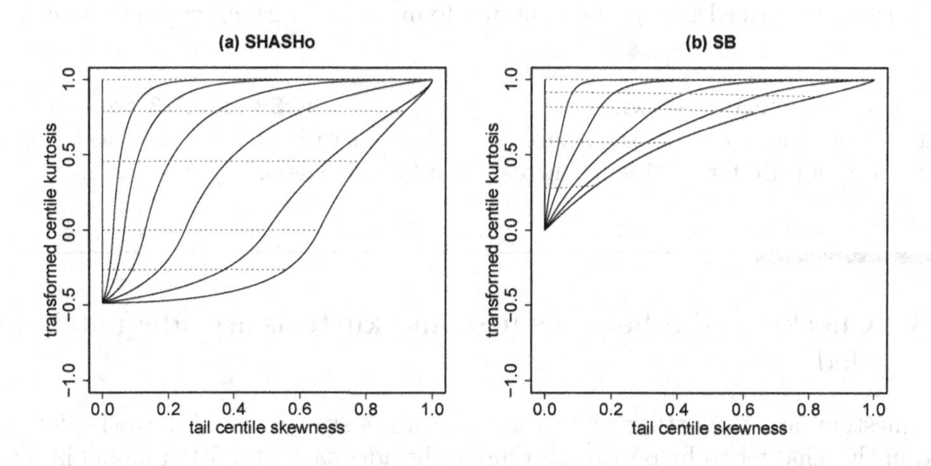

FIGURE 16.7: Transformed centile kurtosis against (positive) tail centile skewness for the (a) SHASHo and (b) SB distributions, showing contours fixing τ and varying ν (dashed lines), and contours fixing ν and varying τ (continuous lines).

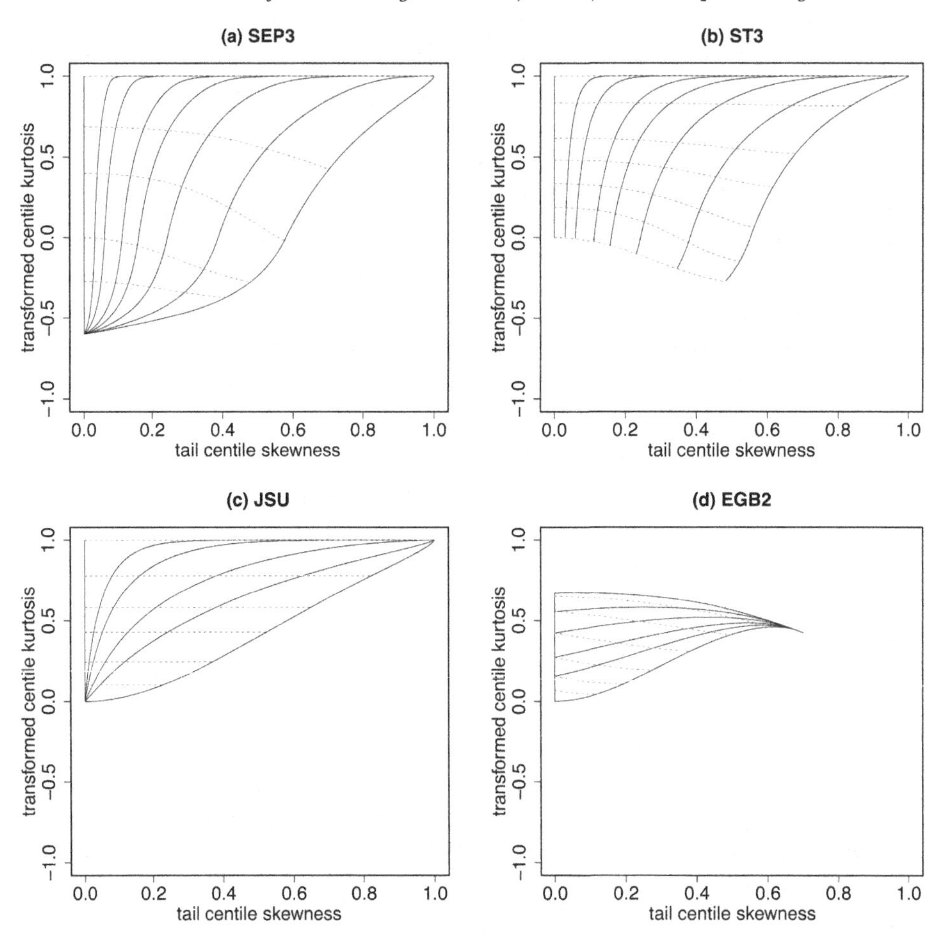

FIGURE 16.8: Transformed centile kurtosis against (positive) tail centile skewness for the (a) SEP3, (b) ST3, (c) JSU, and (d) EGB2 distributions, showing contours fixing τ and varying ν (dashed lines), and contours fixing ν and varying τ (continuous lines).

exhibit a wide range of skewness and kurtosis, while the SB and ST3 are more appropriate for modeling a response variable with high kurtosis and low skewness. The EGB2 is only appropriate for mild leptokurtosis and low skewness.

16.3 Checking whether skewness and kurtosis are adequately modeled

The question here is whether we can use any of the skewness and kurtosis plots developed in this chapter to help us to determine the adequacy of a fitted model in terms of skewness and kurtosis. More generally, especially at an explorative stage of an analysis, we are also interested in finding an appropriate distribution for a response exhibiting skewness and/or kurtosis.

There are two functions in **gamlss** which may help with this:

`checkMomentSK()` for checking the moment skewness and kurtosis, and

`checkCentileSK()` for checking the centile skewness and kurtosis.

Both functions take as argument either a variable (a response variable with no explanatory variables) or a fitted GAMLSS model. In the latter case the residuals of the model are extracted and analyzed. The sample transformed skewness and excess kurtosis of the variable (or residuals) are plotted with the allowable regions of moment or centile skewness and kurtosis of the theoretical distributions. An assessment can then be made on whether the skewness and kurtosis of the variable (or residuals) are adequately fitted or not. However these plots can only tell us about the skewness and/or kurtosis of the variable or residuals. They are not designed for checking the location and scale for the variable or residuals. For information related to location and scale, worm plots (`wp()`) and Q-statistics (`Q.stats()`) are appropriate, which provide information on the location and scale parameter fits as well as skewness and kurtosis, see for example Chapter 12 of Stasinopoulos et al. [2017].

For illustration we use the DAX data first introduced in Section 4.6. For this type of financial data, the main interest lies in the returns `Rdax=-diff(log(dax))`. A histogram of the returns, see Figure 4.7, indicates high kurtosis. The moment skewness for the returns is -0.554 while the moment excess kurtosis is high with a value 6.28. The Jarque-Bera test [Jarque and Bera, 1987] has a value of 3149.64, indicating strong evidence that skewness and/or kurtosis are present in the returns.

16.3.1 Checking moment skewness and kurtosis adequacy

Figure 16.9 shows two different plots created by `checkMomentSK()`. The background of the function is Figure 16.1 reflected about the verical axis at zero, so both negative and positive skewness can be shown. Hence in Figure 16.1 the vertical axis is the transformed moment kurtosis γ_{2t} and the horizontal axis is the transformed moment skewness γ_{1t}. In the middle of the figure there is an additional elliptic region around the zero values of γ_{2t} and γ_{1t}. This region represents a 95% region for $(\hat{\gamma}_{2t}, \hat{\gamma}_{1t})$ based on the Jarque-Bera test, assuming a normal distribution for the variable (or residuals) with $(\gamma_{2t}, \gamma_{1t}) = (0, 0)$. If any $(\hat{\gamma}_{2t}, \hat{\gamma}_{1t})$ falls in this region then we accept the null hypothesis of the normal distribution i.e. the moment skewness and moment excess kurtosis in the variable/residuals are not significantly different from zero values. Values of $(\hat{\gamma}_{2t}, \hat{\gamma}_{1t})$ outside the region result in rejecting the null hypothesis of a normal distribution at the 5% level. This elliptic region depends on the number of observations in the original variable/residuals and therefore will vary for different sample sizes. The values $(\hat{\gamma}_{2t}, \hat{\gamma}_{1t})$ from the variable/residuals produced by the function `checkMomentSK()` is plotted either as character (the default), the variable name or model name from which the residuals were obtained, or as a point. The plots in Figure 16.9(a) and (b) show on the left the skewness and kurtosis plot for the original values of the variable returns, `Rdax`, and on the right the residuals of the fitted model `m1`, respectively. In Figure 16.9(a) point $(\hat{\gamma}_{2t}, \hat{\gamma}_{1t})$, labelled `Rdax`, for the returns falls in the upper left quarter of the plot, indicating that negative skewness and leptokurtosis are present in the `Rdax` variable. The cloud of points around $(\hat{\gamma}_{2t}, \hat{\gamma}_{1t})$ for the `Rdax` variable are 99 values obtained from bootstrapping the `Rdax` values. The cloud gives an indication of the variability of the skewness and kurtosis measures. In this case, since the bootstrap points cross the vertical y-axis, the cloud indicates that skewness maybe is not a problem for the variable Rdax. There is however strong evidence for leptokurtosis. The fitted model m1, using the function fitDist(), resulted in a GT(μ, σ, ν, τ) distribution. Remember that in the GT distribution both ν and τ are kurtosis parameters. Figure 16.9(b) shows the value for $(\hat{\gamma}_{2t}, \hat{\gamma}_{1t})$ for the residuals of the fitted model m1 is within the Jarque-Bera test region and therefore the moment skewness and moment excess kurtosis have been eliminated.

```
dax <- EuStockMarkets[,"DAX"]
Rdax<-diff(log(dax))
m1 <- fitDist(Rdax)
checkMomentSK(Rdax, col.b=gray(.6), pch.b=4); title("(a)")
checkMomentSK(m1, col.b=gray(.6), pch.b=4); title("(b)")
```

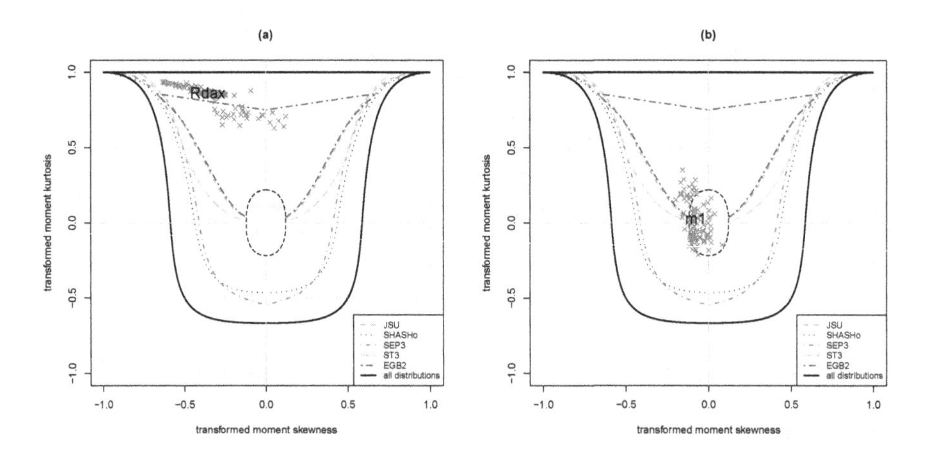

FIGURE 16.9: Transformed moment skewness and kurtosis plots, using the function checkMomentSK() for the 1991 to 1998 DAX returns: (a) the returns, (b) the residuals from the generalized t distribution, GT(μ, σ, ν, τ), fitted to the returns.

Note if a data set includes explanatory variables for a response variable, then the residual skewness and kurtosis for a fitted model should also be investigated using multiple diagnostic plots. For example, Figure 16.9(b) could be split by ranges of an explanatory variable, analogous to the mupliple worm plots (and Q-statistics) used to investigate location, scale, skewness and kurtosis of the residuals as in Stasinopoulos et al. [2017] Chapter 12.

16.3.2 Checking centile skewness and kurtosis adequacy

The centile function checkCentileSK() is similar in nature to checkMomentSK(). It produces a background similar to Figures 16.3 and 16.5, depending on whether 'central' or 'tail' skewness is used, respectively. Both figures are reflected about the y-axis to allow for negative and positive centile skewness. The centile skewness and transformed excess centile kurtosis of the variable/residuals are then plotted together with the cloud of bootstrap values. The plot can be used as a diagnostic tool to check whether skewness and kurtosis have been modeled properly. No elliptical region for testing is provided in the centile skewness and kurtosis plots.

```
checkCentileSK(Rdax, col.b=gray(.6), pch.b=4); title("(a)")
checkCentileSK(Rdax, type="tail", col.b=gray(.6), pch.b=4); title("(b)")
checkCentileSK(m1, col.b=gray(.6), pch.b=4); title("(c)")
checkCentileSK(m1, type="tail", col.b=gray(.6), pch.b=4); title("(d)")
```

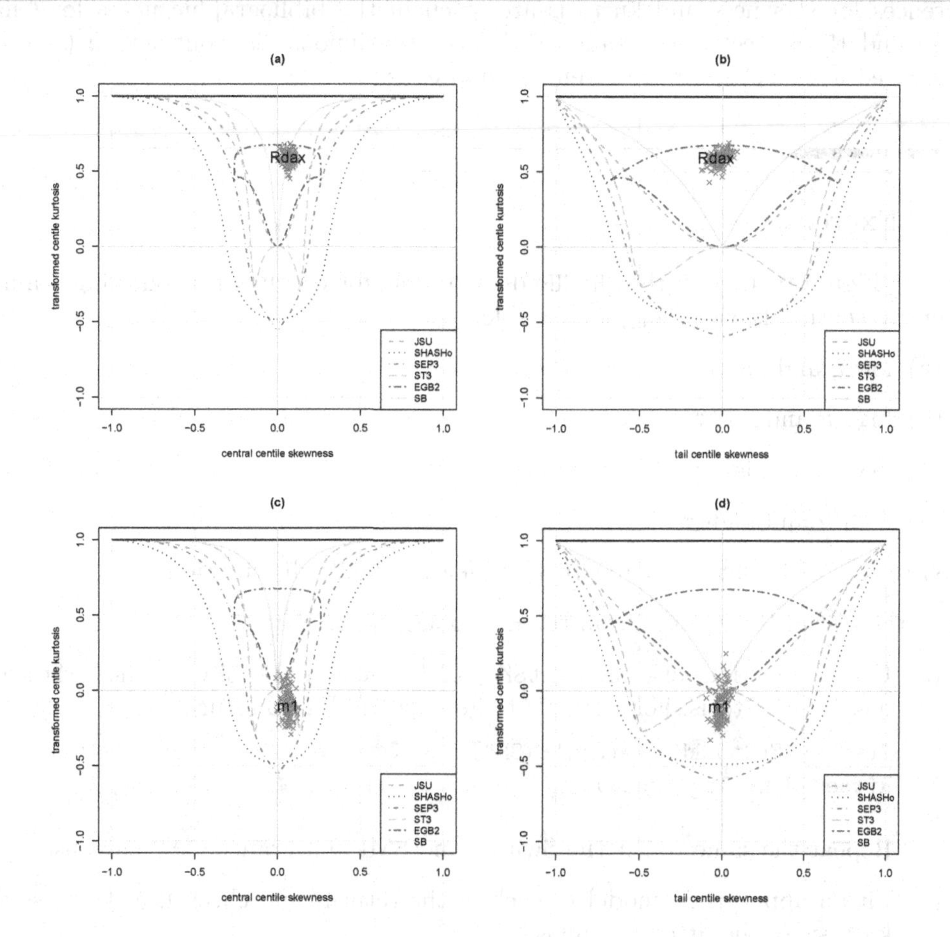

FIGURE 16.10: (a) and (b): Checking the centile skewness and kurtosis of the variable `Rdax`; (c) and (d): checking the residuals of model `m1`. Note that the horizontal axis in (a) and (c) is the 'central' centile skewness while in (b) and (d) it is the 'tail' centile skewness.

Figure 16.9 shows four different plots created by the function `checkCentileSK()`. The horizontal axis in Figures 16.9(a) and (c) shows the 'central' centile skewness while Figures 16.9(b) and (d) shows the 'tail' centile skewness. The top two plots show the actual returns `Rdax`, while the two bottom plots show the residuals of the fitted `GT` distribution model `m1`. The high kurtosis indicated from the top two plots disappears when the model is fitted to the data. Note that when plotting the `Rdax` data using the centile skewness and kurtosis plots, Figure 16.10(a) and (b) show no evidence of 'central' or 'tail' centile skewness, unlike the equivalent moment skewness and kurtosis plot in Figure 16.9(a). This however could be due to 'extreme tail' skewness captured by the moment skewness, but not by the 'tail' centile skewness.

16.4 Bibliographic notes

References for skewness and kurtosis are given in the bibliographic notes for Chapters 14 and 15, respectively. References for the continuous distributions on $(-\infty, \infty)$ investigated in this chapter are given in Chapter 18.

16.5 Exercises

1. Using Figure 16.4(a) for the `SHASHo` distribution, for which of the following is a high central centile skewness $s_{0.25}$ achievable:
   (a) low ν and low τ,
   (b) high ν and low τ,
   (c) low ν and high τ,
   (d) high ν and high τ.
2. Write **R** code to obtain Figure 16.4(a) for the `SHASHo` distribution.
3. For the rest of the returns from the `EuStockMarkets` data:
   (a) Use the functions `checkMomentSK()` and `checkCentileSK()` to check the skewness and kurtosis. For example to get the UK FTSE returns use:

   ```
   ftse <- EuStockMarkets[,"FTSE"]
   Rftse<-diff(log(ftse))
   ```

   Repeat the same to get the Switzerland SMI and France CAC returns.
   (b) Fit an appropriate model to each of the returns and check the skewness and kurtosis of the fitted residuals.
4. This example shows how to check the skewness and kurtosis of several fitted models using the same plot. We use the Munich `rent` data, which come from a survey conducted in April 1993 by Infratest Sozialforschung, in which a random sample of accommodation with new tenancy agreements or increases of rents within the last four years in Munich was selected. The data were analyzed in Stasinopoulos et al. [2017]. There are nine variables in the data set, but for the purpose of this exercise we will use only the following three variables:

   R data file: `rent` in package **gamlss.data** of dimension 1969×9

   var `R` : monthly net rent in Deutsche Marks (DM), i.e. the monthly rent minus calculated or estimated utility cost (response variable)

   `Fl` : floor space in square meters

   `A` : year of construction

(a) Fit different models to the rent data.

```
r1 <- gamlss(R~pb(Fl)+pb(A), data=rent)
r2 <- gamlss(R~pb(Fl)+pb(A), data=rent, family=GA)
r3 <- gamlss(R~pb(Fl)+pb(A), data=rent, family=BCCG)
r4 <- gamlss(R~pb(Fl)+pb(A), sigma.fo=~pb(Fl)+pb(A),data=rent)
r5 <- gamlss(R~pb(Fl)+pb(A), sigma.fo=~pb(Fl)+pb(A),data=rent,
             family=GA)
r6 <- gamlss(R~pb(Fl)+pb(A), sigma.fo=~pb(Fl)+pb(A),data=rent,
             family=BCCG)
```

(b) Inspect the skewness and kurtosis for all models simultaneously:

```
checkMomentSK(r1, boot=T, col.boot="yellow1")
checkMomentSK(r2, add=T, boot=T, col.bootstrap = "turquoise")
checkMomentSK(r3, add=T, boot=T, col.bootstrap = "tan")
checkMomentSK(r4, add=T, boot=T, col.bootstrap = "violet")
checkMomentSK(r5, add=T, boot=T, col.bootstrap = "whitesmoke")
checkMomentSK(r6, add=T, boot=T, col.bootstap = "wheat" )
```

(c) Comment on the results.

(d) Investigate the adequacy of your chosen model using multiple worm plots and Q-statistics (Stasinopoulos et al. [2017], Chapter 12).

17

Heaviness of tails of distributions

CONTENTS

This chapter concentrates on the behaviour of distributional tails. In particular, it:

1. classifies the tails of continuous and discrete distributions; and
2. provides methods for identifying the tail of the distribution of the response variable of a given data set.

This chapter should be of interest when the focus of the analysis is on the tails rather than on the central part of the distribution.

17.1 Introduction

The tail of a distribution is of great importance when the interest is to investigate rare events. There is a vast literature on the subject, some of which is mentioned in the bibliography notes (Section 17.7). Here we will take an unconventionally simplistic (but we believe useful and important) approach to classifying the tails of distributions, by focussing on the functional form of the tails of the probability (density) function $f_Y(y)$ as $y \to \pm\infty$, assuming Y is a continuous random variable with range $(-\infty, \infty)$.

We also consider the tail of the cdf $F_Y(y)$ as $y \to -\infty$ and tail of the survivor function $S_Y(y)$ as $y \to \infty$. When the range of a continuous random variable Y is $(0, \infty)$, only $y \to \infty$ is considered; and when Y is a discrete random variable with unbounded range $\{0, 1, 2, \ldots\}$ or $\{1, 2, \ldots\}$, only $y \to \infty$ is considered.

There are many alternative methods of classifying the tails of distributions, for example, the rigorous classical tail theory [Mikosch, 1999], Mandelbrot's classification of randomness [Mandelbrot, 1997], and Parzen's index [Parzen, 1979]. For further references see Section 17.7 and for a brief introduction see Section 17.6.

A simple advantage of our approach is illustrated by the gamma distribution, where in our classification it is clear how increasing the shape parameter σ affects the tail. In some classifications all gamma distributions are classified as having the same tail heaviness.

It should be clear from the previous chapters that distributions occurring in statistical practice vary considerably. This is because some distributions are symmetrical and some are markedly skew. Some are mesokurtic and some are markedly leptokurtic or platykurtic. In GAMLSS we generally advocate fitting an adequate distribution to the whole of the response variable. Nevertheless there are occasions where this may prove difficult and when the rare events, i.e. the tails of the distribution, are of main interest.

In particular, heavy-tailed response variables are observed in many applications in economics, finance, and natural sciences, where the 'extreme' observations (or outliers) are not mistakes, but an essential part of the distribution. As a result, there are occasions when the tail of the distribution is of primary importance in the statistical analysis. Value at risk (VaR) and Expected shortfall (ES) are well-known concepts in financial analysis and are affected by how the tail of the response distribution behaves. The important point here is the use of a heavy-tailed distribution to model a heavy-tailed response variable results in robust modeling rather than robust estimation. In robust modeling the interest is in both the heavy tail distribution and the regression coefficient for its parameters, as advocated in Lange et al. [1989].

Section 17.2 introduces the basic concepts of tail behaviour and classifies continuous `gamlss.family` distributions into categories according to their tail behaviour. Section 17.3 gives some practical advice on how to determine the tail behaviour of a given response variable. Section 17.4 classifies discrete count `gamlss.family` distributions into categories according to their tail behaviour. Section 17.5 has important Lemmas. Section 17.6 provides a brief introduction to classical tail theory, and Section 17.7 gives a bibliography.

17.2 Types of tails for continuous distributions

Traditionally when we investigate the behaviour of the tail of a continuous distribution we concentrate on the logarithm of the pdf, $\log f_Y(y \mid \theta)$, rather than the pdf, $f_Y(y \mid \theta)$, itself. This is because using a logarithmic scale highlights the tail behaviour. Figure 17.1(a) plots the logarithm of the pdf of the standardized normal, Cauchy, and Laplace distributions. Below -2.5 and above 2.5 the behaviour in the tails of the three distributions varies considerably. The logarithm of the normal distribution pdf is quadratic, the logarithm of the Laplace pdf is linear, while the logarithm of the Cauchy pdf decreases far more slowly than the other two. Hence ordering of the heaviness of the tail of a continuous distribution is clearer when based on the log pdf. Ordering can also be applied to the survivor function of a continuous distribution.

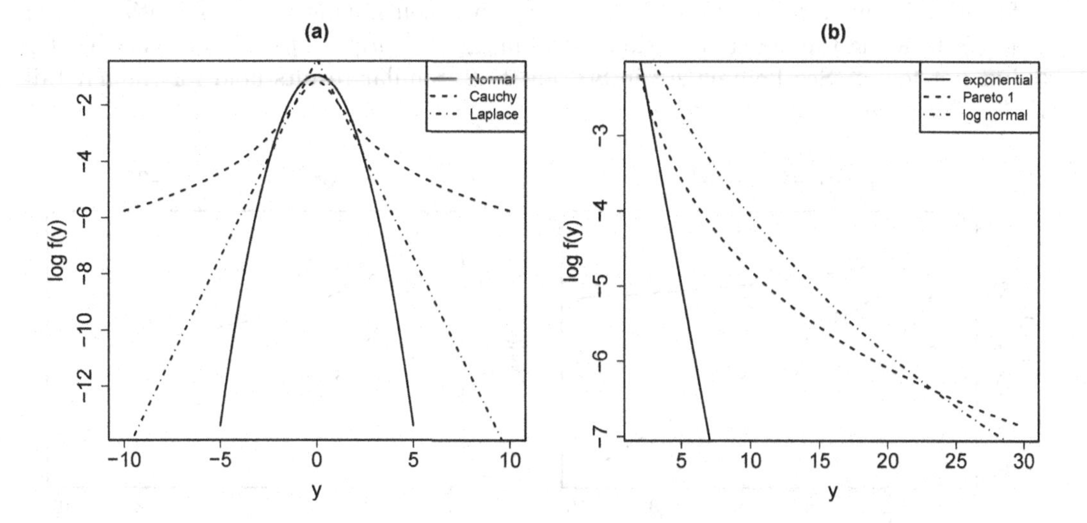

FIGURE 17.1: (a) log pdf of the standardized normal, Cauchy, and Laplace distributions. (b) log pdf of the standardized exponential, Pareto type 1, and log normal distributions.

For continuous distributions defined on $(0, \infty)$ the same idea applies. Figure 17.1 (b) plots the log pdf of the standardized exponential, Pareto type I and log normal distributions. The log pdf of the exponential distribution decreases linearly. Both the Pareto and log normal distributions have heavier tails than the exponential, but the Pareto becomes heavier than the log normal for very high values of y.

Next we define what we mean by a heavier tail distribution.

17.2.1 Definition

If random variables Y_1 and Y_2 have continuous pdfs $f_{Y_1}(y)$ and $f_{Y_2}(y)$ and $\lim_{y\to\infty} f_{Y_1}(y) = \lim_{y\to\infty} f_{Y_2}(y) = 0$, then we use the following (unconventional) definition:

$$Y_2 \text{ has a heavier right tail than } Y_1 \Leftrightarrow \quad \lim_{y\to\infty} [f_{Y_1}(y)/f_{Y_2}(y)] = 0,$$
$$\text{i.e. } f_{Y_1}(y) = o[f_{Y_2}(y)].$$

[Note 'o' means 'order less than'.]

Note that the resulting ordering from $f_Y(y)$ for the right tail heaviness of Y results in the same ordering from $\log f_Y(y)$ where

$$Y_2 \text{ has a heavier right tail than } Y_1 \Leftrightarrow \log[f_{Y_2}(y)] - \log[f_{Y_1}(y)] \to \infty \text{ as } y \to \infty$$

by Lemma A1 in Section 17.1. It also has the same ordering as the ordering from the *survivor* function $S_Y(y) = 1 - F(y)$, where

$$Y_2 \text{ has a heaver right tail than } Y_1 \Leftrightarrow \quad \begin{array}{l} \lim_{y\to\infty} [S_{Y_1}(y)/S_{Y_2}(y)] = 0, \\ \text{i.e. } S_{Y_1}(y) = o[S_{Y_2}(y)] \end{array}$$

if, for y sufficiently large, $F_{Y_1}(y)$ and $F_{Y_2}(y)$ are continuous and differentiable and $f_{Y_2}(y) \neq 0$, while the extra condition that $\lim_{y\to\infty}[f_{Y_1}(y)/f_{Y_2}(y)]$ exists is needed for $\Leftarrow$ but not for $\Rightarrow$. See Lemma A2 in Section 17.2. Similar results hold for the left tail of Y.

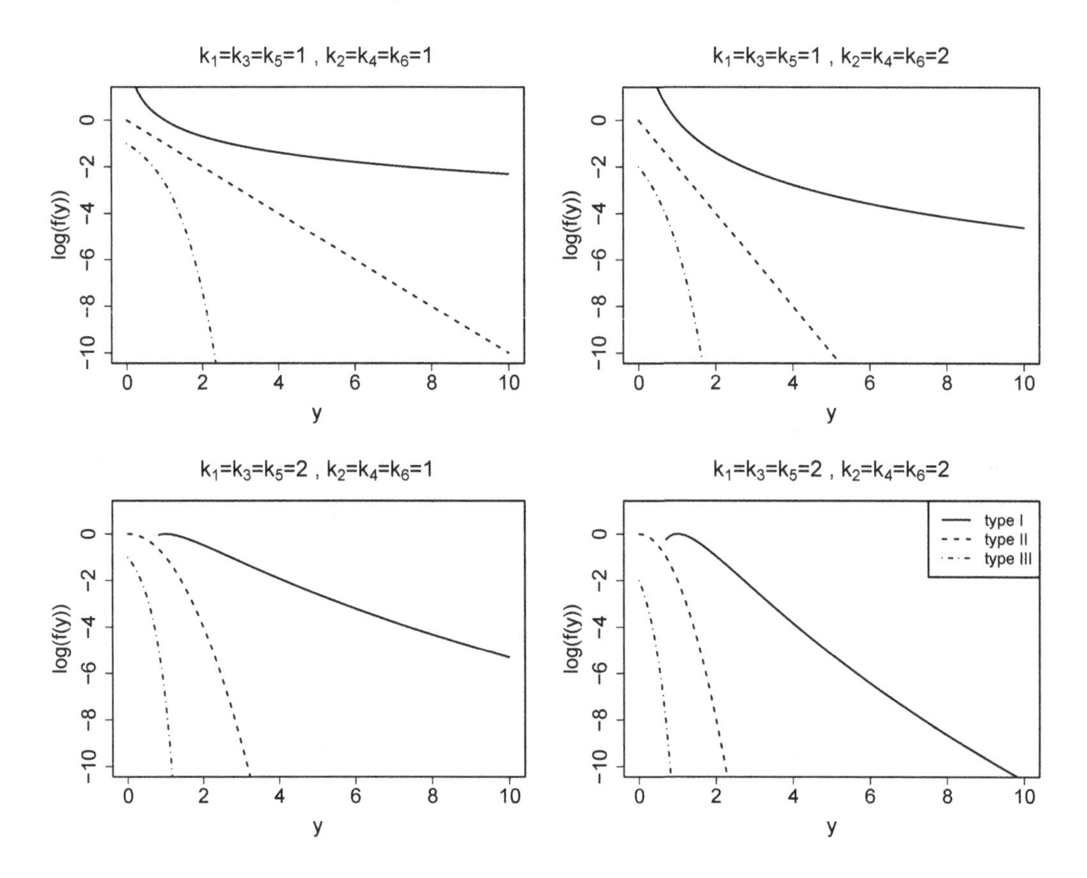

FIGURE 17.2: $\log f_Y(y)$ for type I, II, and III distributions, showing the shape of the tail for $k_1, k_3, k_5 = 1, 2$, and $k_2, k_4, k_6 = 1, 2$. Smaller values of the k's result in heavier tails.

17.2.2 Three main types of tails for the pdf

There are three main asymptotic forms for $\log f_Y(y)$ for a tail of Y, i.e. as $y \to \infty$ (for the right tail) or as $y \to -\infty$ (for the left tail):

$$\log f_Y(y) \sim \begin{cases} \text{Type I} & -k_2 \, (\log|y|)^{k_1} \\ \text{Type II} & -k_4 \, |y|^{k_3} \\ \text{Type III} & -k_6 \, e^{k_5|y|} \end{cases} \tag{17.1}$$

in decreasing order of heaviness of the pdf tail, i.e. type I has the heaviest and type III the lightest tails. Equation (17.1) defines our tail types.

[Note, in equation (17.1), '$\sim$' means 'asymptotically equivalent to', where $h_1(y) \sim h_2(y)$ as $y \to \infty \Leftrightarrow \lim_{y\to\infty}[h_1(y)/h_2(y)] = 1$.]

Note that if $k_1 = 1$ then $k_2 \geq 1$, while if $k_1 > 1$ then $k_2 > 0$. Also $k_j > 0$ for $j = 3, 4, 5, 6$. The right tail of $f_Y(y)$ must be asymptotically lighter than $|y|^{-1}$, (assuming $f_Y(y)$ is continuous and monotonically decreasing for y sufficiently large and $\lim_{y\to\infty} f_Y(y) = 0$), otherwise the integral of the tail would be infinite and $f_Y(y)$ could not be a proper pdf; resulting in $k_1 \geq 1$. Similarly for the left tail.

For type I, decreasing k_1 results in a heavier tail, while decreasing k_2 for fixed k_1 results in a heavier tail. Similarly for type II, with (k_3, k_4) replacing (k_1, k_2) and for type III, with (k_5, k_6) replacing (k_1, k_2). Important special cases are $k_1 = 1$, $k_1 = 2$, $k_3 = 1$, and $k_3 = 2$. Figure 17.2 shows tail behaviour of the three different types for $k_1, k_3, k_5 = 1, 2$, and $k_2, k_4, k_6 = 1, 2$. A higher value of $\log f_Y(y)$ for large y indicates a heavier tail.

17.2.3 Classification tables for tails of continuous distributions

Tables 17.1, 17.2, and 17.3 provide a summary of the tail types for many important continuous distributions on $(-\infty, \infty)$ and $(0, \infty)$, respectively. Many of these distributions have important special cases; see the Reference Guide (Part III). The parameterizations of the distributions (column 3 of Tables 17.1, 17.2, and 17.3) are those used by the **gamlss.dist** package. Note that `CAUCHY`, `SB` and `LA` in Table 17.1, and `logTF` and `logWEI` in Table 17.3 are not currently explicitly available in **gamlss.dist**. However `CAUCHY`$(\mu, \sigma) \equiv$ `TF`$(\mu, \sigma, 1)$; `LA`$(\mu, \sigma) \equiv$ `PE2`$(\mu, \sigma, 1)$; and `logTF`(μ, σ, ν) and `logWEI`(μ, σ) can be obtained by transformation from `TF`(μ, σ, ν) and `WEI`(μ, σ), respectively, see Section 5.6.1. Some distributions are parameterized in two different ways, for example `JSU` and `JSUo`. For many distributions the left and right tails have the same asymptotic form for $\log f_Y(y)$; if not, the relevant tail is specified in the table. Some distributions have different tail forms dependent on a condition on one (or more) parameters, see e.g. the generalized gamma, `GG`(μ, σ, ν), distribution, which has different tail forms for $\nu < 0$ and $\nu > 0$. Note, for example, that all distribution tails with $k_1 = 1$ are heavier than those with $k_1 = 2$. Within the $k_1 = 1$ group a smaller k_2 has the heavier tail. Note from Table 17.1 that the stable (`SB`) distribution and t family (`TF`) distribution with degrees of freedom $0 < \nu < 2$, have the same range for k_2. Pdf tails with $k_3 > 0$ can be:

- heavier than the Laplace (two-sided exponential) if $0 < k_3 < 1$,
- lighter than the Laplace, but heavier than the normal if $1 < k_3 < 2$,
- lighter than the normal if $k_3 > 2$.

TABLE 17.1: Asymptotic form of $\log f_Y(y)$ as $y \to -\infty$ (left tail) or as $y \to \infty$ (right tail) for continuous distributions on $(-\infty, \infty)$ using (17.1). Also the asymptotic form of $\log F_Y(y)$ as $y \to -\infty$ (left tail), and $\log S_Y(y)$ as $y \to \infty$ (right tail) using (17.2). Note $c_0^2 = \Gamma(\frac{1}{\nu})\,[\Gamma(\frac{3}{\nu})]^{-1}$; exp = exponential, gen = generalized.

Value of k_1–k_6	Distribution	**gamlss** name	Condition	Value of k_1–k_6	Parameter range
$k_1 = 1$	Cauchy	`CAUCHY`(μ, σ)		$k_2 = 2$	
	generalized t	`GT`(μ, σ, ν, τ)		$k_2 = \nu\tau + 1$	$\nu > 0, \tau > 0$
	stable	`SB`(μ, σ, ν, τ)		$k_2 = \tau + 1$	$0 < \tau < 2$
	t family	`TF`(μ, σ, ν)		$k_2 = \nu + 1$	$\nu > 0$
$k_1 = 2$	Johnson's SU	`JSU`(μ, σ, ν, τ)		$k_2 = 0.5\tau^2$	$\tau > 0$
	Johnson's SU original	`JSUo`(μ, σ, ν, τ)		$k_2 = 0.5\tau^2$	$\tau > 0$
$k_3 > 0$	power exp	`PE`(μ, σ, ν)		$k_3 = \nu,\ k_4 = (c_0\,\sigma)^{-\nu}$	$\sigma > 0,\ \nu > 0$
	power exp 2	`PE2`(μ, σ, ν)		$k_3 = \nu,\ k_4 = \sigma^{-\nu}$	$\sigma > 0,\ \nu > 0$
	sinh-arcsinh original	`SHASHo`(μ, σ, ν, τ)		$k_3 = 2\tau$	$\sigma > 0,\ \nu > 0,\ \tau > 0$
			right tail	$k_4 = e^{-2\nu} 2^{2\tau-3}\, \sigma^{-2\tau}$	
			left tail	$k_4 = e^{2\nu} 2^{2\tau-3}\, \sigma^{-2\tau}$	
	sinh-arcsinh	`SHASH`(μ, σ, ν, τ)	right tail	$k_3 = 2\tau,\ k_4 = 2^{2\tau-3}\sigma^{-2\tau}$	$\sigma > 0,\ \tau > 0$
			left tail	$k_3 = 2\nu,\ k_4 = 2^{2\nu-3}\sigma^{-2\nu}$	$\sigma > 0,\ \nu > 0$
$k_3 = 1$	exp gen beta 2	`EGB2`(μ, σ, ν, τ)	right tail	$k_4 = \tau\,\sigma^{-1}$	$\sigma > 0,\ \tau > 0$
			left tail	$k_4 = \nu\,\sigma^{-1}$	$\sigma > 0,\ \tau > 0$
	Gumbel	`GU`(μ, σ)	left tail	$k_4 = \sigma^{-1}$	$\sigma > 0$
	Laplace	`LA`(μ, σ)		$k_4 = \sigma^{-1}$	$\sigma > 0$
	logistic	`LO`(μ, σ)		$k_4 = \sigma^{-1}$	$\sigma > 0$
	reverse Gumbel	`RG`(μ, σ)	right tail	$k_4 = \sigma^{-1}$	$\sigma > 0$
$k_3 = 2$	normal	`NO`(μ, σ)		$k_4 = 0.5\sigma^{-2}$	$\sigma > 0$
$k_5 > 0$	Gumbel	`GU`(μ, σ)	right tail	$k_5 = \sigma^{-1},\ k_6 = e^{-\mu/\sigma}$	$-\infty < \mu < \infty,\ \sigma > 0$
	reverse Gumbel	`RG`(μ, σ)	left tail	$k_5 = \sigma^{-1},\ k_6 = e^{\mu/\sigma}$	$-\infty < \mu < \infty,\ \sigma > 0$

TABLE 17.2: (continued from Table 17.1) Azzalini and spliced type distributions. Asymptotic form of $\log f_Y(y)$ as $y \to -\infty$ (left tail) or as $y \to \infty$ (right tail) for continuous distributions on $(-\infty, \infty)$ using (17.1). Also the asymptotic form of $\log F_Y(y)$ as $y \to -\infty$ (left tail), and $\log S_Y(y)$ as $y \to \infty$ (right tail) using (17.2).

Value of k_1–k_6	Distribution	**gamlss** name	Condition	Value of k_1–k_6	Parameter range
$k_1 = 1$	Skew t type 1	ST1(μ, σ, ν, τ)	right tail	$k_2 = \tau + 1$	$\nu \geq 0, \tau > 0$
				$k_2 = 2\tau + 1$	$\nu < 0, \tau > 0$
			left tail	$k_2 = 2\tau + 1,$	$\nu > 0, \tau > 0$
				$k_2 = \tau + 1,$	$\nu \leq 0, \tau > 0$
	Skew t type 2	ST2(μ, σ, ν, τ)		$k_2 = \tau + 1$	$\tau > 0$
	Skew t type 3	ST3(μ, σ, ν, τ)		$k_2 = \tau + 1$	$\tau > 0$
	Skew t type 4	ST4(μ, σ, ν, τ)	right tail	$k_2 = \tau + 1$	$\tau > 0$
			left tail	$k_2 = \nu + 1$	$\nu > 0$
$k_3 > 0$	skew exp power 1	SEP1(μ, σ, ν, τ)		$k_3 = \tau,$	$\sigma > 0, \tau > 0$
			right tail	$k_4 = \tau^{-1}\sigma^{-\tau},$	$\nu \geq 0$
				$k_4 = \tau^{-1}\sigma^{-\tau}(1 + \lvert\nu\rvert^{\tau}),$	$\nu < 0$
			left tail	$k_4 = \tau^{-1}\sigma^{-\tau}(1 + \lvert\nu\rvert^{\tau}),$	$\nu \geq 0$
				$k_4 = \tau^{-1}\sigma^{-\tau},$	$\nu < 0$
	skew power exp 2	SEP2(μ, σ, ν, τ)		$k_3 = \tau,$	$\sigma > 0, \tau > 0$
			right tail	$k_4 = \tau^{-1}\sigma^{-\tau},$	$\nu \geq 0$
				$k_4 = \tau^{-1}\sigma^{-\tau}(1 + \nu^2),$	$\nu < 0$
			left tail	$k_4 = \tau^{-1}\sigma^{-\tau}(1 + \nu^2),$	$\nu \geq 0$
				$k_4 = \tau^{-1}\sigma^{-\tau},$	$\nu < 0$
	skew exp power 3	SEP3(μ, σ, ν, τ)		$k_3 = \tau$	$\sigma > 0, \nu > 0, \tau > 0$
			right tail	$k_4 = 0.5(\sigma\,\nu)^{-\tau}$	
			left tail	$k_4 = 0.5\sigma^{-\tau}\nu^{\tau}$	
	skew exp power 4	SEP4(μ, σ, ν, τ)	right tail	$k_3 = \tau,\ k_4 = \sigma^{-\tau}$	$\sigma > 0, \tau > 0$
			left tail	$k_3 = \nu,\ k_4 = \sigma^{-\nu}$	$\sigma > 0, \tau > 0$

TABLE 17.3: Asymptotic form of $\log f_Y(y)$ as $y \to \infty$ (right tail) for continuous distributions on $(0, \infty)$ using (17.1), also the asymptotic form of $\log S_Y(y)$ as $y \to \infty$ (right tail) using (17.2).
Note $c_1^2 = \Gamma\left(\frac{1}{\tau}\right)\left[\Gamma\left(\frac{3}{\tau}\right)\right]^{-1}$ and $c_2 = \left[K_{\nu+1}\left(\frac{1}{\sigma^2}\right)\right]\left[K_{\nu}\left(\frac{1}{\sigma^2}\right)\right]^{-1}$ where $K_\lambda(\cdot)$ is a modified Bessel function of the second kind.
exp = exponential, gen = generalized; [a]PARETO1o(μ, σ) has range $Y \geq \mu$; [b]logWEI(μ, σ) has range $Y \geq 1$

Value of k_1–k_6	Distribution	**gamlss** name	Condition	Value of k_1–k_6	Parameter range
$k_1 = 1$	Box-Cox Cole Green	BCCG(μ,σ,ν)	$\nu < 0$	$k_2 = \lvert\nu\rvert + 1$	
	Box-Cox power exp	BCPE(μ,σ,ν,τ)	$\nu < 0$	$k_2 = \lvert\nu\rvert + 1$	
	Box-Cox t	BCT(μ,σ,ν,τ)	$\nu \leq 0$	$k_2 = \lvert\nu\rvert + 1$	
			$\nu > 0$	$k_2 = \nu\tau + 1$	$\tau > 0$
	gen beta type 2	GB2(μ,σ,ν,τ)		$k_2 = \sigma\tau + 1$	$\sigma > 0$, $\tau > 0$
	gen gamma	GG(μ,σ,ν)	$\nu < 0$	$k_2 = \left(\sigma^2\lvert\nu\rvert\right)^{-1} + 1$	$\sigma > 0$
	inverse gamma	IGAMMA(μ,σ)		$k_2 = \sigma^{-2} + 1$	$\sigma > 0$
	log t	logTF(μ,σ,ν)		$k_2 = 1$	
	Pareto type 1[a]	PARETO1o(μ,σ)		$k_2 = \sigma + 1$	$\sigma > 0$
$k_1 = 2$	Box-Cox Cole Green	BCCG(μ,σ,ν)	$\nu = 0$	$k_2 = 0.5\sigma^{-2}$	$\sigma > 0$
	log normal	LOGNO(μ,σ)		$k_2 = 0.5\sigma^{-2}$	$\sigma > 0$
$k_1 \geq 1$	Box-Cox power exp	BCPE(μ,σ,ν,τ)	$\nu = 0$, $\tau > 1$	$k_1 = \tau$, $k_2 = (c_1\sigma)^{-\tau}$	$\sigma > 0$
			$\nu = 0$, $\tau = 1$	$k_1 = 1$, $k_2 = 1 + (\sqrt{2}/\sigma)$	$\sigma > 0$
			$\nu = 0$, $\tau < 1$	$k_1 = 1$, $k_2 = 1$	$\sigma > 0$
	log Weibull[b]	logWEI(μ,σ)	$\sigma > 1$	$k_1 = \sigma$, $k_2 = \mu^{-\sigma}$	$\mu > 0$
			$\sigma = 1$	$k_1 = 1$, $k_2 = \mu^{-1} + 1$	$\mu > 0$
			$\sigma < 1$	$k_1 = 1$, $k_2 = 1$	
$k_3 > 0$	Box-Cox Cole Green	BCCG(μ,σ,ν)	$\nu > 0$	$k_3 = 2\nu$, $k_4 = \left[2\mu^{2\nu}\sigma^2\nu^2\right]^{-1}$	$\mu > 0$, $\sigma > 0$
	Box-Cox power exp	BCPE(μ,σ,ν,τ)	$\nu > 0$	$k_3 = \nu\tau$, $k_4 = \left[c_1\mu^{\nu}\sigma\nu\right]^{-\tau}$	$\mu > 0$, $\sigma > 0$, $\tau > 0$
	gen gamma	GG(μ,σ,ν)	$\nu > 0$	$k_3 = \nu$, $k_4 = \left[\mu^{\nu}\sigma^2\nu^2\right]^{-1}$	$\mu > 0$, $\sigma > 0$
	Weibull	WEI(μ,σ)		$k_3 = \sigma$, $k_4 = \mu^{-\sigma}$	$\mu > 0$, $\sigma > 0$
$k_3 = 1$	exponential	EXP(μ)		$k_4 = \mu^{-1}$	$\mu > 0$
	gamma	GA(μ,σ)		$k_4 = \mu^{-1}\sigma^{-2}$	$\mu > 0$, $\sigma > 0$
	gen inverse Gaussian	GIG(μ,σ,ν)		$k_4 = 0.5c_2\mu^{-1}\sigma^{-2}$	$\mu > 0$, $\sigma > 0$
	inverse Gaussian	IG(μ,σ)		$k_4 = 0.5\mu^{-2}\sigma^{-2}$	$\mu > 0$, $\sigma > 0$

It should also be noted that although the tails of two distributions with the same combination of k_1 and k_2 values are not necessarily equally heavy, a reduction in k_2, no matter how small, for either distribution will make it the heavier tail distribution. Similar results apply by replacing (k_1, k_2) by (k_3, k_4) or (k_5, k_6). Hence the important point is that the k values are dominant in determining the heaviness of the tail of the distribution[a].

We define four categories of distribution tails in Tables 17.1, 17.2, and 17.3: light tails ($k_3 \geq 1$ or $k_5 > 0$), heavy tails, i.e. heavier than any exponential distribution but lighter than any of our 'Paretian type' tail ($k_1 > 1$ and $0 < k_3 < 1$), our 'Paretian type' tail ($k_1 = 1$ and $k_2 > 1$), and 'heavier than any Paretian type' tail ($k_1 = k_2 = 1$). Our four categories correspond approximately to mild, slow, wild (pre or proper), and extreme randomness of Mandelbrot [1997], as indicated in Table 17.4. (The light and heavy tails in Table 17.4 are classical tail theory light and heavy tails, as defined in Section 17.6. However our 'Paretian type' tails in Table 17.4 may not always be regularly varying distribution tails, although all distribution tails in Tables 17.1, 17.2, and 17.3 with $k_1 = 1$ and $k_2 > 1$ have regularly varying distribution tails.)

TABLE 17.4: Mandelbrot's classification of randomness.

Our type of tail	Conditions	approximate Mandelbrot's randomness
light	$k_3 \geq 1$ or $k_5 > 0$	mild
heavy	$k_1 > 1$ or $0 < k_3 < 1$	slow
'Paretian type'	$k_1 = 1$ and $k_2 > 1$	wild
'heavier than Paretian'	$k_1 = k_2 = 1$	extreme

One property of our definition of a heavier tail, given in Section 17.2.1, is that location shifting and scaling of the random variable Y can affect the tail heaviness of the distribution. In particular:

(i) for Type I distributions neither k_1 nor k_2 is affected by location shifting or scaling,

(ii) for Type II distributions k_4 is affected by scaling (but not location shifting), while k_3 is not affected by location shifting or scaling,

(iii) for Type III distributions k_6 is affected by both location shifting and scaling, while k_5 is affected by scaling (but not location shifting).

An advantage of our approach is illustrated by the gamma ($\texttt{GA}(\mu, \sigma)$) distribution. For fixed mean μ, the tail is heavier, by our definition, as the shape parameter σ increases, since $k_3 = 1$ and $k_4 = (\mu\sigma^2)^{-1}$. Note a smaller k_4 has a heavier tail. In some classifications all gamma distributions have the same tail heaviness.

Another example is given by comparing the $\texttt{GA}(\mu, \sigma)$ and inverse Gaussian ($\texttt{IG}(\mu, \sigma_1)$) distributions, where $\sigma_1 = \sigma/\sqrt{\mu}$, which have the same mean μ and the same variance $\sigma^2\mu^2$. $\texttt{IG}(\mu, \sigma_1)$ has the heavier tail, by our definition, with $k_3 = 1$ and $k_4 = (2\mu^2\sigma_1^2)^{-1} = (2\mu\sigma^2)^{-1}$, while $\texttt{GA}(\mu, \sigma)$ has $k_3 = 1$ and $k_4 = (\mu\sigma^2)^{-1}$, so $\texttt{IG}(\mu, \sigma_1)$ has essentially asymptotically an exponential tail with twice the mean of $\texttt{GA}(\mu, \sigma)$. In

[a] If it is required to distinguish between the two distributions with the same k values, the second-order terms of $\log f_Y(y)$ can be compared.

some classifications all gamma and inverse Gaussian distributions have the same tail heaviness.

For tail heaviness which is not affected by location shifting and scaling, all Type III distributions should be treated as having equally heavy tails, and Type II distributions with the same k_3 and different values of k_4 should be treated as having equally heavy tails.

17.2.4 Types of tails for the survivor function

Care needs to be taken when switching from the tail of a pdf to the tail of its survivor function (or its cdf). However, from Lemma A3, if $\log f_Y(y) \sim g(y)$ as $y \to \infty$, then $\log S_Y(y) \sim g(y)$ as $y \to \infty$, assuming $\log g'(y) = o(g(y))$ as $y \to \infty$ and $\lim_{y\to\infty} g(y) = \infty$. Also, from Lemma A4, if $\log f_Y(y) \sim -k_2 \log y$ as $y \to \infty$ for $k_2 > 1$, then $\log S_Y(y) \sim -(k_2 - 1)\log y$ as $y \to \infty$.

Hence (using also Corollary A3), as $y \to \infty$, the asymptotic form of $\log S_Y(y)$ is the same as that of $\log f_Y(y)$ for the three types of tails defined in (17.1), except

(i) when $k_1 = 1$ and $k_2 > 1$, k_2 is reduced by 1 for $\log S_Y(y)$;

(ii) when $k_1 = 1$ and $k_2 = 1$ then the specific asymptotic form depends on the particular distribution, as $y \to \infty$.

Hence if the asymptotic form of $\log f_Y(y)$ is given by (17.1), then

$$\log S_Y(y) \sim \begin{cases} \text{Type I} & \begin{cases} -(k_2-1)\log|y|, & \text{if } k_1 = 1 \text{ and } k_2 > 1 \\ -k_2(\log|y|)^{k_1}, & \text{if } k_1 > 1 \text{ and } k_2 > 0 \end{cases} \\ \text{Type II} & -k_4|y|^{k_3}, \quad \text{if } k_3 > 0 \text{ and } k_4 > 0 \\ \text{Type III} & -k_6 e^{k_5|y|}, \quad \text{if } k_5 > 0 \text{ and } k_6 > 0 \end{cases} \tag{17.2}$$

and if $k_1 = 1$ and $k_2 = 1$ in (17.2) then see (ii) above.

Also for the lower tail as $y \to -\infty$, (17.2) holds with $S_Y(y)$ replaced by $F_Y(y)$.

Hence equation (17.2) is correct for *all* the distributions listed in Tables 17.1, 17.2, 17.3, and 17.8, so the tables can be applied to provide the asymptotic form for the left or right tail of $\log S_Y(y)$, for the different distributions using (17.2). For example if $Y \sim \texttt{TF}(\mu, \sigma, \nu)$ then, from Table 17.1, $k_1 = 1$ and $k_2 = \nu + 1$, so $\log f_Y(y) \sim -(\nu + 1)\log|y|$ as $|y| \to \infty$, and hence from (17.2), $\log S_Y(y) \sim -\nu \log|y|$ as $y \to \infty$, and $\log F(y) \sim -\nu \log|y|$ as $y \to -\infty$.

When $k_1 = k_2 = 1$ in Table 17.3 then the specific asymptotic forms for $\log S_Y(y)$ as $y \to \infty$ are given in Table 17.5. Similarly for $\log F_Y(y)$ as $y \to -\infty$. Note that in Table 17.5, $\log S_Y(y) = o(\log|y|)$ as $y \to \infty$.

Note that the distributions in Table 17.5 have extremely heavy right tails, especially `logTF`, and `BCT` (with $\nu = 0$). All of their right tails are heavier than *any* Pareto distribution tail. The results in Table 17.5 were obtained from exponential transformation where $Y = \exp(Z)$, and so $\log S_Y(y) \sim \log[S_Z(z)]$, where $z = \log y$, as $y \to \infty$. For example if $Z \sim \texttt{TF}(\mu, \sigma, \nu)$ and $Y = \exp(Z)$ then $Y \sim \log \texttt{TF}(\mu, \sigma, \nu)$ and $S_Y(y) = S_Z(z)$, where $z = \log y$ and so $\log S_Y(y) \sim \log[S_Z(z)]$ as $y \to \infty$.

TABLE 17.5: Asymptotic form of $\log S_Y(y)$ as $y \to \infty$, when $k_1 = k_2 = 1$.

Distribution	Condition	Asymptotic form of $\log S_Y(y)$
`logTF`		$-\nu \log(\log y)$
`BCT`	$\nu = 0$	$-\tau \log(\log y)$
`logWEI`	$0 < \sigma < 1$	$-\mu^{-\sigma}(\log y)^{\sigma}$
`BCPE`	$\nu = 0,\ 0 < \tau < 1$	$-(c_1\sigma)^{-\tau}(\log y)^{\tau}$

Note that increasingly heavy right-tailed distributions can be obtained by repeated exponential transformation. For example if $Z \sim$ `logTF`(μ, σ, ν) and $Y = \exp(Z)$ then $Y \sim$ `loglogTF`(μ, σ, ν) and the asymptotic form of $\log S_Y(y)$ is $-\nu \log(\log(\log y))$ as $y \to \infty$.

Similarly lighter right tails can be obtained by log transformation. For example if $Z \sim$ `GA`(μ, σ) and $Y = \log Z$ then $Y \sim$ `expGA`(μ, σ) and Y has a Type III right tail, while Z has a Type II right tail.

This is of practical importance when we are looking for an appropriate continuous heavy-tailed distribution. If for example our observed response variable is heavily right-tailed and we fail to find an appropriate distribution for its tail, we can look for an appropriate distribution tail for $\log(y)$ and apply an exponential transform to this distribution, using `gen.Family(, "log")`. See the example in Section 17.3.4.

Note that the log Weibull (`logWEI`), Weibull (`WEI`), and Gumbel (`GU`) distributions have $\log S_Y(y)$ right tails in exactly (i.e. not asymptotically) the forms $-k_2(\log y)^{k_1}$, $-k_4 y^{k_3}$, and $-k_6 e^{k_5 y}$, respectively. Hence the tails of three distributions together provide a very flexible range of tail heaviness.

17.3 Methods for choosing the appropriate tail

The substantive practical implications of ordering of distribution tails is in the development and selection of statistical distributions with tails appropriate for observations on a response variable. The emphasis here is moved from the main bulk of the data in the middle, to the tail of the distribution. This is particularly true, for example, for measures of market risk such as Value-at-Risk (VaR), which is heavily used by financial institutions. Similarly, in the environmental context, heavy rain which results in flooding is of particular interest. The choice of an appropriate tail is particularly important when the probability of rare events is needed, i.e. the expected value above a specified quantile, as is the case with the use of the expected shortfall (ES) in insurance. Underestimation of the VaR or ES can have important consequences, as was evidenced by the global financial crisis of 2008.

A main feature of all the methods we examine below is that lower and middle values of the response variable are discarded and emphasis is placed on the tail. One would need a reasonable amount of data to do that. Ideally GAMLSS could provide a distribution which fits the response variable well throughout its range, but this may not always be achievable. We shall use four different methods for exploring the tails of the response variable. All of them apply to a single response variable (no explanatory variables), but the last two can be extended to regression models:

method 1 the log survival function plot;

method 2 the log-log survival function plot;

method 3 the fit of appropriate truncated distributions to the tail of the data;

method 4 the fit of appropriate distributions to the whole data with emphasis on the tail.

Methods 1 and 2 use the *empirical log survival function*, sometimes called the *complementary cumulative distribution function* (CCDF):

$$S_E(y) = 1 - F_E(y)$$

where $F_E(y)$ is the empirical cumulative distribution function (ecdf) introduced in Sections 1.3.1 and 10.1. Method 1 plots $\log S_E(y)$ against $\log y$, while method 2 plots $\log[-\log S_E(y)]$ against each of $\log(\log y)$, $\log y$, and y. Method 1 is also called the 'log-log plot' and is associated with the investigation of the power law applied on the tail of a distribution and the estimation of its power parameter, see for example Gillespie [2015]. Methods 3 and 4 use straightforward GAMLSS modeling and are recommended for providing potentially more accurate estimates of the tail distribution parameters. All four methods are explained further in Sections 17.3.1 to 17.3.4.

To demonstrate the methods, the total USA box office film revenue, which was recorded for 4031 films from 1988-1999, is used. Film revenues are highly skewed, in such a way that a small number of large revenue films coexist alongside considerably greater numbers of smaller revenue films. Moreover, the skewed nature of this distribution appears to be an empirical regularity, with Pokorny and Sedgwick [2010] dating this phenomenon back to at least the 1930s, making it an early example of a mass market long tail. For demonstration purposes we will use the `film90` data extensively analyzed in Stasinopoulos et al. [2017]. In order to get the total film revenue we add the box office opening week revenue to the revenues after the first week. A histogram of the total revenues is shown in Figure 17.3(a).

R data file: `film90` in package **gamlss.data** of dimension 4031×4.
variables
`lnosc` : log of the number of screens in which the film was played
`lboopen` : log of box office opening week revenues
`lborev1` : log of box office revenues after the first week
`dist` : factor indicating whether the distributor of the film was an "Independent" or a "Major" distributor
purpose: to demonstrate the choice of the distribution for the right tail.

17.3.1 Exploratory method 1: log-survival

From (17.2) the upper tail of $\log S_Y(y)$ is asymptotically, as $y \to \infty$, in the form:

$$\begin{array}{ll} \text{Type I} & -\{k_2 - \mathrm{I}(k_1 = 1)\}(\log y)^{k_1} \\ \text{Type II} & -k_4 e^{k_3 \log y} \\ \text{Type III} & -k_6 e^{k_5 e^{\log y}} \end{array} \qquad (17.3)$$

for $k_1 = 1$ with $k_2 > 1$, (i.e. excluding the case $k_1 = k_2 = 1$), and $k_1 > 1$ with $k_2 > 0$, and $k_j > 0$ for $j = 3,4,5,6$. $\mathrm{I}(\cdot)$ is the indicator function, (taking value 1 if the condition in $(\cdot)$ is true, and 0 otherwise). [Note when $k_1 = k_2 = 1$, $\log S_Y(y)$ can be heavier than in (17.3). For example, for $\texttt{logWEI}(\mu,\sigma)$ with $0 < \sigma < 1$, from Table 17.5, $\log S_Y(y) \sim -\mu^{-\sigma}(\log y)^\sigma$; and for $\texttt{logTF}(\mu,\sigma,\nu)$, $\log S_Y(y) \sim -\nu \log(\log y)$, as $y \to \infty$.]

In method 1 (the classical 'log-log plot'), the empirical log survival function $\log S_E(y) = \log(1 - (i/n + 1))$ is plotted against $\log y_{(i)}$, where $y_{(i)}$ is the ith ordered value of y in the sample. This plot can be used to investigate the appropriateness of the tail form of a theoretical survival function $S_Y(y)$. While plotting $\log S_E(y)$ for the whole data can be informative, typically we include only a certain percentage, e.g. 10%, or 20%, i.e. the right (or left) tail of the empirical survival function.

From equation (17.3), if the upper tail of the empirical log survival function $\log S_E(y)$ is plotted against $\log y$,

- a *linear* plot indicates $k_1 = 1$ (e.g. a Pareto upper tail);
- a *quadratic* plot indicates $k_1 = 2$ (e.g. a log normal upper tail);
- an *exponential* plot indicates a type II distribution, i.e. $k_3 > 0$ (e.g. a Weibull upper tail).

Method 1 is ideally suited to identifying a Pareto type 1, $\texttt{PARETO1o}(\mu,\sigma)$, right tail, for which $\log S_Y(y) = \sigma \log \mu - \sigma \log y$, which has *exactly* a linear relationship between $\log S_Y(y)$ and $\log y$, rather than an asymptotic one. The log Weibull distribution allows *any* (positive) power relationship between $\log S_Y(y)$ and $\log y$ since $\log S_Y(y) \sim -\mu^{-\sigma}(\log y)^\sigma$. Linear, quadratic, and exponential curves can be fitted to the empirical log survival function and the coefficients used as rough estimates of the relevant k parameters. This method is widely used in the literature (e.g. Mandelbrot [1997]) to estimate the parameter σ of a $\texttt{PARETO1o}(\mu,\sigma)$ distribution, but it is not very reliable. For example, the absolute value of the estimate of the slope of the linear line fit in the log-log plot is taken as an estimate of the parameter σ of the $\texttt{PARETO1o}(\mu,\sigma)$ distribution. Note that this corresponds to the Type I tail where $k_1 = 1$ and $k_2 = \sigma + 1$ from Table 17.3. Gillespie [2015] comments that 'estimating the power law exponent on a log-log plot, while appealing, is a very poor technique for fitting these types of models', and advocates MLE instead. See Method 3 in Section 17.3.3.

The **gamlss** function `logSurv()` is designed for exploring method 1. It plots the empirical log survival function $\log S_E(y)$ against $\log y$ for a specified percentage of cases, lying in the tail, and fits linear, quadratic, and exponential curves to the points of the plot.

```
data(film90)
y <-with(film90,  exp(lboopen)+exp(lborev1))/1000000
truehist(y, main="(a)", col=gray(.8))
logSurv(y, prob=0.90, tail="right", col=gray(.6), title="(b)")
```

Figure 17.3(b) shows the log survival plot for the largest 10% of revenues (i.e. the 403 most profitable films, specified by `prob=0.90`) together with fitted linear, quadratic, and exponential functions. The linear fit appears inadequate, hence $k_1 = 1$ (e.g. a Pareto distribution) appears inappropriate. The quadratic and exponential fit adequately, sug-

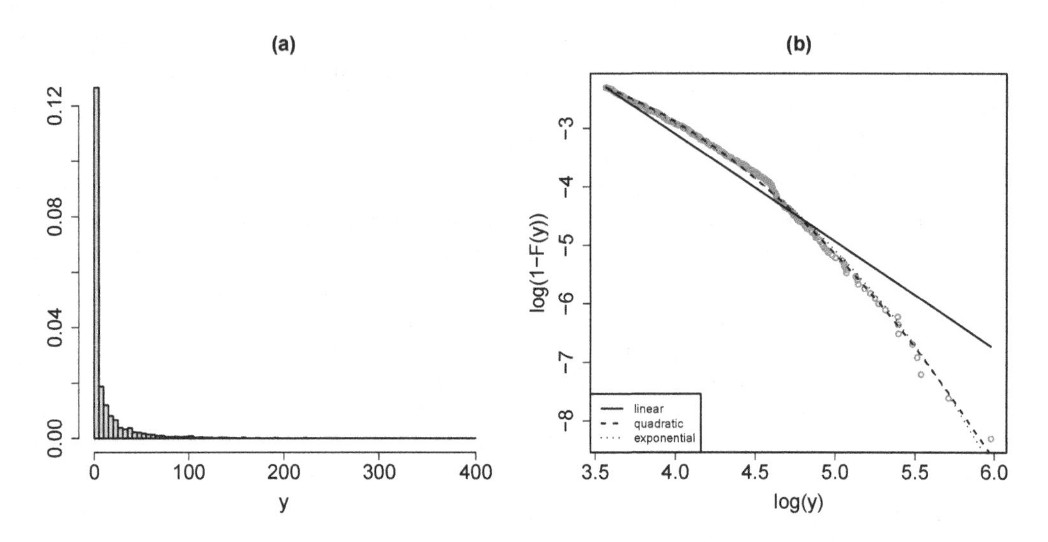

FIGURE 17.3: Film revenues data: (a) histogram, (b) plot of exploratory method 1, using the right 10% tail of revenues.

gesting $k_1 = 2$ or $k_3 > 0$ may be appropriate (see Table 17.3 for the relevant distributions). Note that method 1 is only appropriate to investigate rather heavy tails and, unlike method 2, is not so good at discriminating between the three different tail types.

17.3.2 Exploratory method 2: log-log-survival plot

Method 2 uses the log-log empirical survival function. The inspiration came from Klein and Moeschberger [2003, p410], where it was used in a survival analysis context. Corresponding to (17.2), the upper tail of $\log[-\log S_Y(y)]$ is asymptotically, as $y \to \infty$, in the form:

$$\begin{array}{ll} \textsf{Type I} & \log\{k_2 - \mathrm{I}(k_1 = 1)\} + k_1 \log(\log|y|) \\ \textsf{Type II} & \log k_4 + k_3 \log|y| \\ \textsf{Type III} & \log k_6 + k_5|y| \end{array} \tag{17.4}$$

for $k_1 = 1$ with $k_2 > 1$, (i.e. excluding the case $k_1 = k_2 = 1$), and $k_1 > 1$ with $k_2 > 0$, and $k_j > 0$, for $j = 3, 4, 5, 6$.

Hence a plot of the empirical $\log\{-\log[S_E(y)]\}$ against $\log(\log y)$ or $\log y$ or y will be asymptotically linear in each case. See Figure 17.4 for an example. We call this the log-log-survival plot, see Klein and Moeschberger [2003, p411], Figure 12.2.

Table 17.6 summarizes the relationship between the different types of tails and model terms needed to be fitted (except when $k_1 = k_2 = 1$). The corresponding sample plot, Figure 17.4, can be used to investigate the tail form of $S_Y(y)$. Note from Table 17.6 that method 2 provides rough estimates for all the parameters involved. For example, for Type I tail the fitted constant term $\hat{\beta}_0$ provides the estimate $\hat{k}_2 = \exp(\hat{\beta}_0) + \mathrm{I}(k_1 = 1)$, while the fitted slope $\hat{\beta}_1$ provides an estimate for k_1. Note however that for reasonable estimates, a large sample size (from the tail) may be required, especially in the Type I case. However Method 3 in Section 17.3.3 should provide more reliable parameter estimates for a particular distribution tail.

Method 2 is ideally suited to identifying a `logWEI`(μ, σ), `WEI`(μ, σ) or `GU`(μ, σ) right tail, as they have $\log[-\log S_Y(y)]$ given *exactly* by $-\sigma \log \mu + \sigma \log(\log y)$, $-\sigma \log \mu + \sigma \log y$ and $-\mu/\sigma + y/\sigma$, respectively, i.e. exactly linear in $\log(\log y)$, $\log y$, and y, respectively. Hence, the estimated intercepts and slopes in Figure 17.4(a), (b), and (c) can be used to provide rough estimates for the parameters of the `logWEI`, `WEI` and `GU` distribution tails, respectively, using Table 17.6 together with Tables 17.1 and 17.3. However method 3 provides a better way to estimate the parameters.

An advantage of method 2 is that the response variable, the empirical $\log[-\log S_E(y)]$, is the same in all types in Table 17.6 . This allows straightforward comparison of the three fitted regressions. This is implemented in the `loglogSurv()` function, which fits (for values of y above a particular centile of the tail) the empirical $\log[-\log S_E(y)]$ against $\log(\log y)$, $\log y$, and y, respectively and chooses the best fit according to the residual sum of squares. The functions `loglogSurv1()`, `loglogSurv2()`, and `loglogSurv3()` fit the equivalent model for type I, II, and III tail, respectively. For example, to fit a model for Type I tail, we set the empirical $\log[-\log S_E(y)]$ and $\log(\log y)$ as the response and explanatory variable in a linear regression fit, while we assume an approximate (but incorrect) normal distribution for the error term.

TABLE 17.6: Possible relationships for $\log[-\log S_Y(y)]$ for method 2.

	$\log[-\log S_Y(y)]$	Linear term
Type I	$\log\{k_2 - \mathrm{I}(k_1 = 1)\} + k_1 \log(\log y)$	$\log(\log y)$
Type II	$\log k_4 + k_3 \log y$	$\log y$
Type III	$\log k_6 + k_5 y$	y

Method 2 was applied to the largest 10% of film revenues (y). Figure 17.4 (a), (b), and (c) plot, for the film revenue data, $\log[-\log S_E(y)]$ against $\log(\log y)$, $\log y$, and y, respectively (exploratory method 2). Figure 17.4(b) for Type II provides the best linear fit (error sum of squares equal 0.06823, see Table 17.7), with estimates $\hat{k}_3 = 0.5599$ and $\hat{k}_4 = \exp(-8.90338) = 0.0001359$. This suggests that, for example, a Weibull (`WEI`), or Box-Cox power exponential (`BCPE`) tail may be appropriate. See Table 17.3. It is important to note, however, that Figure 17.4(a) and its error sum of squares is affected by changing the scale of y, and hence a particular scaling of y may fit a `logWEI` distribution tail particularly well. The reason is that `logWEI` is not a scale family distribution. Alternatively the `BCPE` is a scale family distribution and can have similar tails to `logWEI`.

[Note also that, although Figure 17.4(a) provides a reasonably linear fit, its slope is 2.38 indicating $\hat{k}_1 = 2.38$. This is far from a Pareto tail ($k_1 = 1$), although it could possibly be a log Weibull tail or a log normal tail ($k_1 = 2$).]

```
m1 <- loglogSurv1(y, prob=0.90,col=gray(.6), title="(a) TYPE I",
                  cex.axis=1.8, cex.lab=1.8)

## coefficients -2.223148 2.383676
## error sum of squares 0.2370167

m2 <- loglogSurv2(y, prob=0.90, col=gray(.6), title="(b) TYPE II",
                  cex.axis=1.8, cex.lab=1.8)
```

```
## coefficients -1.145337 0.5543395
## error sum of squares 0.06194498

m3 <- loglogSurv3(y, prob=0.90, col=gray(.6), title="(c) TYPE III",
                  cex.axis=1.8, cex.lab=1.8)

## coefficients 0.7523683 0.005623397
## error sum of squares 2.744493
```

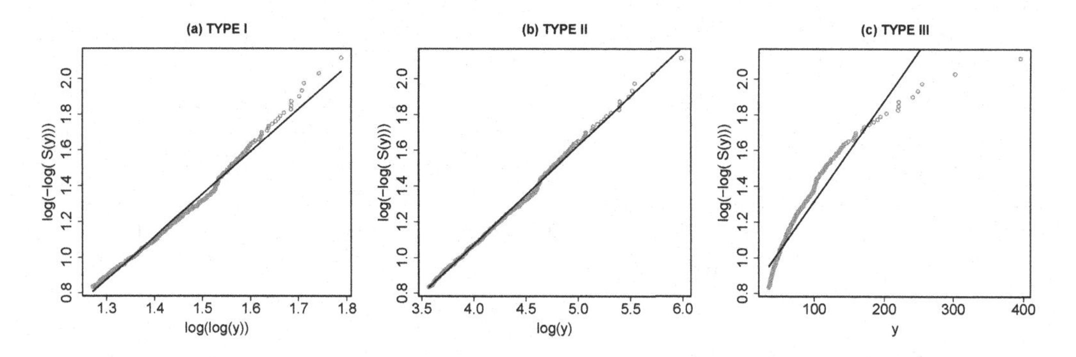

FIGURE 17.4: Exploratory method 2 applied to the film revenues data: plot of empirical $\log[-\log S_E(y)]$ against (a) $\log(\log y)$, (b) $\log y$, and (c) y.

TABLE 17.7: Estimated coefficients from exploratory method 2.

	Intercept	Slope	Error SS
Type I	-2.223	+2.384	0.2370
Type II	-1.145	+0.554	0.0619
Type III	+0.752	+0.006	2.7444

17.3.3 Exploratory method 3: truncated distribution fitting

Method 3 relies on investigating the tail of a distribution by fitting a relevant truncated distribution. Truncation of a distribution is discussed in Section 13.7, while practical examples are given in Sections 3.4.3, 5.6.2 and 6.3.2. Note a truncated distribution can provide a very flexible model for the tail.

Univariate example

Below we fit the truncated log normal, Weibull, and Box-Cox power exponential distributions to the largest 10% of the data. (Note that any other sensible distribution can also be fitted.) The Weibull seems to fit slightly better. Figure 17.5(a) provides a worm plot for the residuals from the truncated Weibull fit to the largest 10% of revenues, indicating a reasonable fit in the upper tail. The estimated Weibull parameters are $\hat{\mu} = 17,222,442$ and $\hat{\sigma} = 0.6258$.

```
library(gamlss.tr)
m1 <- fitTail(y, family=WEI, percentage=10, mu.start=200)
m2 <- fitTail(y, family=LOGNO, percentage=10)
m3 <- fitTail(y, family=BCPE, percentage=10)
```

```
AIC(m1,m2,m3)

##    df      AIC
## m1  2 3696.006
## m2  2 3697.444
## m3  4 3698.751
```

Sequential fits of the truncated Weibull distribution to the largest r revenues, for $r = 8, 9, \ldots, 403$, were followed by a plot of the parameter estimates $\hat{\mu}$ and $\hat{\sigma}$ against $(404 - r)$, shown in Figure 17.5(b). This shows that the fitted parameters $\hat{\mu}$ and $\hat{\sigma}$ are relatively stable until the last 100 observations in the tail are reached (on the right of Figure 17.5(b)). The sequential fitting plot is analogous to a Hill plot [Hill, 1975].

```
wp(m1); title("(a)")
mm<-fitTailAll(y, family=WEI, mu.start=200)
plot(mm, main="(b)")
```

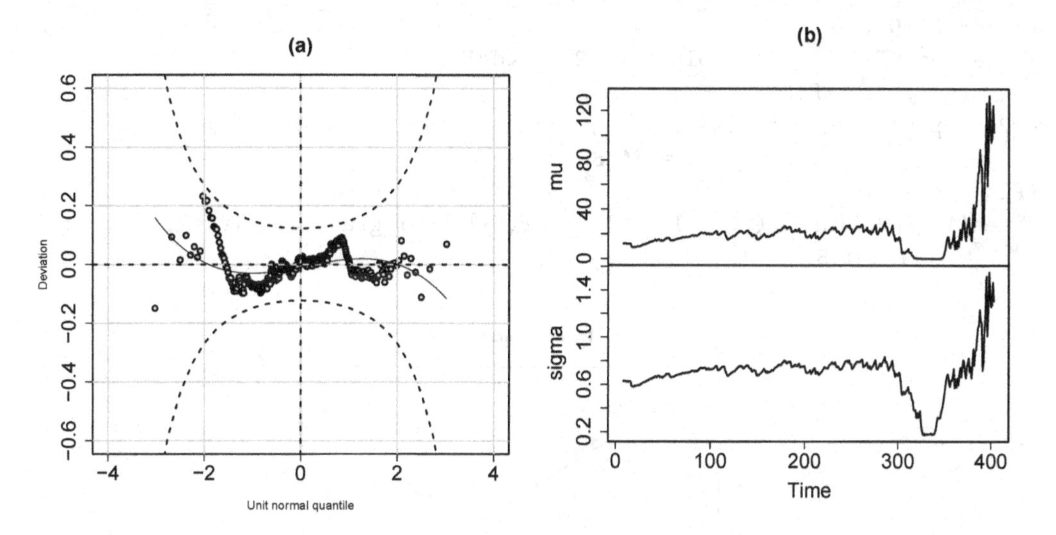

FIGURE 17.5: Exploratory method 3, truncated Weibull: (a) worm plot, (b) sequential plot of $\hat{\mu}$ and $\hat{\sigma}$.

The code below demonstrates the fitting of a truncated PARETO1o distribution to the largest 10% of response variable values. As a PARETO1o(μ, σ), truncated below y_{90} (the 90% sample centile), has exactly a PARETO1o(y_{90}, σ) distribution, it can be fitted directly using PARETO1o, by fixing the value of μ at y_{90}. Then we compare the PARETO1o fit to the corresponding WEI, LOGNO, and BCPE fits above.

```
y90 <- quantile(y, .90)
 Y <-  y[y>y90]
t1 <- gamlss( Y~1, family=PARETO1o, mu.start=y90, trace=FALSE)
fitted(t1, "sigma")[1]+1

##      90%
## 2.707086

AIC(m1,m2,m3,t1)
```

```
##    df      AIC
## m1  2 3696.006
## m2  2 3697.444
## m3  4 3698.751
## t1  1 3723.435
```

The Weibull (`m1`), log normal (`m2`) and Box-Cox power exponential (`m3`) fit equally well, and much better than the Pareto type 1 (`t1`), as predicted by methods 1 and 2.

The code below demonstrates plots of the fitted tail log survival functions against $\log(y)$, for the `PARETO1o` and `WEI3` distributions, as shown in Figure 17.6. This highlights the failure of the Pareto type 1 fitted log survival function (the line) to fit the log empirical survival function in the right tail 10% of the response variable values.

```
logSurv0(y, col="gray", title="PARETO1o and WEI3 fits")
PT<- pPARETO1o(Y, mu=y90,    sigma=fitted(t1, "sigma"),
               lower.tail = FALSE)
PT  <- PT*0.1
lines(log(PT)~ log(Y), col=gray(.2), lwd=1.5)
gen.trun(y90, "WEI")
PP2 <- pWEItr(Y,mu=fitted(m1)[1], sigma=fitted(m1,"sigma")[1],
               lower.tail = FALSE)
PP2<- PP2*0.1
lines(log(PP2)[order(Y)]~ log(Y)[order(Y)], col=gray(.3), lty=2,
        lwd=1.5)
```

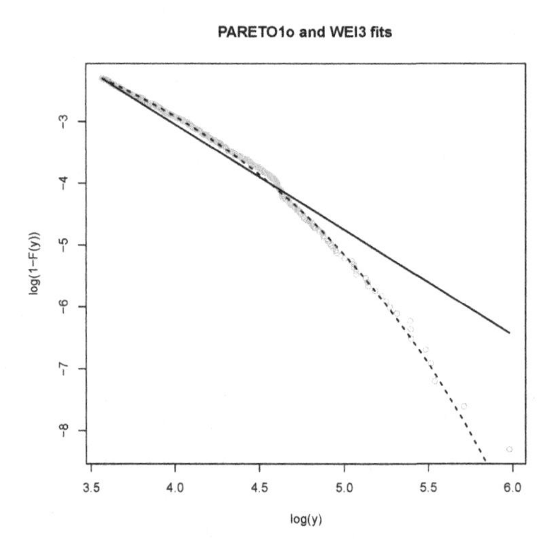

FIGURE 17.6: Exploratory method 3: fitted log survival function against $\log(y)$ from the truncated Weibull and truncated Pareto type 1.

In conclusion, methods 1 and 3 suggest $k_1 = 2$ (e.g. log normal) or $k_3 > 0$ (e.g. Weibull or BCPE), while method 2 just suggests $k_3 > 0$.

Regression example

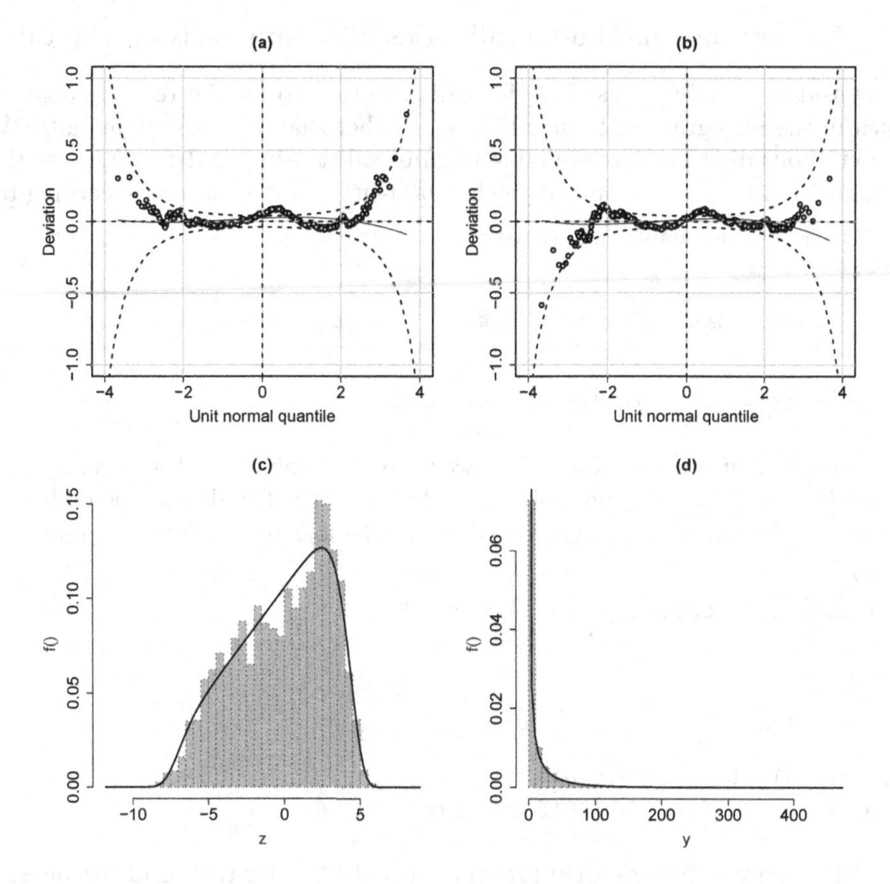

FIGURE 17.7: Exploratory method 4 applied to the film revenues data: (a) worm plot for the fitted BCPEo distribution applied to y, (b) worm plot for the fitted logSEP1 distribution applied to y (or equivalently the fitted SEP1 distribution applied to $\log y$), (c) the fitted SEP1 model fitted to $\log y$, and (d) the fitted logSEP1 model fitted to y.

One of the main concerns when a parametric distribution is fitted within a regression set-up is whether the fitted distribution fits well both in the center but also in the tail(s) of the distribution. In a discussion on the paper "Beyond mean regression" [Kneib, 2013], Rigby et al. [2013] propose a method for fitting the upper tail of the response variable distribution using the following three steps:

- Find the α quantile by fitting an appropriate quantile or GAMLSS model, from which we identify the observations in the upper tail of the distribution.
- Fit a suitable truncated distribution to the tail data.
- Obtain the β quantile of the truncated distribution, which corresponds to the $\tau = \alpha + \beta(1-\alpha)$ quantile of the original data.

An illustration of this approach is given in Exercise 8. More research is needed in this area.

17.3.4 Exploratory method 4: all cases with emphasis on the tail

Let us consider what happens if we fit a distribution to all the revenue observations as shown in the histogram in Figure 17.5(a), (rather than only the right tail). We first fit all continuous distributions available in **gamlss.dist** using `fitDist()`, store the best fitting model in `f1`, and then output the five best-fitting distributions according to AIC.

```
f1 <- fitDist(y, ncpus=4, parallel="snow" )
f1$fits[1:5]

##    BCPEo      GIG  SHASHo2   SHASHo      WEI
## 18450.23 18463.18 18763.89 18763.89 18980.62

wp(f1, ylim.all=1); title("(a)")
```

The `BCPEo` distribution is selected. The worm plot of model `f1` (i.e. `BCPEo`) shown in Figure 17.7(a) indicates that possibly the right tail of the fitted model is underestimating the true right tail of the revenues. We now take logs of the data and refit.

```
z<-log(y)
f2 <- fitDist(z, ncpus=4, parallel="snow")
f2$fits[1:5]

##     SEP1    SHASH     SEP2   SHASHo  SHASHo2
## 19812.91 19819.10 19832.29 19834.93 19834.94

wp(f2, ylim.all=1); title("(b)")
f3 <- histDist(z, family="SEP1", nbins=30, main="(c)")
```

The `SEP1` is suggested here as an appropriate distribution for the log of revenues, shown in Figure 17.7(c), indicating that `logSEP1` is appropriate for the original revenues. So we generate a `logSEP1` distribution and fit it to the revenues, as shown in Figure 17.7(d), with the corresponding worm plot in Figure 17.7(b).

```
gen.Family("SEP1", "log")

## A  log  family of distributions from SEP1 has been generated
##  and saved under the names:
##  dlogSEP1 plogSEP1 qlogSEP1 rlogSEP1 logSEP1

f4 <- histDist(y, family="logSEP1", nbins=30, main="(d)")
AIC(f1, f4)

##    df      AIC
## f1  4 18450.23
## f4  4 18453.26
```

Even though the preference of the AIC is for the `BCPEo` distribution, the `logSEP1` has fitted the right tail better according to the worm plot of Figure 17.7(b), and is a better choice if the right tail of the distribution is of primary interest.

17.4 Classification table for tails of discrete distributions

Table 17.8 provides a summary of the right tail types for many important discrete count distributions with range $\{0,1,2,\ldots\}$ or $\{1,2,\ldots\}$. From classification (17.1), k_1 and k_2 are associated with a Type I probability function tail, while k_3 and k_4 are associated with a Type II tail. No distributions in Table 17.8 have a Type III tail. However DPO and PO have a Type IV tail, given by

- Type IV $\log f_Y(y) \sim -k_8(y\log y)^{k_7}$ as $y \to \infty$.

Note that $ky^{k_7} < k_8(y\log y)^{k_7} < ky^{k_7+\delta}$, for sufficiently large y and any fixed $\delta > 0$ and $k > 0$. Any Type IV tail is lighter than any Type II tail with $k_3 = k_7$, but is heavier than any Type II tail with $k_3 = k_7 + \delta$ for any fixed $\delta > 0$. Hence in Table 17.8 all distributions with $k_7 = 1$ are lighter than all distributions with $k_3 = 1$.

More flexible discrete count distribution right tails can be obtained by discretizing continuous distributions on $(0, \infty)$, see Section 7.3.4. A discretized continuous distribution will have the same tail heaviness as the original continuous distribution, both for the probability function in (17.1) and the survival function in (17.2) and Table 17.5. For example, a discretized log normal distribution has a Type I right tail with $k_1 = 2$. A discretized log Weibull distribution has a more flexible Type I right tail with any $k_1 \geq 1$, in (17.1), see Table 17.3, while a discretized Weibull distribution has a Type II right tail with any $k_3 > 0$ in (17.1), see Table 17.3 and a discretized Gumbel distribution has a Type III right tail with any $k_5 > 0$ in (17.1), see Table 17.1. No discrete distributions in Table 17.8 have these tails.

17.5 Appendix A

Lemma 17.1 (Lemma A1).
Let the random variables Y_1 and Y_2 have pdfs $f_{Y_1}(y)$ and $f_{Y_2}(y)$, respectively, then

$$f_{Y_1}(y) = o\left[f_{Y_2}(y)\right] \text{ as } y \to \infty \Leftrightarrow \left[\log f_{Y_2}(y) - \log f_{Y_1}(y)\right] \to \infty \text{ as } y \to \infty$$

A similar result for the left tail is obtained by replacing $y \to \infty$ by $y \to -\infty$.

Proof A1.

$$\begin{aligned} f_{Y_1}(y) = o\left[f_{Y_2}(y)\right] \quad & \text{as } y \to \infty \\ \Leftrightarrow \quad \lim_{y\to\infty}\left[\frac{f_{Y_1}(y)}{f_{Y_2}(y)}\right] & = 0 \\ \Leftrightarrow \quad \lim_{y\to\infty}\left[\log\frac{f_{Y_2}(y)}{f_{Y_1}(y)}\right] & = \infty \end{aligned} \tag{17.5}$$

□

TABLE 17.8: Asymptotic form of the log of the probability function, $\log f_Y(y)$, as $y \to \infty$, for discrete count distributions, using equation (17.1). Also the asymptotic form of $\log S_Y(y)$ as $y \to \infty$, using equation (17.2). Note $b = K_{\nu+1}(1/\sigma)/K_\nu(1/\sigma)$.

Value of k_1–k_6	Distribution	**gamlss** name	Value of k_1–k_6	Parameter range
$k_1 = 1$	Zipf	`ZIPF`(μ)	$k_2 = 1 + \mu$	$\mu > 0$
	discrete Burr XII	`DBURR12`(μ, σ, ν)	$k_2 = 1 + \sigma\nu$	$\sigma > 0,\ 0 < \nu < \infty$
	beta negative binomial	`BNB`(μ, σ, ν)	$k_2 = 2 + \sigma^{-1}$	$\mu > 0,\ \sigma > 0,\ \nu > 0$
	Waring	`WARING`(μ, σ)	$k_2 = 2 + \sigma^{-1}$	$\mu > 0,\ \sigma > 0$
	Yule	`YULE`(μ)	$k_2 = 2 + \mu^{-1}$	$\mu > 0$
$k_3 = 1$	PSGIG	`PSGIG`(μ, σ, ν, τ)	$k_4 = \log\left[1 + \frac{b}{2\mu\sigma(1-\tau)}\right]$	$\mu > 0,\ \sigma > 0,\ -\infty < \nu < \infty,$ $0 < \tau < 1$
	Delaporte	`DEL`(μ, σ, ν)	$k_4 = \log\left[1 + \frac{1}{\mu\sigma(1-\nu)}\right]$	$\mu > 0,\ \sigma > 0,\ 0 < \nu < 1$
	Sichel	`SI`(μ, σ, ν)	$k_4 = \log\left(1 + \frac{1}{2\mu\sigma}\right)$	$\mu > 0,\ \sigma > 0,\ -\infty < \nu < \infty$
	Sichel	`SICHEL`(μ, σ, ν)	$k_4 = \log\left(1 + \frac{b}{2\mu\sigma}\right)$	$\mu > 0,\ \sigma > 0,\ -\infty < \nu < \infty$
	Poisson inverse Gaussian	`PIG`(μ, σ)	$k_4 = \log\left(1 + \frac{1}{2\mu\sigma}\right)$	$\mu > 0,\ \sigma > 0$
	negative binomial	`NBI`(μ, σ)	$k_4 = \log\left(1 + \frac{1}{\mu\sigma}\right)$	$\mu > 0,\ \sigma > 0$
	negative binomial	`NBII`(μ, σ)	$k_4 = \log\left(1 + \frac{1}{\sigma}\right)$	$\mu > 0,\ \sigma > 0$
	negative binomial family	`NBF`(μ, σ, ν)	$k_4 = \log\left(1 + \frac{1}{\mu^{\nu-1}\sigma}\right)$	$\mu > 0,\ \sigma > 0,\ \nu > 0$
	geometric	`GEOM`(μ)	$k_4 = \log\left(1 + \frac{1}{\mu}\right)$	$\mu > 0$
	geometric	`GEOMo`(μ)	$k_4 = -\log(1 - \mu)$	$0 < \mu < 1$
	generalized Poisson	`GPO`(μ, σ)	$k_4 = \log[1 + (\mu\sigma)^{-1}] - (1 + \mu\sigma)^{-1}$	$\mu > 0,\ \sigma > 0$
	logarithmic	`LG`(μ)	$k_4 = -\log \mu$	$0 < \mu < 1$
$k_7 = 1$	double Poisson	`DPO`(μ, σ)	$k_8 = \sigma^{-1}$	$\mu > 0, \sigma > 0$
	Poisson	`PO`(μ)	$k_8 = 1$	$\mu > 0$

Lemma 17.2 (Lemma A2).
Let random variables Y_1 *and* Y_2 *have pdfs* $f_{Y_1}(y)$ *and* $f_{Y_2}(y)$*, cdfs* $F_{Y_1}(y)$ *and* $F_{Y_2}(y)$ *and survivor functions* $S_{Y_1}(y)$ *and* $S_{Y_2}(y)$*, respectively, and* $\lim_{y\to\infty} f_{Y_1}(y) - \lim_{y\to\infty} f_{Y_2}(y) = 0$*, then*

$$f_{Y_1}(y) = o\left[f_{Y_2}(y)\right] \text{ as } y \to \infty \quad \Leftrightarrow S_{Y_1}(y) = o\left[S_{Y_2}(y)\right] \text{ as } y \to \infty$$

if, for y *sufficiently large,* $F_{Y_1}(y)$ *and* $F_{Y_2}(y)$ *are continuous and differentiable and* $f_{Y_2}(y) \neq 0$*, and, if* $\lim_{y\to\infty} S_{Y_1}(y) = \lim_{y\to\infty} S_{Y_2}(y) = 0$*, while the extra condition that* $\lim\limits_{y\to\infty} \left[f_{Y_1}(y)/f_{Y_2}(y)\right]$ *exists is needed for* $\Leftarrow$*, but not for* $\Rightarrow$*.*

Proof A2. $f_{Y_1}(y) = o\left[f_{Y_2}(y)\right] \quad \text{as} \quad y \to \infty$

$$\Leftrightarrow \quad \lim_{y\to\infty} \frac{f_{Y_1}(y)}{f_{Y_2}(y)} = 0 \tag{17.6}$$

$$\Leftrightarrow \quad \lim_{y\to\infty} \frac{S_{Y_1}(y)}{S_{Y_2}(y)} = 0 \quad \text{using l'Hôpital's rule} \tag{17.7}$$

$$\text{i.e.} \quad S_{Y_1}(y) = o\left[S_{Y_2}(y)\right]$$

The proof follows similarly for the left tail as $y \to -\infty$ giving

$$f_{Y_1}(y) = o\left[f_{Y_2}(y)\right] \text{ as } y \to -\infty \Leftrightarrow F_{Y_1}(y) = o\left[F_{Y_2}(y)\right] \text{ as } y \to -\infty.$$

□

Lemma 17.3 (Lemma A3).
Let $\log f_Y(y) \sim -g(y)$ *as* $y \to \infty$*, then* $\log S_Y(y) \sim -g(y)$ *as* $y \to \infty$*, if* $\log g'(y) = o(g(y))$ *as* $y \to \infty$*, and* $\lim_{y\to\infty} g(y) = \infty$*.*

Proof A3. If $\log f_Y(y) \sim -g(y)$ as $y \to \infty$, then

$$\lim_{y\to\infty} \left[\frac{\log f_Y(y)}{-g(y)}\right] = 1.$$

$\therefore$ For any fixed $\epsilon > 0$, if y is sufficiently large,

$$\begin{aligned}
(1-\epsilon) \quad &< \quad \frac{\log f_Y(y)}{-g(y)} \quad < \quad (1+\epsilon) \\
-(1+\epsilon)g(y) \quad &< \quad \log f_Y(y) \quad < \quad -(1-\epsilon)g(y) \\
\therefore \exp[-(1+\epsilon)g(y)] \quad &< \quad f_Y(y) \quad < \quad \exp[-(1-\epsilon)g(y)].
\end{aligned}$$

If $\log g'(y) = o(g(y))$ as $y \to \infty$, then $\lim_{y\to\infty}[\epsilon g(y) + \log g'(y)] = \lim_{y\to\infty}[\epsilon g(y)] = \infty$, and $\lim_{y\to\infty}[-\epsilon g(y) + \log g'(y)] = \lim_{y\to\infty}[-\epsilon g(y)] = -\infty$.

Hence $\lim_{y\to\infty} \{g'(y)\exp[\epsilon g(y)]\} = \lim_{y\to\infty} \{\exp[\epsilon g(y) + \log g'(y)]\} = \infty$, and $\lim_{y\to\infty} \{g'(y)\exp[-\epsilon g(y)]\} = \lim_{y\to\infty} \{\exp[-\epsilon g(y) + \log g'(y)]\} = 0$.

Hence for y sufficiently large

$$\begin{aligned}\exp[-(1+\epsilon)g(y)] &> g'(y)\exp[-\epsilon g(y)]\exp[-(1+\epsilon)g(y)]\\ &= g'(y)\exp[-(1+2\epsilon)g(y)],\end{aligned}$$

and

$$\begin{aligned}\exp[-(1-\epsilon)g(y)] &< g'(y)\exp[\epsilon g(y)]\exp[-(1-\epsilon)g(y)]\\ &= g'(y)\exp[-(1-2\epsilon)g(y)],\end{aligned}$$

and hence $g'(y)\exp[-(1+2\epsilon)g(y)] < f_Y(y) < g'(y)\exp[-(1-2\epsilon)g(y)]$.

Hence for y sufficiently large

$$\begin{array}{ccccc} g'(y)\exp[-(1+2\epsilon)g(y)] & < & f_Y(y) & < & g'(y)\exp[-(1-2\epsilon)g(y)] \\ \int_y^\infty g'(t)\exp[-(1+2\epsilon)g(t)]dt & < & \int_y^\infty f_Y(t)dt & < & \int_y^\infty g'(t)\exp[-(1-2\epsilon)g(t)]dt \\ \dfrac{1}{(1+2\epsilon)}\exp[-(1+2\epsilon)g(y)] & < & S_Y(y) & < & \dfrac{1}{(1-2\epsilon)}\exp[-(1-2\epsilon)g(y)] \\ -(1+2\epsilon)g(y)-\log(1+2\epsilon) & < & \log S_Y(y) & < & -(1-2\epsilon)g(y)-\log(1-2\epsilon) \\ -(1+4\epsilon)g(y) & < & \log S_Y(y) & < & -(1-4\epsilon)g(y), \end{array}$$

since for y sufficiently large: $-2\epsilon g(y) < -\log(1+2\epsilon)$ and $2\epsilon g(y) > -\log(1-2\epsilon)$, since $\lim_{y\to\infty} g(y) = \infty$.

Hence for y sufficiently large.

$$(1+4\epsilon) > \frac{\log S_Y(y)}{-g(y)} > (1-4\epsilon). \tag{17.8}$$

Hence for any fixed $\delta > 0$ then let $\epsilon = \delta/4$ in (17.8) giving

$$1-\delta < \frac{\log S_Y(y)}{-g(y)} < 1+\delta,$$

for y sufficiently large.

$\therefore \lim_{y\to\infty}\left[\dfrac{\log S_Y(y)}{-g(y)}\right] = 1$

$\therefore \log S_Y(y) \sim -g(y)$ as $y \to \infty$. □

Corollary 17.1 (Corollary A3).

(a) If $\log f_Y(y) \sim k_2(\log y)^{k_1}$ *as* $y \to \infty$, *then* $\log S_Y(y) \sim k_2(\log y)^{k_1}$ *as* $y \to \infty$, *provided* $k_1 > 1$ *and* $k_2 > 0$.

(b) If $\log f_Y(y) \sim k_4(\log y)^{k_3}$ *as* $y \to \infty$, *then* $\log S_Y(y) \sim k_4(\log y)^{k_3}$ *as* $y \to \infty$, *provided* $k_3 > 0$ *and* $k_4 > 0$.

(c) *If* $\log f_Y(y) \sim k_6 (\log y)^{k_5 y}$ *as* $y \to \infty$, *then* $\log S_Y(y) \sim k_6 (\log y)^{k_5 y}$ *as* $y \to \infty$, *provided* $k_5 > 0$ *and* $k_6 > 0$.

.

Proof. Using Lemma A3:

(a) $g(y) = k_2 (\log y)^{k_1}$, so $g'(y) = \dfrac{k_1 k_2}{y} (\log y)^{k_1 - 1}$ and
$\log g'(y) = -\log y + (k_1 - 1) \log(\log y) + \log(k_1 k_2)$.
$\therefore \log g'(y) = o(g(y))$ since $k_1 > 1$. Hence Lemma A3 applies.

(b) $g(y) = k_4 y^{k_3}$, so $g'(y) = k_3 k_4 y^{k_3 - 1}$ and $\log g'(y) = (k_3 - 1) \log y + \log(k_1 k_2)$. Hence Lemma A3 applies.
$\therefore \log g'(y) = o(g(y))$.

(c) $g(y) = k_6 e^{k_5 y}$, so $g'(y) = k_5 k_6 e^{k_5 y}$ and $\log g'(y) = k_5 y + \log(k_5 k_6)$.
$\therefore \log g'(y) = o(g(y))$. Hence Lemma A3 applies.

□

Lemma 17.4 (Lemma A4).
If $\log f_Y(y) \sim -k \log y$ *as* $y \to \infty$,
then $\log S_Y(y) \sim -(k-1) \log y$ *as* $y \to \infty$, *provided* $k > 1$.

Proof. If $\log f_Y(y) \sim -k \log y$ as $y \to \infty$ with $k > 1$, then

$$\lim_{y \to \infty} \left[\frac{\log f_Y(y)}{-k \log y} \right] = 1.$$

$\therefore$ For any fixed ϵ with $0 < \epsilon < (k-1)/k$, there is a value of y sufficiently large that

$$\begin{array}{rcccl}
1 - \epsilon & < & \dfrac{\log f_Y(y)}{-k \log y} & < & 1 + \epsilon \\
-(1+\epsilon) k \log y & < & \log f_Y(y) & < & -(1-\epsilon) k \log y \\
y^{-(1+\epsilon)k} & < & f_Y(y) & < & y^{-(1-\epsilon)k} \\
\int_y^\infty t^{-(1+\epsilon)k} dt & < & \int_y^\infty f_Y(t) dt & < & \int_y^\infty t^{-(1-\epsilon)k} dt \\
\dfrac{1}{k - 1 + \epsilon k} y^{-(k-1+\epsilon k)} & < & S_Y(y) & < & \dfrac{1}{k - 1 - \epsilon k} y^{-(k-1-\epsilon k)}
\end{array}$$

provided $k > 1/(1-\epsilon)$, which holds since $0 < \epsilon < (k-1)/k$. Hence
$-(k - 1 + \epsilon k) \log y - \log(k - 1 + \epsilon k) < \log S_Y(y) < -(k - 1 - \epsilon k) \log y - \log(k - 1 - \epsilon k)$.

But for y sufficiently large, then $k - 1 + \epsilon k > y^{-\epsilon k}$ and $k - 1 - \epsilon k < y^{\epsilon k}$, hence $-\log(k - 1 + \epsilon k) > -\epsilon k \log y$ and $-\log(k - 1 - \epsilon k) < \epsilon k \log y$, hence

$$-(k - 1 + 2\epsilon k) \log y \quad < \quad \log S_Y(y) \quad < \quad -(k - 1 - 2\epsilon k) \log y.$$

Hence for y sufficiently large.

$$1 + \frac{2\epsilon k}{(k-1)} \quad > \quad \frac{\log S_Y(y)}{-(k-1) \log y} \quad > \quad 1 - \frac{2\epsilon k}{(k-1)}. \tag{17.9}$$

Hence for any fixed δ with $0 < \delta < 2$, let $\epsilon = \dfrac{\delta(k-1)}{2k}$ in (17.9), giving

$$1 - \delta \quad > \quad \frac{\log S_Y(y)}{-(k-1)\log y} \quad > \quad 1 + \delta.$$

for y sufficiently large.

$\therefore \lim_{y\to\infty} \left[\dfrac{\log S_Y(y)}{-(k-1)\log y} \right] = 1.$

$\therefore \log S_Y(y) \sim -(k-1)\log y.$

□

17.6 Appendix B: introduction to classical tail theory

A rigorous analysis of classical distribution tail theory is given by Foss et al. [2013], providing the classical approach to classifying distribution tails. Foss et al. [2013] p7-8, p14, p35, provide definitions for a sequence of nested (i.e. subsetted) distribution tail classes. It appears that:
regularly varying $\subset$ subexponential $\subset$ long tailed $\subset$ heavy-tailed,
where $\subset$ means 'a subset of'. Pareto distributions are in the class of regulaly varying distributions. There are distribution tails heavier than the regularly varying type, and there are tails lighter than heavy-tailed called 'light-tailed'.

Foss et al. [2013, p8] state that subexponentially will be 'satisfied by all heavy-tailed distributions likely to be encountered in practice'.

A distribution on $(-\infty, \infty)$ is heavy right-tailed

$$\begin{aligned} \Leftrightarrow \quad & \textstyle\int_{-\infty}^{\infty} e^{\lambda y} f_Y(y)dy = \infty \text{ for all } \lambda > 0, \\ & \text{i.e. the moment generating function } M_Y(\lambda) \text{ is infinite for all } \lambda > 0. \\ \Leftrightarrow \quad & \lim_{y\to\infty} \sup[S_Y(y)e^{\lambda y}] = \infty \text{ for all } \lambda > 0. \end{aligned} \tag{17.10}$$

A distribution on $(-\infty, \infty)$ is light-tailed

$$\begin{aligned} \Leftrightarrow \quad & \textstyle\int_{-\infty}^{\infty} e^{\lambda y} f_Y(y)dy < \infty \text{ for some } \lambda > 0, \\ & \text{i.e. the moment generating function } M_Y(\lambda) \text{ is finite for some } \lambda > 0. \end{aligned} \tag{17.11}$$

A distribution on $(-\infty, \infty)$ is regularly varying at infinity with index $-\alpha < 0$
$\Leftrightarrow S_Y(cy) \sim c^{-\alpha} S_Y(y)$ as $y \to \infty$
$\Leftrightarrow S(y) = y^{-\alpha} l(y)$
for some slowly varying function l, i.e. $\lim\limits_{y\to\infty} \left[\dfrac{l(ay)}{l(y)}\right] = 1.$

If a pdf $f_Y(y)$ is a regularly varying function given, for $\alpha > 0$ by $f_Y(y) = y^{-(\alpha+1)} l(y)$ where $l(y)$ is a slowly varying function, then $\lim\limits_{y\to\infty} \left[\dfrac{y f_Y(y)}{S_Y(y)}\right] = \alpha$ [Mikosch,

1999, p11], Proposition 1.3.2(c), an important result. Hence $S_Y(y) \sim \frac{y f_Y(y)}{\alpha}$ and so $\log S_Y(y) \sim \log y + \log f_Y(y)$, as $y \to \infty$.

An alternative definition of a heavier right tail, (which is not equivalent to the definition given in Section 17.2.1), is given by the following:
Let Y_1 and Y_2 have cdf $F(y)$ and $F_2(y)$ respectively, then
Y_2 has a heavier right tail than Y_1

$\Leftrightarrow$ Y_2 is more van Zwet skew to the right than Y_1 throughout the right tail, denoted $F_1 <_{2R} F_2$, (see Section 14.5),

$\Leftrightarrow$ a QQ plot of Y_2 against Y_1 is strictly convex (but not anywhere linear) throughout the right tail, i.e. a plot of y_{2p} against y_{1p} for $p \in (p_0, 1)$, is strictly convex (but not anywhere linear), for $p \in (p_o, 1)$, where p_0 is a fixed value with $0 < p_0 < 1$, and $y_{jp} = F_j^{-1}(p)$, for $j = 1, 2$.

$\Leftrightarrow \frac{f_1(y_{1p})}{f_2(y_{2p})}$ is increasing with p for $p \in (p_0, 1)$

$\Leftrightarrow \frac{f_1'(y_{1p})}{[f_1(y_{1p})]^2} - \frac{f_2'(y_{2p})}{[f_2(y_{2p})]^2} > 0$ for $p \in (p_0, 1)$,
since $\frac{d}{dp}[\log f(y_p)] = f'(y_p)/[f(y_p)]^2$.

A related alternative classification of distribution right tails was given by Parzen [1979, p16] based on Parzen's right tail index α_o where

$$\alpha_o = -\lim_{p \to 1} \left\{ \frac{(1-p) f_Y'(y_p)}{[f_Y(y_p)]^2} \right\},$$

where a higher value of α_o indicates a heavier right tail.

From Balanda and MacGillivray [1990, p26], if Y_1 and Y_2 have cdf $F_1(y)$ and $F_2(y)$, and indices α_{o1} and α_{o2}, respectively, then $\alpha_{o2} \geq \alpha_{o1} \Leftrightarrow F_1 \leq_{2R} F_2$. (However note $\alpha_{o2} > \alpha_{o1} \not\Leftrightarrow F_1 <_{2R} F_2$.)

17.7 Bibliographic notes

A rigorous analysis of classical distribution tail theory is given by Embrechts et al. [1997], Mikosch [1999], Resnick [1997, 2007] and Foss et al. [2013], providing the classical approach to classifying distribution tails. An alternative classification of distribution right tails was given by Parzen [1979], based on Parzen's right tail index. Schuster [1984] and Alzaid and Al-Osh [1989] further developed Parzen's index.

There are several **R** packages dealing with tails of distributions. Most of them are concerned with estimating the threshold and/or the power law parameters. For example **poweRlaw** [Gillespie, 2015] fits power laws and other heavy-tailed distributions. The package **laeken** [Alfons and Templ, 2013] performs robust Pareto tail modeling. The package **extremefit** [Durrieu et al., 2018] estimates the extreme quantiles and probabilities of rare events by adjusting the tail of the distribution function over a threshold with a Pareto distribution. The package **ercv** [del Castillo et al., 2017] provides methodology for the analysis of extreme values and multiple threshold tests for a generalized Pareto distribution, together with an automatic threshold selection algorithm. The **ecdfHT** package [Nolan, 2016] can be used to plot the ecdf for heavy-tailed data. It splits the data into three different categories and the ecdf in the three categories are calculated differently.

17.8 Exercises

1. (a) $\texttt{NBI}(\mu, \sigma)$ and $\texttt{PIG}(\mu, \sigma)$ have the same mean and the same variance. Show that $\texttt{PIG}(\mu, \sigma)$ has a heavier right tail than $\texttt{NBI}(\mu, \sigma)$.

   (b) $\texttt{NBI}(\mu, \sigma)$ and $\texttt{DEL}(\mu, \sigma_1, \nu)$ where $\sigma_1 = \sigma/(1-\nu)^2$ have the same mean and the same variance. Show that $\texttt{DEL}(\mu, \sigma, \nu)$ has a heavier right tail than $\texttt{NBI}(\mu, \sigma)$.

   [Hint: use their values of k_3 and k_4 from Table 17.8.]

2. Using Tables 17.1, 17.2, 17.3 and 17.5, find a continuous distribution having a right tail with each of the following types:

   (a) Type I with

      (i) $k_1 = 1$ and $k_2 = 2$, (ii) $k_1 = 2$ and $k_2 = 2$, (iii) $k_1 = 3$ and $k_2 = 3$

      (iv) $k_1 = 1$ and $k_2 = 1$
      with asymptotic form $\log S_Y(y) \sim -\log(\log y)$ as $y \to \infty$

      (v) $k_1 = 1$ and $k_2 = 1$, with asymptotic form $\log S_Y(y) \sim -(\log y)^{1/2}$ as $y \to \infty$

   (b) Type II with

      (i) $k_3 = 1$ and $k_4 = 2$, (ii) $k_3 = 0.5$ and $k_4 = 1$, (iii) $k_3 = 2$ and $k_4 = 2$

   (c) Type III with $k_5 = 1$ and $k_6 = 1$

3. Discrete distributions with each of the Types and k values in question 2 can be obtained by discretizing the corresponding continuous distribution. However use Table 17.8 to find the explicit discrete distributions, i.e.

   (a) for which Types and k values in question 2 is there a discrete probability function in Table 17.8?

   (b) for each yes answer in (a), give a discrete distribution from Table 17.8.

4. Continuous distributions on $(-\infty, \infty)$: in each case below, show that the tails of each pdf are of the type specified below in (17.1), with the k values specified below. [Note that, from Corollary A3, this implies that the right tail of the survivor function and the left tail of the cdf have the same types in (17.2) with the same k values specified.]

(a) Show that the normal distribution, $\texttt{NO}(\mu, \sigma)$, has Type II tails with $k_3 = 2$ and $k_4 = 0.5\sigma^{-2}$.

(b) Show that the t family distribution, $\texttt{TF}(\mu, \sigma, \nu)$, has Type I tails, with $k_1 = 1$ and $k_2 = \nu + 1$.

(c) Show that the power exponential type 2 distribution, $\texttt{PE2}(\mu, \sigma, \nu)$, has Type II tails, with $k_3 = \nu$ and $k_4 = \sigma^{-\nu}$.

(d) Show that the Gumbel distribution, $\texttt{GU}(\mu, \sigma)$, has Type III right tail with $k_5 = \sigma^{-1}$ and $k_6 = e^{-\mu/\sigma}$, and a Type II left tail with $k_3 = 1$ and $k_4 = \sigma^{-1}$.

5. Continuous distributions on $(0, \infty)$: in each case below, show that the right tail of the pdf is of the type specified below in (17.1), with the k values specified below. [Note that, from Corollary A3, this imples that the right tail of the survivor function has the same type in (17.2), with the same k values specified.]

(a) Show that the Pareto type 2 distribution, $\texttt{PARETO2o}(\mu, \sigma)$, has a Type I right tail with $k_1 = 1$ and $k_2 = \sigma + 1$.

(b) Show that the log normal distribution, $\texttt{LOGNO}(\mu, \sigma)$, has a Type I right tail with $k_1 = 2$ and $k_2 = 0.5\sigma^{-2}$.

(c) Show that the log t family distribution, $\texttt{logTF}(\mu, \sigma, \nu)$, has a Type I right tail with $k_1 = 1$ and $k_2 = 1$, and $\log S_Y(y) \sim -\nu \log(\log y)$.

6. Discrete distributions: in each case show that the right tail of the probability function is of the type specified in (17.1) with the k values specified below. [Note that, from Corollary A3, this imples that the right tail of the survivor function has the same type in (17.2), with the same k values specified.]

(a) Show that the Poisson distribution, $\texttt{PO}(\mu)$, has a Type IV right tail with $k_7 = 1$ and $k_8 = 1$.

[Hint: Stirling's approximation gives $\Gamma(y+1) \sim (2\pi)^{1/2} y^{y+1/2} e^{-y}$ as $y \to \infty$, Johnson et al. [2005, p7], equation (1.30).]

(b) If Y has a Yule distribution, $\texttt{YULE}(\mu)$, show that

$$f_Y(y) \sim (1 + \mu^{-1})\Gamma(2 + \mu^{-1}) y^{-(2+\mu^{-1})} \quad \text{as } y \to \infty.$$

[Hint: $\Gamma(y+a)/\Gamma(y+b) \sim y^{a-b}$ as $y \to \infty$, Johnson et al. [2005] p8, equation (1.32).]

Hence show that $\texttt{YULE}(\mu)$ has a Type I right tail with $k_1 = 1$ and $k_2 = 2 + \mu^{-1}$.

(c) Show that the Poisson inverse Gaussian distribution, $\texttt{PIG}(\mu, \sigma)$, has a Type II right tail with $k_3 = 1$ and $k_4 = \log[1 + (2\mu\sigma)^{-1}]$.

[Hint: The modified Bessel function of the second kind, $K_\lambda(t)$, has asymptotic form $K_\lambda(t) \sim (2\lambda/t)^\lambda \lambda^{-1/2} e^{-\lambda} (\pi/2)^{1/2}$ as $\lambda \to \infty$, from Jørgensen [1982, p171], equation (A.10), who obtained the result from Ismail [1977].]

7. The generalized extreme value distribution, denoted here by $\texttt{GEV}(\mu, \sigma, \nu)$, has cdf given by

$$F_Y(y) = \exp\left\{ - \left[1 + \frac{\nu(y-\mu)}{\sigma} \right]^{-1/\nu} \right\},$$

where $-\infty < \mu < \infty$, $\sigma > 0$ and $-\infty < \nu < \infty$, where $y > \mu - \sigma/\nu$ if $\nu > 0$, and $y < \mu - \sigma/\nu$ if $\nu < 0$, and $-\infty < y < \infty$ if $\nu = 0$, Johnson et al. [1995, p2-4].

   (a) For $\nu > 0$ in $\texttt{GEV}(\mu, \sigma, \nu)$, the distribution is called Fréchet type distribution.

      (i) Show that $\log S_Y(y) \sim -\nu^{-1} \log y$ as $y \to \infty$, and also that $\log f_Y(y) \sim -(1 + \nu^{-1}) \log y$ as $y \to \infty$.

      [Hint: $1 - e^{-x} \sim x$ as $x \to 0$.]

      (ii) Hence show that for $\nu > 0$, $\texttt{GEV}(\mu, \sigma, \nu)$ has a Type I right tail with $k_1 = 1$ and $k_2 = 1 + \nu^{-1}$.

      (iii) Show that when $\nu > 0$ in $\texttt{GEV}(\mu, \sigma, \nu)$ its distribution is a shifted log reverse Gumbel distribution, by showing that for $\nu > 0$, if $Z \sim \texttt{RG}(\log(\sigma/\nu), \nu)$ then $Y = e^Z + \mu - \sigma/\nu \sim \texttt{GEV}(\mu, \sigma, \nu)$.

   (b) For $\nu < 0$ in $\texttt{GEV}(\mu, \sigma, \nu)$:

      (i) Show that $\log f_Y(y)$ and $\log F_Y(y) \sim -\left(\frac{|\nu||y|}{\sigma}\right)^{1/|\nu|}$ as $y \to -\infty$.

      (ii) Hence show that for $\nu < 0$, $\texttt{GEV}(\mu, \sigma, \nu)$ has a Type II left tail with $k_3 = |\nu|^{-1}$ and $k_4 = (|\nu|/\sigma)^{1/|\nu|}$.

      (iii) Show that when $\nu < 0$ in $\texttt{GEV}(\mu, \sigma, \nu)$ its distribution is a shifted reverse Weibull distribution, or a shifted reverse log Gumbel distribution, by showing that for $\nu < 0$, if $Z \sim \texttt{GU}(\log(\sigma/|\nu|, |\nu|)$, then $e^Z \sim \texttt{WEI}(\sigma/|\nu|, |\nu|^{-1})$ and $Y = \mu - \sigma/\nu - e^Z \sim \texttt{GEV}(\mu, \sigma, \nu)$.

   (c) For $\nu = 0$ in $\texttt{GEV}(\mu, \sigma, \nu)$, $F_Y(y) = \exp\left\{ -\exp\left[-\frac{(y-\mu)}{\sigma} \right] \right\}$, which is the cdf of a reverse Gumbel, $\texttt{RG}(\mu, \sigma)$, distribution. Hence $\texttt{GEV}(\mu, \sigma, 0) \equiv \texttt{RG}(\mu, \sigma)$. Hence use Table 17.1 to find the types of the left and right tails of $\texttt{GEV}(\mu, \sigma, 0)$.

8. This is a practical example of how to use method 3 in a regression situation, as described at the end of Section 17.3.3. We will use the `rent99` data set.

R data file: rent99 in package **gamlss.data** of dimension 3082 × 9
source: Professor Thomas Kneib
variables used in our analysis

- `rent` : the monthly net rent per month (in Euro)
- `rentsqm` : the net rent per month per square meter (in Euro).
- `area` : living area in square meters.
- `yearc` : year of construction.
- `location` : quality of location: a factor indicating whether the location is average location, 1, good location, 2, and top location, 3.
- `bath` : quality of bathroom: a factor indicating whether the bath facilities are standard, 0, or premium, 1.
- `kitchen` : Quality of kitchen: 0 standard 1 premium (factor).
- `cheating` : central heating: a factor 0 without central heating, 1 with central heating.
- `district` : District in Munich.

purpose: to demonstrate the fitting of a tail in a continuous response variable

We will investigate the upper tail of `rent` against `area` alone.

(a) First we fit a 0.80 (denoted α) smooth quantile curve for `rent` against `area` using the **R** package `cobs` with automatic smoothing parameter selection.

```
data(rent99)
plot(rent~area, data=rent99, col=grey(.70),  pch = 15,
     cex = 0.5, xlab="Area", ylab="Rent")
library(cobs) # bring cobs to fit 80% quantile
c2 <- cobs(rent99$area, rent99$rent, lambda=-1, tau=.80)
lines(fitted(c2)[order(rent99$area)]~
    rent99$area[order(rent99$area)],col=gray(.1), lw=2)
```

(b) Select the cases above the 80% quantile curve for rent and plot them:

```
tail <- subset(data.frame(rent99,fv=fitted(c2)),
                fitted(c2)<rent)
dim(rent99)
dim(tail)
plot(rent~area, data=tail, col=grey(.70),  pch = 15,
     cex = 0.5, xlab="Area", ylab="Rent")
lines(tail$fv[order(tail$area)]~tail$area[order(tail$area)],
      col="red", lw=2)
```

(c) Create suitable truncated distributions for `GU`, `WEI3` and `LOGNO`. Note that the point of truncation varies for each observation, so use the argument `varying=TRUE`, and it is the 80% fitted quantile curve `par=tail$fv`. For example:

```
library(gamlss.tr)
gen.trun(par=tail$fv, family="GU", name="tail", type="left",
         varying=TRUE)
```

(d) Fit the three different distributions and select the best according to GAIC.

Note that the number of cycles in the fit, `n.cyc`, has to be increased considerably. Also because of obvious heterogeneity in the data try a model for the scale parameter σ. For example:

```
t1 <- gamlss(rent~pb(area), data=tail, family=GUtail, n.cyc=500)
#now with different scale parameters
t4 <- gamlss(rent~pb(area), sigma.fo=~pb(area), data=tail,
               family=GUtail, n.cyc=500)
```

Compare the models using AIC. Obtain a worm plot for your chosen model.

```
wp(resid=resid(t1), ylim.all=6)
```

(e) Obtain the $\beta = 0.5, 0.9$, and 0.95 quantile curves obtained from 617 observations (out of 3082) using the fitted a truncated Gumbel model.

```
centilesTail(t1, xvar=tail$area, cent=c(50, 90, 95),pch = 15,
      cex = 0.5, col = gray(0.7),  col.centiles=c(gray(.3)),
      lty.centiles = c(1,2,3), lwd.centiles = 1.6)
```

(f) To which quantiles curves (denoted as τ) of the original rent response variable do each of the $\beta = 0.5, 0.9$, and 0.95 quantile curves in (e) correspond? [Note that $\tau = \alpha + \beta(1 - \alpha)$]

(g) Compare the quantile curves in (e) with simple quantile regression fitted curves, for example

```
c99 <- cobs(rent99$area, rent99$rent, lambda=-1,  tau=.99,
              lambda.lo=0.0001)
lines(fitted(c99)[order(rent99$area)]~
      rent99$area[order(rent99$area)], col="green", lw=2)
```

Part III

Reference guide

In the tables in Part III (Chapters 18 to 23), skewness and excess kurtosis refer to moment skewness ($\alpha_1 = \sqrt{\beta_1}$) and moment excess kurtosis ($\alpha_2 = \beta_2 - 3$), respectively. In the tables, a distribution parameter loosely referred to as a 'skewness parameter' or a 'kurtosis parameter' indicates that it affects the skewness or kurtosis, respectively. A parameter is explicitly referred to as a 'primary true skewness parameter' or 'true skewness parameter' if it has been proven to satisfy the definitions 14.2 or 14.3, respectively, in Section 14.5. A parameter is explicitly referred to as a 'primary true kurtosis parameter' or a 'true kurtosis parameter' if it has been proved to satisfy the definitions 15.1 or 15.2, respectively, in Section 15.5.

For the tables of Part III we use the following abbreviations:

pf	probability function, $\mathrm{P}(Y = y)$
pdf	probability density function, $f_Y(y)$
cdf	cumulative distribution function, $F_Y(y)$
inverse cdf	inverse cumulative distribution function, $y_p = F_Y^{-1}(p)$
PGF	probability generating function, $G_Y(t)$
MGF	moment generating function, $M_Y(t)$
par	parameter
comp	component.

The **R** code for the figures in Part III is given in Appendix A.

18

Continuous distributions on $(-\infty, \infty)$

CONTENTS

This chapter gives summary tables and plots for the explicit **gamlss.dist** continuous distributions with range $(-\infty, \infty)$. These are discussed in Chapter 4.

18.1 Location-scale family of distributions

A continuous random variable Y defined on $(-\infty, \infty)$ is said to have a location-scale family of distributions with location shift parameter θ_1 and scaling parameter θ_2 (for fixed values of all other parameters of the distribution) if

$$Z = \frac{Y - \theta_1}{\theta_2}$$

has a cdf which does not depend on θ_1 or θ_2. Hence

$$F_Y(y) = F_Z\left(\frac{y - \theta_1}{\theta_2}\right)$$

and

$$f_Y(y) = \frac{1}{\theta_2} f_Z\left(\frac{y - \theta_1}{\theta_2}\right) ,$$

so $F_Y(y)$ and $\theta_2 f_Y(y)$ only depend on y, θ_1 and θ_2 through the function $z = (y - \theta_1)/\theta_2$. Note $Y = \theta_1 + \theta_2 Z$.

Example: Let Y have a Gumbel distribution, $Y \sim \texttt{GU}(\mu, \sigma)$, then Y has a location-scale family of distributions with location shift parameter μ and scaling parameter σ, since $F_Y(y) = 1 - \exp\left[-\exp\left(\frac{y-\mu}{\sigma}\right)\right]$. Hence $Z = (Y - \mu)/\sigma$ has cdf $F_Z(z) = 1 - \exp\left[-\exp\left(z\right)\right]$ which does not depend on either μ or σ. Note $Z \sim \texttt{GU}(0, 1)$.

All distributions with range $(-\infty, \infty)$ in **gamlss.dist** are location-scale families of distributions with location shift parameter μ and scaling parameter σ, (for fixed values of all other parameters of the distribution), with the exception of $\texttt{NO2}(\mu, \sigma)$, $\texttt{NOF}(\mu, \sigma, \nu)$, and $\texttt{exGAUS}(\mu, \sigma, \nu)$. Hence for these location-scale family distributions, $Y = \mu + \sigma Z$ where the distribution of Z does not depend on μ or σ. Hence, for fixed parameters ν and τ, $\text{Var}(Y)$ is proportional to σ^2 and $\text{SD}(Y)$ is proportional to σ, provided $\text{Var}(Y)$ is finite.

For the location-scale family distributions plotted in Sections 18.3 and 18.4 we fix $\mu = 0$ and $\sigma = 1$, since changing σ from 1 to say σ_1 merely scales the horizontal axis in the figures by the factor σ_1, and scales the vertical axis by the factor $1/\sigma_1$. After this scaling, changing μ from 0 to say μ_1 simply shifts the horizontal axis by μ_1. (The effect of changing μ and σ can be seen, for example, for the $\texttt{NO}(\mu, \sigma)$ distribution in Figure 18.3.)

18.2 Continuous two-parameter distributions on $(-\infty, \infty)$

18.2.1 Gumbel: GU

The pdf of the Gumbel distribution (or reverse extreme value distribution), denoted by $\texttt{GU}(\mu, \sigma)$, is given by

$$f_Y(y \mid \mu, \sigma) = \frac{1}{\sigma} \exp\left[\left(\frac{y-\mu}{\sigma}\right) - \exp\left(\frac{y-\mu}{\sigma}\right)\right] \tag{18.1}$$

for $-\infty < y < \infty$, where $-\infty < \mu < \infty$ and $\sigma > 0$.

Note if $Y \sim \texttt{GU}(\mu, \sigma)$ and $W = -Y$ then $W \sim \texttt{RG}(-\mu, \sigma)$, from which the results in Table 18.1 were obtained. The Gumbel distribution is appropriate for moderately negatively skewed data.

TABLE 18.1: Gumbel distribution.

$\texttt{GU}(\mu, \sigma)$	
Ranges	
Y	$-\infty < y < \infty$
μ	$-\infty < \mu < \infty$, mode, location shift par.
σ	$0 < \sigma < \infty$, scaling parameter
Distribution measures	
mean	$\mu - C\sigma \approx \mu - 0.57722\sigma$
median	$\mu - 0.36651\sigma$
mode	μ
variance	$\pi^2\sigma^2/6 \approx 1.64493\sigma^2$
skewness	-1.13955
excess kurtosis	2.4
MGF	$e^{\mu t}\Gamma(1 + \sigma t)$, for $\sigma\|t\| < 1$
pdf	$\frac{1}{\sigma}\exp\left[\left(\frac{y-\mu}{\sigma}\right) - \exp\left(\frac{y-\mu}{\sigma}\right)\right]$
cdf	$1 - \exp\left[-\exp\left(\frac{y-\mu}{\sigma}\right)\right]$
inverse cdf	$\mu + \sigma\log[-\log(1-p)]$
Reference	Results derived from $Y = -W$ where $W \sim \texttt{RG}(-\mu, \sigma)$
Note	$C \approx 0.57722$ above is Euler's constant

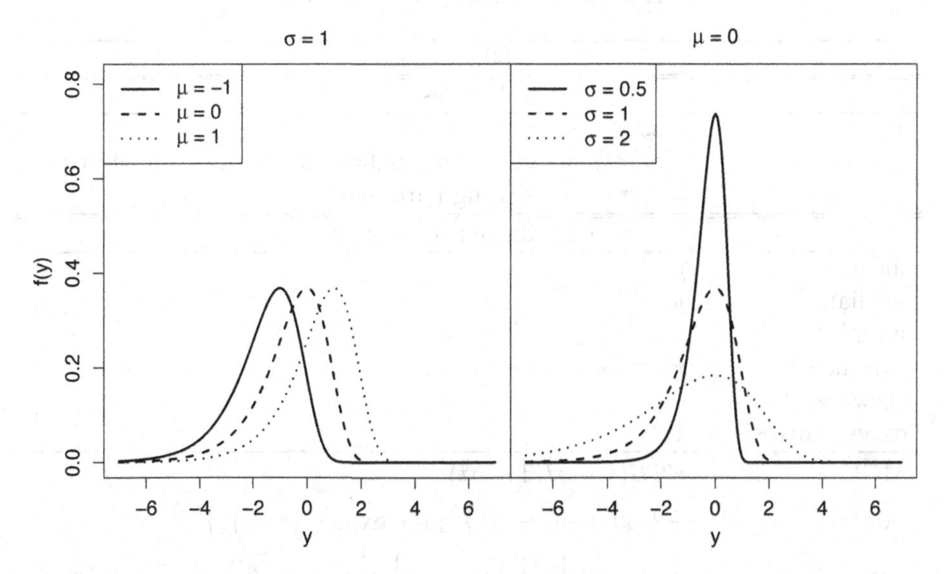

FIGURE 18.1: The Gumbel, $\texttt{GU}(\mu, \sigma)$, distribution with (left) $\sigma = 1$ and $\mu = -1, 0, 1$, and (right) $\mu = 0$ and $\sigma = 0.5, 1, 2$.

18.2.2 Logistic: LO

The pdf of the logistic distribution, denoted by $\texttt{LO}(\mu, \sigma)$, is given by

$$f_Y(y \mid \mu, \sigma) = \frac{1}{\sigma} \left\{ \exp\left[-\left(\frac{y-\mu}{\sigma} \right) \right] \right\} \left\{ 1 + \exp\left[-\left(\frac{y-\mu}{\sigma} \right) \right] \right\}^{-2} \tag{18.2}$$

for $-\infty < y < \infty$, where $-\infty < \mu < \infty$ and $\sigma > 0$. The logistic distribution is symmetric about μ, and is appropriate for a moderately leptokurtic response variable.

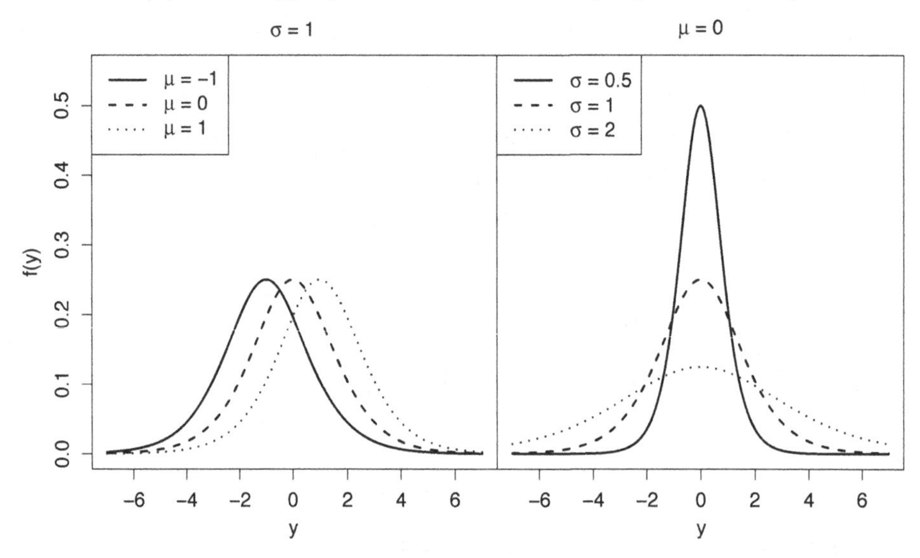

FIGURE 18.2: The logistic, $\texttt{LO}(\mu, \sigma)$, distribution, with (left) $\sigma = 1$ and $\mu = -1, 0, 1$, and (right) $\mu = 0$ and $\sigma = 0.5, 1, 2$.

TABLE 18.2: Logistic distribution.

	$\texttt{LO}(\mu, \sigma)$
	Ranges
Y	$-\infty < y < \infty$
μ	$-\infty < \mu < \infty$, mean, median, mode, location shift par.
σ	$0 < \sigma < \infty$, scaling parameter
	Distribution measures
mean [a]	μ
median	μ
mode	μ
variance [a]	$\pi^2\sigma^2/3$
skewness [a]	0
excess kurtosis [a]	1.2
MGF	$e^{\mu t} B(1 - \sigma t, 1 + \sigma t)$
pdf [a]	$\frac{1}{\sigma} \left\{ \exp\left[-\left(\frac{y-\mu}{\sigma} \right) \right] \right\} \left\{ 1 + \exp\left[-\left(\frac{y-\mu}{\sigma} \right) \right] \right\}^{-2}$
cdf [a]	$\left\{ 1 + \exp\left[-\left(\frac{y-\mu}{\sigma} \right) \right] \right\}^{-1}$
inverse cdf	$\mu + \sigma \log\left(\frac{p}{1-p} \right)$
Reference	[a] Johnson et al. [1995, p115-117] with $\alpha = \mu$ and $\beta = \sigma$

18.2.3 Normal (or Gaussian): NO, NO2

First parameterization, NO

The normal distribution is the default distribution of the `family` argument of the `gamlss()` function. The parameterization used for the normal (or Gaussian) pdf, denoted by $\texttt{NO}(\mu, \sigma)$, is

$$f_Y(y \mid \mu, \sigma) = \frac{1}{\sqrt{2\pi}\sigma} \exp\left[-\frac{(y-\mu)^2}{2\sigma^2}\right] \tag{18.3}$$

for $-\infty < y < \infty$, where $-\infty < \mu < \infty$ and $\sigma > 0$. The $\texttt{NO}(\mu, \sigma)$ distribution is symmetric about μ. Note, in Table 18.3, $\Phi()$ and $\Phi^{-1}()$ are the cdf and inverse cdf of a standard normal, $\texttt{NO}(0, 1)$, distribution.

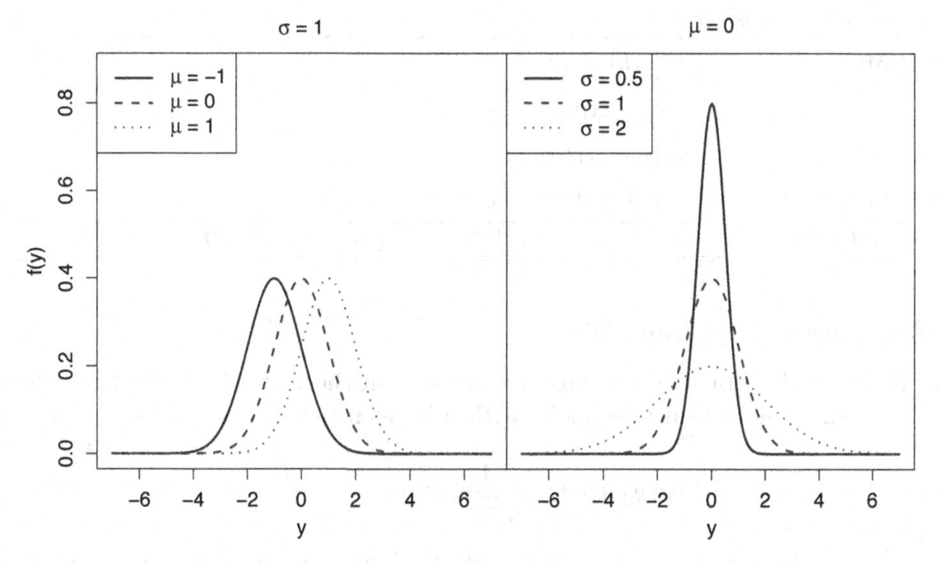

FIGURE 18.3: The normal, $\texttt{NO}(\mu, \sigma)$, distribution, with (left) $\sigma = 1$ and $\mu = -1, 0, 1$ and (right) $\mu = 0$ and $\sigma = 0.5, 1, 2$.

TABLE 18.3: Normal distribution.

NO(μ, σ)	
Ranges	
Y	$-\infty < y < \infty$
μ	$-\infty < \mu < \infty$, mean, median, mode, location shift par.
σ	$0 < \sigma < \infty$, standard deviation, scaling parameter
Distribution measures	
mean	μ
median	μ
mode	μ
variance	σ^2
skewness	0
excess kurtosis	0
MGF	$\exp\left(\mu t + \frac{1}{2}\sigma^2 t^2\right)$
pdf	$\frac{1}{\sqrt{2\pi}\sigma} \exp\left[-\frac{(y-\mu)^2}{2\sigma^2}\right]$
cdf	$\Phi[(y-\mu)/\sigma]$
inverse cdf	$\mu + \sigma z_p$ where $z_p = \Phi^{-1}(p)$
Reference	Johnson et al. [1994] Chapter 13, p80-89.

Second parameterization, NO2

The NO2(μ, σ) distribution is a parameterization of the normal distribution where μ is the mean and σ is the variance of Y, with pdf given by

$$f_Y(y \mid \mu, \sigma) = \frac{1}{\sqrt{2\pi\sigma}} \exp\left[-\frac{(y-\mu)^2}{2\sigma}\right] .$$

The NO2(μ, σ) distribution is symmetric about μ. Note NO2(μ, σ) = NO($\mu, \sigma^{1/2}$).

18.2.4 Reverse Gumbel: RG

The reverse Gumbel distribution, also known as the *type I extreme value distribution*, is a special case of the generalized extreme value distribution, see Johnson et al. [1995] p2, p11-13, and p75-76. The pdf of the reverse Gumbel distribution, denoted by RG(μ, σ), is given by

$$f_Y(y \mid \mu, \sigma) = \frac{1}{\sigma} \exp\left\{-\left(\frac{y-\mu}{\sigma}\right) - \exp\left[-\left(\frac{y-\mu}{\sigma}\right)\right]\right\} \tag{18.4}$$

for $-\infty < y < \infty$, where $-\infty < \mu < \infty$ and $\sigma > 0$.

Note that if $Y \sim$ RG(μ, σ) and $W = -Y$, then $W \sim$ GU($-\mu, \sigma$). The reverse Gumbel distribution is appropriate for moderately positively skewed data.

Since the reverse Gumbel distribution is the type I extreme value distribution, it is the reparameterized limiting distribution of the standardized maximum of a sequence of independent and identically distributed random variables from an 'exponential type distribution', which includes the exponential and gamma.

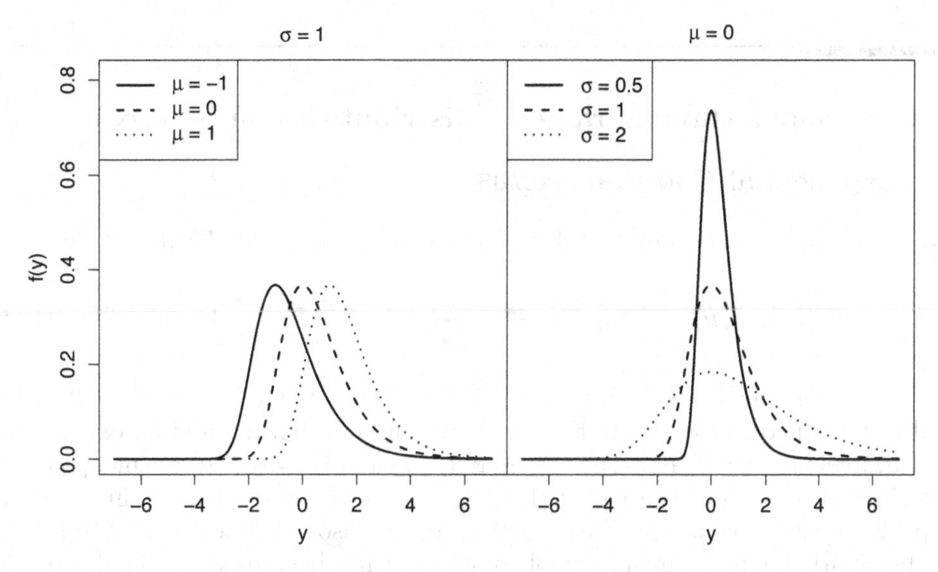

FIGURE 18.4: The reverse Gumbel, $\texttt{RG}(\mu, \sigma)$, distribution with (left) $\sigma = 1$ and $\mu = -1, 0, 1$ and (right) $\mu = 0$ and $\sigma = 0.5, 1, 2$.

TABLE 18.4: Reverse Gumbel distribution.

	$\texttt{RG}(\mu, \sigma)$
	Ranges
Y	$-\infty < y < \infty$
μ	$-\infty < \mu < \infty$, mode, location shift parameter
σ	$0 < \sigma < \infty$, scaling parameter
	Distribution measures
mean	$\mu + C\sigma \approx \mu + 0.57722\sigma$
median	$\mu + 0.36651\sigma$
mode	μ
variance	$\pi^2\sigma^2/6 \approx 1.64493\sigma^2$
skewness	1.13955
excess kurtosis	2.4
MGF	$e^{\mu t}\Gamma(1 - \sigma t)$, for $\sigma\|t\| < 1$
pdf	$\frac{1}{\sigma} \exp\left\{-\left(\frac{y-\mu}{\sigma}\right) - \exp\left[-\left(\frac{y-\mu}{\sigma}\right)\right]\right\}$
cdf	$\exp\left\{-\exp\left[-\left(\frac{y-\mu}{\sigma}\right)\right]\right\}$
inverse cdf	$\mu - \sigma \log[-\log p]$
Reference	Johnson et al. [1995] Chapter 22, p2, p11-13, with $\xi = \mu$ and $\theta = \sigma$.
Note	$C \approx 0.57722$ above is Euler's constant.

18.3 Continuous three-parameter distributions on $(-\infty, \infty)$

18.3.1 Exponential Gaussian: exGAUS

The pdf of the exponential Gaussian distribution, denoted by exGAUS(μ, σ, ν), is

$$f_Y(y \mid \mu, \sigma, \nu) = \frac{1}{\nu} \exp\left[\frac{\mu - y}{\nu} + \frac{\sigma^2}{2\nu^2}\right] \Phi\left(\frac{y-\mu}{\sigma} - \frac{\sigma}{\nu}\right) \tag{18.5}$$

for $-\infty < y < \infty$, where $-\infty < \mu < \infty$, $\sigma > 0$ and $\nu > 0$, and $\Phi(\cdot)$ is the cdf of the standard normal distribution. Note that Y is the sum of a normal and an exponential random variable, i.e. $Y = Y_1 + Y_2$ where $Y_1 \sim$ NO(μ, σ) and $Y_2 \sim$ EXP(ν) are independent. This distribution has also been called the lagged normal distribution, Johnson et al. [1994], p172. See also Davis and Kutner [1976] and Lovison and Schindler [2014]. The exGAUS distribution is popular in psychology where it has been used to model response times.

Note that if $Y \sim$ exGAUS(μ, σ, ν) then $Y_0 = a + bY \sim$ exGAUS$(a + b\mu, b\sigma, b\nu)$. So $Z = Y - \mu \sim$ exGAUS$(0, \sigma, \nu)$. Hence, for fixed σ and ν, the exGAUS(μ, σ, ν) distribution is a location family of distributions with location shift parameter μ. Also exGAUS(μ, σ, ν) is a location-scale family of distributions, but σ is not the scaling parameter. If exGAUS(μ, σ, ν) is reparameterized by setting $\alpha_1 = \sigma + \nu$ and $\alpha_2 = \sigma/\nu$ then the resulting exGAUS2$(\mu, \alpha_1, \alpha_2)$ distribution is a location-scale family of distributions, for fixed α_2, with location shift parameter μ and scaling parameter α_1, since if $Y \sim$ exGAUS2$(\mu, \alpha_1, \alpha_2)$ then $Z = (Y - \mu)/\alpha_1 \sim$ exGAUS2$(0, 1, \alpha_2)$.

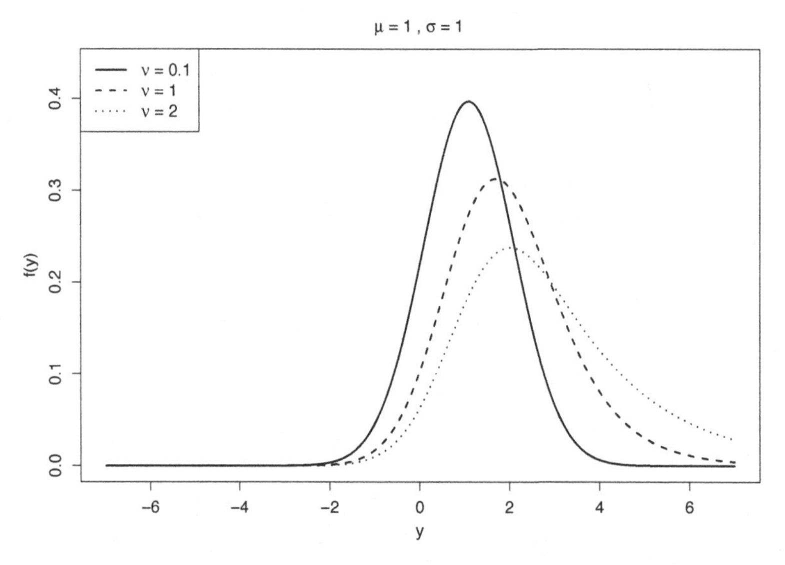

FIGURE 18.5: The exponential Gaussian, exGAUS(μ, σ, ν), distribution, with $\mu = 1$, $\sigma = 1$, and $\nu = 0.1, 1, 2$.

TABLE 18.5: Exponential Gaussian distribution.

exGAUS(μ, σ, ν)	
Ranges	
Y	$-\infty < y < \infty$
μ	$-\infty < \mu < \infty$, mean of normal comp., location shift par.
σ	$0 < \sigma < \infty$, standard deviation of normal component
ν	$0 < \nu < \infty$, mean of exponential component
Distribution measures	
mean [a]	$\mu + \nu$
variance [a]	$\sigma^2 + \nu^2$
skewness [a]	$2\left(1 + \frac{\sigma^2}{\nu^2}\right)^{-1.5}$
excess kurtosis [a]	$6\left(1 + \frac{\sigma^2}{\nu^2}\right)^{-2}$
MGF	$(1 - \nu t)^{-1} \exp(\mu t + \frac{1}{2}\sigma^2 t^2)$
pdf [a]	$\frac{1}{\nu} \exp\left(\frac{\mu - y}{\nu} + \frac{\sigma^2}{2\nu^2}\right) \Phi\left(\frac{y-\mu}{\sigma} - \frac{\sigma}{\nu}\right)$
Reference	[a]Lovison and Schindler [2014]

18.3.2 Normal family of variance-mean relationships: NOF

NOF(μ, σ, ν) defines a normal distribution family with three parameters. The third parameter ν allows the variance of the distribution to be proportional to a power of the mean. The mean of NOF(μ, σ, ν) is μ, while the variance is $\text{Var}(Y) = \sigma^2 \mu^\nu$, so the standard deviation is $\sigma \mu^{\nu/2}$. The pdf of the NOF(μ, σ, ν) distribution is

$$f_Y(y \mid \mu, \sigma, \nu) = \frac{1}{\sqrt{2\pi}\sigma \mu^{\nu/2}} \exp\left[-\frac{(y - \mu)^2}{2\sigma^2 \mu^\nu}\right] \tag{18.6}$$

for $-\infty < y < \infty$, where $\mu > 0$, $\sigma > 0$ and $-\infty < \nu < \infty$. The NOF(μ, σ, ν) distribution is symmetric about μ. Note that if $Y \sim$ NOF(μ, σ, ν), then $Y \sim$ NO$(\mu, \sigma \mu^{\nu/2})$.

NOF(μ, σ, ν) is appropriate for normally distributed regression type models where the variance of the response variable is proportional to a power of the mean. Models of this type are related to the "pseudo likelihood" models of Carroll and Ruppert [1988], but here a proper likelihood is maximized. The ν parameter is usually modeled as a constant, used as a device to model the variance-mean relationship. Note that, due to the high correlation between the σ and ν parameters, the `method=mixed()` and `c.crit=0.0001` method arguments are strongly recommended in the `gamlss()` fitting function to speed the convergence and avoid converging too early. Alternatively a constant ν can be estimated from its profile function, obtained using the **gamlss** function `prof.dev()`.

TABLE 18.6: Normal family (of variance-mean relationships) distribution.

	$\texttt{NOF}(\mu, \sigma, \nu)$
	Ranges
Y	$-\infty < y < \infty$
μ	$0 < \mu < \infty$, mean, median, mode
σ	$0 < \sigma < \infty$
ν	$-\infty < \nu < \infty$
	Distribution measures
mean	μ
median	μ
mode	μ
variance	$\sigma^2 \mu^\nu$
skewness	0
excess kurtosis	0
MGF	$\exp\left(\mu t + \frac{1}{2}\sigma^2 \mu^\nu t^2\right)$
pdf	$\frac{1}{\sqrt{2\pi}\sigma\mu^{\nu/2}} \exp\left[-\frac{(y-\mu)^2}{2\sigma^2\mu^\nu}\right]$
cdf	$\Phi[(y-\mu)/(\sigma\mu^{\nu/2})]$
inverse cdf	$\mu + \sigma\mu^{\nu/2} z_p$ where $z_p = \Phi^{-1}(p)$
Reference	Reparameterize σ to $\sigma\mu^{\nu/2}$ in $\texttt{NO}(\mu, \sigma)$

18.3.3 Power exponential: PE, PE2

First parameterization, $\texttt{PE}(\mu, \sigma, \nu)$

The pdf of the power exponential family distribution, denoted by $\texttt{PE}(\mu, \sigma, \nu)$, is given by

$$f_Y(y \mid \mu, \sigma, \nu) = \frac{\nu \exp\left[-|z|^\nu\right]}{2c\sigma\Gamma\left(\frac{1}{\nu}\right)} \tag{18.7}$$

for $-\infty < y < \infty$, where $-\infty < \mu < \infty$, $\sigma > 0$, $\nu > 0$ and where $z = (y - \mu)/(c\sigma)$ and $c^2 = \Gamma(1/\nu)[\Gamma(3/\nu)]^{-1}$. Note $S = |Z|^\nu \sim \texttt{GA}(\nu^{-1}, \nu^{1/2})$, where $Z = (Y - \mu)/(c\sigma)$. Note $\texttt{PE}(\mu, \sigma, \nu) = \texttt{PE2}(\mu, c\sigma, \nu)$.

In the parameterization (18.7), used by Nelson [1991], $\text{E}(Y) = \mu$ and $\text{Var}(Y) = \sigma^2$. The $\texttt{PE}(\mu, \sigma, \nu)$ distribution includes the Laplace (i.e. two-sided exponential) and normal, $\texttt{NO}(\mu, \sigma)$, distributions as special cases $\nu = 1$ and $\nu = 2$, respectively, while the uniform distribution is the limiting case as $\nu \to \infty$. The $\texttt{PE}(\mu, \sigma, \nu)$ distribution is symmetric about μ, moment leptokurtic for $0 < \nu < 2$, and moment platykurtic for $\nu > 2$.

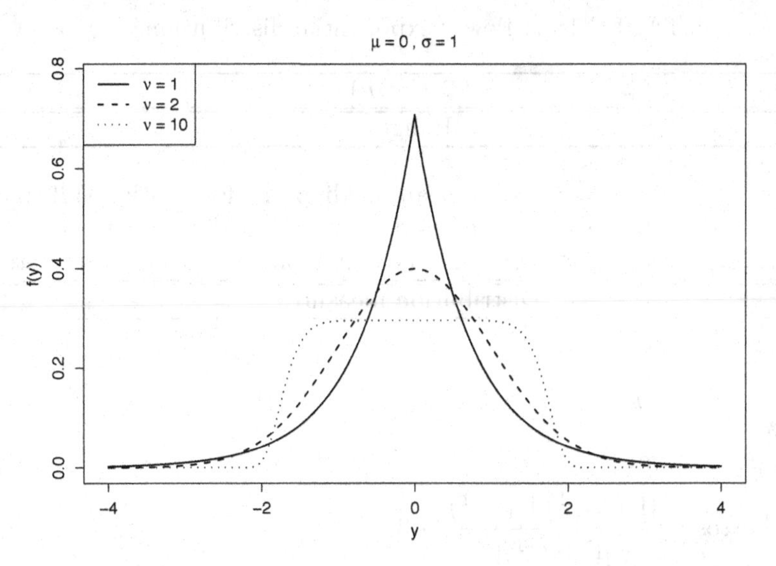

FIGURE 18.6: The power exponential, PE(μ, σ, ν), distribution, with $\mu = 0$, $\sigma = 1$, and $\nu = 1, 2, 10$.

TABLE 18.7: Power exponential distribution.

$\texttt{PE}(\mu, \sigma, \nu)$	
Ranges	
Y	$-\infty < y < \infty$
μ	$-\infty < \mu < \infty$, mean, median, mode, location shift par.
σ	$0 < \sigma < \infty$, standard deviation, scaling parameter
ν	$0 < \nu < \infty$, a primary true moment kurtosis parameter
Distribution measures	
mean	μ
median	μ
mode	μ
variance	σ^2
skewness	0
excess kurtosis	$\frac{\Gamma(5\nu^{-1})\Gamma(\nu^{-1})}{[\Gamma(3\nu^{-1})]^2} - 3$
pdf	$\frac{\nu \exp[-\|z\|^{\nu}]}{2c\sigma\Gamma(\nu^{-1})}$ where $c^2 = \Gamma(\nu^{-1})[\Gamma(3\nu^{-1})]^{-1}$ and $z = (y-\mu)/(c\sigma)$
cdf	$\frac{1}{2}\left[1 + \frac{\gamma(\nu^{-1}, \|z\|^{\nu})}{\Gamma(\nu^{-1})}\text{sign}(y-\mu)\right]$
inverse cdf	$\begin{cases} \mu - c\sigma\left[F_S^{-1}(1-2p)\right]^{1/\nu} & \text{if } p \le 0.5 \\ \mu + c\sigma\left[F_S^{-1}(2p-1)\right]^{1/\nu} & \text{if } p > 0.5 \\ \text{where } S \sim \texttt{GA}(\nu^{-1}, \nu^{1/2}) \end{cases}$
Reference	Reparameterize σ to $c\sigma$ in $\texttt{PE2}(\mu, \sigma, \nu)$
Note	$\gamma(a, x) = \int_0^x t^{a-1}e^{-t}dt$ is the incomplete gamma function

Second parameterization, PE2

An alternative parameterization, the power exponential type 2 distribution, denoted by $\texttt{PE2}(\mu, \sigma, \nu)$, has pdf

$$f_Y(y \mid \mu, \sigma, \nu) = \frac{\nu \exp[-|z|^{\nu}]}{2\sigma\Gamma(\nu^{-1})} \tag{18.8}$$

for $-\infty < y < \infty$, where $-\infty < \mu < \infty$, $\sigma > 0$ and $\nu > 0$ and where $z = (y-\mu)/\sigma$. Note $\texttt{PE2}(\mu, \sigma, \nu) = \texttt{PE}(\mu, \sigma/c, \nu)$.

This is a reparameterization of a version by Subbotin [1923] given in Johnson et al. [1995] Section 24.6, p195-196, equation (24.83).The cdf is given by

$$F_Y(y) = \frac{1}{2}[1 + F_S(s)\text{sign}(z)]$$

where $S = |Z|^{\nu}$ has a gamma, $\texttt{GA}(\nu^{-1}, \nu^{1/2})$, distribution with pdf $f_S(s) = s^{(1/\nu)-1}\exp(-s)/\Gamma(\nu^{-1})$ and $Z = (Y-\mu)/\sigma$, from which the cdf and inverse cdf results in Table 18.8 was obtained.

The $\texttt{PE2}(\mu, \sigma, \nu)$ distribution is symmetric about μ. It includes the (reparameterized) normal and Laplace (i.e. two-sided exponential) distributions as special cases when $\nu = 2$ and $\nu = 1$, respectively, and the uniform distribution as a limiting case as $\nu \to \infty$. The $\texttt{PE2}(\mu, \sigma, \nu)$ distribution is moment leptokurtic for $0 < \nu < 2$ and moment platykurtic for $\nu > 2$.

TABLE 18.8: Second parameterization of power exponential distribution.

	$\texttt{PE2}(\mu, \sigma, \nu)$
	Ranges
Y	$-\infty < y < \infty$
μ	$-\infty < \mu < \infty$, mean, median, mode, location shift par.
σ	$0 < \sigma < \infty$, scaling parameter
ν	$0 < \nu < \infty$, a primary true moment kurtosis parameter
	Distribution measures
mean	μ
median	μ
mode	μ
variance [a]	$\frac{\sigma^2}{c^2}$ where $c^2 = \Gamma(\nu^{-1})[\Gamma(3\nu^{-1})]^{-1}$
skewness	0
excess kurtosis [a]	$\frac{\Gamma(5\nu^{-1})\,\Gamma(\nu^{-1})}{[\Gamma(3\nu^{-1})]^2} - 3$
pdf [a]	$\frac{\nu \exp[-\|z\|^{\nu}]}{2\sigma\Gamma(\nu^{-1})}$
cdf	$\frac{1}{2}\left[1 + \frac{\gamma(\nu^{-1}, \|z\|^{\nu})}{\Gamma(\nu^{-1})} \text{sign}(y-\mu)\right]$ where $z = \frac{y-\mu}{\sigma}$
inverse cdf	$\begin{cases} \mu - \sigma\left[F_S^{-1}(1-2p)\right]^{1/\nu} & \text{if } p \le 0.5 \\ \mu + \sigma\left[F_S^{-1}(2p-1)\right]^{1/\nu} & \text{if } p > 0.5 \\ \text{where } S \sim \texttt{GA}(\nu^{-1}, \nu^{1/2}) & \end{cases}$
Reference	[a] Johnson et al. [1995] Section 24.6, p195-196, equation (24.83), reparameterized by $\theta = \mu$, $\phi = 2^{-1/\nu}\sigma$ and $\delta = 2/\nu$ and hence $\mu = \theta$, $\sigma = \phi 2^{\delta/2}$ and $\nu = 2/\delta$.

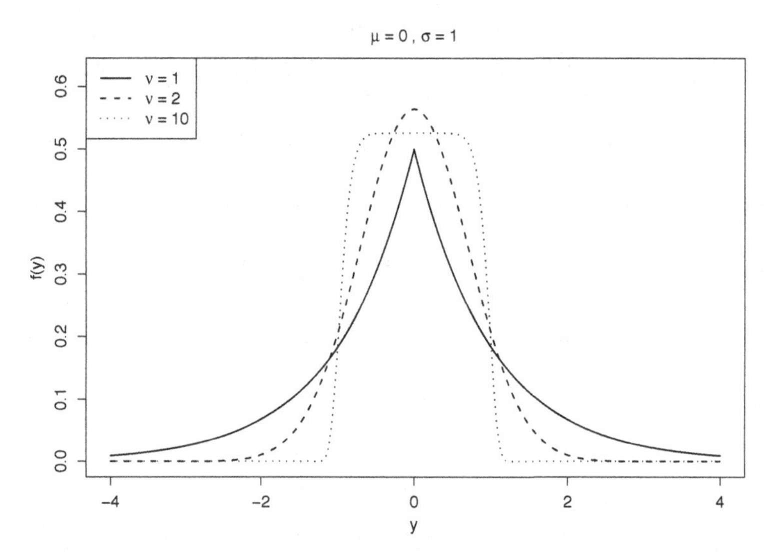

FIGURE 18.7: The power exponential type 2, $\texttt{PE2}(\mu, \sigma, \nu)$, distribution, with $\mu = 0$, $\sigma = 1$, and $\nu = 1, 2, 10$.

18.3.4 Skew normal type 1: SN1

The pdf of a skew normal type 1 distribution, denoted by $\texttt{SN1}(\mu, \sigma, \nu)$, is given by

$$f_Y(y \mid \mu, \sigma, \nu) = \frac{2}{\sigma} \phi(z) \, \Phi(\nu z) \tag{18.9}$$

for $-\infty < y < \infty$, where $-\infty < \mu < \infty$, $\sigma > 0$ and $-\infty < \nu < \infty$, and where $z = (y - \mu)/\sigma$ and $\phi(\cdot)$ and $\Phi(\cdot)$ are the pdf and cdf of a standard normal $\texttt{NO}(0, 1)$ variable, respectively [Azzalini, 1985]. See also Section 13.4. The skew normal type 1 distribution is a special case of the skew exponential power type 1 distribution where $\tau = 2$, i.e. $\texttt{SN1}(\mu, \sigma, \nu) = \texttt{SEP1}(\mu, \sigma, \nu, 2)$.

The $\texttt{SN1}(\mu, \sigma, \nu)$ distribution includes the normal $\texttt{NO}(\mu, \sigma)$ as a special case when $\nu = 0$, the half normal as a limiting case as $\nu \to \infty$ and the reflected half normal (i.e. $Y < \mu$) as a limiting case as $\nu \to -\infty$. The $\texttt{SN1}(\mu, \sigma, \nu)$ distribution is positively moment skewed if $\nu > 0$, and negatively moment skewed if $\nu < 0$.

Note if $Y \sim \texttt{SN1}(\mu, \sigma, \nu)$ then $-Y \sim \texttt{SN1}(-\mu, \sigma, -\nu)$ is a reflection of the distribution of Y about zero. Also $W = (2\mu - Y) \sim \texttt{SN1}(\mu, \sigma, -\nu)$ is a reflection of the distribution of Y about μ.

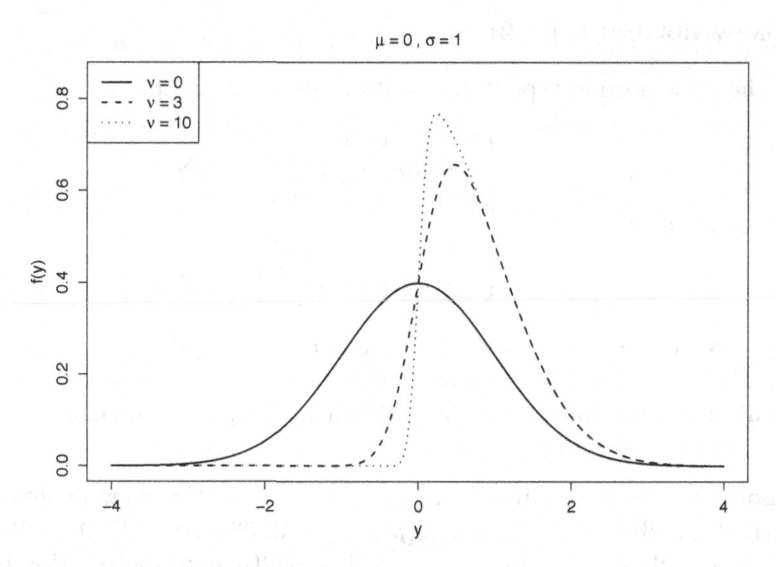

FIGURE 18.8: The skew normal type 1, $\texttt{SN1}(\mu, \sigma, \nu)$, distribution with $\mu = 0$, $\sigma = 1$, and $\nu = 0, 3, 10$. It is symmetric if $\nu = 0$. Changing ν to $-\nu$ reflects the distribution about $y = \mu$, i.e. $y = 0$ above.

TABLE 18.9: Skew normal type 1 distribution.

$\texttt{SN1}(\mu, \sigma, \nu)$	
Ranges	
Y	$-\infty < y < \infty$
μ	$-\infty < \mu < \infty$, location shift parameter
σ	$0 < \sigma < \infty$, scaling parameter
ν	$-\infty < \nu < \infty$, a primary true moment skewness parameter
Distribution measures	
mean	$\mu + \sigma\nu \left[2(1+\nu^2)^{-1}\pi^{-1}\right]^{1/2}$
variance	$\sigma^2 \left[1 - 2\nu^2(1+\nu^2)^{-1}\pi^{-1}\right]$
skewness	$\frac{1}{2}(4-\pi) \left[\frac{\pi}{2}(1+\nu^{-2}) - 1\right]^{-3/2} \text{sign}(\nu)$
excess kurtosis	$2(\pi - 3) \left[\frac{\pi}{2}(1+\nu^{-2}) - 1\right]^{-2}$
MGF	$2\exp\left(\mu t + \frac{1}{2}\sigma^2 t^2\right) \Phi\left(\frac{\sigma\nu t}{\sqrt{1+\nu^2}}\right)$
pdf	$\frac{2}{\sigma}\phi(z)\Phi(\nu z)$ where $z = (y-\mu)/\sigma$
Reference	From Azzalini [1985], p172 and p174, where $\lambda = \nu$ and here $Y = \mu + \sigma Z$

18.3.5 Skew normal type 2: SN2

The pdf of the skew normal type 2 distribution, denoted by SN2(μ, σ, ν), is given by

$$f_Y(y \mid \mu, \sigma \, \nu, \tau) = \begin{cases} \dfrac{c}{\sigma} \exp\left[-\dfrac{1}{2}(\nu z)^2\right] & \text{if } y < \mu \\[2ex] \dfrac{c}{\sigma} \exp\left[-\dfrac{1}{2}\left(\dfrac{z}{\nu}\right)^2\right] & \text{if } y \geq \mu \end{cases} \tag{18.10}$$

for $-\infty < y < \infty$, where $-\infty < \mu < \infty$, $\sigma > 0$ and $\nu > 0$, and where $z = (y-\mu)/\sigma$ and $c = \sqrt{2}\nu/[\sqrt{\pi}(1+\nu^2)]$. This distribution is also called the two-piece normal distribution [Johnson et al., 1994, Section 10.3, p173]. SN2 is a 'scale-spliced' distribution, see Section 13.5.2.

The skew normal type 2 distribution is a special case of the skew exponential power type 3 distribution where $\tau = 2$, i.e. SN2(μ, σ, ν) = SEP3$(\mu, \sigma, \nu, 2)$. Also SN2(μ, σ, ν) is the limiting case of ST3(μ, σ, ν, τ) as $\tau \to \infty$. The SN2(μ, σ, ν) distribution includes the normal NO(μ, σ) as a special case when $\nu = 1$, and is positively moment skewed if $\nu > 1$ and negatively moment skewed if $0 < \nu < 1$.

Note if $Y \sim$ SN2(μ, σ, ν) then $-Y \sim$ SN2$(-\mu, \sigma, \nu^{-1})$ is a reflection of the distribution of Y about zero. Also $W = (2\mu - Y) \sim$ SN2(μ, σ, ν^{-1}) is a reflection of the distribution of Y about μ.

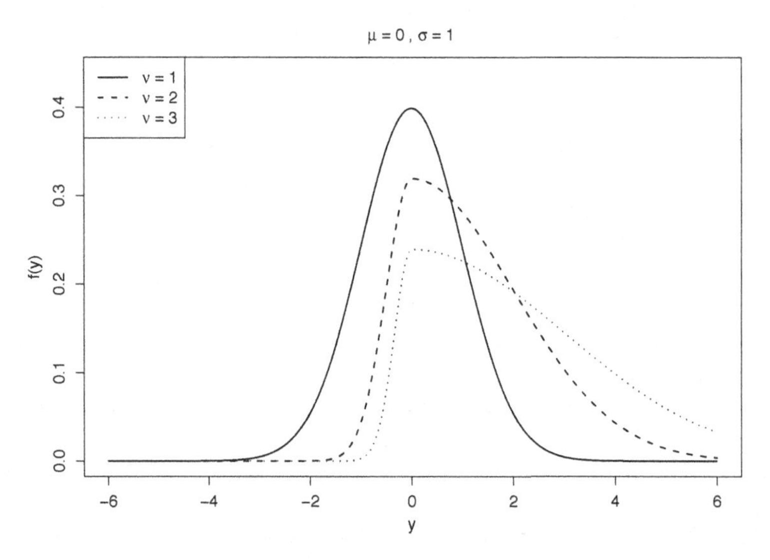

FIGURE 18.9: The skew normal type 2, SN2(μ, σ, ν), distribution with $\mu = 0$, $\sigma = 1$, and $\nu = 1, 2, 3$. It is symmetric if $\nu = 1$. Changing ν to $1/\nu$ reflects the distribution about $y = \mu$, i.e. $y = 0$ above.

TABLE 18.10: Skew normal type 2 distribution.

$\texttt{SN2}(\mu, \sigma, \nu)$	
Ranges	
Y	$-\infty < y < \infty$
μ	$-\infty < \mu < \infty$, mode, location shift parameter
σ	$0 < \sigma < \infty$, scaling parameter
ν	$0 < \nu < \infty$, skewness parameter
Distribution measures	
mean [a]	$\mu + \sigma \mathrm{E}(Z) = \mu + \sigma \frac{\sqrt{2}}{\sqrt{\pi}}(\nu - \nu^{-1})$ where $Z = (Y - \mu)/\sigma$
median [a2]	$\begin{cases} \mu + \frac{\sigma}{\nu}\Phi^{-1}\left(\frac{1+\nu^2}{2}\right) & \text{if } \nu \le 1 \\ \mu + \sigma\nu\Phi^{-1}\left(\frac{3\nu^2 - 1}{4\nu^2}\right) & \text{if } \nu > 1 \end{cases}$
mode	μ
variance [a]	$\sigma^2 \mathrm{Var}(Z) = \sigma^2\{(\nu^2 + \nu^{-2} - 1) - [\mathrm{E}(Z)]^2\}$
skewness [a]	$\begin{cases} \mu_{3Y}/[\mathrm{Var}(Y)]^{1.5} \text{ where} \\ \mu_{3Y} = \sigma^3 \mu_{3Z} = \sigma^3\{\mu'_{3Z} - 3\mathrm{Var}(Z)\mathrm{E}(Z) - [\mathrm{E}(Z)]^3\} \\ \mu'_{3Z} = \mathrm{E}(Z^3) = \frac{2\sqrt{2}(\nu^4 - \nu^{-4})}{\sqrt{\pi}(\nu + \nu^{-1})} \end{cases}$
excess kurtosis [a]	$\begin{cases} \mu_{4Y}/[\mathrm{Var}(Y)]^2 - 3 \text{ where} \\ \mu_{4Y} = \sigma^4 \mu_{4Z} = \sigma^4\{\mu'_{4Z} - 4\mu'_{3Z}\mathrm{E}(Z) + 6\mathrm{Var}(Z)[\mathrm{E}(Z)]^2 \\ \quad +3[E(Z)]^4\} \\ \mu'_{4Z} = \mathrm{E}(Z^4) = 3(\nu^5 + \nu^{-5})/(\nu + \nu^{-1}) \end{cases}$
pdf [a]	$\begin{cases} \frac{c}{\sigma}\exp\left[-\frac{1}{2}(\nu z)^2\right] & \text{if } y < \mu \\ \frac{c}{\sigma}\exp\left[-\frac{1}{2}\left(\frac{z}{\nu}\right)^2\right] & \text{if } y \ge \mu \\ \text{where } z = (y - \mu)/\sigma \text{ and} & c = \frac{\sqrt{2}\nu}{\sqrt{\pi}(1 + \nu^2)} \end{cases}$
cdf [a2]	$\begin{cases} \frac{2\Phi\left[\nu(y - \mu)/\sigma\right]}{(1 + \nu^2)} & \text{if } y < \mu \\ \frac{1}{(1 + \nu^2)}\left\{1 - \nu^2 + 2\nu^2\Phi\left[(y - \mu)/(\sigma\nu)\right]\right\} & \text{if } y \ge \mu \end{cases}$
inverse cdf [a2]	$\begin{cases} \mu + \frac{\sigma}{\nu}\Phi^{-1}\left[\frac{p(1 + \nu^2)}{2}\right] & \text{if } p \le (1 + \nu^2)^{-1} \\ \mu + \sigma\nu\Phi^{-1}\left[\frac{p(1 + \nu^2) - 1 + \nu^2}{2\nu^2}\right] & \text{if } p > (1 + \nu^2)^{-1} \end{cases}$
Note	[a] Set $\tau = 2$ in $\texttt{SEP3}(\mu, \sigma, \nu, \tau)$ [a2] Set $\tau \to \infty$ in $\texttt{ST3}(\mu, \sigma, \nu, \tau)$

18.3.6 t family: TF

The pdf of the t family distribution, denoted by $\texttt{TF}(\mu, \sigma, \nu)$, is given by

$$f_Y(y \mid \mu, \sigma, \nu) = \frac{1}{\sigma B(1/2, \nu/2)\, \nu^{1/2}} \left[1 + \frac{(y-\mu)^2}{\sigma^2 \nu}\right]^{-(\nu+1)/2} \tag{18.11}$$

for $-\infty < y < \infty$, where $-\infty < \mu < \infty$, $\sigma > 0$ and $\nu > 0$, where $B(a,b) = \Gamma(a)\Gamma(b)/\Gamma(a+b)$ is the beta function. Note that $T = (Y - \mu)/\sigma$ has a standard t distribution with ν degrees of freedom, i.e. $T \sim \texttt{TF}(0, 1, \nu) = t_\nu$, with pdf given by Johnson et al. [1995], p363, equation (28.2). Note if $T \sim \texttt{TF}(0, 1, \nu) = t_\nu$ then $W = \nu/(\nu + T^2) \sim \texttt{BEo}(\nu/2, 1/2)$ and $1 - W = T^2/(\nu + T^2) \sim \texttt{BEo}(1/2, \nu/2)$.

The $\texttt{TF}(\mu, \sigma, \nu)$ distribution is symmetric about μ. It is suitable for modeling leptokurtic data, since it has higher kurtosis than the normal distribution. See Lange et al. [1989] for modeling a response variable using the $\texttt{TF}(\mu, \sigma, \nu)$ distribution, including parameter estimation.

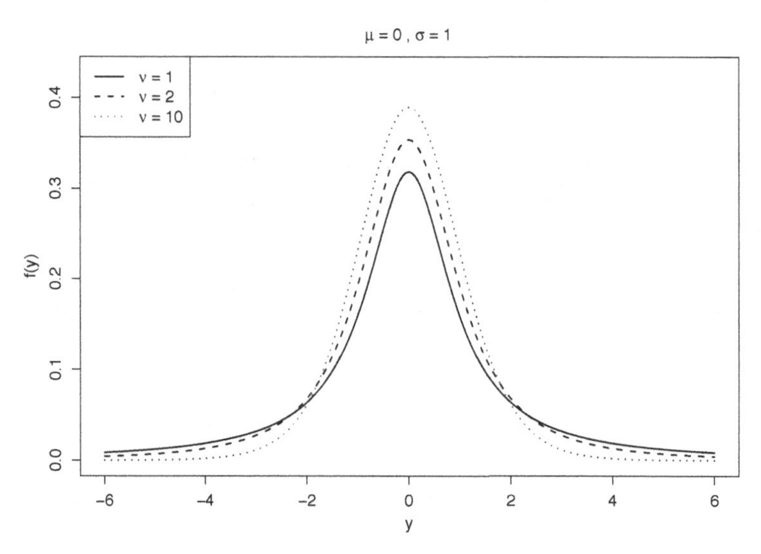

FIGURE 18.10: The t family, $\texttt{TF}(\mu, \sigma, \nu)$, distribution, with $\mu = 0$, $\sigma = 1$, and $\nu = 1, 2, 10$.

TABLE 18.11: t family distribution.

	$\texttt{TF}(\mu, \sigma, \nu)$
	Ranges
Y	$-\infty < y < \infty$
μ	$-\infty < \mu < \infty$, mean, median, mode, location shift parameter
σ	$0 < \sigma < \infty$, scaling parameter
ν	$\begin{cases} 0 < \nu < \infty, \ \nu \text{ is a primary true moment kurtosis parameter} \\ \text{(for } \nu > 4\text{), and the degrees of freedom parameter} \end{cases}$
	Distribution measures
mean	$\begin{cases} \mu & \text{if } \nu > 1, \\ \text{undefined} & \text{if } \nu \leq 1 \end{cases}$
median	μ
mode	μ
variance [a]	$\begin{cases} \dfrac{\sigma^2\nu}{v-2} & \text{if } \nu > 2 \\ \infty & \text{if } 1 < \nu \leq 2 \\ \text{undefined} & \text{if } \nu \leq 1 \end{cases}$
skewness	$\begin{cases} 0 & \text{if } \nu > 3 \\ \text{undefined} & \text{if } \nu \leq 3 \end{cases}$
excess kurtosis [a]	$\begin{cases} \dfrac{6}{v-4} & \text{if } \nu > 4 \\ \infty & \text{if } 2 < \nu \leq 4 \\ \text{undefined} & \text{if } \nu \leq 2 \end{cases}$
pdf [a]	$\dfrac{1}{\sigma B(1/2, \nu/2)\, \nu^{1/2}} \left[1 + \dfrac{(y-\mu)^2}{\sigma^2 \nu}\right]^{-(\nu+1)/2}$
cdf [a]	$0.5 + \dfrac{B(1/2, \nu/2, z^2/[\nu + z^2])}{2B(1/2, \nu/2)} \text{sign}(z)$ where $z = \dfrac{(y-\mu)}{\sigma}$
inverse cdf	$\begin{cases} \mu + \sigma t_{p,\nu} \text{ where } t_{p,\nu} = F_T^{-1}(p) \text{ is the } p \text{ quantile or} \\ 100p \text{ centile valueof } T \sim t_\nu, \text{ i.e. } \text{P}(T < t_{p,\nu}) = p \end{cases}$
Reference	[a] See Johnson et al. [1995], Section 28.2, p363-365.
Note	$B(a, b, x) = \int_o^x t^{a-1}(1-t)^{b-1} dt$ is the incomplete beta function.

18.3.7 t family type 2: TF2

The pdf of the t family type 2 distribution, denoted by $\texttt{TF2}(\mu, \sigma, \nu)$, is

$$f_Y(y \mid \mu, \sigma, \nu) = \frac{1}{\sigma B(1/2, \nu/2)\, (\nu-2)^{1/2}} \left[1 + \frac{(y-\mu)^2}{\sigma^2(\nu-2)}\right]^{-(\nu+1)/2} \tag{18.12}$$

for $-\infty < y < \infty$, where $-\infty < \mu < \infty$, $\sigma > 0$ and $\nu > 2$, where $B(a, b) = \Gamma(a)\Gamma(b)/\Gamma(a+b)$ is the beta function. Note that since $\nu > 2$, the mean $\text{E}(Y) = \mu$ and variance $\text{Var}(Y) = \sigma^2$ of $Y \sim \texttt{TF2}(\mu, \sigma, \nu)$ are always defined and finite.

The $\texttt{TF2}(\mu, \sigma, \nu)$ distribution is symmetric about μ. It is suitable for modeling leptokurtic data, since it has higher kurtosis than the normal distribution.

TABLE 18.12: t family type 2 distribution.

$\texttt{TF2}(\mu, \sigma, \nu)$	
Ranges	
Y	$-\infty < y < \infty$
μ	$-\infty < \mu < \infty$, mean, median, mode, location shift parameter
σ	$0 < \sigma < \infty$, standard deviation, scaling parameter
ν	$\begin{cases} 2 < \nu < \infty, \ \nu \text{ is a primary true moment kurtosis parameter} \\ (\text{for } \nu > 4), \text{ and the degrees of freedom parameter} \end{cases}$
Distribution measures	
mean	μ
median	μ
mode	μ
variance	σ^2
skewness	$\begin{cases} 0 & \text{if } \nu > 3 \\ \text{undefined} & \text{if } 2 < \nu \leq 3 \end{cases}$
excess kurtosis	$\begin{cases} \dfrac{6}{v-4} & \text{if } \nu > 4 \\ \infty & \text{if } 2 < \nu \leq 4 \end{cases}$
pdf	$\dfrac{1}{\sigma B\left(1/2, \nu/2\right)(\nu-2)^{1/2}} \left[1 + \dfrac{(y-\mu)^2}{\sigma^2(\nu-2)}\right]^{-(\nu+1)/2}$
cdf	$0.5 + \dfrac{B\left(1/2, \nu/2, z^2/[\nu+z^2]\right)}{2B(1/2, \nu/2)} \text{sign}(z) \text{ where } z = \dfrac{(y-\mu)\nu^{1/2}}{\sigma(\nu-2)^{1/2}}$
inverse cdf	$\mu + \sigma(\nu-2)^{1/2}\nu^{-1/2} t_{p,\nu}$
Note	Reparameterize σ to $\sigma(\nu-2)^{1/2}\nu^{-1/2}$ in $\texttt{TF}(\mu, \sigma, \nu)$.

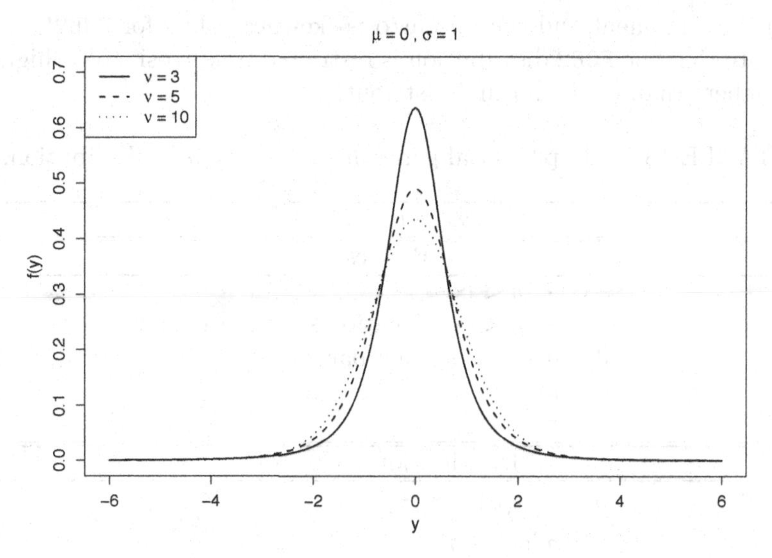

FIGURE 18.11: The t family type 2, TF2(μ, σ, ν), distribution, with $\mu = 0$, $\sigma = 1$, and $\nu = 3, 5, 10$.

18.4 Continuous four-parameter distributions on $(-\infty, \infty)$

18.4.1 Exponential generalized beta type 2: EGB2

The pdf of the exponential generalized beta type 2 distribution, denoted by EGB2(μ, σ, ν, τ), is given by

$$f_Y(y \mid \mu, \sigma, \nu, \tau) = e^{\nu z}\{|\sigma| B(\nu, \tau)[1 + e^z]^{\nu+\tau}\}^{-1} \tag{18.13}$$

for $-\infty < y < \infty$, where $-\infty < \mu < \infty$, $\sigma > 0$, $\nu > 0$ and $\tau > 0$, and where $z = (y - \mu)/\sigma$, [McDonald and Xu, 1995, p141]. Note that McDonald and Xu [1995] appear to allow $\sigma < 0$, however this is unnecessary since EGB2$(\mu, -\sigma, \nu, \tau)$ = EGB2(μ, σ, τ, ν). So we assume $\sigma > 0$ and $|\sigma|$ can be replaced by σ in (18.13).

If $Y \sim$ EGB2(μ, σ, ν, τ), then $-Y \sim$ EGB2$(-\mu, \sigma, \tau, \nu)$ is a reflection of the distribution of Y about zero. Also $W = (2\mu - Y) \sim$ EGB2(μ, σ, τ, ν) is a reflection the distribution of Y about μ, changing the skewness from positive to negative, or vice-versa. The EGB2 distribution is also called the type IV generalized logistic distribution [Johnson et al., 1995, Section 23.10, p142].

If $Y \sim$ EGB2(μ, σ, ν, τ) then $R = \{1+\exp[-(Y-\mu)/\sigma]\}^{-1} \sim$ BEo(ν, τ) from which the cdf in Table 18.13 is obtained. Also $Z = (\tau/\nu)\exp[(Y - \mu)/\sigma] \sim F_{2\nu,2\tau}$, an F distribution, defined in equation (13.36). The EGB2(μ, σ, ν, τ) distribution is symmetric if $\nu = \tau$, and is positively moment skewed if $\nu > \tau$ and negatively moment skewed if $\nu < \tau$. EGB2(μ, σ, ν, τ) is always moment leptokurtic with the normal distribution a limiting case [McDonald, 1996, p 437]. Johnson et al. [1995, p141] indicates that $\sqrt{2/\nu}(Y - \mu)/\sigma$ has a standard normal limiting distribution as $\nu = \tau \to \infty$. Figures 16.2(e), 16.5(d) and 16.8(d) show moment and centile kurtosis-skewness plots for EGB2(μ, σ, ν, τ). The figures indicate that the EGB2 distribution is restricted in kurtosis to be higher (but not extremely higher) than of the normal distribution.

TABLE 18.13: Exponential generalized beta type 2 distribution.

EGB2(μ, σ, ν, τ)	
Ranges	
Y	$-\infty < y < \infty$
μ	$-\infty < \mu < \infty$, location shift parameter
σ	$0 < \sigma < \infty$, scaling parameter
ν	$0 < \nu < \infty$
τ	$0 < \tau < \infty$
Distribution measures	
mean [a2]	$\mu + \sigma[\Psi(\nu) - \Psi(\tau)]$
mode	$\mu + \sigma \log(\nu/\tau)$
variance [a2]	$\sigma^2 \left[\Psi^{(1)}(\nu) + \Psi^{(1)}(\tau)\right]$
skewness [a2]	$\dfrac{\Psi^{(2)}(\nu) - \Psi^{(2)}(\tau)}{[\Psi^{(1)}(\nu) + \Psi^{(1)}(\tau)]^{1.5}}$
excess kurtosis [a2]	$\dfrac{\Psi^{(3)}(\nu) + \Psi^{(3)}(\tau)}{[\Psi^{(1)}(\nu) + \Psi^{(1)}(\tau)]^2}$
MGF [a]	$\dfrac{e^{\mu t} B(\nu + \sigma t, \tau - \sigma t)}{B(\nu, \tau)}$
pdf [a]	$e^{\nu z}\{\lvert\sigma\rvert B(\nu, \tau)[1 + e^z]^{\nu+\tau}\}^{-1}$, where $z = (y - \mu)/\sigma$
cdf	$\dfrac{B(\nu, \tau, c)}{B(\nu, \tau)}$, where $c = \{1 + \exp[-(y - \mu)/\sigma]\}^{-1}$
Reference	[a] McDonald and Xu [1995], p140-141, p150-151, where $\delta = \mu, \sigma = \sigma, p = \nu$ and $q = \tau$. [a2] McDonald [1996], p436-437, where $\delta = \mu, \sigma = \sigma$, $p = \nu$ and $q = \tau$
Note	$\Psi^{(r)}(x) = d^{(r)}\Psi(x)/dx^{(r)}$ is the rth derivative of the psi (or digamma) function $\Psi(x)$

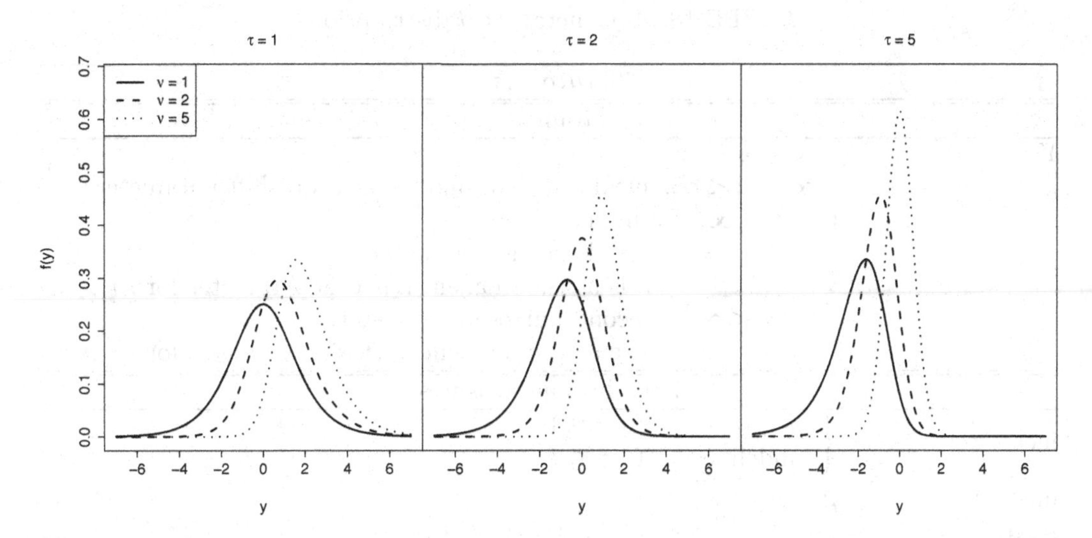

FIGURE 18.12: The exponential generalized beta type 2, $\texttt{EGB2}(\mu, \sigma, \nu, \tau)$, distribution, with $\mu = 0$, $\sigma = 1$, $\nu = 1, 2, 5$, and $\tau = 1, 2, 5$. It is symmetric if $\nu = \tau$. Interchanging ν and τ reflects the distribution about $y = \mu$, i.e. $y = 0$ above.

18.4.2 Generalized t: GT

This pdf of the generalized t distribution, denoted by $\texttt{GT}(\mu, \sigma, \nu, \tau)$, is given by

$$f_Y(y \mid \mu, \sigma\, \nu, \tau) = \tau \left\{ 2\sigma\nu^{1/\tau} B\left(1/\tau, \nu\right) \left[1 + |z|^\tau/\nu\right]^{\nu+(1/\tau)} \right\}^{-1} \tag{18.14}$$

for $-\infty < y < \infty$, where $-\infty < \mu < \infty$, $\sigma > 0$, $\nu > 0$ and $\tau > 0$, and where $z = (y - \mu)/\sigma$, [McDonald and Newey [1988] p430, equation (2.1), where $q = \nu$ and $p = \tau$]. See also Butler et al. [1990] and McDonald [1991]. Note that the t family distribution, $\texttt{TF}(\mu, \sigma, \nu)$, is a special case of the generalized t distribution given by $\texttt{GT}(\mu, \sqrt{2}\sigma, \nu/2, 2)$. The power exponential distribution $\texttt{PE2}(\mu, \sigma, \tau)$ is a limiting distribution of $\texttt{GT}(\mu, \sigma, \nu, \tau)$ as $\nu \to \infty$.

The $\texttt{GT}(\mu, \sigma, \nu, \tau)$ distribution is symmetric about μ and can be moment leptokurtic or platykurtic. It has two kurtosis parameters ν and τ. Parameter τ affects the peakedness around $y = \mu$, with $\tau \leq 1$ resulting in a spike in the pdf at $y = \mu$, and $\tau > 1$ resulting in a flat (i.e. zero derivative) pdf at $y = \mu$. For fixed τ, decreasing ν increases the tail heaviness (and vice-versa), since the pdf of $\texttt{GT}(\mu, \sigma, \nu, \tau)$ has order $O(|y|^{-\nu\tau-1})$ as $|y| \to \infty$, the same order as a t distribution with $\nu\tau$ degrees of freedom, and so the tails are heavier as $\nu\tau$ decreases and are always heavier than the normal distribution.

TABLE 18.14: Generalized t distribution.

$\texttt{GT}(\mu, \sigma, \nu, \tau)$	
Ranges	
Y	$-\infty < y < \infty$
μ	$-\infty < \mu < \infty$, mean, median, mode, location shift parameter
σ	$0 < \sigma < \infty$, scaling parameter
ν	$0 < \nu < \infty$, first kurtosis parameter, ν is a true moment kurtosis parameter for $\nu\tau > 4$
τ	$0 < \tau < \infty$, second kurtosis parameter, τ is a true moment kurtosis parameter for $\nu\tau > 4$
Distribution measures	
mean	$\begin{cases} \mu & \text{if } \nu\tau > 1 \\ \text{undefined} & \text{if } \nu\tau \le 1 \end{cases}$
median	μ
mode	μ
variance [a2]	$\begin{cases} \dfrac{\sigma^2 \nu^{2/\tau} B(3\tau^{-1}, \nu - 2\tau^{-1})}{B(\tau^{-1}, \nu)} & \text{if } \nu\tau > 2 \\ \infty & \text{if } 1 < \nu\tau \le 2 \\ \text{undefined} & \text{if } \nu\tau \le 1 \end{cases}$
skewness	$\begin{cases} 0 & \text{if } \nu\tau > 3 \\ \text{undefined} & \text{if } \nu\tau \le 3 \end{cases}$
excess kurtosis [a2]	$\begin{cases} \dfrac{B(5\tau^{-1}, \nu - 4\tau^{-1}) B(\tau^{-1}, \nu)}{[B(3\tau^{-1}, \nu - 2\tau^{-1})]^2} - 3 & \text{if } \nu\tau > 4 \\ \infty & \text{if } 2 < \nu\tau \le 4 \\ \text{undefined} & \text{if } \nu\tau \le 2 \end{cases}$
pdf [a]	$\tau \left\{ 2\sigma\nu^{1/\tau} B(1/\tau, \nu) \left[1 + \lvert z\rvert^{\tau}/\nu\right]^{\nu + (1/\tau)} \right\}^{-1}$, where $z = (y - \mu)/\sigma$
Reference	[a] McDonald and Newey [1988], p430, equation (2.1), where $q = \nu$ and $p = \tau$. [a2] McDonald [1991], p 274, where $q = \nu$ and $p = \tau$.

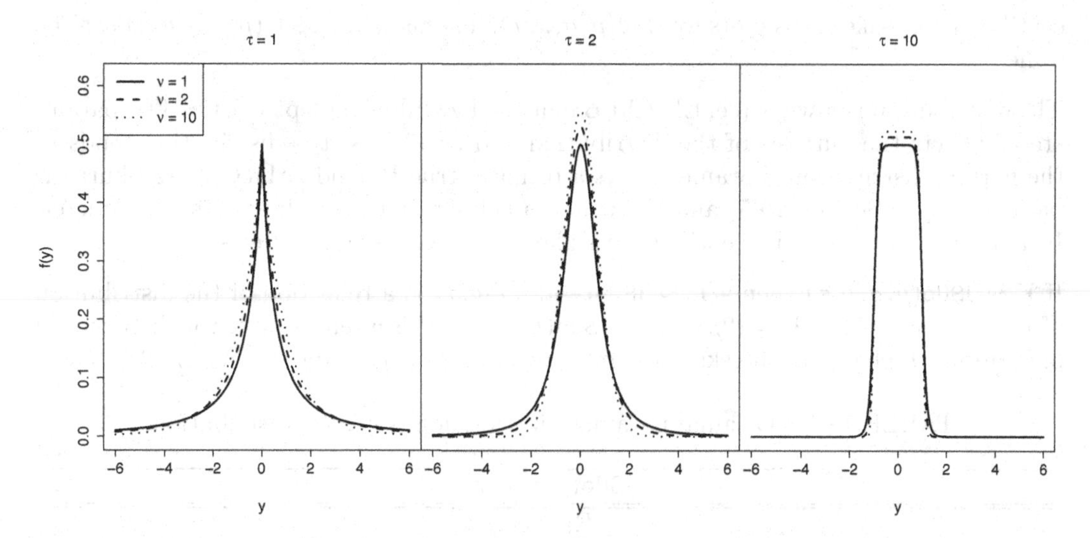

FIGURE 18.13: The generalized t, $\texttt{GT}(\mu, \sigma, \nu, \tau)$, distribution, with $\mu = 0$, $\sigma = 1$, $\nu = 1, 2, 10$, and $\tau = 1, 2, 10$. It is always symmetric.

18.4.3 Johnson's SU: JSUo, JSU

First parameterization, JSUo

This is the original parameterization of the Johnson's S_u distribution [Johnson, 1949] and is denoted by $\texttt{JSUo}(\mu, \sigma, \nu, \tau)$. Its pdf is given by

$$f_Y(y \mid \mu, \sigma\, \nu, \tau) = \frac{\tau}{\sigma(s^2+1)^{1/2}\sqrt{2\pi}} \exp\left[-\frac{1}{2}z^2\right] \tag{18.15}$$

for $-\infty < y < \infty$, where $-\infty < \mu < \infty$, $\sigma > 0$, $-\infty < \nu < \infty$ and $\tau > 0$, and where

$$z = \nu + \tau \sinh^{-1}(s) = \nu + \tau \log\left[s + (s^2+1)^{1/2}\right], \tag{18.16}$$

where $s = (y - \mu)/\sigma$, [Johnson [1949] p162, equation (33) and p152, where $\xi = \mu, \lambda = \sigma, \gamma = \nu, \delta = \tau$ and $x = y, y = s, z = z$]. Hence $s = \sinh\left[(z-\nu)/\tau\right] = \frac{1}{2}\left\{\exp\left[(z-\nu)/\tau\right] - \exp\left[-(z-\nu)/\tau\right]\right\}$ and $y = \mu + \sigma s$. Note that $Z \sim \texttt{NO}(0,1)$, where $Z = \nu + \tau \sinh^{-1}[(Y-\mu)/\sigma]$, from which the results for the cdf, inverse cdf, and median in Table 18.15 are obtained. Also

$$\mathrm{E}\left[\frac{(Y-\mu)^r}{\sigma^r}\right] = \frac{1}{2^r}\sum_{j=0}^{r}(-1)^{r-j}C_j^r \exp\left[\frac{1}{2\tau^2}(r-2j)^2 + \frac{\nu}{\tau}(r-2j)\right] \tag{18.17}$$

for $r = 1, 2, 3, ..,$ where $C_j^r = r!/[j!(r-j)!]$.

The parameter ν affects the skewness of the distribution, which is symmetric if $\nu = 0$, positively (moment, centile, and van Zwet) skew if $\nu < 0$, and negatively skew if $\nu > 0$. Parameter ν is a true van Zwet skewness parameter (see Section 14.5) and hence a true centile skewness parameter and a true moment skewness parameter. Decreasing ν (for fixed τ) increases the moment and centile skewness. Chapter 16 shows moment and centile kurtosis-skewness plots for JSU(μ, σ, ν, τ). The plots for JSUo($\mu, \sigma, -\nu, \tau$) are the same.

The distribution is always (centile and Balanda-MacGillivray) leptokurtic. The parameter τ affects the kurtosis of the distribution and as $\tau \to \infty$ the distribution tends to the normal distribution. Parameter τ is a primary true Balanda-MacGillivray kurtosis parameter (see Section 15.5) and hence it is a primary true centile kurtosis parameter. Increasing τ decreases the centile (and Balanda-MacGillivray) kurtosis.

If $Y \sim$ JSUo(μ, σ, ν, τ) then $-Y \sim$ JSUo($-\mu, \sigma, -\nu, \tau$) is a reflection of the distribution of Y about zero. Also $W = (2\mu - Y) \sim$ JSUo($\mu, \sigma, -\nu, \tau$) is a reflection of the distribution of Y about μ, changing the skewness from positive to negative (or vice-versa).

TABLE 18.15: Original parameterization Johnson's S_u distribution.

JSUo(μ, σ, ν, τ)	
Ranges	
Y	$-\infty < y < \infty$
μ	$-\infty < \mu < \infty$, location shift parameter
σ	$0 < \sigma < \infty$, scaling parameter
ν	$-\infty < \nu < \infty$, true van Zwet skewness parameter
τ	$0 < \tau < \infty$, primary true Balanda-MacGillivray kurtosis parameter
Distribution measures	
mean [a]	$\mu - \sigma\omega^{1/2}\sinh(\nu/\tau)$, where $w = \exp(1/\tau^2)$
median	$\mu - \sigma\sinh(\nu/\tau)$
variance [a]	$\frac{1}{2}\sigma^2(\omega-1)[\omega\cosh(2\nu/\tau)+1]$
skewness [a]	$\begin{cases} \mu_3/[\text{Var}(Y)]^{1.5} \text{ where} \\ \mu_3 = -\frac{1}{4}\sigma^3\omega^{1/2}(\omega-1)^2[\omega(\omega+2)\sinh(3\nu/\tau)+3\sinh(\nu/\tau)] \end{cases}$
excess kurtosis [a]	$\begin{cases} \mu_4/[\text{Var}(Y)]^2 - 3 \text{ where} \\ \mu_4 = \frac{1}{8}\sigma^4(\omega-1)^2[\omega^2(\omega^4+2\omega^3+3\omega^2-3)\cosh(4\nu/\tau) \\ +4\omega^2(\omega+2)\cosh(2\nu/\tau)+3(2\omega+1)] \end{cases}$
cdf	$\Phi(\nu + \tau\sinh^{-1}[(y-\mu)/\sigma])$
inverse cdf	$\mu + \sigma\sinh[(z_p - \nu)/\tau]$ where $z_p = \Phi^{-1}(p)$
Reference	[a] Johnson [1949], p163, equation (37), p152 and p162 where $\xi = \mu, \lambda = \sigma, \gamma = \nu, \delta = \tau$ and $x = y, y = s, z = z$.

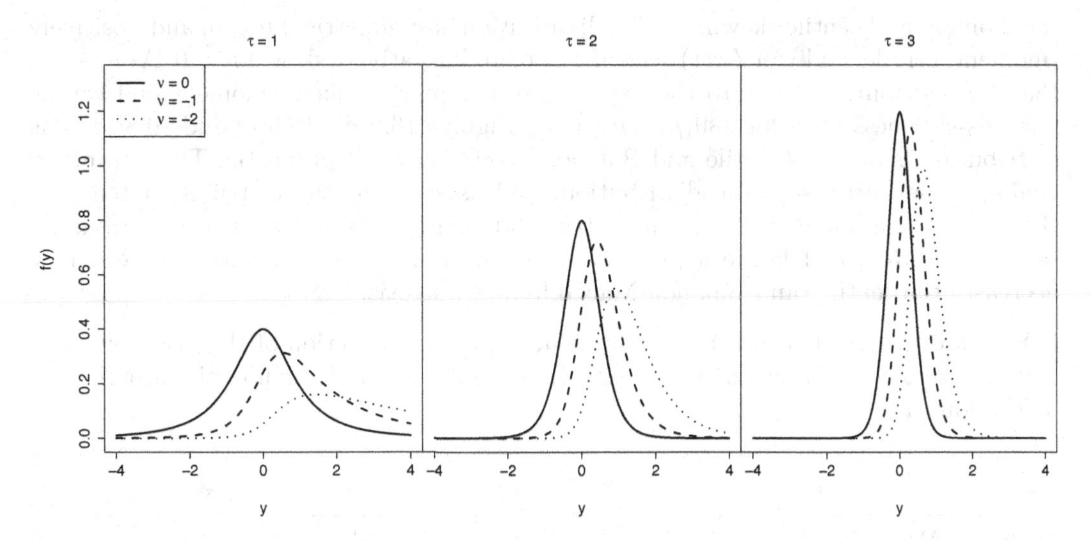

FIGURE 18.14: The original Johnson's S_u, $\texttt{JSUo}(\mu, \sigma, \nu, \tau)$, distribution, with $\mu = 0$, $\sigma = 1$, $\nu = 0, -1, -2$, and $\tau = 1, 2, 3$. It is symmetric if $\nu = 0$. Changing ν to $-\nu$ reflects the distribution about $y = \mu$, i.e. $y = 0$ above.

Second parameterization, JSU

This is a reparameterization of the original Johnson's S_u distribution [Johnson, 1949], so that parameters μ and σ are the mean and the standard deviation of the distribution. The $\texttt{JSU}(\mu, \sigma, \nu, \tau)$ is obtained by reparameterizing $\texttt{JSUo}(\mu_1, \sigma_1, \nu_1, \tau_1)$ to $\mu = \mu_1 - \sigma_1 \omega^{1/2} \sinh(\nu_1/\tau_1)$, $\sigma = \sigma_1/c$, $\nu = -\nu_1$ and $\tau = \tau_1$, where c and ω are defined in equations (18.20) and (18.21) below. Hence $\mu_1 = \mu - c\sigma\omega^{1/2} \sinh(\nu/\tau)$, $\sigma_1 = c\sigma$, $\nu_1 = -\nu$ and $\tau_1 = \tau$.

The pdf of the reparameterized Johnson's S_u, denoted by $\texttt{JSU}(\mu, \sigma, \nu, \tau)$, is given by

$$f_Y(y \mid \mu, \sigma\, \nu, \tau) = \frac{\tau}{c\sigma(s^2+1)^{1/2}\sqrt{2\pi}} \exp\left[-\frac{1}{2}z^2\right] \tag{18.18}$$

for $-\infty < y < \infty$, where $-\infty < \mu < \infty$, $\sigma > 0$, $-\infty < \nu < \infty$, $\tau > 0$, and where

$$z = -\nu + \tau \sinh^{-1}(s) = -\nu + \tau \log\left[s + (s^2+1)^{1/2}\right] \tag{18.19}$$

$$s = \frac{y - \mu + c\sigma w^{1/2} \sinh(\nu/\tau)}{c\sigma}$$

$$c = \left\{\frac{1}{2}(w-1)\left[w \cosh(2\nu/\tau) + 1\right]\right\}^{-1/2} \tag{18.20}$$

$$w = \exp(1/\tau^2)\,. \tag{18.21}$$

Note that $Z \sim \texttt{NO}(0, 1)$, where Z is obtained from (18.19) with y replaced by Y. For $Y \sim \texttt{JSU}(\mu, \sigma, \mu, \tau)$ then $\mathrm{E}(Y) = \mu$ and $\mathrm{Var}(Y) = \sigma^2$.

Parameter ν is a true van Zwet skewness parameter and hence a true centile skewness parameter and a true moment skewness parameter. Increasing ν (for fixed τ) increases the moment and centile skewness. The distribution is symmetric if $\nu = 0$, and positively (moment, centile, and van Zwet) skew if $\nu > 0$, and negatively skew if $\nu < 0$. As $\nu \to \infty$, the JSU distribution tends to the log normal. Chapter 16 shows moment and centile kurtosis-skewness plots for JSU(μ, σ, ν, τ), see Figures 16.2(d), 16.5(c) and 16.8(c). The distribution is always (centile and Balanda-MacGillivray) leptokurtic. The parameter τ affects the kurtosis of the distribution, and as $\tau \to \infty$ the distribution tends to the normal. Parameter τ is a primary true Balanda-MacGillivray kurtosis parameter, see Section 15.5, and hence a primary true centile kurtosis parameter. Increasing τ decreases the centile (and Balanda-MacGillivray) kurtosis.

If $Y \sim$ JSU(μ, σ, ν, τ) then $-Y \sim$ JSU$(-\mu, \sigma, -\nu, \tau)$ is a reflection of the distribution of Y about zero. Also $W = (2\mu - Y) \sim$ JSU$(\mu, \sigma, -\nu, \tau)$ is a reflection of the distribution of Y about μ.

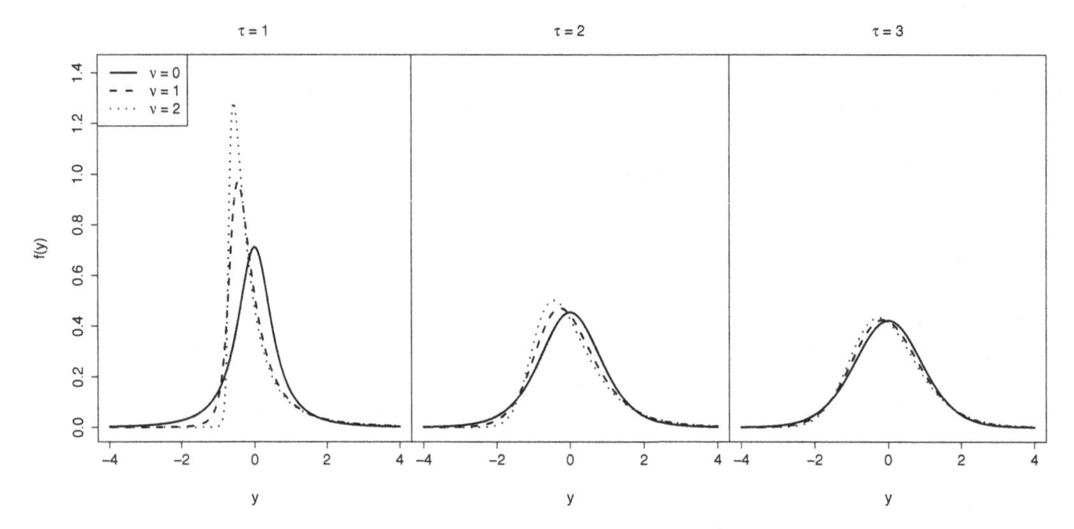

FIGURE 18.15: The reparameterized Johnson S_u, JSU(μ, σ, ν, τ), distribution, with $\mu = 0$, $\sigma = 1$, $\nu = 0, 1, 2$, and $\tau = 1, 2, 3$. It is symmetric if $\nu = 0$. Changing ν to $-\nu$ reflects the distribution about $y = \mu$, i.e. $y = 0$ above.

TABLE 18.16: Second parameterization Johnson's S_u distribution.

JSU(μ, σ, ν, τ)	
Ranges	
Y	$-\infty < y < \infty$
μ	$-\infty < \mu < \infty$, mean, location shift parameter
σ	$0 < \sigma < \infty$, standard deviation, scaling parameter
ν	$-\infty < \nu < \infty$, true van Zwet skewness parameter
τ	$0 < \tau < \infty$, primary true Balanda-MacGillivray kurtosis parameter
Distribution measures	
mean	μ
median	$\begin{cases} \mu_1 + c\sigma \sinh(\nu/\tau) \text{ where} \\ \mu_1 = \mu - c\sigma\omega^{1/2}\sinh(\nu/\tau)\ ,\ \omega = \exp(1/\tau^2) \text{ and } c \text{ is given by (18.20)} \end{cases}$
variance	σ^2
skewness	$\begin{cases} \mu_3/[\text{Var}(Y)]^{1.5} \text{ where} \\ \mu_3 = \frac{1}{4}c^3\sigma^3\omega^{1/2}(\omega-1)^2[\omega(\omega+2)\sinh(3\nu/\tau) + 3\sinh(\nu/\tau)] \end{cases}$
excess kurtosis	$\begin{cases} \mu_4/[\text{Var}(Y)]^2 - 3 \text{ where} \\ \mu_4 = \frac{1}{8}c^4\sigma^4(\omega-1)^2[\omega^2(\omega^4 + 2\omega^3 + 3\omega^2 - 3)\cosh(4\nu/\tau) \\ +4\omega^2(\omega+2)\cosh(2\nu/\tau) + 3(2\omega+1)] \end{cases}$
cdf	$\Phi(-\nu + \tau \sinh^{-1}[(y-\mu_1)/(c\sigma)])$
inverse cdf	$\mu_1 + c\sigma\sinh[(z_p + \nu)/\tau]$ where $z_p = \Phi^{-1}(p)$
Note	Reparameterize $\mu_1, \sigma_1, \nu_1, \tau_1$ in JSUo($\mu_1, \sigma_1, \nu_1, \tau_1$) by letting $\mu_1 = \mu - c\sigma\omega^{1/2}\sinh(\nu/\tau), \sigma_1 = c\sigma, \nu_1 = -\nu$ and $\tau_1 = \tau$ to give JSU(μ, σ, ν, τ).

18.4.4 Normal-exponential-*t*: `NET`

The `NET` is a four-parameter continuous distribution, although in **gamlss** it is used as a two-parameter distribution (μ and σ) with the other two parameters (ν and τ) fixed as constants, by default $\nu = 1.5$ and $\tau = 2$. (These values can be changed by the user.) The `NET` distribution is symmetric about its mean, median, and mode μ. If $Y \sim \texttt{NET}(\mu, \sigma, \nu, \tau)$, then $Z = (Y - \mu)/\sigma$ has a standardized $\texttt{NET}(0, 1, \nu, \tau)$ distribution, with pdf $f_Z(z)$, which is standard normal for $|z| \leq \nu$, exponential with mean $1/\nu$ for $\nu < |z| \leq \tau$, and has t distribution with $(\nu\tau - 1)$ degrees of freedom type tails for $|z| > \tau$, where $z = (y - \mu)/\sigma$. In **gamlss**, μ and σ can be modeled. Parameters ν and τ may be chosen as fixed constants by the user; alternatively estimates of ν and τ can be obtained using the `prof.dev()` function.

The normal-exponential-t distribution, `NET`, was introduced by Rigby and Stasinopoulos [1994] as a robust method of fitting the location and scale parameters of a symmetric distribution as functions of explanatory variables. That is, the distribution is appropriate if the response variable is contaminated and the user wants to robustify the fitting of the mean and variance models. Note that the Huber 'proposal 2' robust estimators of μ and σ [Huber, 1964, 1967], are equivalent to using a normal-exponential distribution for the response variable when fitting μ and a normal-Student-t distribution when fitting σ, and alternating between the fits. (Huber also includes a bias correction for the estimator of σ, which NET does not do.) NET combines these into one distribution. Figure 18.16 (left plot) shows a typical standardized $\texttt{NET}(0, 1, \nu, \tau)$ distribution with $\nu = 1.5$ and $\tau = 2$. The distribution in the interval $(-1.5, 1.5)$ behaves like a normal distribution, in the intervals $(-2, -1.5)$ and $(1.5, 2)$ it behaves like an exponential distribution, while in the intervals $(-\infty, -2)$ and $(2, \infty)$ it behaves like a t distribution. The parameters ν and τ are the breakpoints of the NET distribution and they are *not* modeled in **gamlss**.

The pdf of the $\texttt{NET}(\mu, \sigma, \nu, \tau)$ distribution is given by Rigby and Stasinopoulos [1994] as

$$f_Y(y \mid \mu, \sigma, \nu, \tau) = \frac{c}{\sigma} \begin{cases} \exp\left(-\frac{z^2}{2}\right) & \text{if} \quad |z| \le \nu \\ \exp\left(-\nu|z| + \frac{\nu^2}{2}\right) & \text{if} \quad \nu < |z| \le \tau \\ \exp\left(-\nu\tau \log\left(|z|/\tau\right) - \nu\tau + \frac{\nu^2}{2}\right) & \text{if} \quad |z| > \tau \end{cases} \tag{18.22}$$

for $-\infty < y < \infty$, where $-\infty < \mu < \infty$, $\sigma > 0$, $\nu > 0$, $\tau > \max(\nu, \nu^{-1})$, $z = (y - \mu)/\sigma$ and $c = (c_1 + c_2 + c_3)^{-1}$, where

$$\begin{aligned} c_1 &= \sqrt{2\pi}[2\Phi(\nu) - 1] \\ c_2 &= \frac{2}{\nu} \exp\left\{-\frac{\nu^2}{2}\right\} \\ c_3 &= \frac{2}{\nu(\nu\tau - 1)} \exp\left\{-\nu\tau + \frac{\nu^2}{2}\right\}. \end{aligned} \tag{18.23}$$

The $\texttt{NET}(\mu, \sigma, \nu, \tau)$ is symmetric about μ and is leptokurtic. The pdf of $\texttt{NET}(\mu, \sigma, \nu, \tau)$ has order $O(|y|^{-\nu\tau})$ as $|y| \to \infty$, and so the tails are heavier as $\nu\tau$ decreases. The excess kurtosis is $\gamma_2 = \mu_4/[\text{Var}(Y)]^2 - 3$ where

$$\mu_4 = 2\sigma^4 c \left[\begin{array}{l} 3\sqrt{2\pi}[\Phi(\nu) - 0.5] + \left(\nu + \frac{12}{\nu} + \frac{24}{\nu^3} + \frac{24}{\nu^5}\right) e^{-\nu^2/2} \\ + \left(\frac{\tau^5}{\nu\tau - 5} - \frac{\tau^4}{\nu} - \frac{4\tau^3}{\nu^2} - \frac{12\tau^2}{\nu^3} - \frac{24\tau}{\nu^4} - \frac{24}{\nu^5}\right) e^{-\nu\tau + \nu^2/2} \end{array} \right], \tag{18.24}$$

for $\nu\tau > 5$.

The cdf of $Y \sim \texttt{NET}(\mu, \sigma, \nu, \tau)$ is given by $F_Y(y) = F_Z(z)$ where $Z = (Y - \mu)/\sigma$, $z = (y - \mu)/\sigma$ and

$$F_Z(z) = \begin{cases} \frac{c\tau^{\nu\tau}|z|^{-\nu\tau+1}}{(\nu\tau - 1)} \exp(-\nu\tau + \nu^2/2) & \text{if } z < -\tau \\ \frac{c}{\nu(\nu\tau - 1)} \exp(-\nu\tau + \nu^2/2) + \frac{c}{\nu} \exp(-\nu|z| + \nu^2/2) & \text{if } -\tau \le z < -\nu \\ \frac{c}{\nu(\nu\tau - 1)} \exp(-\nu\tau + \nu^2/2) + \frac{c}{\nu} \exp(-\nu^2/2) & \\ + c\sqrt{2\pi}[\Phi(z) - \Phi(-\nu)] & \text{if } -\nu \le z \le 0 \\ 1 - F_Z(-z) & \text{if } z > 0. \end{cases} \tag{18.25}$$

The inverse cdf of $Y \sim \texttt{NET}(\mu, \sigma, \nu, \tau)$ is given by

$$y_p = \mu + \sigma z_p \text{ where}$$

$$z_p = \begin{cases} -\left(\dfrac{b\nu\tau^{\nu\tau}}{p}\right)^{1/(\nu\tau-1)} & \text{if } p \le \nu\tau b \\ -\dfrac{\nu}{2} + \dfrac{1}{\nu}\log\left[\dfrac{\nu}{c}(p-b)\right] & \text{if } \nu\tau b < p \le b + \dfrac{c}{\nu}\exp\left(-\nu^2/2\right) \\ \Phi^{-1}\left\{\Phi(-\nu) + \dfrac{1}{\sqrt{2\pi}}\left[\dfrac{(p-b)}{c} - \dfrac{1}{\nu}\exp\left(-\nu^2/2\right)\right]\right\} & \\ & \text{if } b + \dfrac{c}{\nu}\exp\left(-\nu^2/2\right) < p \le \dfrac{1}{2} \\ -z_{1-p} & \text{if } p > \dfrac{1}{2}, \end{cases} \tag{18.26}$$

where $b = \dfrac{c}{\nu(\nu\tau - 1)} \exp\left(-\nu\tau + \nu^2/2\right)$.

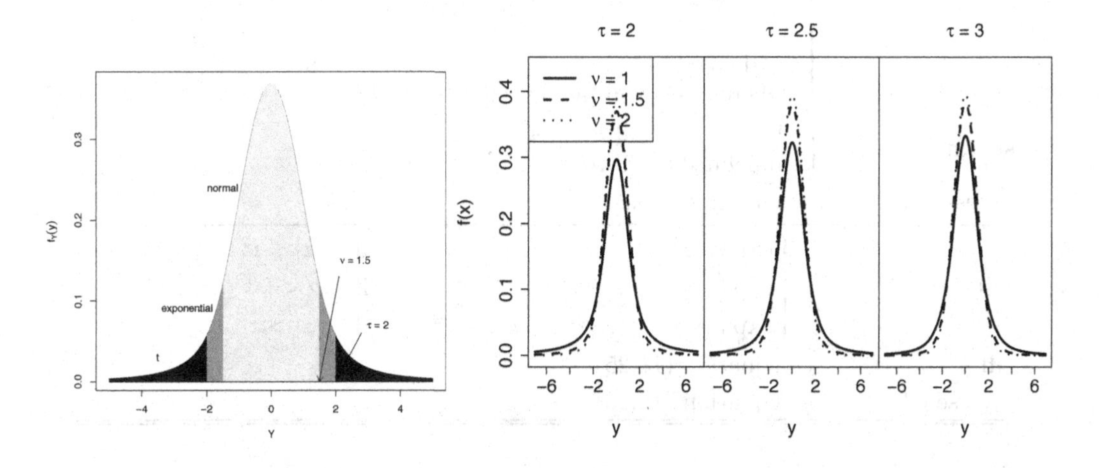

FIGURE 18.16: Left: schematic diagram of the $\texttt{NET}(0, 1, 1.5, 2)$ distribution. Right: the $\texttt{NET}(\mu, \sigma, \nu, \tau)$ distribution with $\mu = 0$, $\sigma = 1$, $\nu = 1, 1.5, 2$, and $\tau = 2, 2.5, 3$. Parameters ν and τ exist to make the distribution fitting robust and cannot be modeled as functions of explanatory variables. The distribution is always symmetric.

TABLE 18.17: Normal-exponential-Student-t distribution.

	$\texttt{NET}(\mu, \sigma, \nu, \tau)$
	Ranges
Y	$-\infty < y < \infty$
μ	$-\infty < \mu < \infty$, mean, median, mode, location shift parameter
σ	$0 < \sigma < \infty$, scaling parameter
ν	$0 < \nu < \infty$, first kurtosis parameter (fixed constant)
τ	$\max(\nu, \nu^{-1}) < \tau < \infty$, second kurtosis parameter (fixed constant)
	Distribution measures
mean	$\begin{cases} \mu & \text{if } \nu\tau > 2 \\ \text{undefined} & \text{if } \nu\tau \le 2 \end{cases}$
median	μ
mode	μ
variance	$\begin{cases} 2\sigma^2 c\Big\{\sqrt{2\pi}\,[\Phi(\nu) - 0.5] & \\ + \left(2/\nu + 2/\nu^3\right)\exp(-\nu^2/2) & \\ + \dfrac{(\nu^2\tau^2 + 4\nu\tau + 6)}{\nu^3(\nu\tau - 3)} \exp(-\nu\tau + \nu^2/2)\Big\} & \text{if } \nu\tau > 3 \\ \infty & \text{if } 2 < \nu\tau \le 3 \\ \text{undefined} & \text{if } \nu\tau \le 2 \\ \text{where } c \text{ is given by (18.23)} & \end{cases}$
skewness	$\begin{cases} 0 & \text{if } \nu\tau > 4 \\ \text{undefined,} & \text{if } \nu\tau \le 4 \end{cases}$
excess kurtosis	see equation (18.24)
pdf	$\dfrac{c}{\sigma}\begin{cases} \exp\left(-z^2/2\right) & \text{if } \quad \lvert z\rvert \le \nu \\ \exp\left(-\nu\lvert z\rvert + \nu^2/2\right) & \text{if } \quad \nu < \lvert z\rvert \le \tau \\ \exp\left[-\nu\tau\log\left(\lvert z\rvert/\tau\right) - \nu\tau + \nu^2/2\right] & \text{if } \quad \lvert z\rvert > \tau \end{cases}$
cdf	see equation (18.25)
inverse cdf	see equation (18.26)

18.4.5 Sinh-arcsinh: SHASH

The pdf of the sinh-arcsinh distribution [Jones, 2005], denoted by $\texttt{SHASH}(\mu, \sigma, \nu, \tau)$, is given by

$$f_Y(y \mid \mu, \sigma\, \nu, \tau) = \frac{c}{\sqrt{2\pi}\sigma(1 + z^2)^{1/2}} \exp\left(-r^2/2\right) \tag{18.27}$$

where

$$\begin{aligned} r &= \tfrac{1}{2}\left\{\exp\left[\tau\sinh^{-1}(z)\right] - \exp\left[-\nu\sinh^{-1}(z)\right]\right\} \\ c &= \tfrac{1}{2}\left\{\tau\exp\left[\tau\sinh^{-1}(z)\right] + \nu\exp\left[-\nu\sinh^{-1}(z)\right]\right\} \end{aligned} \tag{18.28}$$

and $z = (y - \mu)/\sigma$ for $-\infty < y < \infty$, where $-\infty < \mu < \infty$, $\sigma > 0$, $\nu > 0$ and $\tau > 0$. [Note $\sinh^{-1}(z) = \log(u)$ where $u = z + \left(z^2 + 1\right)^{1/2}$. Hence $r = \left(u^\tau - u^{-\nu}\right)/2$.] Note that $R \sim \texttt{NO}(0, 1)$ where R is obtained from (18.28) and $Z = (Y - \mu)/\sigma$. Hence μ is the median of Y (since $y = \mu$ gives $z = 0$, $u = 1$, $\sinh^{-1}(z) = 0$ and hence $r = 0$). Note also R is the normalized quantile residual (or Z score).

The parameter ν controls the left tail heaviness, with the left tail being heavier than the normal distribution if $\nu < 1$ and lighter if $\nu > 1$. Similarly τ controls the right tail. The distribution is symmetric if $\nu = \tau$.

If $Y \sim$ SHASH(μ, σ, ν, τ) then $-Y \sim$ SHASH$(-\mu, \sigma, \tau, \nu)$ is a reflection of the distribution of Y about zero. Also $W = (2\mu - Y) \sim$ SHASH(μ, σ, τ, ν) is a reflection of the distribution of Y about μ.

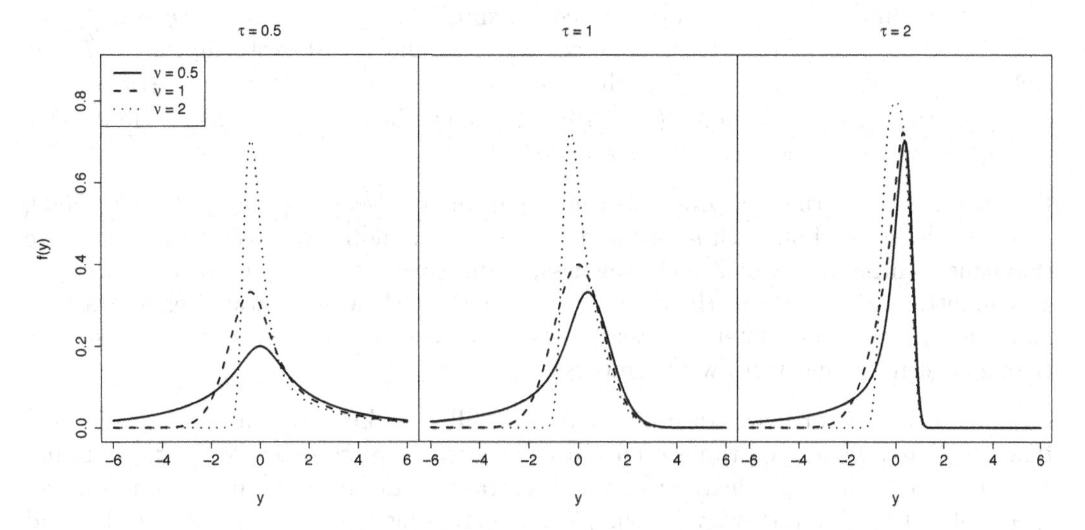

FIGURE 18.17: The sinh-arcsinh, SHASH(μ, σ, ν, τ), distribution, with $\mu = 0$, $\sigma = 1$, $\nu = 0.5, 1, 2$ and $\tau = 0.5, 1, 2$. It is symmetric if $\nu = \tau$. Interchanging ν and τ reflects the distribution about $y = \mu$, i.e. $y = 0$ above.

TABLE 18.18: Sinh-arcsinh distribution.

SHASH(μ, σ, ν, τ)	
Ranges	
Y	$-\infty < y < \infty$
μ	$-\infty < \mu < \infty$, median, location shift parameter
σ	$0 < \sigma < \infty$, scaling parameter
ν	$0 < \nu < \infty$, left tail heaviness parameter
τ	$0 < \tau < \infty$, right tail heaviness parameter
Distribution measures	
median	μ
pdf	$\begin{cases} \frac{c}{\sigma\sqrt{2\pi}(1+z^2)^{1/2}} \exp\left(-r^2/2\right) \\ \text{where } r = \frac{1}{2}\left\{\exp[\tau \sinh^{-1}(z)] - \exp[-\nu \sinh^{-1}(z)]\right\} \text{ and} \\ z = \frac{y-\mu}{\sigma}, \; c = \frac{1}{2}\left\{\tau \exp[\tau \sinh^{-1}(z)] + \nu \exp[-\nu \sinh^{-1}(z)]\right\} \end{cases}$
cdf	$\Phi(r)$
Reference	Jones and Pewsey [2009], page 777 with $(\xi, \eta, \gamma, \delta, x, z)$ replaced by $(\mu, \sigma, \nu, \tau, z, r)$.

18.4.6 Sinh-arcsinh original: SHASHo, SHASHo2

First parameterization, SHASHo

The original sinh-arcsinh distribution, developed by Jones and Pewsey [2009] is denoted by $\texttt{SHASHo}(\mu, \sigma, \nu, \tau)$, with pdf given by

$$f_Y(y \mid \mu, \sigma, \nu, \tau) = \frac{\tau c}{\sigma\sqrt{2\pi}(1+z^2)^{1/2}} \exp\left(-\frac{1}{2}r^2\right) \tag{18.29}$$

where $r = \sinh[\tau \sinh^{-1}(z) - \nu]$, $c = \cosh[\tau \sinh^{-1}(z) - \nu]$ and $z = (y - \mu)/\sigma$ for $-\infty < y < \infty$, $-\infty < \mu < \infty$, $\sigma > 0$, $-\infty < \nu < \infty$ and $\tau > 0$. Note that $c^2 - r^2 = 1$. Note also that $\sinh^{-1}(z) = \log(\mu)$ where $u = z + (z^2+1)^{1/2}$. Hence $z = \frac{1}{2}(u - u^{-1})$. Note also that $R = \sinh[\tau \sinh^{-1}(Z) - \nu)] \sim \texttt{NO}(0,1)$ where $Z = (Y - \mu)/\sigma$. Hence R is the normalized quantile residual (or z-score).

Parameter ν is a true van Zwet skewness parameter, [see Jones and Pewsey, 2009, p763 and Section 14.5], with $\nu > 0$ and $\nu < 0$ corresponding to positive and negative (moment, centile, and van Zwet) skewness, respectively, and $\nu = 0$ corresponding to a symmetrical distribution. Hence it is a true centile skewness parameter and a true moment skewness parameter, see Section 14.5. Increasing ν (for fixed τ) increases the (moment, centile, and van Zwet) skewness.

Parameter τ is a primary true Balanda-MacGillivray kurtosis parameter, see Section 15.5, and hence a primary true centile kurtosis parameter, with $\tau < 1$ and $\tau > 1$ corresponding to heavier and lighter tails than the normal distribution, respectively [Jones and Pewsey, 2009, p762]. Increasing τ decreases the centile (and Balanda-MacGillivray) kurtosis. Figures 16.2(a), 16.4(a), and 16.7(a) show moment and centile kurtosis-skewness plots, for $\texttt{SHASHo}(\mu, \sigma, \nu, \tau)$. If $Y \sim \texttt{SHASHo}(\mu, \sigma, \nu, \tau)$ then $-Y \sim \texttt{SHASHo}(-\mu, \sigma, -\nu, \tau)$ is a reflection of the distribution of Y about zero. Also $W = (2\mu - Y) \sim \texttt{SHASHo}(\mu, \sigma, -\nu, \tau)$ is a reflection of the distribution of Y about μ.

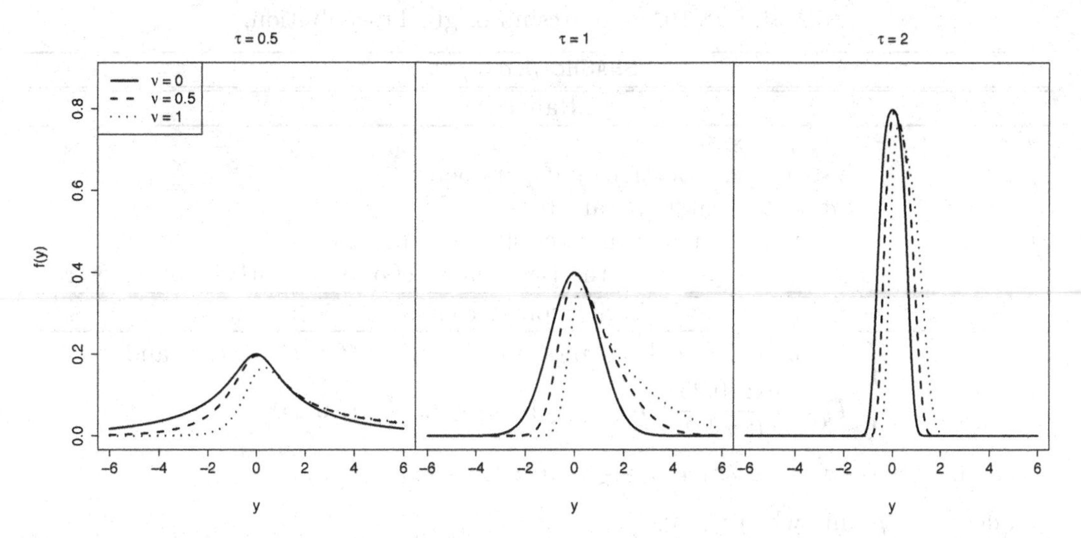

FIGURE 18.18: The original sinh-arcsinh, SHASHo(μ, σ, ν, τ), distribution, with $\mu = 0$, $\sigma = 1$, $\nu = 0, 0.5, 1$, and $\tau = 0.5, 1, 2$. It is symmetric if $\nu = 0$. Changing ν to $-\nu$ reflects the distribution about $y = \mu$, i.e. $y = 0$ above.

TABLE 18.19: Sinh-arcsinh original distribution.

	$\texttt{SHASHo}(\mu, \sigma, \nu, \tau)$
	Ranges
Y	$-\infty < y < \infty$
μ	$-\infty < \mu < \infty$, location shift parameter
σ	$0 < \sigma < \infty$, scaling parameter
ν	$-\infty < \nu < \infty$, true van Zwet skewness parameter
τ	$0 < \tau < \infty$, primary true Balanda-MacGillivray kurtosis parameter
	Distribution measures
mean	$\begin{cases} \mu + \sigma \mathrm{E}(Z) = \mu + \sigma \sinh(\nu/\tau) P_{1/\tau} \text{ where } Z = (Y - \mu)/\sigma \text{ and} \\ P_q = \dfrac{\exp(0.25)}{(8\pi)^{0.5}} \left[K_{(q+1)/2}(0.25) + K_{(q-1)/2}(0.25)\right] \end{cases}$
median	$\mu + \sigma \sinh(\nu/\tau) = \mu + \dfrac{\sigma}{2}\left[\exp(\nu/\tau) - \exp(-\nu/\tau)\right]$
mode	μ only when $\nu = 0$
variance	$\sigma^2 \mathrm{Var}(Z) = \dfrac{\sigma^2}{2}[\cosh(2\nu/\tau) P_{2/\tau} - 1] - \sigma^2[\sinh(\nu/\tau) P_{1/\tau}]^2$
skewness	$\begin{cases} \mu_{3Y}/[\mathrm{Var}(Y)]^{1.5} \text{ where} \\ \mu_{3Y} = \sigma^3 \mu_{3Z} = \sigma^3 \{\mu'_{3Z} - 3\mathrm{Var}(Z)\mathrm{E}(Z) - [\mathrm{E}(Z)]^3\} \\ \mu'_{3Z} = \mathrm{E}(Z^3) = \dfrac{1}{4}[\sinh(3\nu/\tau) P_{3/\tau} - 3\sinh(\nu/\tau) P_{1/\tau}] \end{cases}$
excess kurtosis	$\begin{cases} \mu_{4Y}/[\mathrm{Var}(Y)]^2 - 3 \text{ where} \\ \mu_{4Y} = \sigma^4 \mu_{4Z} = \sigma^4 \{\mu'_{4Z} - 4\mu'_{3Z}\mathrm{E}(Z) + 6\mathrm{Var}(Z)[\mathrm{E}(Z)]^2 + 3[\mathrm{E}(Z)]^4\} \\ \mu'_{4Z} = \mathrm{E}(Z^4) = \dfrac{1}{8}[\cosh(4\nu/\tau) P_{4/\tau} - 4\cosh(2\nu/\tau) P_{2/\tau} + 3] \end{cases}$
pdf	$\begin{cases} \dfrac{\tau c}{\sigma\sqrt{2\pi}(1+z^2)^{1/2}} \exp(-r^2/2) \\ \text{where } r = \sinh[\tau \sinh^{-1}(z) - \nu] \\ \text{and } z = (y - \mu)/\sigma \text{ and } c = \cosh[\tau \sinh^{-1}(z) - \nu] \end{cases}$
cdf	$\Phi(r)$
inverse cdf	$\mu + \sigma \sinh\left\{\dfrac{\nu}{\tau} + \dfrac{1}{\tau}\sinh^{-1}\left[\Phi^{-1}(p)\right]\right\}$
Reference	Jones and Pewsey [2009], page 762-764, with $(\xi, \eta, \epsilon, \delta, x, z)$ replaced by $(\mu, \sigma, \nu, \tau, z, r)$.
Note	$K_\lambda(t)$ is a modified Bessel function of the second kind

Second parameterization, SHASHo2

Jones and Pewsey [2009, p768] suggest reparameterizing $\texttt{SHASHo}(\mu, \sigma, \nu, \tau)$ in order to provide a more orthogonal parameterization, $\texttt{SHASHo2}(\mu, \sigma, \nu, \tau)$, with pdf given by (18.29) with σ replaced by $\sigma\tau$. This solves numerical problems encountered in their original parameterization, i.e. $\texttt{SHASHo}(\mu, \sigma, \nu, \tau)$, when $\tau > 1$. The summary table for $\texttt{SHASHo2}(\mu, \sigma, \nu, \tau)$ is given by replacing σ by $\sigma\tau$ in Table 18.19.

18.4.7 Skew exponential power type 1: SEP1

The pdf of the skew exponential power type 1 distribution, denoted by $\texttt{SEP1}(\mu, \sigma, \nu, \tau)$, is given by

$$f_Y(y \mid \mu, \sigma, \nu, \tau) = \frac{2}{\sigma} f_{Z_1}(z) F_{Z_1}(\nu z) \tag{18.30}$$

for $-\infty < y < \infty$, where $-\infty < \mu < \infty$, $\sigma > 0$, $-\infty < \nu < \infty$ and $\tau > 0$, and where $z = (y - \mu)/\sigma$ and $f_{Z_1}(\cdot)$ and $F_{Z_1}(\cdot)$ are the pdf and cdf of $Z_1 \sim \texttt{PE2}(0, \tau^{1/\tau}, \tau)$, given in Table 18.8. The $\texttt{SEP1}(\mu, \sigma, \nu, \tau)$ distribution was introduced by Azzalini [1986] as his type I distribution. See Section 13.4. The skew normal type 1 distribution, $\texttt{SN1}(\mu, \sigma, \nu)$, is a special case of $\texttt{SEP1}(\mu, \sigma, \nu, \tau)$ given by $\tau = 2$.

The $\texttt{SEP1}(\mu, \sigma, \nu, \tau)$ distribution is symmetric if $\nu = 0$. However from Figures 1 and 2 of Azzalini [1986], for fixed $\tau > 2$, the moment skewness of $\texttt{SEP1}(\mu, \sigma, \nu, \tau)$ can be positive or negative, but is not always positive if $\nu > 0$ (and not always negative if $\nu < 0$) and is not always monotonically increasing with ν. The parameter τ affects the moment kurtosis of the distribution, with increasing τ (for fixed ν) appearing to decrease the moment kurtosis, which can be moment leptokurtosis or platykurtosis. However the shape of the distribution is very flexible.

If $Y \sim \texttt{SEP1}(\mu, \sigma, \nu, \tau)$ then $-Y \sim \texttt{SEP1}(-\mu, \sigma, -\nu, \tau)$ is a reflection of the distribution of Y about zero. Also $W = (2\mu - Y) \sim \texttt{SEP1}(\mu, \sigma, -\nu, \tau)$ is a reflection of the distribution of Y about μ. As $\nu \to \infty$ the $\texttt{SEP1}(\mu, \sigma, \nu, \tau)$ distribution tends to a half power exponential type 2 distribution, i.e. a half $\texttt{PE2}(\mu, \sigma\tau^{1/\tau}, \tau)$, (see Table 18.8). Also $\texttt{SEP1}(\mu, \sigma, 0, \tau) \equiv \texttt{PE2}(\mu, \sigma\tau^{1/\tau}, \tau)$.

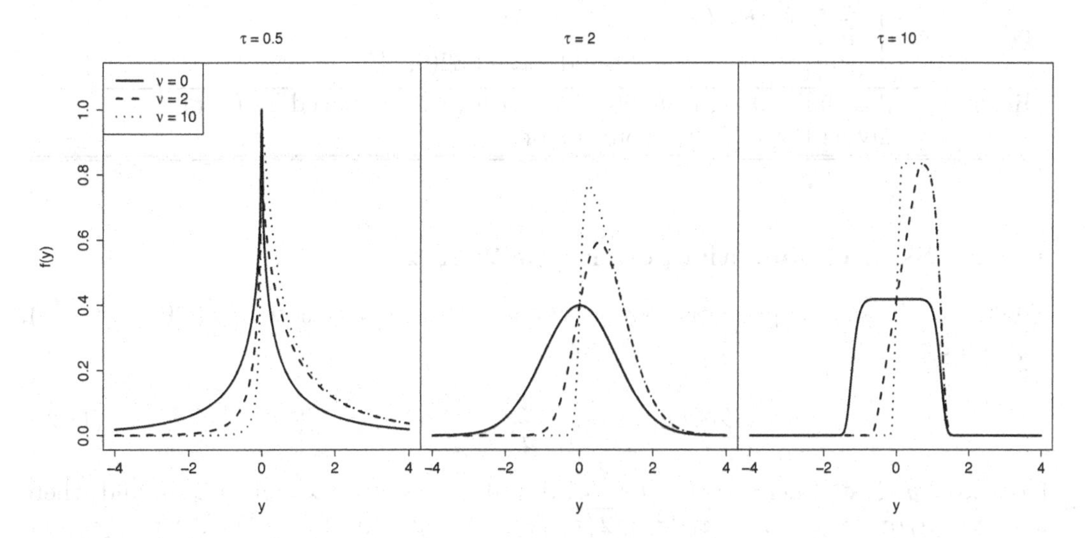

FIGURE 18.19: The skew exponential power type 1, $\texttt{SEP1}(\mu, \sigma, \nu, \tau)$, distribution, with $\mu = 0$, $\sigma = 1$, $\nu = 0, 2, 10$, and $\tau = 0.5, 2, 10$. It is symmetric if $\nu = 0$. Changing ν to $-\nu$ reflects the distribution about $y = \mu$, i.e. $y = 0$ above.

TABLE 18.20: Skew exponential power type 1 distribution.

$\texttt{SEP1}(\mu, \sigma, \nu, \tau)$	
Ranges	
Y	$-\infty < y < \infty$
μ	$-\infty < \mu < \infty$, location shift parameter
σ	$0 < \sigma < \infty$, scaling parameter
ν	$-\infty < \nu < \infty$, skewness parameter
τ	$0 < \tau < \infty$, kurtosis parameter
Distribution measures	
mean	$\begin{cases} \mu + \sigma \text{E}(Z) = \mu + \dfrac{\sigma \text{sign}(\nu) \tau^{1/\tau} \Gamma(2\tau^{-1}) B(\tau^{-1}, 2\tau^{-1}, \nu^\tau/[1+\nu^\tau])}{\Gamma(\tau^{-1}) B(\tau^{-1}, 2\tau^{-1})} \\ \text{where } Z = (Y - \mu)/\sigma \end{cases}$
variance	$\sigma^2 \text{Var}(Z) = \sigma^2 \left\{ \dfrac{\tau^{2/\tau} \Gamma(3\tau^{-1})}{\Gamma(\tau^{-1})} - [\text{E}(Z)]^2 \right\}$
skewness	$\begin{cases} \mu_{3Y}/[\text{Var}(Y)]^{1.5} \text{ where} \\ \mu_{3Y} = \sigma^3 \mu_{3Z} = \sigma^3 \{\mu_{3Z}' - 3\text{Var}(Z)\text{E}(Z) - [\text{E}(Z)]^3\} \\ \mu_{3Z}' = \text{E}(Z^3) = \dfrac{\text{sign}(\nu) \tau^{3/\tau} \Gamma(4\tau^{-1}) B(\tau^{-1}, 4\tau^{-1}, \nu^\tau/[1+\nu^\tau])}{\Gamma(\tau^{-1}) B(\tau^{-1}, 4\tau^{-1})} \end{cases}$
excess kurtosis	$\begin{cases} \mu_{4Y}/[\text{Var}(Y)]^2 - 3 \text{ where} \\ \mu_{4Y} = \sigma^4 \mu_{4Z} = \sigma^4 \{\mu_{4Z}' - 4\mu_{3Z}' \text{E}(Z) + 6\text{Var}(Z)[\text{E}(Z)]^2 + 3[\text{E}(Z)]^4\} \\ \mu_{4Z}' = \text{E}(Z^4) = \dfrac{\tau^{4/\tau} \Gamma(5\tau^{-1})}{\Gamma(\tau^{-1})} \end{cases}$
pdf	$\begin{cases} \dfrac{2}{\sigma} f_{Z_1}(z) F_{Z_1}(\nu z) \\ \text{where } z = (y - \mu)/\sigma \text{ and } Z_1 \sim \texttt{PE2}(0, \tau^{1/\tau}, \tau) \end{cases}$
Reference	Azzalini [1986], page 202-203, with (λ, ω) replaced by (ν, τ) giving the pdf and moments of Z.

18.4.8 Skew exponential power type 2: SEP2

The pdf of the skew exponential power type 2 distribution, denoted by $\texttt{SEP2}(\mu, \sigma, \nu, \tau)$, is given by

$$f_Y(y \mid \mu, \sigma \, \nu, \tau) = \frac{2}{\sigma} f_{Z_1}(z) \, \Phi(\omega) \tag{18.31}$$

for $-\infty < y < \infty$, where $-\infty < \mu < \infty$, $\sigma > 0$, $-\infty < \nu < \infty$, and $\tau > 0$, and where $z = (y - \mu)/\sigma$; $\omega = \text{sign}(z)|z|^{\tau/2} \nu \sqrt{2/\tau}$; $f_{Z_1}(\cdot)$ is the pdf of $Z_1 \sim \texttt{PE2}(0, \tau^{1/\tau}, \tau)$ and $\Phi(\cdot)$ is the standard normal cdf.

This distribution was introduced by Azzalini [1986] as his type II distribution and developed by DiCiccio and Monti [2004]. See Section 13.4. The $\texttt{SEP2}(\mu, \sigma, \nu, \tau)$ distribution is symmetric if $\nu = 0$. However from Figures 3 and 4 of Azzalini [1986], for fixed $\tau > 2$, the moment skewness of $\texttt{SEP2}(\mu, \sigma, \nu, \tau)$ is not always positive if $\nu > 0$ (and not always negative if $\nu < 0$) and is not always monotonically increasing with ν. The parameter τ

affects the moment kurtosis of the distribution, with increasing τ (for fixed ν) appearing to decrease the moment kurtosis, which can be moment leptokurtosis or platykurtosis. However the shape of the distribution is very flexible.

If $Y \sim$ SEP2(μ, σ, ν, τ) then $-Y \sim$ SEP2$(-\mu, \sigma, -\nu, \tau)$ is a reflection of the distribution of Y about zero. Also $W = (2\mu - Y) \sim$ SEP2$(\mu, \sigma, -\nu, \tau)$ is a reflection of the distribution of Y about μ. We have $\text{E}(Y) = \mu + \sigma \text{E}(Z)$ where $Z = (Y - \mu)/\sigma$ and

$$\text{E}(Z) = \frac{2\tau^{1/\tau}\nu}{\sqrt{\pi}\Gamma\left(\tau^{-1}\right)\left(1+\nu^2\right)^{(2/\tau)+0.5}} \sum_{n=0}^{\infty} \frac{\Gamma\left(2\tau^{-1}+n+0.5\right)}{(2n+1)!!}\left(\frac{2\nu^2}{1+\nu^2}\right)^n \tag{18.32}$$

where $(2n+1)!! = 1.3.5\ldots(2n+1)$, and $\mu_{3Z}^{'} = \text{E}(Z^3)$ is given by

$$\text{E}(Z^3) = \frac{2\tau^{3/\tau}\nu}{\sqrt{\pi}\Gamma\left(\tau^{-1}\right)\left(1+\nu^2\right)^{(4/\tau)+0.5}} \sum_{n=0}^{\infty} \frac{\Gamma\left(4\tau^{-1}+n+0.5\right)}{(2n+1)!!}\left(\frac{2\nu^2}{1+\nu^2}\right)^n, \tag{18.33}$$

obtained from DiCiccio and Monti [2004, p440].

The SEP2$(\mu, \sigma, \nu, 2)$ is the skew normal type 1, SN1(μ, σ, ν), distribution, [Azzalini, 1985], while the SEP2$(\mu, \sigma, 0, 2)$ is the normal distribution, NO(μ, σ). As $\nu \to \infty$ the SEP2(μ, σ, ν, τ) distribution tends to a half power exponential type 2 distribution, i.e. a half PE2$(\mu, \sigma\tau^{1/\tau}, \tau)$, (see Table 13.2). Also SEP2$(\mu, \sigma, 0, \tau) =$ PE2$(\mu, \sigma\tau^{1/\tau}, \tau)$.

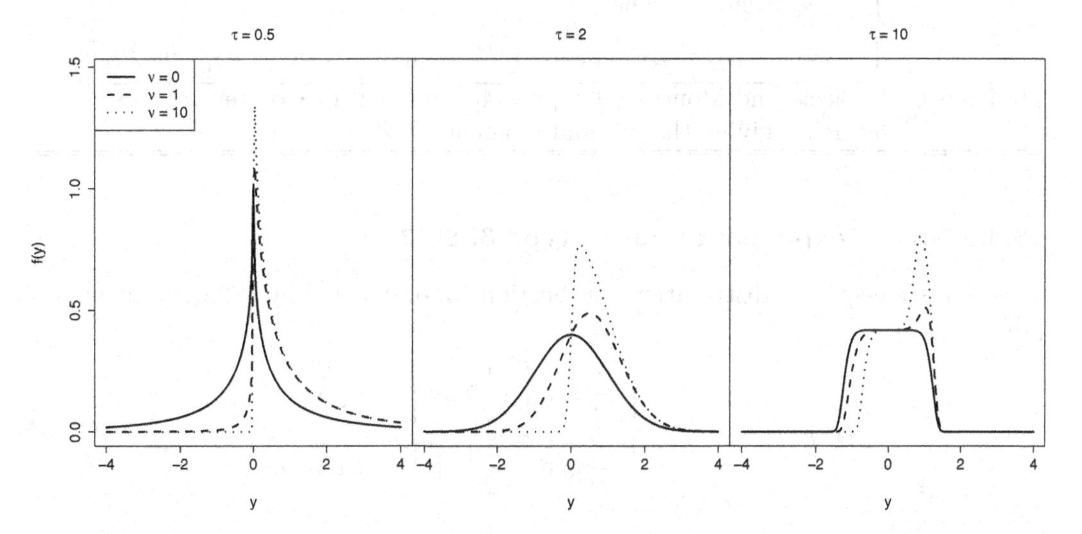

FIGURE 18.20: The skew exponential power type 2, SEP2(μ, σ, ν, τ), distribution with $\mu = 0$, $\sigma = 1$, $\nu = 0, 1, 10$, and $\tau = 0.5, 2, 10$. It is symmetric if $\nu = 0$. Changing ν to $-\nu$ reflects the distribution about $y = \mu$, i.e. $y = 0$ above.

TABLE 18.21: Skew exponential power type 2 distribution.

$\texttt{SEP2}(\mu, \sigma, \nu, \tau)$	
Ranges	
Y	$-\infty < y < \infty$
μ	$-\infty < \mu < \infty$, location shift parameter
σ	$0 < \sigma < \infty$, scaling parameter
ν	$-\infty < \nu < \infty$, skewness parameter
τ	$0 < \tau < \infty$, kurtosis parameter
Distribution measures	
mean	$\begin{cases} \mu + \sigma \text{E}(Z) \\ \text{where } Z = (Y-\mu)/\sigma \text{ and E}(Z) \text{ is given by (18.32)} \end{cases}$
variance	$\sigma^2 \text{Var}(Z) = \sigma^2 \left\{ \frac{\tau^{2/\tau}\Gamma(3\tau^{-1})}{\Gamma(\tau^{-1})} - [E(Z)]^2 \right\}$
skewness	$\begin{cases} \mu_{3Y}/[\text{Var}(Y)]^{1.5} \text{ where} \\ \mu_{3Y} = \sigma^3 \mu_{3Z} = \sigma^3 \{\mu_{3Z}' - 3\text{Var}(Z)\text{E}(Z) - [\text{E}(Z)]^3\} \\ \mu_{3Z}' \text{ is given by (18.33)} \end{cases}$
excess kurtosis	$\begin{cases} \mu_{4Y}/[\text{Var}(Y)]^2 - 3 \text{ where} \\ \mu_{4Y} = \sigma^4 \mu_{4Z} = \sigma^4 \{\mu_{4Z}' - 4\mu_{3Z}'\text{E}(Z) + 6\text{Var}(Z)[\text{E}(Z)]^2 + 3[E(Z)]^4\} \\ \mu_{4Z}' = \text{E}(Z^4) = \tau^{4/\tau}\Gamma(5\tau^{-1})/\Gamma(\tau^{-1}) \end{cases}$
pdf	$\begin{cases} \frac{2}{\sigma} f_{Z_1}(z)\Phi(\omega) \quad \text{where} \\ z = (y-\mu)/\sigma \,, Z_1 \sim \texttt{PE2}(0, \tau^{1/\tau}, \tau) \text{ and } \omega = \text{sign}(z)\|z\|^{\tau/2}\nu\sqrt{2/\tau} \end{cases}$
Reference	DiCiccio and Monti [2004], page 439-440, with (λ, α) replaced by (ν, τ), giving the pdf and moments of Z.

18.4.9 Skew exponential power type 3: SEP3

This is a 'scale-spliced' distribution (see Section 13.5), denoted by $\texttt{SEP3}(\mu, \sigma, \nu, \tau)$, with pdf

$$f_Y(y \mid \mu, \sigma\, \nu, \tau) = \begin{cases} \frac{c}{\sigma} \exp\left[-\frac{1}{2} |\nu z|^\tau \right] & \text{if } y < \mu \\ \frac{c}{\sigma} \exp\left[-\frac{1}{2} \left| \frac{z}{\nu} \right|^\tau \right] & \text{if } y \geq \mu, \end{cases} \tag{18.34}$$

for $-\infty < y < \infty$, where $-\infty < \mu < \infty$, $\sigma > 0$, $\nu > 0$, and $\tau > 0$, and where $z = (y-\mu)/\sigma$ and $c = \nu\tau\left[(1+\nu^2)2^{1/\tau}\Gamma\left(1/\tau\right)\right]^{-1}$, [Fernandez et al., 1995, p 1333]. Note that μ is the mode of Y.

The $\texttt{SEP3}(\mu, \sigma, \nu, \tau)$ distribution is symmetric if $\nu = 0$. It is moment positively skew if $\nu > 1$ and moment negatively skew if $0 < \nu < 1$, [Fernandez et al., 1995, p1334]. The distribution appears to be moment leptokurtic if $\tau < 2$, but can be moment platykurtic or leptokurtic if $\tau > 2$. Figures 16.2(b), 16.5(a) and 16.8(a) show moment and centile kurtosis-skewness plots for the SEP3 distribution.

If $Y \sim \texttt{SEP3}(\mu, \sigma, \nu, \tau)$ then $-Y \sim \texttt{SEP3}(-\mu, \sigma, 1/\nu, \tau)$ is a reflection of the distribution of Y about zero. Also $W = (2\mu - Y) \sim$ SEP3$(\mu, \sigma, 1/\nu, \tau)$ is a reflection of the distribution of Y about μ. The skew normal type 2 (or two-piece normal) distribution, SN2(μ, σ, ν), is the special case SEP3$(\mu, \sigma, \nu, 2)$. The PE2 distribution is a reparameterized special case of SEP3 given by PE2(μ, σ, ν) = SEP3$(\mu, \sigma 2^{-1/\nu}, 1, \nu)$. As $\nu \to \infty$, the SEP3 distribution tends to a half power exponential type 2 distribution, see Table 13.3. Also SEP3$(\mu, \sigma, 1, \tau) \equiv$ PE2$(\mu, \sigma 2^{1/\tau}, \tau)$. The shape of the distribution can be very flexible.

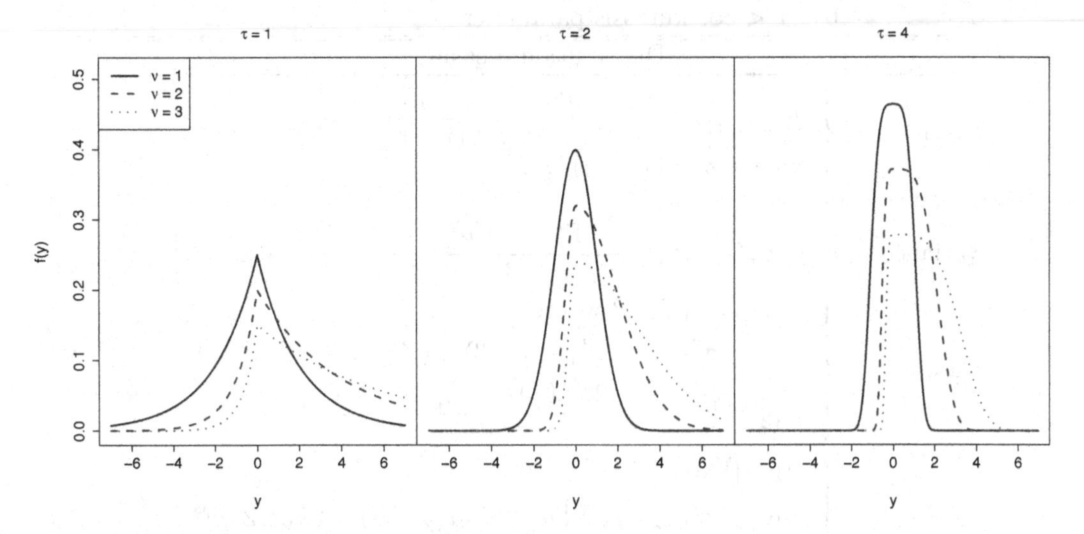

FIGURE 18.21: The skew exponential power type 3, SEP3(μ, σ, ν, τ), distribution, with $\mu = 0$, $\sigma = 1$, $\nu = 1, 2, 3$, and $\tau = 1, 2, 4$. It is symmetric if $\nu = 1$. Changing ν to $1/\nu$ reflects the distribution about $y = \mu$, i.e. $y = 0$ above.

TABLE 18.22: Skew exponential power type 3 distribution.

$\texttt{SEP3}(\mu, \sigma, \nu, \tau)$	
Ranges	
Y	$-\infty < y < \infty$
μ	$-\infty < \mu < \infty$, mode, location shift parameter
σ	$0 < \sigma < \infty$, scaling parameter
ν	$0 < \nu < \infty$, skewness parameter
τ	$0 < \tau < \infty$, kurtosis parameter
Distribution measures	
mean	$\begin{cases} \mu + \sigma E(Z) = \mu + \dfrac{\sigma 2^{1/\tau}\Gamma(2\tau^{-1})(\nu - \nu^{-1})}{\Gamma(\tau^{-1})} \\ \text{where } Z = (Y-\mu)/\sigma \end{cases}$
mode	μ
variance	$\sigma^2 \text{Var}(Z) = \sigma^2 \left\{ \dfrac{2^{2/\tau}\Gamma(3\tau^{-1})(\nu^2 + \nu^{-2} - 1)}{\Gamma(\tau^{-1})} - [\text{E}(Z)]^2 \right\}$
skewness	$\begin{cases} \mu_{3Y}/[\text{Var}(Y)]^{1.5} \text{ where} \\ \mu_{3Y} = \sigma^3 \mu_{3Z} = \sigma^3\{\mu_{3Z}' - 3\text{Var}(Z)\text{E}(Z) - [\text{E}(Z)]^3\} \\ \mu_{3Z}' = \text{E}(Z^3) = \dfrac{2^{3/\tau}\Gamma(4\tau^{-1})(\nu^4 - \nu^{-4})}{\Gamma(\tau^{-1})(\nu + \nu^{-1})} \end{cases}$
excess kurtosis	$\begin{cases} \mu_{4Y}/[\text{Var}(Y)]^2 - 3 \text{ where} \\ \mu_{4Y} = \sigma^4 \mu_{4Z} = \sigma^4 \left\{\mu_{4Z}' - 4\mu_{3Z}'\text{E}(Z) + 6\text{Var}(Z)[\text{E}(Z)]^2 \right. \\ \quad \left. +3[\text{E}(Z)]^4\right\} \\ \mu_{4Z}' = \text{E}(Z^4) = \dfrac{2^{4/\tau}\Gamma(5\tau^{-1})(\nu^5 + \nu^{-5})}{\Gamma(\tau^{-1})(\nu + \nu^{-1})} \end{cases}$
pdf	$\begin{cases} \dfrac{c}{\sigma}\exp\left(-\frac{1}{2}\|\nu z\|^\tau\right) & \text{if } y < \mu \\ \dfrac{c}{\sigma}\exp\left(-\frac{1}{2}\left\|\dfrac{z}{\nu}\right\|^\tau\right) & \text{if } y \ge \mu, \\ \text{where } z = (y-\mu)/\sigma \text{ and} & c = \nu\tau\left[(1+\nu^2)2^{1/\tau}\Gamma(1/\tau)\right]^{-1} \end{cases}$
cdf	$\begin{cases} \dfrac{1}{1+\nu^2}\left[\dfrac{\Gamma(\tau^{-1}, \alpha_1)}{\Gamma(\tau^{-1})}\right] & \text{if } y < \mu \\ 1 - \dfrac{\nu^2\Gamma(\tau^{-1}, \alpha_2)}{(1+\nu^2)\Gamma(\tau^{-1})} & \text{if } y \ge \mu \\ \text{where } \alpha_1 = \dfrac{\nu^2(\mu - y)^\tau}{2\sigma^\tau} \text{ and} & \alpha_2 = \dfrac{(y-\mu)^\tau}{2\sigma^\tau\nu^\tau} \end{cases}$
Reference	Fernandez et al. [1995], p1333, equations (8) and (12), with (γ, q) replaced by (ν, τ) giving the pdf and moments of Z.

18.4.10 Skew exponential power type 4: SEP4

This is a 'shape-spliced' distribution (see Section 13.5), denoted by $\texttt{SEP4}(\mu, \sigma, \nu, \tau)$, with pdf

$$f_Y(y \mid \mu, \sigma, \nu, \tau) = \begin{cases} \dfrac{c}{\sigma} \exp\left[-|z|^{\nu}\right] & \text{if } y < \mu \\ \dfrac{c}{\sigma} \exp\left[-|z|^{\tau}\right] & \text{if } y \geq \mu \end{cases} \tag{18.35}$$

for $-\infty < y < \infty$, where $-\infty < \mu < \infty$, $\sigma > 0$, $\nu > 0$, and $\tau > 0$, and where $z = (y - \mu)/\sigma$ and $c = \left[\Gamma\left(1 + \tau^{-1}\right) + \Gamma\left(1 + \nu^{-1}\right)\right]^{-1}$, [Jones, 2005]. Note that μ is the mode of Y.

Parameters ν and τ affect the left- and right-tail heaviness, respectively, with $0 < \nu < 2$ or $0 < \tau < 2$ a heavier tail than the normal distribution, and $\nu > 2$ or $\tau > 2$ a lighter tail. The $\texttt{SEP4}(\mu, \sigma, \nu, \tau)$ distribution is symmetric if $\nu = \tau$ and $\texttt{SEP4}(\mu, \sigma, \nu, \nu) = \texttt{PE2}(\mu, \sigma, \nu)$. If $Y \sim \texttt{SEP4}(\mu, \sigma, \nu, \tau)$ then $-Y \sim \texttt{SEP4}(-\mu, \sigma, \tau, \nu)$ is a reflection of the distribution of Y about zero. Also $W = (2\mu - Y) \sim \texttt{SEP4}(\mu, \sigma, \tau, \nu)$ is a reflection of the distribution of Y about μ.

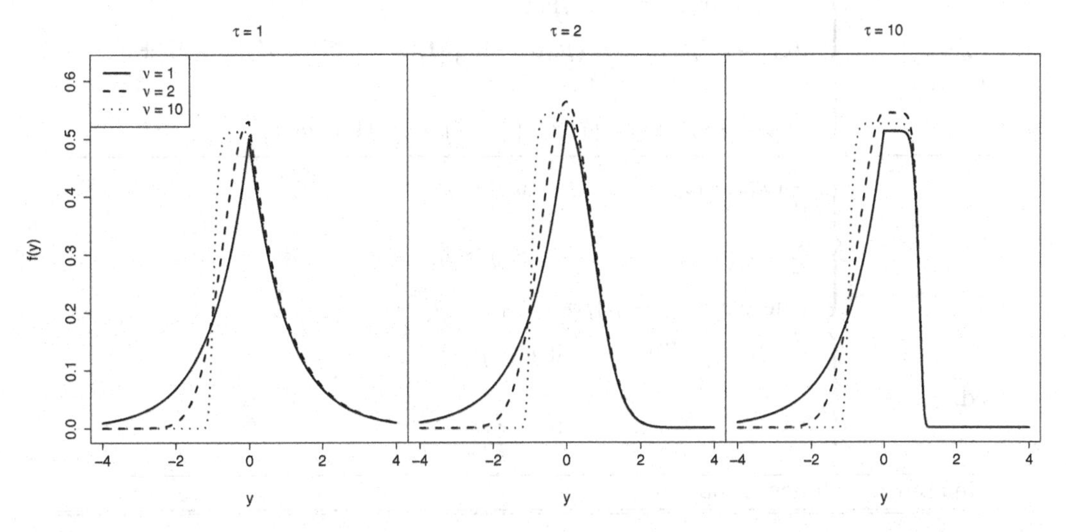

FIGURE 18.22: The skew exponential power type 4, $\texttt{SEP4}(\mu, \sigma, \nu, \tau)$, distribution, with $\mu = 0$, $\sigma = 1$, $\nu = 1, 2, 10$, and $\tau = 1, 2, 10$. It is symmetric if $\nu = \tau$. Interchanging ν and τ reflects the distribution about $y = \mu$, i.e. $y = 0$ above.

TABLE 18.23: Skew exponential power type 4 distribution.

$\texttt{SEP4}(\mu, \sigma, \nu, \tau)$	
Ranges	
Y	$-\infty < y < \infty$
μ	$-\infty < \mu < \infty$, mode, location shift parameter
σ	$0 < \sigma < \infty$, scaling parameter
ν	$0 < \nu < \infty$, left tail heaviness parameter
τ	$0 < \tau < \infty$, right tail heaviness parameter
Distribution measures	
mean	$\begin{cases} \mu + \sigma \text{E}(Z) = \mu + \sigma c \left[\tau^{-1}\Gamma\left(2\tau^{-1}\right) - \nu^{-1}\Gamma\left(2\nu^{-1}\right)\right] \\ \text{where } Z = (Y-\mu)/\sigma \text{ and } c = \left[\Gamma\left(1+\tau^{-1}\right) + \Gamma\left(1+\nu^{-1}\right)\right]^{-1} \end{cases}$
mode	μ
variance	$\sigma^2 \text{Var}(Z) = \sigma^2 \left\{c\left[\tau^{-1}\Gamma\left(3\tau^{-1}\right) + \nu^{-1}\Gamma\left(3\nu^{-1}\right)\right] - [\text{E}(Z)]^2\right\}$
skewness	$\begin{cases} \mu_{3Y}/[\text{Var}(Y)]^{1.5} \text{ where} \\ \mu_{3Y} = \sigma^3 \mu_{3Z} = \sigma^3\{\mu'_{3Z} - 3\text{Var}(Z)\text{E}(Z) - [\text{E}(Z)]^3\} \\ \mu'_{3Z} = \text{E}(Z^3) = c\left[\tau^{-1}\Gamma\left(4\tau^{-1}\right) - \nu^{-1}\Gamma\left(4\nu^{-1}\right)\right] \end{cases}$
excess kurtosis	$\begin{cases} \mu_{4Y}/[\text{Var}(Y)]^2 - 3 \text{ where} \\ \mu_{4Y} = \sigma^4 \mu_{4Z} = \sigma^4\{\mu'_{4Z} - 4\mu'_{3Z}\text{E}(Z) + 6\text{Var}(Z)[\text{E}(Z)]^2 + \\ \qquad 3[\text{E}(Z)]^4\} \\ \mu'_{4Z} = \text{E}(Z^4) = c\left[\tau^{-1}\Gamma\left(5\tau^{-1}\right) + \nu^{-1}\Gamma\left(5\nu^{-1}\right)\right] \end{cases}$
pdf [a]	$\begin{cases} \dfrac{c}{\sigma}\exp\left(-\|z\|^{\nu}\right) & \text{if } y < \mu \\ \dfrac{c}{\sigma}\exp\left(-\|z\|^{\tau}\right) & \text{if } y \ge \mu, \\ \text{where } z = (y-\mu)/\sigma \end{cases}$
cdf	$\begin{cases} \dfrac{c}{\nu}\Gamma(\nu^{-1}, \|z\|^{\nu}) & \text{if } y < \mu \\ 1 - \dfrac{c}{\tau}\Gamma(\tau^{-1}, \|z\|^{\tau}) & \text{if } y \ge \mu \end{cases}$
Reference	[a]Jones [2005]

18.4.11 Skew Student t: SST

Würtz et al. [2006] reparameterized the ST3 distribution, [Fernandez and Steel, 1998], so that in the new parameterization μ is the mean and σ is the standard deviation. They called this the skew Student t distribution, which we denote as SST.

Let $Z_0 \sim \texttt{ST3}(0, 1, \nu, \tau)$ and $Y = \mu + \sigma \left(\frac{Z_0 - m}{s} \right)$, where

$$m = \text{E}(Z_0) = \frac{2\tau^{1/2}(\nu - \nu^{-1})}{(\tau - 1)B(1/2, \tau/2)} \qquad \text{for } \tau > 1 \tag{18.36}$$

$$s^2 = \text{Var}(Z_0) = \frac{\tau}{(\tau - 2)}(\nu^2 + \nu^{-2} - 1) - m^2 \qquad \text{for } \tau > 2\,. \tag{18.37}$$

Hence $Y = \mu_0 + \sigma_0 Z_0$, where $\mu_0 = \mu - \sigma m/s$ and $\sigma_0 = \sigma/s$, and so $Y \sim \texttt{ST3}(\mu_0, \sigma_0, \nu, \tau)$ with $\text{E}(Y) = \mu$ and $\text{Var}(Y) = \sigma^2$ for $\tau > 2$. Let $Y \sim \texttt{SST}(\mu, \sigma, \nu, \tau) = \texttt{ST3}(\mu_0, \sigma_0, \nu, \tau)$, for $\tau > 2$.

Hence the pdf of the skew Student t distribution, denoted by $Y \sim \texttt{SST}(\mu, \sigma, \nu, \tau)$, is given by

$$f_Y(y \mid \mu, \sigma, \nu, \tau) = \begin{cases} \dfrac{c}{\sigma_0}\left(1 + \dfrac{\nu^2 z^2}{\tau}\right)^{-(\tau+1)/2} & \text{if } y < \mu_0 \\ \dfrac{c}{\sigma_0}\left(1 + \dfrac{z^2}{\nu^2 \tau}\right)^{-(\tau+1)/2} & \text{if } y \geq \mu_0, \end{cases}$$

for $-\infty < y < \infty$, where $-\infty < \mu < \infty$, $\sigma > 0, \nu > 0$ and $\tau > 2$ and where $z = (y - \mu_0)/\sigma_0$, $\mu_0 = \mu - \sigma m/s$, $\sigma_0 = \sigma/s$ and $c = 2\nu \left[(1 + \nu^2)B(1/2, \tau/2)\, \tau^{1/2}\right]^{-1}$.

Note that $\text{E}(Y) = \mu$ and $\text{Var}(Y) = \sigma^2$ and the moment based skewness and excess kurtosis of $\texttt{SST}(\mu, \sigma, \nu, \tau)$ are the same as for $\texttt{ST3}(\mu_0, \sigma_0, \nu, \tau)$, and hence the same as for $\texttt{ST3}(0, 1, \nu, \tau)$, depending only on ν and τ. In **gamlss.dist** the default link function for τ is a shifted log link function, $\log(\tau - 2)$, which ensures that τ is always in its valid range, i.e. $\tau > 2$.

Parameter ν mainly affects the skewness, with symmetry when $\nu = 1$, while parameter τ mainly affects the heaviness of the tails, with increasing τ decreasing the heaviness of the tails.

If $Y \sim \texttt{SST}(\mu, \sigma, \nu, \tau)$ then $-Y \sim \texttt{SST}(-\mu, \sigma, 1/\nu, \tau)$ is a reflection of the distribution of Y about zero. Also $W = (2\mu - Y) \sim \texttt{SST}(\mu, \sigma, 1/\nu, \tau)$ is a reflection of the distribution of Y about μ. Also $\texttt{SST}(\mu, \sigma, 1, \tau) = \texttt{TF2}(\mu, \sigma, \tau)$.

TABLE 18.24: Skew Student t distribution.

$\texttt{SST}(\mu, \sigma, \nu, \tau)$	
Ranges	
Y	$-\infty < y < \infty$
μ	$-\infty < \mu < \infty$, mean, location shift parameter
σ	$0 < \sigma < \infty$, standard deviation, scaling parameter
ν	$0 < \nu < \infty$, skewness parameter
τ	$2 < \tau < \infty$, kurtosis parameter
Distribution measures	
mean	μ
median	$\begin{cases} \mu_0 + \dfrac{\sigma_0}{\nu} t_{\alpha_1,\tau} & \text{if } \nu \le 1 \\ \mu_0 + \sigma_0 \nu t_{\alpha_2,\tau} & \text{if } \nu > 1 \\ \text{where } \alpha_1 = \frac{(1+\nu^2)}{4}, \ \alpha_2 = \frac{(3\nu^2-1)}{4\nu^2} & \\ t_{\alpha,\tau} = F_T^{-1}(\alpha) \text{ where } T \sim t_\tau & \end{cases}$
mode	μ_0
variance	σ^2
skewness	equal to skewness of $\texttt{ST3}(0, 1, \nu, \tau)$
excess kurtosis	equal to excess kurtosis of $\texttt{ST3}(0, 1, \nu, \tau)$
pdf	$\begin{cases} \dfrac{c}{\sigma_0}\left[1 + \dfrac{\nu^2 z^2}{\tau}\right]^{-(\tau+1)/2} & \text{if } y < \mu_0 \\ \dfrac{c}{\sigma_0}\left[1 + \dfrac{z^2}{\nu^2 \tau}\right]^{-(\tau+1)/2} & \text{if } y \ge \mu_0, \\ \text{where } z = (y - \mu_0)/\sigma_0 \text{ and} & \\ c = 2\nu\left[(1+\nu^2) B(1/2, \tau/2)\, \tau^{1/2}\right]^{-1} & \end{cases}$
cdf	$\begin{cases} \dfrac{2}{(1+\nu^2)} F_T[\nu(y-\mu_0)/(\sigma_0)] & \text{if } y < \mu_0 \\ \dfrac{1}{(1+\nu^2)}\left\{1 + 2\nu^2\left[F_T\left(\dfrac{y-\mu_0}{\sigma_0 \nu}\right) - 0.5\right]\right\} & \text{if } y \ge \mu_0 \\ \text{where } T \sim t_\tau & \end{cases}$
inverse cdf	$\begin{cases} \mu_0 + \dfrac{\sigma_0}{\nu} t_{\alpha_3,\tau} & \text{if } p \le (1+\nu^2)^{-1} \\ \mu_0 + \sigma_0 \nu t_{\alpha_4,\tau} & \text{if } p > (1+\nu^2)^{-1} \end{cases}$ where $\alpha_3 = \frac{p(1+\nu^2)}{2}$ and $\alpha_4 = \frac{p(1+\nu^2)-1+\nu^2}{2\nu^2}$
Note	μ_0 and σ_0 are defined below equation (18.37)

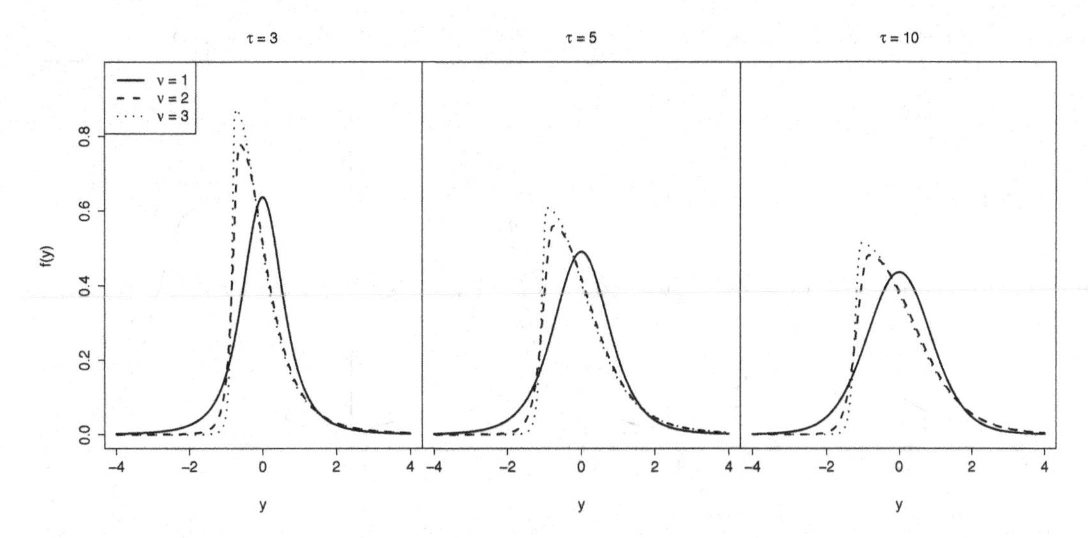

FIGURE 18.23: The skew Student t, $\texttt{SST}(\mu, \sigma, \nu, \tau)$, distribution, with $\mu = 0$, $\sigma = 1$, $\nu = 1, 2, 3$, and $\tau = 3, 5, 10$. It is symmetric if $\nu = 1$. Changing ν and $1/\nu$ reflects the distribution about $y = \mu$, i.e. $y = 0$ above.

18.4.12 Skew t type 1: ST1

The pdf of the skew t type 1 distribution, denoted by $\texttt{ST1}(\mu, \sigma, \nu, \tau)$, is given by

$$f_Y(y \mid \mu, \sigma, \nu, \tau) = \frac{2}{\sigma} f_{Z_1}(z) F_{Z_1}(\nu z) \tag{18.38}$$

for $-\infty < y < \infty$, where $-\infty < \mu < \infty$, $\sigma > 0$, $-\infty < \nu < \infty$ and $\tau > 0$, and where $z = (y - \mu)/\sigma$ and $f_{Z_1}(\cdot)$ and $F_{Z_1}(\cdot)$ are the pdf and cdf of $Z_1 \sim \texttt{TF}(0, 1, \tau) = t_\tau$, the t distribution with $\tau > 0$ degrees of freedom, with τ treated as a continuous parameter. This distribution is in the form of a type I distribution of Azzalini [1986], see Section 13.4. (No summary table is given for ST1.)

If $Y \sim \texttt{ST1}(\mu, \sigma, \nu, \tau)$ then $-Y \sim \texttt{ST1}(-\mu, \sigma, -\nu, \tau)$ is a reflection of the distribution of Y about zero. Also $W = (2\mu - Y) \sim \texttt{ST1}(\mu, \sigma, -\nu, \tau)$ is a reflection of the distribution of Y about μ.

As $\nu \to \infty$, $\texttt{ST1}(\mu, \sigma, \nu, \tau)$ tends to a half t family distribution, i.e. a half $\texttt{TF}(\mu, \sigma, \tau)$, see Table 13.2. Also $\texttt{ST1}(\mu, \sigma, 0, \tau) \equiv \texttt{TF}(\mu, \sigma, \tau)$.

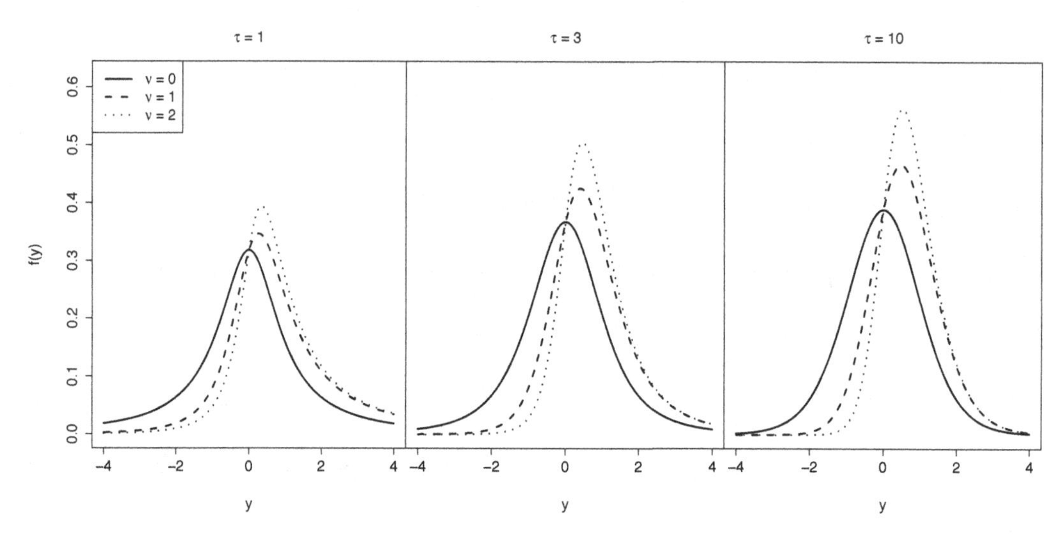

FIGURE 18.24: The skew t type 1, ST1(μ, σ, ν, τ), distribution, with $\mu = 0$, $\sigma = 1$, $\nu = 0, 1, 2$, and $\tau = 1, 3, 10$. It is symmetric if $\nu = 0$. Changing ν and $-\nu$ reflects the distribution about $y = \mu$, i.e. $y = 0$ above.

18.4.13 Skew t type 2: ST2

The pdf of the skew t type 2 distribution, denoted by ST2(μ, σ, ν, τ), is given by

$$f_Y(y \mid \mu, \sigma, \nu, \tau) = \frac{2}{\sigma} f_{Z_1}(z) F_{Z_2}(\omega) \tag{18.39}$$

for $-\infty < y < \infty$, where $-\infty < \mu < \infty$, $\sigma > 0$, $-\infty < \nu < \infty$, and $\tau > 0$, and where $z = (y - \mu)/\sigma$, $\omega = \nu\lambda^{1/2} z$, $\lambda = (\tau + 1)/(\tau + z^2)$, $f_{Z_1}(\cdot)$ is the pdf of $Z_1 \sim \texttt{TF}(0, 1, \tau) = t_\tau$ and $F_{Z_2}(\cdot)$ is the cdf of $Z_2 \sim \texttt{TF}(0, 1, \tau + 1) = t_{\tau+1}$. This distribution is the univariate case of the multivariate skew t distribution introduced by Azzalini and Capitanio [2003, p380, equation (26)]. See Section 13.4.

If $Y \sim \texttt{ST2}(\mu, \sigma, \nu, \tau)$ then $-Y \sim \texttt{ST2}(-\mu, \sigma, -\nu, \tau)$ is a reflection of the distribution of Y about zero. Also $W = (2\mu - Y) \sim \texttt{ST2}(\mu, \sigma, -\nu, \tau)$ is a reflection of the distribution of Y about μ.

As $\nu \to \infty$, ST2(μ, σ, ν, τ) tends to a half t family distribution, i.e. a half TF(μ, σ, τ), see Table 13.2. Also ST2$(\mu, \sigma, 0, \tau)$ = TF(μ, σ, τ).

TABLE 18.25: Skew t type 2 distribution.

ST2(μ, σ, ν, τ)	
Ranges	
Y	$-\infty < y < \infty$
μ	$-\infty < \mu < \infty$, location shift parameter
σ	$0 < \sigma < \infty$, scaling parameter
ν	$-\infty < \nu < \infty$, skewness parameter
τ	$0 < \tau < \infty$, kurtosis parameter
Distribution measures	
mean	$\begin{cases} \mu + \sigma \text{E}(Z) \\ \text{where } Z = (Y-\mu)/\sigma \text{ and} \\ \text{E}(Z) = \dfrac{\nu \tau^{1/2} \Gamma([\tau-1]/2)}{(1+\nu^2)^{1/2} \pi^{1/2} \Gamma(\tau/2)}, \text{ for } \tau > 1 \end{cases}$
variance	$\sigma^2 \text{Var}(Z) = \sigma^2 \left\{ \left(\dfrac{\tau}{\tau-2} \right) - [\text{E}(Z)]^2 \right\}$, for $\tau > 2$
skewness	$\begin{cases} \mu_{3Y}/[\text{Var}(Y)]^{1.5} \text{ where} \\ \mu_{3Y} = \sigma^3 \mu_{3Z} = \sigma^3 \{\mu'_{3Z} - 3\text{Var}(Z)\text{E}(Z) - [\text{E}(Z)]^3\} \\ \mu'_{3Z} = \text{E}(Z^3) = \dfrac{\tau(3-\delta^2)}{(\tau-3)} \text{E}(Z) \text{ for } \tau > 3 \\ \delta = \nu(1+\nu^2)^{-1/2} \end{cases}$
excess kurtosis	$\begin{cases} \mu_{4Y}/[\text{Var}(Y)]^2 - 3 \text{ where} \\ \mu_{4Y} = \sigma^4 \mu_{4Z} = \sigma^4 \{\mu'_{4Z} - 4\mu'_{3Z}\text{E}(Z) + 6\text{Var}(Z)[\text{E}(Z)]^2 \\ \quad +3[\text{E}(Z)]^4\} \\ \mu'_{4Z} = \text{E}(Z^4) = \dfrac{3\tau^2}{(\tau-2)(\tau-4)} \text{ for } \tau > 4 \end{cases}$
pdf	$\begin{cases} \dfrac{2}{\sigma} f_{Z_1}(z) F_{Z_2}(\omega) \\ \text{where } z = (y-\mu)/\sigma,\ \omega = \nu \lambda^{1/2} z,\ \lambda = (\tau+1)/(\tau+z^2) \\ \text{and } Z_1 \sim t_\tau \text{ and } Z_2 \sim t_{\tau+1} \end{cases}$
Reference	Azzalini and Capitanio [2003], p380, equation (26) and p382 with dimension $d = 1$ and $(\xi, \omega, \alpha, \nu)$ and Ω replaced by (μ, σ, ν, τ) and σ^2, respectively, giving the pdf of Y and moments of Z.

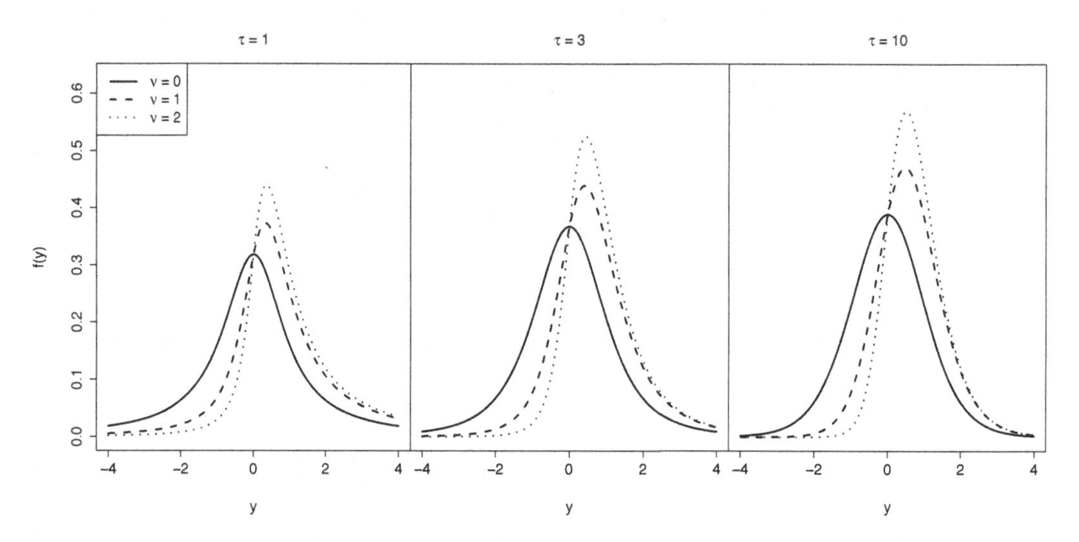

FIGURE 18.25: The skew t type 2, $\texttt{ST2}(\mu, \sigma, \nu, \tau)$, distribution, with $\mu = 0$, $\sigma = 1$, $\nu = 0, 1, 2$, and $\tau = 1, 3, 10$. It is symmetric if $\nu = 0$. Changing ν to $-\nu$ reflects the distribution about $y = \mu$, i.e. $y = 0$ above.

18.4.14 Skew t type 3: `ST3`

This is a 'scale-spliced' distribution (see Section 13.5), denoted by $\texttt{ST3}(\mu, \sigma, \nu, \tau)$, with pdf

$$f_Y(y \mid \mu, \sigma, \nu, \tau) = \begin{cases} \dfrac{c}{\sigma}\left(1 + \dfrac{\nu^2 z^2}{\tau}\right)^{-(\tau+1)/2} & \text{if } y < \mu \\ \dfrac{c}{\sigma}\left(1 + \dfrac{z^2}{\nu^2 \tau}\right)^{-(\tau+1)/2} & \text{if } y \geq \mu \end{cases} \tag{18.40}$$

for $-\infty < y < \infty$, where $-\infty < \mu < \infty$, $\sigma > 0$, $\nu > 0$, and $\tau > 0$, and where $z = (y - \mu)/\sigma$ and $c = 2\nu[(1 + \nu^2)B(1/2, \tau/2)\tau^{1/2}]^{-1}$, [Fernandez and Steel [1998], p362, equation (13), with (γ, ν) replaced by (ν, τ)]. Note that μ is the mode of Y. The moments of Y in Table 18.26 are obtained using Fernandez and Steel [1998], p360, equation (5).

Parameter ν mainly affects the skewness, with symmetry when $\nu = 1$, while parameter τ mainly affects the heaviness of the tails with increasing τ decreasing the heaviness of the tails. The `ST3` distribution appears to be positively skew if $\nu > 1$, negatively skew if $0 < \nu < 1$, and always leptokurtic. Figures 16.2(c), 16.5(b), and 16.8(b) show moment and centile kurtosis-skewness plots for $\texttt{ST3}(\mu, \sigma, \nu, \tau)$.

If $Y \sim \texttt{ST3}(\mu, \sigma, \nu, \tau)$ then $-Y \sim \texttt{ST3}(-\mu, \sigma, 1/\nu, \tau)$ is a reflection of the distribution of Y about zero. Also $W = (2\mu - Y) \sim \texttt{ST3}(\mu, \sigma, 1/\nu, \tau)$ is a reflection of the distribution of Y about μ. As $\nu \to \infty$, $\texttt{ST3}(\mu, \sigma, \nu, \tau)$ tends to a half t family distribution, i.e. a half $\texttt{TF}(\mu, \sigma, \tau)$, see Table 13.3. As $\tau \to \infty$, $\texttt{ST3}(\mu, \sigma, \nu, \tau)$ tends to a skew normal type 2, $\texttt{SN2}(\mu, \sigma, \nu)$, distribution, see Table 13.3. As $\nu \to \infty$ and $\tau \to \infty$, $\texttt{ST3}(\mu, \sigma, \nu, \tau)$ tends to a half normal distribution, i.e. a half $\texttt{NO}(\mu, \sigma)$. Also $\texttt{ST3}(\mu, \sigma, 1, \tau) = \texttt{TF}(\mu, \sigma, \tau)$.

TABLE 18.26: Skew t type 3 distribution.

ST3(μ, σ, ν, τ)	
Ranges	
Y	$-\infty < y < \infty$
μ	$-\infty < \mu < \infty$, mode, location shift parameter
σ	$0 < \sigma < \infty$, scaling parameter
ν	$0 < \nu < \infty$, skewness parameter
τ	$0 < \tau < \infty$, kurtosis parameter
Distribution measures	
mean	$\begin{cases} \mu + \sigma \mathrm{E}(Z) = \mu + \dfrac{2\sigma\tau^{1/2}(\nu - \nu^{-1})}{(\tau-1)B(1/2, \tau/2)}, \text{ for } \tau > 1 \\ \text{where } Z = (Y-\mu)/\sigma \end{cases}$
median	$\begin{cases} \mu + \frac{\sigma}{\nu} t_{\alpha_1,\tau} & \text{if } \nu \le 1 \\ \mu + \sigma\nu t_{\alpha_2,\tau} & \text{if } \nu > 1 \\ \text{where } \alpha_1 = \frac{(1+\nu^2)}{4}, \alpha_2 = \frac{(3\nu^2-1)}{4\nu^2}, & t_{\alpha,\tau} = F_T^{-1}(\alpha) \text{ where } T \sim t_\tau \end{cases}$
mode	μ
variance	$\sigma^2 \mathrm{Var}(Z) = \sigma^2 \left\{ \left(\dfrac{\tau}{\tau-2}\right)(\nu^2 + \nu^{-2} - 1) - [\mathrm{E}(Z)]^2 \right\}$, for $\tau > 2$
skewness	$\begin{cases} \mu_{3Y}/[\mathrm{Var}(Y)]^{1.5} \text{ where} \\ \mu_{3Y} = \sigma^3 \mu_{3Z} = \sigma^3\{\mu'_{3Z} - 3\mathrm{Var}(Z)\mathrm{E}(Z) - [\mathrm{E}(Z)]^3\} \\ \mu'_{3Z} = \mathrm{E}(Z^3) = \dfrac{4\tau^{3/2}(\nu^4 - \nu^{-4})}{(\tau-1)(\tau-3)B(1/2,\tau/2)(\nu+\nu^{-1})}, \text{ for } \tau > 3 \end{cases}$
excess kurtosis	$\begin{cases} \mu_{4Y}/[\mathrm{Var}(Y)]^2 - 3 \text{ where} \\ \mu_{4Y} = \sigma^4 \mu_{4Z} = \sigma^4\{\mu'_{4Z} - 4\mu'_{3Z}\mathrm{E}(Z) + 6\mathrm{Var}(Z)[\mathrm{E}(Z)]^2 + 3[\mathrm{E}(Z)]^4\} \\ \mu'_{4Z} = \mathrm{E}(Z^4) = \dfrac{3\tau^2(\nu^5 + \nu^{-5})}{(\tau-2)(\tau-4)(\nu+\nu^{-1})}, \text{ for } \tau > 4 \end{cases}$
pdf [a]	$\begin{cases} \dfrac{c}{\sigma}\left(1 + \dfrac{\nu^2 z^2}{\tau}\right)^{-(\tau+1)/2} & \text{if } y < \mu \\ \dfrac{c}{\sigma}\left(1 + \dfrac{z^2}{\nu^2\tau}\right)^{-(\tau+1)/2} & \text{if } y \ge \mu \\ \text{where } z = (y-\mu)/\sigma \text{ and} \\ c = 2\nu[(1+\nu^2)B(1/2,\tau/2)\tau^{1/2}]^{-1} \end{cases}$
cdf	$\begin{cases} \dfrac{2}{(1+\nu^2)} F_T[\nu(y-\mu)/\sigma] & \text{if } y < \mu \\ \dfrac{1}{(1+\nu^2)}\left[1 + 2\nu^2\left\{F_T[(y-\mu)/(\sigma\nu)] - \frac{1}{2}\right\}\right] & \text{if } y \ge \mu \end{cases}$
inverse cdf	$\begin{cases} \mu + \dfrac{\sigma}{\nu} t_{\alpha_3,\tau} & \text{if } p \le (1+\nu^2)^{-1} \\ \mu + \sigma\nu t_{\alpha_4,\tau} & \text{if } p > (1+\nu^2)^{-1} \\ \text{where } \alpha_3 = \frac{p(1+\nu^2)}{2}, \alpha_4 = \frac{p(1+\nu^2)-1+\nu^2}{2\nu^2} \end{cases}$
Reference	[a] From Fernandez and Steel [1998], p362, equation (13), with (γ, ν) replaced by (ν, τ).

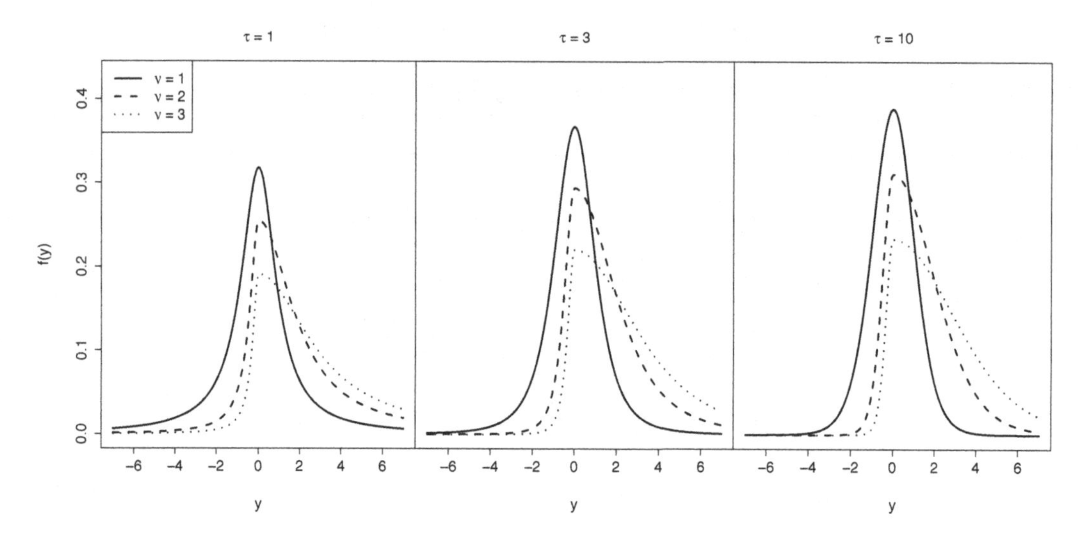

FIGURE 18.26: The skew t type 3, $\texttt{ST3}(\mu, \sigma, \nu, \tau)$, distribution, with $\mu = 0$, $\sigma = 1$, $\nu = 1, 2, 3$, and $\tau = 1, 3, 10$. It is symmetric if $\nu = 1$. Changing ν to $1/\nu$ reflects the distribution about $y = \mu$, i.e. $y = 0$ above.

18.4.15 Skew t type 4: ST4

This is a 'shape-spliced', distribution, (see Section 13.5), denoted by $\texttt{ST4}(\mu, \sigma, \nu, \tau)$, with pdf

$$f_Y(y \mid \mu, \sigma, \nu, \tau) = \begin{cases} \dfrac{c}{\sigma}\left(1 + \dfrac{z^2}{\nu}\right)^{-(\nu+1)/2} & \text{if } y < \mu \\ \dfrac{c}{\sigma}\left(1 + \dfrac{z^2}{\tau}\right)^{-(\tau+1)/2} & \text{if } y \geq \mu \end{cases} \tag{18.41}$$

for $-\infty < y < \infty$, where $-\infty < \mu < \infty$, $\sigma > 0$, $\nu > 0$ and $\tau > 0$, and where $z = (y - \mu)/\sigma$ and $c = 2\left[\nu^{1/2} B\left(1/2, \nu/2\right) + \tau^{1/2} B\left(1/2, \tau/2\right)\right]^{-1}$.

Parameters ν and τ affect the left and right tail heaviness respectively, with increasing ν (τ) decreasing the left (right) tail heaviness, both always heavier than the normal distribution tails. The $\texttt{ST4}(\mu, \sigma, \nu, \tau)$ distribution is symmetric if $\nu = \tau$ and $\texttt{ST4}(\mu, \sigma, \nu, \nu) \equiv \texttt{TF}(\mu, \sigma, \nu)$. If $Y \sim \texttt{ST4}(\mu, \sigma, \nu, \tau)$ then $-Y \sim \texttt{ST4}(-\mu, \sigma, \tau, \nu)$ is a reflection of the distribution of Y about zero. Also $W = (2\mu - Y) \sim \texttt{ST4}(\mu, \sigma, \tau, \nu)$ is a reflection of the distribution of Y about μ.

TABLE 18.27: Skew t type 4 distribution.

ST4(μ, σ, ν, τ)	
Ranges	
Y	$-\infty < y < \infty$
μ	$-\infty < \mu < \infty$, mode, location shift parameter
σ	$0 < \sigma < \infty$, scaling parameter
ν	$0 < \nu < \infty$, left tail heaviness parameter
τ	$0 < \tau < \infty$, right tail heaviness parameter
Distribution measures	
mean	$\begin{cases} \mu + \sigma E(Z) = \mu + \sigma c\left[\tau/(\tau-1) - \nu/(\nu-1)\right] \\ \text{for } \nu > 1 \text{ and } \tau > 1 \text{ where } Z = (Y-\mu)/\sigma\ , c = 2[b_1 + b_2]^{-1} \\ b_1 = \nu^{1/2} B(1/2, \nu/2)\ , \ b_2 = \tau^{1/2} B(1/2, \tau/2) \end{cases}$
median	$\begin{cases} \mu + \sigma t_{\alpha_1, \nu} & \text{if } k \le 1 \\ \mu + \sigma t_{\alpha_2, \tau} & \text{if } k > 1 \\ \text{where } \alpha_1 = \frac{(1+k)}{4}\ , \ \alpha_2 = \frac{(3k-1)}{4k}\ , \ k = b_2/b_1, & \\ t_{\alpha,\tau} = F_T^{-1}(\alpha) \text{ where } T \sim t_\tau & \end{cases}$
mode	μ
Variance	$\begin{cases} \sigma^2 \text{Var}(Z) = \sigma^2\{\text{E}(Z^2) - [\text{E}(Z)]^2\} \text{ where} \\ \text{E}(Z^2) = \dfrac{c\tau b_2}{2(\tau-2)} + \dfrac{c\nu b_1}{2(\nu-2)} \text{ for } \nu > 2 \text{ and } \tau > 2 \end{cases}$
skewness	$\begin{cases} \mu_{3Y}/[\text{Var}(Y)]^{1.5} \text{ where} \\ \mu_{3Y} = \sigma^3 \mu_{3Z} = \sigma^3\{\mu'_{3Z} - 3\text{Var}(Z)\text{E}(Z) - [\text{E}(Z)]^3\} \\ \mu'_{3Z} = \text{E}(Z^3) = 2c\left[\dfrac{\tau^2}{(\tau-1)(\tau-3)} - \dfrac{\nu^2}{(\nu-1)(\nu-3)}\right] \\ \text{for } \nu > 3 \text{ and } \tau > 3 \end{cases}$
excess kurtosis	$\begin{cases} \mu_{4Y}/[\text{Var}(Y)]^2 - 3 \text{ where} \\ \mu_{4Y} = \sigma^4 \mu_{4Z} = \\ \qquad \sigma^4\{\mu'_{4Z} - 4\mu'_{3Z}\text{E}(Z) + 6\text{Var}(Z)[\text{E}(Z)]^2 + 3[\text{E}(Z)]^4\} \\ \mu'_{4Z} = E(Z^4) = 3 + 3c\left[\dfrac{b_2}{(\tau-4)} + \dfrac{b_1}{(\nu-4)}\right] \text{ for } \nu > 4 \text{ and } \tau > 4 \end{cases}$
pdf	$\begin{cases} \dfrac{c}{\sigma}\left(1 + \dfrac{z^2}{\nu}\right)^{-(\nu+1)/2} & \text{if } y < \mu \\ \dfrac{c}{\sigma}\left(1 + \dfrac{z^2}{\tau}\right)^{-(\tau+1)/2} & \text{if } y \ge \mu \\ \text{where } z = (y-\mu)/\sigma & \end{cases}$
cdf	$\begin{cases} \frac{2}{(1+k)} F_{T_1}(z) & \text{if } y < \mu \\ \{1 + 2k\left[F_{T_2}(z) - 0.5\right]\}/(1+k) & \text{if } y \ge \mu \\ \text{where } T_1 \sim t_\nu,\ T_2 \sim t_\tau,\ z = (y-\mu)/\sigma & \end{cases}$
inverse cdf	$\begin{cases} \mu + \sigma t_{\alpha_3, \nu} & \text{if } p \le (1+k)^{-1} \\ \mu + \sigma t_{\alpha_4, \tau} & \text{if } p > (1+k)^{-1} \\ \text{where } \alpha_3 = \frac{p(1+k)}{2}\ , \ \alpha_4 = \frac{p(1+k)-1}{2k} + 0.5 & \end{cases}$

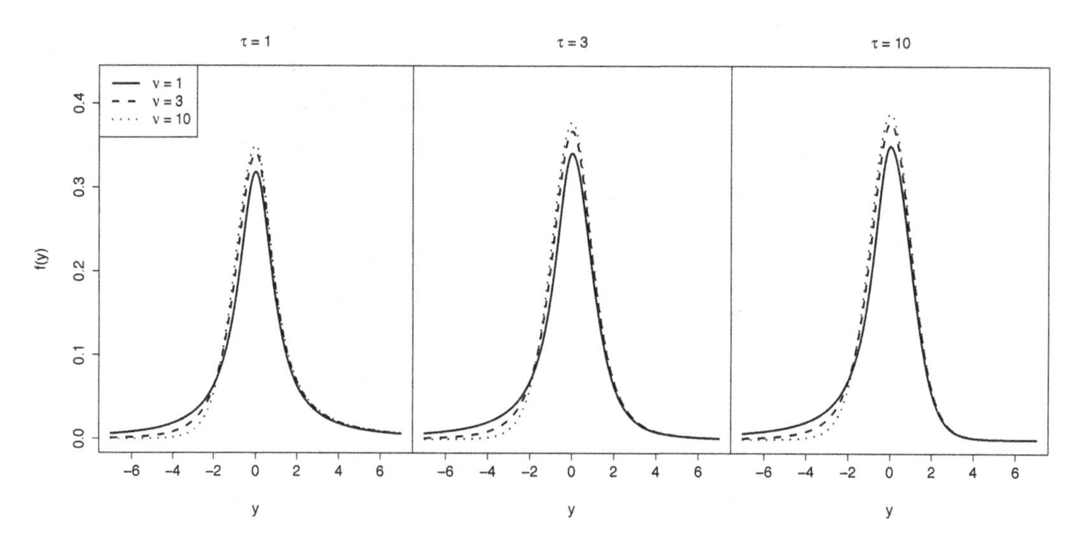

FIGURE 18.27: The skew t type 4, $\texttt{ST4}(\mu, \sigma, \nu, \tau)$, distribution, with $\mu = 0$, $\sigma = 1$, $\nu = 1, 3, 10$, and $\tau = 1, 3, 10$. It is symmetric if $\nu = \tau$. Interchanging ν and τ reflects the distribution about $y = \mu$, i.e. $y = 0$ above.

18.4.16 Skew t type 5: ST5

The pdf of the skew t distribution type 5, denoted by $\texttt{ST5}(\mu, \sigma, \nu, \tau)$, [Jones and Faddy, 2003] is given by

$$f_Y(y \mid \mu, \sigma, \nu, \tau) = \frac{c}{\sigma}\left[1 + \frac{z}{(a+b+z^2)^{1/2}}\right]^{a+1/2}\left[1 - \frac{z}{(a+b+z^2)^{1/2}}\right]^{b+1/2}$$

for $-\infty < y < \infty$, where $-\infty < \mu < \infty$, $\sigma > 0$, $-\infty < \nu < \infty$ and $\tau > 0$, and where $z = (y - \mu)/\sigma$ and

$$\begin{aligned} c &= \left[2^{a+b-1}(a+b)^{1/2}B(a,b)\right]^{-1} \\ \nu &= (a-b)/\left[ab(a+b)\right]^{1/2} \\ \tau &= 2/(a+b)\ . \end{aligned}$$

Hence

$$\begin{aligned} a &= \tau^{-1}[1 + \nu(2\tau + \nu^2)^{-1/2}] \\ b &= \tau^{-1}[1 - \nu(2\tau + \nu^2)^{-1/2}]\ . \end{aligned} \tag{18.42}$$

From Jones and Faddy [2003] p160, equation (2), if $B \sim \texttt{BEo}(a, b)$ then

$$Z = \frac{(a+b)^{1/2}(2B-1)}{2[B(1-B)]^{1/2}} \sim \texttt{ST5}(0, 1, \nu, \tau)\ .$$

Hence $B = \frac{1}{2}\left[1 + Z(a+b+Z^2)^{-1/2}\right] \sim \texttt{BEo}(a, b)$, from which the cdf of Y is obtained. Parameter ν is a skewess parameter with symmetry when $\nu = 0$. The distribution is positively moment skewed if $\nu > 0$ and negatively moment skewed if $\nu < 0$, [Jones and Faddy, 2003, p159]. If $Y \sim \texttt{ST5}(\mu, \sigma, \nu, \tau)$ then $-Y \sim \texttt{ST5}(-\mu, \sigma, -\nu, \tau)$ is a reflection of the distribution of Y about zero. Also $W = (2\mu - Y) \sim \texttt{ST5}(\mu, \sigma, -\nu, \tau)$ is a reflection of the distribution of Y about μ.

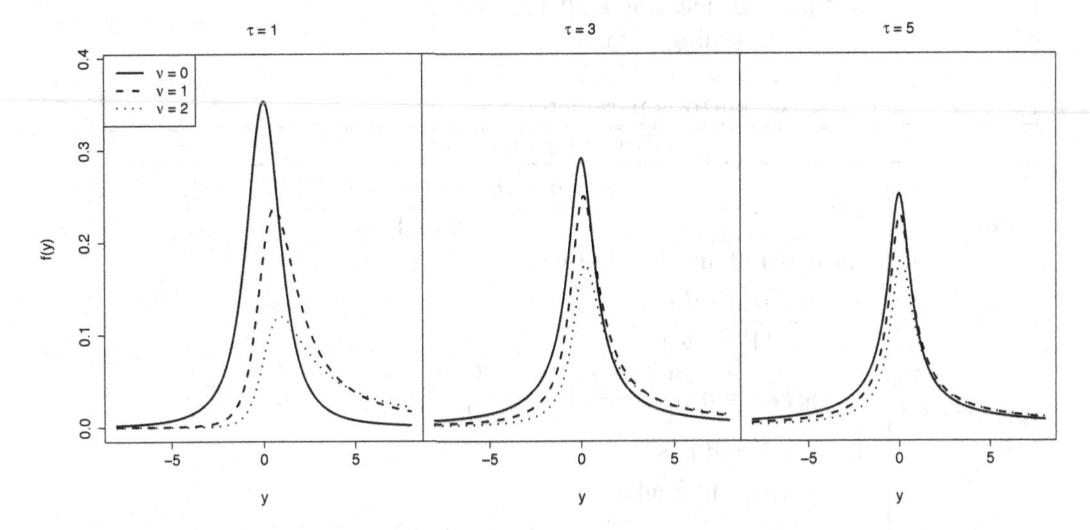

FIGURE 18.28: The skew t type 5, $\texttt{ST5}(\mu, \sigma, \nu, \tau)$, distribution, with $\mu = 0$, $\sigma = 1$, $\nu = 0, 1, 2$, and $\tau = 1, 3, 5$. It is symmetric if $\nu = 0$. Changing ν to $-\nu$ reflects the distribution about $y = \mu$, i.e. $y = 0$ above.

TABLE 18.28: Skew t type 5 distribution.

$\texttt{ST5}(\mu, \sigma, \nu, \tau)$	
Ranges	
Y	$-\infty < y < \infty$
μ	$-\infty < \mu < \infty$, location shift parameter
σ	$0 < \sigma < \infty$, scaling parameter
ν	$-\infty < \nu < \infty$, skewness parameter
τ	$0 < \tau < \infty$, kurtosis parameter
Distribution measures	
mean	$\begin{cases} \mu + \sigma \mathrm{E}(Z) = \mu + \dfrac{\sigma(a+b)^{1/2}(a-b)\Gamma(a-1/2)\Gamma(b-1/2)}{2\Gamma(a)\Gamma(b)} \\ \text{for } a > 1/2 \text{ and } b > 1/2 \text{ where } Z = (Y-\mu)/\sigma \end{cases}$
mode	$\mu + \dfrac{\sigma(a+b)^{1/2}(a-b)}{(2a+1)^{1/2}(2b+1)^{1/2}}$
variance	$\begin{cases} \sigma^2 \mathrm{Var}(Z) = \sigma^2 \left\{ \dfrac{(a+b)[(a-b)^2 + a + b - 2]}{4(a-1)(b-1)} - [\mathrm{E}(Z)]^2 \right\} \\ \text{for } a > 1 \text{ and } b > 1 \end{cases}$
skewness	$\begin{cases} \mu_{3Y}/[\mathrm{Var}(Y)]^{1.5} \text{ where} \\ \mu_{3Y} = \sigma^3 \mu_{3Z} = \sigma^3 \{\mu_{3Z}^{'} - 3\mathrm{Var}(Z)\mathrm{E}(Z) - [\mathrm{E}(Z)]^3\} \\ \mu_{3Z}^{'} = \mathrm{E}(Z^3) = \dfrac{(a+b)^{3/2}\Gamma(a-3/2)\Gamma(b-3/2)}{8\Gamma(a)\Gamma(b)} \times \\ \quad [a^3 + 3a^2 - 7a - b^3 - 3b^2 + 7b + 3ab^2 - 3a^2 b] \\ \quad \text{for } a > 3/2 \text{ and } b > 3/2 \end{cases}$
excess kurtosis	$\begin{cases} \mu_{4Y}/[\mathrm{Var}(Y)]^2 - 3 \text{ where} \\ \mu_{4Y} = \sigma^4 \mu_{4Z} = \sigma^4 \{\mu_{4Z}^{'} - 4\mu_{3Z}^{'}\mathrm{E}(Z) + 6\mathrm{Var}(Z)[\mathrm{E}(Z)]^2 + 3[\mathrm{E}(Z)]^4\} \\ \text{where } \mu_{4Z}^{'} = \mathrm{E}(Z^4) = \dfrac{(a+b)^2}{16(a-1)(a-2)(b-1)(b-2)} \times \\ \quad [a^4 - 2a^3 - a^2 + 2a + b^4 - 2b^3 - b^2 + 2b + \\ \quad 2(a-2)(b-2)(3ab - 2a^2 - 2b^2 - a - b + 3)] \\ \quad \text{for } a > 2 \text{ and } b > 2 \end{cases}$
pdf	$\begin{cases} f_Y(y \mid \mu, \sigma, \nu, \tau) = \dfrac{c}{\sigma} \left[1 + \frac{z}{(a+b+z^2)^{1/2}}\right]^{a+1/2} \left[1 - \frac{z}{(a+b+z^2)^{1/2}}\right]^{b+1/2} \\ \text{where } z = (y-\mu)/\sigma, c = \left[2^{a+b-1}(a+b)^{1/2}B(a,b)\right]^{-1} \\ a = \tau^{-1}[1+d] \text{ and } b = \tau^{-1}[1-d] \text{ where } d = \nu(2\tau + \nu^2)^{-1/2} \end{cases}$
cdf	$\begin{cases} \dfrac{B(a,b,r)}{B(a,b)} \\ \text{where } r = \frac{1}{2}[1 + z(a+b+z^2)^{-1/2}] \text{ and } z = (y-\mu)/\sigma \end{cases}$
Reference	Jones and Faddy [2003], p159, equation (1) reparameterized as on their p164 with (q, p) replaced by (ν, τ), and p162 equations (4b) and (5) giving the pdf, moments and mode of Z, respectively.

19

Continuous distributions on $(0, \infty)$

CONTENTS

This chapter gives summary tables and plots for the explicit **gamlss.dist** continuous distributions with range $(0, \infty)$. These are discussed in Chapter 5. Section 5.6 discuesses creating distributions on $(0, \infty)$ in **gamlss**, either by an inverse log transform or by truncation below zero, from any **gamlss.dist** distribution on $(-\infty, \infty)$.

19.1 Scale family of distributions

A continuous random variable Y defined on $(0, \infty)$, is said to have a scale family of distributions with scaling parameter θ (for fixed values of all other parameters of the distribution) if

$$Z = \frac{Y}{\theta}$$

has a cdf which does not depend on θ. Hence

$$F_Y(y) = F_Z\left(\frac{y}{\theta}\right)$$

and

$$f_Y(y) = \frac{1}{\theta} f_Z\left(\frac{y}{\theta}\right),$$

so $F_Y(y)$ and $\theta f_Y(y)$ only depend on y and θ through the function $z = y/\theta$.

Example: Let Y have a Weibull distribution, $Y \sim$ `WEI`(μ, σ), then Y has a scale family of distributions with scaling parameter μ (for a fixed value of σ), since $F_Y(y) = 1 - \exp\left[-(y/\mu)^\sigma\right]$ and hence $Z = Y/\mu$ has cdf $F_Z(z) = 1 - \exp\left[-(z)^\sigma\right]$ which does not depend on μ. Note $Z \sim$ `WEI`$(1, \sigma)$.

All distributions in **gamlss.dist** with range $(0, \infty)$ are scale families of distributions with scaling parameter μ, except for `GAF`, `IG`, `LNO`, `LOGNO`, and `WEI2`. Hence for these scale family distributions, $Y = \mu Z$ where the distribution of Z does not depend on μ. Hence for fixed parameers other than μ, $\text{E}(Y) = \mu\text{E}(Z)$ and $\text{Var}(Y) = \mu^2\text{Var}(Z)$. Hence, $\text{E}(Y)$ is proportional to μ provided $\text{E}(Z)$ is finite, $\text{Var}(Y)$ is proportional to μ^2, and the standard deviation of Y is proportional to μ, provided $\text{Var}(Z)$ is finite. Note also that $\text{Var}(Y) = [\text{E}(Y)]^2\text{Var}(Z)/[\text{E}(Z)]^2$, so for these scale family distributions the variance-mean relationship is squared, for fixed parameters other than μ. The gamma family, `GAF`(μ, σ, ν), distribution allows for a power variance-mean relationship, since it has mean μ and variance $\sigma^2\mu^\nu$.

For scale family distributions plotted in this chapter we fix $\mu = 1$, since changing μ from 1 to say μ_1 just scales the horizontal axis in the figures by the factor μ_1 and scales the vertical axis by the factor $1/\mu_1$.

19.2 Continuous one-parameter distribution on $(0, \infty)$

19.2.1 Exponential: `EXP`

This is the only one-parameter continuous distribution in the **gamlss.dist** package. The pdf of the exponential distribution, denoted by `EXP`(μ), is given by

$$f_Y(y \mid \mu) = \frac{1}{\mu} \exp\left(-\frac{y}{\mu}\right) \tag{19.1}$$

for $y > 0$, where $\mu > 0$. Hence $\text{E}(Y) = \mu$ and $\text{Var}(Y) = \mu^2$. The exponential distribution is appropriate for moderately positively skewed data see Figure 1.1.

TABLE 19.1: Exponential distribution.

$\texttt{EXP}(\mu)$	
Ranges	
Y	$0 < y < \infty$
μ	$0 < \mu < \infty$, mean, scaling parameter
Distribution measures	
mean	μ
median	$\mu \log 2 \approx 0.69315\mu$
mode	$\to 0$
variance	μ^2
skewness	2
excess kurtosis	6
MGF	$(1 - \mu t)^{-1}$ for $t < 1/\mu$
pdf	$(1/\mu)\exp(-y/\mu)$
cdf	$1 - \exp(-y/\mu)$
inverse cdf	$-\mu \log(1 - p)$
Reference	Johnson et al. [1994] Chapter 19, p494-499, or set $\sigma = 1$ in $\texttt{GA}(\mu, \sigma)$

19.3 Continuous two-parameter distributions on $(0, \infty)$

19.3.1 Gamma: GA

The pdf of the gamma distribution, denoted by $\texttt{GA}(\mu, \sigma)$, is given by

$$f_Y(y \mid \mu, \sigma) = \frac{y^{1/\sigma^2 - 1} e^{-y/(\sigma^2 \mu)}}{(\sigma^2 \mu)^{1/\sigma^2} \Gamma(1/\sigma^2)} \tag{19.2}$$

for $y > 0$, where $\mu > 0$ and $\sigma > 0$. Here $\text{E}(Y) = \mu$ and $\text{Var}(Y) = \sigma^2\mu^2$ and $\text{E}(Y^r) = \mu^r \sigma^{2r} \Gamma\left(1/\sigma^2 + r\right) / \Gamma\left(1/\sigma^2\right)$ for $r > -1/\sigma^2$.

This a reparameterization of Johnson et al. [1994] p343, equation (17.23), obtained by setting $\alpha = 1/\sigma^2$ and $\beta = \mu\sigma^2$. Hence $\mu = \alpha\beta$ and $\sigma^2 = 1/\alpha$.

The gamma distribution is appropriate for positively skewed data.

Parameter σ is a primary true van Zwet skewness parameter, (see van Zwet [1964a,b]). Note also that σ is a true variance parameter (since the variance inscreases monotonically with σ, for fixed μ), and σ is also a primary true moment skewness parameter and a primary true moment kurtosis parameter (since the moment skewness and moment kurtosis increase monotonically with σ, irrespective of the value of μ), see Sections 14.5 and 15.5.

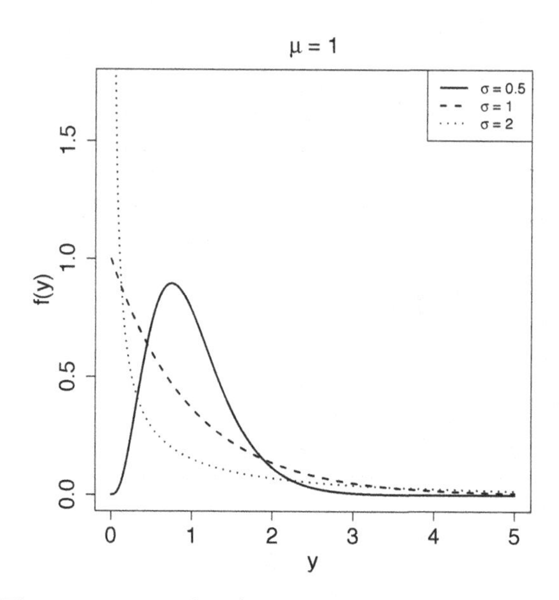

FIGURE 19.1: The gamma, GA(μ, σ), distribution, with $\mu = 1$ and $\sigma = 0.5, 1, 2$.

TABLE 19.2: Gamma distribution.

GA(μ, σ)	
Ranges	
Y	$0 < y < \infty$
μ	$0 < \mu < \infty$, mean, scaling parameter
σ	$0 < \sigma < \infty$, coefficient of variation
Distribution measures	
mean [a]	μ
mode	$\begin{cases} \mu(1-\sigma^2) & \text{if } \sigma < 1 \\ \to 0 & \text{if } \sigma \geq 1 \end{cases}$
variance [a]	$\sigma^2\mu^2$
skewness [a]	2σ
excess kurtosis [a]	$6\sigma^2$
MGF	$(1-\mu\sigma^2 t)^{-1/\sigma^2}$ for $t < (\mu\sigma^2)^{-1}$
pdf [a,a2]	$\dfrac{y^{1/\sigma^2-1} e^{-y/(\sigma^2\mu)}}{(\sigma^2\mu)^{1/\sigma^2}\Gamma(\sigma^{-2})}$
cdf	$\dfrac{\gamma(\sigma^{-2}, y\mu^{-1}\sigma^{-2})}{\Gamma(\sigma^{-2})}$
Reference	[a] Derived from McCullagh and Nelder [1989] p287 reparameterized by $\mu = \mu$ and $\nu = 1/\sigma^2$ [a2] Johnson et al. [1994] p343, equation (17.23) reparameterized by $\alpha = 1/\sigma^2$ and $\beta = \mu\sigma^2$.
Note	$\gamma(a, x) = \int_0^x t^{a-1}e^{-t}dt$, the incomplete gamma function.

19.3.2 Inverse gamma: IGAMMA

The pdf of the inverse gamma distribution, denoted by IGAMMA(μ, σ), is given by

$$f_Y(y \mid \mu, \sigma) = \frac{\mu^{\alpha}(\alpha+1)^{\alpha} y^{-(\alpha+1)}}{\Gamma(\alpha)} \exp\left[-\frac{\mu(\alpha+1)}{y}\right]$$

for $y > 0$, where $\mu > 0$ and $\sigma > 0$ and where $\alpha = 1/\sigma^2$. The inverse gamma distribution is a reparameterized special case of the generalized gamma (GG) distribution given by IGAMMA(μ, σ) = GG($(1+\sigma^2)\mu, \sigma, -1$).

The inverse gamma distribution is appropriate for highly positively skewed data. Note that σ is a true variance parameter (for $\sigma^2 < 1/2$), and a primary true moment skewness parameter (for $\sigma^2 < 1/3$), and a primary true moment kurtosis parameter (for $\sigma^2 < 1/4$), see Sections 14.5 and 15.5.

TABLE 19.3: Inverse gamma distribution.

IGAMMA(μ, σ)	
Ranges	
Y	$0 < y < \infty$
μ	$0 < \mu < \infty$, mode, scaling parameter
σ	$0 < \sigma < \infty$
Distribution measures	
mean [a]	$\begin{cases} \dfrac{(1+\sigma^2)\mu}{(1-\sigma^2)} & \text{if } \sigma^2 < 1 \\ \infty & \text{if } \sigma^2 \geq 1 \end{cases}$
mode [a]	μ
variance [a]	$\begin{cases} \dfrac{(1+\sigma^2)^2\mu^2\sigma^2}{(1-\sigma^2)^2(1-2\sigma^2)} & \text{if } \sigma^2 < 1/2 \\ \infty & \text{if } \sigma^2 \geq 1/2 \end{cases}$
skewness [a]	$\begin{cases} \dfrac{4\sigma(1-2\sigma^2)^{1/2}}{(1-3\sigma^2)} & \text{if } \sigma^2 < 1/3 \\ \infty & \text{if } \sigma^2 \geq 1/3 \end{cases}$
excess kurtosis [a]	$\begin{cases} \dfrac{3\sigma^2(10-22\sigma^2)}{(1-3\sigma^2)(1-4\sigma^2)} & \text{if } \sigma^2 < 1/4 \\ \infty & \text{if } \sigma^2 \geq 1/4 \end{cases}$
pdf [a]	$\dfrac{\mu^{\alpha}(\alpha+1)^{\alpha} y^{-(\alpha+1)}}{\Gamma(\alpha)} \exp\left[-\dfrac{\mu(\alpha+1)}{y}\right]$, where $\alpha = 1/\sigma^2$
cdf [b]	$\dfrac{\Gamma\left[\alpha, \mu(\alpha+1)/y\right]}{\Gamma(\alpha)}$
Note	[a] Set $\mu_1 = (1+\sigma^2)\mu$, $\sigma_1 = \sigma$ and $\nu_1 = -1$ (so $\theta = 1/\sigma_1^2$) in GG(μ_1, σ_1, ν_1). [b] $\Gamma(a, x) = \int_x^{\infty} t^{a-1} e^{-t} dt$, the complement of the incomplete gamma function.

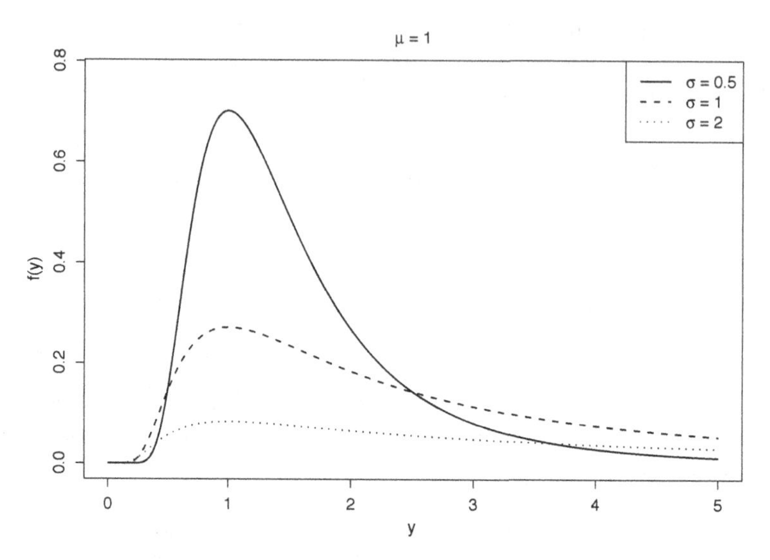

FIGURE 19.2: The inverse gamma, IGAMMA(μ, σ), distribution, with $\mu = 1$ and $\sigma = 0.5, 1, 2$.

19.3.3 Inverse Gaussian: IG

The pdf of the inverse Gaussian distribution, denoted by IG(μ, σ) is given by

$$f_Y(y \mid \mu, \sigma) = \frac{1}{\sqrt{2\pi\sigma^2 y^3}} \exp\left[-\frac{1}{2\mu^2\sigma^2 y}(y-\mu)^2\right] \tag{19.3}$$

for $y > 0$, where $\mu > 0$ and $\sigma > 0$. Hence $\text{E}(Y) = \mu$ and $\text{Var}(Y) = \sigma^2\mu^3$. This is a reparameterization of Johnson et al. [1994, p261 equation (15.4a)], obtained by setting $\sigma^2 = 1/\lambda$. Note that the inverse Gaussian distribution is a reparameterized special case of the generalized inverse Gaussian distribution, given by IG$(\mu, \sigma) =$ GIG$(\mu, \sigma\mu^{1/2}, -0.5)$.

The inverse Gaussian distribution is appropriate for highly positively skewed data.

Note also that if $Y \sim$ IG(μ, σ) then $Y_1 = aY \sim$ IG$(a\mu, a^{-1/2}\sigma)$. Hence IG(μ, σ) is a scale family of distributions, but neither μ nor σ is a scaling parameter. The shape of the IG(μ, σ) distribution depends only on the value of $\sigma^2\mu$, so if $\sigma^2\mu$ is fixed, then changing μ changes the scale but not the shape of the IG(μ, σ) distribution. If σ in the distribution IG(μ, σ) is reparameterized by setting $\alpha = \sigma\mu^{1/2}$, then for fixed α the resulting IG2(μ, α) distribution is a scale family of distributions with scaling parameter μ, since $Z = Y/\mu \sim$ IG2$(1, \alpha)$. Note IG2 is not currently available in **gamlss.dist**.

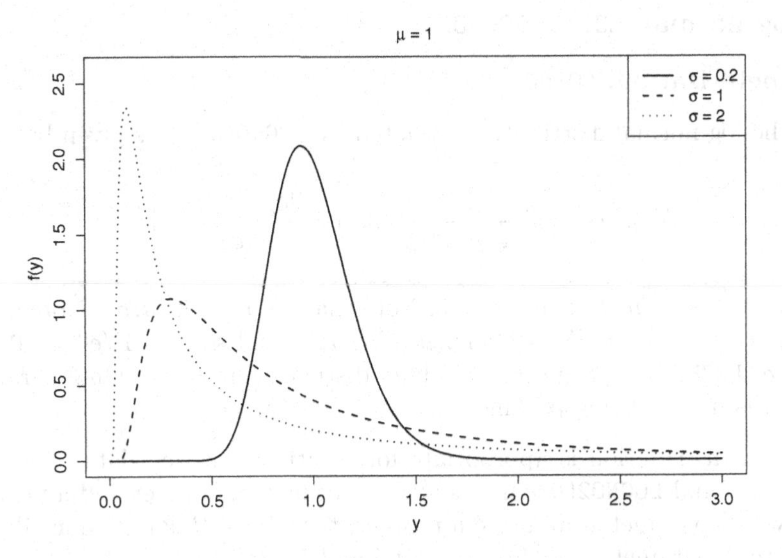

FIGURE 19.3: The inverse Gaussian, $\texttt{IG}(\mu, \sigma)$, distribution, with $\mu = 1$ and $\sigma = 0.2, 1, 2$.

TABLE 19.4: Inverse Gaussian distribution.

	$\texttt{IG}(\mu, \sigma)$
	Ranges
Y	$0 < y < \infty$
μ	$0 < \mu < \infty$, mean, *not* a scaling parameter
σ	$0 < \sigma < \infty$
	Distribution measures
mean [a]	μ
mode [a]	$\dfrac{-3\mu^2\sigma^2 + \mu(9\mu^2\sigma^4 + 4)^{1/2}}{2}$
variance [a]	$\sigma^2\mu^3$
skewness [a]	$3\mu^{1/2}\sigma$
excess kurtosis[a]	$15\mu\sigma^2$
MGF [a]	$\exp\left\{\frac{1}{\mu\sigma^2}\left[1 - \left(1 - 2\mu^2\sigma^2 t\right)^{1/2}\right)\right]\right\}$ for $t < (2\mu^2\sigma^2)^{-1}$
pdf [a]	$(2\pi\sigma^2 y^3)^{-1/2} \exp\left[-\frac{1}{2\mu^2\sigma^2 y}(y-\mu)^2\right]$
cdf [a]	$\Phi\left[(\sigma^2 y)^{-1/2}\left(\frac{y}{\mu} - 1\right)\right] + e^{2(\mu\sigma^2)^{-1}}\Phi\left[-(\sigma^2 y)^{-1/2}\left(\frac{y}{\mu} + 1\right)\right]$
Reference	[a] Johnson et al. [1994] Chapter 15, p261-263 and p268 with equation (15.4a) reparameterized by $\mu = \mu$ and $\lambda = 1/\sigma^2$, or set $\mu_1 = \mu, \sigma_1 = \sigma\mu^{1/2}$ and $\nu_1 = -1/2$, (and so $b = 1$), in $\texttt{GIG}(\mu_1, \sigma_1, \nu_1)$.

19.3.4 Log normal: LOGNO, LOGNO2

First parameterization, LOGNO

The pdf of the log normal distribution, denoted by $\texttt{LOGNO}(\mu, \sigma)$, is given by

$$f_Y(y \mid \mu, \sigma) = \frac{1}{\sqrt{2\pi\sigma^2}} \frac{1}{y} \exp\left[-\frac{(\log y - \mu)^2}{2\sigma^2}\right] \tag{19.4}$$

for $y > 0$, where $-\infty < \mu < \infty$ and $\sigma > 0$. Note that $\log Y \sim \texttt{NO}(\mu, \sigma)$. Note also that if $Y \sim \texttt{LOGNO}(\mu, \sigma)$ then $Y_1 = aY \sim \texttt{LOGNO}(\mu + \log a, \sigma)$ and so $Z = Y/e^{\mu} \sim \texttt{LOGNO}(0, \sigma)$. So, for fixed σ, $\texttt{LOGNO}(\mu, \sigma)$ is a scale family of distributions with scaling parameter e^{μ}. Hence μ itself is not a scaling parameter.

The log normal distribution is appropriate for positively skewed data. Note that, for both $\texttt{LOGNO}(\mu, \sigma)$ and $\texttt{LOGNO2}(\mu, \sigma)$, σ is a true variance parameter and a primary true moment skewness parameter, and also a primary true, both Balanda-MacGillivray and moment, kurtosis parameter, see Sections 14.5 and 15.5.

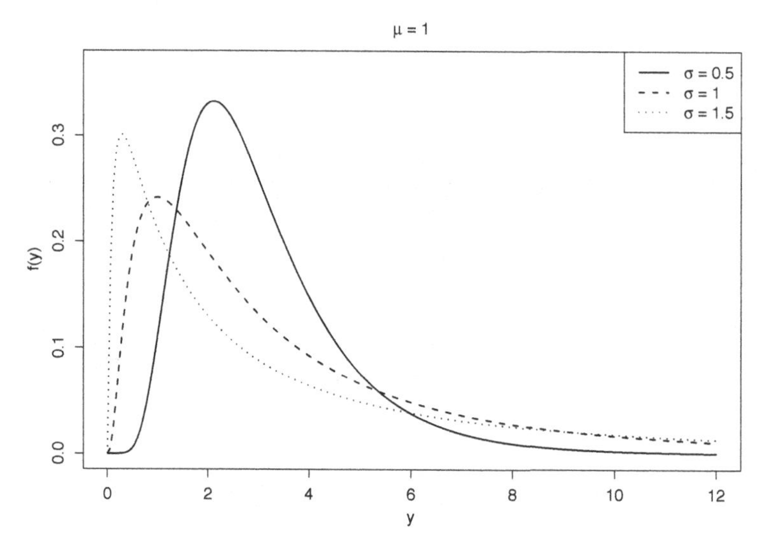

FIGURE 19.4: The log normal, $\texttt{LOGNO}(\mu, \sigma)$, distribution, with $\mu = 1$ and $\sigma = 0.5, 1, 1.5$.

TABLE 19.5: Log normal distribution.

$\texttt{LOGNO}(\mu, \sigma)$	
Ranges	
Y	$0 < y < \infty$
μ	$-\infty < \mu < \infty$, *not* a scaling parameter
σ	$0 < \sigma < \infty$
Distribution measures	
mean [a]	$e^{\mu+\sigma^2/2}$
median [a]	e^{μ}
mode [a]	$e^{\mu-\sigma^2}$
variance [a]	$e^{2\mu+\sigma^2}(e^{\sigma^2}-1)$
skewness [a]	$(e^{\sigma^2}-1)^{0.5}(e^{\sigma^2}+2)$
excess kurtosis [a]	$e^{4\sigma^2}+2e^{3\sigma^2}+3e^{2\sigma^2}-6$
pdf [a2]	$\frac{1}{\sqrt{2\pi\sigma^2}}\frac{1}{y}\exp\left[-\frac{(\log y-\mu)^2}{2\sigma^2}\right]$
cdf	$\Phi\left(\frac{\log y-\mu}{\sigma}\right)$
inverse cdf	$e^{\mu+\sigma z_p}$ where $z_p = \Phi^{-1}(p)$
Reference	[a] Johnson et al. [1994] Chapter 14, p208-213 [a2] Johnson et al. [1994] Chapter 14, p208, equation (14.2) where $\xi = \mu$, $\sigma = \sigma$, $\theta = 0$.

Second parameterization, LOGNO2

The pdf of the second parameterization of the log normal distribution, denoted by $\texttt{LOGNO2}(\mu, \sigma)$, is given by

$$f_Y(y \mid \mu, \sigma) = \frac{1}{\sqrt{2\pi\sigma^2}}\frac{1}{y}\exp\left[-\frac{(\log y - \log \mu)^2}{2\sigma^2}\right] \tag{19.5}$$

for $y > 0$, where $\mu > 0$ and $\sigma > 0$. Note that $\log Y \sim \texttt{NO}(\log \mu, \sigma)$. In this parameterization μ is the median of Y and μ is a scaling parameter, so if $Y \sim \texttt{LOGNO2}(\mu, \sigma)$ then $Z = Y/\mu \sim \texttt{LOGNO2}(0, \sigma)$ which does not depend on μ.

TABLE 19.6: Second parameterization of the log normal distribution.

LOGNO2(μ, σ)	
Ranges	
Y	$0 < y < \infty$
μ	$0 < \mu < \infty$, median, scaling parameter
σ	$0 < \sigma < \infty$
Distribution measures	
mean	$\mu e^{\sigma^2/2}$
median	μ
mode	$\mu e^{-\sigma^2}$
variance	$\mu^2 e^{\sigma^2}(e^{\sigma^2} - 1)$
skewness	$(e^{\sigma^2} - 1)^{0.5}(e^{\sigma^2} + 2)$
excess kurtosis [a]	$e^{4\sigma^2} + 2e^{3\sigma^2} + 3e^{2\sigma^2} - 6$
pdf	$\frac{1}{\sqrt{2\pi\sigma^2}} \frac{1}{y} \exp\left[-\frac{(\log y - \log \mu)^2}{2\sigma^2}\right]$
cdf	$\Phi\left(\frac{\log y - \log \mu}{\sigma}\right)$
inverse cdf	$\mu e^{\sigma z_p}$ where $z_p = \Phi^{-1}(p)$
Reference	Set $\mu_1 = \log \mu$ in LOGNO(μ_1, σ)

19.3.5 Pareto type 1: PARETO1o

The pdf of the Pareto type 1 original distribution, denoted by PARETO1o(μ, σ), is given by

$$f_Y(y \mid \mu, \sigma) = \frac{\sigma \mu^\sigma}{y^{\sigma+1}} \tag{19.6}$$

for $y > \mu$, where $\mu > 0$ and $\sigma > 0$. This is also called the Pareto distribution of the first kind [Johnson et al., 1994, p574]. It is especially important in the discussion of heavy tails, because its survivor function is $S_Y(y) = \mu^\sigma y^{-\sigma}$, for $y > \mu$, i.e. exactly proportional to $y^{-\sigma}$ (see Chapter 17). Note that σ is a true variance parameter (for $\sigma > 2$), a primary true van Zwet skewness parameter and also (for $\sigma > 4$) a primary true moment kurtosis parameter, see Sections 14.5 and 15.5.

TABLE 19.7: Pareto type 1 original distribution.

PARETO1o(μ, σ)	
Ranges	
Y	$\mu < y < \infty$
μ	$0 < \mu < \infty$, scaling parameter (fixed constant)
σ	$0 < \sigma < \infty$
Distribution measures	
mean	$\begin{cases} \dfrac{\mu\sigma}{(\sigma-1)} & \text{if } \sigma > 1 \\ \infty & \text{if } \sigma \le 1 \end{cases}$
median	$\mu 2^{1/\sigma}$
mode	$\to 0$
variance [a]	$\begin{cases} \dfrac{\sigma\mu^2}{(\sigma-1)^2(\sigma-2)} & \text{if } \sigma > 2 \\ \infty & \text{if } \sigma \le 2 \end{cases}$
skewness [a]	$\begin{cases} \dfrac{2(\sigma+1)(\sigma-2)^{1/2}}{(\sigma-3)\sigma^{1/2}} & \text{if } \sigma > 3 \\ \infty & \text{if } \sigma \le 3 \end{cases}$
excess kurtosis[a]	$\begin{cases} \dfrac{3(\sigma-2)(3\sigma^2+\sigma+2)}{\sigma(\sigma-3)(\sigma-4)} - 3 & \text{if } \sigma > 4 \\ \infty & \text{if } \sigma \le 4 \end{cases}$
pdf [a2]	$\dfrac{\sigma\mu^\sigma}{y^{\sigma+1}}$
cdf [a]	$1 - \left(\dfrac{\mu}{y}\right)^\sigma$
inverse cdf	$\mu(1-p)^{-1/\sigma}$
Reference	[a]Johnson et al. [1994] Sections 20.3, 20.4 p574-579. [a2] Johnson et al. [1994], equation (20.3), p574 with $k = \mu$, $a = \sigma$.

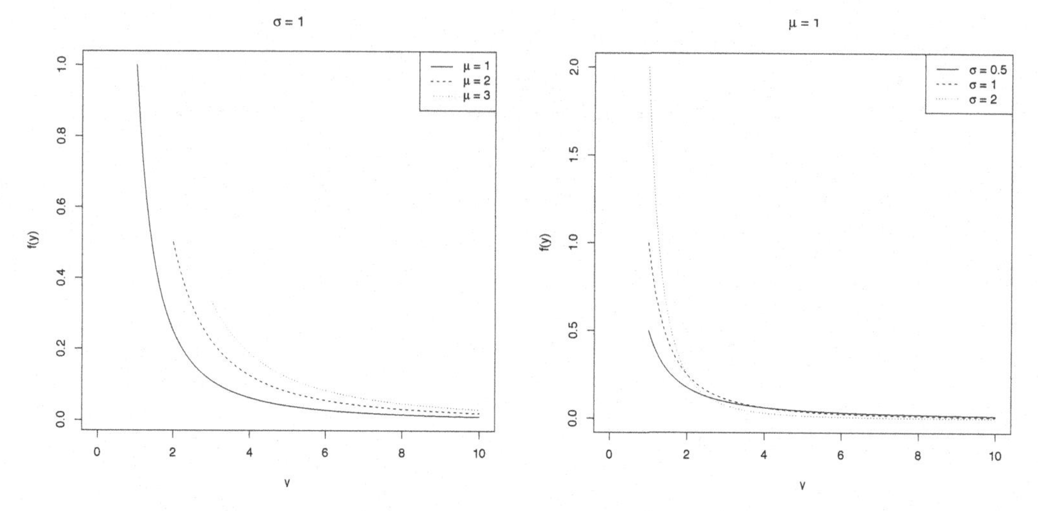

FIGURE 19.5: The Pareto, PARETO1o(μ, σ), distribution, (left) $\sigma = 1$ and $\mu = 1, 2, 3$, and (right) $\mu = 1$ and $\sigma = 0.5, 1, 2$.

19.3.6 Pareto type 2: PARETO2o, PARETO2

First parameterization, PARETO2o(μ, σ)

The pdf of the Pareto type 2 original distribution, denoted by PARETO2o(μ, σ), is

$$f_Y(y \mid \mu, \sigma) = \frac{\sigma \mu^\sigma}{(y + \mu)^{\sigma+1}} \tag{19.7}$$

for $y > 0$, where $\mu > 0$ and $\sigma > 0$. This was called the Pareto distribution of the second kind or Lomax distribution by Johnson et al. [1994, p575], where $C = \mu$ and $a = \sigma$. Note that σ is a true variance parameter (for $\sigma < 2$), a primary true van Zwet skewness parameter, and also (for $\sigma > 4$) a primary true moment kurtosis parameter, see Sections 14.5 and 15.5. Note that if $Y \sim$ PARETO2o(μ, σ) then $Z = Y + \mu \sim$ PARETO1o(μ, σ).

TABLE 19.8: Pareto type 2 original distribution.

PARETO2o(μ, σ)	
Ranges	
Y	$0 < y < \infty$
μ	$0 < \mu < \infty$, scaling parameter
σ	$0 < \sigma < \infty$
Distribution measures	
mean	$\begin{cases} \dfrac{\mu}{(\sigma-1)} & \text{if } \sigma > 1 \\ \infty & \text{if } \sigma \leq 1 \end{cases}$
median	$\mu(2^{1/\sigma} - 1)$
mode	$\to 0$
variance [a]	$\begin{cases} \dfrac{\sigma\mu^2}{(\sigma-1)^2(\sigma-2)} & \text{if } \sigma > 2 \\ \infty & \text{if } \sigma \leq 2 \end{cases}$
skewness [a]	$\begin{cases} \dfrac{2(\sigma+1)(\sigma-2)^{1/2}}{(\sigma-3)\sigma^{1/2}} & \text{if } \sigma > 3 \\ \infty & \text{if } \sigma \leq 3 \end{cases}$
excess kurtosis[a]	$\begin{cases} \dfrac{3(\sigma-2)(3\sigma^2+\sigma+2)}{\sigma(\sigma-3)(\sigma-4)} - 3 & \text{if } \sigma > 4 \\ \infty & \text{if } \sigma \leq 4 \end{cases}$
pdf [a2]	$\dfrac{\sigma\mu^\sigma}{(y+\mu)^{\sigma+1}}$
cdf [a]	$1 - \dfrac{\mu^\sigma}{(y+\mu)^\sigma}$
inverse cdf	$\mu[(1-p)^{-1/\sigma} - 1]$
Reference	[a]Johnson et al. [1994] Sections 20.3, 20.4 p574-579. [a2] Johnson et al. [1994, p574] equation (20.3), p574 with $k = \mu$, $a = \sigma$ and $x = y + \mu$, so $X = Y + \mu$.

Second parameterization, PARETO2(μ, σ)

The pdf of the Pareto type 2 distribution, denoted by PARETO2(μ, σ), is given by

$$f_Y(y \mid \mu, \sigma) = \frac{\sigma^{-1}\mu^{1/\sigma}}{(y+\mu)^{(1/\sigma)+1}} \tag{19.8}$$

for $y > 0$, where $\mu > 0$ and $\sigma > 0$. PARETO2 is given by reparameterizing σ to $1/\sigma$ in PARETO2o, i.e. PARETO2(μ, σ) = PARETO2o$(\mu, 1/\sigma)$. Note that σ is a true variance parameter (for $\sigma < 1/2$), a primary true van Zwet skewness parameter, and (for $\sigma < 1/4$) a primary true moment kurtosis parameter, see Sections 14.5 and 15.5.

TABLE 19.9: Pareto type 2 distribution.

PARETO2(μ, σ)	
Ranges	
Y	$0 < y < \infty$
μ	$0 < \mu < \infty$ scaling parameter
σ	$0 < \sigma < \infty$
Distribution measures	
mean	$\begin{cases} \dfrac{\mu\sigma}{(1-\sigma)} & \text{if } \sigma < 1 \\ \infty & \text{if } \sigma \geq 1 \end{cases}$
median	$\mu(2^{\sigma} - 1)$
mode	$\to 0$
variance	$\begin{cases} \dfrac{\sigma^2\mu^2}{(1-\sigma)^2(1-2\sigma)} & \text{if } \sigma < 1/2 \\ \infty & \text{if } \sigma \geq 1/2 \end{cases}$
skewness	$\begin{cases} \dfrac{2(1+\sigma)(1-2\sigma)^{1/2}}{(1-3\sigma)} & \text{if } \sigma < 1/3 \\ \infty & \text{if } \sigma \geq 1/3 \end{cases}$
excess kurtosis	$\begin{cases} \dfrac{3(1-2\sigma)(2\sigma^2+\sigma+3)}{(1-3\sigma)(1-4\sigma)} - 3 & \text{if } \sigma < 1/4 \\ \infty & \text{if } \sigma \geq 1/4 \end{cases}$
pdf	$\dfrac{\sigma^{-1}\mu^{1/\sigma}}{(y+\mu)^{(1/\sigma)+1}}$
cdf	$1 - \dfrac{\mu^{1/\sigma}}{(y+\mu)^{1/\sigma}}$
inverse cdf	$\mu[(1-p)^{-\sigma} - 1]$
Reference	Set σ to $1/\sigma$ in PARETO2o(μ, σ).

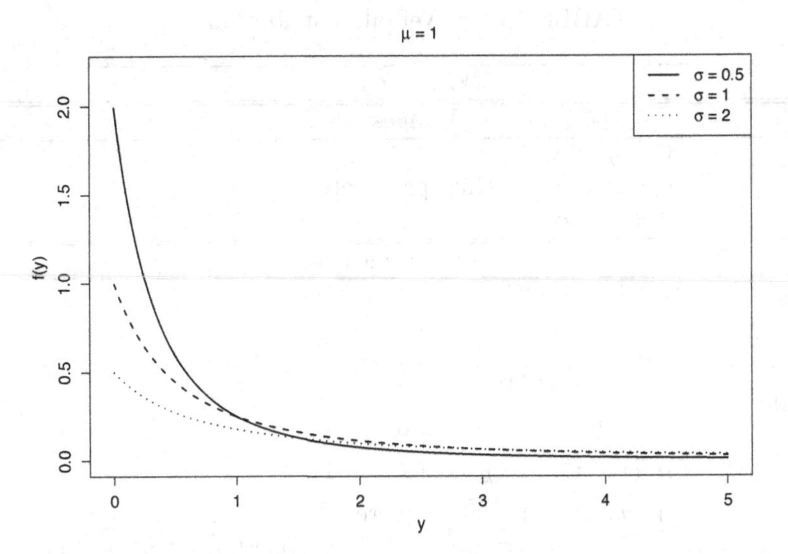

FIGURE 19.6: The Pareto, PARETO2(μ, σ), distribution, with $\mu = 1$ and $\sigma = 0.5, 1, 2$.

19.3.7 Weibull: WEI, WEI2, WEI3

First parameterization, WEI

There are three versions of the two-parameter Weibull distribution implemented in **gamlss**. The first, denoted by WEI(μ, σ), has the following parameterization

$$f_Y(y \mid \mu, \sigma) = \frac{\sigma y^{\sigma-1}}{\mu^{\sigma}} \exp\left[-\left(\frac{y}{\mu}\right)^{\sigma}\right] \tag{19.9}$$

for $y > 0$, where $\mu > 0$ and $\sigma > 0$, see Johnson et al. [1994, p629]. Note that σ is a primary true van Zwet skewness parameter, see Section 14.5. The moment skewness is positive for $\sigma \leq 3.60$ and negative for $\sigma > 3.60$. Note that the exponential distribution, EXP(μ), is a special case of WEI(μ, σ) when $\sigma = 1$.

TABLE 19.10: Weibull distribution.

WEI(μ, σ)	
Ranges	
Y	$0 < y < \infty$
μ	$0 < \mu < \infty$, scaling parameter
σ	$0 < \sigma < \infty$
Distribution measures	
mean	$\mu\Gamma(\sigma^{-1}+1)$
median	$\mu(\log 2)^{1/\sigma}$
mode	$\begin{cases} \mu(1-\sigma^{-1})^{1/\sigma} & \text{if } \sigma > 1 \\ \to 0 & \text{if } \sigma \le 1 \end{cases}$
variance	$\mu^2\{\Gamma(2\sigma^{-1}+1) - [\Gamma(\sigma^{-1}+1)]^2\}$
skewness	$\begin{cases} \mu_3/[\text{Var}(Y)]^{1.5} \quad \text{where} \\ \mu_3 = \mu^3\{\Gamma(3\sigma^{-1}+1) - 3\Gamma(2\sigma^{-1}+1)\Gamma(\sigma^{-1}+1) \\ \quad +2[\Gamma(\sigma^{-1}+1)]^3\} \end{cases}$
excess kurtosis	$\begin{cases} \mu_4/[\text{Var}(Y)]^2 - 3 \quad \text{where} \\ \mu_4 = \mu^4\{\Gamma(4\sigma^{-1}+1) - 4\Gamma(3\sigma^{-1}+1)\Gamma(\sigma^{-1}+1) \\ \quad +6\Gamma(2\sigma^{-1}+1)[\Gamma(\sigma^{-1}+1)]^2 - 3[\Gamma(\sigma^{-1}+1)]^4\} \end{cases}$
pdf [a]	$\dfrac{\sigma y^{\sigma-1}}{\mu^\sigma} \exp\left[-(y/\mu)^\sigma\right]$
cdf	$1 - \exp\left[-(y/\mu)^\sigma\right]$
inverse cdf	$\mu[-\log(1-p)]^{1/\sigma}$
Reference	Johnson et al. [1994] Chapter 21, p628-632. [a] Johnson et al. [1994] equation (21.3), p629, with $\alpha = \mu, c = \sigma$ and $\xi_0 = 0$.

Second parameterization, WEI2

The second parameterization of the Weibull distribution, denoted by WEI2(μ, σ), is defined as

$$f_Y(y \mid \mu, \sigma) = \sigma\mu y^{\sigma-1} \exp\left(-\mu y^\sigma\right) \tag{19.10}$$

for $y > 0$, where $\mu > 0$ and $\sigma > 0$. The parameterization (19.10) gives the usual proportional hazards Weibull model. Note WEI2(μ, σ) = WEI$(\mu^{-1/\sigma}, \sigma)$, and so WEI$(\mu, \sigma)$ = WEI2$(\mu^{-\sigma}, \sigma)$. In this second parameterization, WEI2(μ, σ), of the Weibull distribution, the two parameters μ and σ are highly correlated. As a result the RS method of fitting is very slow and therefore the `CG()` or `mixed()` method of fitting should be used.

TABLE 19.11: Second parameterization of Weibull distribution.

$\texttt{WEI2}(\mu, \sigma)$	
Ranges	
Y	$0 < y < \infty$
μ	$0 < \mu < \infty$, *not* a scaling parameter
σ	$0 < \sigma < \infty$
Distribution measures	
mean	$\mu^{-1/\sigma}\Gamma(\sigma^{-1}+1)$
median	$\mu^{-1/\sigma}(\log 2)^{1/\sigma}$
mode	$\begin{cases} \mu^{-1/\sigma}(1-\sigma^{-1})^{1/\sigma} & \text{if } \sigma > 1 \\ \to 0 & \text{if } \sigma \le 1 \end{cases}$
variance	$\mu^{-2/\sigma}\{\Gamma(2\sigma^{-1}+1) - [\Gamma(\sigma^{-1}+1)]^2\}$
skewness	$\begin{cases} \mu_3/[\text{Var}(Y)]^{1.5} \quad \text{where} \\ \mu_3 = \mu^{-3/\sigma}\{\Gamma(3\sigma^{-1}+1) - 3\Gamma(2\sigma^{-1}+1)\Gamma(\sigma^{-1}+1) \\ \quad +2[\Gamma(\sigma^{-1}+1)]^3\} \end{cases}$
excess kurtosis	$\begin{cases} \mu_4/[\text{Var}(Y)]^2 - 3 \quad \text{where} \\ \mu_4 = \mu^{-4/\sigma}\{\Gamma(4\sigma^{-1}+1) - 4\Gamma(3\sigma^{-1}+1)\Gamma(\sigma^{-1}+1) \\ \quad +6\Gamma(2\sigma^{-1}+1)[\Gamma(\sigma^{-1}+1)]^2 - 3[\Gamma(\sigma^{-1}+1)]^4\} \end{cases}$
pdf	$\mu\sigma y^{\sigma-1}\exp(-\mu y^{\sigma})$
cdf	$1-\exp(-\mu y^{\sigma})$
inverse cdf	$\mu^{-1/\sigma}[-\log(1-p)]^{1/\sigma}$
Note	Set $\mu_1 = \mu^{-1/\sigma}$ in $\texttt{WEI}(\mu_1, \sigma)$.

Third parameterization, WEI3

This is a parameterization of the Weibull distribution where μ is the mean of the distribution. This parameterization of the Weibull distribution, denoted by $\texttt{WEI3}(\mu, \sigma)$, is defined as

$$f_Y(y \mid \mu, \sigma) = \frac{\sigma y^{\sigma-1}}{\beta^{\sigma}} \exp\left[-\left(\frac{y}{\beta}\right)^{\sigma}\right] \tag{19.11}$$

for $y > 0$, where $\mu > 0$ and $\sigma > 0$ and $\beta = \mu[\Gamma(\sigma^{-1}+1)]^{-1}$. The parameterization (19.11) gives the usual accelerated lifetime Weibull model. Note that $\texttt{WEI3}(\mu, \sigma) = \texttt{WEI}(\beta, \sigma)$. Note that the exponential distribution, $\texttt{EXP}(\mu)$, is a special case of $\texttt{WEI}(\mu, \sigma)$ when $\sigma = 1$.

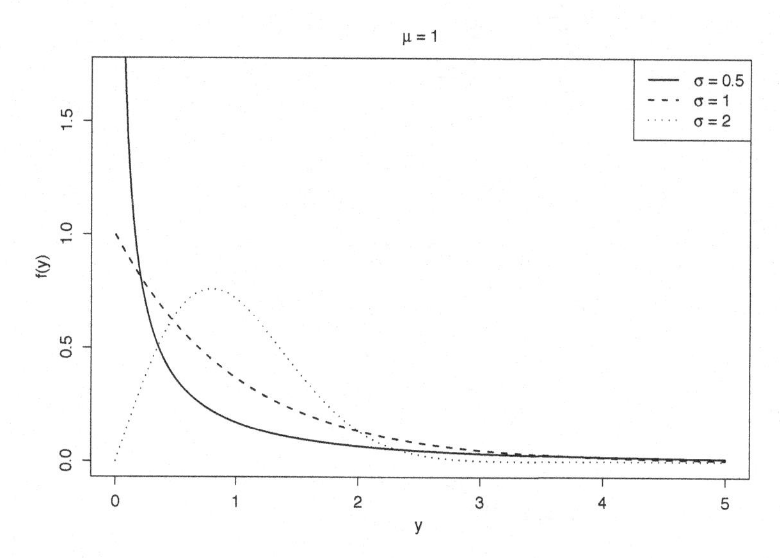

FIGURE 19.7: The third parameterization of Weibull, `WEI3`(μ, σ), distribution, with $\mu = 1$ and $\sigma = 0.5, 1, 2$.

TABLE 19.12: Third parameterization of Weibull distribution.

`WEI3`(μ, σ)	
Ranges	
Y	$0 < y < \infty$
μ	$0 < \mu < \infty$, mean, scaling parameter
σ	$0 < \sigma < \infty$
Distribution measures	
mean	μ
median	$\beta(\log 2)^{1/\sigma}$ where $\beta = \mu[\,\Gamma(\sigma^{-1}+1)\,]^{-1}$
mode	$\begin{cases} \beta(1-\sigma^{-1})^{1/\sigma} & \text{if } \sigma > 1 \\ \to 0, & \text{if } \sigma \le 1 \end{cases}$
variance	$\beta^2\{\Gamma(2\sigma^{-1}+1) - [\Gamma(\sigma^{-1}+1)]^2\}$
skewness	$\begin{cases} \mu_3/[\text{Var}(Y)]^{1.5} \quad \text{where} \\ \mu_3 = \beta^3\{\Gamma(3\sigma^{-1}+1) - 3\Gamma(2\sigma^{-1}+1)\Gamma(\sigma^{-1}+1) \\ \quad +2[\Gamma(\sigma^{-1}+1)]^3\} \end{cases}$
excess kurtosis	$\begin{cases} \mu_4/[\text{Var}(Y)]^2 - 3 \quad \text{where} \\ \mu_4 = \beta^4\{\Gamma(4\sigma^{-1}+1) - 4\Gamma(3\sigma^{-1}+1)\Gamma(\sigma^{-1}+1) \\ \quad +6\Gamma(2\sigma^{-1}+1)[\Gamma(\sigma^{-1}+1)]^2 - 3[\Gamma(\sigma^{-1}+1)]^4\} \end{cases}$
pdf	$\dfrac{\sigma y^{\sigma-1}}{\beta^\sigma} \exp\left[-(y/\beta)^\sigma\right]$
cdf	$1 - \exp\left[-(y/\beta)^\sigma\right]$
inverse cdf	$\beta[\,-\log(1-p)\,]^{1/\sigma}$
Reference	Set μ_1 to β in `WEI`(μ_1, σ), where $\beta = \mu[\Gamma(\sigma^{-1}+1)]^{-1}$.

19.4 Continuous three-parameter distributions on $(0, \infty)$

19.4.1 Box-Cox Cole and Green: BCCG, BCCGo

The Box-Cox Cole and Green distribution is suitable for positively or negatively skewed data. Let $Y > 0$ be a positive random variable having a Box-Cox Cole and Green distribution, denoted by $\texttt{BCCG}(\mu, \sigma, \nu)$, defined through the transformed random variable Z given by

$$Z = \begin{cases} \dfrac{1}{\sigma\nu}\left[\left(\dfrac{Y}{\mu}\right)^{\nu} - 1\right] & \text{if } \nu \neq 0 \\ \dfrac{1}{\sigma}\log\left(\dfrac{Y}{\mu}\right) & \text{if } \nu = 0 \end{cases} \tag{19.12}$$

for $Y > 0$, where $\mu > 0$, $\sigma > 0$ and $-\infty < \nu < \infty$, and where the random variable Z is assumed to follow a truncated standard normal distribution. The condition $Y > 0$ (required for Y^{ν} to be real for all ν) leads to the conditions

$$\begin{aligned} -\frac{1}{\sigma\nu} < Z < \infty \qquad & \text{if } \nu > 0 \\ -\infty < Z < -\frac{1}{\sigma\nu} \qquad & \text{if } \nu < 0\,, \end{aligned}$$

which necessitates the truncated standard normal distribution for Z. Note that

$$Y = \begin{cases} \mu(1+\sigma\nu Z)^{1/\nu} & \text{if } \nu \neq 0 \\ \mu\exp(\sigma Z) & \text{if } \nu = 0\,. \end{cases} \tag{19.13}$$

[See Figure 19.8 for a plot of the relationship between Y and Z for $\mu = 1$, $\sigma = 0.2$ and $\nu = -2$. For this case $-\infty < Z < 2.5$.]

The pdf of $Y \sim \texttt{BCCG}(\mu, \sigma, \nu)$ is given by

$$f_Y(y) = \frac{y^{\nu-1}\exp\left(-\frac{1}{2}z^2\right)}{\mu^{\nu}\sigma\sqrt{2\pi}\,\Phi\left[(\sigma|\nu|)^{-1}\right]} \tag{19.14}$$

for $y > 0$, where $\mu > 0$, $\sigma > 0$ and $-\infty < \nu < \infty$, and where z is given by (19.12). [The distribution $\texttt{BCCGo}(\mu, \sigma, \nu)$ also has pdf given by (19.14). It differs from $\texttt{BCCG}(\mu, \sigma, \nu)$ only in having a default log link function for μ, instead of the default identity link function for μ in $\texttt{BCCG}(\mu, \sigma, \nu)$, in **gamlss.dist**]. Note that $\texttt{LOGNO2}(\mu, \sigma)$ is the special case $\texttt{BCCG}(\mu, \sigma, 0)$.

The exact cdf of $Y \sim \texttt{BCCG}(\mu, \sigma, \nu)$ is given by

$$F_Y(y) = \begin{cases} \dfrac{\Phi(z)}{\Phi\left[(\sigma|\nu|)^{-1}\right]} & \text{if } \nu \leq 0 \\ \dfrac{\Phi(z) - \Phi\left[-(\sigma|\nu|)^{-1}\right]}{\Phi\left[(\sigma|\nu|)^{-1}\right]} & \text{if } \nu > 0 \end{cases}$$

where z is given by (19.12). The term $\Phi\left(-(\sigma|\nu|)^{-1}\right)$ is the truncation probability. If this is negligible, and hence $\Phi\left((\sigma|\nu|)^{-1}\right) \approx 1$, then $F_Y(y) \approx \Phi(z)$. Also if the truncation probability is negligible, Y has median μ. Note if $\nu = 0$ there is no truncation and the truncation probability is zero.

The exact inverse cdf (or quantile) y_p of $Y \sim \texttt{BCCG}(\mu, \sigma, \nu)$, defined by $\mathrm{P}(Y \leq y_p) = p$, is given by

$$y_p = \begin{cases} \mu\left(1 + \sigma\nu z_T\right)^{1/\nu}, & \text{if } \nu \neq 0 \\ \mu \exp(\sigma z_T), & \text{if } \nu = 0 \end{cases}$$

where

$$z_T = \begin{cases} \Phi^{-1}\left\{p\Phi\left[(\sigma|\nu|)^{-1}\right]\right\} & \text{if } \nu \leq 0 \\ \Phi^{-1}\left\{p\Phi\left[(\sigma|\nu|)^{-1}\right] + \Phi\left[-(\sigma|\nu|)^{-1}\right]\right\} & \text{if } \nu > 0 \end{cases}$$

Hence for $\nu \neq 0$, $z_T \approx \Phi^{-1}(p) = z_p$ provided the truncation probability is negligible. The parameterization in (19.12) was originally used by Cole and Green [1992], who assumed a standard normal distribution for Z but also assumed a negligible truncation probability. The $\texttt{BCCG}(\mu, \sigma, \nu)$ is a special case of the $\texttt{BCPE}(\mu, \sigma, \nu, \tau)$, given by setting $\tau = 2$, and hence the results in this section are obtained from Rigby and Stasinopoulos [2004], giving a proper distribution (i.e. not requiring the assumption of a negligible truncation probability).

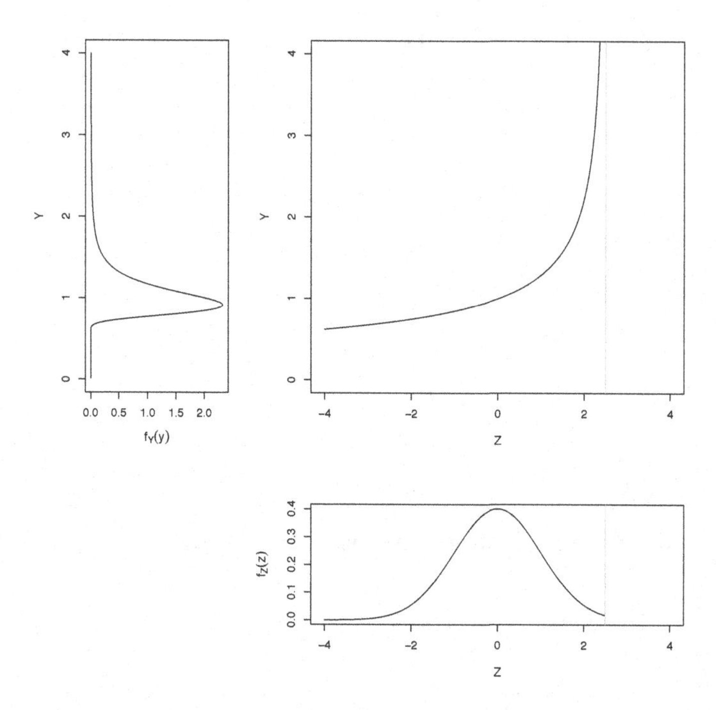

FIGURE 19.8: Relationship between Y and Z for the $\texttt{BCCG}(1, 0.2, -2)$ distribution (top right), $f_Z(z)$ (bottom), and $f_Y(y)$ (top left).

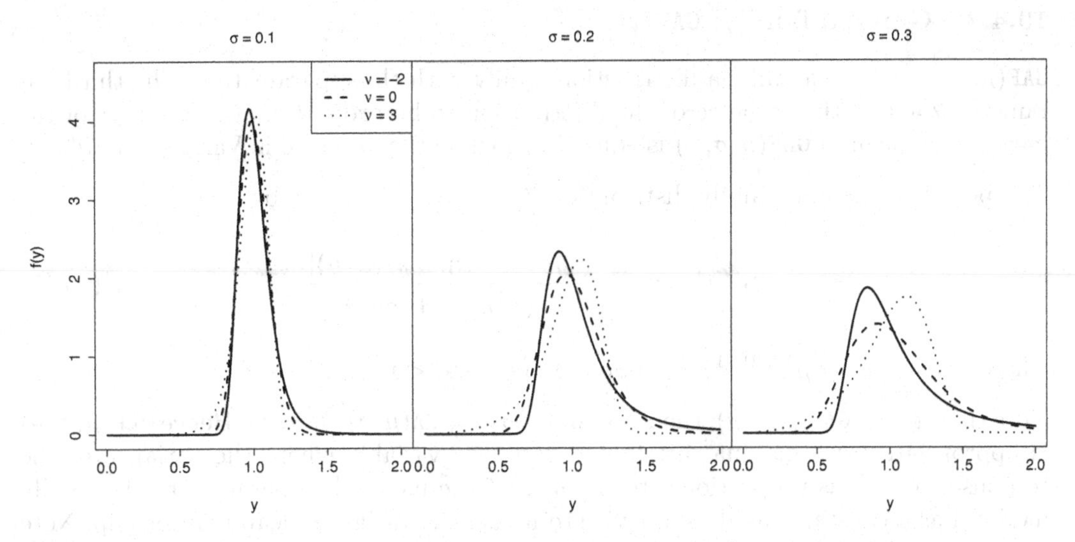

FIGURE 19.9: The Box-Cox Cole and Green, BCCG(μ, σ, ν), distribution, with $\mu = 1$, $\sigma = 0.1, 0.2, 0.3$ and $\nu = -2, 0, 3$.

TABLE 19.13: Box-Cox Cole and Green distribution.

BCCG(μ, σ, ν)	
Ranges	
Y	$0 < y < \infty$
μ	$0 < \mu < \infty$, median[a], scaling parameter
σ	$0 < \sigma < \infty$, approximate coefficient of variation[a,a2]
ν	$-\infty < \nu < \infty$, skewness parameter
Distribution measures	
median [a]	μ
mode [a3]	$\begin{cases} \mu\omega^{1/\nu} & \text{if } \nu \neq 0 \\ \mu e^{-\sigma^2} & \text{if } \nu = 0 \\ \text{where } \omega = \{1 + [1 + 4\sigma^2\nu(\nu - 1)]^{1/2}\}/2 \end{cases}$
pdf	$\dfrac{y^{\nu-1} \exp\left(-\frac{1}{2}z^2\right)}{\mu^\nu \sigma \sqrt{2\pi}\, \Phi\left((\sigma\|\nu\|)^{-1}\right)}$
cdf [a]	$\Phi(z)$ where z is given by (19.12)
inverse cdf[a]	$\begin{cases} \mu(1 + \sigma\nu z_p)^{1/\nu} & \text{if } \nu \neq 0 \\ \mu \exp(\sigma z_p) & \text{if } \nu = 0 \\ \text{where } z_p = \Phi^{-1}(p) \end{cases}$
Reference	Set $\tau = 2$ in BCPE(μ, σ, ν, τ).
Note	[a] Provided $\Phi\left(-(\sigma\|\nu\|)^{-1}\right)$ is negligible, e.g. < 0.0001 for $\sigma\|\nu\| < 0.27$. [a2] Provided ν is reasonably close to 1 ($\nu \geq 0$ and not too large) and σ is small, say $\sigma \leq 0.3$. [a3] If $0 < \nu < 1$ there is a second mode $\to 0$.

19.4.2 Gamma family: GAF

GAF(μ, σ, ν) defines a gamma distribution family with three parameters. The third parameter ν allows the variance of the distribution to be proportional to a power of the mean. The mean of GAF(μ, σ, ν) is equal to μ, while the variance is $\text{Var}(Y) = \sigma^2 \mu^\nu$.

The pdf of the gamma family distribution, GAF(μ, σ, ν), is given by

$$f_Y(y \mid \mu, \sigma, \nu) = \frac{y^{\sigma_1^{-2}-1} \exp[-y/(\sigma_1^2 \mu)]}{(\sigma_1^2 \mu)^{\sigma_1^{-2}} \Gamma(\sigma_1^{-2})} \tag{19.15}$$

where $y > 0$, $\sigma_1 = \sigma \mu^{(\nu/2)-1}$, $\mu > 0$, $\sigma > 0$ and $-\infty < \nu < \infty$.

Note that if $Y \sim$ GAF(μ, σ, ν), then $Y \sim$ GA$(\mu, \sigma_1) =$ GA$(\mu, \sigma \mu^{(\nu/2)-1})$. Hence GAF$(\mu, \sigma, \nu)$ is appropriate for a gamma distributed response variable where the variance of the response variable is proportional to a power of the mean. The parameter ν is usually modeled as a constant, used as a device to model the variance-mean relationship. Note that, due to the high correlation between the σ and ν parameters, the `method=mixed()` and `c.crit=0.0001` arguments are strongly recommended to speed up the convergence, and avoid converging too early.

TABLE 19.14: Gamma family (of variance-mean relationships) distribution.

GAF(μ, σ, ν)	
Ranges	
Y	$0 < y < \infty$
μ	$0 < \mu < \infty$, mean
σ	$0 < \sigma < \infty$
ν	$-\infty < \nu < \infty$
Distribution measures	
mean	μ
mode	$\begin{cases} \mu(1 - \sigma^2 \mu^{\nu-2}) & \text{if } \sigma^2 \mu^{\nu-2} < 1 \\ \to 0 & \text{if } \sigma^2 \mu^{\nu-2} \geq 1 \end{cases}$
variance	$\sigma^2 \mu^\nu$
skewness	$2\sigma \mu^{(\nu/2)-1}$
excess kurtosis	$6\sigma^2 \mu^{\nu-2}$
MGF	$(1 - \sigma^2 \mu^{\nu-1} t)^{-1/(\sigma^2 \mu^{\nu-2})}$, for $t < (\sigma^2 \mu^{\nu-1})^{-1}$
pdf	$\dfrac{y^{\sigma_1^{-2}-1} \exp[-y/(\sigma_1^2 \mu)]}{(\sigma_1^2 \mu)^{\sigma_1^{-2}} \Gamma(\sigma_1^{-2})}$, where $\sigma_1 = \sigma \mu^{(\nu/2)-1}$
cdf	$\dfrac{\gamma(\sigma_1^2, y \mu^{-1} \sigma_1^{-2})}{\Gamma(\sigma_1^{-2})}$
Reference	Set σ_1 to $\sigma \mu^{(\nu/2)-1}$ in GA(μ, σ_1)

19.4.3 Generalized gamma: GG

First parameterization, $\texttt{GG}(\mu, \sigma, \nu)$

The parameterization of the generalized gamma distribution used here and denoted by $\texttt{GG}(\mu,\sigma,\nu)$ was used by Lopatatzidis and Green [2000], with pdf

$$f_Y(y \mid \mu, \sigma, \nu) = \frac{|\nu|\, \theta^\theta z^\theta \exp(-\theta z)}{\Gamma(\theta) y} \tag{19.16}$$

for $y > 0$, where $\mu > 0$, $\sigma > 0$ and $-\infty < \nu < \infty$, $\nu \neq 0$, and where $z = (y/\mu)^\nu$ and $\theta = 1/(\sigma^2\nu^2)$. Note that $Z = (Y/\mu)^\nu \sim \texttt{GA}(1, \sigma\nu)$, from which the results in Table 19.15 are obtained. Note also that the gamma, $\texttt{GA}(\mu, \sigma)$, and Weibull, $\texttt{WEI}(\mu, \sigma)$, distributions are special cases of $\texttt{GG}(\mu, \sigma, \nu)$ given by $\texttt{GG}(\mu, \sigma, 1) = \texttt{GA}(\mu, \sigma)$ and $\texttt{GG}(\mu, \sigma^{-1}, \sigma) = \texttt{WEI}(\mu, \sigma)$, for $\mu > 0$ and $\sigma > 0$. Note also that

$$\text{E}(Y^r) = \frac{\mu^r \Gamma(\theta + r/\nu)}{\theta^{r/\nu} \Gamma(\theta)} \qquad \text{if } \theta > -r/\nu \; .$$

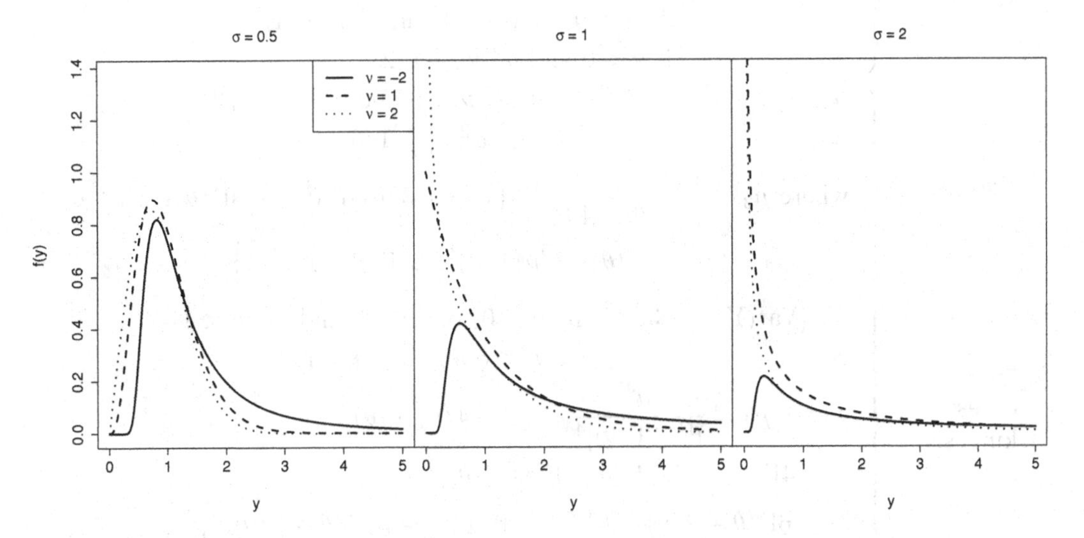

FIGURE 19.10: The generalized gamma, $\texttt{GG}(\mu, \sigma, \nu)$, distribution, with $\mu = 1$, $\sigma = 0.5, 1, 2$, and $\nu = -2, 1, 2$.

TABLE 19.15: Generalized gamma distribution.

$\texttt{GG}(\mu, \sigma, \nu)$	
Ranges	
Y	$0 < y < \infty$
μ	$0 < \mu < \infty$, scaling parameter
σ	$0 < \sigma < \infty$
ν	$-\infty < \nu < \infty, \quad \nu \neq 0$
Distribution measures	
mean	$\begin{cases} \dfrac{\mu\Gamma(\theta + 1/\nu)}{\theta^{1/\nu}\Gamma(\theta)} & \text{if } (\nu > 0) \text{ or } (\nu < 0 \text{ and } \sigma^2\|\nu\| < 1) \\ \infty & \text{if } \nu < 0 \text{ and } \sigma^2\|\nu\| \geq 1 \\ \text{where } \theta = 1/(\sigma^2\nu^2) & \end{cases}$
mode	$\begin{cases} \mu(1 - \sigma^2\nu)^{1/\nu} & \text{if } \sigma^2\nu < 1 \\ \to 0 & \text{if } \sigma^2\nu \geq 1 \end{cases}$
variance	$\begin{cases} \dfrac{\mu^2}{\theta^{2/\nu}[\Gamma(\theta)]^2}\left\{\Gamma(\theta + 2/\nu)\Gamma(\theta) - [\Gamma(\theta + 1/\nu)]^2\right\} & \\ \qquad \text{if } (\nu > 0) \text{ or } (\nu < 0 \text{ and } \sigma^2\|\nu\| < 1/2) & \\ \infty \qquad \text{if } \nu < 0 \text{ and } \sigma^2\|\nu\| \geq 1/2 & \end{cases}$
skewness	$\begin{cases} \mu_3/[\text{Var}(Y)]^{1.5} & \text{if } (\nu > 0) \text{ or } (\nu < 0 \text{ and } \sigma^2\|\nu\| < 1/3) \\ \infty & \text{if } \nu < 0 \text{ and } \sigma^2\|\nu\| \geq 1/3 \\ \text{where } \mu_3 = & \dfrac{\mu^3}{\theta^{3/\nu}[\Gamma(\theta)]^3}\left\{\Gamma(\theta + 3/\nu)[\Gamma(\theta)]^2 - 3\Gamma(\theta + 2/\nu) \times \right. \\ & \left. \Gamma(\theta + 1/\nu)\Gamma(\theta) + 2[\Gamma(\theta + 1/\nu)]^3\right\} \end{cases}$
excess kurtosis	$\begin{cases} \mu_4/[\text{Var}(Y)]^2 - 3 & \text{if } (\nu > 0) \text{ or } (\nu < 0 \text{ and } \sigma^2\|\nu\| < 1/4) \\ \infty & \text{if } \nu < 0 \text{ and } \sigma^2\|\nu\| \geq 1/4 \\ \text{where } \mu_4 = \dfrac{\mu^4}{\theta^{4/\nu}[\Gamma(\theta)]^4}\left\{\Gamma(\theta + 4/\nu)[\Gamma(\theta)]^3 - \right. & \\ \quad 4\Gamma(\theta + 3/\nu)\Gamma(\theta + 1/\nu)[\Gamma(\theta)]^2 + & \\ \quad \left. 6\Gamma(\theta + 2/\nu)[\Gamma(\theta + 1/\nu)]^2\Gamma(\theta) - 3[\Gamma(\theta + 1/\nu)]^4\right\} & \end{cases}$
pdf	$\dfrac{\|\nu\|\,\theta^\theta z^\theta \exp(-\theta z)}{\Gamma(\theta) y}$, where $z = (y/\mu)^\nu$ and $\theta = 1/(\sigma^2\nu^2)$
cdf	$\begin{cases} \gamma[\theta, \theta(y/\mu)^\nu]/\Gamma(\theta) & \text{if } \nu > 0 \\ \Gamma[\theta, \theta(y/\mu)^\nu]/\Gamma(\theta) & \text{if } \nu < 0 \end{cases}$

Second parameterization, $\texttt{GG2}(\mu, \sigma, \nu)$

A second parameterization, given by Johnson et al. [1994] p393, equation (17.128b), denoted here by $\texttt{GG2}(\mu, \sigma, \nu)$, is defined as

$$f_Y(y \mid \mu, \sigma, \nu) = \frac{|\mu|\, y^{\mu\nu - 1}}{\Gamma(\nu)\, \sigma^{\mu\nu}} \exp\left[-\left(\frac{y}{\sigma}\right)^{\mu}\right] \tag{19.17}$$

for $y > 0$, where $-\infty < \mu < \infty$, $\mu \neq 0$, $\sigma > 0$ and $\nu > 0$. The parameterization (19.17) was suggested by Stacy and Mihram [1965], allowing μ to be negative. The moments of $Y \sim \texttt{GG2}(\mu, \sigma, \nu)$ can be obtained from those of $\texttt{GG}(\mu, \sigma, \nu)$, since $\texttt{GG}(\mu, \sigma, \nu) = \texttt{GG2}(\nu, \mu\theta^{-1/\nu}, \theta)$ and $\texttt{GG2}(\mu, \sigma, \nu) = \texttt{GG}(\sigma\nu^{1/\mu}, \left[\mu^2\nu\right]^{-1/2}, \mu)$.

Note that $\texttt{GG2}(\mu, \sigma, \nu)$ is not currently implemented in **gamlss.dist**.

19.4.4 Generalized inverse Gaussian: GIG

The parameterization of the generalized inverse Gaussian distribution, denoted by $\texttt{GIG}(\mu, \sigma, \nu)$, is given by

$$f_Y(y \mid \mu, \sigma, \nu) = \left(\frac{b}{\mu}\right)^{\nu} \left[\frac{y^{\nu - 1}}{2K_\nu\left(\sigma^{-2}\right)}\right] \exp\left[-\frac{1}{2\sigma^2}\left(\frac{b\,y}{\mu} + \frac{\mu}{b\,y}\right)\right] \tag{19.18}$$

for $y > 0$, where $\mu > 0$, $\sigma > 0$ and $-\infty < \nu < \infty$, $b = \left[K_{\nu+1}\left(\sigma^{-2}\right)\right]\left[K_\nu\left(\sigma^{-2}\right)\right]^{-1}$ and $K_\lambda(t)$ is a modified Bessel function of the second kind [Abramowitz and Stegun, 1965], $K_\lambda(t) = \frac{1}{2}\int_0^\infty x^{\lambda-1} \exp\{-\frac{1}{2}t(x + x^{-1})\}\, dx$.

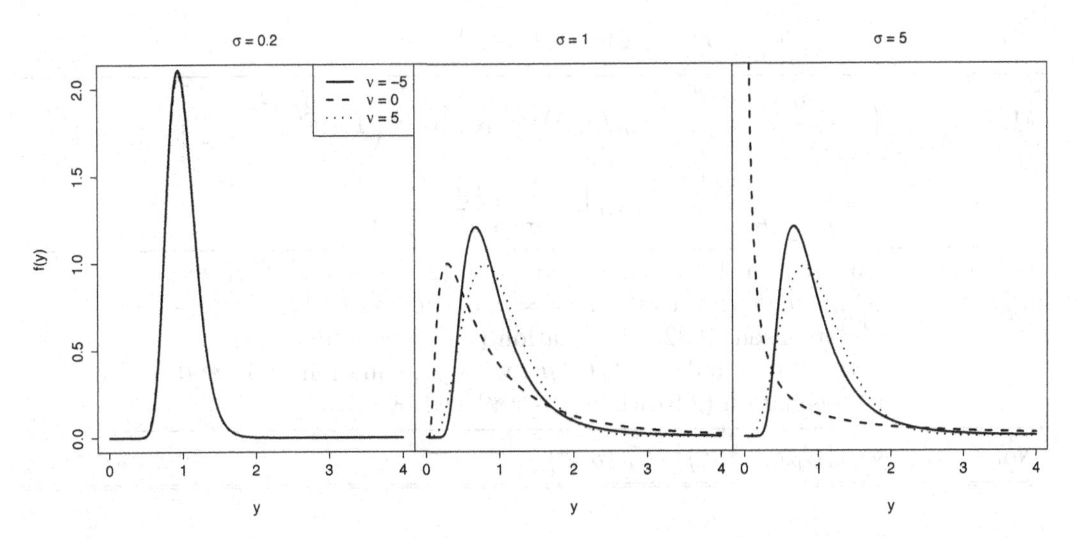

FIGURE 19.11: The generalized inverse Gaussian, $\texttt{GIG}(\mu, \sigma, \nu)$, distribution, with $\mu = 1$, $\sigma = 0.2, 1, 5$ and $\nu = -5, 0, 5$.

$\texttt{GIG}(\mu, \sigma, \nu)$ is a reparameterization of the generalized inverse Gaussian distribution of Jørgensen [1982]. Note also that $\texttt{GIG}(\mu, \sigma, -0.5) = \texttt{IG}(\mu, \sigma\mu^{-1/2})$, a reparameterization of the inverse Gaussian distribution. Note that for $\nu > 0$, $\texttt{GIG}(\mu, \sigma, \nu)$ has limiting distribution $\texttt{GA}(\mu, \nu^{-1/2})$ as $\sigma \to \infty$. (Derived using Jørgensen [1982, p171], equation (17.7): $K_\lambda(t) \sim \Gamma(\lambda) 2^{\lambda-1} t^{-\lambda}$ as $t \to 0$, for $\lambda > 0$.)

TABLE 19.16: Generalized inverse Gaussian distribution.

$\texttt{GIG}(\mu, \sigma, \nu)$	
Ranges	
Y	$0 < y < \infty$
μ	$0 < \mu < \infty$, mean, scaling parameter
σ	$0 < \sigma < \infty$
ν	$-\infty < \nu < \infty$
Distribution measures	
mean [a]	μ
mode [a2]	$\frac{\mu}{b}\{(\nu-1)\sigma^2 + [(\nu-1)^2\sigma^4 + 1]^{1/2}\}$
variance [a]	$\mu^2 \left[\frac{2\sigma^2}{b}(\nu+1) + \frac{1}{b^2} - 1\right]$
skewness [a]	$\begin{cases} \mu_3/[\text{Var}(Y)]^{1.5} \text{ where} \\ \mu_3 = \mu^3 \left[2 - \frac{6\sigma^2}{b}(\nu+1) + \frac{4}{b^2}(\nu+1)(\nu+2)\sigma^4 - \frac{2}{b^2} + \frac{2\sigma^2}{b^3}(\nu+2)\right] \end{cases}$
excess kurtosis [a]	$\begin{cases} k_4/[\text{Var}(Y)]^2 \quad \text{where} \\ k_4 = \mu^4\{-6 + 24\frac{\sigma^2}{b}(\nu+1) + \frac{4}{b^2}[2 - \sigma^4(\nu+1)(7\nu+11)] \\ \quad +4\frac{\sigma^2}{b^3}[2\sigma^4(\nu+1)(\nu+2)(\nu+3) - 4\nu - 5] \\ \quad +\frac{1}{b^4}[4\sigma^4(\nu+2)(\nu+3) - 2]\} \end{cases}$
MGF [a2]	$\left(1 - \frac{2\mu\sigma^2 t}{b}\right)^{-\nu/2} [K_\nu(\sigma^{-2})]^{-1} K_\nu \left[\sigma^{-2}\left(1 - \frac{2\mu\sigma^2 t}{b}\right)^{1/2}\right]$
pdf [a]	$\left(\frac{b}{\mu}\right)^\nu \left[\frac{y^{\nu-1}}{2K_\nu(\sigma^{-2})}\right] \exp\left[-\frac{1}{2\sigma^2}\left(\frac{by}{\mu} + \frac{\mu}{by}\right)\right]$
Reference	[a] Jørgensen [1982] page 6, equation (2.2), reparameterized by $\eta = \mu/b$, $\omega = 1/\sigma^2$ and $\lambda = \nu$, and pages 15-17. [a2] Jørgensen [1982] p1, equation (1.1) reparameterized by $\chi = \mu/(\sigma^2 b)$ and $\psi = b/(\sigma^2\mu)$, with p7, equation (2.6), and p12, equation (2.9) where t is replaced by $-t$.
Note	$b = [K_{\nu+1}(\sigma^{-2})] [K_\nu(\sigma^{-2})]^{-1}$

19.4.5 Log normal (Box-Cox) family: LNO

The `gamlss.family` distribution $\texttt{LNO}(\mu, \sigma, \nu)$ allows the use of the Box-Cox power transformation approach [Box and Cox, 1964], where a transformation is applied to Y in order to remove skewness, given by

$$Z = \begin{cases} (Y^\nu - 1)/\nu & \text{if } \nu \neq 0 \\ \log(Y) & \text{if } \nu = 0 \,. \end{cases} \tag{19.19}$$

The transformed variable Z is then assumed to have a normal, $\texttt{NO}(\mu, \sigma)$, distribution. The resulting distribution of Y is denoted by $\texttt{LNO}(\mu, \sigma, \nu)$ with pdf

$$f_Y(y \mid \mu, \sigma, \nu) = \frac{y^{\nu-1}}{\sqrt{2\pi\sigma^2}} \exp\left[-\frac{(z-\mu)^2}{2\sigma^2}\right] \tag{19.20}$$

for $y > 0$, where $-\infty < \mu < \infty$, $\sigma > 0$ and $-\infty < \nu < \infty$, and where z is defined as in (19.19). Strictly (19.20) is not exactly a proper distribution, since Y is positive and hence Z should have a truncated normal distribution. Hence the integral of $f_Y(y \mid \mu, \sigma, \nu)$ for y from 0 to ∞ is not exactly equal to 1 (and can be far from 1 especially if σ is large relative to μ). This is the *only* distribution in **gamlss.dist** which is not a proper distribution. When $\nu = 0$, this results in the `LOGNO2` distribution (equation (19.5)). The distribution in (19.20) can be fitted for fixed ν only, e.g. $\nu = 0.5$, using the following arguments of `gamlss()`:

```
family = LNO, nu.fix = TRUE, nu.start = 0.5
```

If ν is unknown, it can be estimated from its profile likelihood. Alternatively instead of (19.20), the more orthogonal parameterization given by the `BCCG` distribution in Section 19.4.1 is recommented, and `BCCG` is a proper distribution.

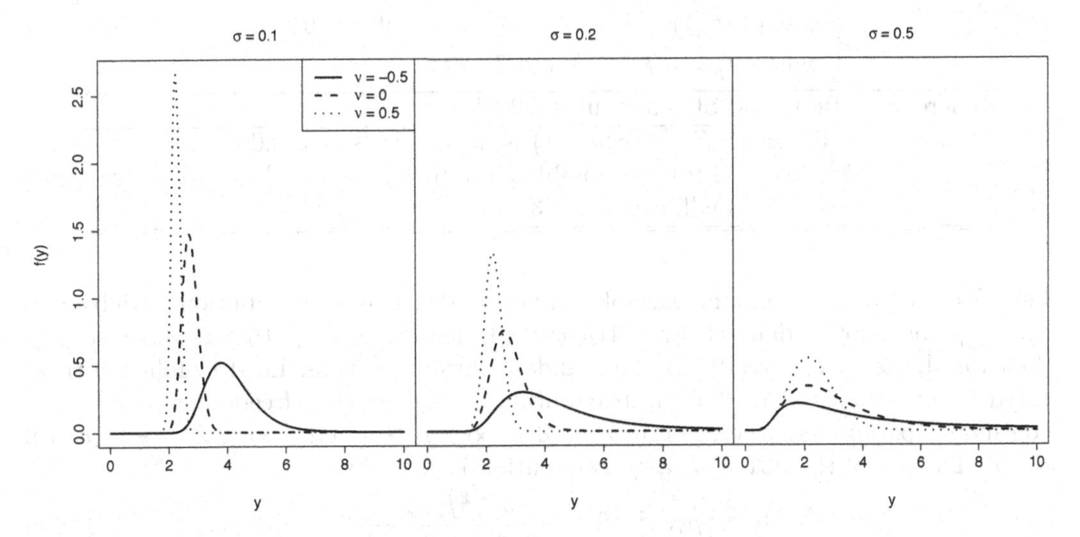

FIGURE 19.12: The log normal family, $\texttt{LNO}(\mu, \sigma, \nu)$, distribution, with $\mu = 1$, $\sigma = 0.1, 0.2, 0.5$, and $\nu = -0.5, 0, 0.5$.

19.5 Continuous four-parameter distributions on $(0, \infty)$

19.5.1 Box-Cox t: BCT, BCTo

The Box-Cox t distribution, $\texttt{BCT}(\mu, \sigma, \nu, \tau)$, has been found to provide a good model for many response variables Y on $(0, \infty)$ when Y is not very close to zero.

TABLE 19.17: Box-Cox t distribution.

$\texttt{BCT}(\mu, \sigma, \nu, \tau)$	
Ranges	
Y	$0 < y < \infty$
μ	$0 < \mu < \infty$, median[a], scaling parameter
σ	$0 < \sigma < \infty$, approximate coefficient of variation[a,a2]
ν	$-\infty < \nu < \infty$, skewness parameter
τ	$0 < \tau < \infty$, kurtosis parameter
Distribution measures	
median[a]	μ
pdf	$\dfrac{y^{\nu-1} f_T(z)}{\mu^\nu \sigma F_T\left[(\sigma\lvert\nu\rvert)^{-1}\right]}$ where $T \sim t_\tau = \texttt{TF}(0, 1, \tau)$
cdf[a]	$F_T(z)$ where $T \sim t_\tau$ and z is given by (19.12)
inverse cdf[a]	$\begin{cases} \mu(1 + \sigma\nu t_{p,\tau})^{1/\nu} & \text{if } \nu \neq 0 \\ \mu \exp(\sigma t_{p,\tau}) & \text{if } \nu = 0 \\ \text{where } t_{p,\tau} = F_T^{-1}(p) \text{ and } T \sim t_\tau \end{cases}$
Reference	Rigby and Stasinopoulos [2006]
Notes	[a] Provided $F_T\left(-(\sigma\lvert\nu\rvert)^{-1}\right)$ is negligible (say < 0.0001). [a2] Provided ν is reasonably close to 1 ($\nu \geq 0$ and not too large), and σ is small, say $\sigma < 0.3$.

Let Y be a positive random variable having a Box-Cox t distribution, [Rigby and Stasinopoulos, 2006], denoted by $\texttt{BCT}(\mu, \sigma, \nu, \tau)$, defined through the transformed random variable Z given by (19.12). The random variable Z is assumed to follow a truncated t_τ, i.e. truncated $\texttt{TF}(0, 1, \tau)$, distribution, with degrees of freedom τ treated as a continuous parameter, where $-1/(\sigma\nu) < Z < \infty$ if $\nu > 0$ and $-\infty < Z < -1/(\sigma\nu)$ if $\nu < 0$. The pdf of the $\texttt{BCT}(\mu, \sigma, \nu, \tau)$ distribution is given by

$$f_Y(y \mid \mu, \sigma, \nu, \tau) = \frac{y^{\nu-1} f_T(z)}{\mu^\nu \sigma F_T\left[(\sigma\lvert\nu\rvert)^{-1}\right]} \tag{19.21}$$

for $y > 0$, where $\mu > 0$, $\sigma > 0$, $-\infty < \nu < \infty$, $\tau > 0$, and z is given by (19.12) and $f_T(\cdot)$ and $F_T(\cdot)$ are the pdf and cdf, respectively, of $T \sim t_\tau = \texttt{TF}(0, 1, \tau)$, see Section 18.3.6, Hence

$$f_T(z) = \frac{1}{B\left(\frac{1}{2}, \frac{\tau}{2}\right) \tau^{1/2}} \left(1 + z^2/\tau\right)^{-(\tau+1)/2}.$$

[The distribution $\texttt{BCTo}(\mu, \sigma, \nu, \tau)$ also has pdf given by (19.21). It differs from

$\texttt{BCT}(\mu, \sigma, \nu, \tau)$ only in having default log link function for μ, instead of the default identity link function for μ in $\texttt{BCT}(\mu, \sigma, \nu, \tau)$ in **gamlss.dist**.] Note that the special case $\texttt{BCT}(\mu, \sigma, 0, \tau)$ gives a log t family distribution. Note that $\texttt{BCCG}(\mu, \sigma, \nu)$ is the limiting distribution of $\texttt{BCT}(\mu, \sigma, \nu, \tau)$ as $\tau \to \infty$.

The exact cdf of Y is given by

$$F_Y(y) = \begin{cases} \dfrac{F_T(z)}{F_T\left[(\sigma|\nu|)^{-1}\right]} & \text{if } \nu \leq 0 \\ \dfrac{F_T(z) - F_T\left[-(\sigma|\nu|)^{-1}\right]}{F_T\left[(\sigma|\nu|)^{-1}\right]} & \text{if } \nu > 0 \end{cases} \tag{19.22}$$

where z is given by (19.12).

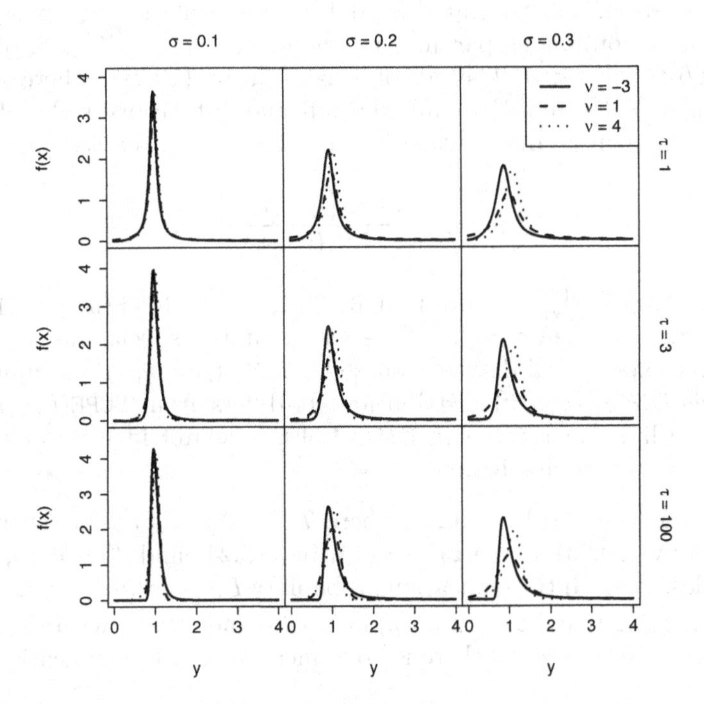

FIGURE 19.13: The Box-Cox t, $\texttt{BCT}(\mu, \sigma, \nu, \tau)$, distribution, with $\mu = 1$, $\sigma = 0.1, 0.2, 0.3$, $\nu = -3, 1, 4$, and $\tau = 1, 3, 100$.

The exact inverse cdf y_p of Y, defined by $P(Y \leq y_p) = p$, is given by

$$y_p = \begin{cases} \mu(1 + \sigma\nu z_T)^{1/\nu}, & \text{if } \nu \neq 0 \\ \mu \exp(\sigma z_T), & \text{if } \nu = 0\ , \end{cases} \tag{19.23}$$

where, [Rigby and Stasinopoulos, 2006],

$$z_T = \begin{cases} F_T^{-1}\left[p F_T\left((\sigma|\nu|)^{-1}\right)\right] & \text{if } \nu \leq 0 \\ F_T^{-1}\left[p F_T\left((\sigma|\nu|)^{-1}\right) + F_T\left(-(\sigma|\nu|)^{-1}\right)\right] & \text{if } \nu > 0\ . \end{cases} \tag{19.24}$$

If the truncation probability $F_T\left(-(\sigma|\nu|)^{-1}\right)$ is negligible, then $F_Y(y) = F_T(z)$ in (19.22) and $z_T = F_T^{-1}(p) = t_{p,\tau}$ in (19.23) and (19.24), so Y has median μ. Note if $\nu = 0$ there is no truncation and the truncation probability is zero. The mean of Y is finite if $\nu < -1$ and also for $\tau > 1/\nu$ if $\nu > 0$. The variance of Y is finite if $\nu < -2$ and also for $\tau > 2/\nu$ if $\nu > 0$.

19.5.2 Box-Cox power exponential: BCPE, BCPEo

The Box-Cox exponential distribution, BCPE(μ, σ, ν, τ), is a very flexible distribution for a response variable Y on $(0, \infty)$, when Y is not very close to 0.

Let Y be a positive random variable having a Box-Cox power exponential distribution [Rigby and Stasinopoulos, 2004], denoted by BCPE(μ, σ, ν, τ), defined through the transformed random variable Z given by (19.12). Z is assumed to follow a truncated standard power exponential, i.e. truncated PE($0, 1, \tau$) distribution, with power parameter $\tau > 0$ treated as a continuous parameter, where $-1/(\sigma\nu) < Z < \infty$ if $\nu > 0$ and $-\infty < Z < -1/(\sigma\nu)$ if $\nu < 0$. The pdf of Y is given by (19.21), where Z is given by (19.12), and where $f_T(\cdot)$ and $F_T(\cdot)$ are the pdf and cdf, respectively, of T having a standard power exponential distribution, i.e. $T \sim$ PE($0, 1, \tau$), see Section 18.3.3. Hence

$$f_T(z) = \frac{\tau \exp\left[-\left|z/c\right|^\tau\right]}{2c\Gamma\left(\tau^{-1}\right)}$$

where $c^2 = \Gamma(\tau^{-1})[\Gamma(3\tau^{-1})]^{-1}$. Note that BCCG($\mu, \sigma, \nu$) = BCPE($\mu, \sigma, \nu, 2$) is a special case of BCPE(μ, σ, ν, τ) when $\tau = 2$. Note also that the special case BCPE($\mu, \sigma, 0, \tau$) given a log power exponential distribution. [The BCPEo(μ, σ, ν, τ) distribution also has the same pdf as BCPE(μ, σ, ν, τ) distribution. It differs from BCPE(μ, σ, ν, τ) only in having default log link function for μ, instead of the default identity link function for μ in BCPE(μ, σ, ν, τ), in **gamlss.dist**.]

The exact cdf of Y is given by (19.22) where $T \sim$ PE($0, 1, \tau$). The exact inverse cdf y_p of Y is given by (19.23) where z_T is given by (19.24) and $T \sim$ PE($0, 1, \tau$), [Rigby and Stasinopoulos, 2004]. If the truncation probability $F_T\left[-(\sigma|\nu|)^{-1}\right]$ is negligible, then $F_Y(y) = F_T(z)$ in (19.22) and $z_T = F_T^{-1}(p)$ in (19.23) and (19.24) where $T \sim$ PE($0, 1, \tau$), so Y has median μ. Note if $\nu = 0$ there is no truncation and the truncation probability is zero.

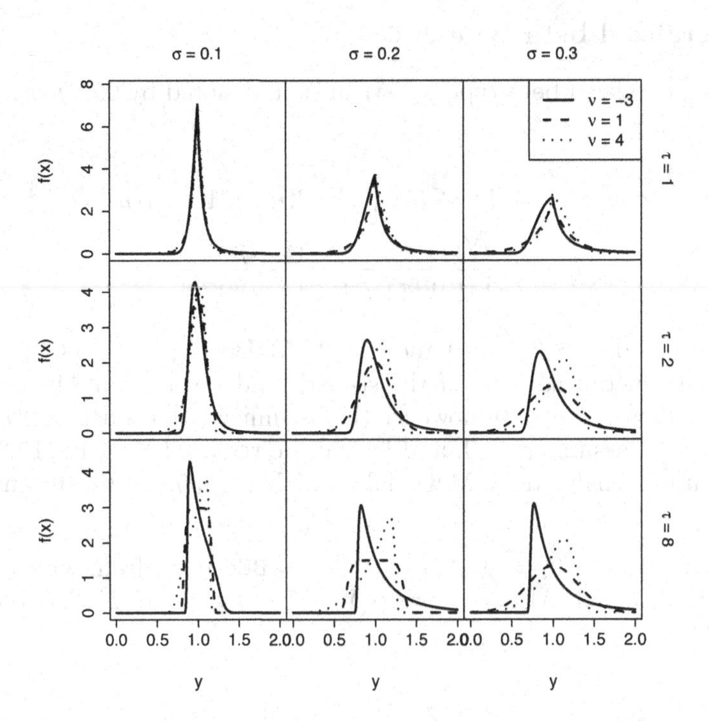

FIGURE 19.14: The Box-Cox power exponential, $\texttt{BCPE}(\mu, \sigma, \nu, \tau)$, distribution, with $\mu = 1$, $\sigma = 0.1, 0.2, 0.3$, $\nu = -3, 1, 4$, and $\tau = 1, 2, 8$.

TABLE 19.18: Box-Cox power exponential distribution.

	$\texttt{BCPE}(\mu, \sigma, \nu, \tau)$
	Ranges
Y	$0 < y < \infty$
μ	$0 < \mu < \infty$, median[a], scaling parameter
σ	$0 < \sigma < \infty$, approximate coefficient of variation[a,a2]
ν	$-\infty < \nu < \infty$, skewness parameter
τ	$0 < \tau < \infty$ kurtosis parameter
	Distribution measures
median [a]	μ
pdf	$\dfrac{y^{\nu-1} f_T(z)}{\mu^\nu \sigma F_T\left[(\sigma\|\nu\|)^{-1}\right]}$ where $T \sim \texttt{PE}(0, 1, \tau)$
cdf [a]	$F_T(z)$ where $T \sim \texttt{PE}(0, 1, \tau)$ and z is given by (19.12)
inverse cdf[a]	$\begin{cases} \mu\left[1 + \sigma\nu F_T^{-1}(p)\right]^{1/\nu} & \text{if } \nu \neq 0 \\ \mu \exp\left[\sigma F_T^{-1}(p)\right] & \text{if } \nu = 0 \\ \text{where } T \sim \texttt{PE}(0, 1, \tau) & \end{cases}$
Reference	Rigby and Stasinopoulos [2004]
Notes	[a] Provided $F_T\left(-(\sigma\|\nu\|)^{-1}\right)$ is neglibible (e.g. say < 0.0001). [a2] Provided ν is reasonably close to 1 ($\nu \geq 0$ and not too large), and σ is small, say $\sigma < 0.3$.

19.5.3 Generalized beta type 2: GB2

This pdf of the generalized beta type 2 distribution, denoted by GB2(μ, σ, ν, τ), is given by

$$\begin{aligned} f_Y(y \mid \mu, \sigma, \nu, \tau) &= |\sigma| \, y^{\sigma\nu - 1} \left\{ \mu^{\sigma\nu} B(\nu, \tau) \left[1 + (y/\mu)^\sigma\right]^{\nu+\tau} \right\}^{-1} \\ &= \frac{\Gamma(\nu+\tau)}{\Gamma(\nu)\Gamma(\tau)} \frac{|\sigma| \, (y/\mu)^{\sigma\nu}}{y \left[1 + (y/\mu)^\sigma\right]^{\nu+\tau}} \end{aligned} \tag{19.25}$$

for $y > 0$, where $\mu > 0$, $\sigma > 0$, $\nu > 0$ and $\tau > 0$, [McDonald [1984] p648, equation (3), who called it the generalized beta of the second kind.] Note that McDonald and Xu [1995] appear to allow for $\sigma < 0$, however this is unnecessary since GB2$(\mu, -\sigma, \nu, \tau) =$ GB2(μ, σ, τ, ν). So we assume $\sigma > 0$, and $|\sigma|$ can be replaced by σ in (19.25). The GB2 distribution is also considered by McDonald and Xu [1995, p136-140] and McDonald [1996, p433-435].

If $Y \sim$ GB2(μ, σ, ν, τ) then $Y_1 = [1 + (Y/\mu)^{-\sigma}]^{-1} \sim$ BEo(ν, τ), from which the cdf of Y in Table 19.19 is obtained. Also $Z = (\tau/\nu)(Y/\mu)^\sigma \sim F_{2\nu, 2\tau}$, an F distribution defined in equation (13.36).

Note

$$\mathrm{E}(Y^r) = \mu^r \frac{B(\nu + r\sigma^{-1}, \tau - r\sigma^{-1})}{B(\nu, \tau)} \qquad \text{for } \tau > r\sigma^{-1}$$

[McDonald and Xu, 1995, p136, equation (2.8)], from which the mean, variance, skewness, and excess kurtosis in Table 19.19 are obtained.

Setting $\sigma = 1$ in (19.25) gives a form of the Pearson type VI distribution:

$$f_Y(y \mid \mu, \nu, \tau) = \frac{\Gamma(\nu+\tau)}{\Gamma(\nu)\Gamma(\tau)} \frac{\mu^\tau y^{\nu-1}}{(y+\mu)^{\nu+\tau}} \,. \tag{19.26}$$

Setting $\nu = 1$ in (19.25) gives the Burr XII (or Singh-Maddala) distribution:

$$f_Y(y \mid \mu, \sigma, \tau) = \frac{\tau\sigma (y/\mu)^\sigma}{y \left[1 + (y/\mu)^\sigma\right]^{\tau+1}} \,. \tag{19.27}$$

Setting $\tau = 1$ in (19.25) gives the Burr III (or Dagum) distribution

$$f_Y(y \mid \mu, \sigma, \nu) = \frac{\nu\sigma (y|\mu)^{\sigma\nu}}{y[1 + (y/\mu)^\sigma]^{\nu+1}} \,.$$

Setting $\sigma = 1$ and $\nu = 1$ in (19.25) gives the Pareto type 2 original distribution, PARETO2o(μ, τ). Setting $\nu = 1$ and $\tau = 1$ in (19.25) gives the log logistic distribution. Other special cases and limiting cases are given in McDonald and Xu [1995, pp136, 139].

TABLE 19.19: Generalized beta type 2 distribution.

$\texttt{GB2}(\mu, \sigma, \nu, \tau)$	
Ranges	
Y	$0 < y < \infty$
μ	$0 < \mu < \infty$, scaling parameter
σ	$0 < \sigma < \infty$
ν	$0 < \nu < \infty$
τ	$0 < \tau < \infty$
Distribution measures	
mean	$\begin{cases} \mu \dfrac{B(\nu+\sigma^{-1}, \tau-\sigma^{-1})}{B(\nu,\tau)} & \text{if } \tau > \sigma^{-1} \\ \infty & \text{if } \tau \le \sigma^{-1} \end{cases}$
mode	$\begin{cases} \mu \left(\dfrac{\sigma\nu - 1}{\sigma\tau + 1}\right)^{1/\sigma} & \text{if } \nu > \sigma^{-1} \\ \to 0 & \text{if } \nu \le \sigma^{-1} \end{cases}$
variance	$\begin{cases} \dfrac{\mu^2\{B(\nu+2\sigma^{-1}, \tau-2\sigma^{-1})B(\nu,\tau) - [B(\nu+\sigma^{-1}, \tau-\sigma^{-1})]^2\}}{[B(\nu,\tau)]^2} & \text{if } \tau > 2\sigma^{-1} \\ \infty & \text{if } \tau \le 2\sigma^{-1} \end{cases}$
skewness	$\begin{cases} \mu_3/[\text{Var}(Y)]^{1.5} & \text{if } \tau > 3\sigma^{-1} \\ \infty & \text{if } \tau \le 3\sigma^{-1} \quad \text{where} \\ \mu_3 = \mu^3 \dfrac{1}{[B(\nu,\tau)]^3} \{B(\nu+3\sigma^{-1}, \tau-3\sigma^{-1})[B(\nu,\tau)]^2 & \\ \quad -3B(\nu+2\sigma^{-1}, \tau-2\sigma^{-1})B(\nu+\sigma^{-1}, \tau-\sigma^{-1})B(\nu,\tau) & \\ \quad +2[B(\nu+\sigma^{-1}, \tau-\sigma^{-1})]^3\} & \end{cases}$
excess kurtosis	$\begin{cases} \mu_4/[\text{Var}(Y)]^2 - 3 & \text{if } \tau > 4\sigma^{-1} \\ \infty & \text{if } \tau \le 4\sigma^{-1} \quad \text{where} \\ \mu_4 = \mu^4 \dfrac{1}{[B(\nu,\tau)]^4} \{B(\nu+4\sigma^{-1}, \tau-4\sigma^{-1})[B(\nu,\tau)]^3 & \\ \quad -4B(\nu+3\sigma^{-1}, \tau-3\sigma^{-1})B(\nu+\sigma^{-1}, \tau-\sigma^{-1})[B(\nu,\tau)]^2 & \\ \quad +6B(\nu+2\sigma^{-1}, \tau-2\sigma^{-1})[B(\nu+\sigma^{-1}, \tau-\sigma^{-1})]^2 B(\nu,\tau) & \\ \quad -3[B(\nu+\sigma^{-1}, \tau-\sigma^{-1})]^4\} & \end{cases}$
pdf[a]	$\lvert\sigma\rvert y^{\sigma\nu-1} \left\{\mu^{\sigma\nu} B(\nu,\tau) \left[1 + (y/\mu)^{\sigma}\right]^{\nu+\tau}\right\}^{-1}$
cdf	$\dfrac{B(\nu,\tau,c)}{B(\nu,\tau)}$ where $c = 1/[1 + (y/\mu)^{-\sigma}]$
Reference	[a] McDonald [1984] p648, equation (3) where $b = \mu$, $a = \sigma$, $p = \nu$ and $q = \tau$.

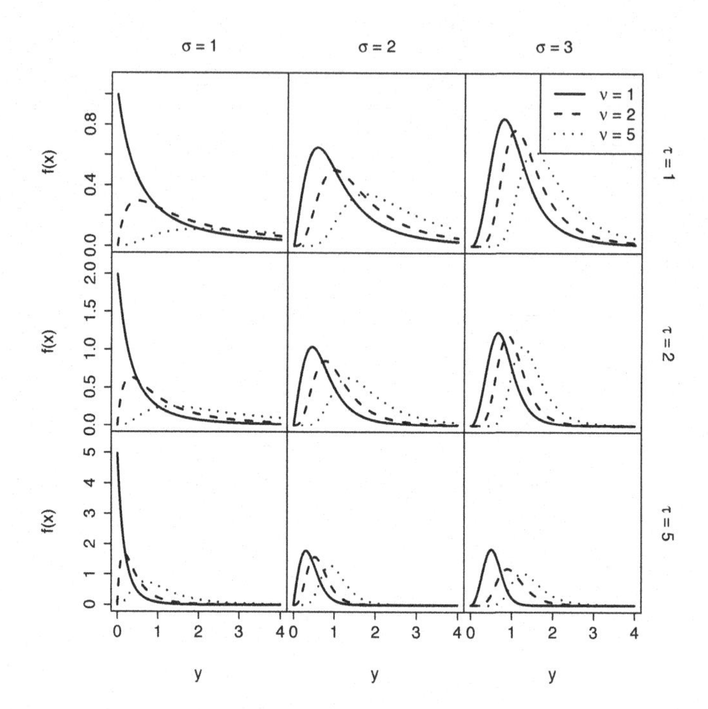

FIGURE 19.15: The generalized beta type 2, GB2(μ, σ, ν, τ), distribution, with $\mu = 1$, $\sigma = 1, 2, 3$, $\nu = 1, 2, 5$, and $\tau = 1, 2, 5$.

20

Mixed distributions on $[0, \infty)$

CONTENTS

This chapter gives summary tables and plots for the explicit **gamlss.dist** mixed distributions with range $[0, \infty)$, which are continuous on $(0, \infty)$ and include a point probability at zero. These are discussed in Section 9.2, which also discusses creating distributions on $[0, \infty)$ in the **gamlss** packages either by generating a zero-adjusted distribution [i.e. adding a point probability at zero to any `gamlss.family` distribution on $(0, \infty)$] or by generalized Tobit model [i.e. censoring below zero any **gamlss.dist** distribution on $(-\infty, \infty)$ to give a point probability at zero].

20.1 Zero-adjusted gamma: ZAGA

The zero-adjusted gamma distribution, ZAGA, is appropriate when the response variable Y takes values from zero to infinity, including exact value zero, i.e. $[0,\infty)$. Here $Y = 0$ with nonzero probability ν, and $Y \sim$ GA(μ, σ) with probability $(1 - \nu)$. The mixed probability function of the zero-adjusted gamma distribution, denoted by ZAGA(μ, σ, ν), is given (informally) by

$$f_Y(y|\mu,\sigma,\nu) = \begin{cases} \nu & \text{if } y = 0 \\ (1-\nu)\left[\dfrac{y^{1/\sigma^2-1}e^{-y/(\sigma^2\mu)}}{(\sigma^2\mu)^{1/\sigma^2}\Gamma(1/\sigma^2)}\right] & \text{if } y > 0 \end{cases} \tag{20.1}$$

for $y \geq 0$, where $\mu > 0$, $\sigma > 0$ and $0 < \nu < 1$.

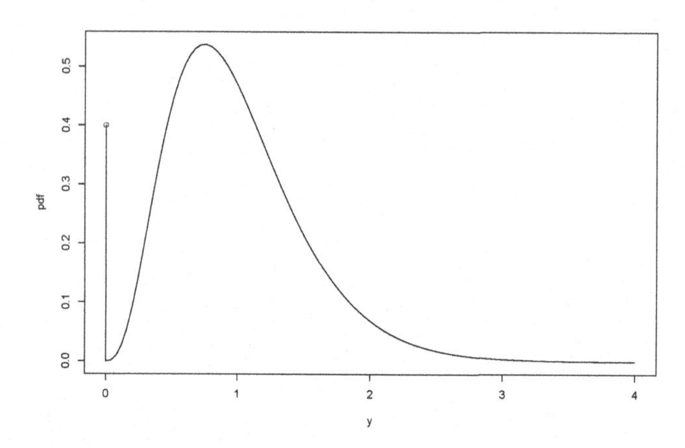

FIGURE 20.1: Zero-adjusted gamma, `ZAGA`(μ, σ, ν), distribution, with $\mu = 1$, $\sigma = 0.5$, and $\nu = 0.4$.

TABLE 20.1: Zero-adjusted gamma distribution.

`ZAGA`(μ, σ, ν)	
Ranges	
Y	$0 \leq y < \infty$,
μ	$0 < \mu < \infty$, mean of gamma component
σ	$0 < \sigma < \infty$, coefficient of variation of gamma component
ν	$0 < \nu < 1$, $\text{P}(Y = 0)$
Distribution measures	
mean	$(1-\nu)\mu$
mode	$\begin{cases} 0 & \text{if } \sigma \geq 1 \\ 0 \text{ and } \mu(1-\sigma^2) & \text{if } \sigma < 1 \end{cases}$
variance	$(1-\nu)\mu^2(\sigma^2 + \nu)$
skewness	$\begin{cases} \mu_3/[\text{Var}(Y)]^{1.5} \text{ where} \\ \mu_3 = \mu^3(1-\nu)\left(2\sigma^4 + 3\nu\sigma^2 + 2\nu^2 - \nu\right) \end{cases}$
excess kurtosis	$\begin{cases} \mu_4/[\text{Var}(Y)]^2 - 3 \text{ where} \\ \mu_4 = \mu^4(1-\nu)\left[6\sigma^6 + 3\sigma^4 + 8\nu\sigma^4 + 6\nu^2\sigma^2 + \nu(1 - 3\nu + 3\nu^2)\right] \end{cases}$
MGF	$\nu + (1-\nu)(1-\mu\sigma^2 t)^{-1/\sigma^2}$ for $t < (\mu\sigma^2)^{-1}$
pdf	$\begin{cases} \nu & \text{if } y = 0 \\ (1-\nu)\left[\dfrac{y^{1/\sigma^2 - 1} e^{-y/(\sigma^2\mu)}}{(\sigma^2\mu)^{1/\sigma^2}\Gamma(1/\sigma^2)}\right] & \text{if } y > 0 \end{cases}$
cdf	$\begin{cases} \nu & \text{if } y = 0 \\ \nu + (1-\nu)\dfrac{\gamma(\sigma^{-2}, y\mu^{-1}\sigma^{-2})}{\Gamma(\sigma^{-2})} & \text{if } y > 0 \end{cases}$
Reference	Obtained from equations (9.1), (9.2), (9.3), and (9.4), where $Y_1 \sim$ `GA`(μ, σ).

20.2 Zero-adjusted inverse Gaussian: ZAIG

The zero-adjusted inverse Gaussian distribution, **ZAIG**, is appropriate when the response variable Y takes values from zero to infinity, including exact value zero, i.e. $[0, \infty)$. Here $Y = 0$ with nonzero probability ν, and $Y \sim$ **IG**(μ, σ) with probability $(1-\nu)$. The mixed probability function of the zero-adjusted inverse Gaussian distribution, denoted by **ZAIG**(μ, σ, ν), is given (informally) by

$$f_Y(y|\mu,\sigma,\nu) = \begin{cases} \nu & \text{if } y = 0 \\ (1-\nu)\dfrac{1}{\sqrt{2\pi\sigma^2 y^3}} \exp\left[-\dfrac{1}{2\mu^2\sigma^2 y}(y-\mu)^2\right] & \text{if } y > 0 \end{cases} \tag{20.2}$$

for $y \geq 0$, where $\mu > 0$, $\sigma > 0$ and $0 < \nu < 1$.

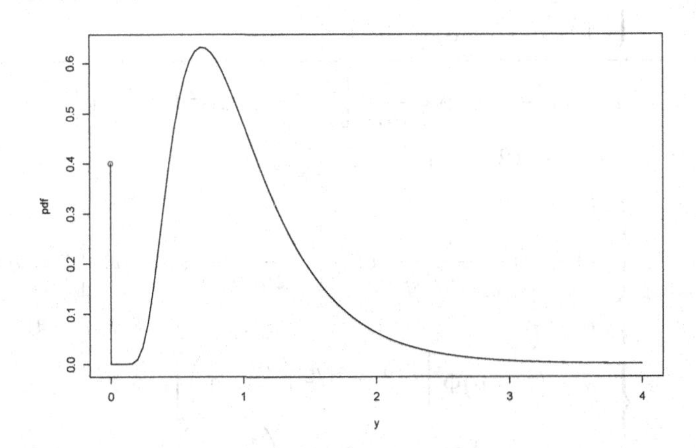

FIGURE 20.2: Zero-adjusted inverse Gaussian, **ZAIG**(μ, σ, ν), distribution, with $\mu = 1$, $\sigma = 0.5$, and $\nu = 0.4$.

TABLE 20.2: Zero-adjusted inverse Gaussian distribution.

	$\texttt{ZAIG}(\mu, \sigma, \nu)$
	Ranges
Y	$0 \leq y < \infty$
μ	$0 < \mu < \infty$, mean of inverse Gaussian component
σ	$0 < \sigma < \infty$
ν	$0 < \nu < 1$, $\quad \text{P}(Y = 0)$
	Distribution measures
mean	$(1-\nu)\mu$
mode	0 and $\left[-3\mu^2\sigma^2 + \mu(9\mu^2\sigma^4 + 4)^{1/2}\right]/2$
variance	$(1-\nu)\mu^2(\nu + \mu\sigma^2)$
skewness	$\begin{cases} \mu_3/[\text{Var}(Y)]^{1.5} \text{ where} \\ \mu_3 = \mu^3(1-\nu)\left[3\mu^2\sigma^4 + 3\mu\sigma^2\nu + 2\nu^2 - \nu\right] \end{cases}$
excess kurtosis	$\begin{cases} \mu_4/[\text{Var}(Y)]^2 - 3 \text{ where} \\ \mu_4 = \mu^4(1-\nu)\left[3\mu^2\sigma^4 + 15\mu^3\sigma^6 + \nu(1 + 12\mu^2\sigma^4) + \right. \\ \left. \nu^2(6\mu\sigma^2 - 3) + 3\nu^3\right] \end{cases}$
MGF	$\nu + (1-\nu)\exp\left\{\frac{1}{\mu\sigma^2}\left[1 - \left(1 - 2\mu^2\sigma^2 t\right)^{1/2}\right]\right\}$ for $t < (2\mu^2\sigma^2)^{-1}$
pdf	$\begin{cases} \nu & \text{if } y = 0 \\ (1-\nu)\frac{1}{\sqrt{2\pi\sigma^2 y^3}} \exp\left[-\frac{1}{2\mu^2\sigma^2 y}(y-\mu)^2\right] & \text{if } y > 0 \end{cases}$
cdf	$\begin{cases} \nu & \text{if } y = 0 \\ \nu + (1-\nu)\Phi\left[(\sigma^2 y)^{-1/2}\left(\frac{y}{\mu} - 1\right)\right] & \\ +e^{2(\mu\sigma^2)^{-1}}\Phi\left[-(\sigma^2 y)^{-1/2}\left(\frac{y}{\mu} - 1\right)\right] & \text{if } y > 0 \end{cases}$
Reference	Obtained from equations (9.1), (9.2), (9.3), and (9.4), where $Y_1 \sim \texttt{IG}(\mu, \sigma)$.

21

Continuous and mixed distributions on $[0, 1]$

CONTENTS

This chapter gives summary tables and plots for the explicit **gamlss.dist** continuous distributions with range $(0, 1)$. These are discussed in Chapter 6, which also discusses creating distributions on $(0, 1)$ in the **gamlss** packages, either by an inverse logit transform, or by truncation below zero and above one, from any `gamlss.family` distribution on $(-\infty, \infty)$, or by truncation above one from any `gamlss.family` distribution on $(0, \infty)$.

This chapter also gives summary tables and plots for the explicit **gamlss.dist** mixed distributions with ranges $[0, 1)$, $(0, 1]$, and $[0, 1]$, which are continuous on $(0, 1)$ and include point probabilities at zero or one, or both, respectively. These are discussed in Section 9.3, which also discusses creating distributions on $[0, 1)$, $(0, 1]$, or $[0, 1]$ in **gamlss**, firstly by generating an 'inflated distribution' [i.e. by adding point probabilities at zero, or one, or both, to any `gamlss.family` continuous distribution on $(0, 1)$], and secondly by a generalized Tobit model [i.e. obtaining the point probabilities at zero, or one, or both, by censoring].

Note that if a random variable Y has any range from a to b, where a and b are known and finite, then $Z = (Y - a)/(b - a)$ has range from 0 to 1 and can be modeled using distributions in this chapter.

21.1 Continuous two-parameter distributions on $(0, 1)$

21.1.1 Beta: BE, BEo

The beta distribution is appropriate when the response variable takes values in a known restricted range, excluding the endpoints of the range. Appropriate standardization can be applied to make the range of the response variable (0,1). Note that the exact values $Y = 0$ and $Y = 1$ are not included in the range and so have zero density under the model.

First parameterization, BEo

The original parameterization of the beta distribution, denoted by $\texttt{BEo}(\mu, \sigma)$, has pdf

$$f_Y(y \mid \mu, \sigma) = \frac{1}{B(\mu, \sigma)} y^{\mu-1}(1-y)^{\sigma-1}$$

for $0 < y < 1$, with parameters $\mu > 0$ and $\sigma > 0$. Here $\text{E}(Y) = \mu/(\mu+\sigma)$ and $\text{Var}(y) = \mu\sigma(\mu+\sigma)^{-2}(\mu+\sigma+1)^{-1}$.

TABLE 21.1: The beta distribution (original parameterization).

	$\texttt{BEo}(\mu, \sigma)$
	Ranges
Y	$0 < y < 1$
μ	$0 < \mu < \infty$
σ	$0 < \sigma < \infty$
	Distribution measures
mean [a]	$\frac{\mu}{\mu+\sigma}$
mode	$\begin{cases} \frac{\mu-1}{\mu+\sigma-2} & \text{if } \mu > 1 \text{ and } \sigma > 1 \\ \to 0 & \text{if } 0 < \mu < 1 \text{ and } \sigma > 1 \\ \to 1 & \text{if } \mu > 1 \text{ and } 0 < \sigma < 1 \\ \to 0 \text{ and } 1 & \text{if } 0 < \mu < 1 \text{ and } 0 < \sigma < 1 \end{cases}$
variance [a]	$\frac{\mu\sigma}{(\mu+\sigma)^2(\mu+\sigma+1)}$
skewness [a]	$\frac{2(\sigma-\mu)(\mu+\sigma+1)^{0.5}}{\mu^{0.5}\sigma^{0.5}(\mu+\sigma+2)}$
excess kurtosis [a]	$\begin{cases} \beta_2 - 3 \text{ where} \\ \beta_2 = \frac{3(\mu+\sigma+1)[2(\mu+\sigma)^2 + \mu\sigma(\mu+\sigma-6)]}{\mu\sigma(\mu+\sigma+2)(\mu+\sigma+3)} \end{cases}$
pdf [a]	$\frac{1}{B(\mu,\sigma)} y^{\mu-1}(1-y)^{\sigma-1}$
cdf	$\frac{B(\mu,\sigma,y)}{B(\mu,\sigma)}$
Reference	[a] Johnson et al. [1995] chapter 25, p210, equation (25.2) and p217, with (p, q) set to (μ, σ)

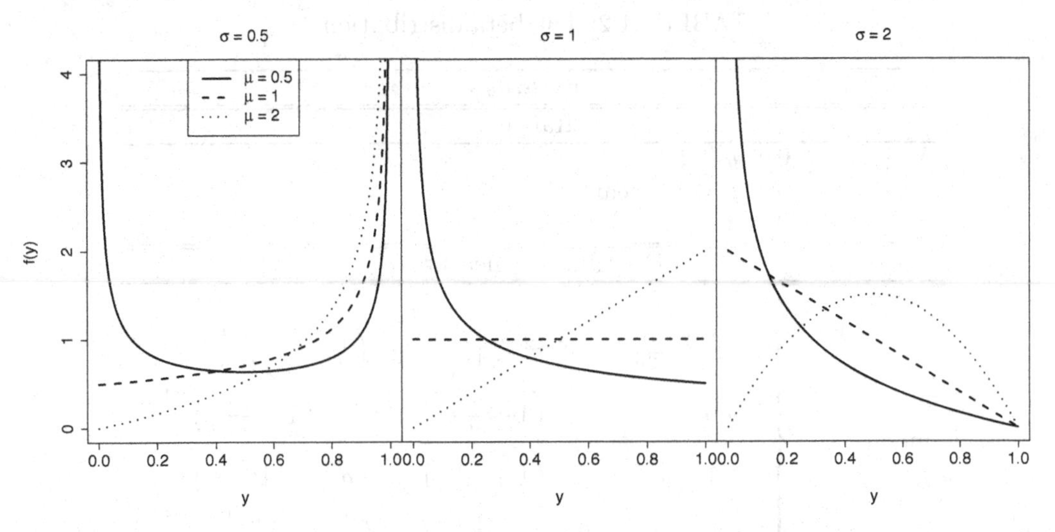

FIGURE 21.1: Beta, BEo(μ, σ), distribution, with $\mu = 0.5, 1, 2$ and $\sigma = 0.5, 1, 2$.

Second parameterization, BE

The pdf of the beta distribution, denoted by BE(μ, σ), is given by

$$f_Y(y \mid \mu, \sigma) = \frac{1}{B(\alpha, \beta)} \, y^{\alpha-1}(1-y)^{\beta-1} \tag{21.1}$$

for $0 < y < 1$, where

$$\begin{aligned} \alpha &= \mu(1-\sigma^2)/\sigma^2 \\ \beta &= (1-\mu)(1-\sigma^2)/\sigma^2 \ , \end{aligned} \tag{21.2}$$

$\alpha > 0$, and $\beta > 0$ and hence $0 < \mu < 1$ and $0 < \sigma < 1$. Note the relationship between parameters (μ,σ) and (α,β) is given by

$$\begin{aligned} \mu &= \alpha/(\alpha+\beta) \\ \sigma &= (\alpha+\beta+1)^{-1/2} \ . \end{aligned}$$

Hence BE(μ, σ) = BEo(α, β). In the parameterization $Y \sim$ BE(μ, σ), the mean and variance of Y are $\mathrm{E}(Y) = \mu$ and $\mathrm{Var}(Y) = \sigma^2\mu(1-\mu)$. Hence $\mathrm{SD}(Y) = \sigma\mu^{0.5}(1-\mu)^{0.5}$ and σ scales the standard deviation of Y.

TABLE 21.2: The beta distribution.

	BE(μ, σ)
	Ranges
Y	$0 < y < 1$
μ	$0 < \mu < 1$, mean
σ	$0 < \sigma < 1$
	Distribution measures
mean	μ
mode	$\begin{cases} \mu + \frac{(2\mu-1)\sigma^2}{1-3\sigma^2} & \text{if } \sigma^2 < \left[1 + \max\left(\frac{1}{\mu}, \frac{1}{1-\mu}\right)\right]^{-1} \\ \to 0 & \text{if } \left(1 + \frac{1}{\mu}\right)^{-1} < \sigma^2 < \left(1 + \frac{1}{1-\mu}\right)^{-1} \\ \to 1 & \text{if } \left(1 + \frac{1}{1-\mu}\right)^{-1} < \sigma^2 < \left(1 + \frac{1}{\mu}\right)^{-1} \\ \to 0 \text{ and } 1 & \text{if } \sigma^2 > \left[1 + \min\left(\frac{1}{\mu}, \frac{1}{1-\mu}\right)\right]^{-1} \end{cases}$
variance	$\sigma^2\mu(1-\mu)$
skewness	$\dfrac{2(1-2\mu)\sigma}{\mu^{0.5}(1-\mu)^{0.5}(1+\sigma^2)}$
excess kurtosis	$\begin{cases} \beta_2 - 3 \text{ where} \\ \beta_2 = \dfrac{6\sigma^2 + 3\mu(1-\mu)(1-7\sigma^2)}{\mu(1-\mu)(1+\sigma^2)(1+2\sigma^2)} \end{cases}$
pdf	$\frac{1}{B(\alpha,\beta)} y^{\alpha-1}(1-y)^{\beta-1}$ where $\alpha = \mu(1-\sigma^2)/\sigma^2$ and $\beta = (1-\mu)(1-\sigma^2)/\sigma^2$
cdf	$B(\alpha, \beta, y)/B(\alpha, \beta)$
Reference	Reparameterize BEo(α, β) by setting $\alpha = \mu(1-\sigma^2)/\sigma^2$ and $\beta = (1-\mu)(1-\sigma^2)/\sigma^2$

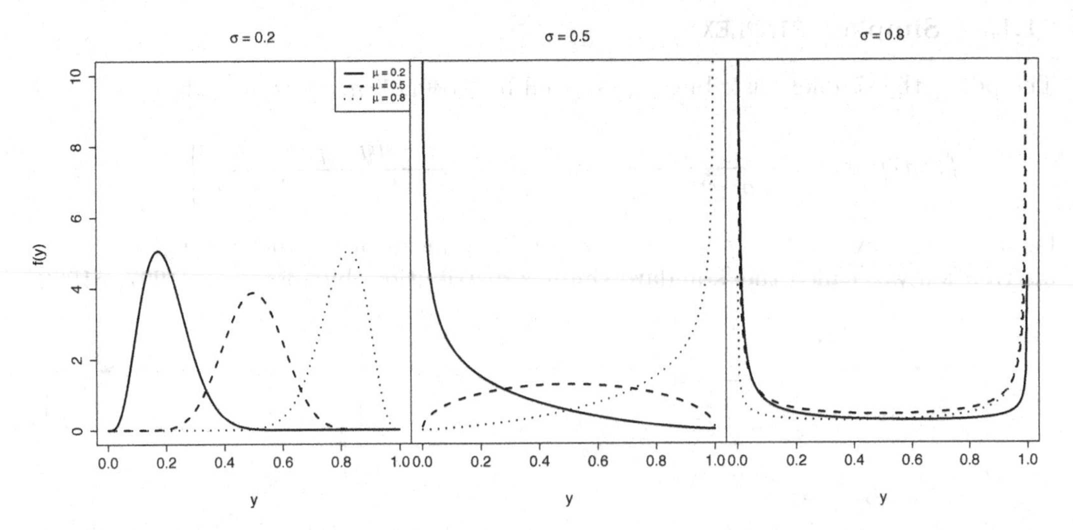

FIGURE 21.2: Beta, BE(μ, σ), distribution, with $\mu = 0.2, 0.5, 0.8$ and $\sigma = 0.2, 0.5, 0.8$

21.1.2 Logit normal: LOGITNO

The pdf of the logit normal distribution, denoted by LOGITNO(μ, σ), is

$$f_Y(y \mid \mu, \sigma) = \frac{1}{\sqrt{2\pi\sigma^2}} \frac{1}{y(1-y)} \exp\left(-\frac{\{\log[y/(1-y)] - \log[\mu/(1-\mu)]\}^2}{2\sigma^2} \right) \quad (21.3)$$

for $0 < y < 1$, where $0 < \mu < 1$ and $\sigma > 0$.

Note that $\log[Y/(1-Y)] \sim$ NO$(\log[\mu/(1-\mu)], \sigma)$ and μ is the median of Y.

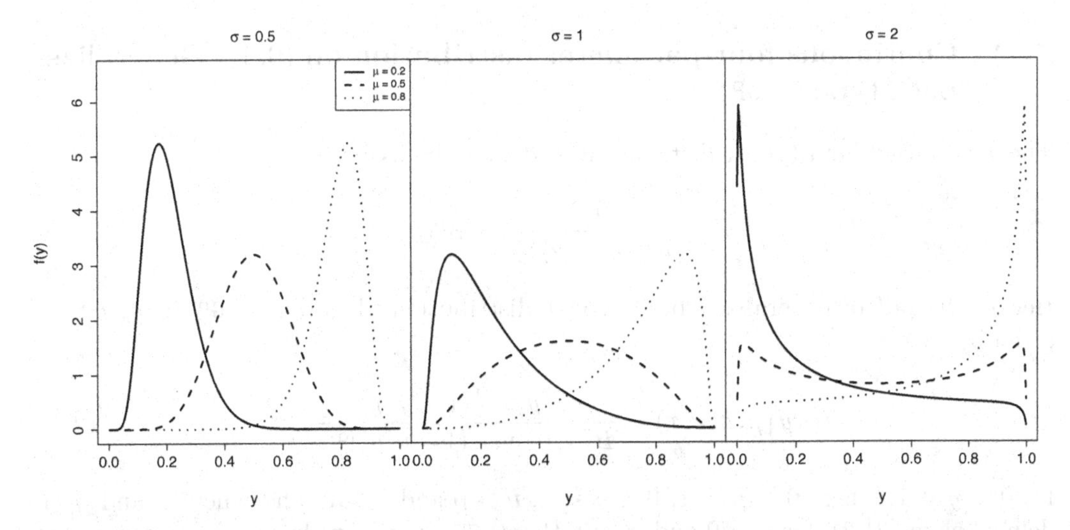

FIGURE 21.3: The logit normal, LOGITNO(μ, σ), distribution, for $\mu = 0.2, 0.5, 0.8$ and $\sigma = 0.5, 1, 2$.

21.1.3 Simplex: SIMPLEX

The pdf of the simplex distribution, denoted by SIMPLEX(μ, σ), is given by

$$f_Y(y \mid \mu, \sigma) = \frac{1}{[2\pi\sigma^2 y^3(1-y)^3]^{1/2}} \exp\left[-\frac{(y-\mu)^2}{2\sigma^2 y(1-y)\mu^2(1-\mu)^2}\right], \tag{21.4}$$

for $0 < y < 1$, where $0 < \mu < 1$ and $\sigma > 0$. The mean of the distribution is μ. This distribution was called the 'standard simplex distribution' by Jørgensen [1997, p199].

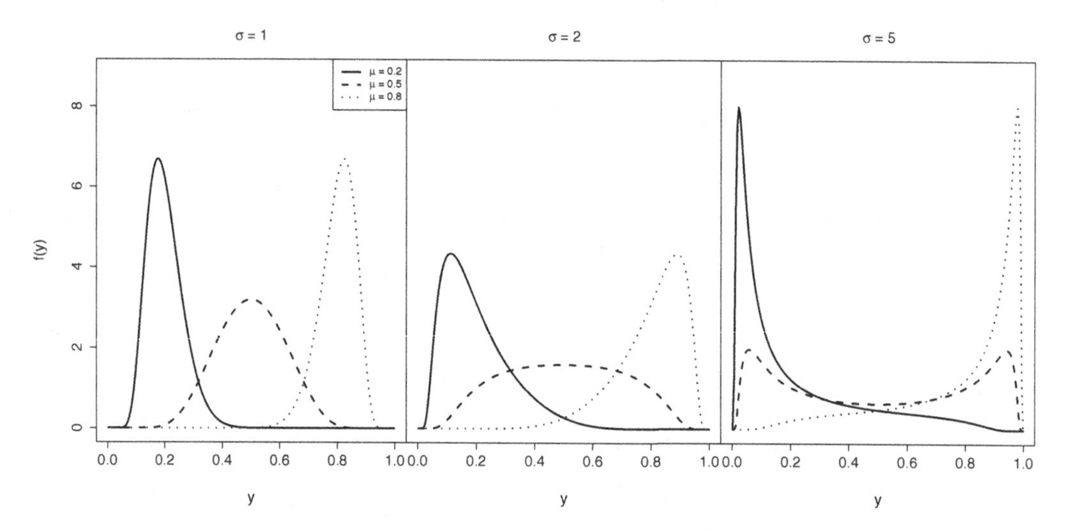

FIGURE 21.4: The simplex, SIMPLEX(μ, σ), distribution, with $\mu = 0.2, 0.5, 0.8$ and $\sigma = 1, 2, 5$

21.2 Continuous four-parameter distribution on $(0, 1)$: Generalized beta type 1, GB1

The generalized beta type 1 distribution is defined by assuming

$$Z = \frac{Y^\tau}{\nu + (1-\nu)Y^\tau} \sim \text{BE}(\mu, \sigma) \ .$$

Hence, the pdf of generalized beta type 1 distribution, denoted by GB1(μ, σ, ν, τ), is given by

$$f_Y(y \mid \mu, \sigma, \nu, \tau) = \frac{\tau \nu^\beta y^{\tau\alpha - 1}(1-y^\tau)^{\beta-1}}{B(\alpha, \beta)[\nu + (1-\nu)y^\tau]^{\alpha+\beta}} \tag{21.5}$$

for $0 < y < 1$, where $0 < \mu < 1$, $0 < \sigma < 1$, $\nu > 0$ and $\tau > 0$, and where α and β are defined as in (21.2), for $\alpha > 0$ and $\beta > 0$. Hence GB1(μ, σ, ν, τ) has adopted parameters $\mu = \alpha/(\alpha + \beta)$, $\sigma = (\alpha + \beta + 1)^{-1/2}$, ν and τ.

For $0 < \nu < 1$, GB1(μ, σ, ν, τ) is a reparameterized submodel, with range $0 < Y < 1$, of the five parameter generalized beta (GB) distribution of McDonald and Xu [1995, equation (2.8)] where $\texttt{GB1}(\mu,\sigma,\nu,\tau) = \texttt{GB}(\tau,\nu^{1/\tau},1-\nu,\alpha,\beta)$. (Note that GB1 is different from the generalized beta of the first kind of McDonald and Xu [1995].) The generalized three-parameter beta (GB3) distribution of Pham-Gia and Duong (1989) and Johnson et al. [1995, p251] is a reparameterized submodel of GB1 where $\tau = 1$, given by $\texttt{G3B}(\alpha_1,\alpha_2,\lambda) = \texttt{GB1}(\alpha_1(\alpha_1+\alpha_2)^{-1},(\alpha_1+\alpha_2+1)^{-1},\lambda^{-1},1)$. The $\texttt{BE}(\mu,\sigma)$ distribution is a submodel of $\texttt{GB1}(\mu,\sigma,\nu,\tau)$ where $\nu = 1$ and $\tau = 1$, i.e. $\texttt{BE}(\mu,\sigma) = \texttt{GB1}(\mu,\sigma,1,1)$.

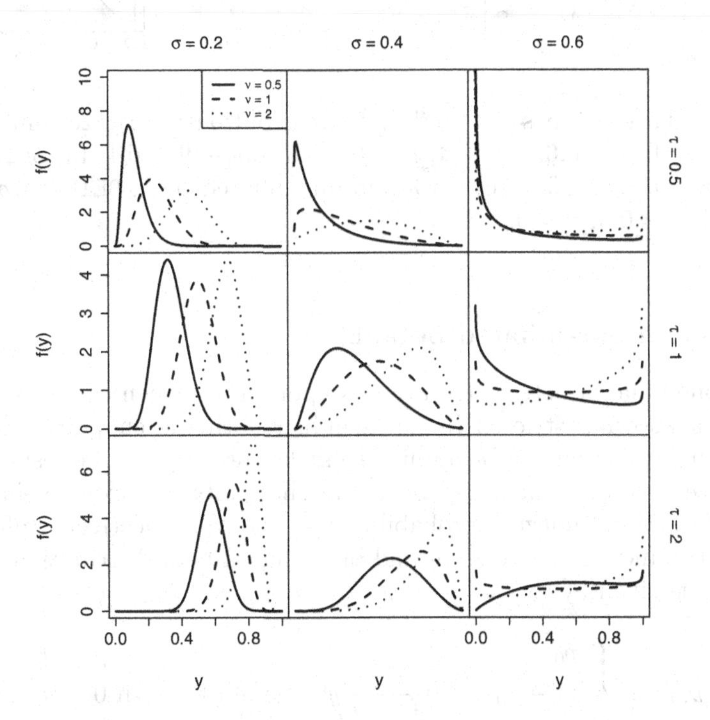

FIGURE 21.5: The generalized beta type 1 distribution, $\texttt{GB1}(\mu,\sigma,\nu,\tau)$, with $\mu = .5$, $\sigma = 0.2, 0.4, 0.6$, $\nu = 0.5, 1, 2$, and $\tau = 0.5, 1, 2$

21.3 Inflated distributions on [0,1), (0,1], and [0,1]

The three types of inflated distributions are shown in Figure 21.6, which is identical to Figure 9.3.

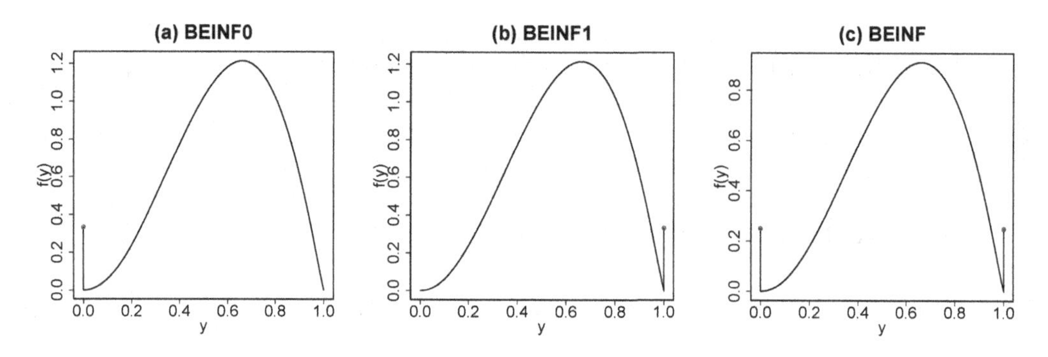

FIGURE 21.6: Three types of inflated distributions: (a) zero-inflated beta, BEINF0(μ, σ, ν), with $\mu = 0.6$, $\sigma = 0.4$, $\nu = 0.5$, (b) one-inflated beta, BEINF1(μ, σ, ν), with $\mu = 0.6$, $\sigma = 0.4$, $\nu = 0.5$, (c) zero- and one-inflated beta, BEINF(μ, σ, ν, τ), with $\mu = 0.6$, $\sigma = 0.4$, $\nu = 0.5$, $\tau = 0.5$.

21.3.1 Zero- and one-inflated beta: BEINF

The zero- and one-inflated beta distribution is appropriate when the response variable takes values in a known restricted range including both the endpoints of the range. Appropriate standardization can be applied to make the range of the response variable [0,1], i.e. from zero to one, including both the endpoints, i.e. exact values 0 and 1. Values zero and one have nonzero probabilities p_0 and p_1 respectively. Informally, the mixed probability function of the zero- and one-inflated beta distribution, denoted by BEINF(μ, σ, ν, τ), is given by

$$f_Y(y \mid \mu, \sigma, \nu, \tau) = \begin{cases} p_0 & \text{if } y = 0 \\ (1 - p_0 - p_1) \frac{1}{B(\alpha,\beta)} \, y^{\alpha-1}(1-y)^{\beta-1} & \text{if } 0 < y < 1 \\ p_1 & \text{if } y = 1 \end{cases} \tag{21.6}$$

for $0 \leq y \leq 1$, where α and β are given by (21.2), $p_0 = \nu(1+\nu+\tau)$ and $p_1 = \tau/(1+\nu+\tau)$ where $\alpha > 0$, $\beta > 0$, $0 < p_0 < 1$, $0 < p_1 < 1$ and $0 < p_0 + p_1 < 1$.

Hence BEINF(μ, σ, ν, τ) has parameters

$$\begin{aligned} \mu &= \alpha/(\alpha + \beta) \\ \sigma &= (\alpha + \beta + 1)^{-1/2} \\ \nu &= \frac{p_0}{1 - p_0 - p_1} \\ \tau &= \frac{p_1}{1 - p_0 - p_1} \, , \end{aligned}$$

and $0 < \mu < 1$, $0 < \sigma < 1$, $\nu > 0$ and $\tau > 0$. Note that $\mathrm{E}(Y) = (\mu + \tau)/(1 + \nu + \tau)$.

TABLE 21.3: The zero- and one-inflated beta distribution.

BEINF(μ, σ, ν, τ)	
Ranges	
Y	$0 \leq y \leq 1$
μ	$0 < \mu < 1$, mean of BE(μ, σ) component
σ	$0 < \sigma < 1$
ν	$0 < \nu < \infty$
τ	$0 < \tau < \infty$
Distribution measures	
mean	$(\mu + \tau)(1 + \nu + \tau)^{-1}$
variance	$\dfrac{\sigma^2 \mu(1-\mu) + \mu^2 + \tau + (\mu+\tau)^2 (1+\nu+\tau)^{-1}}{1+\nu+\tau}$
pdf	$\begin{cases} p_0 & \text{if } y = 0 \\ (1 - p_0 - p_1) \dfrac{1}{B(\alpha, \beta)} y^{\alpha-1}(1-y)^{\beta-1} & \text{if } 0 < y < 1 \\ p_1 & \text{if } y = 1 \\ \text{where} & \\ \alpha = \mu(1-\sigma^2)/\sigma^2, \ \beta = (1-\mu)(1-\sigma^2)/\sigma^2, & \\ p_0 = \nu(1+\nu+\tau)^{-1}, & \\ p_1 = \tau(1+\nu+\tau)^{-1} & \end{cases}$
cdf	$\begin{cases} p_0 & \text{if } y = 0 \\ p_0 + \dfrac{(1-p_0-p_1)B(\alpha,\beta,y)}{B(\alpha,\beta)} & \text{if } 0 < y < 1 \\ 1 & \text{if } y = 1 \end{cases}$

21.3.2 Zero-inflated beta: BEINF0

Informally, the mixed probability function of the zero-inflated beta distribution, denoted by BEINF0(μ, σ, ν), is given by

$$f_Y(y \mid \mu, \sigma, \nu) = \begin{cases} p_0 & \text{if } y = 0 \\ (1 - p_0) \frac{1}{B(\alpha,\beta)} \ y^{\alpha-1}(1-y)^{\beta-1} & \text{if } 0 < y < 1 \end{cases} \tag{21.7}$$

for $0 \leq y < 1$, where α and β are given by (21.2) and $p_0 = \nu/(1+\nu)$, where $\alpha > 0$, $\beta > 0$, $0 < p_0 < 1$. Hence BEINF0(μ, σ, ν) has parameters

$$\begin{aligned} \mu &= \alpha/(\alpha+\beta) \\ \sigma &= (\alpha+\beta+1)^{-1/2} \\ \nu &= \frac{p_0}{1-p_0}, \end{aligned}$$

and $0 < \mu < 1$, $0 < \sigma < 1$ and $\nu > 0$. Note that for BEINF0(μ, σ, ν), $\mathrm{E}(Y) = \mu/(1+\nu)$.

TABLE 21.4: The zero-inflated beta distribution.

BEINF0(μ, σ, ν)	
Ranges	
Y	$0 \leq y < 1$
μ	$0 < \mu < 1$, mean of BE(μ, σ) component
σ	$0 < \sigma < 1$
ν	$0 < \nu < \infty$
Distribution measures	
mean	$\dfrac{\mu}{1+\nu}$
variance	$\dfrac{\sigma^2\mu(1-\mu)+\mu^2+\mu^2(1+\nu)^{-1}}{1+\nu}$
pdf	$\begin{cases} p_0 & \text{if } y=0 \\ (1-p_0)\dfrac{1}{B(\alpha,\beta)}y^{\alpha-1}(1-y)^{\beta-1} & \text{if } 0<y<1 \\ \text{where} & \\ \alpha=\dfrac{\mu(1-\sigma^2)}{\sigma^2},\ \beta=\dfrac{(1-\mu)(1-\sigma^2)}{\sigma^2}, & \\ p_0=\nu(1+\nu)^{-1} & \end{cases}$
cdf	$\begin{cases} p_0 & \text{if } y=0 \\ p_0+\dfrac{(1-p_0)B(\alpha,\beta,y)}{B(\alpha,\beta)} & \text{if } 0<y<1 \\ 1 & \text{if } y=1 \end{cases}$
Reference	Set $\tau=0$ in BEINF(μ, σ, ν, τ)

21.3.3 One-inflated beta: BEINF1

Informally, the mixed probability function of the one-inflated beta distribution, denoted by BEINF1(μ, σ, ν), is given by

$$f_Y(y \mid \mu, \sigma, \nu) = \begin{cases} (1-p_1)\frac{1}{B(\alpha,\beta)}\, y^{\alpha-1}(1-y)^{\beta-1} & \text{if } 0<y<1 \\ p_1 & \text{if } y=1 \end{cases} \tag{21.8}$$

for $0 < y \leq 1$, where α and β are given by (21.2), $p_1 = \nu/(1+\nu)$, where $\alpha > 0$, $\beta > 0$ and $0 < p_1 < 1$. Hence BEINF1(μ, σ, ν) has parameters

$$\begin{aligned} \mu &= \alpha/(\alpha+\beta) \\ \sigma &= (\alpha+\beta+1)^{-1/2} \\ \nu &= \frac{p_1}{1-p_1}\,, \end{aligned}$$

and $0 < \mu < 1$, $0 < \sigma < 1$ and $\nu > 0$. Note that $\mathrm{E}(Y) = (\mu+\nu)/(1+\nu)$.

TABLE 21.5: The one-inflated beta distribution.

BEINF1(μ, σ, ν)	
Ranges	
Y	$0 < y \leq 1$
μ	$0 < \mu < 1$, mean of BE(μ, σ) component
σ	$0 < \sigma < 1$
ν	$0 < \nu < \infty$
Distribution measures	
mean	$\dfrac{\mu + \nu}{1 + \nu}$
variance	$\dfrac{\sigma^2 \mu(1-\mu) + \mu^2 + \nu + (\mu+\nu)^2(1+\nu)^{-1}}{1+\nu}$
pdf	$\begin{cases} (1-p_1)\dfrac{1}{B(\alpha,\beta)} y^{\alpha-1}(1-y)^{\beta-1} & \text{if } 0 < y < 1 \\ p_1 & \text{if } y = 1 \\ \text{where} & \\ \alpha = \dfrac{\mu(1-\sigma^2)}{\sigma^2}, \quad \beta = \dfrac{(1-\mu)(1-\sigma^2)}{\sigma^2} & \\ p_1 = \nu(1+\nu)^{-1} & \end{cases}$
cdf	$\begin{cases} \dfrac{(1-p_1)B(\alpha,\beta,y)}{B(\alpha,\beta)} & \text{if } 0 < y < 1 \\ 1 & \text{if } y = 1 \end{cases}$
Reference	Set $\nu = 0$ and then $\tau = \nu$ in BEINF(μ, σ, ν, τ)

BEZI(μ, σ, ν) and BEOI(μ, σ, ν) are different parameterizations of the BEINF0(μ, σ, ν) and BEINF1(μ, σ, ν) distributions contributed to **gamlss.dist** by Raydonal Ospina, see Ospina and Ferrari [2010] and Section 9.3.2.

22

Discrete count distributions

CONTENTS

This chapter gives summary tables and plots for explicit **gamlss.dist** discrete count distributions. All the distributions in this chapter have range $\{0, 1, 2, 3, \ldots\}$, with the exception of the logarithmic ($\texttt{LG}(\mu)$) and the Zipf ($\texttt{ZIPF}(\mu)$) distributions, which have range $\{1, 2, 3, \ldots\}$. Discrete count distributions are discussed in Chapter 7. Section

7.3.4 discusses creating discrete count distributions by discretizing any continuous `gamlss.family` distribution defined on $(0, \infty)$.

22.1 One-parameter count distributions

22.1.1 Geometric: GEOM, GEOMo

There are two parameterizations of the geometric distribution in the **gamlss.dist** package: GEOM(μ) and GEOMo(μ).

First parameterization, GEOM

The probability function (pf) of the geometric distribution, GEOM(μ), is given by

$$P(Y = y \mid \mu) = \frac{\mu^y}{(\mu+1)^{y+1}} \tag{22.1}$$

for $y = 0, 1, 2, 3, \ldots$, where $\mu > 0$. Figure 22.1 shows the shapes of the GEOM(μ) distribution for $\mu = 1, 3, 5$. The mode of Y is always at zero, and the probabilities always decrease as Y increases. For large values of y (and all y), $P(Y = y \mid \mu) = q \exp\left\{-y \log\left(1 + \mu^{-1}\right)\right\}$ where $q = (\mu+1)^{-1}$, i.e. an exponential right tail.

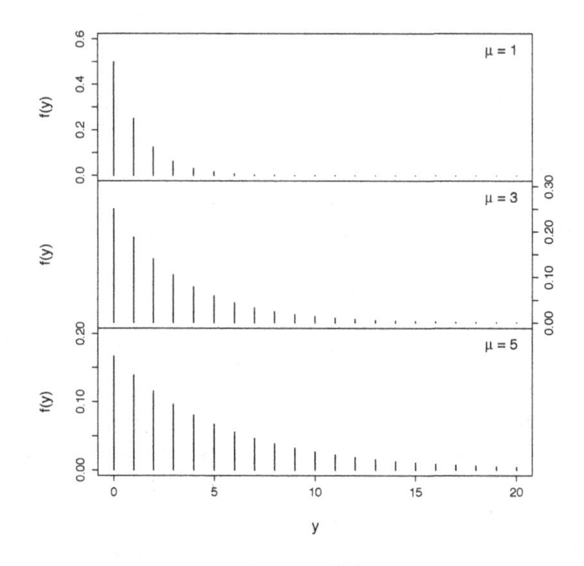

FIGURE 22.1: The geometric, GEOM(μ), distribution, with $\mu = 1, 3, 5$.

TABLE 22.1: Geometric distribution.

GEOM(μ)	
Ranges	
Y	$0, 1, 2, 3 \ldots$
μ	$0 < \mu < \infty$, mean
Distribution measures	
mean	μ
median	$\begin{cases} \lfloor \alpha \rfloor & \text{if } \alpha \text{ is not an integer} \\ \alpha - 1 & \text{if } \alpha \text{ is an integer} \end{cases}$ where $\alpha = \dfrac{-\log 2}{\log[\,\mu\,(\mu+1)^{-1}]}$
mode	0
variance	$\mu + \mu^2$
skewness	$(1+2\mu)\left(\mu+\mu^2\right)^{-0.5}$
excess kurtosis	$6 + (\mu + \mu^2)^{-1}$
PGF	$[1 + \mu(1-t)]^{-1}$ for $0 < t < 1 + \mu^{-1}$
pf	$\dfrac{\mu^y}{(\mu+1)^{y+1}}$
cdf	$1 - \left(\frac{\mu}{\mu+1}\right)^{y+1}$
inverse cdf	$\begin{cases} \lfloor \alpha_p \rfloor & \text{if } \alpha_p \text{ is not an integer} \\ \alpha_p - 1 & \text{if } \alpha_p \text{ is an integer} \end{cases}$ where $\alpha_p = \dfrac{\log(1-p)}{\log[\,\mu(\mu+1)^{-1}]}$
Reference	Set $\sigma = 1$ in NBI(μ, σ)
Notes	$\lfloor \alpha \rfloor$ is the largest integer less than or equal to α, i.e. the floor function.

Second parameterization, GEOMo

For the original parameterization of the geometric distribution, GEOMo(μ), replace μ in (22.1) by $(1-\mu)/\mu$, giving

$$P(Y = y \mid \mu) = (1-\mu)^y \mu$$

for $y = 0, 1, 2, 3, \ldots$, and $0 < \mu < 1$. Hence μ in this case is $P(Y = 0)$. Other characteristics of the GEOMo(μ) distribution are given in Table 22.2.

TABLE 22.2: Geometric distribution (original).

	`GEOMo`(μ)	
	Ranges	
Y	$0, 1, 2, 3 \ldots$	
μ	$0 < \mu < 1,$	
	Distribution measures	
mean	$(1-\mu)\mu^{-1}$	
median	$\begin{cases} \lfloor \alpha \rfloor, & \text{if } \alpha \text{ is not an integer} \\ \alpha - 1, & \text{if } \alpha \text{ is an integer} \end{cases}$	where $\alpha = \dfrac{-\log 2}{\log(1-\mu)}$
mode	0	
variance	$(1-\mu)\,\mu^{-2}$	
skewness	$(2-\mu)\,(1-\mu)^{-0.5}$	
excess kurtosis	$6 + \mu^2\,(1-\mu)^{-1}$	
PGF	$\mu\,[1-(1-\mu)t]^{-1}$ for $0 < t < (1-\mu)^{-1}$	
pf	$\mu(1-\mu)^y$	
cdf	$1-(1-\mu)^{y+1}$	
inverse cdf	$\begin{cases} \lfloor \alpha_p \rfloor, & \text{if } \alpha_p \text{ is not an integer} \\ \alpha_p - 1, & \text{if } \alpha_p \text{ is an integer} \end{cases}$	where $\alpha_p = \dfrac{\log(1-p)}{\log(1-\mu)}$
Reference	Set μ_1 to $(1-\mu)\mu^{-1}$ in `GEOM`(μ_1)	

22.1.2 Logarithmic: LG

The probability function of the logarithmic distribution, denoted by `LG`(μ), is given by

$$P(Y = y \mid \mu) = \frac{\alpha \mu^y}{y} \tag{22.2}$$

for $y = 1, 2, 3, \ldots$, where $\alpha = -\left[\log(1-\mu)\right]^{-1}$ for $0 < \mu < 1$. Note that the range of Y starts from 1. As $y \to \infty$ (and all y), $P(Y = y \mid \mu)$ is proportional to $\exp\{y \log \mu - \log y\}$. Hence $\log P(Y = y \mid \mu) \sim y \log \mu$ as $y \to \infty$. The mode of Y is always at 1 and the probabilities always decrease as Y increases.

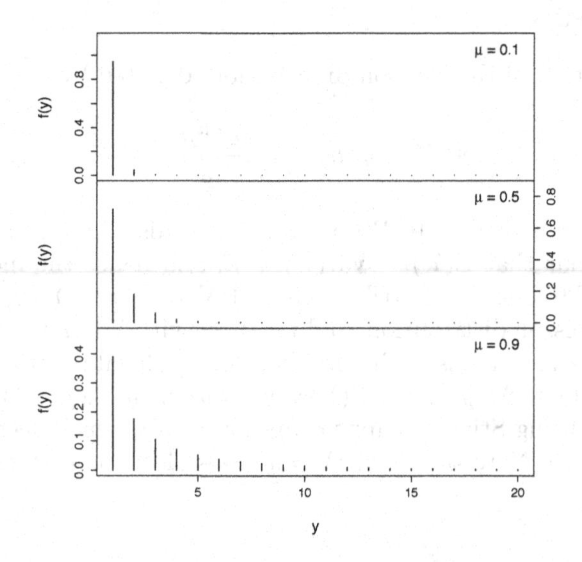

FIGURE 22.2: The logarithmic, $\texttt{LG}(\mu)$, distribution, with $\mu = 0.1, 0.5, 0.9$.

TABLE 22.3: Logarithmic distribution.

$\texttt{LG}(\mu)$	
Ranges	
Y	$1, 2, 3 \ldots$
μ	$0 < \mu < 1$
Distribution measures	
mean[a]	$\alpha\mu(1-\mu)^{-1}$ where $\alpha = -[\log(1-\mu)]^{-1}$
mode[a]	1
variance[a]	$\alpha\mu(1-\alpha\mu)(1-\mu)^{-2}$
skewness[a]	$\mu_3/[\text{Var}(Y)]^{1.5}$ where $\mu_3 = \alpha\mu(1+\mu-3\alpha\mu+2\alpha^2\mu^2)(1-\mu)^{-3}$
excess kurtosis	$k_4/[\text{Var}(Y)]^2$ where $k_4 = \alpha\mu[1+4\mu+\mu^2-\alpha\mu(7+4\mu)+12\alpha^2\mu^2-6\alpha^3\mu^3](1-\mu)^{-4}$
PGF[a]	$\dfrac{\log(1-\mu t)}{\log(1-\mu)}$ for $0 < t < \mu^{-1}$
pf [a]	$\dfrac{\alpha\mu^y}{y}$
Reference	[a] Johnson et al. [2005], Section 7.1, p302-307, parameterized by $\theta = \mu$

22.1.3 Poisson: PO

The probability function of the Poisson distribution, denoted by PO(μ), is given by

$$P(Y = y \mid \mu) = \frac{e^{-\mu}\mu^y}{y!} \tag{22.3}$$

for $y = 0, 1, 2, 3, \ldots$, where $\mu > 0$. Properties of the distribution are given in Table 22.4. Note in particular that $\text{E}(Y) = \text{Var}(Y) = \mu$, and hence the index of dispersion $\text{Var}(Y)/\text{E}(Y)$ is equal to one. For distributions with $\text{Var}(Y) > \text{E}(Y)$ we have overdispersion relative to the Poisson distribution, and for those with $\text{Var}(Y) < \text{E}(Y)$ we have underdispersion. The distribution is positively skew for small values of μ, but almost symmetric for large μ values. As $y \to \infty$, $P(Y = y) \sim q \exp\left[y(\log\mu + 1) - (y + 0.5)\log y\right]$ where $q = e^{-\mu}\sqrt{2\pi}$, (using Stirling's approximation to $y!$), which decreases faster than an exponential $\exp(-y)$. Note that $\log P(Y = y) \sim -y \log y$ as $y \to \infty$.

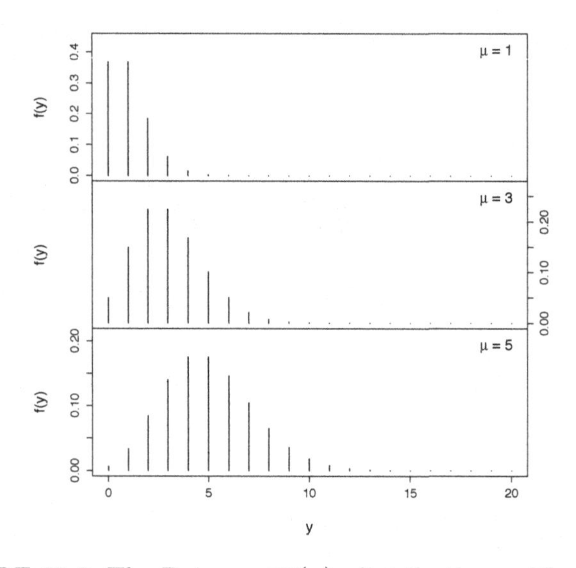

FIGURE 22.3: The Poisson, PO(μ), distribution, with $\mu = 1, 3, 5$.

TABLE 22.4: Poisson distribution.

$\texttt{PO}(\mu)$	
Ranges	
Y	$0, 1, 2, 3 \ldots$
μ	$0 < \mu < \infty$, mean
Distribution measures	
mean	μ
mode	$\begin{cases} \lfloor \mu \rfloor & \text{if } \mu \text{ is not an integer} \\ \mu - 1 \text{ and } \mu & \text{if } \mu \text{ is an integer} \end{cases}$
variance	μ
skewness	$\mu^{-0.5}$
excess kurtosis	μ^{-1}
PGF	$e^{\mu(t-1)}$
pf	$\dfrac{e^{-\mu}\mu^y}{y!}$
cdf	$\dfrac{\Gamma(y+1, \mu)}{\Gamma(y)}$
Reference	Johnson et al. [2005] Sections 4.1, 4.3, 4.4, p156, p161-165, p307
Note	$\lfloor \alpha \rfloor$ is the largest integer less than or equal to α $\Gamma(\alpha, x) = \int_x^\infty t^{\alpha-1} e^{-t} dt$ is the complement of the incomplete gamma function

22.1.4 Yule: YULE

The probability function of the Yule distribution, denoted by $\texttt{YULE}(\mu)$, is given by

$$P(Y = y \mid \mu) = (\mu^{-1} + 1) B(y + 1, \mu^{-1} + 2) \tag{22.4}$$

for $y = 0, 1, 2, 3, \ldots,$ where $\mu > 0$. Note that this parameterization only includes Yule distributions with a finite mean μ. The mode is always at zero, and the probabilities always decrease with increasing Y. The $\texttt{YULE}(\mu)$ distribution is a special case of the $\texttt{WARING}(\mu, \sigma)$ where $\mu = \sigma$. It can be shown (using equation (1.32) of Johnson et al. [2005, p8]) that as $y \to \infty$, $P(Y = y \mid \mu) \sim q y^{-(\mu^{-1}+2)}$ where $q = (\mu^{-1} + 1)\Gamma(\mu^{-1} + 2)$. Hence the $\texttt{YULE}(\mu)$ distribution has a heavy tail, especially for large μ .

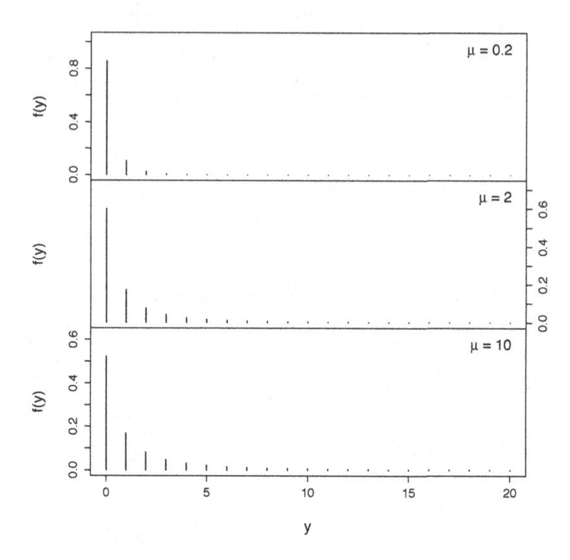

FIGURE 22.4: The Yule, YULE(μ), distribution, with $\mu = 0.2, 2, 10$.

TABLE 22.5: Yule distribution.

YULE(μ)	
Ranges	
Y	$0, 1, 2, 3 \ldots$
μ	$0 < \mu < \infty$, mean
Distribution measures	
mean	μ
mode	0
variance	$\begin{cases} \mu(\mu+1)^2(1-\mu)^{-1} & \text{if } \mu < 1 \\ \infty & \text{if } \mu \geq 1 \end{cases}$
skewness	$\begin{cases} (2\mu+1)^2(1-\mu)^{0.5}(\mu+1)^{-1}(1-2\mu)^{-1}\mu^{-0.5} & \text{if } \mu < 1/2 \\ \infty & \text{if } \mu \geq 1/2 \end{cases}$
excess kurtosis	$\begin{cases} \dfrac{(1+11\mu+18\mu^2-6\mu^3-36\mu^4)}{\mu(\mu+1)(1-2\mu)(1-3\mu)} & \text{if } \mu < 1/3 \\ \infty & \text{if } \mu \geq 1/3 \end{cases}$
PGF	$(\mu+1)(2\mu+1)^{-1}{}_2F_1(1, 1; 3+\mu^{-1}; t)$
pf	$(\mu^{-1}+1)B(y+1, \mu^{-1}+2)$
cdf	$1-(y+1)B(y+1, 2+\mu^{-1})$
Reference	Set $\sigma = \mu$ in WARING(μ, σ)

22.1.5 Zipf: ZIPF

The probability function of the Zipf distribution, denoted by ZIPF(μ), is given by

$$P(Y = y \mid \mu) = \frac{y^{-(\mu+1)}}{\zeta(\mu+1)} \tag{22.5}$$

for $y = 1, 2, 3, \ldots$, where $\mu > 0$ and $\zeta(b) = \Sigma_{i=1}^{\infty} i^{-b}$ is the Riemann zeta function. Note that the range of Y starts from 1. The mode of Y is always at 1, and the probabilities always decrease with increasing Y.

This distribution is also known as the Riemann zeta distribution or the discrete Pareto distribution. as $y \to \infty$ (and all y), $P(Y = y \mid \mu) = qy^{-(\mu+1)}$ where $q = [\zeta(\mu+1)]^{-1}$, so the ZIPF(μ) distribution has a very heavy tail especially for μ close to zero. It is suitable for very heavy-tailed count data, as can be seen from Figure 22.5 where it is plotted for $\mu = 0.1, 0.5, 1$.

The rth raw moment is $E(Y^r) = \zeta(\mu - r + 1)/\zeta(\mu + 1)$, provided that $\mu > r$, [Johnson et al., 2005, p528]. Hence this gives the mean, variance, skewness, and excess kurtosis of Y. The mean increases as μ decreases and is infinite if $\mu \leq 1$.

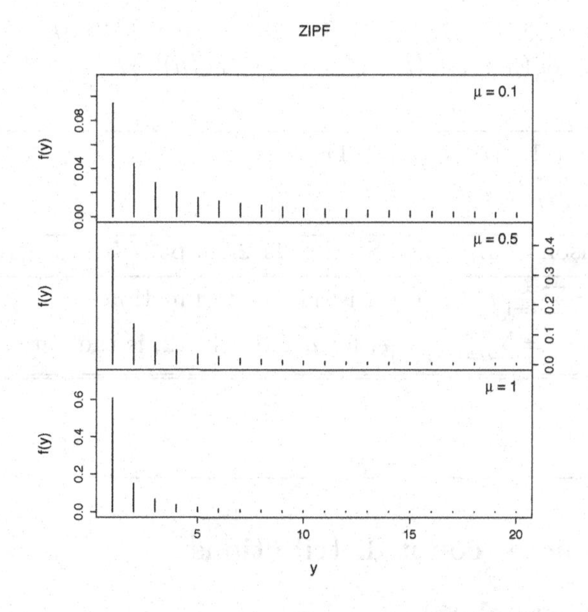

FIGURE 22.5: The ZIPF(μ) distribution, with $\mu = 0.1, 0.5, 1$.

TABLE 22.6: ZIPF distribution.

ZIPF(μ)	
Ranges	
Y	$1, 2, 3 \ldots$
μ	$0 < \mu < \infty$
Distribution measures	
mean	$\begin{cases} \zeta(\mu)/\zeta(\mu+1) & \text{if } \mu > 1 \\ \infty & \text{if } \mu \leq 1 \end{cases}$
mode	1
variance	$\begin{cases} \left\{\zeta(\mu+1)\zeta(\mu-1) - [\zeta(\mu)]^2\right\}/[\zeta(\mu+1)]^2 & \text{if } \mu > 2 \\ \infty & \text{if } \mu \leq 2 \end{cases}$
skewness	$\begin{cases} \mu_3/[\text{Var}(Y)]^{1.5} \quad \text{where} & \\ \mu_3 = \{[\zeta(\mu+1)]^2\zeta(\mu-2) - 3\zeta(\mu+1)\zeta(\mu)\zeta(\mu-1) + & \\ \quad 2[\zeta(\mu)]^3\}/[\zeta(\mu+1)]^3 & \text{if } \mu > 3 \\ \infty & \text{if } \mu \leq 3 \end{cases}$
excess kurtosis	$\begin{cases} \{\mu_4/[\text{Var}(Y)]^2\} - 3 \quad \text{where} & \\ \mu_4 = \{[\zeta(\mu+1)]^3\zeta(\mu-3) - 4[\zeta(\mu+1)]^2\zeta(\mu)\zeta(\mu-2) + & \\ \quad 6\zeta(\mu+1)[\zeta(\mu)]^2\zeta(\mu-1) - 3[\zeta(\mu)]^4\}/[\zeta(\mu+1)]^4 & \text{if } \mu > 4 \\ \infty & \text{if } \mu \leq 4 \end{cases}$
PGF[a]	$t\phi(t, \mu+1, 1)/\phi(1, \mu+1, 1)$
pf [a]	$[y^{(\mu+1)}\zeta(\mu+1)]^{-1}$
Reference	[a]Johnson et al. [2005] Section 11.2.20, p527-528, where $\rho = \mu$
Notes	$\zeta(b) = \sum_{i=1}^{\infty} i^{-b}$ is the Riemann zeta function
	$\phi(a, b, c) = \sum_{i=0}^{\infty} \frac{a^i}{(i+c)^b}$, for $c \neq 0, -1, -2$, is the Lerch function

22.2 Two-parameter count distributions

22.2.1 Double Poisson: DPO

The double Poisson distribution, denoted by DPO(μ, σ), has probability function

$$P(Y = y \mid \mu, \sigma) = c(\mu, \sigma)\sigma^{-1/2}e^{-\mu/\sigma}\left(\frac{\mu}{y}\right)^{y/\sigma}\frac{e^{y/\sigma - y}y^y}{y!} \tag{22.6}$$

for $y = 0, 1, 2, 3, \ldots$, where $\mu > 0$, $\sigma > 0$ and $c(\mu, \sigma)$ is a normalizing constant (ensuring that the distribution probabilities sum to one) given by

$$c(\mu, \sigma) = \left[\sum_{y=0}^{\infty} \sigma^{-1/2}e^{-\mu/\sigma}\left(\frac{\mu}{y}\right)^{y/\sigma}\frac{e^{y/\sigma - y}y^y}{y!}\right]^{-1} \tag{22.7}$$

(Lindsey [1995, p131], reparameterized by $\nu = \mu$ and $\psi = 1/\sigma$).

The double Poisson distribution is a special case of the double exponential family of Efron [1986]. It has approximate mean μ and approximate variance $\sigma\mu$. The $\texttt{DPO}(\mu, \sigma)$ distribution is a $\texttt{PO}(\mu)$ distribution if $\sigma = 1$. It is (approximately) an overdispersed Poisson distribution if $\sigma > 1$ and underdispersed Poisson if $\sigma < 1$. Unlike some other implementations, **gamlss.dist** approximates $c(\mu, \sigma)$ using a finite sum with a very large number of terms ($3 \times \max(y)$) in equation (22.7), rather than a potentially less accurate functional approximation of $c(\mu, \sigma)$. As $y \to \infty$, $\log \mathrm{P}(Y = y \mid \mu, \sigma) \sim -(y \log y)/\sigma$.

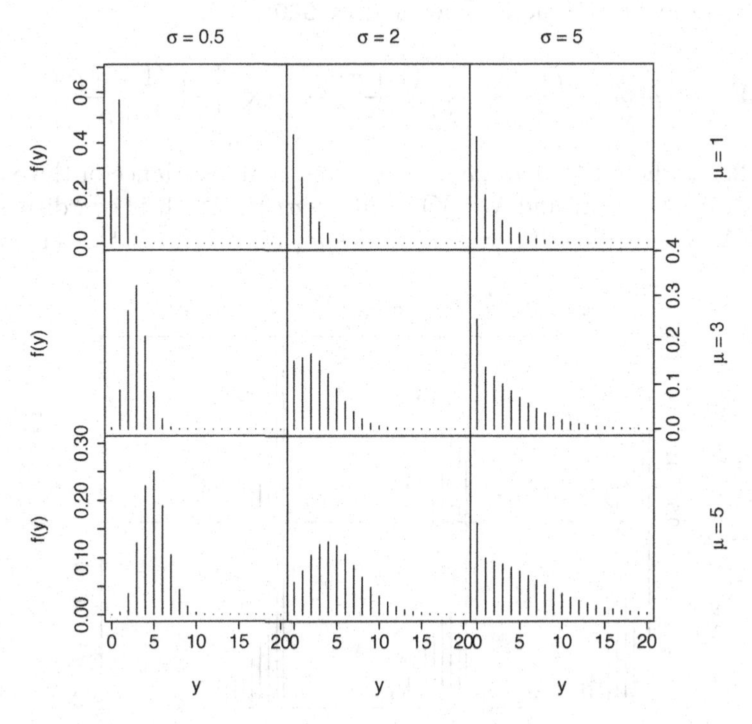

FIGURE 22.6: The double Poisson, $\texttt{DPO}(\mu, \sigma)$, distribution with $\mu = 1, 3, 5$ and $\sigma = .5, 2, 5$.

22.2.2 Generalized Poisson: GPO

The probability function of the generalized Poisson distribution [Consul and Jain, 1973], [Consul, 1989], [Poortema, 1999], [Johnson et al., 2005, p336], is given by

$$\mathrm{P}(Y = y \mid \theta, \lambda) = \frac{\theta(\theta + \lambda y)^{y-1}\, e^{-(\theta+\lambda y)}}{y!} \tag{22.8}$$

for $y = 0, 1, 2, ...$ where $\theta > 0$ and $0 \leq \lambda \leq 1$. The generalized Poisson distribution was derived by Consul and Jain [1973] as an approximation of a generalized negative binomial distribution. In recognition of P. C. Consul's contribution to this distribution, it is sometimes called *Consul's generalized Poisson distribution.* The mean and the variance of the generalized Poisson distribution are given by $\text{E}(Y) = \theta/(1-\lambda)$ and $\text{Var}(Y) = \theta/(1-\lambda)^3$, [Johnson et al., 2005, p337-338].

In **gamlss.dist**, a parameterization of the generalized Poisson distribution which is more suitable for regression modeling is used. This is given by reparameterizing (22.8) by setting

$$\mu = \frac{\theta}{1-\lambda} \qquad \text{and} \qquad \sigma = \frac{\lambda}{\theta}$$

and hence

$$\theta = \frac{\mu}{1+\sigma\mu} \qquad \text{and} \qquad \lambda = \frac{\sigma\mu}{1+\sigma\mu}\,.$$

This gives the probability function, denoted by $\texttt{GPO}(\mu,\sigma)$:

$$\text{P}(Y = y \,|\, \mu, \sigma) = \left(\frac{\mu}{1+\sigma\mu}\right)^y \frac{(1+\sigma y)^{y-1}}{y!} \exp\left[\frac{-\mu\,(1+\sigma y)}{1+\sigma\mu}\right] \tag{22.9}$$

for $y = 0, 1, 2, ...$, where $\mu > 0$ and $\sigma > 0$. The mean and variance of this version of the distribution are $\text{E}(Y) = \mu$ and $\text{Var}(Y) = \mu(1+\sigma\mu)^2$, which is overdispersed Poisson since $\sigma > 0$. As $y \to \infty$, $\log \text{P}(Y = y \,|\, \mu, \sigma) \sim -y\left\{\log\left[1 + (\mu\sigma)^{-1}\right] - (1+\mu\sigma)^{-1}\right\}$.

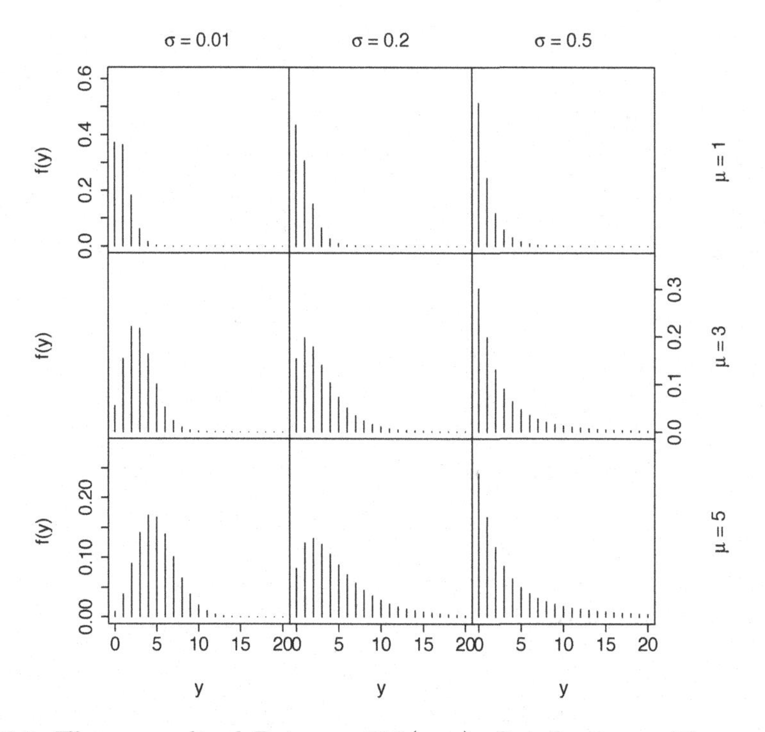

FIGURE 22.7: The generalized Poisson, $\texttt{GPO}(\mu,\sigma)$, distribution, with $\mu = 1, 3, 5$ and $\sigma = 0.01, 0.2, 0.5$.

TABLE 22.7: Generalized Poisson distribution.

	$\texttt{GPO}(\mu, \sigma)$
	Ranges
Y	$0, 1, 2, 3 \ldots$
μ	$0 < \mu < \infty$, mean
σ	$0 < \sigma < \infty$, dispersion parameter
	Distribution measures
mean	μ
variance	$\mu\,(1 + \sigma\mu)^2$
skewness	$(1 + 3\sigma\mu)\,/\mu^{0.5}$
excess kurtosis	$(1 + 10\sigma\mu + 15\sigma^2\mu^2)\,/\mu$
pf	$\left(\frac{\mu}{1+\sigma\mu}\right)^y \frac{(1+\sigma y)^{y-1}}{y!} \exp\left[\frac{-\mu\,(1+\sigma y)}{1+\sigma\mu}\right]$
Reference	Johnson et al. [2005] p336-339 reparameterized by $\theta = \mu/(1+\sigma\mu)$ and $\lambda = \sigma\mu/(1+\sigma\mu)$ and hence $\mu = \theta/(1-\lambda)$ and $\sigma = \lambda/\theta$

22.2.3 Negative binomial: NBI, NBII

Negative binomial type I, NBI The probability function of the negative binomial distribution type I, denoted by $\texttt{NBI}(\mu, \sigma)$, is given by

$$P(Y = y \mid \mu, \sigma) = \frac{\Gamma(y + \frac{1}{\sigma})}{\Gamma(\frac{1}{\sigma})\Gamma(y+1)} \left(\frac{\sigma\mu}{1+\sigma\mu}\right)^y \left(\frac{1}{1+\sigma\mu}\right)^{1/\sigma}$$

for $y = 0, 1, 2, 3 \ldots$, where $\mu > 0$, $\sigma > 0$. The mean and variance are $E(Y) = \mu$ and $\text{Var}(Y) = \mu + \sigma\mu^2$. The above parameterization is equivalent to that used by Anscombe [1950], with the exception that he used $\alpha = 1/\sigma$, [Johnson et al., 2005, p209]. The Poisson, $\texttt{PO}(\mu)$, distribution is the limiting case of $\texttt{NBI}(\mu, \sigma)$ as $\sigma \to 0$.

[Note that the original parameterization of the negative binomial distribution, $\texttt{NBo}(k, \pi)$, is given by setting $\mu = k(1-\pi)/\pi$ and $\sigma = 1/k$, giving

$$P(Y = y \mid \pi) = \frac{\Gamma(y+k)}{\Gamma(y+1)\Gamma(k)}(1-\pi)^y \pi^k$$

for $y = 0, 1, 2, 3 \ldots$, where $k > 0$ and $0 < \pi < 1$, [Johnson et al., 2005, p209]. Hence $\pi = 1/(1+\mu\sigma)$ and $k = 1/\sigma$. When k is an integer, this version has the interpretation of being the distribution of the number of failures until the kth success, in independent Bernoulli trials, where the probability of success in each trial is π. The $\texttt{NBo}(k, \pi)$ distribution is not available in **gamlss.dist**.]

As $y \to \infty$, $P(Y = y \mid \mu, \sigma) \sim q \exp\left[-y \log\left(1 + \frac{1}{\mu\sigma}\right) + (\frac{1}{\sigma} - 1)\log y\right]$ where $q = \left[\Gamma(1/\sigma)(1+\sigma\mu)^{1/\sigma}\right]^{-1}$, essentially an exponential tail. Note $\log P(Y = y \mid \mu, \sigma) \sim -y \log\left(1 + \frac{1}{\mu\sigma}\right)$ as $y \to \infty$.

See NBII(μ, σ) below with variance $\text{Var}(Y) = \mu + \sigma\mu$ for an alternative parameterization. Also see NBF(μ, σ, ν) in Section 22.3.4 with variance $\text{Var}(Y) = \mu + \sigma\mu^\nu$ for a family of reparameterizations of the NBI(μ, σ).

TABLE 22.8: Negative binomial type I distribution.

NBI(μ, σ)		
Ranges		
Y	$0, 1, 2, 3 \ldots$	
μ	$0 < \mu < \infty$, mean	
σ	$0 < \sigma < \infty$, dispersion	
Distribution measures		
mean	μ	
mode	$\lfloor(1-\sigma)\mu\rfloor$	if $(1-\sigma)\mu$ is not an integer and $\sigma < 1$
	$(1-\sigma)\mu - 1$ and $(1-\sigma)\mu$	if $(1-\sigma)\mu$ is an integer and $\sigma < 1$
	0	if $\sigma \geq 1$
variance	$\mu + \sigma\mu^2$	
skewness	$(1+2\mu\sigma)(\mu+\sigma\mu^2)^{-0.5}$	
excess kurtosis	$6\sigma + (\mu+\sigma\mu^2)^{-1}$	
PGF	$[1+\mu\sigma(1-t)]^{-1/\sigma}$ for $0 < t < 1 + (\mu\sigma)^{-1}$	
pf [a]	$\dfrac{\Gamma(y+\sigma^{-1})}{\Gamma(\sigma^{-1})\Gamma(y+1)} \left(\dfrac{\sigma\mu}{1+\sigma\mu}\right)^y \left(\dfrac{1}{1+\sigma\mu}\right)^{1/\sigma}$	
cdf	$1 - \dfrac{B(y+1, \sigma^{-1}, \mu\sigma(1+\mu\sigma)^{-1})}{B(y+1, \sigma^{-1})}$	
Reference	Johnson et al. [2005], Sections 5.1 to 5.5, p209-217, reparameterized by $p = 1/(1+\mu\sigma)$ and $k = 1/\sigma$ and hence $\mu = k(1-p)/p$ and $\sigma = 1/k$ [a] McCullagh and Nelder [1989] p 373 reparameterized by $\alpha = \mu\sigma$ and $k = 1/\sigma$.	
Note	$\lfloor\alpha\rfloor$ is the largest integer less than or equal to α $B(\alpha, \beta, x) = \int_0^x t^{\alpha-1}(1-t)^{\beta-1}\,dt$ is the incomplete beta function	

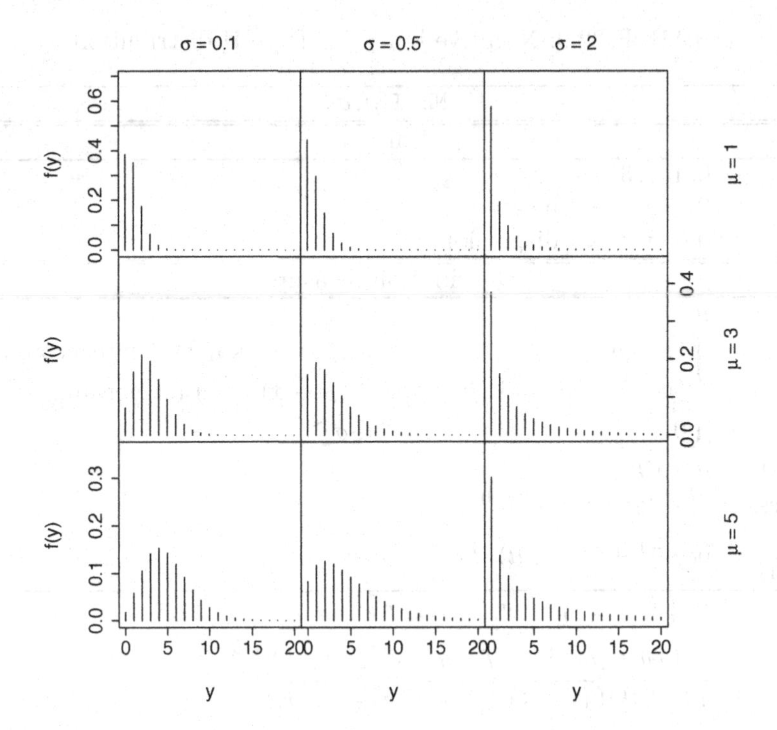

FIGURE 22.8: The negative binomial type I, `NBI`(μ, σ), distribution, with $\mu = 1, 3, 5$ and $\sigma = 0.1, 0.5, 2$.

Negative binomial type II, `NBII`

The probability function of the negative binomial type II distribution, denoted by `NBII`(μ, σ), is given by

$$\mathrm{P}(Y = y \mid \mu, \sigma) = \frac{\Gamma(y + \frac{\mu}{\sigma})\sigma^y}{\Gamma(\frac{\mu}{\sigma})\Gamma(y+1)(1+\sigma)^{y+\mu/\sigma}}$$

for $y = 0, 1, 2, 3, \ldots$, where $\mu > 0$ and $\sigma > 0$.

This parameterization was used by Evans [1953], and is obtained by reparameterizing σ to σ/μ in `NBI`(μ, σ). The important difference between the `NBI` and `NBII` parameterizations is the variance-mean relationship: both distributions have $\mathrm{E}(Y) = \mu$; in `NBI` the variance is $\mathrm{Var}(Y) = \mu(1 + \sigma\mu)$, (i.e. the variance is quadratic in μ), and in `NBII` it is $\mathrm{Var}(Y) = \mu(1 + \sigma)$, (i.e. the variance is linear in μ).

TABLE 22.9: Negative binomial Type II distribution.

NBII(μ, σ)	
Ranges	
Y	$0, 1, 2, 3 \ldots$
μ	$0 < \mu < \infty$, mean
σ	$0 < \sigma < \infty$, dispersion
Distribution measures	
mean	μ
mode	$\begin{cases} \lfloor \mu - \sigma \rfloor & \text{if } (\mu - \sigma) \text{ is not an integer and } \sigma < \mu \\ (\mu - \sigma - 1) \text{ and } (\mu - \sigma) & \text{if } (\mu - \sigma) \text{ is an integer and } \sigma < \mu \\ 0 & \text{if } \sigma \geq \mu \end{cases}$
variance	$\mu + \sigma\mu$
skewness	$(1 + 2\sigma)(\mu + \sigma\mu)^{-0.5}$
excess kurtosis	$6\sigma\mu^{-1} + (\mu + \sigma\mu)^{-1}$
PGF	$[1 + \sigma(1 - t)]^{-\mu/\sigma}$ for $0 < t < 1 + \sigma^{-1}$
pf	$\dfrac{\Gamma(y + \mu\sigma^{-1})}{\Gamma(\mu\sigma^{-1})\Gamma(y + 1)} \left(\dfrac{\sigma}{1 + \sigma}\right)^{y} \left(\dfrac{1}{1 + \sigma}\right)^{\mu/\sigma}$
cdf	$1 - \dfrac{B(y + 1, \mu\sigma^{-1}, \sigma(1 + \sigma)^{-1})}{B(y + 1, \mu\sigma^{-1})}$
Reference	Reparameterize σ_1 to σ/μ in NBI(μ, σ_1)

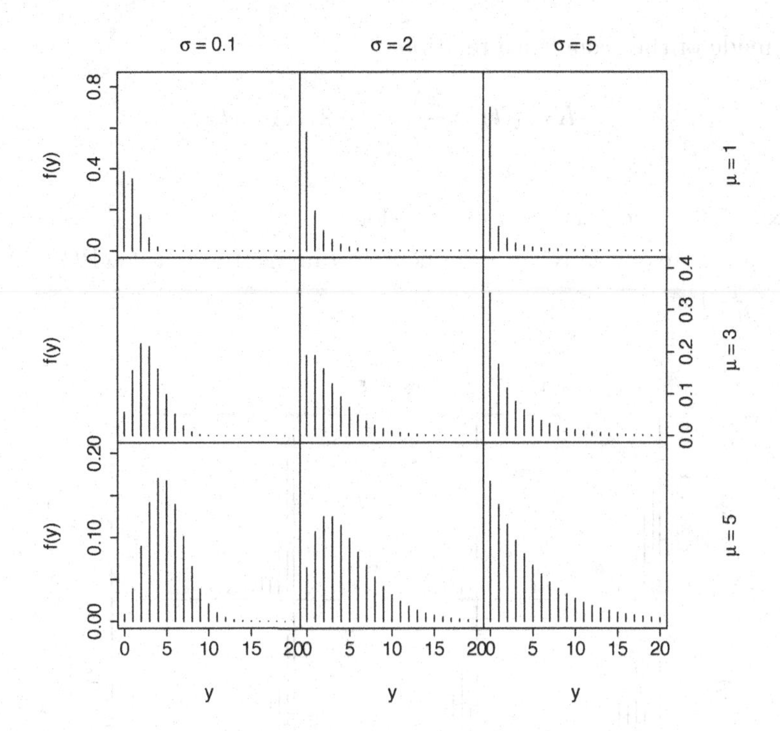

FIGURE 22.9: The negative binomial type II, `NBII`(μ, σ), distribution, with $\mu = 1, 3, 5$ and $\sigma = 0.1, 2, 5$.

22.2.4 Poisson-inverse Gaussian: PIG, PIG2

First parameterization, PIG

The probability function of the Poisson-inverse Gaussian distribution, denoted by `PIG`(μ, σ), is given by

$$P(Y = y \mid \mu, \sigma) = \left(\frac{2\alpha}{\pi}\right)^{1/2} \frac{\mu^y e^{1/\sigma} K_{y-\frac{1}{2}}(\alpha)}{y!(\alpha\sigma)^y} \tag{22.10}$$

for $y = 0, 1, 2, 3, \ldots$, where $\mu > 0$, $\sigma > 0$, $\alpha^2 = \sigma^{-2} + 2\mu\sigma^{-1}$ and $\alpha > 0$ and $K_\lambda(t)$ is the modified Bessel function of the second kind given by $K_\lambda(t) = \frac{1}{2}\int_0^\infty x^{\lambda-1} \exp\left[-\frac{1}{2}t(x + x^{-1})\right] dx$, [Abramowitz and Stegun, 1965, Section 9.6, p 374].

Note that $\sigma = \left[\left(\mu^2 + \alpha^2\right)^{0.5} - \mu\right]^{-1}$. This parameterization was used by Dean et al. [1989], and it is the special case of the **gamlss.dist** distributions `SI`(μ, σ, ν) and `SICHEL` (μ, σ, ν) when $\nu = -\frac{1}{2}$. The Poisson distribution is the limiting case of `PIG`(μ, σ) as $\sigma \to 0$.

The Bessel function $K_\lambda(t)$ in general presents challenges in computation; however, when the order λ of the Bessel function is a half-integer, as it is in (22.10), computations are simplified considerably as

$$K_{-1/2}(t) = K_{1/2}(t) = \sqrt{\frac{\pi}{2t}}\, e^{-t}$$

and use is made of the recurrence relation

$$K_{\lambda+1}(t) = \frac{2\lambda}{t} K_\lambda(t) + K_{\lambda-1}(t) \ .$$

As $y \to \infty$, $\text{P}(Y = y \mid \mu, \sigma) \sim r \exp\left[-y \log\left(1 + \frac{1}{2\mu\sigma}\right) - \frac{3}{2}\log y\right]$, where r does not depend on y, i.e. essentially an exponential tail. Note that $\log \text{P}(Y = y \mid \mu, \sigma) \sim -y \log\left(1 + \frac{1}{2\mu\sigma}\right)$ as $y \to \infty$.

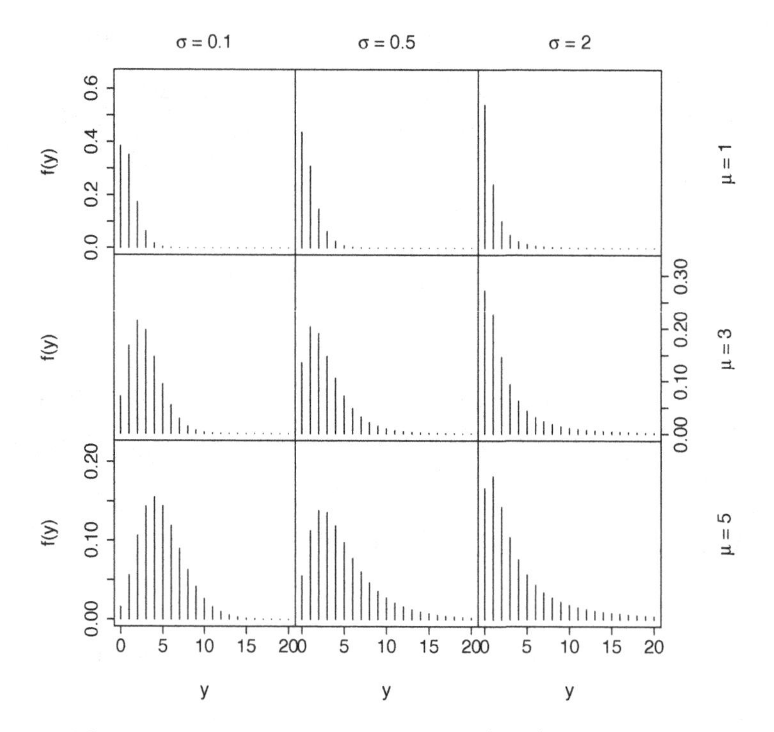

FIGURE 22.10: The Poisson-inverse Gaussian, `PIG`(μ, σ), distribution, with $\mu = 1, 3, 5$ and $\sigma = 0.1, 0.5, 2$.

TABLE 22.10: Poisson-inverse Gaussian distribution.

PIG(μ, σ)	
Ranges	
Y	$0, 1, 2, 3 \ldots$
μ	$0 < \mu < \infty$, mean
σ	$0 < \sigma < \infty$, dispersion
Distribution measures	
mean	μ
variance	$\mu + \sigma\mu^2$
skewness	$(1 + 3\mu\sigma + 3\mu^2\sigma^2)(1 + \mu\sigma)^{-1.5}\mu^{-0.5}$
excess kurtosis	$(1 + 7\mu\sigma + 18\mu^2\sigma^2 + 15\mu^3\sigma^3)(1 + \mu\sigma)^{-2}\mu^{-1}$
PGF[a]	$e^{1/\sigma - q}$ where $q^2 = \sigma^{-2} + 2\mu(1-t)\sigma^{-1}$
pf [a]	$\left(\frac{2\alpha}{\pi}\right)^{1/2} \frac{\mu^y e^{1/\sigma} K_{y-\frac{1}{2}}(\alpha)}{y!(\alpha\sigma)^y}$ where $\alpha^2 = \sigma^{-2} + 2\mu\sigma^{-1}$
Reference	Set $\nu = -\frac{1}{2}$ (and hence $c = 1$) in SICHEL(μ, σ, ν), see also [a] Dean et al. [1989]

Second parameterization, PIG2

In the second parameterization of the Poisson-inverse Gaussian distribution, denoted PIG2(μ, σ), the probability function is given by

$$P(Y = y \mid \mu, \sigma) = \left(\frac{2\sigma}{\pi}\right)^{1/2} \frac{\mu^y e^{1/\alpha} K_{y-\frac{1}{2}}(\sigma)}{y!(\sigma\alpha)^y} \tag{22.11}$$

for $y = 0, 1, 2, 3, \ldots$, where $\mu > 0$, $\sigma > 0$ and $\alpha = \left[(\mu^2 + \sigma^2)^{0.5} - \mu\right]^{-1}$. Note that $\sigma^2 = \alpha^{-2} + 2\mu\alpha^{-1}$. Currently under development in **gamlss.dist**.

This parameterization was given by Stein et al. [1987] and has the advantage that μ and σ are informationally orthogonal (and hence their maximum likelihood estimators are asymptotically uncorrelated). Heller et al. [2018] demonstrate that this results in (22.11) having more robust estimation of the model for the mean parameter μ, under misspecification of the model for σ. Note however that (22.11) has the disadvantage of a more complicated variance-mean relationship.

The PIG2 parameterization (22.11) is given by interchanging σ and α in (22.10), i.e. in (22.10) σ is reparameterized to $[(\mu^2 + \sigma^2)^{0.5} - \mu]^{-1}$ giving

$$\texttt{PIG2}(\mu, \sigma) = \texttt{PIG}(\mu, [(\mu^2 + \sigma^2)^{0.5} - \mu]^{-1})$$

and

$$\texttt{PIG}(\mu, \sigma) = \texttt{PIG2}(\mu, \sqrt{\sigma^{-2} + 2\mu\sigma^{-1}}) \; .$$

The Poisson is the limiting distribution of PIG2(μ, σ) as $\sigma \to \infty$ and the variance of PIG2(μ, σ) increases with decreasing σ. As $y \to \infty$, $P(Y = y \mid \mu, \sigma) \sim$

$r \exp\left[-y \log\left(1+\frac{1}{2\mu\alpha}\right) - \frac{3}{2}\log y\right]$, where r does not depend on y, i.e. essentially an exponential tail. Note that $\log P(Y=y|\mu,\sigma) \sim -y\log\left(1+\frac{1}{2\mu\alpha}\right)$ as $y \to \infty$.

TABLE 22.11: Second parameterization of Poisson-inverse Gaussian distribution.

	PIG2(μ,σ)
	Ranges
Y	$0, 1, 2, 3 \ldots$
μ	$0 < \mu < \infty$, mean
σ	$0 < \sigma < \infty$, dispersion
	Distribution measures
mean	μ
variance	$\mu + \alpha\mu^2 = \mu + \dfrac{\mu^2}{(\mu^2+\sigma^2)^{0.5} - \mu}$ where $\alpha = [(\mu^2+\sigma^2)^{0.5} - \mu]^{-1}$
skewness	$(1 + 3\mu\alpha + 3\mu^2\alpha^2)(1+\mu\alpha)^{-1.5}\mu^{-0.5}$
excess kurtosis	$(1 + 7\mu\alpha + 18\mu^2\alpha^2 + 15\mu^3\alpha^3)(1+\mu\alpha)^{-2}\mu^{-1}$
PGF[a]	$e^{1/\alpha - q}$ where $q^2 = \alpha^{-2} + 2\mu(1-t)\alpha^{-1}$
pf [a]	$\left(\dfrac{2\sigma}{\pi}\right)^{1/2} \dfrac{\mu^y e^{1/\alpha} K_{y-\frac{1}{2}}(\sigma)}{y!(\sigma\alpha)^y}$
Reference	[a] Stein et al. [1987], equation (2.1) with (ξ,α) replaced by (μ,σ)

22.2.5 Waring: WARING

The probability function of the Waring distribution (also called the beta geometric), denoted by WARING(μ,σ), is given by

$$P(Y = y \,|\, \mu,\sigma) = \frac{B(y + \mu\sigma^{-1}, \sigma^{-1}+2)}{B(\mu\sigma^{-1}, \sigma^{-1}+1)} \tag{22.12}$$

for $y = 0, 1, 2, 3, \ldots$, where $\mu > 0$ and $\sigma > 0$. Note that this parameterization of WARING(μ,σ) only includes Waring distributions with finite mean μ. The WARING(μ,σ) distribution is a reparameterization of the distribution given in Wimmer and Altmann [1999, p643], where $b = \sigma^{-1}+1$ and $n = \mu\sigma^{-1}$. Hence $\mu = n(b-1)^{-1}$ and $\sigma = (b-1)^{-1}$. It is also a reparameterization of the distribution given in equation 16.131 in Johnson et al. [2005, p290], where $\alpha = \mu\sigma^{-1}$ and $c = (\mu+1)\,\sigma^{-1}+1$.

The WARING(μ,σ) distribution is a special case of the beta negative binomial, BNB(μ,σ,ν), distribution where $\nu = 1$. It can be derived as a beta mixture of geometric distributions, by assuming $Y \,|\, \pi \sim$ GEOMo(π), where $\pi \sim$ BEo(b,n), $b = \sigma^{-1}+1$ and $n = \mu\sigma^{-1}$. Hence the WARING(μ,σ) distribution can be considered as an overdispersed geometric distribution.

As $y \to \infty$,

$$P(Y = y \mid \mu, \sigma) \sim q y^{-(\sigma^{-1}+2)}$$

where

$$q = (\sigma^{-1} + 1)\,\Gamma\left([\mu + 1]\,\sigma^{-1} + 1\right)/\Gamma\left(\mu\sigma^{-1}\right)$$

and hence the `WARING`(μ, σ) has a heavy right tail, especially for large σ.

TABLE 22.12: Waring distribution.

`WARING`(μ, σ)	
Ranges	
Y	$0, 1, 2, 3 \ldots$
μ	$0 < \mu < \infty$, mean
σ	$0 < \sigma < \infty$
Distribution measures	
mean	μ
mode	0
variance	$\begin{cases} \mu(\mu+1)(1+\sigma)(1-\sigma)^{-1} & \text{if } \sigma < 1 \\ \infty & \text{if } \sigma \geq 1 \end{cases}$
skewness	$\begin{cases} \dfrac{(2\mu+1)(1+2\sigma)(1-\sigma)^{1/2}}{\mu^{1/2}(\mu+1)^{1/2}(1+\sigma)^{1/2}(1-2\sigma)} & \text{if } \sigma < 1/2 \\ \infty & \text{if } \sigma \geq 1/2 \end{cases}$
excess kurtosis	$\begin{cases} \beta_2 - 3 & \text{if } \sigma < 1/3 \\ \infty & \text{if } \sigma \geq 1/3 \end{cases}$ where $\beta_2 = \dfrac{(1-\sigma)\left\{1 + 7\sigma + 6\sigma^2 + \mu(\mu+1)[9 + 21\sigma + 18\sigma^2]\right\}}{\mu(\mu+1)(1+\sigma)(1-2\sigma)(1-3\sigma)}$
PGF [a]	$\dfrac{(\sigma+1)}{(\mu+\sigma+1)}\, {}_2F_1(\mu\sigma^{-1}, 1; [\mu + 2\sigma + 1]\sigma^{-1}; t)$
pf [a,a2]	$\dfrac{B(y + \mu\sigma^{-1}, \sigma^{-1} + 2)}{B(\mu\sigma^{-1}, \sigma^{-1} + 1)}$
cdf [a2]	$1 - \dfrac{\Gamma(y + \mu\sigma^{-1} + 1)\Gamma([\mu+1]\sigma^{-1} + 1)}{\Gamma(y + [\mu+1]\sigma^{-1} + 2)\Gamma(\mu\sigma^{-1})}$
Reference	[a] Wimmer and Altmann [1999], p643, reparameterized by $b = \sigma^{-1} + 1$ and $n = \mu\sigma^{-1}$ [a2] `http://reference.wolfram.com/language/ref/WaringYuleDistribution.html` reparameterized by $\alpha = \sigma^{-1} + 1$ and $\beta = \mu\sigma^{-1}$. Set $\nu = 1$ in `BNB`(μ, σ, ν)

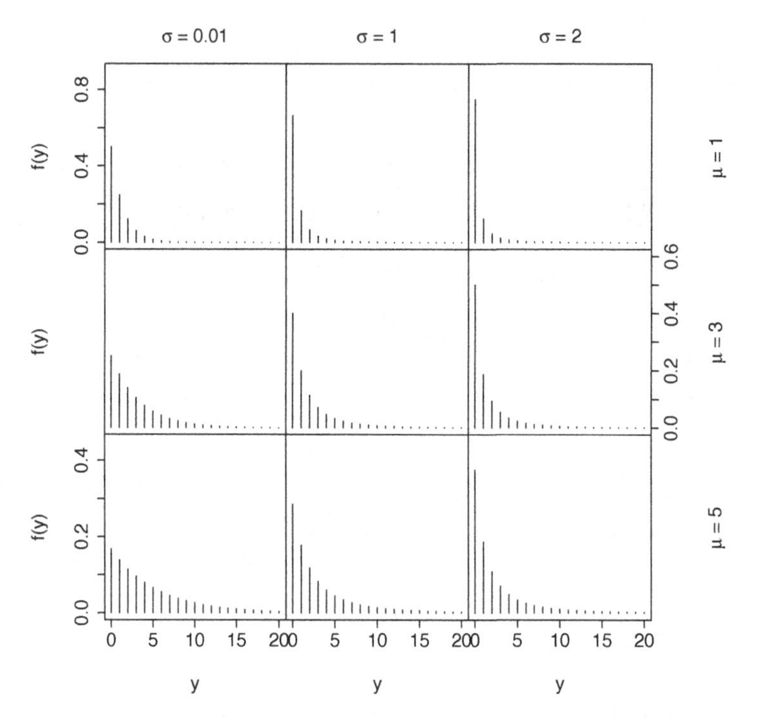

FIGURE 22.11: The WARING(μ, σ), distribution, with $\mu = 1, 3, 5$ and $\sigma = 0.01, 1, 2$.

22.2.6 Zero-adjusted logarithmic: ZALG

Let $Y = 0$ with probability σ, and $Y = Y_1$ where $Y_1 \sim$ LG(μ), (logarithmic distribution), with probability $(1 - \sigma)$. Then Y has a zero-adjusted logarithmic distribution, denoted by ZALG(μ, σ), with probability function given by

$$P(Y = y \mid \mu, \sigma) = \begin{cases} \sigma & \text{if } y = 0 \\ (1 - \sigma)(\alpha\mu^y)/y & \text{if } y = 1, 2, 3, \ldots \end{cases} \tag{22.13}$$

where $\alpha = -\left[\log(1 - \mu)\right]^{-1}$ for $0 < \mu < 1$ and $0 < \sigma < 1$, see Johnson et al. [2005, p355]. For large (and all) y, $P(Y = y \mid \mu, \sigma) = (1-\sigma)\alpha \exp\left(y \log \mu - \log y\right)$, so $\log P(Y = y \mid \mu, \sigma) \sim y \log \mu$ as $y \to \infty$.

TABLE 22.13: Zero-adjusted logarithmic distribution.

$\texttt{ZALG}(\mu, \sigma)$	
Ranges	
Y	$0, 1, 2, 3 \ldots$
μ	$0 < \mu < 1$
σ	$0 < \sigma < 1, \ \sigma = \text{P}(Y = 0)$
Distribution measures	
mean	$(1-\sigma)\, \alpha\mu\, (1-\mu)^{-1}$ where $\alpha = -[\log(1-\mu)]^{-1}$
mode	$\begin{cases} 0 & \text{if } \sigma > \frac{\alpha\mu}{(1+\alpha\mu)} \\ 0 \text{ and } 1 & \text{if } \sigma = \frac{\alpha\mu}{(1+\alpha\mu)} \\ 1 & \text{if } \sigma < \frac{\alpha\mu}{(1+\alpha\mu)} \end{cases}$
variance	$(1-\sigma)\alpha\mu[1-(1-\sigma)\alpha\mu](1-\mu)^{-2}$
skewness	$\begin{cases} \mu_3/[\text{Var}(Y)]^{1.5} \text{ where} \\ \mu_3 = (1-\sigma)\alpha\mu[1+\mu-3(1-\sigma)\alpha\mu+2(1-\sigma)^2\alpha^2\mu^2](1-\mu)^{-3} \end{cases}$
excess kurtosis	$\begin{cases} k_4/[\text{Var}(Y)]^2 \text{ where} \\ k_4 = (1-\sigma)\alpha\mu[1+4\mu+\mu^2-(1-\sigma)\alpha\mu(7+4\mu)+ \\ 12(1-\sigma)^2\alpha^2\mu^2-6(1-\sigma)^3\alpha^3\mu^3](1-\mu)^{-4} \end{cases}$
PGF	$\sigma + (1-\sigma)\dfrac{\log(1-\mu t)}{\log(1-\mu)}$ for $0 < t < \mu^{-1}$
pf	$\begin{cases} \sigma & \text{if } y = 0 \\ (1-\sigma)\alpha\mu^y/y & \text{if } y = 1, 2, 3 \ldots \end{cases}$

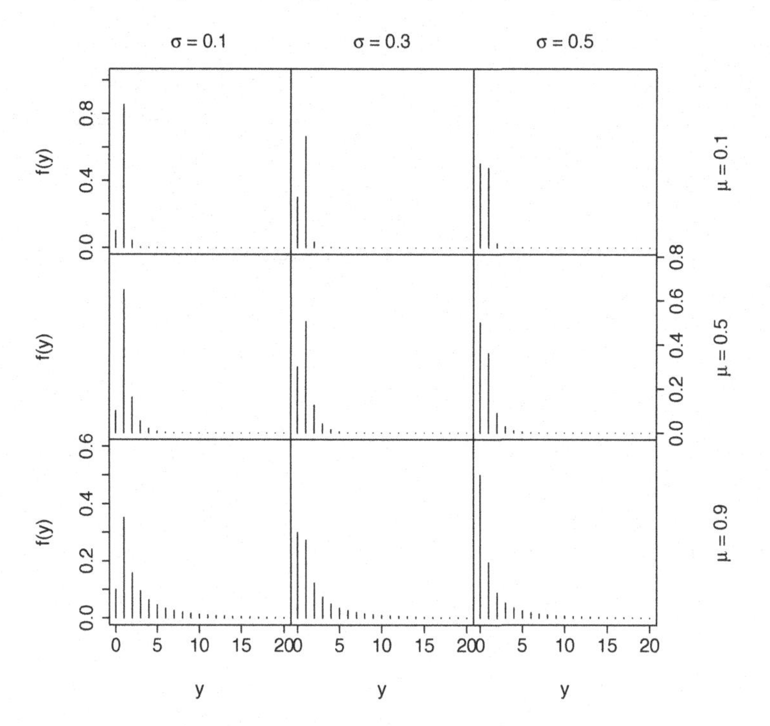

FIGURE 22.12: The zero-adjusted logarithmic, ZALG(μ, σ), distribution, with $\mu = 0.1, 0.5, 0.9$ and $\sigma = 0.1, 0.3, 0.5$.

22.2.7 Zero-adjusted Poisson: ZAP

Let $Y = 0$ with probability σ, and $Y = Y_1$ where $Y_1 \sim$ POtr(μ) with probability $(1-\sigma)$, where POtr(μ) is a Poisson distribution truncated at zero. Then Y has a zero-adjusted Poisson distribution, denoted by ZAP(μ, σ), with probability function

$$P(Y = y \mid \mu, \sigma) = \begin{cases} \sigma & \text{if } y = 0 \\ (ce^{-\mu}\mu^y)/y! & \text{if } y = 1, 2, 3, \ldots \end{cases} \tag{22.14}$$

for $\mu > 0$ and $0 < \sigma < 1$, where $c = (1-\sigma)/(1-e^{-\mu})$.

As $y \to \infty$, $P(Y = y \mid \mu, \sigma) \sim q \exp[y(\log \mu + 1) - (y + 0.5)\log y]$. where $q = ce^{-\mu}\sqrt{2\pi}$, so $\log P(Y = y \mid \mu, \sigma) \sim -y \log y$, as $y \to \infty$.

TABLE 22.14: Zero-adjusted Poisson distribution.

	ZAP(μ, σ)
	Ranges
Y	$0, 1, 2, 3 \ldots$
μ	$0 < \mu < \infty$, mean of Poisson component before truncation at 0
σ	$0 < \sigma < 1$, $\sigma = P(Y = 0)$
	Distribution measures
mean [a,a2]	$c\mu$ where $c = (1-\sigma)/(1-e^{-\mu})$
variance [a,a2]	$c\mu + c\mu^2 - c^2\mu^2$
skewness [a,a2]	$\begin{cases} \mu_3/[\text{Var}(Y)]^{1.5} \\ \text{where } \mu_3 = c\mu[1 + 3\mu(1-c) + \mu^2(1 - 3c + 2c^2)] \end{cases}$
excess kurtosis [a,a2]	$\begin{cases} \mu_4/[\text{Var}(Y)]^2 - 3 \\ \text{where } \mu_4 = c\mu[1 + \mu(7-4c) + 6\mu^2(1 - 2c + c^2) + \\ \mu^3(1 - 4c + 6c^2 - 3c^3)] \end{cases}$
PGF [a,a2]	$(1-c) + ce^{\mu(t-1)}$
pf [a,a2]	$\begin{cases} \sigma & \text{if } y = 0 \\ (ce^{-\mu}\mu^y)/y! & \text{if } y = 1, 2, 3, \ldots \end{cases}$
cdf [a2]	$\begin{cases} \sigma & \text{if } y = 0 \\ \sigma + c\left[\frac{\Gamma(y+1,\mu)}{\Gamma(y)} - e^{-\mu}\right] & \text{if } y = 1, 2, 3, \ldots \end{cases}$
Reference	[a] let $\sigma \to 0$ in ZANBI(μ, σ, ν) and then set ν to σ [a2] obtained from equations (7.11), (7.12), (7.13), and (7.14), where $Y_2 \sim$ PO(μ)

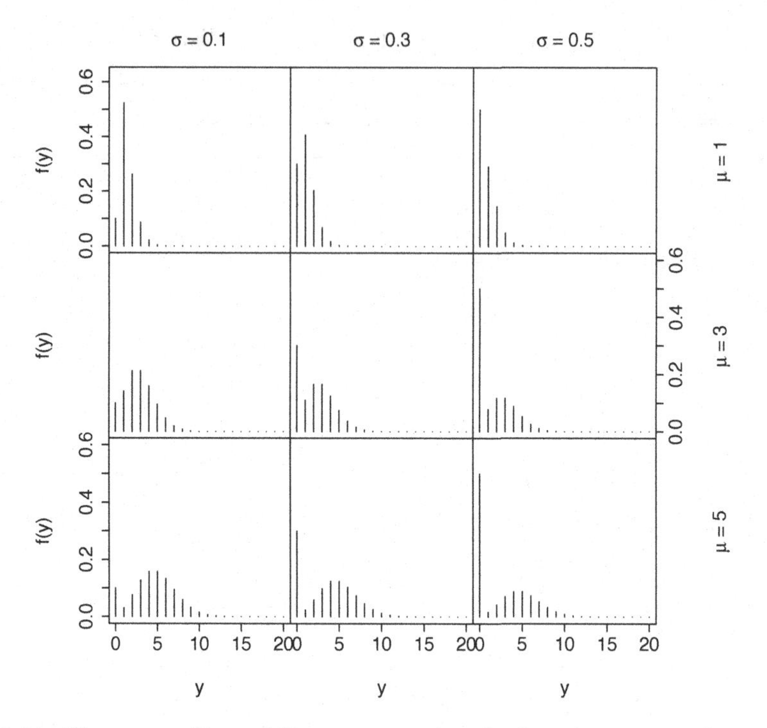

FIGURE 22.13: The zero-adjusted Poisson, ZAP(μ, σ), distribution, with $\mu = 1, 3, 5$ and $\sigma = 0.1, 0.3, 0.5$.

22.2.8 Zero-adjusted Zipf: ZAZIPF

Let $Y = 0$ with probability σ, and $Y = Y_1$ with probability $(1-\sigma)$, where $Y_1 \sim$ ZIPF(μ). Then Y has a zero-adjusted Zipf distribution, denoted by ZAZIPF(μ, σ), with probability function

$$P(Y = y \mid \mu, \sigma) = \begin{cases} \sigma & \text{if } y = 0 \\ (1-\sigma) y^{-(\mu+1)} / \zeta(\mu+1) & \text{if } y = 1, 2, 3, \ldots \end{cases} \tag{22.15}$$

for $\mu > 0$ and $0 < \sigma < 1$.

TABLE 22.15: Zero-adjusted Zipf distribution.

ZAZIPF(μ, σ)	
Ranges	
Y	$0, 1, 2, 3 \ldots$
μ	$0 < \mu < \infty$
σ	$0 < \sigma < 1, \ \sigma = P(Y = 0)$
Distribution measures	
mean[a]	$\begin{cases} b(1-\sigma) & \text{if } \mu > 1 \\ \infty & \text{if } \mu \leq 1 \end{cases}$ where $b = \zeta(\mu)/\zeta(\mu+1)$
mode	$\begin{cases} 0 & \text{if } \sigma > [1+\zeta(\mu+1)]^{-1} \\ 0 \text{ and } 1 & \text{if } \sigma = [1+\zeta(\mu+1)]^{-1} \\ 1 & \text{if } \sigma < [1+\zeta(\mu+1)]^{-1} \end{cases}$
variance[a]	$\begin{cases} (1-\sigma)[\zeta(\mu-1)/\zeta(\mu+1)] - (1-\sigma)^2 b^2 & \text{if } \mu > 2 \\ \infty & \text{if } \mu \leq 2 \end{cases}$
skewness[a]	$\begin{cases} \mu_3/[\text{Var}(Y)]^{1.5} \text{ where} & \\ \mu_3 = \dfrac{(1-\sigma)\zeta(\mu-2) - 3b(1-\sigma)^2\zeta(\mu-1)}{\zeta(\mu+1)} + & \\ \quad 2b^3(1-\sigma)^3 & \text{if } \mu > 3 \\ \infty & \text{if } \mu \leq 3 \end{cases}$
excess kurtosis [a]	$\begin{cases} \mu_4/[\text{Var}(Y)]^2 - 3 \text{ where} & \\ \mu_4 = \{[(1-\sigma)\zeta(\mu-3) - 4b(1-\sigma)^2\zeta(\mu-2) + & \\ \quad 6b^2(1-\sigma)^3\zeta(\mu-1)]/\zeta(\mu+1)\} - 3b^4(1-\sigma)^4 & \text{if } \mu > 4 \\ \infty & \text{if } \mu \leq 4 \end{cases}$
PGF	$\sigma + (1-\sigma)t\phi(t, \mu+1, 1)/\phi(1, \mu+1, 1)$
pf	$\begin{cases} \sigma & \text{if } y = 0 \\ (1-\sigma)y^{-(\mu+1)}/\zeta(\mu+1) & \text{if } y = 1, 2, 3, \ldots \end{cases}$
Reference	[a] Obtained using equation (2.1)
Notes	$\zeta(b) = \sum_{i=1}^{\infty} i^{-b}$ is the Riemann zeta function $\phi(a, b, c) = \sum_{i=0}^{\infty} \frac{a^i}{(i+c)^b}$, for $c \neq 0, -1, -2$, is the Lerch function

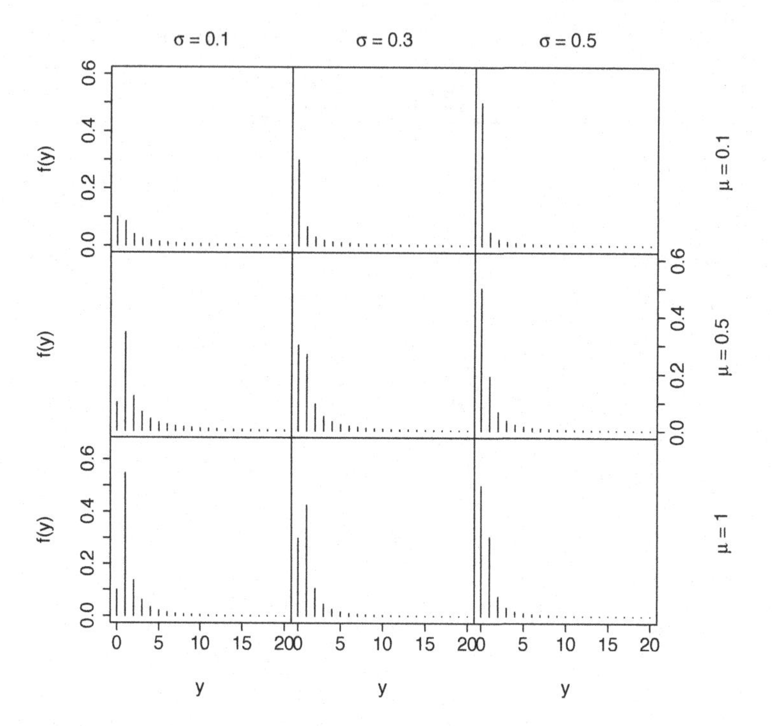

FIGURE 22.14: The zero-adjusted Zipf, ZAZIPF(μ, σ), distribution, with $\mu = 0.1, 0.5, 1$ and $\sigma = 0.1, 0.3, 0.5$.

22.2.9 Zero-inflated Poisson: ZIP, ZIP2

First parameterization, ZIP

Let $Y = 0$ with probability σ, and $Y = Y_1$ with probability $(1 - \sigma)$, where $Y_1 \sim$ PO(μ). Then Y has a zero-inflated Poisson distribution, denoted by ZIP(μ, σ), given by

$$P(Y = y \mid \mu, \sigma) = \begin{cases} \sigma + (1 - \sigma)e^{-\mu} & \text{if} \quad y = 0 \\ (1 - \sigma)e^{-\mu}\mu^y/y! & \text{if} \quad y = 1, 2, 3, \ldots \end{cases} \tag{22.16}$$

for $\mu > 0$ and $0 < \sigma < 1$. See Johnson et al. [2005, p193] for this parameterization, which was also used by Lambert [1992]. As $y \to \infty$,

$$P(Y = y|\mu, \sigma) \sim q \exp\left[y\left(\log \mu + 1\right) - (y + 0.5)\log y\right]$$

where $q = (1 - \sigma)\ e^{-\mu}\sqrt{2\pi}$, so, as $y \to \infty$,

$$\log P(Y = y \mid \mu, \sigma) \sim -y \log y\ .$$

TABLE 22.16: Zero-inflated Poisson distribution.

	$\texttt{ZIP}(\mu, \sigma)$
	Ranges
Y	$0, 1, 2, 3 \ldots$
μ	$0 < \mu < \infty$, mean of Poisson component
σ	$0 < \sigma < 1$, inflation probability at zero
	Distribution measures
mean[a,a2]	$(1-\sigma)\mu$
variance[a,a2]	$\mu(1-\sigma)(1+\mu\sigma)$
skewness[a,a2]	$\mu_3/[\text{Var}(Y)]^{1.5}$ where $\mu_3 = \mu(1-\sigma)[1 + 3\mu\sigma + \mu^2\sigma(2\sigma - 1)]$
excess kurtosis[a,a2]	$k_4/[\text{Var}(Y)]^2$ where $k_4 = \mu(1-\sigma)[1 + 7\mu\sigma - 6\mu^2\sigma + 12\mu^2\sigma^2 + \mu^3\sigma(1 - 6\sigma + 6\sigma^2)]$
PGF[a,a2]	$\sigma + (1-\sigma)e^{\mu(t-1)}$
pf [a]	$\begin{cases} \sigma + (1-\sigma)e^{-\mu} & \text{if } y = 0 \\ (1-\sigma)e^{-\mu}\mu^y/y! & \text{if } y = 1, 2, 3, \ldots \end{cases}$
cdf[a,a2]	$\sigma + \dfrac{(1-\sigma)\Gamma(y+1, \mu)}{\Gamma(y)}$
Reference	[a] let $\sigma \to 0$ in $\texttt{ZINBI}(\mu, \sigma, \nu)$ and then set ν to σ [a2] obtained from equations (7.5), (7.6), (7.7), and (7.8), where $Y_1 \sim \texttt{PO}(\mu)$

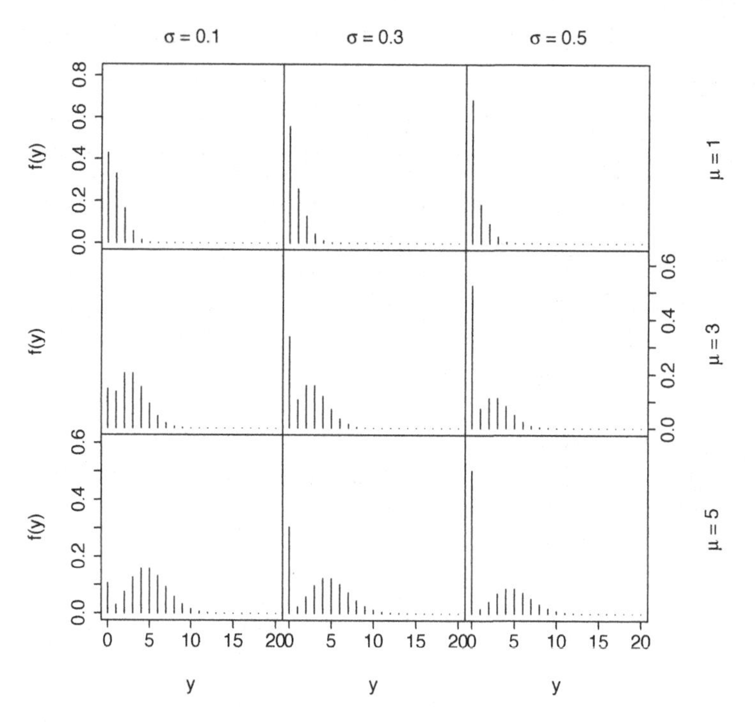

FIGURE 22.15: The zero-inflated Poisson, `ZIP`(μ, σ), distribution, with $\mu = 0.5, 1, 5$ and $\sigma = 0.1, 0.5$.

Second parameterization, `ZIP2`

A different parameterization of the zero-inflated Poisson distribution, denoted by `ZIP2`(μ, σ), has a probability function given by

$$P(Y = y \mid \mu, \sigma) = \begin{cases} \sigma + (1-\sigma)e^{-\left(\frac{\mu}{1-\sigma}\right)} & \text{if} \quad y = 0 \\ \dfrac{\mu^y}{y!(1-\sigma)^{y-1}} e^{-\left(\frac{\mu}{1-\sigma}\right)} & \text{if} \quad y = 1, 2, 3, \ldots \end{cases} \tag{22.17}$$

for $\mu > 0$ and $0 < \sigma < 1$. The `ZIP2`(μ, σ) distribution is given by reparameterizing μ to $\mu/(1-\sigma)$ in the `ZIP`(μ, σ) distribution. This has the advantage that $E(Y) = \mu$.

As $y \to \infty$,

$$P(Y = y|\mu, \sigma) \sim q \exp\left[y\left(\log(\mu/(1-\sigma)) + 1\right) - (y + 0.5)\log y\right] ,$$

where $q = \sqrt{2\pi}(1-\sigma)\ e^{-\mu/(1-\sigma)}$, so $\log P(Y = y \mid \mu, \sigma) \sim -y \log y \quad \text{as } y \to \infty$.

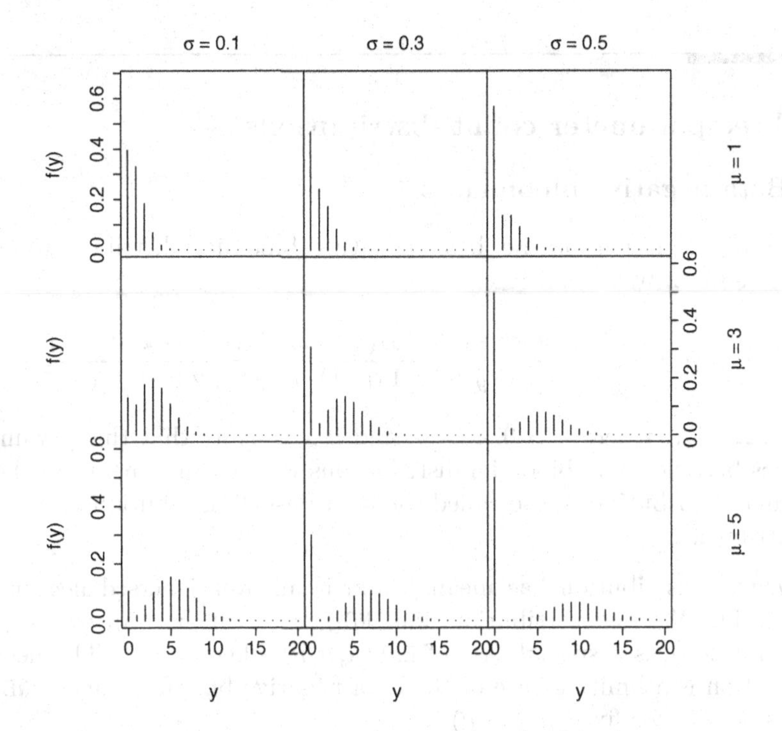

FIGURE 22.16: The zero-inflated Poisson, ZIP2(μ, σ), distribution type 2, with $\mu = 1, 3, 5$ and $\sigma = 0.1, 0.3, 0.5$.

TABLE 22.17: Zero-inflated Poisson distribution type 2.

ZIP2(μ, σ)	
Ranges	
Y	$0, 1, 2, 3 \ldots$
μ	$0 < \mu < \infty$, mean
σ	$0 < \sigma < 1$, inflation probability at zero
Distribution measures	
mean	μ
variance	$\mu\left[1 + \mu\sigma/(1-\sigma)\right]$
skewness	$\mu_3/[\text{Var}(Y)]^{1.5}$ where $\mu_3 = \mu\left[1 + \frac{3\mu\sigma}{(1-\sigma)} + \frac{\mu^2\sigma(2\sigma-1)}{(1-\sigma)^2}\right]$
excess kurtosis	$k_4/[\text{Var}(Y)]^2$ where $k_4 = \mu[1 + \frac{7\mu\sigma}{(1-\sigma)} + \frac{6\mu^2\sigma(2\sigma-1)}{(1-\sigma)^2} + \frac{\mu^3\sigma}{(1-\sigma)^3}(1 - 6\sigma + 6\sigma^2)]$
PGF[a,b]	$\sigma + (1-\sigma)e^{\mu(t-1)/(1-\sigma)}$
pf[a]	$\begin{cases} \sigma + (1-\sigma)e^{-\left(\frac{\mu}{1-\sigma}\right)} & \text{if } y = 0 \\ \frac{\mu^y}{y!(1-\sigma)^{y-1}} e^{-\left(\frac{\mu}{1-\sigma}\right)} & \text{if } y = 1, 2, 3, \ldots \end{cases}$
cdf[b]	$\sigma + (1-\sigma)\Gamma(y+1, \frac{\mu}{1-\sigma})/\Gamma(y)$
Reference	Reparameterize μ_1 to $\mu/(1-\sigma)$ in ZIP(μ_1, σ)

22.3 Three-parameter count distributions

22.3.1 Beta negative binomial: BNB

The probability function of the beta negative binomial distribution, denoted by $\texttt{BNB}(\mu,\sigma,\nu)$, is given by

$$\text{P}(Y=y\,|\,\mu,\sigma,\nu)=\frac{\Gamma(y+\nu^{-1})\ \ B(y+\mu\sigma^{-1}\nu,\sigma^{-1}+\nu^{-1}+1)}{\Gamma(y+1)\ \ \Gamma(\nu^{-1})\ \ B(\mu\sigma^{-1}\nu,\sigma^{-1}+1)} \tag{22.18}$$

for $y=0,1,2,3,\ldots$, where $\mu>0$, $\sigma>0$, and $\nu>0$. Note that this parameterization only includes beta negative binomial distributions with a finite mean μ. The beta negative binomial distribution is also called the beta Pascal distribution or the generalized Waring distribution.

The $\texttt{BNB}(\mu,\sigma,\nu)$ distribution has mean μ and is an overdispersed negative binomial distribution. The Waring distribution, $\texttt{WARING}(\mu,\sigma)$, (which is an overdispersed geometric distribution), is a special case of $\texttt{BNB}(\mu,\sigma,\nu)$ where $\nu=1$. The negative binomial distribution is a limiting case of the beta negative binomial, since $\texttt{BNB}(\mu,\sigma,\nu)\to\texttt{NBI}(\mu,\nu)$ as $\sigma\to 0$ (for fixed μ and ν).

The $\texttt{BNB}(\mu,\sigma,\nu)$ is a reparameterization of the distribution given in Wimmer and Altmann [1999, p19], where $m=\sigma^{-1}+1$ and $n=\mu\sigma^{-1}\nu$ and $k=\nu^{-1}$. Hence $\mu=kn(m-1)^{-1}$, $\sigma=(m-1)^{-1}$ and $\nu=1/k$. It can be derived as a beta mixture of negative binomial distributions, by assuming $y\,|\,\pi\sim\texttt{NBo}(k,\pi)$ where $\pi\sim\texttt{BEo}(m,n)$. See Section 22.2.3 for $\texttt{NBo}(k,\pi)$. Hence the $\texttt{BNB}(\mu,\sigma,\nu)$ can be considered as an overdispersed negative binomial distribution.

As $y\to\infty$, $\text{P}(Y=y\,|\,\mu,\sigma,\nu)\sim qy^{-(\sigma^{-1}+2)}$ where

$$q=\Gamma\left([\mu\nu+1]\sigma^{-1}+1\right)/[B(\sigma^{-1}+1,\nu^{-1})\Gamma(\mu\sigma^{-1}\nu)]$$

and hence the $\texttt{BNB}(\mu,\sigma,\nu)$ distribution has a heavy right tail, especially for large σ.

The probability function (22.18) and the mean, variance, skewness, and kurtosis in Table 22.18 can be obtained from Johnson et al. [2005, p259, 263] setting $\alpha=-\mu\nu\sigma^{-1}$, $b=(\mu\nu+1)\sigma^{-1}$ and $n=-\nu^{-1}$. The equivalence of their probability function to (22.18) is shown using their equation (1.24).

To interpret the parameters of $\texttt{BNB}(\mu,\sigma,\nu)$, μ is the mean, σ is a right tail heaviness parameter (and increasing σ increases the variance, for finite σ, i.e. $\sigma<1$), and ν increases the variance (for $\nu^2>\sigma/\mu$ and $\sigma<1$), while the variance is infinite for $\sigma\geq 1$.

TABLE 22.18: Beta negative binomial distribution.

BNB(μ, σ, ν)	
Ranges	
Y	$0, 1, 2, 3 \ldots$
μ	$0 < \mu < \infty$, mean
σ	$0 < \sigma < \infty$
ν	$0 < \nu < \infty$
Distribution measures	
mean	μ
variance	$\begin{cases} \mu(1+\mu\nu)(1+\sigma\nu^{-1})(1-\sigma)^{-1} & \text{if } \sigma < 1 \\ \infty & \text{if } \sigma \geq 1 \end{cases}$
skewness	$\begin{cases} \dfrac{(2\mu\nu+1)(1+2\sigma\nu^{-1})(1-\sigma)^{1/2}}{\mu^{1/2}(\mu\nu+1)^{1/2}(1+\sigma\nu^{-1})^{1/2}(1-2\sigma)} & \text{if } \sigma < 1/2 \\ \infty & \text{if } \sigma \geq 1/2 \end{cases}$
excess kurtosis	$\beta_2 - 3$ where $\beta_2 = \begin{cases} (1-\sigma)\left[1+\sigma+6\sigma\nu^{-1}+6\sigma^2\nu^{-2}+ \right. \\ \left. 3\mu(\mu\nu+1)(1+2\nu+\sigma\nu^{-1}+6\sigma+6\sigma^2\nu^{-1})\right] \times \\ \left[\mu(\mu\nu+1)(1+\sigma\nu^{-1})(1-2\sigma)(1-3\sigma)\right]^{-1} & \text{if } \sigma < 1/3 \\ \infty & \text{if } \sigma \geq 1/3 \end{cases}$
PGF[a]	$\dfrac{{}_2F_1(\nu^{-1}, \mu\sigma^{-1}\nu; \mu\sigma^{-1}\nu+\sigma^{-1}+\nu^{-1}+1; t)}{{}_2F_1(\nu^{-1}, \mu\sigma^{-1}\nu; \mu\sigma^{-1}\nu+\sigma^{-1}+\nu^{-1}+1; 1)}$
pf[a]	$\dfrac{\Gamma(y+\nu^{-1})B(y+\mu\sigma^{-1}\nu, \sigma^{-1}+\nu^{-1}+1)}{\Gamma(y+1)\Gamma(\nu^{-1})B(\mu\sigma^{-1}\nu, \sigma^{-1}+1)}$
Reference	[a] Wimmer and Altmann [1999] p19, reparameterized by $m = \sigma^{-1}+1$, $n = \mu\sigma^{-1}\nu$ and $k = \nu^{-1}$ and hence $\mu = kn(m-1)^{-1}$, $\sigma = (m-1)^{-1}$ and $\nu = 1/k$

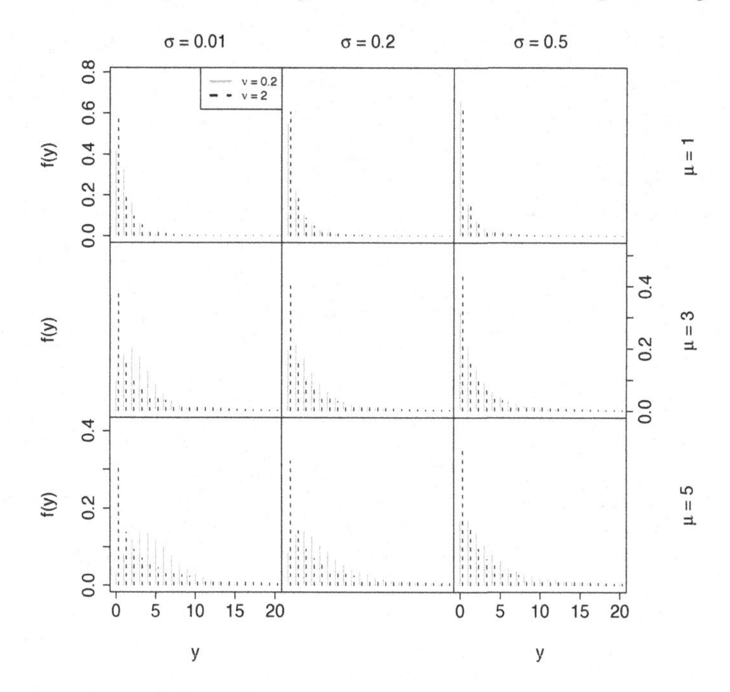

FIGURE 22.17: The beta negative binomial, BNB(μ, σ, ν), distribution, with $\mu = 1, 3, 5$, $\sigma = 0.01, 0.2, 0.5$, and $\nu = 0.2, 2$.

22.3.2 Discrete Burr XII: DBURR12

The discrete Burr XII distribution, denoted by DBURR12(μ, σ, ν), has probability function given by

$$P(Y = y \mid \mu, \sigma, \nu) = \left[1 + \left(\frac{y}{\mu}\right)^{\sigma}\right]^{-\nu} - \left[1 + \left(\frac{y+1}{\mu}\right)^{\sigma}\right]^{-\nu} \tag{22.19}$$

for $y = 0, 1, 2, 3, \ldots$, where $\mu > 0$, $\sigma > 0$ and $\nu > 0$.

This distribution was investigated by Para and Jan [2016], who used parameterization $\gamma = \mu$, $c = \sigma$ and $\beta = \exp(-\nu)$, so $\nu = -\log(\beta)$ and hence (22.19) can be written in the form

$$P(Y = y \mid \mu, \sigma, \beta) = \beta^{\log\left[1+\left(\frac{y}{\mu}\right)^{\sigma}\right]} - \beta^{\log\left[1+\left(\frac{y+1}{\mu}\right)^{\sigma}\right]} \tag{22.20}$$

since $\beta^{\log \alpha} = \exp\left[(\log \alpha)(\log \beta)\right] = \alpha^{\log \beta}$.

Note parameters μ and ν in DBURR12(μ, σ, ν) can be informationally highly correlated, so arguments `method=mixed(10,100)` and `c.crit=0.0001` are highly recommended in the `gamlss()` fitting function, in order to speed convergence and avoid converging too early. As $y \to \infty$, $\log P(Y = y|\mu, \sigma, \nu) \sim -(1 + \sigma\nu) \log y$, so DBURR12$(\mu, \sigma, \nu)$ can have a very heavy tail if $\sigma\nu$ is close to 0.

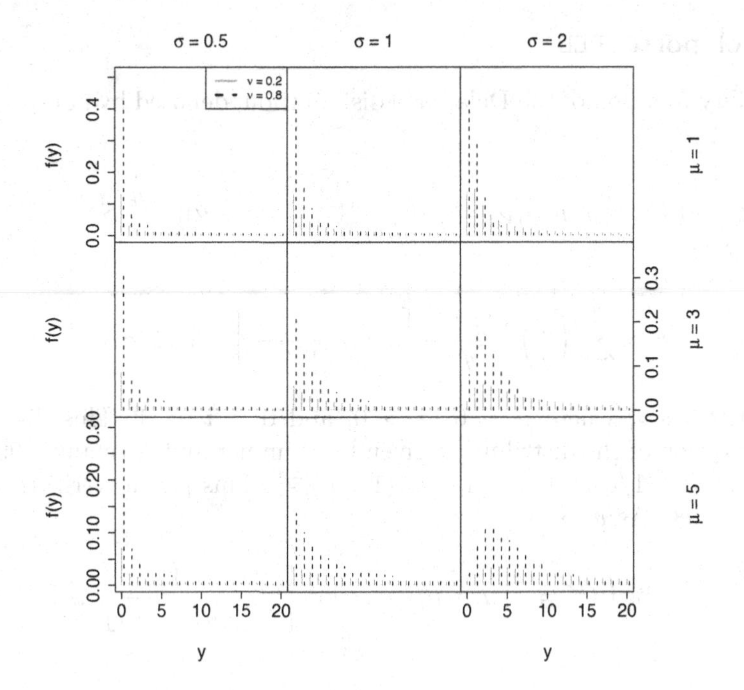

FIGURE 22.18: The discrete Burr XII, DBURR12(μ, σ, ν), distribution, with $\mu = 1, 3, 5$, $\sigma = 0.5, 1, 2$, and $\nu = 0.2, 0.8$.

TABLE 22.19: Discrete Burr XII distribution.

	DBURR12(μ, σ, ν)
	Ranges
Y	$0, 1, 2, 3 \ldots$
μ	$0 < \mu < \infty$
σ	$0 < \sigma < \infty$
ν	$0 < \nu < \infty$
	Distribution measures
mean	$\sum_{y=1}^{\infty} \left[1 + \left(\frac{y}{\mu}\right)^{\sigma}\right]^{-\nu}$, for $\sigma\nu > 1$
pf[a]	$\left[1 + \left(\frac{y}{\mu}\right)^{\sigma}\right]^{-\nu} - \left[1 + \left(\frac{y+1}{\mu}\right)^{\sigma}\right]^{-\nu}$
cdf	$1 - \left[1 + \left(\frac{y+1}{\mu}\right)^{\sigma}\right]^{-\nu}$
inverse cdf	$\left\lceil \mu \left\{ \exp\left[-\frac{\log(1-p)}{\nu}\right] - 1 \right\}^{1/\sigma} - 1 \right\rceil$
Reference	Para and Jan [2016] with parameters (γ, c, β) replaced by $(\mu, \sigma, \exp(-\nu))$.
Note	$\lceil x \rceil$ is the ceiling function, i.e. the smallest integer greater than or equal to x

22.3.3 Delaporte: DEL

The probability function of the Delaporte distribution, denoted by $\texttt{DEL}(\mu, \sigma, \nu)$, is given by

$$P(Y = y \mid \mu, \sigma, \nu) = \frac{e^{-\mu\nu}}{\Gamma(1/\sigma)} \left[1 + \mu\sigma(1-\nu)\right]^{-1/\sigma} S \tag{22.21}$$

where

$$S = \sum_{j=0}^{y} \binom{y}{j} \frac{\mu^y \nu^{y-j}}{y!} \left[\mu + \frac{1}{\sigma(1-\nu)}\right]^{-j} \Gamma\left(1/\sigma + j\right)$$

for $y = 0, 1, 2, 3, \ldots$, where $\mu > 0$, $\sigma > 0$, and $0 < \nu < 1$. This distribution is a reparameterization of the distribution given by Wimmer and Altmann [1999, p515-516] where $\alpha = \mu\nu$, $k = 1/\sigma$ and $\rho = [1 + \mu\sigma(1-\nu)]^{-1}$. This parameterization is given by Rigby et al. [2008]. As $y \to \infty$,

$$\log P(Y = y \mid \mu, \sigma, \nu) \sim -y \log \left[1 + \frac{1}{\mu\sigma(1-\nu)}\right] .$$

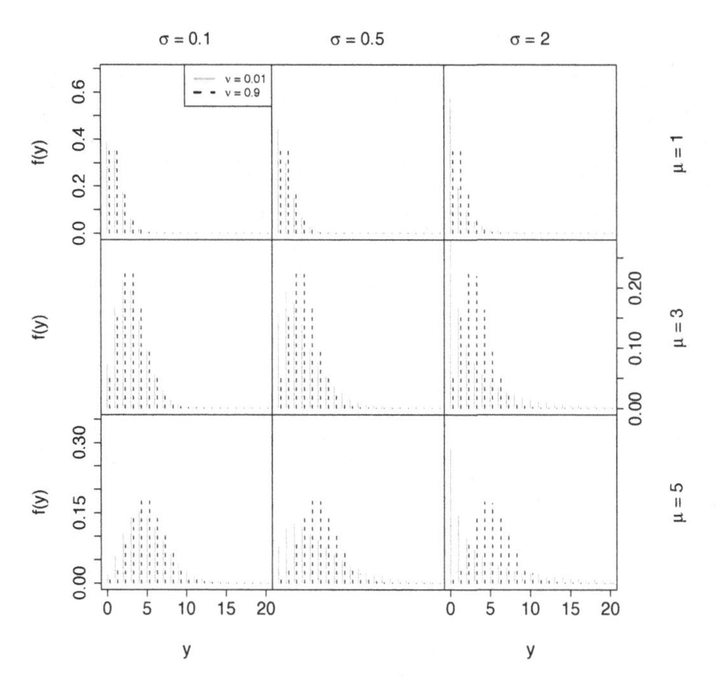

FIGURE 22.19: The Delaporte, $\texttt{DEL}(\mu, \sigma, \nu)$, distribution, with $\mu = 1, 3, 5$, $\sigma = 0.1, 0.5, 2$, and $\nu = 0.01, 0.9$.

TABLE 22.20: Delaporte distribution.

DEL(μ,σ,ν)	
Ranges	
Y	$0,1,2,3\ldots$
μ	$0<\mu<\infty$, mean
σ	$0<\sigma<\infty$
ν	$0<\nu<1$
Distribution measures	
mean	μ
variance	$\mu+\mu^2\sigma(1-\nu)^2$
skewness	$\mu_3/[\text{Var}(Y)]^{1.5}$ where $\mu_3=\mu[1+3\mu\sigma(1-\nu)^2+2\mu^2\sigma^2(1-\nu)^3]$
excess kurtosis	$\begin{cases} k_4/[\text{Var}(Y)]^2 \text{ where} \\ k_4=\mu[1+7\mu\sigma(1-\nu)^2+12\mu^2\sigma^2(1-\nu)^3+6\mu^3\sigma^3(1-\nu)^4] \end{cases}$
PGF	$e^{\mu\nu(t-1)}[1+\mu\sigma(1-\nu)(1-t)]^{-1/\sigma}$
pf	$\frac{\exp(-\mu\nu)}{\Gamma(1/\sigma)}[1+\mu\sigma(1-\nu)]^{-1/\sigma}\ S$ where
	$S=\sum_{j=0}^{y}\binom{y}{j}\frac{\mu^y\nu^{y-j}}{y!}\left[\mu+\frac{1}{\sigma(1-\nu)}\right]^{-j}\Gamma(1/\sigma+j)$
Reference	Rigby et al. [2008]

22.3.4 Negative binomial family: NBF

The probability function of the negative binomial family distribution, denoted NBF(μ,σ,ν), is given by

$$P(Y=y\,|\,\mu,\sigma,\nu)=\frac{\Gamma\left(y+\sigma^{-1}\mu^{2-\nu}\right)\sigma^y\mu^{y(\nu-1)}}{\Gamma\left(\sigma^{-1}\mu^{2-\nu}\right)\Gamma\left(y+1\right)\left(1+\sigma\mu^{\nu-1}\right)^{\sigma^{-1}\mu^{2-\nu}+y}} \qquad (22.22)$$

for $y=0,1,2,3,\ldots$, where $\mu>0$, $\sigma>0$ and $\nu>0$.

This family of reparameterizations of the negative binomial distribution is obtained by reparameterizing σ to $\sigma\mu^{\nu-2}$ in NBI(μ,σ). The variance of $Y\sim$ NBF(μ,σ,ν) is $\text{Var}(Y)=\mu+\sigma\mu^\nu$. Hence ν is the power in the variance-mean relationship.

TABLE 22.21: Negative binomial family distribution.

	NBF(μ, σ, ν)
	Ranges
Y	$0, 1, 2, 3 \ldots$
μ	$0 < \mu < \infty$ mean
σ	$0 < \sigma < \infty$ dispersion
ν	$0 < \nu < \infty$ power parameter in variance-mean relationship
	Distribution measures
mean	μ
variance	$\mu + \sigma\mu^{\nu}$
skewness	$(1 + 2\sigma\mu^{\nu-1})(\mu + \sigma\mu^{\nu})^{-0.5}$
excess kurtosis	$6\sigma\mu^{\nu-2} + (\mu + \sigma\mu^{\nu})^{-1}$
PGF	$\left[1 + \sigma\mu^{\nu-1}(1-t)\right]^{-1/(\sigma\mu^{\nu-2})}$ for $0 < t < 1 + (\sigma\mu^{\nu-1})^{-1}$
pf	$\dfrac{\Gamma\left(y + \sigma^{-1}\mu^{2-\nu}\right)\sigma^{y}\mu^{y(\nu-1)}}{\Gamma\left(\sigma^{-1}\mu^{2-\nu}\right)\Gamma\left(y+1\right)\left(1 + \sigma\mu^{\nu-1}\right)^{\sigma^{-1}\mu^{2-\nu}+y}}$
cdf	$1 - \dfrac{B\left(y+1, \sigma^{-1}\mu^{2-\nu}, \sigma\mu^{\nu-1}(1+\sigma\mu^{\nu-1})^{-1}\right)}{B\left(y+1, \sigma^{-1}\mu^{2-\nu}\right)}$
Reference	Reparameterized σ_1 to $\sigma\mu^{\nu-2}$ in NBI(μ, σ_1)

22.3.5 Sichel: SICHEL, SI

First parameterization, SICHEL

This parameterization of the Sichel distribution, [Rigby et al., 2008], denoted by SICHEL(μ, σ, ν), has probability function

$$P(Y = y \mid \mu, \sigma, \nu) = \frac{(\mu/b)^{y} K_{y+\nu}(\alpha)}{y!\,(\alpha\sigma)^{y+\nu}\, K_{\nu}\left(1/\sigma\right)} \tag{22.23}$$

for $y = 0, 1, 2, 3, \ldots$, where $\mu > 0$, $\sigma > 0$ and $-\infty < \nu < \infty$, and

$$\alpha^2 = \sigma^{-2} + 2\mu(b\sigma)^{-1}\,, \quad b = \frac{K_{\nu+1}(1/\sigma)}{K_{\nu}(1/\sigma)}$$

and $K_{\lambda}(t)$ is the modified Bessel function of the second kind [Abramowitz and Stegun, 1965]. Note

$$\sigma = \left[\left(\mu^2/b^2 + \alpha^2\right)^{0.5} - \mu/b\right]^{-1}.$$

The SICHEL(μ, σ, ν) distribution is a reparameterization of the SI(μ, σ, ν) distribution given by setting μ to μ/b, so that the mean of SICHEL(μ, σ, ν) is μ. As $\sigma \to 0$, SICHEL$(\mu, \sigma, \nu) \to$ PO(μ). For $\nu > 0$, as $\sigma \to \infty$ SICHEL$(\mu, \sigma, \nu) \to$ NBI(μ, ν^{-1}). For $\nu = 0$, as $\sigma \to \infty$, SICHEL$(\mu, \sigma, \nu) \to$ ZALG(μ_1, σ_1) where $\mu_1 = (2\mu \log\sigma)/(1 + 2\mu\log\sigma)$

and $\sigma_1 = 1 - [\log(1 + 2\mu \log \sigma)] \,/\, (2 \log \sigma)$. This tends to a degenerate point probability 1 at $y = 0$ in the limit as $\sigma \to \infty$.

As $y \to \infty$, $P(Y = y \mid \mu, \sigma, \nu) \sim q \exp\left[-y \log\left(1 + \frac{b}{2\mu\sigma}\right) - (1 - \nu) \log y\right]$ where q does not depend on y, i.e. essentially an exponential tail. The tail becomes heavier as $2\mu\sigma/b$ increases.

TABLE 22.22: Sichel distribution, first parameterization.

SICHEL(μ, σ, ν)	
Ranges	
Y	$0, 1, 2, 3 \ldots$
μ	$0 < \mu < \infty$, mean
σ	$0 < \sigma < \infty$
ν	$-\infty < \nu < \infty$
Distribution measures	
mean	μ
variance	$\mu + \mu^2 g_1$ where $g_1 = \dfrac{2\sigma(\nu+1)}{b} + \dfrac{1}{b^2} - 1$
skewness	$\mu_3/[\text{Var}(Y)]^{1.5}$ where $\mu_3 = \mu + 3\mu^2 g_1 + \mu^3(g_2 - 3g_1)$ where $g_2 = \dfrac{2\sigma(\nu+2)}{b^3} + \dfrac{[4\sigma^2(\nu+1)(\nu+2)+1]}{b^2} - 1$
excess kurtosis	$k_4/[\text{Var}(Y)]^2$ where $k_4 = \mu + 7\mu^2 g_1 + 6\mu^3(g_2 - 3g_1) + \mu^4(g_3 - 4g_2 + 6g_1 - 3g_1^2)$ and where $g_3 = [1 + 4\sigma^2(\nu+2)(\nu+3)]/b^4 + [8\sigma^3(\nu+1)(\nu+2)(\nu+3) + 4\sigma(\nu+2)]/b^3 - 1$
PGF	$\dfrac{K_\nu(q)}{(q\sigma)^\nu K_\nu(1/\sigma)}$ where $q^2 = \sigma^{-2} + 2\mu(1-t)(b\sigma)^{-1}$
pf	$\dfrac{(\mu/b)^y K_{y+\nu}(\alpha)}{y!\,(\alpha\sigma)^{y+\nu} K_\nu(1/\sigma)}$ where $\alpha^2 = \sigma^{-2} + 2\mu(b\sigma)^{-1}$
Reference	Rigby et al. [2008]
Notes	$b = K_{\nu+1}(1/\sigma)/K_\nu(1/\sigma)$ $K_\lambda(t)$ is the modified Bessel function of the second kind

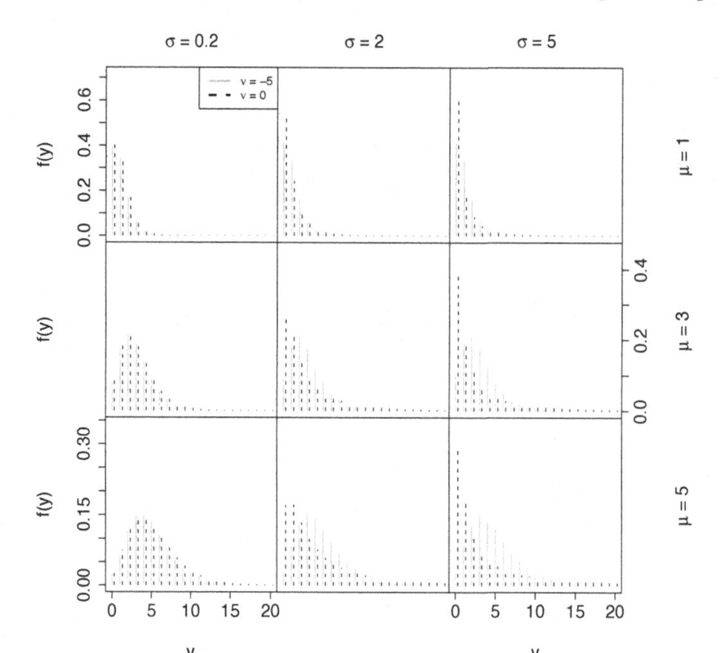

FIGURE 22.20: The Sichel, SICHEL(μ, σ), distribution, with $\mu = 1, 3, 5$, $\sigma = 0.2, 2, 5$, and $\nu = -5, 0$.

Second parameterization, SI

The probability function of the second parameterization of the Sichel distribution, denoted by SI(μ, σ, ν), is given by

$$P(Y = y \mid \mu, \sigma, \nu) = \frac{\mu^y K_{y+\nu}(\alpha)}{y!(\alpha\sigma)^{y+\nu} K_\nu(1/\sigma)} \tag{22.24}$$

for $y = 0, 1, 2, 3, \ldots$, where $\mu > 0$, $\sigma > 0$, $-\infty < \nu < \infty$, and where $\alpha^2 = \sigma^{-2} + 2\mu\sigma^{-1}$, and $K_\lambda(t)$ is the modified Bessel function of the second kind [Abramowitz and Stegun, 1965]. Note that $\sigma = [(\mu^2 + \alpha^2)^{0.5} - \mu]^{-1}$. Note that the above parameterization (22.24) is different from that of Stein et al. [1987, Section 2.1], who use the above probability function but treat μ, α, and ν as the parameters.

As $\sigma \to 0$, SI$(\mu, \sigma, \nu) \to$ PO(μ). Following Stein et al. [1987], for $\nu = 0$, then as $\sigma \to \infty$, SI$(\mu, \sigma, \nu) \to$ ZALG(μ_1, σ_1) where $\mu_1 = 2\mu\sigma/(1 + 2\mu\sigma)$ and $\sigma_1 = 1 - [\log(1 + 2\mu\sigma)]/[2 \log \sigma]$. This tends to a degenerate ZALG$(1, 0.5)$ in the limit as $\sigma \to \infty$.

As $y \to \infty$, $P(Y = y \mid \mu, \sigma, \nu) \sim q \exp\left[-y \log(1 + \frac{1}{2\mu\sigma}) - (1 - \nu) \log y\right]$, where q does not depend on y, i.e. essentially an exponential tail. The tail becomes heavier as $2\mu\sigma$ increases.

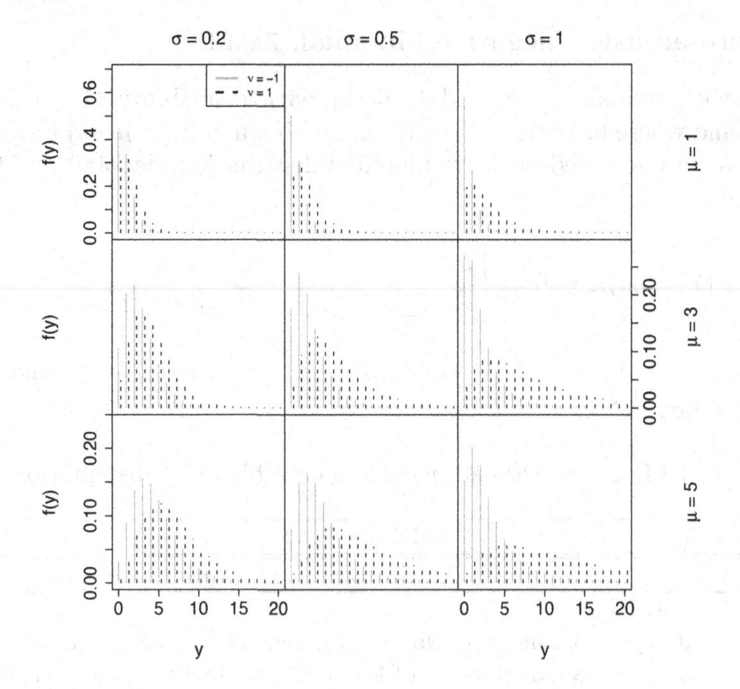

FIGURE 22.21: The Sichel, $\texttt{SI}(\mu, \sigma, \nu)$, distribution, with $\mu = 1, 3, 5$, $\sigma = 0.2, 0.5, 1$, and $\nu = -1, 1$.

TABLE 22.23: Sichel distribution, second parameterization.

	$\texttt{SI}(\mu, \sigma, \nu)$	
	Ranges	
Y	$0, 1, 2, 3 \ldots$	
μ	$0 < \mu < \infty$	
σ	$0 < \sigma < \infty$	
ν	$-\infty < \nu < \infty$	
	Distribution measures	
mean	$b\mu$	
variance	$b\mu + b^2\mu^2 g_1$	
skewness	$\mu_3/[\text{Var}(Y)]^{1.5}$ where $\mu_3 = b\mu + 3b^2\mu^2 g_1 + b^3\mu^3(g_2 - 3g_1)$	
excess kurtosis	$\begin{cases} k_4/[\text{Var}(Y)]^2 \text{ where} \\ k_4 = b\mu + 7b^2\mu^2 g_1 + 6b^3\mu^3(g_2 - 3g_1) \\ \quad +b^4\mu^4(g_3 - 4g_2 + 6g_1 - 3g_1^2) \end{cases}$	
PGF	$\dfrac{K_\nu(q)}{(q\sigma)^\nu K_\nu(1/\sigma)}$	where $q^2 = \sigma^{-2} + 2\mu(1-t)\sigma^{-1}$
pf	$\dfrac{\mu^y K_{y+\nu}(\alpha)}{y!(\alpha\sigma)^{y+\nu} K_\nu(1/\sigma)}$	where $\alpha^2 = \sigma^{-2} + 2\mu\sigma^{-1}$
Reference	Reparameterize μ_1 to $b\mu$ in $\texttt{SICHEL}(\mu_1, \sigma, \nu)$	
Note	Formulae for b, g_1, g_2, and g_3 are given in Table 22.22	

22.3.6 Zero-adjusted negative binomial: ZANBI

Let $Y = 0$ with probability ν, and $Y = Y_1$ with probability $(1-\nu)$, where $Y_1 \sim$ NBItr(μ, σ) and where NBItr(μ, σ) is a negative binomial (NBI(μ, σ)) truncated at zero. Then Y has a zero-adjusted negative binomial distribution, denoted by ZANBI(μ, σ, ν), with probability function

$$P(Y = y \mid \mu, \sigma, \nu) = \begin{cases} \nu & \text{if } \; y = 0 \\ c P(Y_2 = y \mid \mu, \sigma) & \text{if } \; y = 1, 2, 3, \ldots \end{cases} \tag{22.25}$$

where $\mu > 0$, $\sigma > 0$, $0 < \nu < 1$, $Y_2 \sim$ NBI(μ, σ), $c = (1-\nu)/(1-p_0)$ and $p_0 = P(Y_2 = 0 \mid \mu, \sigma) = (1 + \mu\sigma)^{-1/\sigma}$.

TABLE 22.24: Zero-adjusted negative binomial distribution.

ZANBI(μ, σ, ν)	
Ranges	
Y	$0, 1, 2, 3 \ldots$
μ	$0 < \mu < \infty$, mean of NBI component before zero truncation
σ	$0 < \sigma < \infty$, dispersion of NBI component before zero truncation
ν	$0 < \nu < 1$, $\nu = P(Y = 0)$
Distribution measures	
mean [a,a2]	$c\mu$ where $c = \dfrac{(1-\nu)}{[1 - (1+\mu\sigma)^{-1/\sigma}]}$
variance [a,a2]	$c\mu + c\mu^2(1 + \sigma - c)$
skewness[a,a2]	$\begin{cases} \mu_3/[\text{Var}(Y)]^{1.5} \; \text{ where} \\ \mu_3 = c\mu\left[1 + 3\mu(1+\sigma-c) + \mu^2\left(1 + 3\sigma + 2\sigma^2 - 3c - 3c\sigma + 2c^2\right)\right] \end{cases}$
excess kurtosis [a,a2]	$\begin{cases} \mu_4/[\text{Var}(Y)]^2 - 3 \; \text{ where} \\ \mu_4 = c\mu\,[1 + \mu(7 + 7\sigma - 4c) + \\ \quad 6\mu^2\left(1 + 3\sigma + 2\sigma^2 - 2c - 2c\sigma + c^2\right) + \\ \quad \mu^3\left(1 - 4c + 6c^2 - 3c^3 + 6\sigma + 11\sigma^2 + \right. \\ \quad \left. 6\sigma^3 - 12c\sigma + 6c^2\sigma - 8c\sigma^2\right)\,] \end{cases}$
PGF [a]	$(1-c) + c[1 + \mu\sigma(1-t)]^{-1/\sigma}$ for $0 < t < 1 + (\mu\sigma)^{-1}$
pf [a]	$\begin{cases} \nu & \text{if } \; y = 0 \\ c P(Y_2 = y \mid \mu, \sigma) & \text{if } \; y = 1, 2, 3, \ldots \end{cases}$ where $Y_2 \sim$ NBI(μ, σ)
cdf [a]	$\nu + c\left[1 - \dfrac{B(y+1, \sigma^{-1}, \mu\sigma(1+\mu\sigma)^{-1})}{B(y+1, \sigma^{-1})} - (1+\mu\sigma)^{-1/\sigma}\right]$
Reference	[a] obtained from equations (7.11), (7.12), (7.13), and (7.14), where $Y_2 \sim$ NBI(μ, σ). [a2] set $\nu = (1-c)$ in moments of ZINBI(μ, σ, ν)

22.3.7 Zero-adjusted Poisson-inverse Gaussian: ZAPIG

Let $Y = 0$ with probability ν, and $Y = Y_1$ with probability $(1-\nu)$, where $Y_1 \sim$ $\texttt{PIGtr}(\mu,\sigma)$ and where $\texttt{PIGtr}(\mu,\sigma)$ is a Poisson-inverse Gaussian ($\texttt{PIG}(\mu,\sigma)$) truncated at zero. Then Y has a zero-adjusted Poisson-inverse Gaussian distribution, denoted by $\texttt{ZAPIG}(\mu,\sigma,\nu)$, with probability function

$$P(Y = y \mid \mu,\sigma,\nu) = \begin{cases} \nu & \text{if } \ y = 0 \\ cP(Y_2 = y \mid \mu,\sigma) & \text{if } \ y = 1,2,3,\ldots \end{cases} \tag{22.26}$$

for $y = 0,1,2,3,\ldots$, where $\mu > 0$, $\sigma > 0$, $0 < \nu < 1$, $Y_2 \sim \texttt{PIG}(\mu,\sigma)$, $c = (1-\nu)/(1-p_0)$ and $p_0 = P(Y_2 = 0 \mid \mu,\sigma) = \exp(1/\sigma - \alpha)$, where $\alpha^2 = \sigma^{-2} + 2\mu\sigma^{-1}$ and $\alpha > 0$.

TABLE 22.25: Zero-adjusted Poisson-inverse Gaussian.

$\texttt{ZAPIG}(\mu,\sigma,\nu)$	
Ranges	
Y	$0,1,2,3\ldots$
μ	$0 < \mu < \infty$, mean of $\texttt{PIG}$ component before zero truncation
σ	$0 < \sigma < \infty$, dispersion of $\texttt{PIG}$ component before zero truncation
ν	$0 < \nu < 1$, $\nu = P(Y = 0)$
Distribution measures	
mean [a,a2]	$c\mu$ where $c = (1-\nu)/[1-\exp(1/\sigma - \alpha)]$ and $\alpha^2 = \sigma^{-2} + 2\mu\sigma^{-1}$
variance [a,a2]	$c\mu + c\mu^2(1+\sigma-c)$
skewness [a,a2]	$\begin{cases} \mu_3/[\text{Var}(Y)]^{1.5} \quad \text{where} \\ \mu_3 = c\mu\,[1 + 3\mu(1+\sigma-c) + \\ \qquad \mu^2\,(1+3\sigma+3\sigma^2-3c-3c\sigma+2c^2)\,] \end{cases}$
excess kurtosis [a,a2]	$\begin{cases} k_4/[\text{Var}(Y)]^2 \text{ where} \\ k_4 = \; c\mu\,[1 + 7\mu(1+\sigma-c) + \\ \qquad 6\mu^2\,(1+3\sigma+3\sigma^2-3c-3c\sigma+2c^2) \\ \qquad +\mu^3\,(1-7c+12c^2-6c^3+6\sigma+15\sigma^2 + \\ \qquad 15\sigma^3 - 18c\sigma + 12c^2\sigma - 15c\sigma^2)\,] \end{cases}$
PGF [a]	$(1-c) + c\exp(1/\sigma - q)$ where $q^2 = \sigma^{-2} + 2\mu(1-t)\sigma^{-1}$
pf [a]	$\begin{cases} \nu & \text{if } \ y = 0 \\ cP(Y_2 = y \mid \mu,\sigma) & \text{if } \ y = 1,2,3,\ldots \end{cases}$ where $Y_2 \sim \texttt{PIG}(\mu,\sigma)$.
Reference	[a] obtained from equations (7.11), (7.12), and (7.14), where $Y_2 \sim \texttt{PIG}(\mu,\sigma)$. [a2] set $\nu = (1-c)$ in moments of $\texttt{ZIPIG}(\mu,\sigma,\nu)$

22.3.8 Zero-inflated negative binomial: ZINBI

Let $Y = 0$ with probability ν and $Y = Y_1$ with probability $(1-\nu)$, where $Y_1 \sim \texttt{NBI}(\mu,\sigma)$. Then Y has a zero-inflated negative binomial distribution, denoted by $\texttt{ZINBI}(\mu,\sigma,\nu)$, with probability function

$$P(Y = y \mid \mu, \sigma, \nu) = \begin{cases} \nu + (1-\nu)\,P(Y_1 = 0 \mid \mu, \sigma) & \text{if } y = 0 \\ (1-\nu)P(Y_1 = y \mid \mu, \sigma) & \text{if } y = 1, 2, 3, \ldots \end{cases} \tag{22.27}$$

for $\mu > 0$, $\sigma > 0$, $0 < \nu < 1$, where $Y_1 \sim \texttt{NBI}(\mu, \sigma)$ and so $P(Y_1 = 0 \mid \mu, \sigma) = (1 + \mu\sigma)^{-1/\sigma}$. The mean is given by $E(Y) = (1-\nu)\,\mu$ and the variance by $\text{Var}(Y) = \mu\,(1-\nu) + \mu^2\,(1-\nu)\,(\sigma+\nu)$. Hence $\text{Var}(Y) = E(Y) + [E(Y)]^2\,(\sigma+\nu)/(1-\nu)$.

TABLE 22.26: Zero-inflated negative binomial distribution.

$\texttt{ZINBI}(\mu, \sigma, \nu)$	
Ranges	
Y	$0, 1, 2, 3 \ldots$
μ	$0 < \mu < \infty$, mean of the negative binomial component
σ	$0 < \sigma < \infty$, dispersion of the negative binomial component
ν	$0 < \nu < 1$, inflation probability at zero
Distribution measures	
mean [a]	$(1-\nu)\mu$
variance [a]	$\mu(1-\nu) + \mu^2(1-\nu)(\sigma+\nu)$
skewness [a]	$\begin{cases} \mu_3/[\text{Var}(Y)]^{1.5} \quad \text{where} \\ \mu_3 = \mu(1-\nu)\left[1 + 3\mu(\sigma+\nu) + \mu^2\left(2\sigma^2 + 3\sigma\nu + 2\nu^2 - \nu\right)\right] \end{cases}$
excess kurtosis [a]	$\begin{cases} \mu_4/[\text{Var}(Y)]^2 - 3 \quad \text{where} \\ \mu_4 = \mu(1-\nu)\left[1 + \mu(3 + 7\sigma + 4\nu) + 6\mu^2\left(\sigma + 2\sigma^2 + 2\sigma\nu + \nu^2\right) + \right. \\ \left. \mu^3\left(3\sigma^2 + 6\sigma^3 + 6\sigma\nu^2 + 8\sigma^2\nu + \nu - 3\nu^2 + 3\nu^3\right)\right] \end{cases}$
PGF [a]	$\nu + (1-\nu)[1 + \mu\sigma(1-t)]^{-1/\sigma}$ for $0 < t < 1 + (\mu\sigma)^{-1}$
pf [a]	$\begin{cases} \nu + (1-\nu)\,P(Y_1 = 0 \mid \mu, \sigma) & \text{if } y = 0 \\ (1-\nu)P(Y_1 = y \mid \mu, \sigma) & \text{if } y = 1, 2, 3, \ldots \end{cases}$ where $Y_1 \sim \texttt{NBI}(\mu, \sigma)$
cdf [a]	$1 - \dfrac{(1-\nu)B(y+1, \sigma^{-1}, \mu\sigma(1+\mu\sigma)^{-1})}{B(y+1, \sigma^{-1})}$
Reference	[a]Obtained from equations (7.5), (7.6), (7.7) and (7.8), where $Y_1 \sim \texttt{NBI}(\mu, \sigma)$

22.3.9 Zero-inflated Poisson-inverse Gaussian: ZIPIG

Let $Y = 0$ with probability ν, and $Y = Y_1$ with probability $(1-\nu)$, where $Y_1 \sim \texttt{PIG}(\mu, \sigma)$. Then Y has a zero-inflated Poisson-inverse Gaussian distribution, denoted by ZIPIG(μ, σ, ν), with probability function

$$P(Y = y \mid \mu, \sigma, \nu) = \begin{cases} \nu + (1-\nu)\,P(Y_1 = 0 \mid \mu, \sigma) & \text{if} \quad y = 0 \\ (1-\nu)\,P(Y_1 = y \mid \mu, \sigma) & \text{if} \quad y = 1, 2, 3, \ldots \end{cases} \tag{22.28}$$

for $\mu > 0$, $\sigma > 0$ and $0 < \nu < 1$, where $Y_1 \sim$ PIG(μ, σ) and so $P(Y_1 = 0|\mu, \sigma) = \exp(1/\sigma - \alpha)$, where $\alpha^2 = \sigma^{-2} + 2\mu\sigma^{-1}$ and $\alpha > 0$. The mean of Y is given by $E(Y) = (1-\nu)\,\mu$ and the variance by $\text{Var}(Y) = \mu(1-\nu) + \mu^2(1-\nu)(\sigma+\nu)$. Hence $\text{Var}(Y) = E(Y) + [E(Y)]^2\,(\sigma+\nu)/(1-\nu)$.

TABLE 22.27: Zero-inflated Poisson-inverse Gaussian distribution.

ZIPIG(μ, σ, ν)	
Ranges	
Y	$0, 1, 2, 3 \ldots$
μ	$0 < \mu < \infty$, mean of the PIG component
σ	$0 < \sigma < \infty$, dispersion of the PIG component
ν	$0 < \nu < 1$, inflation probability at zero
Distribution measures	
mean[a]	$(1-\nu)\mu$
variance[a]	$\mu(1-\nu) + \mu^2(1-\nu)(\sigma+\nu)$
skewness[a]	$\begin{cases} \mu_3/[\text{Var}(Y)]^{1.5} \quad \text{where} \\ \mu_3 = \mu(1-\nu)[1 + 3\mu(\sigma+\nu) + \mu^2(3\sigma^2 + 3\sigma\nu + 2\nu^2 - \nu)] \end{cases}$
excess kurtosis [a]	$\begin{cases} k_4/[\text{Var}(Y)]^2 \quad \text{where} \\ k_4 = \mu(1-\nu)\left[1 + 7\mu(\sigma+\nu) + 6\mu^2\left(3\sigma^2 + 3\sigma\nu + 2\nu^2 - \nu\right) + \right. \\ \qquad \left. \mu^3\left(\nu - 6\nu^2 + 6\nu^3 + 15\sigma^3 - 6\sigma\nu + 12\sigma\nu^2 + 15\sigma^2\nu\right)\right] \end{cases}$
PGF[a]	$\nu + (1-\nu)e^{(1/\sigma)-q}$ where $q^2 = \sigma^{-2} + 2\mu(1-t)\sigma^{-1}$
pf[a]	$\begin{cases} \nu + (1-\nu)\,P(Y_1 = 0 \mid \mu, \sigma) & \text{if} \quad y = 0 \\ (1-\nu)\,P(Y_1 = y \mid \mu, \sigma) & \text{if} \quad y = 1, 2, 3, \ldots \end{cases}$ where $Y_1 \sim$ PIG(μ, σ)
Reference	[a]Obtained from equations (7.5), (7.6), and (7.8), $Y_1 \sim$ PIG(μ, σ)

22.4 Four-parameter count distributions

22.4.1 Poisson shifted generalized inverse Gaussian: PSGIG

The Poisson shifted generalized inverse Gaussian distribution, [Rigby et al., 2008], denoted by PSGIG(μ, σ, ν, τ), has probability function given by

$$P(Y = y \mid \mu, \sigma, \nu, \tau) \quad = \quad \frac{e^{-\mu\tau}T}{K_\nu(1/\sigma)} \tag{22.29}$$

where $y = 0, 1, 2, 3, \ldots, \mu > 0, \sigma > 0, -\infty < \nu < \infty, 0 < \tau < 1$,

$$T = \sum_{j=0}^{y} \binom{y}{j} \frac{\mu^y \tau^{y-j} K_{\nu+j}(\delta)}{y! d^j (\delta\sigma)^{\nu+j}}, \tag{22.30}$$

$d = b/(1-\tau)$, $b = K_{\nu+1}(1/\sigma)/K_\nu(1/\sigma)$ and $\delta^2 = \sigma^{-2} + 2\mu(d\sigma)^{-1}$. Note that

$$\sigma = \left[\left(\frac{\mu^2}{d^2} + \delta^2 \right)^{1/2} - \frac{\mu}{d} \right]^{-1}.$$

As $y \to \infty$, $\log \mathrm{P}(Y = y \mid \mu, \sigma, \nu, \tau) \sim -y \log \left[1 + \frac{b}{2\mu\sigma(1-\tau)}\right]$. Note that the distribution is currently under development in **gamlss.dist**.

TABLE 22.28: Poisson shifted generalized inverse Gaussian distribution.

PSGIG(μ, σ, ν, τ)	
Ranges	
Y	$0, 1, 2, 3 \ldots$
μ	$0 < \mu < \infty$, mean
σ	$0 < \sigma < \infty$
ν	$-\infty < \nu < \infty$
τ	$0 < \tau < 1$
Distribution measures	
mean	μ
variance	$\mu + (1-\tau)^2 \mu^2 g_1$ where $g_1 = 2\sigma(\nu+1)b^{-1} + b^{-2} - 1$
skewness	$\begin{cases} \mu_3/[\mathrm{Var}(Y)]^{1.5} \text{ where} \\ \mu_3 = \mu + 3(1-\tau)^2\mu^2 g_1 + (1-\tau)^3\mu^3(g_2 - 3g_1) \\ g_2 = 2\sigma(\nu+2)b^{-3} + [4\sigma^2(\nu+1)(\nu+2)+1]b^{-2} - 1 \end{cases}$
excess kurtosis	$\begin{cases} k_4/[\mathrm{Var}(Y)]^2 \text{ where} \\ k_4 = \mu + 7(1-\tau)^2\mu^2 g_1 + 6(1-\tau)^3\mu^3(g_2 - 3g_1) + \\ \quad (1-\tau)^4\mu^4(g_3 - 4g_2 + 6g_1 - 3g_1^2) \\ g_3 = [1 + 4\sigma^2(\nu+2)(\nu+3)]b^{-4} + \\ \quad [8\sigma^3(\nu+1)(\nu+2)(\nu+3) + 4\sigma(\nu+2)]b^{-3} - 1 \end{cases}$
PGF	$\dfrac{\exp[\mu\tau(t-1)]K_\nu(r)}{(r\sigma)^\nu K_\nu(1/\sigma)}$ where $r^2 = \sigma^{-2} + 2\mu(1-t)(d\sigma)^{-1}$ and $d = b/(1-\tau)$
pf	$\dfrac{\exp(-\mu t)T}{K_\nu(1/\sigma)}$ where $T = \sum_{j=0}^{y} \binom{y}{j} \dfrac{\mu^y \tau^{y-j} K_{\nu+j}(\delta)}{y! d^j (\delta\sigma)^{\nu+j}}$ where $\delta^2 = \sigma^{-2} + 2\mu(d\sigma)^{-1}$
Reference	Rigby et al. [2008]
Notes	$b = K_{\nu+1}(1/\sigma)/K_\nu(1/\sigma)$ $K_\lambda(t)$ is the modified Bessel function of the second kind

22.4.2 Zero-adjusted beta negative binomial: ZABNB

Let $Y = 0$ with probability τ, and $Y = Y_1$ with probability $(1-\tau)$, where $Y_1 \sim$ $\texttt{BNBtr}(\mu, \sigma, \nu)$, i.e. the beta negative binomial ($\texttt{BNB}(\mu, \sigma, \tau)$) distribution truncated at zero. Then Y has a zero-adjusted beta negative binomial distribution, denoted by $\texttt{ZABNB}(\mu, \sigma, \nu, \tau)$, with probability function

$$P(Y = y \mid \mu, \sigma, \nu, \tau) = \begin{cases} \tau & \text{if } y = 0 \\ cP(Y_2 = y \mid \mu, \sigma, \nu) & \text{if } y = 1, 2, 3, \ldots \end{cases} \tag{22.31}$$

where $\mu > 0$, $\sigma > 0$, $\nu > 0$, $0 < \tau < 1$, $Y_2 \sim \texttt{BNB}(\mu, \sigma, \nu)$ and $c = (1-\tau)/(1-p_0)$ where

$$p_0 = P(Y_2 = 0 \mid \mu, \sigma, \nu) = \frac{B(\mu\sigma^{-1}\nu, \sigma^{-1} + \nu^{-1} + 1)}{B(\mu\sigma^{-1}\nu, \sigma^{-1} + 1)} .$$

The moments of Y can be obtained from the moments of Y_2 using equation (7.12), from which the mean, variance, skewness, and excess kurtosis of Y can be found. In particular the mean of Y is $E(Y) = c\mu$.

TABLE 22.29: Zero-adjusted beta negative binomial.

$\texttt{ZABNB}(\mu, \sigma, \nu, \tau)$	
Ranges	
Y	$0, 1, 2, 3 \ldots$
μ	$0 < \mu < \infty$, mean of BNB component before zero truncation
σ	$0 < \sigma < \infty$
ν	$0 < \nu < \infty$
τ	$0 < \tau < 1$, $\tau = P(Y = 0)$
Distribution measures	
mean [a,a2]	$c\mu$ where $c = (1-\tau)/(1-p_0)$ and $p_0 = P(Y_2 = 0 \mid \mu, \sigma, \nu)$ where $Y_2 \sim \texttt{BNB}(\mu, \sigma, \nu)$
variance [a,a2]	$\begin{cases} c\mu(1+\mu\nu)(1+\sigma\nu^{-1})(1-\sigma)^{-1} + c(1-c)\mu^2 & \text{for } \sigma < 1 \\ \infty & \text{for } \sigma \geq 1 \end{cases}$
pf [a]	$\begin{cases} \tau & \text{if } y = 0 \\ cP(Y_2 = y \mid \mu, \sigma, \nu) & \text{if } y = 1, 2, 3, \ldots \end{cases}$ where $Y_2 \sim \texttt{BNB}(\mu, \sigma, \nu)$
Reference	[a] Obtained from equations (7.11) and (7.12). where $Y_2 \sim \texttt{BNB}(\mu, \sigma, \nu)$. [a2] set $\nu = (1-c)$ in moments of $\texttt{ZIBNB}(\mu, \sigma, \nu)$

22.4.3 Zero-adjusted Sichel: ZASICHEL

Let $Y = 0$ with probability τ, and $Y = Y_1$ with probability $(1-\tau)$, where $Y_1 \sim$ $\texttt{SICHELtr}(\mu, \sigma, \nu)$, i.e. $\texttt{SICHEL}(\mu, \sigma, \nu)$ truncated at zero. Then Y has a zero-adjusted

Sichel distribution, denoted by ZASICHEL(μ,σ,ν,τ), with probability function

$$P(Y=y\,|\,\mu,\sigma,\nu)=\begin{cases}\tau & \text{if } y=0\\ cP(Y_2=y\,|\,\mu,\sigma,\nu) & \text{if } y=1,2,3,\ldots\end{cases} \tag{22.32}$$

where $\mu>0$, $\sigma>0$, $-\infty<\nu<\infty$, $0<\tau<1$, $Y_2\sim$ SICHEL(μ,σ,ν) and $c=(1-\tau)/(1-p_0)$ where

$$p_0=P(Y_2=0\,|\,\mu,\sigma,\nu)=\frac{K_\nu(\alpha)}{(\alpha\sigma)^\nu K_\nu(1/\sigma)}$$

and $\alpha^2=\sigma^{-2}+2\mu(b\sigma)^{-1}$, $b=K_{\nu+1}(1/\sigma)/K_\nu(1/\sigma)$.

TABLE 22.30: Zero-adjusted Sichel distribution.

ZASICHEL(μ,σ,ν,τ)	
Ranges	
Y	$0,1,2,3\ldots$
μ	$0<\mu<\infty$, mean of SICHEL component before zero truncation
σ	$0<\sigma<\infty$
ν	$-\infty<\nu<\infty$
τ	$0<\tau<1$, $\tau=P(Y=0)$
Distribution measures	
mean [a,a2]	$c\mu$ where $c=(1-\tau)/(1-p_0)$ and $p_0=P(Y_2=0)$ where $Y_2\sim$ SICHEL(μ,σ,ν)
variance [a,a2]	$c\mu+c^2\mu^2h_1$ where $h_1=c^{-1}\left[2\sigma(\nu+1)b^{-1}+b^{-2}\right]-1$
skewness [a,a2]	$\begin{cases}\mu_3/[\text{Var}(Y)]^{1.5} \text{ where}\\ \mu_3=c\mu+3c^2\mu^2h_1+c^3\mu^3(h_2-3h_1-1)\\ h_2=c^{-2}\left\{2\sigma(\nu+2)b^{-3}+\left[4\sigma^2(\nu+1)(\nu+1)+1\right]b^{-2}\right\}\end{cases}$
excess kurtosis [a,a2]	$\begin{cases}k_4/[\text{Var}(Y)]^2 \text{ where}\\ k_4=c\mu+7c^2\mu^2h_1+6c^3\mu^3(h_2-3h_1-1)+\\ \quad c^4\mu^4\left(h_3-4h_2+6h_1-3h_1^2+3\right)\\ h_3=c^{-3}\left\{b^{-4}\left[1+4\sigma^2(\nu+2)(\nu+3)\right]+\right.\\ \quad \left.b^{-3}\left[8\sigma^3(\nu+1)(\nu+2)(\nu+3)+4\sigma(\nu+2)\right]\right\}\end{cases}$
PGF [a]	$(1-c)+\dfrac{cK_\nu(q)}{(q\sigma)^\nu K_\nu(1/\sigma)}$ where $q^2=\sigma^{-2}+2\mu(1-\tau)(b\sigma)^{-1}$
pf [a]	$\begin{cases}\tau & \text{if } y=0\\ cP(Y_2=y\,|\,\mu,\sigma,\nu) & \text{if } y=1,2,3,\ldots\end{cases}$ where $Y_2\sim$ SICHEL(μ,σ,ν)
Reference	[a] Obtained from equations (7.11), (7.12), and (7.14) where $Y_2\sim$ SICHEL(μ,σ,ν). [a2] set $\tau=(1-c)$ in moments of ZISICHEL(μ,σ,ν,τ).
Notes	b and $K_\lambda(t)$ are defined in Table 22.22.

22.4.4 Zero-inflated beta negative binomial: ZIBNB

Let $Y = 0$ with probability τ, and $Y = Y_1$ with probability $(1-\tau)$ where $Y_1 \sim$ $\texttt{BNB}(\mu, \sigma, \nu)$. Then Y has a zero-inflated beta negative binomial distribution, denoted by $\texttt{ZIBNB}(\mu, \sigma, \nu, \tau)$, with probability function

$$P(Y = y \,|\, \mu, \sigma, \nu, \tau) = \begin{cases} \tau + (1-\tau)\,P(Y_1 = 0 \,|\, \mu, \sigma, \nu) & \text{if } y = 0 \\ (1-\tau)P(Y_1 = y \,|\, \mu, \sigma, \nu) & \text{if } y = 1, 2, 3, \ldots \end{cases} \tag{22.33}$$

for $\mu > 0$, $\sigma > 0$, $\nu > 0$, $0 < \tau < 1$ and $Y_1 \sim \texttt{BNB}(\mu, \sigma, \nu)$. The moments of Y can be obtained from the moments of Y_1, using equations (7.6), from which the mean, variance, skewness, and excess kurtosis of Y can be found. In particular the mean of Y is $E(Y) = (1-\tau)\mu$.

TABLE 22.31: Zero-inflated beta negative binomial distribution.

$\texttt{ZIBNB}(\mu, \sigma, \nu, \tau)$	
Ranges	
Y	$0, 1, 2, 3 \ldots$
μ	$0 < \mu < \infty$, mean of the BNB component
σ	$0 < \sigma < \infty$
ν	$0 < \nu < \infty$
τ	$0 < \tau < 1$, inflation probability at zero
Distribution measures	
mean	$(1-\tau)\,\mu$
variance	$(1-\tau)\mu(1+\mu\nu)(1+\sigma\nu^{-1})(1-\sigma)^{-1} + \tau(1-\tau)\mu^2$
pf	$\begin{cases} \tau + (1-\tau)\,P(Y_1 = 0 \,|\, \mu, \sigma, \nu) & \text{if } y = 0 \\ (1-\tau)P(Y_1 = y \,|\, \mu, \sigma, \nu) & \text{if } y = 1, 2, 3, \ldots \end{cases}$ where $Y_1 \sim \texttt{BNB}(\mu, \sigma, \nu)$
Reference	Obtained from equations (7.5) and (7.6), where $Y_1 \sim \texttt{BNB}(\mu, \sigma, \nu)$

22.4.5 Zero-inflated Sichel: ZISICHEL

Let $Y = 0$ with probability τ, and $Y = Y_1$, with probability $(1-\tau)$ where $Y_1 \sim \texttt{SICHEL}(\mu, \sigma, \nu)$. Then Y has a zero-inflated Sichel distribution, denoted by $\texttt{ZISICHEL}(\mu, \sigma, \nu, \tau)$, with probability function

$$P(Y = y \,|\, \mu, \sigma, \nu, \tau) = \begin{cases} \tau + (1-\tau)\,P(Y_1 = 0 \,|\, \mu, \sigma, \nu) & \text{if } y = 0 \\ (1-\tau)P(Y_1 = y \,|\, \mu, \sigma, \nu) & \text{if } y = 1, 2, 3, \ldots \end{cases} \tag{22.34}$$

for $\mu > 0$, $\sigma > 0$, $-\infty < \nu < \infty$ and $0 < \tau < 1$, and $Y_1 \sim \texttt{SICHEL}(\mu, \sigma, \nu)$.

TABLE 22.32: Zero-inflated Sichel distribution.

ZISICHEL(μ, σ, ν, τ)	
Ranges	
Y	$0, 1, 2, 3 \ldots$
μ	$0 < \mu < \infty$, mean of the SICHEL component
σ	$0 < \sigma < \infty$
ν	$-\infty < \nu < \infty$
τ	$0 < \tau < 1$, inflation probability at zero
Distribution measures	
mean[a]	$(1-\tau)\,\mu$
variance[a]	$\begin{cases} (1-\tau)\,\mu + (1-\tau)^2\,\mu^2 h_1 \\ \text{where } h_1 = (1-\tau)^{-1}\left[2\sigma\,(\nu+1)b^{-1} + b^{-2}\right] - 1 \end{cases}$
skewness[a]	$\begin{cases} \mu_3/\left[\text{Var}(Y)\right]^{1.5} \text{ where} \\ \mu_3 = (1-\tau)\,\mu + 3\,(1-\tau)^2\,\mu^2 h_1 + (1-\tau)^3\,\mu^3\,(h_2 - 3h_1 - 1) \\ h_2 = (1-\tau)^{-2}\left\{2\sigma\,(\nu+2)/b^3 + \left[4\sigma^2\,(\nu+1)\,(\nu+2)+1\right]/b^2\right\} \end{cases}$
excess kurtosis [a]	$\begin{cases} k_4/\left[\text{Var}(Y)\right]^2 \text{ where} \\ k_4 = (1-\tau)\,\mu + 7\,(1-\tau)^2\,\mu^2 h_1 + 6\,(1-\tau)^3\,\mu^3\,(h_2 - 3h_1 - 1) + \\ \qquad (1-\tau)^4\,\mu^4\,(h_3 - 4h_2 + 6h_1 - 3h_1^2 + 3) \\ h_3 = (1-\tau)^{-3}\left\{\left[1 + 4\sigma^2\,(\nu+2)\,(\nu+3)\right]/b^4 + \right. \\ \qquad \left. \left[8\sigma^3\,(\nu+1)\,(\nu+2)\,(\nu+3) + 4\sigma\,(\nu+2)\right]/b^3\right\} \end{cases}$
PGF[a]	$\tau + (1-\tau)\dfrac{K_\nu(q)}{(q\sigma)^\nu K_\nu(1/\sigma)}$ where $q^2 = \sigma^{-2} + 2\mu(1-t)(b\sigma)^{-1}$
pf[a]	$\begin{cases} \tau + (1-\tau)\,\text{P}(Y_1 = 0\,\|\,\mu, \sigma, \nu) & \text{if } y = 0 \\ (1-\tau)\text{P}(Y_1 = y\,\|\,\mu, \sigma, \nu) & \text{if } y = 1, 2, 3, \ldots \end{cases}$ where $Y_1 \sim$ SICHEL(μ, σ, ν)
Reference	[a]Obtained from equations (7.5), (7.6) and (7.8), where $Y_1 \sim$ SICHEL(μ, σ, ν)
Note	b and $K_\nu(\cdot)$ are defined in Table 22.22.

23

Binomial type distributions and multinomial distributions

CONTENTS

This chapter gives summary tables and plots for the explicit **gamlss.dist** discrete binomial type distributions with range $\{0, 1, \ldots, n\}$, in which it is assumed throughout that n is a known positive integer. These are discussed in Chapter 8. This chapter also covers the explicit **gamlss.dist** multinomial distributions.

23.1 One-parameter binomial type distribution

23.1.1 Binomial: `BI`

The probability function of the binomial distribution, denoted as `BI`(n, μ), is given by

$$\begin{aligned} P(Y = y|n, \mu) &= \binom{n}{y} \mu^y (1-\mu)^{n-y} \\ &= \frac{n!}{y!(n-y)!} \mu^y (1-\mu)^{n-y} \end{aligned} \tag{23.1}$$

for $y = 0, 1, 2, \ldots, n$, where $0 < \mu < 1$ and n is a known positive integer.

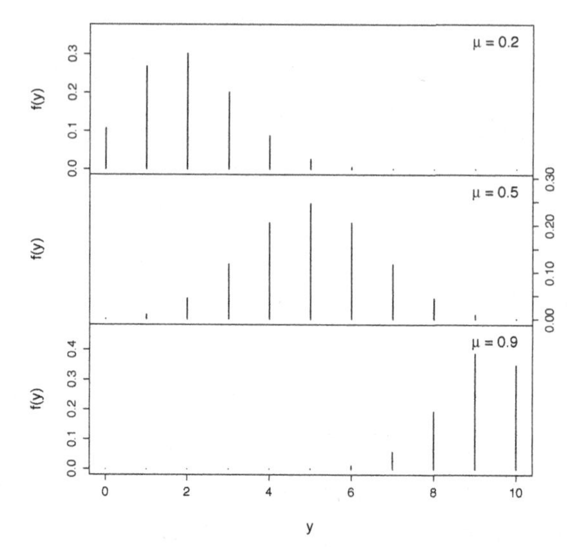

FIGURE 23.1: The binomial, BI(n, μ), distribution, with $\mu = 0.2, 0.5, 0.9$ and binomial denominator $n = 10$.

TABLE 23.1: Binomial distribution.

	BI(n, μ)
	Ranges
Y	$0, 1, 2, \ldots, n$ where n is a known positive integer
μ	$0 < \mu < 1$
	Distribution measures
mean	$n\mu$
mode	$\begin{cases} \lfloor (n+1)\mu \rfloor & \text{if } (n+1)\mu \text{ is not an integer} \\ 0 & \text{if } \mu < 1/(n+1) \\ n & \text{if } \mu > n/(n+1) \\ (n+1)\mu - 1 \text{ and } (n+1)\mu, & \text{if } (n+1)\mu \text{ is an integer} \end{cases}$
variance	$n\mu(1-\mu)$
skewness	$\dfrac{(1-2\mu)}{[n\mu(1-\mu)]^{0.5}}$
excess kurtosis	$\dfrac{1-6\mu(1-\mu)}{n\mu(1-\mu)}$
PGF	$(1-\mu+\mu t)^n$
pf	$\binom{n}{y}\mu^y(1-\mu)^{n-y}$
cdf	$1 - \dfrac{B(y+1, n-y, \mu)}{B(y+1, n-y)}$
Reference	Johnson et al. [2005] p108, 110, 112, 113, with $p = \mu$
Note	$\lfloor x \rfloor$ is the floor function or integer part of x

23.2 Two-parameter binomial type distributions

23.2.1 Beta binomial: BB

The probability function of the beta binomial distribution, denoted as $\texttt{BB}(n, \mu, \sigma)$, is given by

$$P(Y = y|n, \mu, \sigma) = \binom{n}{y} \frac{B(y + \mu/\sigma, n + (1-\mu)/\sigma - y)}{B(\mu/\sigma, (1-\mu)/\sigma)} \tag{23.2}$$
$$= \frac{\Gamma(n+1)}{\Gamma(y+1)\Gamma(n-y+1)} \frac{\Gamma(\frac{1}{\sigma})\Gamma(y+\frac{\mu}{\sigma})\Gamma(n+\frac{(1-\mu)}{\sigma}-y)}{\Gamma(n+\frac{1}{\sigma})\Gamma(\frac{\mu}{\sigma})\Gamma(\frac{1-\mu}{\sigma})}$$

for $y = 0, 1, \ldots, n$, where $0 < \mu < 1$, $\sigma > 0$, and n is a known positive integer.

The binomial $\texttt{BI}(n, \mu)$ distribution is the limiting distribution of $\texttt{BB}(n, \mu, \sigma)$ as $\sigma \to 0$. For $\mu = 0.5$ and $\sigma = 0.5$, $\texttt{BB}(n, \mu, \sigma)$ is a discrete uniform distribution. The probability function (23.2) and the mean, variance, skewness, and kurtosis in Table 23.2 can be obtained from Johnson et al. [2005], p259 equation (6.31) and p263, by setting $a = -\mu/\sigma$ and $b = -(1-\mu)/\sigma$. The equivalence of their probability function to (23.2) is shown using their equation (1.24).

The $\texttt{BB}(n, \mu, \sigma)$ distribution can be derived as a beta mixture of binomial distributions, by assuming $Y|\pi \sim \texttt{BI}(n, \pi)$ where $\pi \sim \texttt{BEo}(\frac{\mu}{\sigma}, \frac{1-\mu}{\sigma})$. Hence the $\texttt{BB}(n, \mu, \sigma)$ is an overdispersed binomial distribution.

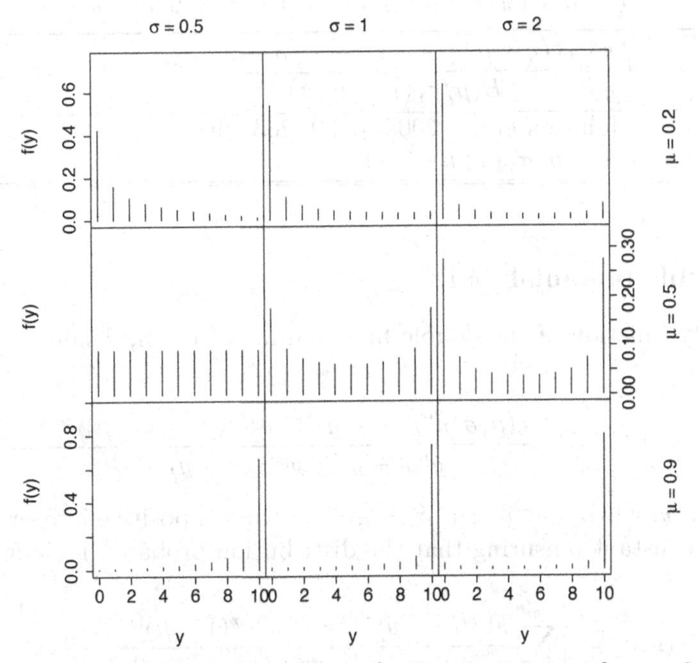

FIGURE 23.2: The beta binomial, $\texttt{BB}(n, \mu, \sigma)$, distribution, with $\mu = 0.2, 0.5, 0.9$, $\sigma = 0.5, 1, 2$ and binomial denominator $n = 10$.

TABLE 23.2: Beta binomial distribution.

	BB(n, μ, σ)
	Ranges
Y	$0, 1, 2, \ldots, n$ where n is a known positive integer
μ	$0 < \mu < 1$
σ	$0 < \sigma < \infty$
	Distribution measures
mean[a]	$n\mu$
mode	$\begin{cases} \lfloor \frac{(n+1)(\mu-\sigma)}{1-2\sigma} \rfloor & \text{if } \sigma < \min(\sigma_0, \sigma_1) \\ 0 & \text{if } \sigma_0 < \sigma < \sigma_1 \\ n & \text{if } \sigma_1 < \sigma < \sigma_0 \\ 0 \text{ and } n & \text{if } \sigma > \max(\sigma_0, \sigma_1) \\ \text{where} & \\ \sigma_0 = [(n+1)\mu - 1]/(n-1) & \\ \sigma_1 = [(n+1)(1-\mu) - 1]/(n-1) & \end{cases}$
variance[a]	$n\mu(1-\mu)\left[1 + \frac{(n-1)\sigma}{(1+\sigma)}\right]$
skewness[a]	$\left[\frac{(1+\sigma)}{n\mu(1-\mu)(1+n\sigma)}\right]^{0.5} \frac{(1-2\mu)(1+2n\sigma)}{(1+2\sigma)}$
excess kurtosis[a]	$\begin{cases} \beta_2 - 3 \text{ where} \\ \beta_2 = \frac{(1+\sigma)}{n\mu(1-\mu)(1+n\sigma)(1+2\sigma)(1+3\sigma)} [(1+6n\sigma- \\ \quad \sigma + 6n^2\sigma^2) - 3\mu(1-\mu)(6n^2\sigma^2 + 6n\sigma - n^2\sigma - n + 2)] \end{cases}$
pf[a]	$\binom{n}{y} \frac{B(y + \mu/\sigma, n + (1-\mu)/\sigma - y)}{B(\mu/\sigma, (1-\mu)/\sigma)}$
Reference	[a]Johnson et al. [2005] p259, 263 with $a = -\mu/\sigma$ and $b = -(1-\mu)/\sigma$

23.2.2 Double binomial: DBI

The probability function of the double binomial distribution, denoted by DBI(n, μ, σ), is given by

$$P(Y = y \mid n, \mu, \sigma) = \frac{c(\mu, \sigma) n! y^y (n-y)^{n-y} n^{n/\sigma} \mu^{y/\sigma} (1-\mu)^{(n-y)/\sigma}}{y!(n-y)! n^n y^{y/\sigma} (n-y)^{(n-y)/\sigma}} \tag{23.3}$$

for $y = 0, 1, \ldots, n$, where $0 < \mu < 1$, $\sigma > 0$, n is a known positive integer, and $c(\mu, \sigma)$ is a normalizing constant (ensuring that the distribution probabilities sum to one) given by

$$c(\mu, \sigma) = \left[\sum_{y=0}^{n} \frac{n! y^y (n-y)^{n-y} n^{n/\sigma} \mu^{y/\sigma} (1-\mu)^{(n-y)/\sigma}}{y!(n-y)! n^n y^{y/\sigma} (n-y)^{(n-y)/\sigma}}\right]^{-1} \tag{23.4}$$

obtained from Lindsey [1995, p131], reparameterized by $\pi = \mu$ and $\psi = 1/\sigma$.

The $\texttt{DBI}(n, \mu, \sigma)$ distribution is a special case of the double exponential family of Efron [1986]. It has approximate mean $n\mu$ and approximate variance $n\sigma\mu(1-\mu)$. The $\texttt{DBI}(n, \mu, \sigma)$ distribution is a binomial ($\texttt{BI}(n, \mu)$) distribution if $\sigma = 1$. It is (approximately) an overdispersed binomial if $\sigma > 1$ and underdispersed if $\sigma < 1$. Unlike some other implementations, **gamlss.dist** calculates $c(\mu, \sigma)$ using a finite sum with a very large number of terms in equation (23.4), rather than a potentially less accurate functional approximation of $c(\mu, \sigma)$.

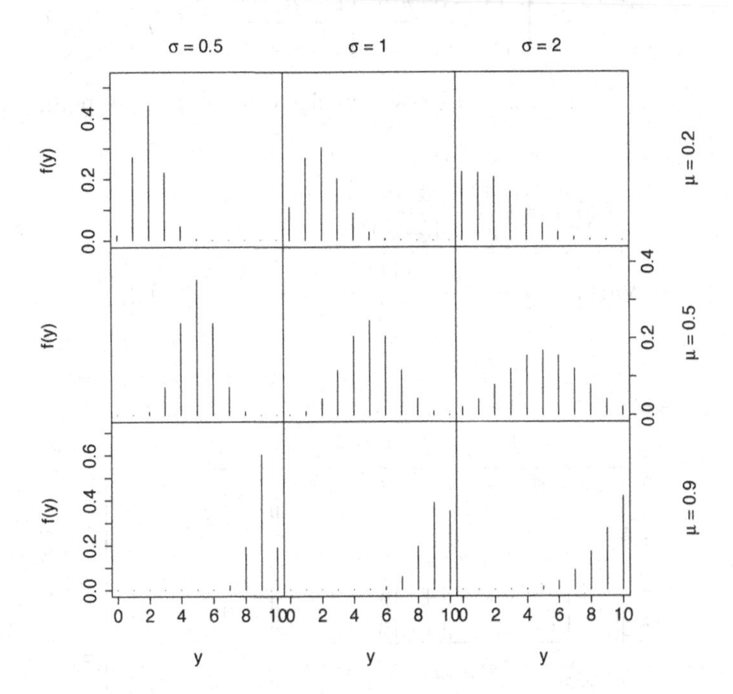

FIGURE 23.3: The double binomial, $\texttt{DBI}(n, \mu, \sigma)$, distribution, with $\mu = 0.2, 0.5, 0.9$, $\sigma = .5, 1, 2$, and binomial denominator $n = 10$.

TABLE 23.3: Double binomial distribution.

$\texttt{DBI}(n, \mu, \sigma)$	
Ranges	
Y	$0, 1, 2, \ldots, n$ where n is a known positive integer
μ	$0 < \mu < 1$
σ	$0 < \sigma < \infty$
Distribution measures	
mean[a]	$n\mu$
variance[a]	$n\sigma\mu(1-\mu)$
pf[a2]	$\dfrac{c(\mu, \sigma) n! y^y (n-y)^{n-y} n^{n/\sigma} \mu^{y/\sigma} (1-\mu)^{(n-y)/\sigma}}{y!(n-y)! n^n y^{y/\sigma} (n-y)^{(n-y)/\sigma}}$ where $c(\mu, \sigma)$ is given by equation (23.4)
Reference	[a2] Lindsey [1995] p131, where $\pi = \mu$ and $\psi = 1/\sigma$
Note	[a] approximate

23.2.3 Zero-adjusted binomial: ZABI

Let $Y = 0$ with probability σ, and $Y = Y_1$ with probability $(1-\sigma)$, where $Y_1 \sim$ `BItr`(n, μ) a binomial distribution truncated at zero. Then Y has a zero-adjusted (or altered) binomial distribution, denoted by `ZABI`(n, μ, σ), given by

$$\mathrm{P}(Y = y \mid n, \mu, \sigma) = \begin{cases} \sigma & \text{if } y = 0 \\ \dfrac{(1-\sigma)\, n!\mu^y\, (1-\mu)^{n-y}}{\left[1-(1-\mu)^n\right] y!\,(n-y)!} & \text{if } y = 1, 2, 3, \ldots \end{cases} \tag{23.5}$$

for $0 < \mu < 1$, $0 < \sigma < 1$ and n is a known positive integer. The mean and variance of Y are given by

$$\mathrm{E}(Y) = \frac{(1-\sigma)\, n\mu}{\left[1-(1-\mu)^n\right]}$$

$$\mathrm{Var}(Y) = \frac{n\mu\,(1-\sigma)\,(1-\mu+n\mu)}{\left[1-(1-\mu)^n\right]} - \left[\mathrm{E}\,(Y)\right]^2 .$$

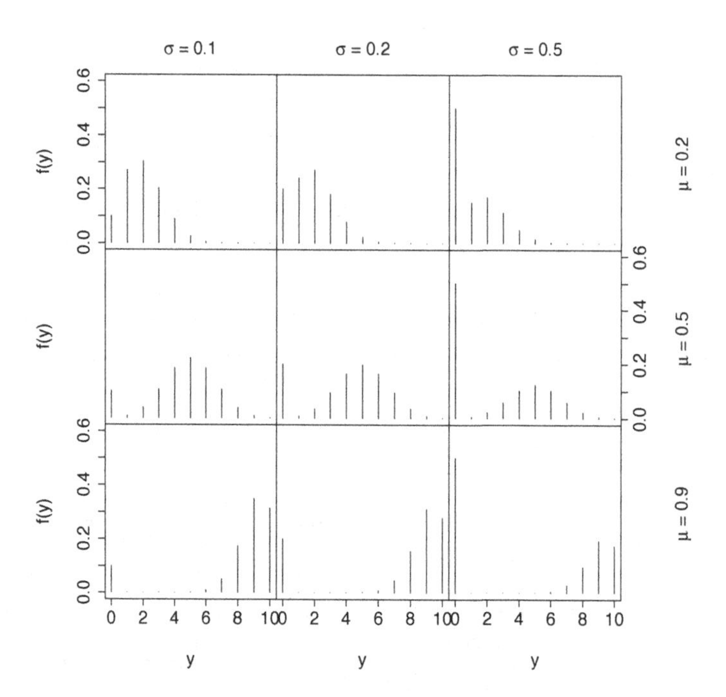

FIGURE 23.4: The zero-adjusted binomial, `ZABI`(n, μ, σ), distribution, with $\mu = 0.2, 0.5, 0.9$, $\sigma = 0.1, 0.2, 0.5$, and binomial denominator $n = 10$.

23.2.4 Zero-inflated binomial: ZIBI

Let $Y = 0$ with probability σ, and $Y = Y_1$ with probability $(1-\sigma)$, where $Y_1 \sim \texttt{BI}(n, \mu)$, then Y has a zero-inflated binomial distribution, denoted by $\texttt{ZIBI}(n, \mu, \sigma)$, given by

$$P(Y = y \mid n, \mu, \sigma) \begin{cases} \sigma + (1-\sigma)(1-\mu)^n & \text{if } y = 0 \\ \dfrac{(1-\sigma)\, n!\mu^y (1-\mu)^{n-y}}{y!\,(n-y)!} & \text{if } y = 1, 2, 3, \ldots \end{cases} \tag{23.6}$$

for $0 < \mu < 1$, $0 < \sigma < 1$ and n is a known positive integer. The mean and variance of Y are given by

$$\begin{aligned} \text{E}(Y) &= (1-\sigma)\, n\mu \\ \text{Var}(Y) &= n\mu\,(1-\sigma)\,[1-\mu+n\mu\sigma] \ . \end{aligned}$$

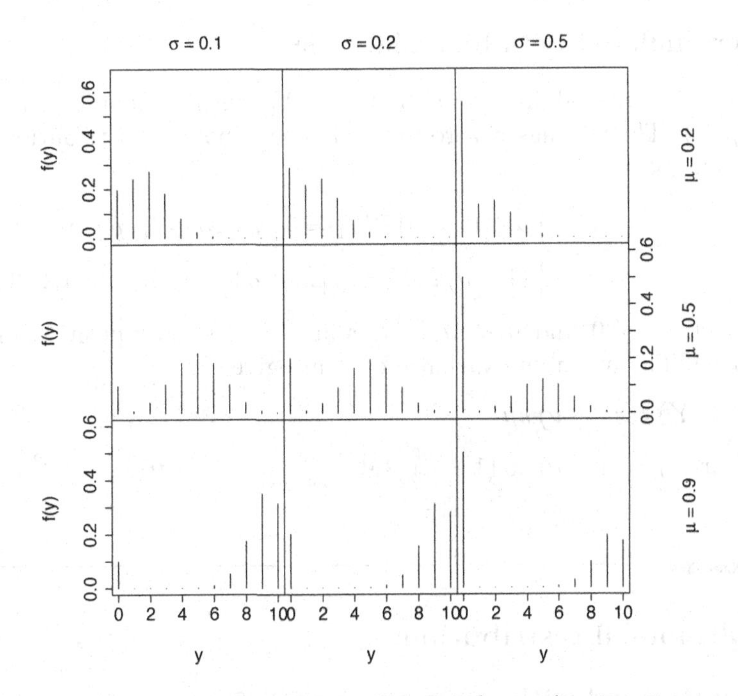

FIGURE 23.5: The zero-inflated binomial, $\texttt{ZIBI}(n, \mu, \sigma)$, distribution, with $\mu = 0.2, 0.5, 0.9$, $\sigma = 0.1, 0.2, 0.5$, and binomial denominator $n = 10$.

23.3 Three-parameter binomial type distributions

23.3.1 Zero-adjusted beta binomial: ZABB

Let $Y = 0$ with probability ν, and $Y = Y_1$ with probability $(1-\nu)$, where $Y_1 \sim \texttt{BBtr}(n, \mu, \sigma)$ a beta binomial distribution truncated at zero. Then Y has a zero-adjusted (or altered) beta binomial distribution, denoted by ZABB(n,μ,σ,ν), given by

$$P(Y=y\,|\,n,\mu,\sigma,\nu)=\begin{cases}\nu & \text{if } y=0\\ \dfrac{(1-\nu)\,P(Y_2=y\,|\,n,\mu,\sigma)}{1-P(Y_2=0\,|\,n,\mu,\sigma)} & \text{if } y=1,2,3,\ldots\end{cases} \tag{23.7}$$

for $0<\mu<1$, $\sigma>0$ and $0<\nu<1$ and n is a known positive integer, where $Y_2 \sim$ BB(n,μ,σ). The mean and variance of Y are given by

$$E(Y)=\frac{(1-\nu)\,n\mu}{[1-P(Y_2=0\,|\,n,\mu,\sigma)]}$$

$$\text{Var}(Y)=\frac{(1-\nu)\left\{n\mu\,(1-\mu)\left[1+\frac{\sigma}{1+\sigma}\,(n-1)\right]-n^2\mu^2\right\}}{[1-P(Y_2=0\,|\,n,\mu,\sigma)]}-[E(Y)]^2\ .$$

23.3.2 Zero-inflated beta binomial: ZIBB

Let $Y=0$ with probability ν, and $Y=Y_1$ with probability $(1-\nu)$ where $Y_1 \sim$ BB(n,μ,σ). Then Y has a zero-inflated beta binomial distribution, denoted by ZIBB(n,μ,σ,ν), given by

$$P(Y=y\,|\,n,\mu,\sigma,\nu)=\begin{cases}\nu+(1-\nu)\,P(Y_1=0\,|\,n,\mu,\sigma) & \text{if } y=0\\ (1-\nu)\,P(Y_1=y\,|\,n,\mu,\sigma) & \text{if } y=1,2,3,\ldots\end{cases} \tag{23.8}$$

for $0<\mu<1$, $\sigma>0$ and $0<\nu<1$, and n is a known positive integer, where $Y_1 \sim$ BB(n,μ,σ). The mean and variance of Y are given by

$$\begin{aligned}E(Y)&=(1-\nu)\,n\mu\\ \text{Var}(Y)&=(1-\nu)\,n\mu\,(1-\mu)\left[1+\tfrac{\sigma}{1+\sigma}\,(n-1)\right]+\nu\,(1-\nu)\,n^2\mu^2\ .\end{aligned}$$

23.4 Multinomial distributions

23.4.1 Multinomial with five categories: MN5

Let Y be a categorical variable with five possible categories (or levels) labelled '0', '1', '2', '3' and '4'.

The probability function of the multinomial distribution with 5 categories, denoted by MN5(μ,σ,ν,τ), is given by

$$P(Y=y\,|\,n,\mu,\sigma,\nu,\tau)=\begin{cases}p_0 & \text{if } y=0\\ p_1 & \text{if } y=1\\ p_2 & \text{if } y=2\\ p_3 & \text{if } y=3\\ p_4 & \text{if } y=4,\end{cases} \tag{23.9}$$

where $p_0 = \mu/s$, $p_1 = \sigma/s$, $p_2 = \nu/s$, $p_3 = \tau/s$, $p_4 = 1/s$, and $s = (1 + \mu + \sigma + \nu + \tau)$ and so $\sum_{j=0}^{4} p_j = 1$.

Hence $\mu = p_0/p_4$, $\sigma = p_1/p_4$, $\nu = p_2/p_4$, $\tau = p_3/p_4$, and $\mu > 0$, $\sigma > 0$, $\nu > 0$ and $\tau > 0$. The default link functions for μ, σ, ν, and τ in `MN5`(μ, σ, ν, τ) in **gamlss.dist** are all logarithmic.

23.4.2 Multinomial with four categories: `MN4`

The probability function of the multinomial distribution with four categories, denoted by `MN4`(μ, σ, ν), is given by setting $p_3 = 0$ in (23.9) for `MN5`(μ, σ, ν, τ), hence $\tau = 0$, and then relabelling '4' to '3', and p_4 to p_3.

23.4.3 Multinomial with three categories: `MN3`

The probability function of the multinomial distribution with three categories, denoted by `MN3`(μ, σ), is given by setting $p_2 = p_3 = 0$ in (23.9) for `MN5`(μ, σ, ν, τ), hence $\nu = \tau = 0$, and then relabelling '4' to '2', and p_4 to p_2.

A

R *code for plots of the reference guide*

CONTENTS

The code for most of the plots presented in the reference guide are given in the next sections.

A.1 Continuous distributions on $(-\infty, \infty)$

```
contR_2_12("GU",mu=c(-1,0,1),sigma=c(0.5,1,2),cex.axis=0.9,cex.all=0.9)
contR_2_12("LO",mu=c(-1,0,1),sigma=c(0.5,1,2),cex.axis=0.9,cex.all=0.9)
contR_2_12("NO",mu=c(-1,0,1),sigma=c(0.5,1,2),cex.axis=0.9,cex.all=0.9)
contR_2_12("RG",mu=c(-1,0,1),sigma=c(0.5,1,2),cex.axis=0.9,cex.all=0.9)
contR_3_11("exGAUS",mu=1,sigma=1,nu=c(0.1,1,2),cex.axis=1.1,cex.all=1.3)
contR_3_11("PE",mu=0,sigma=1,nu=c(1,2,10),maxy=4,cex.axis=1.1,
          cex.all=1.3)
contR_3_11("PE2",mu=0,sigma=1,nu=c(1,2,10),maxy=4,cex.axis=1.1,
          cex.all=1.3)
contR_3_11("SN1",mu=0,sigma=1,nu=c(0,3,10),maxy=4,cex.axis=1.1,
          cex.all=1.3)
contR_3_11("SN2",mu=0,sigma=1,nu=c(1,2,3),maxy=6,cex.axis=1.1,
          cex.all=1.3)
contR_3_11("TF",mu=0,sigma=1,nu=c(1,2,10),maxy=6,cex.axis=1.1,
          cex.all=1.3)
contR_3_11("TF2",mu=0,sigma=1,nu=c(3,5,10),maxy=6,cex.axis=1.1,
          cex.all=1.3)
contR_4_13("EGB2",mu=0,sigma=1,nu=c(1,2,5),tau=c(1,2,5), maxy=7,
          cex.axis=1.4)
contR_4_13("GT",mu=0,sigma=1,nu=c(1,2,10),tau=c(1,2,10),maxy=6,
          cex.axis=1.4)
contR_4_13("JSUo",mu=0,sigma=1,nu=c(0,-1,-2),tau=c(1,2,3),maxy=4,
          cex.axis=1.4)
```

```
contR_4_13("JSU",mu=0,sigma=1,nu=c(0,1,2),tau=c(1,2,3),maxy=4,
           cex.axis=1.4)
contR_4_13("NET",mu=0,sigma=1,nu=c(1,1.5,2),tau=c(2,2.5,3))
contR_4_13("SHASH",mu=0,sigma=1,nu=c(0.5,1,2),tau=c(0.5,1,2),
           maxy=6,cex.axis=1.4)
contR_4_13("SHASHo",mu=0,sigma=1,nu=c(0,0.5,1),tau=c(0.5,1,2),
           maxy=6,cex.axis=1.4)
contR_4_13("SEP1",mu=0,sigma=1,nu=c(0,2,10),tau=c(0.5,2,10),
           maxy=4,cex.axis=1.4)
contR_4_13("SEP2",mu=0,sigma=1,nu=c(0,1,10),tau=c(0.5,2,10),
           maxy=4,cex.axis=1.4)
contR_4_13("SEP3",mu=0,sigma=1,nu=c(1,2,3),tau=c(1,2,4),maxy=7,
           cex.axis=1.4)
contR_4_13("SEP4",mu=0,sigma=1,nu=c(1,2,10),tau=c(1,2,10),
           maxy=4,cex.axis=1.4)
contR_4_13("SST",mu=0,sigma=1,nu=c(1,2,3),tau=c(3,5,10),maxy=4,
           cex.axis=1.4)
contR_4_13("ST1",mu=0,sigma=1,nu=c(0,1,2),tau=c(1,3,10),maxy=4,
           cex.axis=1.4)
contR_4_13("ST2",mu=0,sigma=1,nu=c(0,1,2),tau=c(1,3,10),maxy=4,
           cex.axis=1.4)
contR_4_13("ST3",mu=0,sigma=1,nu=c(1,2,3),tau=c(1,3,10),maxy=7,
           cex.axis=1.4)
contR_4_13("ST4",mu=0,sigma=1,nu=c(1,3,10),tau=c(1,3,10),maxy=7,
           cex.axis=1.4)
contR_4_13("ST5",mu=0,sigma=1,nu=c(0,1,2),tau=c(1,3,5),maxy=8,
           cex.axis=1.4)
```

A.2 Continuous distributions on $(0, \infty)$

```
contRplus_2_11("IGAMMA",mu=1,sigma=c(0.5,1,2),maxy=5)
contRplus_2_11("IG",mu=1,sigma=c(0.2,1,2),maxy=3)
contRplus_2_11("LOGNO",mu=1,sigma=c(0.5,1,1.5),maxy=12)
contRplus_2_11("PARETO2",mu=1,sigma=c(0.5,1,2),maxy=5)
contRplus_2_11("WEI3",mu=1,sigma=c(0.5,1,2),maxy=5,
               y.axis.lim=0.08)
contRplus_3_13("BCCG",mu=1,sigma=c(0.1,0.2,0.3),nu=c(-2,0,3),
               maxy=2,cex.axis=1.4)
contRplus_3_13("GG",mu=1,sigma=c(0.5,1,2),nu=c(-2,1,2),maxy=5,
               y.axis.lim=0.03,cex.axis=1.4)
contRplus_3_13("GIG",mu=1,sigma=c(0.2,1,5),nu=c(-5,0,5),
               y.axis.lim=0.2,cex.axis=1.4)
contRplus_3_13("LNO",mu=1,sigma=c(0.5,1,5),nu=c(-1,0,2),
               maxy=6,y.axis.lim=0.08,cex.axis=1.4)
contRplus_4_33("BCT", mu=1, sigma=c(.1,.2,.3), nu=c(-3,1,4),
```

```
            tau=c(1,3,100))
contRplus_4_33("BCPE", mu=1, sigma=c(.1,.2,.3), nu=c(-3,1,4),
            tau=c(1,2,8),maxy=2)
contRplus_4_33("GB2", mu=1, sigma=c(1,2,3), nu=c(1,2,5),
            tau=c(1,2,5),y.axis.lim=1.1)
```

A.3 Mixed distributions on [0,∞), including 0

```
plotZAGA(mu=1, sigma=.5, nu=.4, from=0,to=4,main="")
plotZAIG(mu=1, sigma=.5, nu=.4, from=0,to=4, main="")
```

A.4 Continuous distributions on $(0, 1)$ excluding 0 and 1

```
contR01_2_13("BEo", mu=c(0.5,1,2), sigma=c(0.5,1,2),
            maxYlim = 5)
contR01_2_13("BE", mu=c(0.2,0.5,0.8), sigma=c(0.2,0.5,0.8),
            maxYlim = 10)
title("BE")
contR01_2_13("LOGITNO", mu=c(0.2,0.5,0.8), sigma=c(0.5,1,2),
            maxYlim = 10)
title("LOGITNO")
contR01_2_13("SIMPLEX", mu=c(0.2,0.5,0.8), sigma=c(1,2,5),
            maxYlim = 10)
title("SIMPLEX")
contR01_4_33("GB1", mu=0.5, sigma=c(0.2,0.4,0.6),
            nu=c(0.5,1,2), tau=c(0.5,1,2), maxYlim=10)
```

A.5 Mixed distributions on 0 to 1

```
plotBEINF0(mu=0.6, sigma=.4, nu=0.5,ylab="f(y)", xlab="y");
title("(a) BEINF0", cex.main=2)
plotBEINF1(mu=0.6, sigma=.4, nu=0.5, ylab="f(y)", xlab="y");
title("(b) BEINF1", cex.main=2)
plotBEINF(mu=0.6, sigma=.4, nu=0.5,tau=0.5,ylab="f(y)",
        xlab="y");title("(c) BEINF", cex.main=2)
```

A.6 Binomial type distributions and multinomial distributions

```
binom_1_31("BI",bd=10,mu=c(.2,.5,.9), maxy =20)
binom_2_33("BB", bd=10, mu=c(.2,.5,.9), sigma=c(0.5, 1, 2))
binom_2_33("DBI", bd=10, mu=c(.2,.5,.9), sigma=c(0.5, 1, 2))
binom_2_33("ZABI", bd=10,  mu=c(.2,.5,.9), sigma=c(0.1, 0.2, 0.5))
binom_2_33("ZIBI", bd=10, mu=c(.2,.5,.9), sigma=c(0.1, 0.2, 0.5))
```

Bibliography

M. Abramowitz and I. A. Stegun. *Handbook of Mathematical Functions: with Formulas, Graphs, and Mathematical Tables.* Dover, New York, 1965.

N. Ahmadi, S. A. Chung, A. Gibbs, and C. M. Shapiro. The Berlin questionnaire for sleep apnea in a sleep clinic population: relationship to polysomnographic measurement of respiratory disturbance. *Sleep and Breathing*, 12(1):39–45, 2008. doi: 10.1007/s11325-007-0125-y.

M. A. Aitkin. *Statistical Inference: An Integrated Bayesian/Likelihood Approach.* Chapman & Hall/CRC, Boca Raton, 2010.

M. A. Aitkin, B. Francis, J. Hinde, and R. Darnell. *Statistical Modelling in R.* Oxford University Press, Oxford, 2009.

H. Akaike. A new look at the statistical model identification. *IEEE Transactions on Automatic Control*, 19(6):716–723, 1974.

H. Akaike. Information measures and model selection. *Bulletin of the International Statistical Institute*, 50(1):277–290, 1983.

A. Alfons and M. Templ. Estimation of social exclusion indicators from complex surveys: The R package laeken. *Journal of Statistical Software*, 54(15):1–25, 2013. URL `http://www.jstatsoft.org/v54/i15/`.

M. M. Ali. Stochastic ordering and kurtosis measure. *Journal of the American Statistical Association*, 69:543–545, 1974.

A. Alimadad and M. Salibian-Barrera. An outlier-robust fit for generalized additive models with applications to outbreak detection. *Journal of the American Statistical Association*, 106(494):719–731, 2011.

M. A. Aljarrah, C. Lee, and F. Famoye. On generating t-x family of distributions using quantile functions. *Journal of Statistical Distributions and Applications*, 1(1):1–17, 2014.

A. Alzaatreh, F. Famoye, and C. Lee. Weibull-Pareto distribution and its applications. *Communications in Statistics: Theory and Methods*, 42(9):1673–1691, 2013.

A. A. Alzaid and M. A. Al-Osh. Ordering probability distributions by tail behavior. *Statistics and Probability Letters*, 8:185–188, 1989.

D. F. Andrews, P. J. Bickel, F. R. Hampel, P. J. Huber, W. H. Rogers, and J. W. Tukey. Robust estimation of location: Survey and advances. Technical report, Princeton University Press, Princeton, NJ, 1972.

F. J. Anscombe. Sampling theory of the negative binomial and logarithmic series approximations. *Biometrika*, 37:358–382, 1950.

A. Azzalini. A class of distributions which includes the normal ones. *Scandinavian Journal of Statistics*, 12:171–178, 1985.

A. Azzalini. Further results on a class of distributions which includes the normal ones. *Statistica*, 46:199:208, 1986.

A. Azzalini. *sn: The Skew-Normal and Related Distributions such as the Skew-t*. Università di Padova, Italia, 2018. URL `http://azzalini.stat.unipd.it/SN`. R package version 1.5-3.

A. Azzalini and A. Capitanio. Distributions generated by perturbation of symmetry with emphasis on a multivariate skew t-distribution. *Journal of the Royal Statistical Society: Series B*, 65:367–389, 2003.

N. Balakrishnan and V. B. Nevzorov. *A Primer on Statistical Distributions*. John Wiley & Sons, Hoboken, New Jersey, 2003.

K. P. Balanda and H. L. MacGillivray. Kurtosis: A critical review. *The American Statistician*, 42(2):111–119, 1988.

K. P. Balanda and H. L. MacGillivray. Kurtosis and spread. *Canadian Journal of Statistics*, 18(1):17–30, 1990.

O. E. Barndorff-Nielsen and B. Jørgensen. Some parametric models on the simplex. *Journal of Multivariate Analysis*, 39(1):106–116, 1991.

V. Barnett. *Comparative Statistical Inference*. John Wiley & Sons, 3rd edition, 1999.

J. Bernoulli. *Ars conjectandi (the art of conjecture)*. Thurneysen Brothers, Basel, 1713.

T. Bollerslev. Generalized autoregressive conditional heteroskedasticity. *Journal of Econometrics*, 31(3):307–327, 1986.

A. Bowman, E. Crawford, G. Alexander, and R. W. Bowman. rpanel: Simple interactive controls for R functions using the tcltk package. *Journal of Statistical Software*, 17 (9):1–18, 2007.

G. E. P. Box. Robustness in the strategy of scientific model building. *Robustness in Statistics*, 1:201–236, 1979.

G. E. P. Box and D. R. Cox. An analysis of transformations (with discussion). *Journal of the Royal Statistical Society: Series B*, 26:211–252, 1964.

G. E. P. Box and G. C. Tiao. *Bayesian Inference in Statistical Analysis*. John Wiley & Sons, New York, 1973.

R.J. Butler, J. B. McDonald, R.D. Nelson, and S.B. White. Robust and partially adaptive estimation of regression models. *The Review of Economics and Statistics*, 72(2):321–327, 1990.

A. C. Cameron and P. K. Trivedi. *Regression Analysis of Count Data.* Cambridge University Press, 2nd edition, 2013.

E. Cantoni and E. Ronchetti. Robust inference for generalized linear models. *Journal of the American Statistical Association*, 96(455):1022–1030, 2001.

R. J. Carroll and D. Ruppert. Robust estimation in heteroscedastic linear models. *Annals of Statistics*, 10:429–441, 1982.

R. J. Carroll and D. Ruppert. *Transformations and Weighting in Regression.* Chapman & Hall, London, 1988.

G. Claeskens and N. L. Hjort. The focused information criterion. *Journal of the American Statistical Association*, 98:900–916, 2003.

G. Claeskens and N. L. Hjort. *Model Selection and Model Averaging.* Cambridge University Press, 2008.

T. J. Cole and P. J. Green. Smoothing reference centile curves: The LMS method and penalized likelihood. *Statistics in Medicine.*, 11:1305–1319, 1992.

P. C. Consul. *Generalized Poisson Distributions.* Marcel Dekker, New York, 1989.

P. C. Consul and G. C. Jain. A generalization of the Poisson distribution. *Technometrics*, 15(4):791–799, 1973.

G. M. Cordeiro and M. de Castro. A new family of generalized distributions. *Journal of Statistical Computation and Simulation*, 81(7):883–898, 2011.

G. M. Cordeiro, M. Alizadeh, T. G. Ramires, and E. M. M. Ortega. The generalized odd half-Cauchy family of distributions: Properties and applications. *Communications in Statistics - Theory and Methods*, 46(11):5685–5705, 2017.

D. R. Cox and D. V. Hinkley. *Theoretical Statistics.* Chapman & Hall/CRC, 1979.

H. Cramér. *Mathematical Methods of Statistics (PMS-9).* Princeton University Press, 1946.

F. Cribari-Neto and A. Zeileis. Beta regression in R. *Journal of Statistical Software*, 34(2):1–24, 2010.

C. Croux, I. Gijbels, and I. Prosdocimi. Robust estimation of mean and dispersion functions in extended generalized additive models. *Biometrics*, 68(1):31–44, 2012.

M. Davidian and R. J. Carroll. A note on extended quasi-likelihood. *Journal of the Royal Statistical Society*, 50:74–82, 1988.

G. C. Jr. Davis and M. H. Kutner. The lagged normal family of probability density functions applied to indicator-dilution curves. *Biometrics*, 32:669–675, 1976.

P. de Jong and G. Z. Heller. *Generalized Linear Models for Insurance Data.* Cambridge University Press, 2008.

C. Dean, J. F. Lawless, and G. E. Willmot. A mixed Poisson-inverse-Gaussian regression model. *Canadian Journal of Statistics*, 17(2):171–181, 1989.

P. Deb and P.K. Trivedi. Demand for medical care by the elderly: A finite mixture approach. *Journal of Applied Econometrics*, 12:313–336, 1997.

J. del Castillo, D. Moriña Soler, and I. Serra. *ercv: Fitting Tails by the Empirical Residual Coefficient of Variation*, 2017. URL `https://CRAN.R-project.org/package=ercv`. R package version 1.0.0.

M. L. Delignette-Muller and C. Dutang. fitdistrplus: An R package for fitting distributions. *Journal of Statistical Software*, 64(4):1–34, 2015. URL `http://www.jstatsoft.org/v64/i04/`.

M. Denuit, X. Maréchal, S. Pitrebois, and J.-F. Walhin. *Actuarial Modelling of Claim Counts: Risk Classification, Credibility and Bonus-Malus Systems.* John Wiley & Sons, 2007.

J. C. Diaz Zapata. *ZOIP: ZOIP Distribution, ZOIP Regression, ZOIP Mixed Regression*, 2017. URL `https://CRAN.R-project.org/package=ZOIP`. R package version 0.1.

T. J. DiCiccio and A. C. Monti. Inferential aspects of the skew exponential power distribution. *Journal of the American Statistical Association*, 99:439–450, 2004.

S. Dossou-Gbété and D. Mizère. An overview of probability models for statistical modelling of count data. *Monografías del Seminario Matemático García de Galdeano*, 33:237–244, 2006.

N. Duan, W. G. Manning, C. N. Morris, and J. P. Newhouse. A comparison of alternative models for the demand for medical care. *Journal of Business and Economic Statistics*, 1(2):115–126, 1983.

P. K. Dunn. *Tweedie: Evaluation of Tweedie Exponential Family Models*, 2017. R package version 2.3.0.

G. Durrieu, I. Grama, K. Jaunatre, Q.-K. Pham, and J.-M. Tricot. extremefit: A package for extreme quantiles. *Journal of Statistical Software*, 87(12):1–20, 2018. doi: 10.18637/jss.v087.i12.

B. Efron. Double exponential families and their use in generalized linear regression. *Journal of the American Statistical Association*, 81:709–721, 1986.

B. Efron and R.J. Tibshirani. *An Introduction to the Bootstrap.* Chapman & Hall, New York, 1993.

P. Embrechts, C. Kluppelberg, and T. Mikosch. *Modelling Extremal Events for Insurance and Finance.* Springer, Berlin, 1997.

R. F. Engle. Autoregressive conditional heteroscedasticity with estimates of the variance of United Kingdom inflation. *Econometrica: Journal of the Econometric Society*, pages 987–1007, 1982.

D. G. Enki, A. Noufaily, P. Farrington, P. Garthwaite, N. Andrews, and A. Charlett. Taylor's power law and the statistical modelling of infectious disease surveillance data. *Journal of the Royal Statistical Society: Series A*, 180(1):45–72, 2017.

D. A. Evans. Experimental evidence concerning contagious distributions in ecology. *Biometrika*, 40:186–211, 1953.

C. Fernandez and M. F. J. Steel. On Bayesian modelling of fat tails and skewness. *Journal of the American Statistical Association*, 93:359–371, 1998.

C. Fernandez, J. Osiewalski, and M. J. F. Steel. Modeling and inference with v-spherical distributions. *Journal of the American Statistical Association*, 90:1331–1340, 1995.

S. Ferrari and F. Cribari-Neto. Beta regression for modelling rates and proportions. *Journal of Applied Statistics*, 31(7):799–815, 2004.

C. Forbes, M. Evans, N. Hastings, and B. Peacock. *Statistical Distributions.* John Wiley & Sons, Hoboken, New Jersey, 4th edition, 2011.

S. Foss, D. Korshunov, and S. Zachary. *An Introduction to Heavy-tailed and Subexponential Distributions.* Springer Series in Operations Research and Financial Engineering. Springer, New York, 2nd edition, 2013.

A.M. Fredriks, S. van Buuren, R.J.F. Burgmeijer, J.F. Meulmeester, R.J. Beuker, E. Brugman, M.J. Roede, S.P. Verloove-Vanhorick, and J. M. Wit. Continuing positive secular change in The Netherlands, 1955-1997. *Pediatric Research*, 47:316–323, 2000a.

A.M. Fredriks, S. van Buuren, J.M. Wit, and S. P. Verloove-Vanhorick. Body index measurements in 1996-7 compared with 1980. *Archives of Childhood Diseases*, 82: 107–112, 2000b.

A. Gelman, J. B. Carlin, H. S. Stern, D. B. Dunson, A. Vehtari, and D. B. Rubin. *Bayesian Data Analysis.* Chapman & Hall/CRC, Boca Raton, 3rd edition, 2013.

J. F. Gibbons and S. Mylroie. Estimation of impurity profiles in ion-implanted amorphous targets using joined half-Gaussian distributions. *Applied Physics Letters*, 22: 568–569, 1973.

R. Gilchrist. Regression models for data with a non-zero probability of a zero response. *Communications in Statistics: Theory and Methods*, 29:1987–2003, 2000.

W.R. Gilks, S. Richardson, and D.J. Spiegelhalter. *Markov Chain Monte Carlo in Practice.* Chapman & Hall/CRC, Boca Raton, 1996.

J. Gill. *Bayesian Methods: A Social and Behavioral Sciences Approach.* Chapman & Hall/CRC, London, 3rd edition, 2014.

C. S. Gillespie. Fitting heavy tailed distributions: The poweRlaw package. *Journal of Statistical Software*, 64(2):1–16, 2015. URL `http://www.jstatsoft.org/v64/i02/`.

C. Gourieroux, A. Monfort, and A. Trognon. Pseudo maximum likelihood methods: Theory. *Econometrica: Journal of the Econometric Society*, 52(3):681–700, 1984.

M. Greenwood and G. U. Yule. An inquiry into the nature of frequency distributions representative of multiple happenings with particular reference to the occurrence of multiple attacks of disease or of repeated accidents. *Journal of the Royal Statistical Society*, 83(2):255–279, 1920.

R. A. Groeneveld and G. Meeden. Measuring skewness and kurtosis. *The Statistician*, 33:391–399, 1984.

E. J. Gumbel. *Statistics of Extremes.* Columbia University Press, 1958.

P. L. Gupta, R. C. Gupta, and R. C. Tripathi. Analysis of zero-adjusted count data. *Computational Statistics and Data Analysis*, 23(2):207–218, 1996.

R. C. Gupta, P. L. Gupta, and R. D. Gupta. Modeling failure time data by Lehman alternatives. *Communications in Statistics: Theory and Methods*, 27(4):887–904, 1998.

F. A. Haight. *Handbook of the Poisson Distribution.* John Wiley & Sons, New York, 1967.

D. B. Hall. Zero-inflated Poisson and binomial regression with random effects: a case study. *Biometrics*, 56(4):1030–1039, 2000.

F. Hampel. *Contributions to the theory of robust estimation.* PhD thesis, University of California, Berkeley, 1968.

F. Hampel. The influence curve and its role in robust estimation. *Journal of the American Statistical Association*, 69:383–393, 1974.

F. R. Hampel, E. M. Ronchetti, P. J. Rousseeuw, and W. A. Stahel. *Robust Statistics: The Approach Based on Influence Functions.* John Wiley & Sons, New York, 1986.

D. J. Hand, F. Daly, A. D. Lunn, K. J. McConway, and E. Ostrowski. *A Handbook of Small Data Sets.* Chapman & Hall, London, 1994.

B. Hansen. Autoregressive conditional density estimation. *International Economic Review*, 35:705–730, 1994.

W. Härdle. *Smoothing Techniques, with Implementation in S.* Springer-Verlag, New York, 1990.

F. E. Harrell. Plasma retinol and beta-carotene dataset. `http://biostat.mc.vanderbilt.edu/wiki/pub/Main/DataSets/plasma.html`, 2002. [Accessed 3 March 2019].

T. J. Hastie, R. J. Tibshirani, and J. Friedman. *The Elements of Statistical Learning: Data Mining, Inference and Prediction.* Springer, New York, 2nd edition, 2009.

X. He and P. Ng. COBS: Qualitative constrained smoothing via linear programming. *Computational Statistics*, 14:315–337, 1999.

D. C. Heilbron. Zero-altered and other regression models for count data with added zeros. *Biometrical Journal*, 36(5):531–547, 1994.

L. Held and D. Sabanés Bové. *Applied Statistical Inference: Likelihood and Bayes.* Springer, New York, 2014.

G. Heller, M.D. Stasinopoulos, and R.A. Rigby. The zero-adjusted inverse Gaussian distribution as a model for insurance claims. In J. Hinde, editor, *Proceedings of the 21st International Workshop on Statistical Modelling*, pages 109–121. Galway, Ireland, 2006.

G. Z. Heller, D.-L. Couturier, and S. R. Heritier. Beyond mean modelling: Bias due to misspecification of dispersion in Poisson-inverse Gaussian regression. *Biometrical Journal*, 61(2):333–342, 2018. doi: 10.1002/bimj.201700218.

A. Henningsen. *censReg: Censored Regression (Tobit) Models*, 2017. URL `https://CRAN.R-project.org/package=censReg`. R package version 0.5-26.

B. M. Hill. A simple general approach to inference about the tail of a distribution. *Annals of Statistics*, 3:1163–1174, 1975.

J. Hinde. Compound Poisson regression models. In R. Gilchrist, editor, *GLIM 82, Proceedings of the International Conference on Generalised Linear Models*, pages 109–121. Springer, New York, 1982.

B. Hofner, A. Mayr, and M. Schmid. gamboostLSS: An R package for model building and variable selection in the GAMLSS framework. *Journal of Statistical Software*, 74(1):1–31, 2016.

A. Hossain. *General Methods for Analysing Bounded Propotion Data*, 2017. PhD Thesis, London Metropolitan University.

A. Hossain, R.A. Rigby, D.M. Stasinopoulos, and M. Enea. Modelling a proportion response variable using generalized additive models for location, scale and shape. In *Proceedings of the 30th International Workshop on Statistical Modelling*, volume 2, pages 182–186, 2015.

A. Hossain, R.A. Rigby, D.M. Stasinopoulos, and M. Enea. Centile estimation for a proportion response variable. *Statistics in Medicine*, 35(6):895–904, 2016a. URL `http://dx.doi.org/10.1002/sim.6748`.

A. Hossain, R.A. Rigby, D.M. Stasinopoulos, and M. Enea. A flexible approach for modelling a proportion response variable: Loss given default. In *Proceedings of the 31st International Workshop on Statistical Modelling*, pages 127–132, 2016b.

P. J. Huber. Robust estimation of a location parameter. *Annals of Mathematical Statistics*, 35:73–101, 1964.

P. J. Huber. The behavior of maximum likelihood estimates under nonstandard conditions. In *Proceedings of the Fifth Berkeley Symposium on Mathematical Statistics and Probability*, volume 1, pages 221–233, 1967.

P. J. Huber. *Robust Statistics*. John Wiley & Sons, New York, 1981.

M. E. H. Ismail. Integral representations and complete monotonicity of various quotients of Bessel functions. *Canadian Journal of Mathematics*, 29(6):1198–1207, 1977.

S. Jackman. *pscl: Classes and Methods for R Developed in the Political Science Computational Laboratory*. United States Studies Centre, University of Sydney, Sydney, New South Wales, Australia, 2017. URL `https://github.com/atahk/pscl/`. R package version 1.5.2.

C. M. Jarque and A. K. Bera. A test for normality of observations and regression residuals. *International Statistical Review*, 55(2):163–172, 1987.

N. L. Johnson. Systems of frequency curves generated by methods of translation. *Biometrika*, 36:149–176, 1949.

N. L. Johnson, S. Kotz, and N. Balakrishnan. *Continuous Univariate Distributions*, volume 1. John Wiley & Sons, New York, 2nd edition, 1994.

N. L. Johnson, S. Kotz, and N. Balakrishnan. *Continuous Univariate Distributions*, volume 2. John Wiley & Sons, New York, 2nd edition, 1995.

N. L. Johnson, A. W. Kemp, and S. Kotz. *Univariate Discrete Distributions.* John Wiley & Sons, New York, 3rd edition, 2005.

M. Jones. On families of distributions with shape parameters. *International Statistical Review*, 83(2):175–192, 2015.

M. C. Jones. In discussion of Rigby, R. A. and Stasinopoulos, D. M. (2005) Generalized additive models for location, scale and shape. *Applied Statistics*, 54:507–554, 2005.

M. C. Jones and M. J. Faddy. A skew extension of the t distribution, with applications. *Journal of the Royal Statistical Society: Series B*, 65:159–174, 2003.

M. C. Jones and A. Pewsey. Sinh-arcsinh distributions. *Biometrika*, 96:761–780, 2009.

B. Jørgensen. *Statistical properties of the generalized inverse Gaussian distribution.* Lecture Notes in Statistics No. 9. Springer-Verlag, New York, 1982.

B. Jørgensen. *The Theory of Dispersion Models.* Chapman & Hall, London, 1997.

D. Karlis and E. Xekalaki. The polygonal distribution. In B. C. Arnold, N. Balakrishnan, J. M. Sarabia, and R. Minguez, editors, *Advances in Mathematical and Statistical Modeling.* Springer Science & Business Media, 2009.

R. Kieschnick and B. D. McCullough. Regression analysis of variates observed on (0, 1): percentages, proportions and fractions. *Statistical Modelling*, 3(3):193–213, 2003.

C. Kleiber and A. Zeileis. Visualizing count data regressions using rootograms. *The American Statistician*, 70(3):296–303, 2016.

J. P. Klein and M. L. Moeschberger. *Survival Analysis: Techniques for Censored and Truncated Data.* Springer Science & Business Media, 2nd edition, 2003.

T. Kneib. Beyond mean regression (with discussion and rejoinder). *Statistical Modelling*, 13(4):275–303, 2013.

R. Koenker. Quantile regression: 40 years on. *Annual Review of Economics*, 9:155–176, 2017.

R. Koenker, P. Ng, and S. Portnoy. Quantile smoothing splines. *Biometrika*, 81(4): 673–680, 1994. doi: 10.1093/biomet/81.4.673.

S. Kotz and J. R. van Dorp. *Beyond Beta: Other Continuous Families of Distributions with Bounded Support and Applications.* World Scientific, Singapore, 2004.

K. Krishnamoorthy. *Handbook of Statistical Distributions with Applications.* Chapman & Hall/CRC, Boca Raton, 2nd edition, 2016.

D. Lambert. Zero-inflated Poisson regression, with an application to defects in manufacturing. *Technometrics*, 34(1):1–14, 1992.

P. Lambert and J. K. Lindsey. Analysing financial returns using regression models based on non-symmetric stable distributions. *Applied Statistics*, 48:409–424, 1999.

K. L. Lange, R. J. A. Little, and J. M. G. Taylor. Robust statistical modelling using the t distribution. *Journal of the American Statistical Association*, 84:881–896, 1989.

C. Lee, F. Famoye, and A. Y. Alzaatreh. Methods for generating families of univariate continuous distributions in the recent decades. *Wiley Interdisciplinary Reviews: Computational Statistics*, 5(3):219–238, 2013.

P. Li, M. Qi, X. Zhang, and X. Zhao. Further investigation of parametric loss given default modeling. *Economics Working Paper, Office of the Comptroller of the Currency*, pages 1–41, 2014.

J. K. Lindsey. *Modelling Frequency and Count Data.* Clarendon Press, Oxford, 1995.

J. K. Lindsey. *Parametric Statistical Inference.* Clarendon Press, Oxford, 1996.

J. K. Lindsey. On the use of correction for overdispersion. *Applied Statistics*, 48: 553–561, 1999.

A. Lopatatzidis and P. J. Green. Private Communication, 2000.

G. Lovison and C. Schindler. Separate regression modelling of the Gaussian and exponential components of an EMG response from respiratory physiology. In T. Kneib, F. Sobotka, J. Fahrenholz, and H. Irmer, editors, *Proceedings of the 29th International Workshop on Statistical Modelling*, pages 189–194, 2014.

H. L. MacGillivray. Skewness and asymmetry: measures and orderings. *Annals of Statistics*, 14(3):994–1011, 1986.

H. L. MacGillivray and K. P. Balanda. The relationships between skewness and kurtosis. *Australian Journal of Statistics*, 30(3):319–337, 1988.

B. B. Mandelbrot. *Fractals and Scaling in Finance: Discontinuity, Concentration, Risk.* Springer, New York, 1997.

G. Marra and R. Radice. *GJRM: Generalised Joint Regression Modelling*, 2019. URL `https://CRAN.R-project.org/package=GJRM`. R package version 0.2.0.

B. H. McArdle and M. J. Anderson. Variance heterogeneity, transformations, and models of species abundance: a cautionary tale. *Canadian Journal of Fisheries and Aquatic Sciences*, 61(7):1294–1302, 2004.

P. McCullagh and J. A. Nelder. *Generalized Linear Models.* Chapman & Hall, London, 2nd edition, 1989.

J. B. McDonald. Some generalized functions for the size distributions of income. *Econometrica*, 52:647–663, 1984.

J. B. McDonald. Parametric models for partially adaptive estimation with skewed and leptokurtic residuals. *Economic Letters*, 37:273–278, 1991.

J. B. McDonald. Probability distributions for financial models. In G. S. Maddala and C. R. Rao, editors, *Handbook of Statistics*, volume 14, pages 427–460. Elsevier Science, 1996.

J. B. McDonald and W. K. Newey. Partially adaptive estimation of regression models via the generalized t distribution. *Econometric Theory*, 4:428–457, 1988.

J. B. McDonald and Y. J. Xu. A generalisation of the beta distribution with applications. *Journal of Econometrics*, 66:133–152, 1995.

T. Mikosch. Regular variation, subexponentiality and their applications in probability theory. Technical report, University of Groningen, 1999.

R. B. Millar. *Maximum Likelihood Estimation and Inference: With Examples in R, SAS and ADMB*. John Wiley & Sons, 2011.

Y. Min and A. Agresti. Modeling nonnegative data with clumping at zero: a survey. *Journal of the Iranian Statistical Society*, 1(1):7–33, 2002.

G. S. Mudholkar and D. K. Srivastava. Exponentiated Weibull family for analyzing bathtub failure-rate data. *IEEE Transactions on Reliability*, 42(2):299–302, 1993.

J. Mullahy. Specification and testing of some modified count data models. *Journal of Econometrics*, 33(3):341–365, 1986.

S. Nadarajah and R. Rocha. Newdistns: An R package for new families of distributions. *Journal of Statistical Software*, 69(10):1–32, 2016.

S. Nadarajah, V. G. Cancho, and E. M. M. Ortega. The geometric exponential Poisson distribution. *Statistical Methods and Application*, 22:355–380, 2013.

T. Nakagawa and S. Osaki. The discrete Weibull distribution. *IEEE Transactions on Reliability*, 24(5):300–301, 1975.

A. K. Nandi and D. Mämpel. An extension of the generalized Gaussian distribution to include asymmetry. *Journal of the Franklin Institute*, 332:67–75, 1995.

J. A. Nelder and Y. Lee. Likelihood, quasi-likelihood and pseudolikelihood: Some comparisons. *Journal of the Royal Statistical Society: Series B*, 54:273–284, 1992.

J. A. Nelder and D. Pregibon. An extended quasi-likelihood function. *Biometrika*, 74: 221–232, 1987.

J. A. Nelder and R. W. M. Wedderburn. Generalized linear models. *Journal of the Royal Statistical Society: Series A*, 135:370–384, 1972.

D. B. Nelson. Conditional heteroskedasticity in asset returns: a new approach. *Econometrica*, 59:347–370, 1991.

W. K. Newey and D. McFadden. Large sample estimation and hypothesis testing. *Handbook of Econometrics*, 4:2111–2245, 1994.

J. P. Nolan. Numerical calculation of stable densities and distribution functions. *Communications in Statistics. Stochastic Models*, 13(4):759–774, 2007.

J. P. Nolan. *ecdfHT: Empirical CDF for Heavy Tailed Data*, 2016. URL `https://CRAN.R-project.org/package=ecdfHT`. R package version 0.1.1.

P. Np and M. Maechler. A fast and efficient implementation on qualitatively constrained quantile smoothing splines. *Statistical Modelling*, 7:315–328, 2007.

R. Ospina and S. L. P. Ferrari. Inflated beta distributions. *Statistical Papers*, 51: 111–126, 2010.

R. Ospina and S. L. P. Ferrari. A general class of zero-or-one inflated beta regression models. *Computational Statistics and Data Analysis*, 56:1609–1623, 2012.

B. A. Para and T. R. Jan. On discrete three parameter Burr type XII and discrete Lomax distributions and their applications to model count data from medical science. *Biometrics and Biostatistics International Journal*, 4(2):1–15, 2016.

E. Parzen. Nonparametric statistical data modeling (with discussion). *Journal of the American statistical association*, 74(365):105–121, 1979.

N. A. T. Payandeh and S. Mohammadpour. A k-inflated negative binomial mixture regression model: Application to rate–making systems. *Asia-Pacific Journal of Risk and Insurance*, 12(2):1–31, 2018.

K. Pearson. Contributions to the mathematical theory of evolution ii. skew variation in homogeneous material. *Philosophical Transactions of the Royal Society of London. Series B*, 186:343–414, 1895.

W. F. Perks. On some experiments in the graduation of mortality statistics. *Journal of the Institute of Actuaries*, 58:12–57, 1932.

S. D. Poisson. *Recherches sur la probabilité des jugements en matière criminelle et en matière civile precédées des règles générales du calcul des probabilités.* Bachelier, Paris, 1837.

M. Pokorny and J. Sedgwick. Profitability trends in Hollywood: 1929 to 1999: somebody must know something. *Economic History Review*, 63:56–84, 2010.

K. Poortema. On modelling overdispersion of counts. *Statistica Neerlandica*, 53(1): 5–20, 1999.

F. Proschan. Theoretical explanation of observed decreasing failure rate. *Technometrics*, 5(3):375–383, 1963.

P.H. Quanjer, S. Stanojevic, J. Stocks, G.L. Hall, K.V.V. Prasad, T.J. Cole, M. Rosenthal, R. Perez-Padilla, J.L. Hankinson, E. Falaschetti, et al. Changes in the FEV_1/FVC ratio during childhood and adolescence: an intercontinental study. *European Respiratory Journal*, 36(6):1391, 2010.

C. P. Quesenberry and C. Hales. Concentration bands for uniformity plots. *Journal of Statistical Computation and Simulation*, 11:41–53, 1980.

A. E. Raftery. Approximate Bayes factors and accounting for model uncertainty in generalized linear models. *Biometrika*, 83:251–266, 1996.

A. E. Raftery. Bayes Factors and BIC, comment on: A critique of the Bayesian Information Criterion for Model Selection. *Sociological Methods and Research*, 27:411–427, 1999.

T. G. Ramires, L. R. Nakamura, A. J. Righetto, R. R. Pescim, J. Mazuchelli, R. A. Rigby, and D. M. Stasinopoulos. Validation of stepwise-based procedure in GAMLSS. *submitted*, 2019.

S. I. Resnick. Heavy tail modeling and teletraffic data: special invited paper. *Annals of Statistics*, 25(5):1805–1869, 1997.

S. I. Resnick. *Heavy-tail Phenomena: Probabilistic and Statistical Modeling*. Springer Science & Business Media, New York, 2007.

R. A. Rigby and D. M. Stasinopoulos. Robust fitting of an additive model for variance heterogeneity. In R. Dutter and W. Grossmann, editors, *COMPSTAT: Proceedings in Computational Statistics*, pages 263–268. Physica, Heidelberg, 1994.

R. A. Rigby and D. M. Stasinopoulos. Construction of reference centiles using mean and dispersion additive models. *Statistician*, 49:41–50, 2000.

R. A. Rigby and D. M. Stasinopoulos. Smooth centile curves for skew and kurtotic data modelled using the Box Cox power exponential distribution. *Statistics in Medicine*, 23:3053–3076, 2004.

R. A. Rigby and D. M. Stasinopoulos. Generalized additive models for location, scale and shape (with discussion). *Applied Statistics*, 54:507–554, 2005.

R. A. Rigby and D. M. Stasinopoulos. Using the Box-Cox t distribution in GAMLSS to model skewness and kurtosis. *Statistical Modelling*, 6(3):209, 2006.

R. A. Rigby and D. M. Stasinopoulos. Automatic smoothing parameter selection in GAMLSS with an application to centile estimation. *Statistical Methods in Medical Research*, 23(4):318–332, 2013.

R. A. Rigby, D. M. Stasinopoulos, and C. Akantziliotou. A framework for modelling overdispersed count data, including the Poisson-shifted generalized inverse Gaussian distribution. *Computational Statistics and Data Analysis*, 53(2):381–393, 2008. ISSN 0167-9473.

R. A. Rigby, D. M. Stasinopoulos, and V. Voudouris. A comparison of GAMLSS with quantile regression, in discussion of Kneib, T. (2013) Beyond mean regression. *Statistical Modelling*, 13(4):335–348, 2013.

B. D. Ripley. *Pattern Recognition and Neural Networks*. Cambridge University Press, Cambridge, 1996.

J. L. Rosenberger and M. Gasko. Comparing location estimators: Trimmed means, medians and trimean. In D. C. Hoaglin, F. Mosteller, and J. W. Tukey, editors, *Understanding Robust and Exploratory Data Analysis*, pages 297–338. John Wiley, New York, 1983.

A. Routhier-Labadie, V. Ouellet, W. Bellemare, D. Richard, L. Lakhal-Chaieb, E. Turcotte, and A. C. Carpentier. Outdoor temperature, age, sex, body mass index, and diabetic status determine the prevalence, mass, and glucose-uptake activity of 18F-FDG-detected BAT in humans. *The Journal of Clinical Endocrinology and Metabolism*, 96(1):192–199, 2011. doi: 10.1210/jc.2010-0989.

P. Royston and E. M. Wright. Goodness-of-fit statistics for age-specific reference intervals. *Statistics in Medicine*, 19:2943–2962, 2000.

A. Saei, J. Ward, and C. A. McGilchrist. Threshold models in a methadone programme evaluation. *Statistics in Medicine*, 15(20):2253–2260, 1996.

E. F. Schuster. Classification of probability laws by tail behavior. *Journal of the American Statistical Association*, 79(388):936–939, 1984.

G. E. Schwarz. Estimating the dimension of a model. *Annals of Statistics*, 6(2):461–464, 1978.

H. S. Sichel. On a family of discrete distributions particularly suited to represent long-tailed frequency data. In N. F. Laubscher, editor, *Proceedings of the Third Symposium on Mathematical Statistics*, pages 75–83. Pretoria, South Africa: Council of Scientific and Industrial Research, 1971.

B. W. Silverman. *Density Estimation for Statistics and Data Analysis.* Chapman & Hall / CRC, Boca Raton, 1986.

S. D. Silvey. *Statistical Inference.* Chapman & Hall/CRC, Boca Raton, 1975.

J. G. Skellam. A probability distribution derived from the binomial distribution by regarding the probability of success as variable between the sets of trials. *Journal of the Royal Statistical Society: Series B*, 10(2):257–261, 1948.

P. J. Smith. A recursive formulation of the old problem of obtaining moments from cumulants and vice versa. *The American Statistician*, 49(2):217–218, 1995.

R. L. Smith and J. C. Naylor. A comparison of maximum likelihood and Bayesian estimators for the three-parameter Weibull distribution. *Applied Statistics*, 36:358–369, 1987.

E. W. Stacy and G. A. Mihram. Parameter estimation for a generalized gamma distribution. *Technometrics*, 7(3):349–358, 1965.

S. Stadlmann. *distreg.vis: Framework for the Visualization of Distributional Regression Models*, 2019. URL `https://CRAN.R-project.org/package=distreg.vis`. R package version 1.1.0.

S. Stahl. The evolution of the normal distribution. *Mathematics Magazine*, 79(2): 96–113, 2006.

S. Stanojevic, A. Wade, J. Stocks, J. Hankinson, A. L. Coates, H. Pan, M. Rosenthal, M. Corey, P. Lebecque, and T. J. Cole. Reference ranges for spirometry across all ages: a new approach. *American Journal of Respiratory and Critical Care Medicine*, 177:253–260, 2008.

D. M. Stasinopoulos. Contribution to the discussion of the paper by Lee and Nelder, Double hierarchical generalized linear models. *Applied Statistics*, 55:171–172, 2006.

D. M. Stasinopoulos, R. A. Rigby, and L. Fahrmeir. Modelling rental guide data using mean and dispersion additive models. *Statistician*, 49:479–493, 2000.

D. M. Stasinopoulos, R. A. Rigby, G. Z. Heller, V. Voudouris, and F. De Bastiani. *Flexible Regression and Smoothing: Using GAMLSS in R*. Chapman & Hall/CRC, Boca Raton, 2017.

D. M. Stasinopoulos, R. A. Rigby, and F. De Bastiani. GAMLSS: a distributional regression approach. *Statistical Modelling*, 18(3-4):248–273, 2018.

M. Stasinopoulos and S. Mohammadpour. *gamlss.countKinf: Generating and Fitting K-inflated 'Discrete gamlss.family' Distributions*, 2018. URL `http://www.gamlss.org/`. R package version 3.5.1.

Statistical Society of Canada. Obstructive sleep apnea. `https://ssc.ca/en/case-study/obstructive-sleep-apnea`, 2006. [Accessed 9 August 2010].

Statistical Society of Canada. Determinants of the presence and volume of brown fat in humans. `https://ssc.ca/en/case-study/determinants-presence-and-volume-brown-fat-human`, 2011. [Accessed 13 February 2019].

G. Z. Stein, W. Zucchini, and J. M. Juritz. Parameter estimation of the Sichel distribution and its multivariate extension. *Journal of the American Statistical Association*, 82:938–944, 1987.

M. T. Subbotin. On the law of frequency of errors. *Mathematicheskii Sbornik*, 31: 296–301, 1923.

P. R. Tadikamalla and N. L. Johnson. Systems of frequency curves generated by transformations of logistic variables. *Biometrika*, 69:461–465, 1982.

M. H. Tahir and S. Nadarajah. Parameter induction in continuous univariate distributions: Well-established G families. *Anais da Academia Brasileira de Ciências*, 87(2): 539–568, 2015.

R. N. Tamura and S. S. Young. A stabilized moment estimator for the beta binomial distribution. *Biometrics*, 43:813–824, 1987.

P. Theodossiou. Financial data and the skewed generalized *t* distribution. *Management Science*, 44:1650–1661, 1998.

J. Tobin. Estimation of relationships for limited dependent variables. *Econometrica: Journal of the Econometric Society*, 26:24–36, 1958.

E.N.C. Tong, C. Mues, and L. Thomas. A zero-adjusted gamma model for mortgage loan loss given default. *International Journal of Forecasting*, 29(4):548–562, 2013.

J. M. Tukey. Some graphic and semigraphic displays. In *Statistical Papers in Honor of George W. Snedecor*, volume V, pages 293–316. TA Bancroft, 1972.

M. C. K. Tweedie. An index which distinguishes between some important exponential families. In *Statistics: Applications and New Directions: Proceedings of the Indian Statistical Institute Golden Jubilee International Conference*, pages 579–604, 1984.

N. Umlauf, N. Klein, and A. Zeileis. BAMLSS: Bayesian additive models for location, scale and shape (and beyond). *Journal of Computational and Graphical Statistics*, 27:612–627, 2017. URL `http://dx.doi.org/10.1080/10618600.2017.1407325`.

S. van Buuren and M. Fredriks. Worm plot: a simple diagnostic device for modelling growth reference curves. *Statistics in Medicine*, 20:1259–1277, 2001.

W. R. van Zwet. Convex transformations: a new approach to skewness and kurtosis. *Statistica Neerlandica*, 18:433–441, 1964a.

W. R. van Zwet. *Convex Transformations of Random Variables. Mathematica Centre Tracts 7.* Mathematisch Centrum, Amsterdam, 1964b.

W. N. Venables and B. D. Ripley. *S Programming.* Springer, New York, 2000.

D. J. Venzon and S. H. Moolgavkar. A method for computing profile-likelihood-based confidence intervals. *Applied Statistics*, 37:87–94, 1988.

L. von Bortkiewicz. *Das Gesetz der Kleinen Zahlen.* BG Teubner, 1898.

A. Wald. Note on the consistency of the maximum likelihood estimate. *The Annals of Mathematical Statistics*, 20(4):595–601, 1949.

M. P. Wand and M. C. Jones. *Kernel Smoothing.* Chapman & Hall, Essen, Germany, 1999.

R. W. M. Wedderburn. Quasi-likelihood functions, generalized linear models and the Gauss-Newton method. *Biometrika*, 61:439–447, 1974.

H. White. Maximum likelihood estimation of misspecified models. *Econometrica: Journal of the Econometric Society*, 50(1):1–25, 1982.

Multicentre Growth Reference Study Group WHO. *WHO Child Growth Standards: Length/height-for-age, weight-for-age, weight-for-length, weight-for-height and body mass index-for-age: Methods and development.* Geneva: World Health Organization, 2006.

Multicentre Growth Reference Study Group WHO. *WHO Child Growth Standards: Head circumference-for-age, arm circumference-for-age, triceps circumference-for-age and subscapular skinford-for-age: Methods and development.* Geneva: World Health Organization, 2007.

Multicentre Growth Reference Study Group WHO. *WHO Child Growth Standards: Growth velocity based on weight, length and head circumference: Methods and development.* Geneva: World Health Organization, 2009.

C. B. Williams. Some applications of the logarithmic series and the index of diversity to ecological problems. *Journal of Ecology*, 32:1–44, 1944.

G. Wimmer and G. Altmann. *Thesaurus of Univariate Discrete Probability Distributions.* Stamm Verlag, Essen, Germany, 1999.

R. Winkelmann. *Econometric Analysis of Count Data.* Springer Verlag, Berlin, 3rd edition, 2008.

S. N. Wood. *Core Statistics*, volume 6. Cambridge University Press, 2015.

S. N. Wood. *Generalized Additive Models. An Introduction with R.* Chapman & Hall/CRC, Boca Raton, 2nd edition, 2017.

S. N. Wood, N. Pya, and B. Säfken. Smoothing parameter and model selection for general smooth models. *Journal of the American Statistical Association*, 111(516): 1548–1563, 2017.

D. Wuertz and M. Maechler. *stabledist: Stable Distribution Functions*, 2016. URL `https://CRAN.R-project.org/package=stabledist`. R package version 0.7-1.

D. Würtz, Y. Chalabi, and L. Luksan. Parameter estimation of ARMA models with GARCH/APARCH errors. An R and SPlus software implementation. *Journal of Statistical Software*, 55:28–33, 2006.

Y. Yang and D. Simpson. Unified computational methods for regression analysis of zero-inflated and bound-inflated data. *Computational Statistics and Data Analysis*, 54(6):1525–1534, 2010.

T. W. Yee. *VGAM: Vector Generalized Linear and Additive Models*, 2019. URL `https://CRAN.R-project.org/package=VGAM`. R package version 1.1-1.

A. Zeileis, C. Kleiber, and S. Jackman. Regression models for count data in R. *Journal of Statistical Software*, 27(8):1–25, 2008. URL `http://www.jstatsoft.org/v27/i08/`.

P. Zhang, Z. Qiu, and C. Shi. simplexreg: An R package for regression analysis of proportional data using the simplex distribution. *Journal of Statistical Software*, 71 (11):1–21, 2016.

A. Ziegler. *Generalized Estimating Equations.* Springer Science and Business Media, New York, 2011.

Index

Printed in the United States
by Baker & Taylor Publisher Services